Formulas from Geometry

Formulas for Area (A), Perimeter (P), Circumference (C), and Volume (V):

Square

$A = s^2$

$P = 4s$

Rectangle

$A = lw$

$P = 2l + 2w$

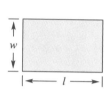

Circle

$A = \pi r^2$

$C = 2\pi r$

Triangle

$A = \dfrac{1}{2}bh$

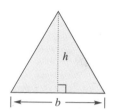

Trapezoid

$A = \dfrac{1}{2}h(b_1 + b_2)$

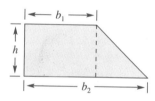

Parallelogram

$A = bh$

$P = 2a + 2b$

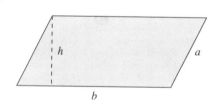

Pythagorean Theorem

$a^2 + b^2 = c^2$

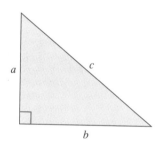

Cube

$V = s^3$

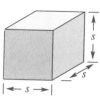

Rectangular Solid

$V = lwh$

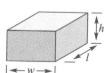

Circular Cylinder

$V = \pi r^2 h$

Sphere

$V = \dfrac{4}{3}\pi r^3$

Intermediate Algebra

Second Edition

Roland E. Larson

Robert P. Hostetler

The Pennsylvania State University
The Behrend College

with the assistance of
David E. Heyd
The Pennsylvania State University
The Behrend College

D. C. Heath and Company
Lexington, Massachusetts Toronto

Address editorial correspondence to:
D. C. Heath and Company
125 Spring Street
Lexington, MA 02173

Acquisitions Editors: Ann Marie Jones, Charles Hartford
Managing Editor: Catherine B. Cantin
Development Editor: Emily Keaton
Production Editor (art): Rachel D'Angelo Wimberly
Marketing Manager: Christine Hoag
Designer: Henry Rachlin
Interior Photo Researchers: Derek Wing and Billie Porter
Production Coordinator: Lisa Merrill
Composition: Meridian Creative Group
Art: Folium, Inc.; Meridian Creative Group; Patrice Rossi; Illustrious, Inc.
Cover Photo Researcher: Linda Finigan

Trademark Acknowledgments: TI is a registered trademark of Texas Instruments, Inc.
Casio is a registered trademark of Casio, Inc. Sharp is a registered trademark of Sharp
Electronics Corp.

Published simultaneously in Canada.

Printed in the United States of America.

International Standard Book Number: 0-669-39615-X

Library of Congress Catalog Number: 95-75643

10 9 8 7 6 5 4 3 2

Preface

The primary goals of *Intermediate Algebra,* Second Edition, are to encourage students to develop their proficiency in algebra and to show how algebra is a modern modeling language for real-life problems.

New to the Second Edition

In the Second Edition, all text elements were considered for revision, and many new examples, exercises, and applications were added to the text. Following are the major changes in the Second Edition.

Improved Coverage   The new Second Edition was designed to be flexible with respect to the order of coverage of core algebra topics, adapting easily to a wide variety of course syllabi and teaching styles. This text begins with Prerequisites, a review chapter. All or part of this material may be covered or it can be omitted. Graphing is now introduced in Chapter 2, earlier than in the previous edition. The use of graphs encourages visualization to offer an opportunity for more conceptual understanding, strengthens graph-reading skills, and supports a smoother transition to math courses students may take in the future. Throughout the text, greater emphasis is given to geometry, collecting and interpreting data and statistics, updated data analysis, and creating models, as well as to the NCTM Standards and Addenda and the AMATYC Guidelines.

Problem Solving   A general problem-solving process for applied problems is stressed throughout the text: form a verbal model, label terms, create a mathematical model, solve, and check the answer in the original statement of the problem (see page 74). This problem-solving process helps students understand the problem, organize their work, and develop facility with verbal, analytical, graphical, and numerical approaches to problem solving. Students are also reminded of specific problem-solving strategies (see page 74) that are reinforced throughout the text in the exercises (see Exercises 31–34 on page 365 and Exercise 125 on page 651).

Exercises   The exercise sets were completely revised and expanded—by nearly 40%—for the Second Edition. These comprehensive exercise sets offer students ample opportunity to practice algebraic techniques (see pages 280–282 and 387–389) and develop their conceptual and critical-thinking skills (see Exercises 79 and 80 on page 144, Exercises 55 and 56 on page 178, Exercise 117 on page 216, Exercises 89 and 90 on page 294, Exercise 41 on page 461, and Exercise 104 on page 573). The broad range of computational, conceptual, and applied problems in each exercise set is carefully graded to provide a smooth transition from routine to more challenging problems. Section and review exercises—as well as mid-chapter quizzes (see page 155) and chapter tests (see page 196)—consistently encourage student mastery of algebraic skills and concepts through practice and self-assessment.

Technology   Recognizing that graphing technology is becoming increasingly available, the Second Edition offers the opportunity to use graphing utilities throughout, but without requiring their use. This is achieved through a combination of features, including—at point of use—discovery opportunities that require scientific or graphing calculators (see pages 401 and 578), graphing utility instructions (see pages 306 and 383), and clearly labeled exercises that require the use of a graphing utility (see Exercise 64 on page 431 and Exercises 67–78 on pages 505 and 506).

Group Activities   Each section ends with a Group Activity. This exercise reinforces students' understanding by exploring mathematical concepts in a variety of ways: You Be the Instructor, Extending the Concept, Problem Solving, Exploring with Technology, and Communicating Mathematically. Some Group Activities encourage interpretation or discovery of mathematical concepts and results (see pages 202, 513, and 568); some provide opportunities for problem posing and error analysis (see pages 101, 261, 402, and 580); and others reinforce methods of interpreting and constructing mathematical models, tables, and graphs (see pages 386, 442, 459, and 602). Designed to be completed in class or as homework assignments, the Group Activities give students the opportunity to work cooperatively as they think, talk, and write about mathematics.

Data Analysis/Modeling   Throughout the Second Edition, students are offered more opportunities to collect and interpret data, make conjectures, and construct mathematical models. Students are exposed to combining mathematical models to make related models (see Exercise 101 on page 206 and Exercise 39 on page 263); encouraged to use mathematical models to make predictions and estimates from real data (see Exercise 40 on page 143, Exercise 43 on page 312, and Exercise 95 on page 446); invited to compare two or more models or compare actual data with a model (see Exercise 103 on page 573 and Exercise 97 on page 625); and asked to use curve-fitting techniques to write their own models from data (see Exercise 97 on page 446, Exercises 37–39 on page 540, and Exercises 39–41 on page 553). This edition encourages greater use of charts, tables, scatter plots, and graphs to summarize, analyze, and interpret data.

Applications   To emphasize for students the connection between mathematical concepts and real-world situations, up-to-date, real-life applications are integrated throughout the text. Appearing as examples (see pages 141, 362, 501, and 602), exercises (see Exercise 74 on page 134, Exercise 98 on page 190, Exercise 41 on page 515, and Exercise 116 on page 614), group activities (see pages 74 and 525), and projects (see pages 315 and 556), these applications help students validate the material they are learning and offer them frequent opportunities to use and review their problem-solving skills. A wide range of disciplines is represented by the applications—including such areas as physics, chemistry, electronics, the social sciences, biology, and business—as well as the career interviews, covering areas such as insurance, real estate, architecture, engineering, graphic arts, business, education, scuba diving, biochemistry, and economics.

Connections   In addition to highlighting the connections between algebra and areas outside mathematics through real-world applications, this text also emphasizes the connections between algebra and other branches of mathematics, such as probability (see pages 273 and 668), geometry (see page 50), logic (see Appendix A), and statistics (see Appendix B). Too, many examples and exercises throughout the text reinforce the connections among graphical, numerical, and algebraic representations of important algebraic concepts (see Exercises 27 and 28 on page 414).

There are many other new features of the Second Edition as well, including Discovery, Chapter Opening Applications, Study Tips, Historical Notes, Mid-Chapter Quizzes, Chapter Summaries, Career Interviews, and Chapter Projects. These and other features of the Second Edition are described in greater detail on the following pages.

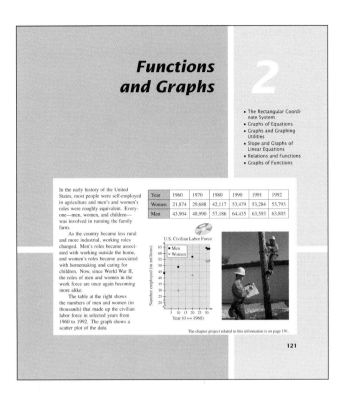

Chapter Opener

Each chapter opens with a look at a real-life application that is explored in depth in the Chapter Project at the end of the chapter. Real data is manipulated using graphical, numerical, and algebraic techniques. In addition, a list of the section titles shows students how the topics fit into the overall development of algebra.

Section Outline

Each section begins with a list of the major topics covered in that section. These topics are also the subsection titles and can be used for easy reference and review by students.

Historical Notes

To help students understand that algebra has a past, historical notes featuring mathematical artifacts or mathematicians and their work are included in each chapter.

Notes

Notes anticipate students' needs by offering additional insight, pointing out common errors, and describing generalizations.

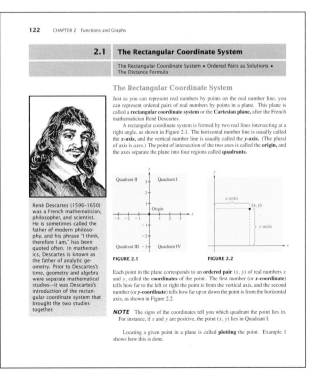

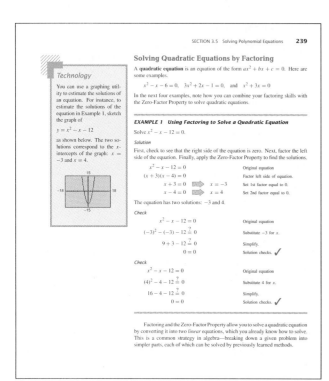

Technology

Instructions for using graphing utilities appear in the margin at point of use. They offer convenient reference for users of graphing technology and can easily be omitted if desired. Additionally, problems in the Exercise Sets that require a graphing utility have been identified with a graphing calculator icon.

Study Tips

Study Tips appear in the margin at point of use. They offer students specific, helpful, and insightful suggestions for studying algebra. "How to Study Algebra" on page xxvi and "Reading and Writing About Mathematics" on page xxix outline a general plan designed to improve student study skills.

Problem Solving

The text provides ample opportunity for students to develop their problem-solving skills. They are taught the following approach to solving applied problems: (1) Construct a verbal model; (2) Label variable and constant terms; (3) Construct an algebraic model; (4) Using the model, solve the problem; and (5) Check the answer in the original statement of the problem. This process has wide applicability, and it is used with verbal, analytical, graphical, and numerical approaches to problem solving. In the Second Edition, there is increased emphasis on identifying units of measure and checking solutions, and many solutions were rewritten with explanations and additional help in the form of comments adjacent to the computation. There is also increased use of color to emphasize and clarify the solution steps.

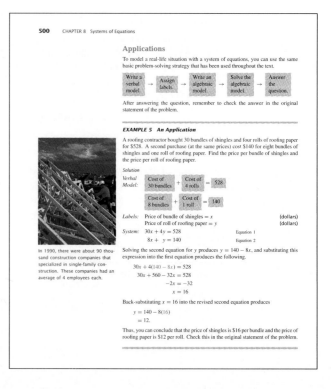

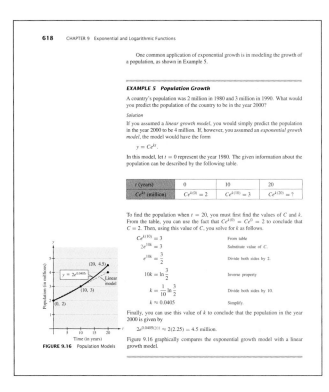

Applications

Real-life applications are integrated throughout the text in examples and exercises. These applications offer students constant review of problem-solving skills and emphasize the relevance of the mathematics. Many of the applications use recent, real data, and all are titled for easy reference. Photographs with captions throughout the text also encourage students to see the link between mathematics and real life.

Examples

Each of the text examples was carefully chosen to illustrate a particular mathematical concept, problem-solving approach, or computational technique, and to enhance students' understanding. The examples in the text cover a wide variety of problem types, including computational, real-life applications (many with real data), and those requiring the use of graphing technology. Each example is titled for easy reference, and real-life applications are labeled. Many examples include side comments in color, which clarify the key steps of the solution process.

Discovery

Throughout the text, Discovery notes encourage active participation by students, often taking advantage of the power of technology (graphing calculators and scientific calculators) to explore mathematical concepts and discover mathematical patterns. Using a variety of approaches, including visualization, verification, pattern recognition, and modeling, students develop an intuitive understanding of algebraic topics.

Definitions and Rules

All of the important rules, formulas, guidelines, properties, definitions, and summaries are highlighted for emphasis. Each is also titled for easy reference.

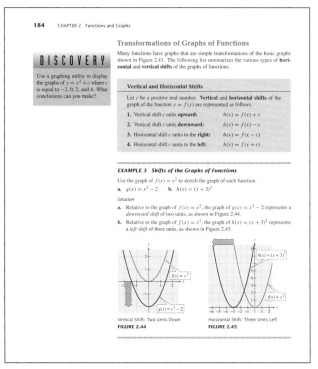

SECTION 2.2 Graphs of Equations **143**

In Exercises 17–24, sketch the graph of the equation.

17. $y = 3x$
18. $y = \frac{1}{2}x$
19. $y = 2x - 3$
20. $y = -x + 2$
21. $y = x^2 - 1$
22. $y = -x^2$
23. $y = |x| - 1$
24. $y = |x - 1|$

In Exercises 25–28, find the x- and y-intercepts (if any) of the graph of the equation.

25. $x + 2y = 10$
26. $3x - 2y + 12 = 0$
27. $y = (x + 5)(x - 5)$
28. $y = (x + 1)^2$

In Exercises 29–38, sketch the graph of the equation and show the coordinates of three solution points.

29. $y = 3 - x$
30. $y = x - 3$
31. $y = 4$
32. $x = -6$
33. $4x + y = 3$
34. $y - 2x = -4$
35. $y = x^2 - 4$
36. $y = 1 - x^2$
37. $y = |x + 2|$
38. $y = |x| + 2$

39. *Using a Graph* The force F (in pounds) to stretch a spring x inches from its natural length is given by
$$F = \frac{4}{3}x, \quad 0 \le x \le 12.$$

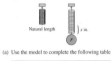

(a) Use the model to complete the following table.

x	0	3	6	9	12
F					

(b) Sketch the graph of the model.

(c) Use the graph to determine how the length of the spring changes each time the force is doubled. Explain your reasoning.

40. *Comparing Data with a Model* The number of farms in the United States with milk cows has been decreasing. The number of farms N (in thousands) for 1984 through 1991 is given in the table.

t	4	5	6	7	8	9	10	11
N	282	269	249	228	217	204	194	182

A model for this data is
$$N = -14.5t + 337.1$$
where t is time in years, with $t = 0$ representing 1980. (Source: U.S. Department of Agriculture)

(a) Sketch the graph of the model and plot the data in the table on the same graph.

(b) How well does the model represent the data? Explain your reasoning.

(c) Use the model to predict the number of farms with milk cows in 1994.

(d) Explain why this model may not be accurate in the future.

41. *Misleading Graphs* Graphs can help you visualize relationships between two variables, but they can also be misused to imply results that are not correct. The two graphs below represent the *same* data points. Which graph is misleading, and why?

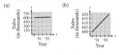

42. *Exploration* Sketch the graphs of $y = x^2 + 1$ and $y = -(x^2 + 1)$ on the same set of coordinate axes. Explain how the graph of an equation changes when the expression for y is multiplied by -1. Justify your answer by giving additional examples.

SECTION 2.4 Slope and Graphs of Linear Equations **167**

In Exercises 95–98, use a graphing utility to graph the three equations on the same viewing rectangle. Describe the relationships among the graphs. (Use the *square* setting so the slopes of the lines appear visually correct.)

95.
$y_1 = 3x$
$y_2 = -3x$
$y_3 = \frac{1}{3}x$

96.
$y_1 = \frac{3}{4}x$
$y_2 = -\frac{4}{3}x$
$y_3 = \frac{4}{3}$

97.
$y_1 = \frac{1}{4}x$
$y_2 = \frac{1}{4}x - 2$
$y_3 = \frac{1}{4}x + 3$

98.
$y_1 = 2x$
$y_2 = 2x - 5$
$y_3 = 2x + \frac{3}{2}$

99. *Graphical Estimation* The graph shows the earnings per share of common stock for Johnson & Johnson for the years 1987 through 1993. Use the slope of each segment to determine the year when earnings (a) decreased most rapidly and (b) increased most rapidly. (Source: *Johnson & Johnson 1993 Annual Report*)

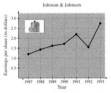

Johnson & Johnson

100. *Road Grade* When driving down a mountain road, you notice warning signs indicating that it is a "12% grade." This means that the slope of the road is $-\frac{12}{100}$. Over a stretch of road, your elevation drops by 2000 feet. What is the horizontal change in your position?

101. *Graphical Estimation* The graph gives the declared dividend per share of common stock for Emerson Electric Company for the years 1987 through 1993. Use the slope of each segment to determine the year when the dividend increased most rapidly. (Source: Emerson Electric Company)

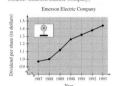

102. *Height of an Attic* The slope, or pitch, of a roof is such that it rises (or falls) 3 feet for every 4 feet of horizontal distance. Determine the maximum height in the attic of the house if the house is 30 feet wide.

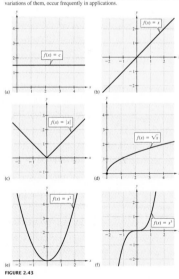

Graphics

Visualization is a critical problem-solving skill. To encourage the development of this ability, the text has numerous figures in examples, exercises, and answers to odd-numbered exercises. Included are graphs of equations and functions, geometric figures, displays of statistical information, scatter plots, and numerous screen outputs from graphing technology. All graphs of equations and functions, computer- or calculator-generated for accuracy, are designed to resemble students' actual screen outputs as closely as possible. Graphics are also used to emphasize graphical interpretation, comparison, and estimation.

SECTION 2.6 Graphs of Functions **183**

Graphs of Basic Functions

To become good at sketching the graphs of functions, it helps to be familiar with the graphs of some basic functions. The functions shown in Figure 2.43, and variations of them, occur frequently in applications.

NOTE Try using a graphing utility to verify the graphs at the right. The names of these functions are as follows.

(a) Constant function
(b) Identity function
(c) Absolute value function
(d) Square root function
(e) Squaring function
(f) Cubing function

(a) $f(x) = c$

(b) $f(x) = x$

(c) $f(x) = |x|$

(d) $f(x) = \sqrt{x}$

(e) $f(x) = x^2$

(f) $f(x) = x^3$

FIGURE 2.43

Group Activities   Communicating Mathematically

Translating a Formula   Use the information provided in the following statement to write a mathematical formula for the 10-second pulse count.

"The target heart rate is the heartbeat rate a person should have during aerobic exercise to get the full benefit of the exercise for cardiovascular conditioning. . . . Using the American College of Sports Medicine Method to calculate one's target heart rate, an individual should subtract his or her age from 220, then multiply by the desired intensity level (as a percent—sedentary persons may want to use 60% and highly fit individuals may want to use 85 to 95%) of the workout. Then divide the answer by 6 for a 10-second pulse count. (The 10-second pulse count is useful for checking whether the target heart rate is being achieved during the workout. One can easily check one's pulse—at the wrist or side of the neck—counting the number of beats in 10 seconds.)" (Source: Aerobic Fitness Association of America)

Use the formula you have found to find your own 10-second pulse count.

Group Activities   Extending the Concept

Using Inequalities   Try the following activity. One person picks a point with whole number coordinates on a grid like the one at left without revealing the coordinates. A second person writes the equation of a line passing through the grid region. The first person graphs the line on the grid and indicates whether the secret point lies above, below, or on the line. Continue writing and graphing lines until the second person is able to guess the coordinates of the secret point. Switch roles and try again. What is the fewest number of turns your team required to guess the point?

Group Activities

The Group Activities that appear at the end of sections reinforce students' understanding by approaching mathematical concepts in a variety of ways: Communicating Mathematically, You Be the Instructor, Extending the Concept, Problem Solving, and Exploring with Technology. Designed to be completed as group projects in class or as homework assignments, the Group Activities give students opportunities for interactive learning and to think, talk, and write about mathematics.

Group Activities   Problem Solving

Fitting a Quadratic Model   The data in the table represents the United States government's annual net receipts y (in billions of dollars) from individual income taxes for the year x from 1990 through 1992, where $x = 0$ corresponds to 1990.   (Source: U.S. Department of the Treasury)

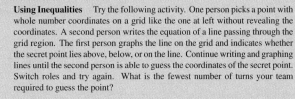

x	0	1	2
y	467	468	476

Use a system of three linear equations to find a quadratic model that fits the data. According to your model, what were the annual net receipts from individual income taxes in 1993? The actual annual net receipts for 1993 were $510 billion. How does the value obtained from your quadratic model compare? Suppose you had been involved in planning the 1993 federal budget and had used this model to estimate how much federal income could be expected from 1993 individual income taxes. When you review the actual 1993 tax receipts and see that the model wasn't completely accurate, how do you evaluate the model's prediction performance? Are you satisfied with it? Why or why not?

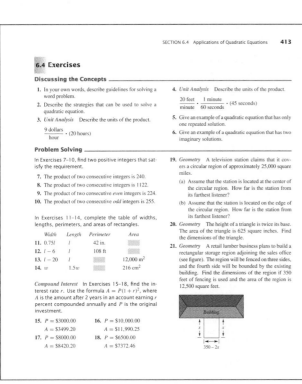

Exercises

In the completely revised and expanded exercise sets of the Second Edition, problems are now grouped into four categories: Discussing the Concepts, Problem Solving, Reviewing the Major Concepts, and Additional Problem Solving. To accommodate a variety of teaching and learning styles, the exercise sets offer numerous computational, conceptual, and applied problems, including multi-part, exploration and discovery, writing, estimation, numeracy, geometry, and challenging exercises, as well as real-life applications, mathematical modeling, graphical comparisons, data interpretation and analysis, fitting a line to data, and exercises that require graphing technology. Applications are labeled for easy reference. Designed to build competence, skill, and understanding, each part of the exercise set is graded in difficulty to allow students to gain confidence as they progress. Detailed solutions to all odd-numbered exercises are given in the Student Solutions Guide, and answers to all odd-numbered exercises appear in the back of the text.

Geometry

Geometric formulas and concepts are reviewed throughout the text. For reference, common formulas are listed inside the back cover of this text.

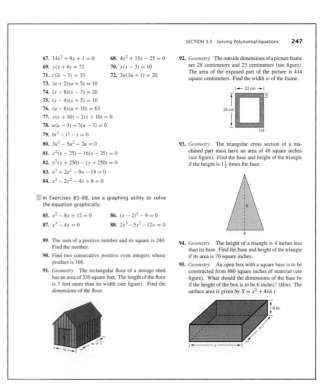

CAREER INTERVIEW

Lisa M. Deitemeyer

Civil Engineer

Johnson-Brittain & Associates, Inc.

Tucson, AZ 85701

Johnson-Brittain does highway design work, primarily for the Arizona Department of Transportation. I am responsible for drainage design of roadways and intersections. It is important that water properly drain off the road surface to avoid flooding problems. One strategy for removing excess water is to use a pipe drainage system that empties into a retention pond. When designing a pipe system and choosing pipe size, I use the equation $V = Q/A$ to find the velocity V of water moving at flow rate Q (volume per unit time) through a given pipe of cross-sectional area A. Finding the water velocity is very important. If it is too fast, erosion can occur in the retention pond. If it is too slow, sedimentation can clog the pipe. As you can see, algebra is very important to my work. I am always solving for different variables that are needed for drainage design.

Career Interviews

Appearing in each chapter, Career Interviews with people who use algebra in their jobs help students understand that algebra is a modern, problem-solving language.

Math Matters

Each chapter contains a Math Matters feature that engages student interest by discussing an historical note or mathematical problem. For those features that pose a question, the answers appear in the back of the text.

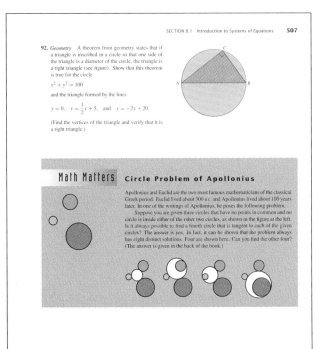

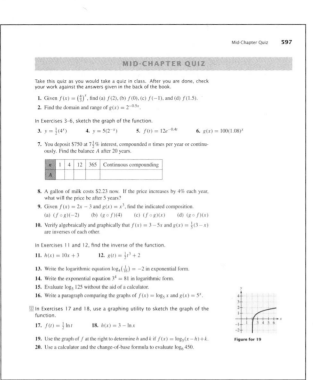

MID-CHAPTER QUIZ

Take this quiz as you would take a quiz in class. After you are done, check your work against the answers given in the back of the book.

1. Given $f(x) = \left(\frac{4}{3}\right)^x$, find (a) $f(2)$, (b) $f(0)$, (c) $f(-1)$, and (d) $f(1.5)$.
2. Find the domain and range of $g(x) = 2^{-0.5x}$.

In Exercises 3–6, sketch the graph of the function.

3. $y = \frac{1}{2}(4^x)$ 4. $y = 5(2^{-x})$ 5. $f(t) = 12e^{-0.4t}$ 6. $g(x) = 100(1.08)^x$

7. You deposit \$750 at $7\frac{1}{2}\%$ interest, compounded n times per year or continuously. Find the balance A after 20 years.

n	1	4	12	365	Continuous compounding
A					

8. A gallon of milk costs \$2.23 now. If the price increases by 4% each year, what will the price be after 5 years?
9. Given $f(x) = 2x - 3$ and $g(x) = x^3$, find the indicated composition.
   (a) $(f \circ g)(-2)$ (b) $(g \circ f)(4)$ (c) $(f \circ g)(x)$ (d) $(g \circ f)(x)$
10. Verify algebraically and graphically that $f(x) = 3 - 5x$ and $g(x) = \frac{1}{5}(3 - x)$ are inverses of each other.

In Exercises 11 and 12, find the inverse of the function.

11. $h(x) = 10x + 3$ 12. $g(t) = \frac{1}{3}t^3 + 2$

13. Write the logarithmic equation $\log_4\left(\frac{1}{16}\right) = -2$ in exponential form.
14. Write the exponential equation $3^4 = 81$ in logarithmic form.
15. Evaluate $\log_5 125$ without the aid of a calculator.
16. Write a paragraph comparing the graphs of $f(x) = \log_5 x$ and $g(x) = 5^x$.

In Exercises 17 and 18, use a graphing utility to sketch the graph of the function.

17. $f(t) = \frac{1}{2}\ln t$ 18. $h(x) = 3 - \ln x$

19. Use the graph of f at the right to determine h and k if $f(x) = \log_5(x - h) + k$.

Figure for 19

20. Use a calculator and the change-of-base formula to evaluate $\log_6 450$.

Mid-Chapter Quizzes

Each chapter contains a Mid-Chapter Quiz with answers in the back of the text. This feature allows the student to perform a self-assessment midway through the chapter.

Chapter Project

Chapter Projects, referenced in the chapter opener, are engaging applications that use real data, graphs, and modeling to enhance students' understanding of mathematical concepts. Designed as individual or group projects, they offer additional opportunities to think, discuss, and write about mathematics. Many projects include research assignments that give students the opportunity to collect, analyze, and interpret their own data. Each Chapter Project is also available in an interactive, multimedia, CD-ROM format.

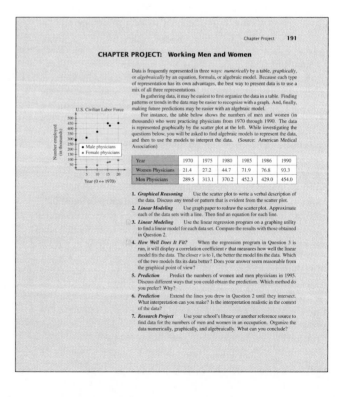

CHAPTER PROJECT: Working Men and Women

Data is frequently represented in three ways: *numerically* by a table, *graphically*, or *algebraically* by an equation, formula, or algebraic model. Because each type of representation has its own advantages, the best way to present data is to use a mix of all three representations.

In gathering data, it may be easiest to first organize the data in a table. Finding patterns or trends in the data may be easier to recognize with a graph. And, finally, making future predictions may be easier with an algebraic model.

For instance, the table below shows the numbers of men and women (in thousands) who were practicing physicians from 1970 through 1990. The data is represented graphically by the scatter plot at the left. While investigating the questions below, you will be asked to find algebraic models to represent the data, and then to use the models to interpret the data. (Source: American Medical Association)

Year	1970	1975	1980	1985	1986	1990
Women Physicians	21.4	27.2	44.7	71.9	76.8	93.3
Men Physicians	289.5	313.1	370.2	452.3	429.0	454.0

1. *Graphical Reasoning*   Use the scatter plot to write a verbal description of the data. Discuss any trend or pattern that is evident from the scatter plot.
2. *Linear Modeling*   Use graph paper to redraw the scatter plot. Approximate each of the data sets with a line. Then find an equation for each line.
3. *Linear Modeling*   Use the linear regression program on a graphing utility to find a linear model for each data set. Compare the results with those obtained in Question 2.
4. *How Well Does It Fit?*   When the regression program in Question 3 is run, it will display a correlation coefficient r that measures how well the linear model fits the data. The closer r is to 1, the better the model fits the data. Which of the two models fits its data better? Does your answer seem reasonable from the graphical point of view?
5. *Prediction*   Predict the numbers of women and men physicians in 1995. Discuss different ways that you could obtain the prediction. Which method do you prefer? Why?
6. *Prediction*   Extend the lines you drew in Question 2 until they intersect. What interpretation can you make? Is the interpretation realistic in the context of the data?
7. *Research Project*   Use your school's library or another reference source to find data for the numbers of men and women in an occupation. Organize the data numerically, graphically, and algebraically. What can you conclude?

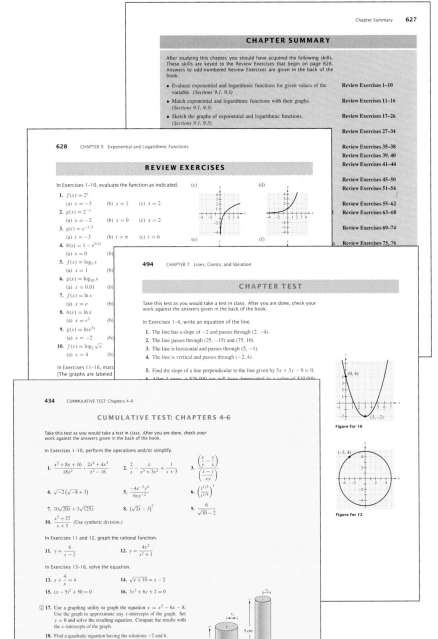

Chapter Summary

The Chapter Summary reviews the skills covered in the chapter. Section references for the major topics make this an effective study tool, and correlation to the review exercises offers guided practice.

Review Exercises

The Review Exercises at the end of each chapter offer the student an opportunity for additional practice. Each set of review exercises includes both computational and applied problems covering a wide range of topics.

Chapter Test

Chapter Tests allow students to assess their own level of success.

Cumulative Tests

The Cumulative Tests that appear after Chapters 3, 6, and 9 help students judge their mastery of previously covered material, as well as reinforce the knowledge students have been accumulating throughout the text—preparing them for other exams and for future courses.

Supplements

Intermediate Algebra, Second Edition, by Larson and Hostetler, is accompanied by a comprehensive supplements package. All items are keyed to the text.

Printed Resources

Student Solutions Guide by Gerry C. Fitch, Louisiana State University
- Detailed, step-by-step solutions to all odd-numbered section exercises (except Discussing the Concept) and review exercises
- Detailed, step-by-step solutions to all Mid-Chapter Quiz, Chapter Test, and Cumulative Test questions

Study Guide by Jay Wiestling, Palomar College
- Section summaries
- Additional examples with solutions
- Starter exercises with answers

Graphing Technology Keystroke Guide: Algebra by Benjamin N. Levy
- Keystroke instructions for Texas Instruments, Sharp, Casio, and Hewlett-Packard graphing calculators
- Examples with step-by-step solutions
- Extensive graphics screen output
- Technology tips

Instructor's Annotated Edition
- Includes the entire student edition of the text, with the student answers section
- Instructor's Answers section: answers to all Discussing the Concepts exercises, all remaining even-numbered exercises, and all Discovery Boxes, Technology Boxes, Group Activities, and Chapter Projects
- Specific teaching strategies and suggestions
- Hints for implementing Group Activities
- Common student error annotations
- Additional examples, exercises, class activities, historical notes, and group activities

Test Item File and Instructor's Resource Guide
- Printed test bank with approximately 4000 test items (multiple-choice, open-ended, and writing) coded by level of difficulty
- Technology-required test items coded for easy reference
- Bank of chapter test forms with answer keys
- Two final exams
- Transparency masters
- Notes to the Instructor, which includes information on standardized tests such as the Texas Academic Skills Program (TASP), Florida College Level

Academic Skills Test (CLAST), and the California State University Entry Level Mathematics (ELM) Examination and provides a list of skills covered by the test and the corresponding section(s) in the text where the topic can be found, as well as notes on contemporary instructional strategies such as alternative assessment and cooperative learning

Media Resources

Tutor (IBM, Macintosh)
- Extensive additional practice

Videotapes by Dana Mosely
- Comprehensive coverage keyed to the text by section
- Detailed explanation of important concepts
- Numerous examples and applications, often illustrated via computer-generated animations
- Discussion of study skills
- For media resource centers; by popular demand, also available for student purchase

D.C. Heath Interactive Math Series CD-ROM Projects
- Real-life applications in an interactive, multimedia CD-ROM format
- IBM PC for Windows; Macintosh
- See page xvi for a description.

Computerized Testing
- Test-generating software for both IBM and Macintosh computers
- Approximately 4000 test items
- Also available as a printed test bank

CD-ROM Projects
for Intermediate Algebra, Second Edition

To accommodate a variety of teaching and learning styles, a series of real-life applications is available in a multimedia, interactive CD-ROM format. Suitable for individual or group assignments, these projects reinforce a variety of mathematical concepts. For each text chapter project is a CD-ROM project, allowing students to explore interactively questions that expand upon the topic and goals of the text project. Students have the opportunity to discover the nature of data sets through exploration, using a combination of graphical, numerical, and algebraic approaches in a guided learning environment. Throughout the text, you will notice a CD-ROM icon that reminds you of the availability of this multimedia software in conjunction with the chapter projects.

These multimedia projects broaden the scope of the text's chapter projects by offering additional opportunities for finding patterns and drawing conclusions, covering related topics and concepts, and providing practice with interpreting graphs, charts, and tables. The multimedia format provides access to extensive real data sets and facilitates hands-on data manipulation for practicing data analysis and modeling techniques. In addition, the projects include animations, color photographs, and audio enhancements.

Each multimedia project is presented in four parts: Introduction, Data, Exploration, and Exercises. The Introduction explains the goals of the project and the background of the project topic. The Data section presents all of the data that may be manipulated in the context of the project in a format that is appropriate to the placement in the text; additional history or pertinent facts may often be found in this section. The Exploration section enables students to manipulate data and discover certain facts about or patterns within the data. For example, the Mass Transportation project allows students to use graphs to find patterns and interactively experiment with placing a line on a scatter plot of actual data to approximate a best fitting line. The Exercises section is a set of questions designed to guide the student to the types of discoveries that may be made from exploration of the data. For example, with the Mass Transportation project students are asked to interpret slopes and y-intercepts, consider predictions, and compare various models.

The CD-ROM Projects for *Intermediate Algebra*, Second Edition, are available for use with multimedia Macintosh or IBM with Windows computers. They cover the following topics:

Chapter P	Playing the Stock Market	**Chapter 6**	Gravitation
Chapter 1	Animal Voices and Hearing	**Chapter 7**	Transportation
Chapter 2	Job Comparisons	**Chapter 8**	Retail Sales of Companies
Chapter 3	Volume of a Box	**Chapter 9**	Half-Life and Radioactivity
Chapter 4	Parachutes and Ratios	**Chapter 10**	Mortgages and Finance

Acknowledgments

We would like to thank the many people who have helped us at various stages of this project to prepare the text and supplements package. Their encouragement, criticisms, and suggestions have been invaluable to us.

Second Edition Advisory Panel: Mary Jean Brod, University of Montana; Kenneth Johnston, Hinds Community College; Beverly Michael, University of Pittsburgh.

Second Edition Reviewers: Cynthia Fleck, Wright State University; Lisa Grenier, Pima College—Downtown; Brenda Lackey, University of Tennessee at Martin; Wanda Long, St. Charles County Community College; Judith Marwick, Prairie State College; Jon Odell, Richland Community College; John Squires, Cleveland State Community College; Pat Stanley, Ball State University.

Second Edition Survey Respondents: Over 160 professors took time to respond to an Algebra Survey. We appreciate your comments.

Career Interviews: Our thanks to Laura Balaoro, Jacquelyn Bick, Dean R. Brookie, Mary Kay Brown, Alfred A. Campos, Melanie Cansler, Lisa M. Deitemeyer, Bernie Khoo, Peggy Murray, Lt. Robert Orr, and Richard S. Schroeder for their help in creating the career interviews. We appreciate their time and effort.

Thanks to all of the people at D. C. Heath and Company who worked with us in the development and production of the text, especially Charles Hartford and Ann Marie Jones, Mathematics Acquisitions Editors; Cathy Cantin, Managing Editor; Emily Keaton, Developmental Editor; Carolyn Johnson, Editorial Associate; Karen Carter and Rachel Wimberly, Production Editors; Henry Rachlin, Designer; Gary Crespo, Art Editor; Lisa Merrill, Production Coordinator; and Billie L. Porter, Photo Researcher.

We would also like to thank the staff at Larson Texts, Inc., who assisted with proofreading the manuscript; preparing and proofreading the art package; and checking and typesetting the supplements.

On a personal level, we are grateful to our wives, Deanna Gilbert Larson and Eloise Hostetler, for their love, patience, and support. Also, a special thanks goes to R. Scott O'Neil.

If you have suggestions for improving the text, please feel free to write to us. Over the past two decades, we have received many useful comments from both instructors and students, and we value these very much.

Roland E. Larson
Robert P. Hostetler

Contents

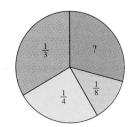
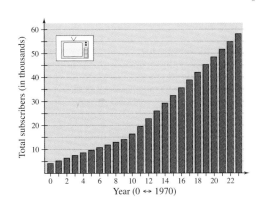

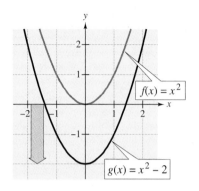

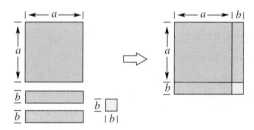

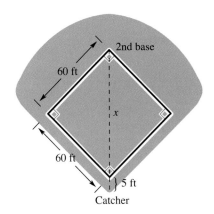

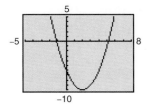

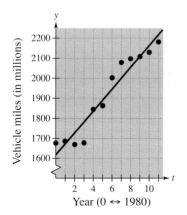

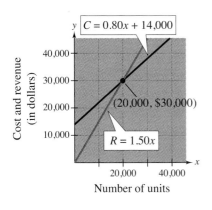

Chapter 8 Systems of Equations 495

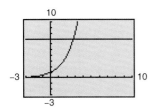

Chapter 9 Exponential and Logarithmic Functions 561

To the Student

- *How to Study Algebra*
- *Reading and Writing About Mathematics*
- *What Is Algebra?*

How to Study Algebra

After years of teaching and guiding students through algebra courses, we have compiled the following list of suggestions for studying algebra. These study tips may take some time and effort—but they work!

Making a Plan   Make your own course plan right now! Determine the number of hours you need to spend on algebra each week. Write your plans on your calendar or some other schedule planner, and then *stick to your plan.*

Preparing for Class   Before attending class, read the portion of the text that is to be covered. This takes a lot of self-discipline, but it pays off. By going to class prepared, you will be able to benefit much more from your instructor's presentation. Algebra, like most other technical subjects, is easier to understand the second or third time you hear it.

Attending Class   Attend every class. Arrive on time with your text, a pen or pencil, paper for notes, and your calendar.

Participating in Class   As you are reading the text before class, write down any questions that you have about the material. Then, ask your instructor during class.

Taking Notes   Take notes in class, especially on definitions, examples, concepts, and rules. Then, as soon after class as possible, read through your notes, adding any explanations that are necessary to make your notes understandable *to you*.

Doing the Homework   Learning algebra is like learning to play the piano or learning to play basketball. You cannot become skilled by just watching someone else do it. You must also do it yourself. A general guideline is to spend two to four hours of study outside of class for each hour in class. When working exercises, your ultimate goal is to be able to solve the problems accurately and quickly. When you start a new exercise set, however, understanding is much more important than speed.

Finding a Study Partner   When you get stuck on a problem, it may help to try to work with someone else. Even if you feel you are giving more help than you are getting, you will find that an excellent way to learn is by teaching others.

Working in a Group   Agree on what you have to do and make a plan. Listen to each other's ideas, and try to build on them. Ask for help when you need it; give help when asked. Finish the project together. Discuss what you did well together and what you could do differently next time.

Building a Math Library   Start building a library of books that can help you with this and future math courses. You might consider using the *Study and Solutions Guide* that accompanies the text. Also, since you will probably be taking other math courses after you finish this course, we suggest that you keep the text. It will be a valuable reference book. Tutorial software and videos available with this text will also be valuable additions to your mathematics library.

Keeping Up with the Work   Don't let yourself fall behind in the course. If you think that you are having trouble, seek help immediately. Ask your instructor, attend your school's tutoring services, talk with your study partner, use additional study aids such as videos or software tutorials—but do something. If you are having trouble with the material in one chapter of your algebra text, there is a good chance that you will also have trouble in later chapters.

Getting Stuck   *Everyone* who has ever taken a math course has had this experience: You are working on a problem and cannot see how to solve it, or you have solved it but your answer does not agree with the answer given in the back of the book. People have different approaches to this sort of problem. You might ask for help, take a break to clear your thoughts, sleep on it, rework the problem, or reread the section in the text. The point is, try not to get frustrated or spend too much time on a single problem.

Assessing Your Progress   In the middle of each chapter is a *Mid-Chapter Quiz*. Take the quiz as you would if you were in class, then check your answers in the back of the text.

Checking Your Work   One of the nice things about algebra is that you don't have to wonder whether your solution is correct. You can tell whether it is correct by checking it in the original statement of the problem. If, in addition to your "solving skills," you work on your "checking skills," you should find your test scores improving.

Preparing for Exams   Cramming for algebra exams seldom works. If you have kept up with the work and followed the suggestions given here, you should be almost ready for the exam. At the end of each chapter, we have included three features that should help as a final preparation. Read the *Chapter Summary*, work the *Review Exercises,* and set aside an hour to take the sample *Chapter Test.*

Taking Exams   Most instructors suggest that you do *not* study right up to the minute you are taking a test. This tends to make people anxious. The best cure for anxiousness during tests is to prepare well before taking the test. Once the test has begun, read the directions carefully, and try to work at a reasonable pace. (You might want to read the entire test first, then work the problems in the order with which you feel most comfortable.) Hurrying tends to cause people to make careless errors. If you finish early, take a few moments to clear your thoughts and then take time to go over your work.

Learning from Mistakes   When you get an exam back, be sure to go over any errors that you might have made. Don't be too quick to pass off an error as just a "dumb mistake." Take advantage of any mistakes by hunting for ways to continually improve your test-taking abilities.

Reading and Writing About Mathematics

Reading a Mathematics Textbook

The following suggestions can help you read your textbook most effectively.

- Before each class, read the portion of the text that will be covered during class. Read the material carefully, and keep a pen or pencil and your calculator nearby and ready to use. Work the examples *before* reading the solution, and try the Discovery activities.

- It may help to take notes as you read, paying particular attention to terms, special symbols, and new ideas. Don't expect to read a mathematics textbook as quickly as you would a novel or magazine. A mathematics textbook takes a little more time and your full attention.

- See how the text is organized. Notice that the major parts of a section are listed at the beginning of each section. These are the key concepts and objectives for that section. Notice that important terms are highlighted in boldface type. Make sure you understand this vocabulary. Study the side comments next to solution steps that show how to proceed from one step to the next. Note the Study Tips in the margin.

- Especially when reading definitions, consider every sentence, word, and symbol carefully. It is helpful to read these more than once—first to get the general idea of the statement and then a second time for details, such as under what conditions the statement is true.

- Ask questions in class based on what you have read.

- After class, reread the text. Take your time. Have a pen or pencil and calculator in hand. Make sure that you now understand any material that was unclear during your first reading.

- Be patient. The ability to read mathematics (or any technical material) is a skill that will be useful to you in this course, in other mathematics courses, and for most jobs.

Writing About Mathematics

Mathematics is a language—a way of communicating ideas using symbols. As a mathematics student, you will use the mathematics language in many different ways. You will show how a problem can be solved in a logical series of steps. You may also write about mathematical ideas and what they mean in the real world. Being able to think about a problem in a logical way is a useful skill. Here are some suggestions to help you write about mathematics.

- When asked to discuss or explain a mathematical finding, begin by making a short list of important points.
- Put the list of points in order, and see if you have left out anything. Imagine that you are explaining your answer to a friend. Would he or she understand it?
- Use complete statements and explanations when writing mathematics for others to read. (This is just like using complete sentences in English.)
- If you have difficulty writing a solution, look at the solutions in the text to see if you can use one as a model.
- You may find it easier to begin the solution to a problem by describing the problem in your own words.
- It may help to give an example or counterexample as part of your explanation.
- List any assumptions and define all variables used in your writing.
- When writing a solution to a problem, try to explain or justify your reasoning for *all* steps.
- Use a graph or a table of data if you think it will make your answer easier to understand.
- Ask yourself if you have covered all the possible outcomes of the situation.
- Check your written answers to make sure that they are organized logically, have detailed steps, and are convincing.
- Be neat.
- If research is involved, list your sources fully and accurately.
- The time you spend learning to write about mathematics is worth it. The ability to write a logical mathematical argument or explanation is a skill that will be useful in your algebra course, in other mathematics courses, and in the workplace.

What Is Algebra?

To some, algebra is manipulating symbols or performing mathematical operations with letters instead of numbers. To others, it is factoring, solving equations, or solving word problems. And to still others, algebra is a mathematical language that can be used to model real-world problems. In fact, algebra is all of these!

As you study this text, it is helpful to view algebra from the "big picture"—to see how the various rules, operations, and strategies fit together.

The rules of arithmetic form the foundation of algebra. These rules are generalized through the use of symbols and letters to form the basic rules of algebra, which are used to *rewrite* algebraic expressions and equations in new, more useful forms. The ability to rewrite algebraic expressions and equations is the common skill involved in the three major components of algebra—*simplifying* algebraic expressions, *solving* algebraic equations, and *graphing* algebraic functions. The following chart shows how this college algebra text fits into the "big picture" of algebra.

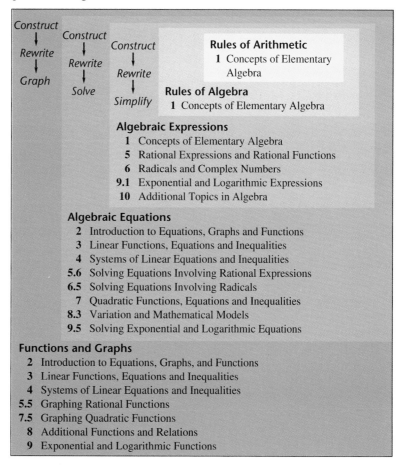

Construct
↓
Rewrite
↓
Graph

Construct
↓
Rewrite
↓
Solve

Construct
↓
Rewrite
↓
Simplify

Rules of Arithmetic
  1 Concepts of Elementary Algebra

Rules of Algebra
  1 Concepts of Elementary Algebra

Algebraic Expressions
  1 Concepts of Elementary Algebra
  5 Rational Expressions and Rational Functions
  6 Radicals and Complex Numbers
  9.1 Exponential and Logarithmic Expressions
  10 Additional Topics in Algebra

Algebraic Equations
  2 Introduction to Equations, Graphs and Functions
  3 Linear Functions, Equations and Inequalities
  4 Systems of Linear Equations and Inequalities
  5.6 Solving Equations Involving Rational Expressions
  6.5 Solving Equations Involving Radicals
  7 Quadratic Functions, Equations and Inequalities
  8.3 Variation and Mathematical Models
  9.5 Solving Exponential and Logarithmic Equations

Functions and Graphs
  2 Introduction to Equations, Graphs, and Functions
  3 Linear Functions, Equations and Inequalities
  4 Systems of Linear Equations and Inequalities
  5.5 Graphing Rational Functions
  7.5 Graphing Quadratic Functions
  8 Additional Functions and Relations
  9 Exponential and Logarithmic Functions

Prerequisites: Fundamental Concepts of Algebra

P

- Real Numbers: Order and Absolute Value
- Operations with Real Numbers
- Properties of Real Numbers
- Algebraic Expressions
- Constructing Algebraic Expressions

The New York Stock Exchange is the largest marketplace in the United States for buying and selling stocks and bonds. Analysts use the trends of the market as one of the predictors of the future economy of the country.

Stock prices listed by the New York Stock Exchange are *records of transactions that have occurred*—they are not necessarily prices at which you can buy stocks. Prices of stocks are negotiable—the actual price can be any amount on which a seller and buyer agree.

The table and double-line graph at the right show the high and low trading prices per share for Micron Technology, Inc. for the week of October 17, 1994. From the graph you can see that the greatest difference between the daily high and low prices occurred on Wednesday. (Source: *Wall Street Journal*)

Day	Monday	Tuesday	Wednesday	Thursday	Friday
High	35	$34\frac{3}{8}$	$34\frac{7}{8}$	36	$35\frac{7}{8}$
Low	$33\frac{1}{2}$	$33\frac{3}{8}$	$32\frac{3}{4}$	$34\frac{1}{2}$	$34\frac{5}{8}$

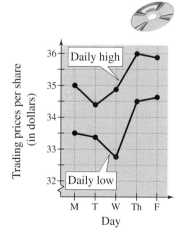

The chapter project related to this information is on page 51.

| **P.1** | **Real Numbers: Order and Absolute Value** |

Sets and Real Numbers ▪ The Real Number Line ▪
Distance on the Real Line ▪ Absolute Value

Sets and Real Numbers

This chapter reviews the basic definitions, operations, and rules that form the fundamental concepts of algebra. The chapter begins with real numbers and their representation on the real number line. Sections P.2 and P.3 review operations and properties of real numbers, and Sections P.4 and P.5 review algebraic expressions.

The formal term that is used in mathematics to talk about a collection of objects is the word **set.** For instance, the set

$$\{1, 2, 3\} \qquad \text{A set with three members}$$

contains the three numbers 1, 2, and 3. Note that a pair of braces { } is used to list the members of the set. Parentheses ( ) and brackets [ ] are used to represent other concepts.

The set of numbers that is used in arithmetic is called the set of **real numbers.** The term *real* distinguishes real numbers from *imaginary* or *complex* numbers—a type of number that you will study later in this text.

If all members of a set *A* are also members of a set *B*, then *A* is a **subset** of *B*. One of the most commonly used subsets of real numbers is the set of **natural numbers** or **positive integers**

$$\{1, 2, 3, 4, \ldots\}. \qquad \text{The set of positive integers}$$

Note that the three dots indicate that the pattern continues. For instance, the set also contains the numbers 5, 6, 7, and so on.

Positive integers can be used to describe many quantities that you encounter in everyday life—for instance, you might be taking four classes this term, or you might be paying 240 dollars a month for rent. But even in everyday life, positive integers cannot describe some concepts accurately. For instance, you could have a zero balance in your checking account, or the temperature could be $-10°$ (10 degrees below zero). To describe such quantities you need to expand the set of positive integers to include **zero** and the **negative integers.** The expanded set is called the set of **integers.**

$$\underbrace{\{\ldots, -3, -2, -1,}_{\text{Negative integers}} \overset{\text{Zero}}{0,} \underbrace{1, 2, 3, \ldots\}}_{\text{Positive integers}} \qquad \text{The set of integers}$$

The set of integers is also a *subset* of the set of real numbers.

STUDY TIP

In this text, whenever a mathematical term is introduced, the word will appear in boldface type. Be sure you understand the meaning of each new word—it is important that each word become part of your mathematical vocabulary.

Technology

You can use a calculator to round decimals. For instance, to round 0.2846 to three decimal places on a scientific calculator, enter

| FIX | | 3 | | .2846 | | = |

or, on a *TI-82* graphing calculator, enter

round (.2846, 3) | ENTER |.

Without using a calculator, round 0.38174 to four decimal places. Verify your answer with a calculator.

Even with the set of integers, there are still many quantities in everyday life that you cannot describe accurately. The costs of many items are not in whole-dollar amounts, but in parts of dollars, such as $1.19 or $39.98. You might work $8\frac{1}{2}$ hours, or you might miss the first *half* of a movie. To describe such quantities, you can expand the set of integers to include **fractions.** The expanded set is called the set of **rational numbers.** Formally, a real number is **rational** if it can be written as the ratio p/q of two integers, where $q \neq 0$ (the symbol $\neq$ means *does not equal*). For instance,

$$2 = \frac{2}{1}, \quad \frac{1}{3} = 0.333\ldots, \quad \frac{1}{8} = 0.125, \quad \text{and} \quad \frac{125}{111} = 1.126126\ldots$$

are rational numbers. A real number that cannot be written as a ratio of two integers is **irrational.** For instance, the numbers

$$\sqrt{2} = 1.4142135\ldots \quad \text{and} \quad \pi = 3.1415926\ldots$$

are irrational. The decimal representation of a rational number is either *terminating* or *repeating.* For instance, the decimal representation of $\frac{1}{4} = 0.25$ is terminating, and the decimal representation of

$$\frac{4}{11} = 0.363636\ldots = 0.\overline{36}$$

is repeating. (The line over "36" indicates which digits repeat.)

The decimal representation of an irrational number neither terminates nor repeats. When you perform calculations using decimal representations of non-terminating decimals, you usually use a decimal approximation that has been **rounded** to a certain number of decimal places. For instance, rounded to four decimal places, the decimal approximations of $\frac{2}{3}$ and π are

$$\frac{2}{3} \approx 0.6667 \quad \text{and} \quad \pi \approx 3.1416.$$

NOTE The rounding rule used in this text is to round up if the succeeding digit is 5 or more and round down if the succeeding digit is 4 or less. For example, to one decimal place, 7.35 would *round up* to 7.4. Similarly, to two decimal places, 2.364 would *round down* to 2.36.

The symbol $\approx$ means **is approximately equal to.** Figure P.1 shows several commonly used subsets of real numbers and their relationships to each other.

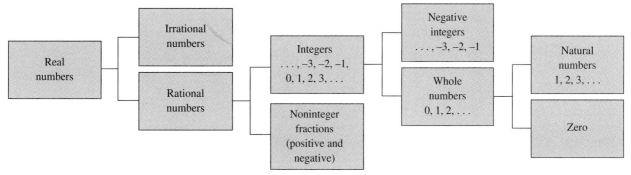

FIGURE P.1

The Real Number Line

The picture that represents the real numbers is called the **real number line.** It consists of a horizontal line with a point (the **origin**) labeled as 0. Numbers to the left of 0 are **negative** and numbers to the right of 0 are **positive,** as shown in Figure P.2.

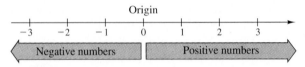

FIGURE P.2   The Real Number Line

The real number zero is neither positive nor negative. Thus, to describe a real number that might be positive *or* zero, you can use the term **nonnegative real number.**

Each point on the real number line corresponds to exactly one real number, and each real number corresponds to exactly one point on the real number line, as shown in Figure P.3. When you draw the point (on the real number line) that corresponds to a real number, you are **plotting** the real number.

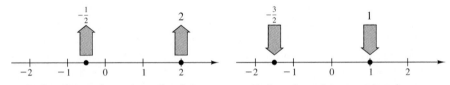

Each point on the real number line corresponds to a real number
FIGURE P.3

Each real number corresponds to a point on the real number line.

EXAMPLE 1   Plotting Points on the Real Number Line

Plot the points that represent the real numbers $-\frac{5}{3}$, 2.3, and $\frac{9}{4}$.

Solution

All three points are shown in Figure P.4.

a.   The point representing the real number $-\frac{5}{3} = -1.666\ldots$ lies between -2 and -1 on the real number line.

b.   The point representing the real number 2.3 lies between 2 and 3 on the real number line.

c.   The point representing the real number $\frac{9}{4} = 2.25$ lies between 2 and 3 on the real number line. Note that the point representing $\frac{9}{4}$ lies slightly to the left of the point representing 2.3.

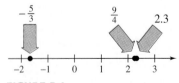

FIGURE P.4

The real number line provides a way of comparing any two real numbers. For instance, if you choose any two (different) numbers on the real number line, one of the numbers must be to the left of the other. You can describe this by saying that the number to the left is **less than** the number to the right, or that the number to the right is **greater than** the number to the left, as shown in Figure P.5.

$$a < b$$

FIGURE P.5   *a* is to the left of *b*.

Order on the Real Number Line

If the real number *a* lies to the left of the real number *b* on the real number line, then *a* is **less than** *b*, which is written as

$$a < b.$$

This relationship can also be described by saying that *b* is **greater than** *a* and writing $b > a$. The symbol $a \le b$ means that *a* is **less than or equal to** *b*, and the symbol $b \ge a$ means that *b* is **greater than or equal to** *a*. The symbols $<, >, \le,$ and $\ge$ are called **inequality symbols.**

When asked to **order** two numbers, you are simply being asked to say which of the two numbers is greater.

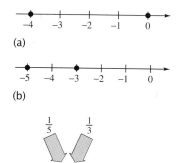

(a)

(b)

(c)

FIGURE P.6

EXAMPLE 2   *Ordering Real Numbers*

Place the correct inequality symbol ($<$ or $>$) between the two numbers.

a. $-4 \quad\rule{1cm}{0.4pt}\quad 0$ **b.** $-3 \quad\rule{1cm}{0.4pt}\quad -5$ **c.** $\dfrac{1}{5} \quad\rule{1cm}{0.4pt}\quad \dfrac{1}{3}$

Solution

a.   Because -4 lies to the left of 0 on the real number line, you can say that -4 is *less than* 0, and write $-4 < 0$, as shown in Figure P.6(a).

b.   Because -3 lies to the right of -5 on the real number line, you can say that -3 is *greater than* -5, and write $-3 > -5$, as shown in Figure P.6(b).

c.   Because $\frac{1}{5}$ lies to the left of $\frac{1}{3}$ on the real number line, you can say that $\frac{1}{5}$ is *less than* $\frac{1}{3}$, and write $\frac{1}{5} < \frac{1}{3}$, as shown in Figure P.6(c).

Distance on the Real Line

Once you know how to represent real numbers as points on the real number line, it is natural to talk about the **distance between two real numbers.** Specifically, if a and b are two real numbers such that $a \leq b$, then the distance between a and b is defined to be $b - a$.

Distance Between Two Real Numbers

If a and b are two real numbers such that $a \leq b$, then the **distance between a and b** is given by

(Distance between a and b) $= b - a$.

Note from this definition that if $a = b$, the distance between a and b is zero. If $a \neq b$, the distance between a and b is positive.

EXAMPLE 3   Finding the Distance Between Two Real Numbers

Find the distance between each pair of real numbers.

a. -2 and 3 **b.** 0 and 4 **c.** -4 and 0

Solution

a. The distance between -2 and 3 is

$$3 - (-2) = 3 + 2 = 5 \qquad \text{Distance between } -2 \text{ and } 3$$

as shown in Figure P.7(a).

b. The distance between 0 and 4 is

$$4 - 0 = 4 \qquad \text{Distance between } 0 \text{ and } 4$$

as shown in Figure P.7(b).

c. The distance between -4 and 0 is

$$0 - (-4) = 0 + 4 = 4 \qquad \text{Distance between } -4 \text{ and } 0$$

as shown in Figure P.7(c).

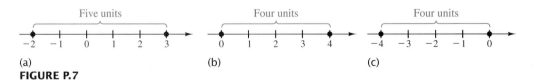

(a)   (b)   (c)

FIGURE P.7

Absolute Value

The distance between a real number a and 0 (the origin) is called the **absolute value** of a. Absolute value is denoted by double vertical bars, $|\ \ |$. For example,

$$|5| = \text{"distance between 5 and 0"} = 5$$

and

$$|-8| = \text{"distance between } -8 \text{ and 0"} = 8.$$

NOTE   Be sure you see from this definition that the absolute value of a real number is never negative. For instance, if $a = -3$, then $|-3| = -(-3) = 3$. Moreover, the only real number whose absolute value is zero is 0. That is, $|0| = 0$.

Absolute Value of a Real Number

The **absolute value** of a real number a is the distance between a and 0 on the real number line.

1. If $a > 0$, then $|a| = a - 0 = a$.

2. If $a = 0$, then $|a| = 0 - 0 = 0$.

3. If $a < 0$, then $|a| = 0 - a = -a$.

Two real numbers are called **opposites** of each other if they lie the same distance from, but on opposite sides of, 0. For instance, -2 is the opposite of 2. Because *opposite* numbers lie the same distance from 0 on the real number line, they have the same absolute value. Thus, $|5| = 5$ and $|-5| = 5$. The use of a negative sign to denote the *opposite* of a number gives meaning to expressions such as $-(-3)$ and $-|-3|$, as follows.

$$-(-3) = (\text{opposite of } -3) = 3$$
$$-|-3| = (\text{opposite of } |-3|) = -3$$

EXAMPLE 4   Finding Absolute Values

a. $|-10| = 10$ The absolute value of -10 is 10.

b. $\left|\dfrac{3}{4}\right| = \dfrac{3}{4}$ The absolute value of $\frac{3}{4}$ is $\frac{3}{4}$.

c. $|-3.2| = 3.2$ The absolute value of -3.2 is 3.2.

d. $-|-6| = -(6) = -6$ The opposite of $|-6|$ is -6.

Note that part (d) does not contradict the fact that the absolute value of a number cannot be negative. The expression $-|-6|$ calls for the *opposite* of an absolute value; hence, it must be negative.

For any two real numbers a and b, exactly one of the following must be true:

$$a < b, \quad a = b, \quad \text{or} \quad a > b.$$

This property of real numbers is called the **Law of Trichotomy.** In words, this property tells you that if a and b are any two real numbers, then a is less than b, a is equal to b, or a is greater than b.

EXAMPLE 5   *Comparing Real Numbers*

Place the correct symbol ($<$, $>$, or $=$) between each pair of real numbers.

a. $|-2|$ ▢ 1 **b.** -4 ▢ $-|-4|$ **c.** $|12|$ ▢ $|-15|$

Solution

a. $|-2| > 1$, because $|-2| = 2$ and 2 is greater than 1.

b. $-4 = -|-4|$, because both numbers are equal to -4.

c. $|12| < |-15|$, because $|12| = 12$ and $|-15| = 15$, and 12 is less than 15.

When the distance between the two real numbers a and b was defined to be $b - a$, we specified that a was less than or equal to b. Using absolute value, you can generalize this definition. That is, if a and b are *any* two real numbers, then the distance between a and b is given by

(Distance between a and b) $= |a - b|$.

For instance, the distance between -2 and 1 is given by

$$|-2 - 1| = |-3| = 3. \qquad \text{\small Distance between } -2 \text{ and } 1$$

Group Activities   E x t e n d i n g   t h e   C o n c e p t

Comparing Real Numbers   For each of the following lists of real numbers, arrange the entries in increasing order. Compare your lists with those of others in your group and resolve any differences.

a. $3, -\sqrt{7}, \sqrt{2}, \pi, -3, 1, \frac{22}{7}, -\frac{15}{7}$

b. $\left|-\frac{3}{5}\right|, -\left|\frac{3}{5}\right|, \left|-\frac{5}{3}\right|, -\left|\frac{5}{3}\right|, \left|-\frac{3}{4}\right|, -\left|\frac{3}{4}\right|, \left|-\frac{4}{5}\right|, -\left|\frac{4}{5}\right|$

P.1 Exercises

Discussing the Concepts

1. Describe the difference between the set of natural numbers and the set of integers.

2. Describe the difference between the rational numbers 0.15 and $0.\overline{15}$.

3. Is there a difference between saying that a real number is positive and saying that a real number is nonnegative? Explain your answer.

4. Which lies farther from -4: -8 or 6? Explain your reasoning.

5. If you are given two real numbers a and b, how can you tell which is greater?

6. *True or False?* Because $|a|$ is a nonnegative real number for any real number a, there exists no real number a such that $|a| = -a$. Explain.

Problem Solving

7. Which of the real numbers in the set are (a) natural numbers, (b) integers, (c) rational numbers, and (d) irrational numbers?

$$\left\{-10,\ -\sqrt{5},\ -\tfrac{2}{3},\ -\tfrac{1}{4},\ 0,\ \tfrac{5}{8},\ 1,\ \sqrt{3},\ 4,\ 2\pi,\ 6\right\}$$

8. Which of the real numbers in the set are (a) natural numbers, (b) integers, (c) rational numbers, and (d) irrational numbers?

$$\left\{-\tfrac{7}{2},\ -\sqrt{6},\ -\pi,\ -\tfrac{3}{8},\ 0,\ \sqrt{15},\ \tfrac{10}{3},\ 8,\ 245\right\}$$

In Exercises 9–12, list all members of the set.

9. The integers between -5.8 and 3.2

10. The even integers between -2.1 and 10.5

11. The odd integers between 0 and 3π

12. All prime numbers between 0 and 25

In Exercises 13 and 14, locate the real numbers on the real number line.

13. (a) 3 (b) $\tfrac{5}{2}$ (c) $-\tfrac{7}{2}$ (d) -5.2

14. (a) 8 (b) $\tfrac{4}{3}$ (c) -6.75 (d) $-\tfrac{9}{2}$

In Exercises 15–20, plot the numbers and place the correct inequality symbol ($<$ or $>$) between them.

15. 2 ___ 5

16. 8 ___ 3

17. -7 ___ -2

18. -2 ___ -5

19. $-\tfrac{2}{3}$ ___ $-\tfrac{10}{3}$

20. $\tfrac{11}{4}$ ___ π

In Exercises 21–24, approximate the numbers and order them.

21.

22.

23.

24.
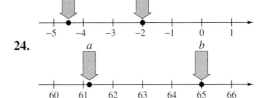

In Exercises 25–30, find the distance between the numbers.

25. 4 and 10

26. 75 and 20

27. 18 and -32

28. -54 and 32

29. -35 and 0

30. 0 and 35

In Exercises 31–34, evaluate the expression.

31. $-|3.5|$

32. $|-6|$

33. $-|-25|$

34. $-\left|\tfrac{3}{4}\right|$

In Exercises 35–38, place the correct symbol (<, >, or =) between the two real numbers.

35. $|-6|$ ▒ $|2|$ **36.** $|150|$ ▒ $|-310|$
37. $-|-16.8|$ ▒ $-|16.8|$
38. $|-\frac{3}{4}|$ ▒ $-|\frac{4}{5}|$

In Exercises 39–42, find the opposite and absolute value of the number.

39. 14 **40.** -22.5
41. $-\frac{5}{4}$ **42.** π

In Exercises 43–46, plot the number and its opposite on the real number line.

43. -3 **44.** 3.5
45. $\frac{5}{3}$ **46.** $-\frac{3}{4}$

In Exercises 47–50, write the statement using inequality notation.

47. x is negative. **48.** y is more than 25.
49. The price p is less than $225.
50. The Dow Jones Average A will exceed 4000.

Additional Problem Solving

In Exercises 51–58, plot the numbers and place the correct inequality symbol (< or >) between them.

51. -5 ▒ 2 **52.** -8 ▒ 3
53. $\frac{1}{3}$ ▒ $\frac{1}{4}$ **54.** $\frac{4}{5}$ ▒ 1
55. $-\frac{5}{8}$ ▒ $\frac{1}{2}$ **56.** $-\frac{3}{2}$ ▒ $-\frac{5}{2}$
57. $-\frac{5}{3}$ ▒ $-\frac{3}{2}$ **58.** $-\pi$ ▒ -3

In Exercises 59–64, find the distance between the numbers.

59. -12 and 7 **60.** 14 and -6
61. -8 and 0 **62.** 0 and 125
63. -6 and -45 **64.** -300 and -110

In Exercises 65–72, evaluate the expression.

65. $|18.6|$ **66.** $|-14|$
67. $-|16|$ **68.** $-|\frac{3}{8}|$
69. $-|-85|$ **70.** $-|-36.5|$
71. $|-\pi|$ **72.** $|0|$

In Exercises 73–76, place the correct symbol (<, >, or =) between the numbers.

73. $|-2|$ ▒ $|2|$ **74.** $|47|$ ▒ $|-27|$
75. $-|12.5|$ ▒ $|-25|$
76. $-|-\frac{7}{3}|$ ▒ $-|\frac{1}{3}|$

In Exercises 77–82, find the opposite and absolute value of the number.

77. 34 **78.** 225
79. -160 **80.** -52
81. $-\frac{3}{11}$ **82.** $\frac{5}{32}$

In Exercises 83–86, plot the number and its opposite.

83. 7 **84.** $\frac{7}{4}$
85. $-\frac{3}{5}$ **86.** -4.25

In Exercises 87–90, write the statement using inequality notation.

87. x is nonnegative. **88.** u is at least 16.
89. z is greater than 2 and no more than 10.
90. The tire pressure p is at least 30 pounds per square inch and no more than 35 pounds per square inch.

True or False? In Exercises 91–94, decide whether the statement is true or false. If the statement is false, give an example of a real number that makes the statement false.

91. Every integer is a rational number.
92. Every real number is either rational or irrational.
93. Every rational number is an integer.
94. The absolute value of every real number is positive.

| **P.2** | **Operations with Real Numbers** |

Operations with Real Numbers ▪ Positive Integer Exponents ▪
Order of Operations ▪ Calculators and Order of Operations

Operations with Real Numbers

This section reviews the four basic operations of arithmetic: addition, subtraction, multiplication, and division.

Addition of Two Real Numbers

1. To add two real numbers with *like signs,* add their absolute values and attach the common sign to the result.

2. To add two real numbers with *unlike signs*, subtract the smaller absolute value from the greater absolute value and attach the sign of the number with the greater absolute value.

The result of adding two real numbers is the **sum** of the two numbers, and the two real numbers are the **terms** of the sum.

EXAMPLE 1 Adding Integers and Decimals

a. $-84 + 14 = -(84 - 14) = -70$
b. $-138 + (-62) = -(138 + 62) = -200$
c. $3.2 + (-0.4) = +(3.2 - 0.4) = 2.8$
d. $58.06 + 24.7 = 82.76$

Subtraction of one real number from another can be described as *adding the opposite* of the second number to the first number.

Subtraction of Two Real Numbers

To **subtract** the real number b from the real number a, add the opposite of b to a. That is,

$$a - b = a + (-b).$$

The result of this subtraction is the **difference** of a and b.

EXAMPLE 2   Subtracting Integers and Decimals

a. $6 - 14 = 6 + (-14) = -(14 - 6) = -8$

b. $-25 - (-27) = -25 + 27 = 2$

c. $-13.8 - 7.02 = -13.8 + (-7.02) = -(13.8 + 7.02) = -20.82$

Addition and Subtraction of Fractions

1. *Like Denominators:* The sum and difference of two fractions with like denominators are as follows.

$$\frac{a}{c} + \frac{b}{c} = \frac{a+b}{c} \qquad \text{Sum}$$

$$\frac{a}{c} - \frac{b}{c} = \frac{a-b}{c} \qquad \text{Difference}$$

2. *Unlike Denominators:* To add or subtract two fractions with unlike denominators, first rewrite the fractions so that they have the same denominator and apply the first rule, *or* apply the following rules.

$$\frac{a}{c} + \frac{b}{d} = \frac{ad+bc}{cd} \qquad \text{Sum}$$

$$\frac{a}{c} - \frac{b}{d} = \frac{ad-bc}{cd} \qquad \text{Difference}$$

NOTE The **least common denominator** of two or more fractions is the least common multiple of their denominators. For instance, the least common denominator of $\frac{1}{24}$ and $\frac{5}{18}$ is 72.

EXAMPLE 3   Adding and Subtracting Fractions

a. $\dfrac{5}{17} + \dfrac{9}{17} = \dfrac{5+9}{17}$ Add numerators.

$\qquad\qquad = \dfrac{14}{17}$ Simplify.

b. $\dfrac{3}{8} - \dfrac{5}{12} = \dfrac{3(3)}{8(3)} - \dfrac{5(2)}{12(2)}$ Least common denominator is 24.

$\qquad\qquad = \dfrac{9}{24} - \dfrac{10}{24}$ Simplify.

$\qquad\qquad = \dfrac{9-10}{24}$ Subtract numerators.

$\qquad\qquad = -\dfrac{1}{24}$ Simplify.

NOTE In Example 3(b), the fractions are subtracted by the least common denominator method. Try reworking the problem using the rule

$$\frac{a}{c} - \frac{b}{d} = \frac{ad-bc}{cd}.$$

EXAMPLE 4   Adding Mixed Numbers

$$1\frac{4}{5} + \frac{11}{7} = \frac{9}{5} + \frac{11}{7}$$ Write $1\frac{4}{5}$ as $\frac{9}{5}$.

$$= \frac{9(7)}{5(7)} + \frac{11(5)}{7(5)}$$ Least common denominator is 35.

$$= \frac{63}{35} + \frac{55}{35}$$ Simplify.

$$= \frac{63 + 55}{35}$$ Add numerators.

$$= \frac{118}{35}$$ Simplify.

Multiplication of two real numbers can be described as repeated addition. For instance, 7×3 can be described as $3 + 3 + 3 + 3 + 3 + 3 + 3$. Multiplication is denoted in a variety of ways. For instance,

$$7 \times 3, \quad 7 \cdot 3, \quad 7(3), \quad \text{and} \quad (7)(3)$$

all denote the product "7 times 3."

Multiplication of Two Real Numbers

The result of multiplying two real numbers is their **product,** and each of the two numbers is a **factor** of the product.

1. To multiply two real numbers with *like signs,* find the product of their absolute values. The product is *positive*.

2. To multiply two real numbers with *unlike signs,* find the product of their absolute values, and attach a minus sign. The product is *negative*.

The product of zero and any other number is zero. For instance, the product of 0 and 4 is 0.

STUDY TIP

To find the product of more than two numbers, first find the product of their absolute values. If there is an *even* number of negative factors, as in Example 5(c), the product is positive. If there is an *odd* number of negative factors, the product is negative.

EXAMPLE 5   Multiplying Integers

a. $-6 \cdot 9 = -54$ The product is negative.

b. $(-5)(-7) = 35$ The product is positive.

c. $5(-3)(-4)(7) = 420$ The product is positive.

Multiplication of Two Fractions

The product of the two fractions a/c and b/d is given by

$$\frac{a}{c} \cdot \frac{b}{d} = \frac{ab}{cd}.$$

STUDY TIP

When operating with fractions, you should check to see whether your answers can be reduced by dividing out factors that are common to the numerator and denominator. For instance, the fraction $\frac{4}{6}$ can be written in reduced form as

$$\frac{4}{6} = \frac{2 \cdot 2}{2 \cdot 3} = \frac{2}{3}.$$

EXAMPLE 6   Multiplying Fractions

$$\left(-\frac{3}{8}\right)\left(\frac{11}{6}\right) = -\frac{3(11)}{8(6)}$$ Multiply numerators and denominators.

$$= -\frac{3(11)}{8(2)(3)}$$ Factor and reduce.

$$= -\frac{11}{16}$$ Simplify.

Division of Two Real Numbers

To divide the real number a by the nonzero real number b, multiply a by the **reciprocal** of b. That is,

$$a \div b = a \cdot \frac{1}{b}.$$

The result of dividing two real numbers is the **quotient** of the numbers. The number a is the **dividend** and the number b is the **divisor.** Using the symbols a/b or $\frac{a}{b}$, a is the **numerator** and b is the **denominator.**

NOTE   In the definition at the right, be sure you see that *division by zero* is not defined.

EXAMPLE 7   Division of Real Numbers

a. $-30 \div 5 = -30 \cdot \frac{1}{5} = -\frac{30}{5} = -\frac{6 \cdot 5}{5} = -6$

b. $-\frac{9}{14} \div -\frac{1}{3} = -\frac{9}{14}\left(-\frac{3}{1}\right) = \frac{27}{14}$

c. $\frac{5}{16} \div 2\frac{3}{4} = \frac{5}{16} \div \frac{11}{4} = \frac{5}{16} \cdot \frac{4}{11} = \frac{5(4)}{16(11)} = \frac{5}{44}$

d. $\frac{-2/3}{3/5} = -\frac{2}{3} \div \frac{3}{5} = -\frac{2}{3} \cdot \frac{5}{3} = -\frac{10}{9}$

Positive Integer Exponents

Just as multiplication by a positive integer can be described as repeated addition, *repeated multiplication* can be written in what is called **exponential form.** Here is an example.

Repeated Multiplication   *Exponential Form*

$$\underbrace{7 \cdot 7 \cdot 7 \cdot 7}_{\text{4 factors of 7}} = 7^4$$

Exponential Notation

Let n be a positive integer and let a be a real number. Then the product of n factors of a is given by

$$a^n = \underbrace{a \cdot a \cdot a \cdots a}_{n \text{ factors}}.$$

In the exponential form a^n, a is the **base** and n is the **exponent.** Writing the exponential form a^n is called **"raising a to the nth power."**

When a number is raised to the *first* power, you usually do not write the exponent 1. For instance, you usually write 5 rather than 5^1. Raising a number to the *second* power is called **squaring** the number. Raising a number to the *third* power is called **cubing** the number.

EXAMPLE 8 Evaluating Exponential Expressions

a. $(-3)^4 = (-3)(-3)(-3)(-3) = 81$ Negative sign is part of the base.

b. $-3^4 = -(3)(3)(3)(3) = -81$ Negative sign is not part of the base.

c. $\left(\frac{2}{5}\right)^3 = \left(\frac{2}{5}\right)\left(\frac{2}{5}\right)\left(\frac{2}{5}\right) = \frac{8}{125}$

d. $(-2)^5 = (-2)(-2)(-2)(-2)(-2) = -32$

e. $(-2)^6 = (-2)(-2)(-2)(-2)(-2)(-2) = 64$

NOTE In parts (a) and (b) of Example 8, be sure you see the distinction between the expressions $(-3)^4$ and -3^4. Here is a similar example.

$$(-5)^2 = (-5)(-5) = 25$$ Negative sign is part of the base.

$$-5^2 = -(5)(5) = -25$$ Negative sign is not part of the base.

Order of Operations

One of your goals as you study this book is to learn to communicate about algebra by reading and writing information about numbers. One way to help avoid confusion when you are communicating algebraic ideas is to establish an **order of operations.** This is done by giving priorities to different operations. First priority is given to exponents, second priority is given to multiplication and division, and third priority is given to addition and subtraction. To distinguish between operations with the same priority, use the *Left-to-Right Rule.*

NOTE The order of operation for multiplication applies when multiplication is written with the symbols $\times$ or $\cdot$. When multiplication is implied by parentheses, it has a higher priority than the Left-to-Right Rule. For instance,

$$8 \div 4(2) = 8 \div 8 = 1$$

but

$$8 \div 4 \cdot 2 = 2 \cdot 2 = 4.$$

Order of Operations

To evaluate an expression involving more than one operation, use the following order.

1. First do operations that occur within symbols of grouping.

2. Then evaluate powers.

3. Then do multiplications and divisions from left to right.

4. Finally, do additions and subtractions from left to right.

EXAMPLE 9   Using Order of Operations

a. $20 - 2 \cdot 3^2 = 20 - 2 \cdot 9 = 20 - 18 = 2$

b. $5 - 6 - 2 = (5 - 6) - 2 = -1 - 2 = -3$

c. $-4^2 = -(4^2) = -16$

d. $8 \div 2 \div 2 = (8 \div 2) \div 2 = 4 \div 2 = 2$

When you want to change the established order of operations, you must use parentheses or other grouping symbols.

EXAMPLE 10   Order of Operations

a. $7 - 3(4 - 2) = 7 - 3(2) = 7 - 6 = 1$

b. $(4 - 5) - (3 - 6) = (-1) - (-3) = -1 + 3 = 2$

c. $4 - 3(2)^3 = 4 - 3(8) = 4 - 24 = -20$

d. $1 - [4 - (5 - 3)] = 1 - (4 - 2) = 1 - 2 = -1$

Calculators and Order of Operations

This text includes several examples and exercises that use a calculator. As each new calculator application occurs, we will give general instructions for using a calculator. These instructions, however, may not agree precisely with the steps required by *your* calculator, so be sure you are familiar with the use of the keys on your own calculator.

For each of the calculator examples in the text, we will give two possible keystroke sequences: one for a standard *scientific* calculator, and one for a *graphing* calculator.

Next to fingers, the abacus is the oldest calculating device known. It can be used to add, subtract, multiply, divide, and calculate square and cube roots. Different forms of the abacus were used by Egyptians, Greeks, Romans, Hindus, and Chinese.

EXAMPLE 11   *Evaluating Expressions on a Calculator*

a.   To evaluate the expression $7 - (5 \cdot 3)$, use the following keystrokes.

Keystrokes	Display	
7 − (5 × 3) =	−8	Scientific
7 − (5 × 3) ENTER	−8	Graphing

b.   To evaluate the expression $24 \div 2^3$, use the following keystrokes.

Keystrokes	Display	
24 ÷ 2 y^x 3 =	3	Scientific
24 ÷ 2 ∧ 3 ENTER	3	Graphing

c.   To evaluate the expression $(24 \div 2)^3$, use the following keystrokes.

Keystrokes	Display	
(24 ÷ 2) y^x 3 =	1728	Scientific
(24 ÷ 2) ∧ 3 ENTER	1728	Graphing

d.   To evaluate the expression $5/(4 + 3 \cdot 2)$, use the following keystrokes.

Keystrokes	Display	
5 ÷ (4 + 3 × 2) =	0.5	Scientific
5 ÷ (4 + 3 × 2) ENTER	0.5	Graphing

Some calculators follow the established order of operations and some don't. To see whether yours does, try entering

$$30 - 5 \times 4.$$

If your calculator follows the established order of operations, it will display 10. If it doesn't, it will display 100.

EXAMPLE 12   Evaluating Expressions on a Calculator

a.   To evaluate the expression $-4 - 5$, use the following keystrokes.

Keystrokes	Display	
4 [+/-] [−] 5 [=]	-9	Scientific
[(-)] 4 [−] 5 [ENTER]	-9	Graphing

NOTE   Be sure you see the difference between the change sign key [+/-] and the subtraction key [−] on a scientific calculator. Also notice the difference between the minus key [(-)] and the subtraction key [−] on a graphing calculator.

b.   To evaluate the expression $-3^2 + 4$, use the following keystrokes.

Keystrokes	Display	
3 [x^2] [+/-] [+] 4 [=]	-5	Scientific
[(-)] 3 [x^2] [+] 4 [ENTER]	-5	Graphing

c.   To evaluate the expression $(-3)^2 + 4$, use the following keystrokes.

Keystrokes	Display	
3 [+/-] [x^2] [+] 4 [=]	13	Scientific
[(] [(-)] 3 [)] [x^2] [+] 4 [ENTER]	13	Graphing

Group Activities   Communicating Mathematically

Evaluating Expressions   Decide which of the following expressions are equal to 27 when you follow the standard order of operations.

a. $40 - 10 + 3$ **b.** $5^2 + \frac{1}{2} \cdot 4$

c. $8 \cdot 3 + 30 \div 2$ **d.** $75 \div 2 + 1 + 2$

e. $9 \cdot 4 - 18 \div 2$ **f.** $7 \cdot 4 - 4 - 5$

For the expressions that are not equal to 27, see if you can discover a way to insert grouping symbols (parentheses, brackets, and absolute value symbols) that makes the expression equal to 27. Discuss the value of grouping symbols to mathematical communication.

P.2 Exercises

Discussing the Concepts

1. Can the sum of two real numbers be less than either number? If so, give an example.

2. Explain how to subtract one real number from another.

3. Is the following statement true? Explain.

$$\frac{2}{3} + \frac{3}{2} = \frac{2+3}{3+2} = 1$$

4. In your own words, describe the rules for determining the sign of the product or quotient of two real numbers.

5. Is it true that $3 \cdot 4^2 = 12^2$? Explain your reasoning.

6. In your own words, describe the established order of operations. Without these priorities, explain why the expression $6 - 5 - 2$ would be ambiguous.

Problem Solving

In Exercises 7–20, evaluate the expression.

7. $13 + 32$

8. $16 + 84$

9. $-13 + 32$

10. $16 + (-84)$

11. $-7 - 15$

12. $-5 + (-52)$

13. $\dfrac{3}{4} - \dfrac{1}{4}$

14. $\dfrac{5}{6} + \dfrac{7}{6}$

15. $\dfrac{5}{8} + \dfrac{1}{4} - \dfrac{5}{6}$

16. $\dfrac{3}{11} + \dfrac{-5}{2}$

17. $5\frac{3}{4} + 7\frac{3}{8}$

18. $8\frac{1}{2} - 24\frac{2}{3}$

19. $5.8 - 6.2 + 1.1 - 4.7 - 9.2$

20. $46.08 - 35.1 - 16.25 + 14.78$

In Exercises 21 and 22, write the expression as a repeated addition problem.

21. $4(5)$

22. $6(-2)$

In Exercises 23 and 24, write the expression as a multiplication problem.

23. $(-15) + (-15) + (-15) + (-15)$

24. $9 + 9 + 9 + 9$

In Exercises 25–28, evaluate the product without using a calculator.

25. $5(-6)$

26. $-7(3)$

27. $\left(-\frac{5}{8}\right)\left(-\frac{4}{5}\right)$

28. $\frac{2}{3}\left(-\frac{18}{5}\right)\left(-\frac{5}{6}\right)$

In Exercises 29–34, evaluate the expression using a calculator. Round the result to two decimal places.

29. $\dfrac{-18}{-3}$

30. $-\dfrac{30}{-15}$

31. $-\dfrac{4}{5} \div \dfrac{8}{25}$

32. $\dfrac{-11/12}{5/24}$

33. $5\frac{3}{4} \div 2\frac{1}{8}$

34. $-3\frac{5}{6} \div -2\frac{2}{3}$

In Exercises 35 and 36, write the expression as a repeated multiplication problem.

35. $(-3)^4$

36. $\left(\frac{2}{3}\right)^3$

In Exercises 37 and 38, write the expression using exponential notation.

37. $(-5)(-5)(-5)(-5)$

38. $-(5 \times 5 \times 5 \times 5 \times 5 \times 5)$

In Exercises 39–42, evaluate the expression.

39. $(-4)^2$

40. $(-3)^6$

41. -4^2

42. -3^6

In Exercises 43–46, evaluate the expression.

43. $16 - 5(6 - 10)$

44. $72 - 8(6^2 \div 9)$

45. $\dfrac{3^2 - 5}{12} - 3\frac{1}{6}$

46. $\dfrac{3}{2} - \left(\dfrac{1}{3} \div \dfrac{5}{6}\right)$

In Exercises 47–50, evaluate the expression using a calculator. Round the result to two decimal places.

47. $\dfrac{25.5}{6.325}$

48. $300(1.09)^{10}$

49. $\dfrac{(1.5)^{15}}{3}$

50. $\dfrac{3.94}{58.35}$

51. *Savings Plan*   You save $50 a month for 18 years. How much do you set aside during the 18 years?

52. *Compound Interest*   If the money in Exercise 51 is deposited in a savings account earning 9% interest compounded monthly, the total amount in the fund after 18 years will be

$$50\left[\left(1+\dfrac{0.09}{12}\right)^{216}-1\right]\left(1+\dfrac{12}{0.09}\right).$$

Use a calculator to determine this amount.

Additional Problem Solving

In Exercises 53–68, evaluate the expression.

53. $13+(-32)$

54. $-16+84$

55. $-13+(-8)$

56. $-22-6$

57. $4-16+(-8)$

58. $-15+(-6)+32$

59. $\dfrac{3}{8}+\dfrac{7}{8}$

60. $\dfrac{5}{9}-\dfrac{1}{9}$

61. $\dfrac{3}{5}+\left(-\dfrac{1}{2}\right)$

62. $\dfrac{5}{6}-\dfrac{3}{4}$

63. $85-|-25|$

64. $-36+|-8|$

65. $|-16.25|-54.78$

66. $-(-11.35)+|34.65|$

67. $-\left|-15\tfrac{2}{3}\right|-12\tfrac{1}{3}$

68. $-|-15.667|-12.333$

In Exercises 69 and 70, write the expression as a repeated addition problem.

69. $3(-4)$

70. $5(2)$

In Exercises 71 and 72, write the expression as a multiplication problem.

71. $\tfrac{1}{4}+\tfrac{1}{4}+\tfrac{1}{4}+\tfrac{1}{4}+\tfrac{1}{4}+\tfrac{1}{4}$

72. $\left(-\tfrac{5}{22}\right)+\left(-\tfrac{5}{22}\right)+\left(-\tfrac{5}{22}\right)$

In Exercises 73–78, evaluate the product without using a calculator.

73. $(-8)(-6)$

74. $4(-2.5)$

75. $6.3(5.1)$

76. $(-4.4)(-3.2)$

77. $\left(\dfrac{10}{13}\right)\left(-\dfrac{3}{5}\right)$

78. $-\dfrac{9}{8}\left(\dfrac{16}{27}\right)\left(\dfrac{1}{2}\right)$

In Exercises 79–84, evaluate the expression using a calculator. Round the result to two decimal places.

79. $\dfrac{-48}{16}$

80. $-\dfrac{27}{-9}$

81. $-\dfrac{-1/3}{5/6}$

82. $\dfrac{8}{15}\div\dfrac{32}{5}$

83. $4\tfrac{1}{8}\div3\tfrac{3}{2}$

84. $26\tfrac{2}{3}\div10\tfrac{5}{6}$

In Exercises 85–88, write the expression as a repeated multiplication problem.

85. 4^{3}

86. $\left(\tfrac{2}{3}\right)^{4}$

87. $\left(-\tfrac{4}{5}\right)^{6}$

88. $(-6)^{5}$

In Exercises 89–92, write the expression using exponential notation. In exponential notation, is it necessary to use parentheses to clearly convey the expression's meaning? Explain your reasoning.

89. $\left(\tfrac{5}{8}\right)\times\left(\tfrac{5}{8}\right)\times\left(\tfrac{5}{8}\right)\times\left(\tfrac{5}{8}\right)$

90. $-(7\times7\times7)$

91. $(-4)(-4)(-4)(-4)(-4)(-4)$

92. $(-7)\times(-7)\times(-7)$

In Exercises 93–98, evaluate the expression.

93. $(-4)^{4}$

94. $(-7)^{2}$

95. -4^{4}

96. -7^{2}

97. $\left(\tfrac{4}{5}\right)^{3}$

98. $-\left(\tfrac{2}{3}\right)^{4}$

In Exercises 99–104, evaluate the expression.

99. $5^3 + |-14 + 4|$

100. $|-36| + 7(12 - 16)$

101. $45 + 3(16 \div 4)$

102. $|(-2)^5| - (25 + 7)$

103. $0.2(6 - 10)^3 + 85$

104. $\dfrac{5^3 - 50}{-15} + 27$

In Exercises 105–110, evaluate the expression using a calculator. Round the result to two decimal places.

105. $5.6[13 - 2.5(-6.3)]$

106. $35(1032 - 4650)$

107. $5^6 - 3(400)$

108. $5(100 - 3.6^4) \div 4.1$

109. $\dfrac{500}{(1.055)^{20}}$

110. $\dfrac{265.45}{25.6}$

111. *Balance in an Account*   During one month, you made the following transactions in your checking account.

Initial Balance:	$2618.68
Deposit:	$1236.45
Withdrawal:	$25.62
Withdrawal:	$455.00
Withdrawal:	$125.00
Withdrawal:	$715.95

Find the balance at the end of the month. (Disregard any interest that may have been earned.)

112. *Company Profit*   The midyear financial statement of a company showed a profit of $1,415,322.62. At the close of the year the financial statement showed a profit for the year of $916,489.26. Find the profit of the company for the second 6 months of the year.

Circle Graphs   In Exercises 113 and 114, find the unknown fractional part of the circle graph. What property of pie graphs did you use to solve the problem?

113.

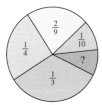

114.

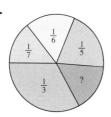

115. *Organizing Data*   On Monday you purchased $500 worth of stock. The value of the stock during the remainder of the week is shown in the bar graph.

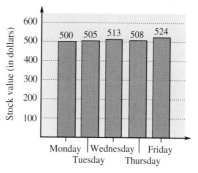

(a) Use the graph to complete the table.

Day	Daily Gain or Loss
Tuesday	?
Wednesday	?
Thursday	?
Friday	?

(b) Find the sum of the daily gains and losses. Interpret the result in the context of the problem. How could you determine this sum from the graph?

116. *Organizing Data*   The annual profits for a company (in millions of dollars) are shown in the bar graph. Use the graph to create a table that shows the yearly gains and losses.

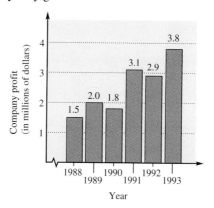

Geometry In Exercises 117 and 118, answer the questions using the information that a bale of hay in the form of a rectangular solid has dimensions 14 inches by 18 inches by 42 inches and weighs approximately 50 pounds (see figure).

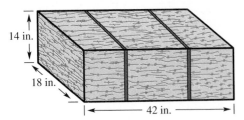

14 in.

18 in.

42 in.

117. Find the volume of a bale of hay in cubic feet if the volume of a rectangular solid is the product of its length, width, and height. (Use the fact that 1728 cubic inches = 1 cubic foot.)

118. Approximate the number of bales in a ton of hay. Then, approximate the volume of a stack of baled hay that weighs 12 tons.

True or False? In Exercises 119–122, decide whether the statement is true or false. Explain your reasoning.

119. The reciprocal of every nonzero integer is an integer.

120. The reciprocal of every nonzero rational number is a rational number.

121. If a negative real number is raised to the 12th power, the result will be positive.

122. If a negative real number is raised to the 11th power, the result will be positive.

Math Matters   Creating Numbers from Fours

It is possible to create each whole number from 1 to 100 using *exactly* four 4's and the following calculator keys.

The repeating decimal $0.\overline{4} = 0.4444\ldots$ is also allowed. (To enter $0.4444\ldots$ into your calculator, divide 4 by 9, and store the result for later use.) Note that the symbol ! is used to represent $n!$, as in $4! = 4 \cdot 3 \cdot 2 \cdot 1$. Here are ways to represent the first 10 whole numbers. How many others can you find?

$1 = 44 \div 44$

$2 = (4 \div 4) + (4 \div 4)$

$3 = (4 + 4 + 4) \div 4$

$4 = (4! - 4 - 4) \div 4$

$5 = [(4 \times 4) + 4] \div 4$

$6 = 4 + [(4 + 4) \div 4]$

$7 = 4 + 4 - (4 \div 4)$

$8 = 4 + 4 + 4 - 4$

$9 = 4 + 4 + (4 \div 4)$

$10 = (4 \div .\overline{4}) + (4 \div 4)$

The Greeks thought that the number 4 represented harmony and justice. This made the number 4 lucky, as did the fact that the Greeks believed there were four elements: earth, air, fire, and water. The color of the number four was green, and it was represented by the astrological sign of Cancer the Crab.

P.3	**Properties of Real Numbers**

Basic Properties of Real Numbers ▪
Additional Properties of Real Numbers

Basic Properties of Real Numbers

The following list summarizes the basic properties of addition and multiplication. Although the examples involve real numbers, you will learn later that these properties can also be applied to algebraic expressions.

NOTE Why are the operations of subtraction and division not listed at the right? It is because they fail to possess many of these properties. For instance, subtraction and division are not commutative or associative. To see this, consider the following.

- $4 - 3 \neq 3 - 4$
- $15 \div 5 \neq 5 \div 15$
- $8 - (6 - 2) \neq (8 - 6) - 2$
- $20 \div (4 \div 2) \neq (20 \div 4) \div 2$

Properties of Real Numbers

Let a, b, and c represent real numbers, variables, or algebraic expressions.

Property	Example
Commutative Property of Addition: $a + b = b + a$	$3 + 5 = 5 + 3$
Commutative Property of Multiplication: $ab = ba$	$2 \cdot 7 = 7 \cdot 2$
Associative Property of Addition: $(a + b) + c = a + (b + c)$	$(4 + 2) + 3 = 4 + (2 + 3)$
Associative Property of Multiplication: $(ab)c = a(bc)$	$(2 \cdot 5) \cdot 7 = 2 \cdot (5 \cdot 7)$
Distributive Property: $a(b + c) = ab + ac$ $(a + b)c = ac + bc$	$4(7 + 3) = 4 \cdot 7 + 4 \cdot 3$ $(2 + 5)3 = 2 \cdot 3 + 5 \cdot 3$
Additive Identity Property: $a + 0 = a$	$4 + 0 = 4$
Multiplicative Identity Property: $a \cdot 1 = 1 \cdot a = a$	$-5 \cdot 1 = 1 \cdot (-5) = -5$
Additive Inverse Property: $a + (-a) = 0$	$3 + (-3) = 0$
Multiplicative Inverse Property: $a \cdot \dfrac{1}{a} = 1, \; a \neq 0$	$5 \cdot \dfrac{1}{5} = 1$

EXAMPLE 1   Identifying Properties of Real Numbers

a. $4(a + 3) = 4 \cdot a + 4 \cdot 3$ Distributive Property

b. $6 \cdot \dfrac{1}{6} = 1$ Multiplicative Inverse Property

c. $-3 + (2 + b) = (-3 + 2) + b$ Associative Property of Addition

d. $(b + 8) + 0 = b + 8$ Additive Identity Property

The properties of real numbers make up the third component of what is called a **mathematical system.** These three components are a set of numbers (Section 1), operations with the set of numbers (Section 2), and properties of the operations with the numbers (Section 3).

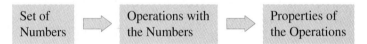

Be sure you see that the properties of real numbers can be applied to variables and algebraic expressions as well as to simple numbers.

EXAMPLE 2   Using the Properties of Real Numbers

Complete each statement using the specified property of real numbers.

a. Multiplicative Identity Property: $(4a)1 = $

b. Associative Property of Addition: $(b + 8) + 3 = $

c. Additive Inverse Property: $0 = 5c + $

d. Distributive Property: $4 \cdot b + 4 \cdot 5 = $

Solution

a. By the Multiplicative Identity Property, $(4a)1 = 4a$.

b. By the Associative Property of Addition, $(b + 8) + 3 = b + (8 + 3)$.

c. By the Additive Inverse Property, $0 = 5c + (-5c)$.

d. By the Distributive Property, $4 \cdot b + 4 \cdot 5 = 4(b + 5)$.

NOTE   As practice to make sure you understand each property of real numbers, try stating the properties in words. For instance, the Associative Property of Addition can be worded as follows: *When three real numbers are added, it makes no difference which two are added first.*

Additional Properties of Real Numbers

Once you have determined the basic properties (or *axioms*) of a mathematical system, you can go on to develop other properties. These additional properties are **theorems,** and the formal arguments that justify the theorems are **proofs.**

Additional Properties of Real Numbers

Let a, b, and c be real numbers, variables, or algebraic expressions.

Properties of Equality

Addition Property of Equality: If $a = b$, then $a + c = b + c$.

Multiplication Property of Equality: If $a = b$, then $ac = bc$.

Cancellation Property of Addition: If $a + c = b + c$, then $a = b$.

Cancellation Property of Multiplication: If $ac = bc$ and $c \neq 0$, then $a = b$.

Properties of Zero

Multiplication Property of Zero: $0 \cdot a = 0$

Division Property of Zero: $\dfrac{0}{a} = 0, \quad a \neq 0$

Division by Zero is Undefined: $\dfrac{a}{0}$ is undefined.

Properties of Negation

Multiplication by -1: $(-1)a = -a, \ (-1)(-a) = a$

Placement of Minus Signs: $(-a)(b) = -(ab) = (a)(-b)$

Product of Two Opposites: $(-a)(-b) = ab$

STUDY TIP

When you are using properties of real numbers in actual applications, the process is usually less formal than it would appear from the list on this page. For instance, the steps shown at the right are less formal than those shown in Examples 5 and 6 on page 27. The importance of the properties is that they can be used to justify the steps of a solution. They do not always need to be listed for *every* step of the solution.

In Section 1.1, you will see that the Properties of Equality are useful for solving equations, as shown below. Note that the Addition and Multiplication Properties of Equality can be used to subtract or divide both sides of an equation by the same nonzero quantity.

$5x + 4 = -2x + 18$	Original equation
$5x + 4 - 4 = -2x + 18 - 4$	Subtract 4 from both sides.
$5x = -2x + 14$	Simplify.
$5x + 2x = -2x + 2x + 14$	Add $2x$ to both sides.
$7x = 14$	Simplify.
$\dfrac{7x}{7} = \dfrac{14}{7}$	Divide both sides by 7.
$x = 2$	Simplify.

Each of the additional properties in the list can be proved using the basic properties of real numbers. Examples 3 and 4 illustrate such proofs.

EXAMPLE 3 *Proof of the Cancellation Property of Addition*

Prove that if $a + c = b + c$, then $a = b$.

Solution

Notice how each step is justified from the previous step by means of a property of real numbers.

$a + c = b + c$	Given equation
$(a + c) + (-c) = (b + c) + (-c)$	Addition Property of Equality
$a + [c + (-c)] = b + [c + (-c)]$	Associative Property of Addition
$a + 0 = b + 0$	Additive Inverse Property
$a = b$	Additive Identity Property

EXAMPLE 4 *Proof of a Property of Negation*

Prove that $(-1)a = -a$.

Solution

At first glance, it is a little difficult to see what you are asked to prove. However, a good way to start is to carefully consider the definitions of each of the three numbers in the equation.

$$a = \text{given real number}$$
$$-1 = \text{the additive inverse of } 1$$
$$-a = \text{the additive inverse of } a$$

Now, by showing that $(-1)a$ has the same properties as the additive inverse of a, you will be showing that $(-1)a$ must be the additive inverse of a.

$(-1)a + a = (-1)a + (1)(a)$	Multiplicative Identity Property
$= (-1 + 1)a$	Distributive Property
$= (0)a$	Additive Inverse Property
$= 0$	Multiplication Property of Zero

Because $(-1)a + a = 0$, you can use the fact that $-a + a = 0$ to conclude that $(-1)a + a = -a + a$. From this, you can complete the proof as follows.

$(-1)a + a = -a + a$	Shown in first part of proof
$(-1)a = -a$	Cancellation Property of Addition

The list of additional properties of real numbers forms a very important part of algebra. Knowing the names of the properties is not especially important, but knowing how to use each property is extremely important. The next two examples show how several of the properties can be used to solve equations. (You will study these techniques in detail in Section 1.1.)

EXAMPLE 5 *Applying the Properties of Real Numbers*

$b + 2 = 6$	Given equation
$(b + 2) + (-2) = 6 + (-2)$	Addition Property of Equality
$b + [2 + (-2)] = 6 - 2$	Associative Property of Addition
$b + 0 = 4$	Additive Inverse Property
$b = 4$	Additive Identity Property

EXAMPLE 6 *Applying the Properties of Real Numbers*

$3a = 9$	Given equation
$\left(\dfrac{1}{3}\right)(3a) = \left(\dfrac{1}{3}\right)(9)$	Multiplication Property of Equality
$\left(\dfrac{1}{3} \cdot 3\right)(a) = 3$	Associative Property of Multiplication
$(1)(a) = 3$	Multiplicative Inverse Property
$a = 3$	Multiplicative Identity Property

Group Activities   You Be the Instructor

Error Analysis   One of your students argues that the following statement is true because of the Associative Property.

$$(7 - 3) + 4 = 7 - (3 + 4)$$

Discuss why this statement is incorrect. Can you explain how to alter the statement so that the Associative Property may be correctly applied?

P.3 Exercises

Discussing the Concepts

1. In your own words, give a verbal description of the Commutative Property of Addition.

2. What is the additive inverse of a real number? Give an example of the Additive Inverse Property.

3. What is the multiplicative inverse of a real number? Give an example of the Multiplicative Inverse Property.

4. Does every real number have a multiplicative inverse? Explain.

5. State the Multiplication Property of Zero.

6. Explain how the Addition Property of Equality can be used to allow you to subtract the same number from both sides of an equation.

Problem Solving

In Exercises 7–18, name the property of real numbers that justifies the statement.

7. $(10 + 8) + 3 = 10 + (8 + 3)$

8. $7(9 + 15) = 7 \cdot 9 + 7 \cdot 15$

9. $6(-10) = -10(6)$ **10.** $2(6 \cdot 3) = (2 \cdot 6)3$

11. $2x - 2x = 0$ **12.** $3x + 0 = 3x$

13. $1 \cdot (5t) = 5t$ **14.** $4 \cdot \frac{1}{4} = 1$

15. $3(2 + x) = 3 \cdot 2 + 3x$

16. $4 + (3 - x) = (4 + 3) - x$

17. $10(2x) = (10 \cdot 2)x$ **18.** $(x+1) - (x+1) = 0$

In Exercises 19–24, complete the statement.

19. Associative Property of Multiplication:

$3(6y) = $

20. Commutative Property of Addition:

$10 + (-6) = $

21. Distributive Property:

$5(6 + z) = $

22. Additive Inverse Property:

$3x + $ $= 0$

23. Multiplicative Identity Property:

$(x + 8) \cdot 1 = $

24. Additive Identity Property:

$(8x) + 0 = $

In Exercises 25–28, write (a) the additive inverse and (b) the multiplicative inverse of the quantity.

25. 10   **26.** -52

27. $x + 1$ **28.** $y - 4$

In Exercises 29–32, rewrite the expression using the Associative Property of Addition or Multiplication.

29. $(x + 5) - 3$ **30.** $(z + 6) + 10$

31. $3(4 \cdot 5)$ **32.** $(10 \cdot 8) \cdot 5$

In Exercises 33–36, rewrite the expression using the Distributive Property.

33. $20(2 + 5)$ **34.** $6(2x + 5)$

35. $(x + 6)(-2)$ **36.** $(z - 10)(12)$

Mental Math In Exercises 37 and 38, use the Distributive Property to perform the required arithmetic mentally. For example, suppose you work in an industry where the wage is $14 per hour with "time and one-half" for overtime. Thus, your hourly wage for overtime is

$14(1.5) = 14\left(1 + \frac{1}{2}\right) = 14 + 7 = \$21.$

37. $16(1.75) = 16\left(2 - \frac{1}{4}\right)$

38. $12(19.75) = 12\left(20 - \frac{1}{4}\right)$

In Exercises 39 and 40, identify the property of real numbers that justifies each rewriting of the equation.

39.
$$x + 5 = 3$$
$$(x + 5) + (-5) = 3 + (-5)$$
$$x + [5 + (-5)] = 3 - 5$$
$$x + 0 = -2$$
$$x = -2$$

40.
$$3x + 4 = 10$$
$$(3x + 4) + (-4) = 10 + (-4)$$
$$3x + [4 + (-4)] = 6$$
$$3x + 0 = 6$$
$$3x = 6$$
$$\frac{1}{3}(3x) = \frac{1}{3}(6)$$
$$\left(\frac{1}{3} \cdot 3\right)x = 2$$
$$1 \cdot x = 2$$
$$x = 2$$

41. Use the figure to fill in the blanks. What property is being demonstrated?

$$\boxed{}\ (\ \boxed{}\ +\ \boxed{}\ )=\ \boxed{}\ +\ \boxed{}$$

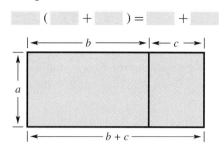

42. Use the figure to fill in the blanks. What property is being demonstrated?

$$\boxed{}\ (\ \boxed{}\ -\ \boxed{}\ )=\ \boxed{}\ -\ \boxed{}$$

Additional Problem Solving

In Exercises 43–62, name the property of real numbers that justifies the statement.

43. $3 + (-5) = -5 + 3$

44. $-5(7) = 7(-5)$

45. $5(2a) = (5 \cdot 2)a$

46. $5 + 0 = 5$

47. $7 \cdot 1 = 7$

48. $3x - 3x = 0$

49. $3x + 0 = 3x$

50. $8y \cdot 1 = 8y$

51. $10x \cdot \dfrac{1}{10x} = 1$

52. $\dfrac{1}{y} \cdot y = 1$

53. $25 - 25 = 0$

54. $25 + 35 = 35 + 25$

55. $(5 + 10)(8) = 8(5 + 10)$

56. $3(6 + b) = 3 \cdot 6 + 3 \cdot b$

57. $3 + (12 - 9) = (3 + 12) - 9$

58. $(8 - 5)(10) = 8 \cdot 10 - 5 \cdot 10$

59. $(16 + 8) - 5 = 16 + (8 - 5)$

60. $6(x + 3) = 6 \cdot x + 6 \cdot 3$

61. $(6 + x) - m = 6 + (x - m)$

62. $(-4 \cdot 10) \cdot 8 = -4(10 \cdot 8)$

In Exercises 63–68, complete the statement.

63. Commutative Property of Multiplication:

$$15(-3) = \boxed{}$$

64. Associative Property of Addition:

$$6 + (5 - y) = \boxed{}$$

65. Distributive Property:

$$-3(4 - x) = \boxed{}$$

66. Distributive Property:

$$(8 - y)(4) = \boxed{}$$

67. Commutative Property of Addition:

$$25 + (-x) = \boxed{}$$

68. Additive Inverse Property:

$$13x - 13x = \boxed{}$$

In Exercises 69–72, write (a) the additive inverse and (b) the multiplicative inverse of the quantity.

69. -16 **70.** 18

71. $6z$ **72.** $2y$

In Exercises 73–76, rewrite the expression using the Associative Property of Addition or Multiplication.

73. $3 + (8 - x)$ **74.** $(x + 4) + 5$

75. $6(2y)$ **76.** $8(3x)$

In Exercises 77–80, rewrite the expression using the Distributive Property.

77. $-8(3 - 5)$ **78.** $-3(4 - 8)$

79. $5(3x - 4)$ **80.** $(x + 2)5$

In Exercises 81 and 82, identify the property of real numbers that justifies each rewriting of the equation.

81.
$$x - 8 = 20$$
$$(x - 8) + 8 = 20 + 8$$
$$x + (-8 + 8) = 28$$
$$x + 0 = 28$$
$$x = 28$$

82.
$$2x - 5 = 6$$
$$(2x - 5) + 5 = 6 + 5$$
$$2x + (-5 + 5) = 11$$
$$2x + 0 = 11$$
$$2x = 11$$
$$\frac{1}{2}(2x) = \frac{1}{2}(11)$$
$$\left(\frac{1}{2} \cdot 2\right)x = \frac{11}{2}$$
$$1 \cdot x = \frac{11}{2}$$
$$x = \frac{11}{2}$$

83. Prove that if $ac = bc$ and $c \neq 0$, then $a = b$.

84. Prove that $(-1)(-a) = a$.

Interpreting a Graph In Exercises 85–88, the dividend paid per share by the General Electric Company for the years 1989 through 1993 is approximated by the model

Dividend per share $= -0.93 + 0.66x$.

In this model, the dividend per share is measured in dollars and x represents the earnings per share in dollars (see figure). (Source: General Electric Company)

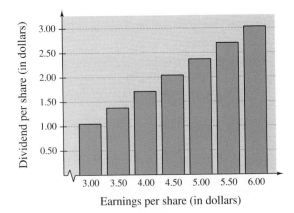

85. Use the graph to approximate the increase in the dividend per share if the earnings per share increases by $0.50.

86. Use the graph to approximate the increase in the dividend per share if the earnings per share increases by $1.00.

87. In 1989, General Electric's earnings per share was $3.88. Use the model to approximate the dividend per share.

88. In 1993, General Electric's earnings per share was $5.18. Use the model to approximate the dividend per share.

89. *Exploration* Suppose you define a new mathematical operation using the symbol $\odot$. This operation is defined as $a \odot b = 2 \cdot a + b$. Give examples to show that neither the Commutative Property nor the Associative Property is true for this operation.

MID-CHAPTER QUIZ

Take this quiz as you would take a quiz in class. After you are done, check your work against the answers given in the back of the book.

In Exercises 1 and 2, show each real number as a point on the real number line and place the correct inequality symbol ($<$ or $>$) between the real numbers.

1. -4.5 ▪▪▪ -6 **2.** $\frac{3}{4}$ ▪▪▪ $\frac{3}{2}$

In Exercises 3 and 4, evaluate the expression.

3. $|-3.2|$ **4.** $-|5.75|$

In Exercises 5 and 6, find the distance between the two real numbers.

5. -15 and 7 **6.** -10.5 and -6.75

In Exercises 7–14, evaluate the expression. Write fractions in reduced form.

7. $32 + (-18)$ **8.** $-10 - 12$

9. $\dfrac{3}{4} + \dfrac{7}{4}$ **10.** $2 + \dfrac{2}{3} - \dfrac{1}{6}$

11. $\left| 4 - 1\dfrac{3}{8} \right|$ **12.** $\left(-\dfrac{4}{5} \right)\left(\dfrac{15}{32} \right)$

13. $\dfrac{7}{12} \div \dfrac{5}{6}$ **14.** $\left(-\dfrac{3}{2} \right)^3$

In Exercises 15 and 16, evaluate the expression and round your answer to two decimal places. (A calculator may be useful.)

15. $\dfrac{-130.45}{-45.2}$ **16.** $(3.8)^4$

In Exercises 17 and 18, name the property of real numbers that justifies the given statement.

17. (a) $8(u - 5) = 8 \cdot u - 8 \cdot 5$ (b) $10x - 10x = 0$

18. (a) $(7 + y) - z = 7 + (y - z)$ (b) $2x \cdot 1 = 2x$

19. You deposit $30 in a retirement account twice each month. How much will you deposit in the account in 5 years?

20. Determine the unknown fractional part of the circle graph at the right. Explain how you were able to make this determination.

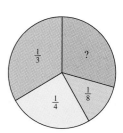

Figure for 20

P.4	**Algebraic Expressions**
	Algebraic Expressions ▪ Properties of Exponents ▪ Simplifying Algebraic Expressions ▪ Evaluating Algebraic Expressions

Algebraic Expressions

One of the basic characteristics of algebra is the use of letters (or combinations of letters) to represent numbers. The letters used to represent the numbers are called **variables,** and combinations of letters and numbers are called **algebraic expressions.** Here are some examples.

$$3x, \quad x + 2, \quad \frac{x}{x^2 + 1}, \quad 2x - 3y$$

Algebraic Expression

A collection of letters (called **variables**) and real numbers (called **constants**) combined using the operations of addition, subtraction, multiplication, and division is called an **algebraic expression.**

The **terms** of an algebraic expression are those parts that are separated by *addition.* For example, the algebraic expression $x^2 - 3x + 6$ has three terms: x^2, $-3x$, and 6. Note that $-3x$ is a term, rather than $3x$, because

$$x^2 - 3x + 6 = x^2 + (-3x) + 6. \qquad \text{Terms are separated by addition.}$$

The terms x^2 and $-3x$ are the **variable terms** of the expression, and 6 is the **constant term.** The numerical factor of a variable term is called the **coefficient.** For instance, the coefficient of the variable term $-3x$ is -3, and the coefficient of the variable term x^2 is 1.

EXAMPLE 1   Identifying the Terms of an Algebraic Expression

Algebraic Expression	*Terms*
a. $5x - \frac{1}{3}$	$5x, \quad -\frac{1}{3}$
b. $4y + 6x - 9$	$4y, \quad 6x, \quad -9$
c. $-x^3 + 4x^2 - 5x + 1$	$-x^3, \quad 4x^2, \quad -5x, \quad 1$
d. $\frac{1}{2} + 5x^4$	$\frac{1}{2}, \quad 5x^4$

Properties of Exponents

When multiplying two exponential expressions that have the *same* base, you add exponents. To see why this is true, consider the product $a^3 \cdot a^2$. Because the first expression represents $a \cdot a \cdot a$ and the second represents $a \cdot a$, it follows that the product of the two expressions represents $a \cdot a \cdot a \cdot a \cdot a$, as follows.

$$a^3 \cdot a^2 = \underbrace{(a \cdot a \cdot a)}_{\substack{\text{Three} \\ \text{factors}}} \cdot \underbrace{(a \cdot a)}_{\substack{\text{Two} \\ \text{factors}}} = \underbrace{(a \cdot a \cdot a \cdot a \cdot a)}_{\substack{\text{Five} \\ \text{factors}}} = a^{3+2} = a^5$$

This and one other property of exponents are summarized below.

Properties of Exponents

Let m and n be positive integers, and let a and b be real numbers, variables, or algebraic expressions.

1. To multiply two exponential expressions that have the same base, add the exponents.

$$a^m \cdot a^n = a^{m+n}$$

2. To raise a product to a power, raise each factor to the power and multiply the results.

$$(ab)^m = a^m b^m$$

The next two examples illustrate the use of these rules.

EXAMPLE 2 Applying the Properties of Exponents

a. $-3^2 \cdot 3^4 \cdot 3 = -(3^2 \cdot 3^4 \cdot 3^1) = -3^7$
b. $b^3 b^2 b = b^{3+2+1} = b^6$
c. $4^2 x^2 \cdot x = (4^2)(x^2 \cdot x) = 16x^3$
d. $(-3x^2 y)(5xy)(2y^2) = (-3)(5)(2)(x^2 \cdot x)(y \cdot y \cdot y^2) = -30x^3 y^4$

EXAMPLE 3 Applying the Properties of Exponents

a. $(3x)^3 = 3^3 \cdot x^3 = 27x^3$
b. $(-x^2)(x^2) = (-1)(x^2 \cdot x^2) = (-1)x^{2+2} = -x^4$
c. $(-x)^2(x^2) = x^2 \cdot x^2 = x^{2+2} = x^4$
d. $3x^2 \cdot (-5x)^3 = 3 \cdot x^2 \cdot (-5)^3 \cdot x^3 = 3(-125)x^{2+3} = -375x^5$

NOTE The rules at the right extend to three or more factors. For example,

$$a^m \cdot a^n \cdot a^k = a^{m+n+k}$$

and

$$(abc)^m = a^m b^m c^m.$$

Simplifying Algebraic Expressions

In an algebraic expression, two terms are said to be **like terms** if they are both constant terms or if they have the same variable factors. For example, the terms $4x$ and $-2x$ are like terms because they have the same variable factor. Similarly, $2x^2y$, $-x^2y$, and $\frac{1}{2}(x^2y)$ are like terms. Note that $4x^2y$ and $-x^2y^2$ are not like terms because their variable factors are different.

One of the most common uses of the basic rules of algebra is to rewrite an algebraic expression in a simpler form. One way to **simplify** an algebraic expression is to remove symbols of grouping and combine like terms.

EXAMPLE 4   Combining Like Terms

a. $2x + 3x - 4 = (2 + 3)x - 4$ Distributive Property

$\qquad\qquad\qquad = 5x - 4$ Simplify.

b. $-3 + 5 + 2y - 7y = (-3 + 5) + (2 - 7)y$ Distributive Property

$\qquad\qquad\qquad\qquad = 2 - 5y$ Simplify.

c. $5x + 3y - 4x = 3y + 5x - 4x$ Commutative Property

$\qquad\qquad\qquad = 3y + (5x - 4x)$ Associative Property

$\qquad\qquad\qquad = 3y + (5 - 4)x$ Distributive Property

$\qquad\qquad\qquad = 3y + x$ Simplify.

As you gain experience with the rules of algebra, you may want to combine some of the steps in your work. For instance, you might feel comfortable listing only the following steps to solve Example 4(c).

$5x + 3y - 4x = 3y + (5x - 4x)$ Group like terms.

$\qquad\qquad\quad = 3y + x$ Combine like terms.

EXAMPLE 5   Combining Like Terms

a. $7x + 7y - 4x - y = (7x - 4x) + (7y - y)$ Group like terms.

$\qquad\qquad\qquad\quad = 3x + 6y$ Combine like terms.

b. $2x^2 + 3x - 5x^2 - x = (2x^2 - 5x^2) + (3x - x)$ Group like terms.

$\qquad\qquad\qquad\qquad = -3x^2 + 2x$ Combine like terms.

c. $3xy^2 - 4x^2y^2 + 2xy^2 + (xy)^2 = (3xy^2 + 2xy^2) + (-4x^2y^2 + x^2y^2)$

$\qquad\qquad\qquad\qquad\qquad\qquad = 5xy^2 - 3x^2y^2$

When removing symbols of grouping, remove the innermost symbols first and combine like terms. Then repeat the process for any remaining symbols of grouping.

EXAMPLE 6 *Removing Symbols of Grouping*

a. $3(x - 5) - (2x - 7) = 3x - 15 - 2x + 7$ Distributive Property

$\qquad\qquad\qquad\quad = (3x - 2x) + (-15 + 7)$ Group like terms.

$\qquad\qquad\qquad\quad = x - 8$ Combine like terms.

b. $-4(x^2 + 4) + x^2(x + 4) = -4x^2 - 16 + x^3 + 4x^2$ Distributive Property

$\qquad\qquad\qquad\qquad\quad = x^3 + (4x^2 - 4x^2) - 16$ Group like terms.

$\qquad\qquad\qquad\qquad\quad = x^3 + 0 - 16$ Simplify.

$\qquad\qquad\qquad\qquad\quad = x^3 - 16$ Simplify.

EXAMPLE 7 *Removing Symbols of Grouping*

a. $5x - 2x[3 + 2(x - 7)] = 5x - 2x(3 + 2x - 14)$ Distributive Property

$\qquad\qquad\qquad\qquad\quad = 5x - 2x(2x - 11)$ Combine like terms.

$\qquad\qquad\qquad\qquad\quad = 5x - 4x^2 + 22x$ Distributive Property

$\qquad\qquad\qquad\qquad\quad = -4x^2 + 27x$ Combine like terms.

b. $-3x(5x^4) + (2x)^5 = -15x^5 + (2^5)(x^5)$ Properties of exponents

$\qquad\qquad\qquad\qquad = -15x^5 + 32x^5$ Simplify.

$\qquad\qquad\qquad\qquad = 17x^5$ Combine like terms.

A set of parentheses preceded by a *minus* sign can be removed by changing the sign of each term inside the parentheses. For instance,

$$3x - (2x - 7) = 3x - 2x + 7.$$

This is equivalent to using the Distributive Property with a multiplier of -1. That is,

$$3x - (2x - 7) = 3x + (-1)(2x - 7) = 3x - 2x + 7.$$

A set of parentheses preceded by a *plus* sign can be removed without changing the sign of each term inside the parentheses. For instance,

$$3x + (2x - 7) = 3x + 2x - 7.$$

PROGRAMMING

You can use a graphing utility to evaluate an expression using the program below. The following program may be entered into a *TI-82* calculator. Programs for other calculator models may be found in the Appendix.

```
PROGRAM: EVALUATE
: Lbl 1
: Disp ''ENTER X''
: Input X
: Disp Y₁
: Goto 1
```

Now enter the equation $y = 3x + 2$ into your utility by the following keystrokes. $\boxed{Y=}$
3 $\boxed{X, T, \theta}$ $\boxed{+}$ 2. Then run the program using several x-values. Create a table showing the values of the expression for several values of x.

Evaluating Algebraic Expressions

To **evaluate** an algebraic expression, substitute numerical values for each of the variables in the expression. Here are some examples.

Expression	Value of Variable	Substitute	Value of Expression
$3x + 2$	$x = 2$	$3(2) + 2$	$6 + 2 = 8$
$4x^2 + 2x - 1$	$x = -1$	$4(-1)^2 + 2(-1) - 1$	$4 - 2 - 1 = 1$
$2x(x + 4)$	$x = -2$	$2(-2)(-2 + 4)$	$2(-2)(2) = -8$

Note that you must substitute for *each* occurrence of the variable.

EXAMPLE 8   Evaluating Algebraic Expressions

Evaluate each algebraic expression when $x = -2$ and $y = 5$.

a. $2y - 3x$ **b.** $5 + x^2$ **c.** $5 - x^2$

Solution

a. When $x = -2$ and $y = 5$, the expression $2y - 3x$ has a value of
$$2(5) - 3(-2) = 10 + 6 = 16.$$

b. When $x = -2$, the expression $5 + x^2$ has a value of
$$5 + (-2)^2 = 5 + 4 = 9.$$

c. When $x = -2$, the expression $5 - x^2$ has a value of
$$5 - (-2)^2 = 5 - 4 = 1.$$

EXAMPLE 9   Evaluating Algebraic Expressions

Evaluate each algebraic expression when $x = 2$ and $y = -1$.

a. $y - x$ **b.** $|y - x|$

Solution

a. When $x = 2$ and $y = -1$, the expression $y - x$ has a value of
$$(-1) - (2) = -1 - 2 = -3.$$

b. When $x = 2$ and $y = -1$, the expression $|y - x|$ has a value of
$$|(-1) - (2)| = |-3| = 3.$$

EXAMPLE 10   Using a Mathematical Model

From 1980 to 1991, the average hourly wage for miners in the United States can be modeled by

$$\text{Hourly wage} = -0.045t^2 + 0.8t + 9.2, \qquad 0 \le t \le 11$$

where $t = 0$ represents 1980. Create a table that shows the average hourly wages for these years.   (Source: U.S. Bureau of Labor Statistics)

Solution

Year	1980	1981	1982	1983	1984	1985
t	0	1	2	3	4	5
Hourly Wage	$9.20	$9.96	$10.62	$11.20	$11.68	$12.08

Year	1986	1987	1988	1989	1990	1991
t	6	7	8	9	10	11
Hourly Wage	$12.38	$12.60	$12.72	$12.76	$12.70	$12.56

In 1991, the mining industry in the United States employed about 750,000 people.

Group Activities

Exploring with Technology

Using a Graphing Utility   The *TI-82* calculator can evaluate an algebraic expression at several x-values and display the results in a table. For instance, to evaluate $2x^2 - 3x + 2$ when x is 0, 1, 2, 3, 4, and 5, you can use the following steps.

| TblSet | TblMin=0, ΔTbl=1

| Y= | $Y_1 = 2X^2 - 3X + 2$

| TABLE |

X	Y₁	
0	2	
1	1	
2	4	
3	11	
4	22	
5	37	
6	56	

The results are shown at the left. Try creating another table that shows the values of $-4x^2 + 5x - 8$ when x is 0, 1, 2, 3, 4, 5, and 6.

P.4 Exercises

Discussing the Concepts

1. Explain the difference between terms and factors in an algebraic expression.

2. In your own words, explain how to combine like terms. Give an example.

3. Explain how the Distributive Property can be used to simplify the expression $5x + 3x$.

4. Describe a procedure for removing nested symbols of grouping.

5. Explain how to rewrite $[x - (3 \cdot 4)] \div 5$ without using symbols of grouping.

6. Is it possible to evaluate $(x + 2)/(y - 3)$ when $x = 5$ and $y = 3$? Explain.

Problem Solving

In Exercises 7–10, identify the terms of the expression.

7. $10x + 5$

8. $25z^3 - 4.8z^2$

9. $\dfrac{3}{t^2} - \dfrac{4}{t} + 6$

10. $4x^2 - 3y^2 - 5x + 2y$

In Exercises 11 and 12, identify the coefficient.

11. $5y^3$

12. $-8.4x$

In Exercises 13–22, identify the rule of algebra that is illustrated by the equation.

13. $4 - 3x = -3x + 4$

14. $(10 + x) - y = 10 + (x - y)$

15. $-5(2x) = (-5 \cdot 2)x$

16. $(x - 2)(3) = 3(x - 2)$

17. $(x + 5) \cdot \dfrac{1}{x + 5} = 1$

18. $(x^2 + 1) - (x^2 + 1) = 0$

19. $5(y^3 + 3) = 5y^3 + 5 \cdot 3$

20. $8x + 0 = 8x$

21. $(16t^4) \cdot 1 = 16t^4$

22. $-32(u^2 - 3u) = -32u^2 - (-32)(3u)$

In Exercises 23 and 24, write in exponential notation.

23. $(-5x)(-5x)(-5x)(-5x)$

24. $(y \cdot y \cdot y)(y \cdot y \cdot y \cdot y)$

In Exercises 25–28, simplify the expression.

25. $x^5 \cdot x^7$

26. $(-4x)^3$

27. $(2xy)(3x^2y^3)$

28. $(-5a^2b^3)(2ab^4)$

In Exercises 29–32, simplify the expression by combining like terms.

29. $3x - 2y + 5x + 20y$

30. $-2a + \frac{1}{3}b - 7a - b$

31. $2uv + 5u^2v^2 - uv - (uv)^2$

32. $-5y^3 + 3y - 6y^2 + 8y^3 + y - 4$

In Exercises 33–36, use the Distributive Property to rewrite the expression.

33. $4(2x^2 + x - 3)$

34. $-5(-x^2 + 2y + 1)$

35. $4x - 8$

36. $9x + 18y$

In Exercises 37–40, simplify the expression.

37. $10(x - 3) + 2x - 5$

38. $5(a + 6) - 4(a^2 - 2a - 1)$

39. $2[3(b - 5) - (b^2 + b + 3)]$

40. $4[5 - 3(x^2 + 10)]$

In Exercises 41 and 42, evaluate the expression when $x = -3$.

41. $-x^2 + 2x + 5$

42. $-3x^2 - 8x + 1$

In Exercises 43–46, evaluate the expression for the specified values of the variable(s). If it is not possible, state the reason.

Expression	*Values*
43. $5 - 3x$	(a) $x = \frac{2}{3}$ (b) $x = 5$
44. $2x^2 + 5x - 3$	(a) $x = \frac{1}{2}$ (b) $x = -3$
45. $\dfrac{x}{x - y}$	(a) $x = 0, y = 10$
	(b) $x = 4, y = 4$
46. $|y - x|$	(a) $x = 2, y = 5$
	(b) $x = -2, y = -2$

Geometry   In Exercises 47 and 48, find an expression for the area of the figure. Then evaluate the expression for the given value of the variable.

47. $b = 15$ **48.** $h = 12$

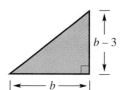

Additional Problem Solving

In Exercises 49–52, identify the terms of the expression.

49. $-3y^2 + 2y - 8$ **50.** $-16t^2 + 48$

51. $x^2 - 2.5x - \dfrac{1}{x}$ **52.** $14u^2 + 25uv - 3v^2$

In Exercises 53 and 54, identify the coefficient.

53. $-\frac{3}{4}t^2$ **54.** $4x^6$

In Exercises 55–60, identify the rule of algebra that is illustrated by the equation.

55. $3[2(x + 4)] = (3 \cdot 2)(x + 4)$

56. $(3x + 5) - (3x + 5) = 0$

57. $(z + 10) \cdot \dfrac{1}{z + 10} = 1$

58. $-4(y - 25) = -4y - (-4)(25)$

59. $(x + 8) - 4 = x + (8 - 4)$

60. $1 \cdot (x^2 + 2) = x^2 + 2$

In Exercises 61–72, use the given property to rewrite the expression.

61. Distributive Property:

$5(x + 6) = $

62. Commutative Property of Multiplication:

$5(x + 6) = $

63. Distributive Property:

$6x + 6 = $

64. Commutative Property of Addition:

$6x + 6 = $

65. Commutative Property of Multiplication:

$6(xy) = $

66. Associative Property of Multiplication:

$6(xy) = $

67. Additive Identity Property:

$3 + 0 = $

68. Commutative Property of Addition:

$3 + 0 = $

69. Additive Inverse Property:

$4 + (-4) = $

70. Commutative Property of Addition:

$4 + (-4) = $

71. Associative Property of Addition:

$(3 + 6) + (-9) = $

72. Additive Inverse Property:

$9 + (-9) = $

In Exercises 73 and 74, rewrite the expression as a repeated multiplication problem.

73. $(-2x)^3$ **74.** $z^2 \cdot z^5$

Geometry In Exercises 75 and 76, write an expression for the area of the region. Then simplify the expression.

75.

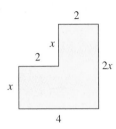

76.

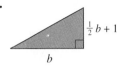

In Exercises 77–82, simplify the expression.

77. $(-4x)^2$

78. $u^3 \cdot u^5 \cdot u$

79. $3^3 y^4 \cdot y^2$

80. $6^2 x^3 \cdot x^5$

81. $(3uv)^2(-6u^3 v)$

82. $(10x^2 y)^3 (2x^4 y)$

In Exercises 83–88, simplify the expression by combining like terms.

83. $3x + 4x$

84. $-2x^2 + 4x^2$

85. $9y - 5y + 4y$

86. $8y + 7y - y$

87. $3\left(\dfrac{1}{x}\right) - \dfrac{1}{x} + 8$

88. $8z^2 + \dfrac{3}{2}z - \dfrac{5}{2}z^2 + 10$

In Exercises 89–92, use the Distributive Property to rewrite the expression.

89. $-3(6y^2 - y - 2)$

90. $8(z^3 - 4z^2 + 2)$

91. $ab + 2a$

92. $3bx^2 - 6bx$

In Exercises 93–98, simplify the expression.

93. $-3(y^2 + 3y - 1) + 2(y - 5)$

94. $3(x^2 + 1) + x^2 - 6$

95. $2x(5x^2) - (x^3 + 5)$

96. $x(x^2 + 3) - 3(x + 4)$

97. $y^2(y + 1) + y(y^2 + 1)$

98. $2ab(b^2 - 3) - ab(b^2 + 2)$

In Exercises 99–106, evaluate the expression for the specified values of the variable(s). If it is not possible, state the reason.

	Expression	*Values*
99.	$10 - |x|$	(a) $x = 3$ (b) $x = -3$
100.	$\frac{3}{2}x - 2$	(a) $x = 6$ (b) $x = -3$
101.	$\dfrac{x}{x^2 + 1}$	(a) $x = 0$ (b) $x = 3$
102.	$5 - \dfrac{3}{x}$	(a) $x = 0$ (b) $x = -6$
103.	$3x + 2y$	(a) $x = 1, y = 5$ (b) $x = -6, y = -9$
104.	$x^2 - xy + y^2$	(a) $x = 2, y = -1$ (b) $x = -3, y = -2$
105.	rt	(a) $r = 40, t = 5\frac{1}{4}$ (b) $r = 35, t = 4$
106.	Prt	(a) $P = \$5000, r = 0.085, t = 10$ (b) $P = \$750, r = 0.07, t = 3$

107. *Creating a Table*

(a) Complete the table by evaluating $2x - 5$.

x	-1	0	1	2	3	4
$2x - 5$						

(b) How much does the value of the expression increase for each one-unit increase in x?

(c) Use the results of (a) and (b) to guess the increase in $\frac{3}{4}x + 5$ for each one-unit increase in x.

108. *Creating a Table*

(a) Complete the table by evaluating $3x + 2$.

x	-1	0	1	2	3	4
$3x + 2$						

(b) How much does the value of the expression increase for each one-unit increase in x?

(c) Use the results of (a) and (b) to guess the increase in $\frac{2}{3}x - 4$ for each one-unit increase in x.

Using a Model　In Exercises 109 and 110, use the following model, which approximates the annual sales (in millions of dollars) of exercise equipment in the United States from 1983 to 1992 (see figure).

Sales $= 10.38t^2 - 14.97t + 935.42$

In this formula, $t = 0$ represents 1980.　(Source: National Sporting Goods Association)

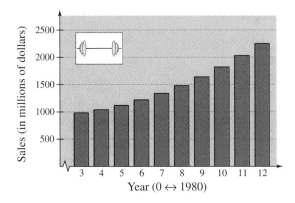

109. Graphically approximate the sales of exercise equipment in 1988. Then use the model to confirm your estimate algebraically.

110. Graphically approximate the sales of exercise equipment in 1991. Then use the model to confirm your estimate algebraically.

111. *Geometry*　The area of the trapezoid shown in the figure, with parallel bases of lengths b_1 and b_2 and height h, is given by $A = \frac{1}{2}(b_1 + b_2)h$. Use the Distributive Property to show that the area can also be expressed as

$$A = b_1 h + \frac{1}{2}(b_2 - b_1)h.$$

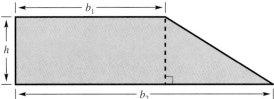

112. *Geometry*　Use both formulas given in Exercise 111 to find the area of a trapezoid with $b_1 = 7$, $b_2 = 12$, and $h = 3$.

113. *Geometry*　The roof shown in the figure is made up of two trapezoids and two triangles. Find the total area of the roof.

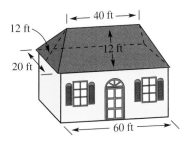

C A R E E R　 I N T E R V I E W

Alfred A. Campos

**Senior Underwriter
Property and Marine
Department**

Chubb & Son, Inc.

Boston, MA 02111

Chubb is a Fortune 500, multinational, U. S.-based property and casualty insurance provider. Corporations requiring insurance consult with one of our agents, who then prepares a proposal for coverage. As an underwriter, I analyze and review these potential insureds for risk acceptability; write the terms of the insurance contract; and determine the premium to be charged. To determine a premium, I must know the *exposure,* or dollar value of assets to be insured, and use an appropriate *all-risk rate* (as a percent) made of three separate insurance rates for fire, extended coverage, and any other peril. The expression I use to calculate the dollar amount of a premium is

$$\frac{\text{all-risk rate} \times \text{exposure}}{100}.$$

| P.5 | **Constructing Algebraic Expressions** |

Translating Phrases ▪ Hidden Operations

Translating Phrases

In this section, you will study ways to *construct* algebraic expressions. When you translate a verbal sentence or phrase into an algebraic expression, it helps to watch for key words and phrases that indicate the four different operations of arithmetic.

Translating Key Words and Phrases

Key Words and Phrases	Verbal Description	Algebraic Expression
Addition: Sum, plus, greater than, increased by, more than, exceeds, total of	The sum of 5 and x Seven more than y	$5 + x$ $y + 7$
Subtraction: Difference, minus, less than, decreased by, subtracted from, reduced by, the remainder	b is subtracted from 4. Three less than z	$4 - b$ $z - 3$
Multiplication: Product, multiplied by, twice, times, percent of	Two times x	$2x$
Division: Quotient, divided by, ratio, per	The ratio of x and 8	$\dfrac{x}{8}$

EXAMPLE 1   Translating Verbal Phrases

a. *Verbal Description:* Seven more than three times x
   Algebraic Expression: $3x + 7$

b. *Verbal Description:* Four less than the product of 6 and n
   Algebraic Expression: $6n - 4$

EXAMPLE 2 Translating Verbal Phrases

a. *Verbal Description:* Eight added to the product of 2 and n
    Algebraic Expression: $2n + 8$

b. *Verbal Description:* Four times the sum of y and 9
    Algebraic Expression: $4(y + 9)$

c. *Verbal Description:* Three times the ratio of x and 7
    Algebraic Expression: $3\left(\dfrac{x}{7}\right)$

In Examples 1 and 2, the verbal description specified the name of the variable. In most real-life situations, however, the variables are not specified and it is your task to assign variables to the *appropriate* quantities.

EXAMPLE 3 Translating Verbal Phrases

a. *Verbal Description:* The sum of 7 and a number
    Label: The number $= x$
    Algebraic Expression: $7 + x$

b. *Verbal Description:* Four decreased by the product of 2 and a number
    Label: The number $= x$
    Algebraic Expression: $4 - 2x$

A good way to learn algebra is to do it forwards and backwards. For instance, the next example translates algebraic expressions into verbal form. Keep in mind that other key words could be used to describe the operations in each expression.

EXAMPLE 4 Translating Expressions into Verbal Phrases

Without using a variable, write a verbal description for each.

a. $5x - 10$ b. $\dfrac{3 + x}{4}$

Solution

a. 10 less than the product of 5 and a number

b. The sum of 3 and some number, all divided by 4

Hidden Operations

When verbal phrases are translated into algebraic expressions, products are often overlooked. Watch for hidden products in the next two examples.

EXAMPLE 5 Discovering Hidden Products

A cash register contains x quarters. Write an expression for this amount of money in dollars.

Solution

Verbal Model:

Value of coin	$\cdot$	Number of coins

Labels: Value of coin $= 0.25$ (dollars per quarter)
              Number of coins $= x$ (quarters)

Expression: $0.25x$

EXAMPLE 6 Discovering Hidden Products

A cash register contains n nickels and d dimes. Write an expression for this amount of money in cents.

Solution

Verbal Model:

Value of nickel	$\cdot$	Number of nickels	$+$	Value of dime	$\cdot$	Number of dimes

Labels: Value of nickel $= 5$ (cents per nickel)
              Number of nickels $= n$ (nickels)
              Value of dime $= 10$ (cents per dime)
              Number of dimes $= d$ (dimes)

Expression: $5n + 10d$

In Example 6, the final expression $5n + 10d$ is measured in cents. This makes "sense" in the following way.

$$\frac{5 \text{ cents}}{\text{nickel}} \cdot n \text{ nickels} + \frac{10 \text{ cents}}{\text{dime}} \cdot d \text{ dimes}$$

Note that the nickels and dimes "cancel," leaving cents as the unit of measure for each term. This technique is called *unit analysis*, and it can be very helpful in determining the final unit of measure.

EXAMPLE 7 *Discovering Hidden Products*

Write an expression showing how far a person can ride a bicycle in t hours if the person travels at a constant rate of 12 miles per hour.

Solution

For this problem, use the formula (distance) = (rate)(time).

Verbal Model: | Rate | $\cdot$ | Time |

Labels: Rate = 12   (miles per hour)
              Time = t (hours)

Expression: $12t$

NOTE Using unit analysis, you can see that the expression in Example 7 has *miles* as its unit of measure.

$$12 \, \frac{\text{miles}}{\text{hour}} \cdot t \ \text{hours}$$

When translating verbal phrases involving percents, be sure you write the percent *in decimal form.* For instance, 25% should be written as 0.25. Remember that when you find a percent of a number, you multiply. For instance, 25% of 78 is given by

$$0.25(78) = 19.5.\qquad \text{25\% of 78}$$

EXAMPLE 8 *Discovering Hidden Operations*

A person adds k liters of a fluid containing 55% antifreeze to a car radiator. Write an expression that indicates how much antifreeze was added.

Solution

Verbal Model: | Percent antifreeze | $\cdot$ | Number of liters |

Labels: Percent of antifreeze = 0.55   (in decimal form)
              Number of liters = k (liters)

Expression: $0.55k$

Note that the algebraic expression uses the decimal form of 55%. That is, you compute with 0.55 rather than 55%.

EXAMPLE 9   Discovering Hidden Operations

You paid x dollars for an automobile. Your total cost included a 7% sales tax. Write an expression for the total cost of the automobile.

Solution

Verbal Model: Cost of automobile + Tax

Labels:
Cost of automobile $= x$ (dollars)
Sales tax rate $= 0.07$ (in decimal form)
Sales tax $= 0.07x$ (dollars)

Expression: $x + 0.07x = (1 + 0.07)x$

$$= 1.07x$$

Hidden operations are often involved when labels are assigned to *two* unknown quantities. For example, suppose two numbers add up to 18 and one of the numbers is assigned the variable x. What expression can you use to represent the second number? Let's try a specific case first, then apply it to a general case.

Specific case: If the first number is 7, the second number is $18 - 7 = 11$.

General case: If the first number is x, the second number is $18 - x$.

The strategy of using a *specific* case to help to determine the general case is often helpful in applications. Observe the use of this strategy in the next example.

EXAMPLE 10   Using Specific Cases to Model General Cases

a. A person's weekly salary is d dollars. Write an expression for the person's annual salary.

b. A person's annual salary is y dollars. Write an expression for the person's monthly salary.

Solution

a. *Specific Case:* If the weekly salary is $300, the annual salary is 52(300) dollars.
General Case: If the weekly salary is d dollars, the annual salary is $52 \cdot d$ or $52d$ dollars.

b. *Specific Case:* If the annual salary is $24,000, the monthly salary is $24,000 \div 12$ dollars.
General Case: If the annual salary is y dollars, the monthly salary is $y \div 12$ or $y/12$ dollars.

In mathematics it is useful to know how to represent certain types of integers algebraically. For instance, consider the set {2, 4, 6, 8, . . .} of *even* integers. What algebraic symbol could you use to denote an unknown, but even, integer? Because every even integer has 2 as a factor,

$$2 = 2 \cdot 1, \quad 4 = 2 \cdot 2, \quad 6 = 2 \cdot 3, \quad 8 = 2 \cdot 4, \ldots$$

it follows that any integer n multiplied by 2 is sure to be the *even* number $2n$. Moreover, if $2n$ is even, then $2n - 1$ and $2n + 1$ are sure to be *odd* integers. For example, choose $n = 5$. Then $2n = 2 \cdot 5 = 10$ is even, whereas $2n - 1 = 10 - 1 = 9$ and $2n + 1 = 10 + 1 = 11$ are both odd.

Two integers are called **consecutive integers** if they differ by 1. Hence, for any integer n, its next two larger consecutive integers are $n + 1$ and $(n + 1) + 1$ or $n + 2$. Thus, you can denote three consecutive integers by n, $n + 1$, and $n + 2$. These results are summarized below.

Labels for Integers

Let n represent an integer. Then even integers, odd integers, and consecutive integers can be represented as follows.

1. $2n$ denotes an *even* integer for $n = 1, 2, 3, \ldots$.

2. $2n - 1$ and $2n + 1$ denote *odd* integers for $n = 1, 2, 3, \ldots$.

3. $\{n, n + 1, n + 2, \ldots\}$ denotes a set of *consecutive* integers.

Group Activities   Extending the Concept

Avoiding Ambiguity   When you write a verbal model or construct an algebraic expression, watch out for statements that might be ambiguous. For instance, the statement

"The sum of 4 and x divided by 5"

could mean

$$\frac{4 + x}{5} \quad \text{or} \quad 4 + \frac{x}{5}.$$

Write up at least three other statements like this that can be interpreted in two different ways. Try them out with other people in your group to see whether they are truly ambiguous.

P.5 Exercises

Discussing the Concepts

1. The phrase *reduced by* indicates what operation?
2. The word *ratio* indicates what operation?
3. Translate the expression $3n + 2$ into a verbal phrase.
4. Which are equivalent to the expression $4x$?
   (a) x multiplied by 4
   (b) x increased by 4
   (c) the product of x and 4
   (d) the ratio of 4 and x

5. When each phrase below is translated into an algebraic expression, is order important? Explain.
   (a) y is multiplied by 5
   (b) 5 is decreased by y
   (c) y divided by 5
   (d) the sum of 5 and y
6. If n is an integer, what type of integers are $2n - 1$ and $2n + 1$? Explain.

Problem Solving

In Exercises 7–14, translate the statement into an algebraic expression.

7. The sum of 8 and a number n
8. The sum of 12 and twice a number n
9. Fifteen decreased by three times a number n
10. Seven-fifths of a number n
11. The quotient of a number x and 6
12. The product of a number y and 10 is decreased by 35.
13. The sum of 3 and four times x, all divided by 8
14. The absolute value of the quotient of y and 4

In Exercises 15–20, write a verbal description of the algebraic expression without using the variable.

15. $x - 5$
16. $2y + 3$
17. $\dfrac{z}{2}$
18. $\dfrac{4}{5}x$
19. $\dfrac{x - 2}{3}$
20. $-6(a + 5)$

In Exercises 21–26, write an algebraic expression that represents the specified quantity in the verbal statement and simplify if possible.

21. The amount of money (in dollars) represented by n quarters

22. The amount of money (in cents) represented by m dimes and n quarters
23. The distance traveled in t hours at an average speed of 55 miles per hour
24. The average rate of speed when traveling 360 miles in t hours
25. The amount of antifreeze in a cooling system containing y gallons of coolant that is 45% antifreeze
26. The amount of sales tax on a purchase valued at L dollars if the tax rate is 6%

Geometry In Exercises 27 and 28, write an expression for the area of the region.

27.

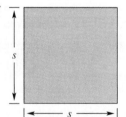

28.

29. Write an expression for the sum of three consecutive integers, the first of which is n.
30. Write an expression for the sum of two consecutive even integers, the first of which is $2n$.

Additional Problem Solving

In Exercises 31–44, translate the statement into an algebraic expression.

31. Five more than a number n

32. The total of 25 and three times a number n

33. Six less than a number n

34. Four times a number n minus 3

35. One-third of a number n

36. Forty percent of the cost C

37. Thirty percent of the list price L

38. The ratio of y and 3

39. The sum of x and 5 is divided by 10.

40. The number c is quadrupled and the product is increased by 10.

41. The product of three and the square of a number x, all decreased by 4

42. The sum of 10 and one-fourth the square of x

43. The absolute value of the difference between a number n and 5

44. Eight times the ratio of 3 and 5

In Exercises 45–56, write a verbal description of the algebraic expression without using the variable.

45. $t - 2$ **46.** $y + 50$

47. $3x + 2$ **48.** $4x - 5$

49. $8(x - 5)$ **50.** $-3(x + 2)$

51. $\dfrac{y}{8}$ **52.** $\dfrac{4x}{5}$

53. $\dfrac{x + 10}{3}$ **54.** $25 + \dfrac{x}{6}$

55. $x(x + 7)$ **56.** $x^2 + 2$

In Exercises 57–72, write an algebraic expression that represents the specified quantity in the verbal statement and simplify if possible.

57. The amount of money (in dollars) represented by x nickels

58. The amount of money (in dollars) represented by m dimes

59. The amount of money (in cents) represented by m nickels and n dimes

60. The amount of money (in cents) represented by x dimes and y quarters

61. The time to travel 100 miles at an average speed of r miles per hour

62. The distance traveled in 5 hours at an average speed of r miles per hour

63. The amount of wage tax due for a taxable income of I dollars that is taxed at the rate of 1.25%

64. The amount of water in q quarts of a food product that is 65% water

65. The sale price of a coat that has a list price of L dollars if the sale is a "20% off" sale

66. The total cost for a family to stay one night at a campground if the charge is $18 for the parents plus $3 for each of the n children

67. The total hourly wage for an employee when the base pay is $8.25 per hour plus 60 cents for each of q units produced per hour

68. The total hourly wage for an employee when the base pay is $11.65 per hour plus 80 cents for each of q units produced per hour

69. The auto repair bill if the cost for parts was $69.50 and there were t hours of labor at $32 per hour

70. The cost per acre of 50 acres of land that was sold for P dollars

71. The length of a rectangle if the length is 6 feet more than the width

72. The area of the rectangle shown in the figure

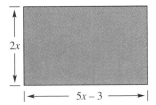

Geometry In Exercises 73–76, write expressions for the perimeter and area of the region. Then simplify the expressions.

73.
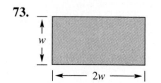

74.
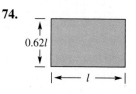

75.
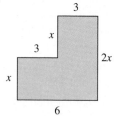

76.
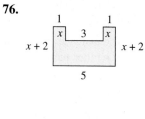

77. Write an expression for the sum of a number n and three times the number.

78. Write an expression for the sum of two consecutive integers, the first of which is n.

79. Write an expression for the sum of two consecutive odd integers, the first of which is $2n + 1$.

80. Write an expression for the product of two consecutive *even* integers divided by 4.

81. *Geometry* Write an expression that represents the area of the top of the billiard table in the figure. What is the unit of measure for the area?

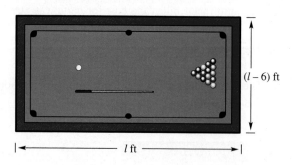

82. *Geometry* Write an expression that represents the area of the advertising banner (see figure). What are the units of measure for the area?

83. *Finding a Pattern* Complete the table. The third row contains the differences between consecutive entries of the second row. Describe the pattern of the third row.

n	0	1	2	3	4	5
$5n - 3$						
Differences						

84. *Finding a Pattern* Complete the table. The third row contains the differences between consecutive entries of the second row. Describe the pattern of the third row.

n	0	1	2	3	4	5
$3n + 1$						
Differences						

85. *Finding a Pattern* Using the results of Exercises 83 and 84, guess the third-row difference that would result in a similar table if the algebraic expression were $an + b$.

86. *Think About It* Find a and b so that the expression $an + b$ would yield the following table.

n	0	1	2	3	4	5
$an + b$	3	7	11	15	19	23

CHAPTER PROJECT: Playing the Market

Although the price of a stock is listed as the price per share, stocks are most commonly sold in lots of 100. If you buy 100 shares of stock at $42\frac{1}{2}$ dollars per share, your total bill is $100\left(42\frac{1}{2}\right) = \$4250$.

Investigate the questions using the tables below, which give the daily high and low prices per stock for the week of October 17, 1994 for three companies traded on the New York Stock Exchange. (Source: *Wall Street Journal*)

1. What was the difference between Monday's low price and Friday's low price for Cypress Semiconductor Corporation?

2. On Wednesday you bought 100 shares of each of the stocks listed in the table at the low price. Which of the following expressions can be used to calculate the amount of money you spent?

    (a) $17\frac{5}{8} + 54\frac{1}{4} + 67\frac{1}{4}$

    (b) $100\left(17\frac{5}{8}\right) + 100\left(54\frac{1}{4}\right) + 100\left(67\frac{1}{4}\right)$

    (c) $100\left(17\frac{5}{8} + 54\frac{1}{4} + 67\frac{1}{4}\right)$

3. Create a double-line graph for the high and low stock prices for Motorola, Inc. Which days had lows that were higher than Monday and Tuesday's high of $54\frac{1}{2}$? Explain how this is represented on your graph.

4. What strategy would you want to use to make the most money when buying and selling stocks? Explain.

5. *Playing the Market*   Divide the class into groups of two or three people and stake each group to $20,000 in play money. For four weeks, each group is to "invest" their money in stocks listed on the New York Stock Exchange. Buy and sell orders must be placed with the investment banker (someone chosen from the class) using the closing price of the day. After four weeks, determine which group made the most money (or lost the least). Discuss how their strategy was the same or different from those of other groups.

Cypress Semiconductor

Day	High	Low
Monday	$17\frac{5}{8}$	$16\frac{3}{4}$
Tuesday	18	17
Wednesday	$18\frac{1}{2}$	$17\frac{5}{8}$
Thursday	$18\frac{1}{4}$	$17\frac{3}{4}$
Friday	$18\frac{1}{4}$	$17\frac{7}{8}$

Motorola, Incorporated

Day	High	Low
Monday	$54\frac{1}{2}$	$53\frac{3}{4}$
Tuesday	$54\frac{1}{2}$	$53\frac{3}{4}$
Wednesday	$56\frac{1}{8}$	$54\frac{1}{4}$
Thursday	$56\frac{7}{8}$	$55\frac{1}{2}$
Friday	$56\frac{3}{8}$	$55\frac{5}{8}$

Texas Instruments

Day	High	Low
Monday	$69\frac{3}{4}$	$68\frac{1}{4}$
Tuesday	$68\frac{3}{4}$	$67\frac{3}{4}$
Wednesday	70	$67\frac{1}{4}$
Thursday	$70\frac{3}{4}$	$69\frac{5}{8}$
Friday	$70\frac{5}{8}$	$69\frac{1}{4}$

CHAPTER SUMMARY

After studying this chapter, you should have acquired the following skills. These skills are keyed to the Review Exercises that begin on page 53. Answers to odd-numbered Review Exercises are given in the back of the book.

- Plot real numbers on a number line and compare them by using inequality symbols. *(Section P.1)*

  Review Exercises 1– 4

- Evaluate expressions involving absolute value. *(Section P.1)*

  Review Exercises 5–8

- Evaluate expressions containing operations with real numbers. *(Section P.2)*

  Review Exercises 9–40, 71, 72

- Name the property of real numbers that justifies a statement. *(Section P.3)*

  Review Exercises 41– 46

- Identify the rule of algebra that is illustrated by an equation. *(Section P.4)*

  Review Exercises 47–50

- Expand expressions using the Distributive Property. *(Sections P.3, P.4)*

  Review Exercises 51–56

- Simplify expressions by removing symbols of grouping. *(Section P.4)*

  Review Exercises 57– 64

- Simplify expressions by applying the properties of exponents. *(Section P.4)*

  Review Exercises 65–70

- Evaluate expressions for specified values of the variable(s). *(Section P.4)*

  Review Exercises 73–76

- Use geometry to write expressions for the perimeters and areas of regions. *(Sections P.4, P.5)*

  Review Exercises 77– 80

- Interpret graphs representing real-life data. *(Section P.2)*

  Review Exercises 81–84

- Translate real-life situations into arithmetic expressions and solve. *(Section P.2)*

  Review Exercises 85, 86

- Perform calculator experiments and interpret the results. *(Section P.2)*

  Review Exercises 87, 88

- Translate verbal phrases into algebraic expressions. *(Section P.5)*

  Review Exercises 89–94, 99–104

- Translate algebraic expressions into verbal phrases. *(Section P.5)*

  Review Exercises 95–98

REVIEW EXERCISES

In Exercises 1–4, plot the real numbers on a number line and place the correct inequality symbol (< or >) between them.

1. $-\frac{1}{8}$ ___ 3

2. -2 ___ -8

3. $-\frac{8}{5}$ ___ $-\frac{2}{5}$

4. 8.4 ___ $-\pi$

In Exercises 5–8, evaluate the expression.

5. $|-7.2|$

6. $|1.6|$

7. $-|-7.2|$

8. $|-3.6|$

In Exercises 9–40, evaluate the expression. If it is not possible, state the reason. Write all fractions in reduced form.

9. $15 + (-4)$

10. $-12 + 3$

11. $340 - 115 + 5$

12. $-154 + 86 - 240$

13. $|-96| - |134|$

14. $16(3000)$

15. $120(-5)(7)$

16. $(-16)(-15)(-4)$

17. $\frac{-56}{-4}$

18. $\frac{85}{0}$

19. $\frac{45 - |-45|}{2}$

20. $\frac{288}{2 \cdot 3 \cdot 3}$

21. $\frac{4}{21} + \frac{7}{21}$

22. $\frac{21}{16} - \frac{13}{16}$

23. $-\frac{5}{6} + 1$

24. $\frac{21}{32} + \frac{11}{24}$

25. $8\frac{3}{4} - 6\frac{5}{8}$

26. $-2\frac{9}{10} + 5\frac{3}{20}$

27. $\frac{3}{8} \cdot \frac{-2}{15}$

28. $\frac{5}{21} \cdot \frac{21}{5}$

29. $-\frac{7}{15} \div -\frac{7}{30}$

30. $-\frac{2}{3} \div \frac{4}{15}$

31. $\frac{(3/4) - (1/2)}{5/8}$

32. $\frac{-(4/5) - (3/2)}{7/10}$

33. $\frac{6.25}{1.25}$

34. $-35 + 26.5 + 13.75$

35. $(-6)^3$

36. $-(-3)^4$

37. $\frac{3}{6^2}$

38. $-4(25 - 4^2)$

39. $120 - (5^2 \cdot 4)$

40. $45 - 45 \div 3^2$

In Exercises 41–46, name the property of real numbers that justifies the statement.

41. $13 - 13 = 0$

42. $7\left(\frac{1}{7}\right) = 1$

43. $7(9 + 3) = 7 \cdot 9 + 7 \cdot 3$

44. $15(4) = 4(15)$

45. $5 + (4 - y) = (5 + 4) - y$

46. $6(4z) = (6 \cdot 4)z$

In Exercises 47–50, identify the rule of algebra that is illustrated by the equation.

47. $(u - v)(2) = 2(u - v)$

48. $(x + y) + 0 = x + y$

49. $ab \cdot \frac{1}{ab} = 1$

50. $x(yz) = (xy)z$

In Exercises 51–56, expand the expression by using the Distributive Property.

51. $\frac{2}{3}(6s - 12t)$

52. $-5(2x - 4y)$

53. $-y(3y - 10)$

54. $x(3x + 4y)$

55. $-(-u + 3v)$

56. $(5 - 3j)(-4)$

In Exercises 57–64, simplify the expression.

57. $5(x - 4) + 10$

58. $15 - 7(z + 2)$

59. $3x - (y - 2x)$

60. $30x - (10x + 80)$

61. $-2(1 - 20u) + 2(1 - 10u)$

62. $9(11 - 2y) - 6(11 - 2y)$

63. $3[b + 5(b - a)]$

64. $-2t[8 - (6 - t)] + 5t$

In Exercises 65–70, simplify the expression.

65. $x^2 \cdot x^3 \cdot x$

66. $y^3(-2y^2)$

67. $(xy)(-3x^2y^3)$

68. $3uv(-2uv^2)^2$

69. $(-2a^2)^3(8a)$

70. $2(a-b)^4(a-b)^2$

71. Perform the indicated operations and simplify.

$$6 \cdot 10^3 + 9 \cdot 10^2 + 1 \cdot 10^1$$

72. Perform the indicated operations and simplify.

$$5 \cdot 10^3 + 4 \cdot 10^2 + 9 \cdot 10^1$$

In Exercises 73–76, evaluate the expression for the specified values of the variable(s).

Expression	*Values*
73. $x^2 - 2x - 3$	(a) $x = 3$ (b) $x = 0$
74. $x^3 - 125$	(a) $x = 5$ (b) $x = -2$
75. $3x^2 - y(x+1)$	(a) $x = 2, y = -1$ (b) $x = -1, y = 20$
76. $\dfrac{x}{y+2}$	(a) $x = 0, y = 3$ (b) $x = 5, y = 2$

Geometry   In Exercises 77–80, write expressions for the perimeter and area of the region, then simplify.

77.

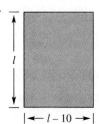

78.

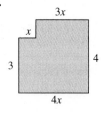

79.

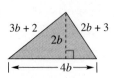

80.

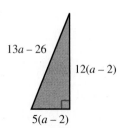

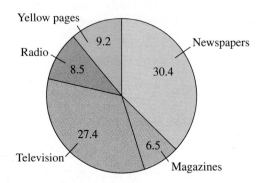

81. Determine the combined expenditures for advertising by the five media.

82. What is the difference in expenditures between television and radio?

Graphical Interpretation   In Exercises 83 and 84, use the figure, which shows the average weekly overtime for production workers in August for the years 1982, 1986, 1990, and 1994.   (Source: U.S. Department of Labor)

Weekly Overtime for Production Workers

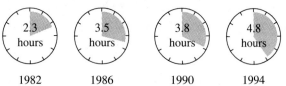

83. Determine the increase in average weekly overtime during the 4 years from 1990 to 1994.

84. During which 4-year period did the average weekly overtime increase the least?

85. *Total Charge*   You purchase a product and make a down payment of $239 plus nine monthly payments of $45 each. What is the total amount you pay for the product?

86. *Total Charge*   You purchase a product and make a down payment of $387 plus 12 monthly payments of $68 each. What is the total amount you pay for the product?

87. *Calculator Experiment*   Enter any number between 0 and 1 in a calculator. Take the square root of the number. Then take the square root of the result, and keep continuing this process. What number does the calculator display seem to be approaching?

88. *Calculator Experiment*   Use either a scientific or graphing calculator to calculate 12^4 in two ways.

12 $\boxed{y^x}$ 4 $\boxed{=}$ Scientific

12 $\boxed{x^2}$ $\boxed{x^2}$ Scientific

12 $\boxed{\wedge}$ 4 $\boxed{\text{ENTER}}$ Graphing

12 $\boxed{x^2}$ $\boxed{x^2}$ $\boxed{\text{ENTER}}$ Graphing

Why do these two methods give the same result?

In Exercises 89–94, translate the phrase into an algebraic expression. (Let n represent the arbitrary real number.)

89. Two hundred decreased by three times a number

90. One hundred increased by the product of 15 and a number

91. The sum of the square of a number and 49

92. The absolute value of the sum of a number and 10

93. The absolute value of the quotient of a number and 5

94. The sum of a number and the square of the number

In Exercises 95–98, write a verbal description of the algebraic expression without using the variable.

95. $2y + 7$ **96.** $5u - 3$

97. $\dfrac{x - 5}{4}$ **98.** $-3(a - 10)$

In Exercises 99–104, write an algebraic expression that represents the quantity given by the verbal statement.

99. The amount of income tax on a taxable income of I dollars when the tax rate is 18%

100. The distance traveled in 8 hours at an average speed of r miles per hour

101. The area of a rectangle whose length is l inches and whose width is five units less than its length

102. The sum of three consecutive odd integers, the first of which is $2n + 1$

103. The cost of 30 acres of land if the price per acre is p dollars

104. The time to copy z pages if the copy rate is 8 pages per minute

CHAPTER TEST

Take this test as you would take a test in class. After you are done, check your work against the answers given in the back of the book.

1. Place the correct symbol ($<$ or $>$) between the numbers.

    (a) $-\frac{5}{2}$ _____ $-|-3|$ (b) $-\frac{2}{3}$ _____ $-\frac{3}{2}$

2. Find the distance between -6.2 and 5.7.

In Exercises 3–8, evaluate the expression.

3. $-2(225 - 150)$

4. $\frac{2}{3} + \left(-\frac{7}{6}\right)$

5. $\left(-\frac{7}{16}\right)\left(-\frac{8}{21}\right)$

6. $\frac{5}{18} \div \frac{15}{8}$

7. $\left(-\frac{3}{5}\right)^3$

8. $\frac{4^2 - 6}{5} + 13$

9. Name the property of real numbers demonstrated by the equation.

    (a) $(-3 \cdot 5) \cdot 6 = -3(5 \cdot 6)$ (b) $3y \cdot \dfrac{1}{3y} = 1$

10. Rewrite the expression $5(2x - 3)$ using the Distributive Property.

In Exercises 11–14, simplify the expression.

11. $(3x^2 y)(-xy)^2$

12. $3x^2 - 2x - 5x^2 + 7x - 1$

13. $a(5a - 4) - 2(2a^2 - 2a)$

14. $4t - [3t - (10t + 7)]$

15. Explain the meaning of "evaluating an expression." Evaluate the expression $4 - (x + 1)^2$ for the value of x.

    (a) $x = -1$ (b) $x = 3$

16. Translate the following statement into an algebraic expression.

    "The product of a number n and 5 is decreased by 8."

17. Write algebraic expressions for the perimeter and area of the rectangle shown at the right. Then simplify the expressions.

18. Write an algebraic expression for the sum of two consecutive even integers, the first of which is $2n$.

19. It is necessary to cut a 144-foot rope into nine pieces of equal length. What is the length of each piece?

20. A *cord* of wood is a pile 4 feet high, 4 feet wide, and 8 feet long. The volume of a rectangular solid is its length times its width times its height. Find the number of cubic feet in 5 cords of wood.

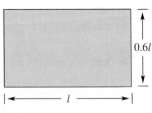

$0.6l$

Figure for 17

Linear Equations and Inequalities

1

- Linear Equations
- Linear Equations and Problem Solving
- Business and Scientific Problems
- Linear Inequalities
- Absolute Value Equations and Inequalities

Sounds are *emitted* by vibrations, which produce sound waves that travel through air or water. Sounds are *heard* when the sound waves are received by the ear, which transmits messages to the brain.

Sound waves are measured by their amplitude (loud or soft) and frequency (high or low). Frequencies are measured in hertz, with 1 hertz equaling one cycle or vibration per second.

The frequency ranges that different types of animals can emit and receive vary. The table and graph at the right show the frequency ranges in which humans and dolphins emit and receive sound. Notice that the ranges for dolphins are higher than those for humans. Thus, although dophins can hear most human sounds, humans cannot hear most dolphin sounds.

Animal	Frequency Emittance	Frequency Reception
Human	$85 \leq f \leq 1100$	$20 \leq f \leq 20{,}000$
Dolphin	$7000 \leq f \leq 120{,}000$	$150 \leq f \leq 150{,}000$

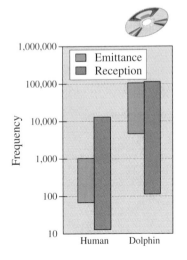

The chapter project related to this information is on page 115.

1.1	**Linear Equations**

Introduction ▪ Solving Linear Equations in Standard Form ▪
Solving Linear Equations in Nonstandard Form

Introduction

An **equation** is a statement that equates two mathematical expressions. Some
examples are

$$x = 4, \quad 4x + 3 = 15, \quad 2x - 8 = 2(x - 4), \quad \text{and} \quad x^2 - 16 = 0.$$

Solving an equation involving x means finding all values of x for which the
equation is true. Such values are **solutions** and are said to **satisfy** the equation.
For instance, 3 is a solution of $4x + 3 = 15$ because $4(3) + 3 = 15$ is a true
statement.

The **solution set** of an equation is the set of all solutions of the equation.
Sometimes, an equation will have the set of all real numbers as its solution set.
Such an equation is an **identity.** For instance, the equation

$$2x - 8 = 2(x - 4) \qquad \text{Identity}$$

is an identity because the equation is true for all real values of x. Try values such
as 0, 1, -2, and 5 in this equation to see that each one is a solution.

An equation whose solution set is not the entire set of real numbers is called
a **conditional equation.** For instance, the equation

$$x^2 - 16 = 0 \qquad \text{Conditional equation}$$

is a conditional equation because it has only two solutions, 4 and -4. Example 1
shows how to **check** whether a given value is a solution.

NOTE When checking a solution, we suggest you write a
question mark over the equal
sign to indicate that you are un-
certain whether the "equation"
is true.

EXAMPLE 1   *Checking a Solution of an Equation*

Decide whether -3 is a solution of $-3x - 5 = 4x + 16$.

Solution

$$-3x - 5 = 4x + 16 \qquad \text{Original equation}$$

$$-3(-3) - 5 \stackrel{?}{=} 4(-3) + 16 \qquad \text{Substitute } -3 \text{ for } x.$$

$$9 - 5 \stackrel{?}{=} -12 + 16 \qquad \text{Simplify.}$$

$$4 = 4 \qquad \text{Solution checks.} ✓$$

Because both sides turn out to be the same number, you can conclude that -3 is
a solution of the original equation. Try checking to see whether -2 is a solution.
You should be able to decide that it is not.

In the late 1800s, a movement was begun to identify a complete set of axioms for each branch of mathematics from which all other propositions could be deduced. In 1931 Kurt Gödel, a faculty member at the University of Vienna, showed that this goal was unattainable. He proved that a complete set of axioms could never be identified for a branch of mathematics such that all of its propositions could be proven or disproven on the basis of those axioms. Although this closed one avenue of research, Gödel also pointed out new directions for the future.

It is helpful to think of an equation as having two sides that are "in balance." Consequently, when you try to solve an equation, you must be careful to maintain that balance by performing the same operation(s) on both sides.

Two equations that have the same set of solutions are **equivalent.** For instance, the equations $x = 3$ and $x - 3 = 0$ are equivalent because both have only one solution—the number 3. When any one of the four techniques in the following list is applied to an equation, the resulting equation is equivalent to the original equation.

Forming Equivalent Equations: Properties of Equality

A given equation can be transformed into an *equivalent equation* using one or more of the following procedures.

	Original Equation	*Equivalent Equation*
1. *Simplify Each Side:* Remove symbols of grouping, combine like terms, or reduce fractions on one or both sides of the equation.	$3x - x = 8$	$2x = 8$
2. *Apply the Addition Property of Equality:* Add (or subtract) the same quantity to (from) *both* sides of the equation.	$x - 3 = 5$	$x = 8$
3. *Apply the Multiplication Property of Equality:* Multiply (or divide) *both* sides of the equation by the same nonzero quantity.	$3x = 9$	$x = 3$
4. *Interchange Sides:* Interchange the two sides of the equation.	$7 = x$	$x = 7$

When solving an equation, you can use any of the four techniques for forming equivalent equations to eliminate terms or factors in the equation. For example, to solve the equation $x + 4 = 2$, you need to get rid of the term $+4$ on the left side. This is accomplished by subtracting 4 from both sides.

$x + 4 = 2$	Original equation
$x + 4 - 4 = 2 - 4$	Subtract 4 from both sides.
$x + 0 = -2$	Combine like terms.
$x = -2$	Simplify.

Although this solution involved subtracting 4 from both sides, you could just as easily have added -4 to both sides. Both techniques are legitimate—which one you decide to use is a matter of personal preference.

Solving Linear Equations in Standard Form

The most common type of equation in one variable is a **linear equation.**

Definition of Linear Equation

A **linear equation** in one variable x is an equation that can be written in the standard form

$$ax + b = c$$

where a, b, and c are real numbers with $a \neq 0$.

A linear equation in one variable is also called a **first-degree equation** because its variable has an implied exponent of 1. Some examples of linear equations in the standard form $ax + b = c$ are $3x + 2 = 0$ and $5x - 4 = -6$.

Remember that to *solve* an equation in x means that you are to find the values of x that satisfy the equation. For a linear equation in the standard form $ax + b = c$, the goal is to **isolate** x by rewriting the standard equation in the form

$$x = \text{(a number)}.$$

Beginning with the original equation, you write a sequence of equivalent equations, each having the same solution as the original equation.

EXAMPLE 2   Solving a Linear Equation in Standard Form

$4x - 2 = 10$	Original equation
$4x - 2 + 2 = 10 + 2$	Add 2 to both sides.
$4x = 12$	Combine like terms.
$\dfrac{4x}{4} = \dfrac{12}{4}$	Divide both sides by 4.
$x = 3$	Simplify.

It appears that the solution is 3. You can check this as follows.

Check

$4x - 2 = 10$	Original equation
$4(3) - 2 \overset{?}{=} 10$	Substitute 3 for x.
$12 - 2 \overset{?}{=} 10$	Simplify.
$10 = 10$ ✓	Solution checks. ✓

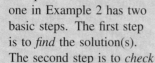

STUDY TIP

Be sure you see that solving an equation such as the one in Example 2 has two basic steps. The first step is to *find* the solution(s). The second step is to *check* that each solution you find actually satisfies the original equation. You can improve your accuracy in algebra by developing the habit of checking each solution.

You know that 3 is a solution of the equation in Example 2, but at this point you might be asking, "How can I be sure that the equation does not have other solutions?" The answer is that a linear equation in one variable always has *exactly* *one* solution. You can prove this with the following three steps.

NOTE   Equations in one variable that are not linear may have two or more solutions. For instance, the nonlinear equation $x^2 = 4$ has two solutions: 2 and -2.

$$ax + b = c \qquad\qquad \text{Original equation, with } a \neq 0$$

$$ax + b - b = c - b \qquad\qquad \text{Subtract } b \text{ from both sides.}$$

$$ax = c - b \qquad\qquad \text{Combine like terms.}$$

$$\frac{ax}{a} = \frac{c - b}{a} \qquad\qquad \text{Divide both sides by } a.$$

$$x = \frac{c - b}{a} \qquad\qquad \text{Simplify.}$$

It is clear that the last equation has only one solution, $x = (c - b)/a$. Moreover, because the last equation is equivalent to the given equation, you can conclude that every linear equation in one variable has exactly one solution.

EXAMPLE 3   Solving a Linear Equation in Standard Form

Solve the equation $2x - 3 = -5$.

Solution

$$2x - 3 = -5 \qquad\qquad \text{Original equation}$$

$$2x - 3 + 3 = -5 + 3 \qquad\qquad \text{Add 3 to both sides.}$$

$$2x = -2 \qquad\qquad \text{Combine like terms.}$$

$$\frac{2x}{2} = \frac{-2}{2} \qquad\qquad \text{Divide both sides by 2.}$$

$$x = -1 \qquad\qquad \text{Simplify.}$$

The solution is -1. Check this in the original equation.

Technology

A graphing utility can be used to check solutions of equations. For instance, here are the steps that will evaluate $2x - 3$ when $x = -1$ in Example 3 on a *TI-82*.

- Use [Y=] key to store the expression as Y_1.
- [2nd] [QUIT]
- Store -1 in X.
  -1 [STO▷] [X, T, θ] [ENTER]
- Display Y_1
  [Y-vars] [ENTER] [ENTER]
  and then press
  [ENTER] again.

Solve the equation $3(x - 2) + 5 = 10$. Check the solution using a graphing utility.

As you gain experience in solving linear equations, you will probably find that you can perform some of the solution steps in your head. For instance, you might solve the equation given in Example 3 by performing two of the steps mentally, and writing only the following steps.

$$2x - 3 = -5 \qquad\qquad \text{Original equation}$$

$$2x = -2 \qquad\qquad \text{Add 3 to both sides.}$$

$$x = -1 \qquad\qquad \text{Divide both sides by 2.}$$

Remember, however, that you should not skip the final step—checking your solution. You may find your calculator useful for checking solutions.

Solving Linear Equations in Nonstandard Form

Linear equations often occur in nonstandard forms that contain symbols of grouping or like terms that are not combined. Here are some examples.

$$x + 2 = 2x - 6, \quad 6(y - 1) + 4y = 3(7y + 1), \quad \frac{x}{18} + \frac{3x}{4} = 2$$

The next three examples show how to solve these linear equations.

EXAMPLE 4 Solving a Linear Equation in Nonstandard Form

$x + 2 = 2x - 6$	Original equation
$-2x + x + 2 = -2x + 2x - 6$	Add $-2x$ to both sides.
$-x + 2 = -6$	Combine like terms.
$-x + 2 - 2 = -6 - 2$	Subtract 2 from both sides.
$-x = -8$	Combine like terms.
$(-1)(-x) = (-1)(-8)$	Multiply both sides by -1.
$x = 8$	Simplify.

The solution is 8. Check this in the original equation.

In most cases, it helps to remove symbols of grouping as a first step to solving an equation. This is illustrated in Example 5.

EXAMPLE 5 Solving a Linear Equation in Nonstandard Form

$6(y - 1) + 4y = 3(7y + 1)$	Original equation
$6y - 6 + 4y = 21y + 3$	Distributive Property
$10y - 6 = 21y + 3$	Combine like terms.
$10y - 21y - 6 = 21y - 21y + 3$	Subtract $21y$ from both sides.
$-11y - 6 = 3$	Combine like terms.
$-11y - 6 + 6 = 3 + 6$	Add 6 to both sides.
$-11y = 9$	Combine like terms.
$\dfrac{-11y}{-11} = \dfrac{9}{-11}$	Divide both sides by -11.
$y = -\dfrac{9}{11}$	Simplify.

The solution is $-\frac{9}{11}$. Check this in the original equation.

If a linear equation contains fractions, we suggest that you first *clear the equation of fractions* by multiplying both sides of the equation by the least common denominator (LCD) of the fractions.

EXAMPLE 6   Solving a Linear Equation That Contains Fractions

Solve the equation $\dfrac{x}{18} + \dfrac{3x}{4} = 2$.

Solution

$$\dfrac{x}{18} + \dfrac{3x}{4} = 2 \qquad \text{Original equation}$$

$$36\left(\dfrac{x}{18} + \dfrac{3x}{4}\right) = 36(2) \qquad \text{Multiply both sides by LCD of 36.}$$

$$36 \cdot \dfrac{x}{18} + 36 \cdot \dfrac{3x}{4} = 36(2) \qquad \text{Distributive Property}$$

$$2x + 27x = 72 \qquad \text{Simplify.}$$

$$29x = 72 \qquad \text{Combine like terms.}$$

$$\dfrac{29x}{29} = \dfrac{72}{29} \qquad \text{Divide both sides by 29.}$$

$$x = \dfrac{72}{29} \qquad \text{Simplify.}$$

The solution is $\frac{72}{29}$. Check this in the original equation.

The next example shows how to solve a linear equation involving decimals. The procedure is basically the same, but the arithmetic can be messier.

EXAMPLE 7   Solving a Linear Equation Involving Decimals

NOTE   A different approach to Example 7 would be to begin by multiplying both sides of the equation by 100. This would clear the equation of decimals to produce

$12x + 9(5000 - x) = 51,300.$

Try solving this equation to see that you obtain the same solution.

$$0.12x + 0.09(5000 - x) = 513 \qquad \text{Original equation}$$

$$0.12x + 450 - 0.09x = 513 \qquad \text{Distributive Property}$$

$$(0.12x - 0.09x) + 450 = 513 \qquad \text{Group like terms.}$$

$$0.03x + 450 - 450 = 513 - 450 \qquad \text{Subtract 450 from both sides.}$$

$$0.03x = 63 \qquad \text{Combine like terms.}$$

$$\dfrac{0.03x}{0.03} = \dfrac{63}{0.03} \qquad \text{Divide both sides by 0.03.}$$

$$x = 2100 \qquad \text{Simplify.}$$

The solution is 2100. Check this in the original equation.

When solving equations that are in nonstandard form, you should be aware that it is possible for the equation to have *no solution*. This possibility is illustrated in Example 8.

EXAMPLE 8   A Linear Equation with No Solution

Solve the equation $2x - 4 = 2(x - 3)$.

Solution

$$2x - 4 = 2(x - 3)$$ Original equation

$$2x - 4 = 2x - 6$$ Distributive Property

$$-4 = -6$$ Subtract $2x$ from both sides.

Because the last equation has no solution, you can conclude that the original equation also has no solution.

Group Activities   Extending the Concept

Analyzing and Interpreting Equations   Classify each of the following equations as being an identity, being a conditional equation, or having no solution. Compare your conclusions with those of the rest of your group, and discuss the reasons for each conclusion.

a. $2x - 3 = -4 + 2x$

b. $x + 0.05x = 37.75$

c. $5x(3 + x) = 15x + 5x^2$

Discuss possible realistic situations in which the equations you classified as an identity and a conditional equation might apply. Write a brief description of these situations and explain how the equation could be used.

1.1 Exercises

Discussing the Concepts

1. Explain the difference between a conditional equation and an identity.

2. Explain how you can decide whether a real number is a solution of an equation.

3. Explain the difference between evaluating an expression and solving an equation.

4. Give the standard form of a linear equation. Why is a linear equation called a first-degree equation?

5. What is meant by equivalent equations? Give an example of two equivalent equations.

6. In your own words, describe the steps used to transform an equation into an equivalent equation.

Problem Solving

In Exercises 7–10, decide whether the values of the variable are solutions of the equation.

Equation	Values	
7. $3x - 7 = 2$	(a) $x = 0$	(b) $x = 3$
8. $5x + 9 = 4$	(a) $x = -1$	(b) $x = 2$
9. $\frac{1}{4}x = 3$	(a) $x = -4$	(b) $x = 12$
10. $3(y + 2) = y - 5$	(a) $y = -\frac{3}{2}$	(b) $y = -5.5$

In Exercises 11–14, identify the equation as a conditional equation, an identity, or an equation with no solution.

11. $3(x - 1) = 3x$

12. $2x + 8 = 6x$

13. $5(x + 3) = 2x + 3(x + 5)$

14. $\frac{2}{3}x + 4 = \frac{1}{3}x + 12$

In Exercises 15–18, determine whether the equation is linear. If it is not, state why.

15. $3x + 4 = 10$

16. $x^2 + 3 = 8$

17. $\frac{4}{x} - 3 = 5$

18. $3(x - 2) = 4x$

In Exercises 19–22, solve the equation in two ways. Then explain which way you prefer.

19. $2(x - 4) = 6$

20. $5(x + 1) = 6$

21. $\frac{1}{2}(x + 3) = 7$

22. $\frac{2}{3}(x - 4) = 8$

In Exercises 23 and 24, justify each step of the solution.

23.
$$3x + 15 = 0$$
$$3x + 15 - 15 = 0 - 15$$
$$3x = -15$$
$$\frac{3x}{3} = \frac{-15}{3}$$
$$x = -5$$

24.
$$7x - 21 = 0$$
$$7x - 21 + 21 = 0 + 21$$
$$7x = 21$$
$$\frac{7x}{7} = \frac{21}{7}$$
$$x = 3$$

In Exercises 25–40, solve the equation and check the result. (If it is not possible, state the reason.)

25. $3x = 12$

26. $8z - 10 = 0$

27. $7 - 8x = 13x$

28. $2s - 16 = 34s$

29. $-8t + 7 = -8t$

30. $4x = -12x$

31. $8(x - 8) = 24$

32. $6(x + 2) = 30$

33. $12y = 6(y + 1)$

34. $8x - 3(x - 2) = 12$

35. $t - \frac{2}{5} = \frac{3}{2}$

36. $z + \frac{1}{15} = -\frac{3}{10}$

37. $\frac{t}{5} - \frac{t}{2} = 1$

38. $\frac{t}{6} + \frac{t}{8} = 1$

39. $1.2(x - 3) = 10.8$

40. $6.5(1 - 2x) = 13$

In Exercises 41 and 42, solve the equation with the aid of a calculator. Round the solution to two decimal places.

41. $1.234x + 3 = 7.805$

42. $2x + \dfrac{1}{8.6} = \dfrac{3}{2}$

43. *Writing a Model*   The sum of two consecutive integers is 251. Find the integers.

44. *Writing a Model*   The bill for the repair of your car was \$210. The cost for parts was \$162. The cost for labor was \$32 per hour. How many hours did the repair work take?

Reviewing the Major Concepts

In Exercises 45–50, evaluate the expression.

45. $-360 + 120$

46. $5(57 - 33)$

47. $-\frac{4}{15} \times \frac{15}{16}$

48. $\frac{3}{8} \div \frac{5}{16}$

49. $(12 - 15)^3$

50. $\left(\frac{5}{8}\right)^2$

Additional Problem Solving

In Exercises 51–54, decide whether the values of the variable are solutions of the equation.

	Equation		Values
51.	$x + 8 = 3x$	(a) $x = 4$	(b) $x = -4$
52.	$10x - 3 = 7x$	(a) $x = 0$	(b) $x = -1$
53.	$3x + 3 = 2(x - 4)$	(a) $x = -11$	(b) $x = 5$
54.	$7x - 1 = 5(x + 5)$	(a) $x = 2$	(b) $x = 13$

In Exercises 55 and 56, justify each step of the solution.

55.
$$-2x + 5 = 12$$
$$-2x + 5 - 5 = 12 - 5$$
$$-2x = 7$$
$$\frac{-2x}{-2} = \frac{7}{-2}$$
$$x = -\frac{7}{2}$$

56.
$$25 - 3x = 10$$
$$25 - 3x + 3x = 10 + 3x$$
$$25 = 10 + 3x$$
$$25 - 10 = 10 + 3x - 10$$
$$15 = 3x$$
$$\frac{15}{3} = \frac{3x}{3}$$
$$5 = x$$

In Exercises 57–86, solve the equation and check the result. (If it is not possible, state the reason.)

57. $6y = 42$

58. $-14x = 28$

59. $6x + 4 = 0$

60. $3 - 2y = 5$

61. $23x - 4 = 42$

62. $15x - 18 = 27$

63. $4y - 3 = 4y$

64. $24 - 2x = x$

65. $8 - 5t = 20 + t$

66. $3y + 14 = y + 20$

67. $15t = 0$

68. $6a + 2 = 6a$

69. $5 - (2y - 4) = 15$

70. $26 - (3x - 10) = 6$

71. $-4(t + 2) = 0$

72. $8(z - 8) = 0$

73. $12(x + 3) = 7(x + 3)$

74. $-25(x - 100) = 16(x - 100)$

75. $2(x + 7) - 9 = 5(x - 4)$

76. $6[x - (5x - 7)] = 4 - 5x$

77. $\dfrac{u}{5} = 10$

78. $-\dfrac{z}{2} = 7$

79. $\frac{1}{3}x + 1 = \frac{1}{12}x - 4$

80. $\frac{1}{9}x + \frac{1}{3} = \frac{11}{18}$

81. $\dfrac{t + 4}{14} = \dfrac{2}{7}$

82. $\dfrac{11x}{6} + \dfrac{1}{3} = 0$

83. $\dfrac{25 - 4u}{3} = \dfrac{5u + 12}{4} + 6$

84. $\dfrac{8 - 3x}{4} - 4 = \dfrac{x}{6}$

85. $0.3x + 1.5 = 8.4$

86. $16.3 - 0.2x = 7.1$

In Exercises 87 and 88, solve the equation using a calculator. Round the solution to two decimal places.

87. $\dfrac{x}{10.625} = 2.850$ **88.** $325x - 4125 = 612$

89. *Finding a Pattern*   The length of a rectangle is t times its width. Thus, the perimeter P is given by $P = 2w + 2(tw)$, where w is the width of the rectangle. The perimeter of the rectangle is 1200 meters.

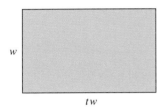

w

tw

(a) Complete the table of lengths, widths, and areas of the rectangle for the specified values of t.

t	1	1.5	2	3	4	5
Width						
Length						
Area						

(b) Use the table to write a short paragraph describing the relationship among the width, length, and area of a rectangle that has a *fixed* perimeter.

90. *Maximum Height of a Fountain*   Consider the fountain shown in the figure. The initial velocity of the stream of water is 48 feet per second. The velocity v of the water at any time t (in seconds) is then given by $v = 48 - 32t$. Find the time when the maximum height is obtained. Explain your reasoning.

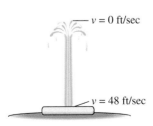

$v = 0$ ft/sec

$v = 48$ ft/sec

91. *Maximum Height of an Object*   The velocity v of an object projected vertically with an initial velocity of 64 feet per second is given by $v = 64 - 32t$, where t is time in seconds. When does the object reach its maximum height? Explain.

92. *Work Rate*   Two people can complete a task in t hours, where t must satisfy the equation

$$\frac{t}{10} + \frac{t}{15} = 1.$$

Find the required time t.

93. *Creating a Bar Graph*   The total amount of tuition and fees y (in millions of dollars) collected by colleges and universities in the United States from 1985 to 1990 can be approximated by the model

$$y = 8166.4 + 2526.2t, \quad 5 \le t \le 10$$

where $t = 5$ represents 1985. Create a bar graph that shows the amounts from 1985 through 1990. What can you conclude?   (Source: U.S. National Center for Education Statistics)

94. *Using a Model*   The average annual expenditures per student y for primary and secondary public schools in the United States from 1980 to 1991 can be approximated by the model

$$y = 2151.0 + 274.1t$$

where $t = 0$ represents 1980 (see figure). According to this model, during which year did the expenditures reach $4000? Explain how to answer the question graphically and algebraically.   (Source: National Education Association)

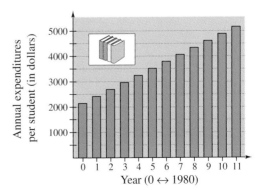

Year (0 ↔ 1980)

Using a Two-Part Model   In Exercises 95 and 96, use the following *two-part* model, which approximates the number of cable television subscribers in the United States from 1970 to 1993.

$$y = 4193 + 1099.4t, \qquad 0 \le t \le 9$$
$$y = -15{,}731 + 3210t, \qquad 10 \le t \le 23$$

In this model, y represents the number of subscribers (in thousands) and t represents the year, with $t = 0$ corresponding to 1970 (see figure).   (Source: Corporation for Public Broadcasting)

95. During which year were there 10,789.4 (thousand) cable television subscribers?

96. During which year were there 42,049 (thousand) cable television subscribers?

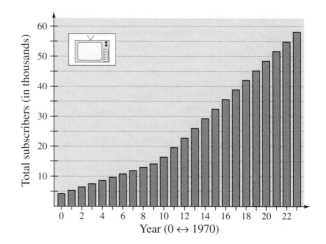

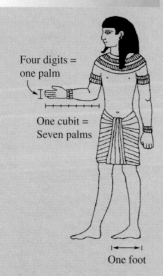

Math Matters   Miles or Kilometers

To convert from miles per hour to kilometers per hour, multiply miles per hour by 1.609344.

  Miles per hour × 1.609344 = kilometers per hour

To convert from kilometers per hour to miles per hour, multiply kilometers per hour by 0.621371.

  Kilometers per hour × 0.621371 = miles per hour

For a quick approximation that you can perform in your head, use the following rules.

1. *Quick Approximation:* To convert miles per hour to kilometers per hour, multiply by 8 and divide by 5. For instance, 50 miles per hour is approximately $(50 \times 8) \div 5 = 400 \div 5 = 80$ kilometers per hour.

2. *Quick Approximation:* To convert kilometers per hour to miles per hour, multiply by 5 and divide by 8. For instance, 88 kilometers per hour is approximately $(88 \times 5) \div 8 = 440 \div 8 = 55$ miles per hour.

In ancient Egypt, the units of measure were related to parts of the body. The *cubit* was the distance from the tip of a man's elbow to the tip of his middle finger. The *digit* was the width of one finger. These units fit together like this: four digits are equal to one *palm*, and seven palms are equal to one cubit.

1.2	**Linear Equations and Problem Solving**
	Mathematical Modeling ▪ Percent Problems ▪ Ratios and Unit Prices ▪ Solving Proportions

Mathematical Modeling

In this section you will see how algebra can be used to solve problems that occur in real-life situations. This process is called **mathematical modeling,** and its basic steps are shown below.

Verbal Description ⇨ Verbal Model ⇨ Assign Labels ⇨ Algebraic Equation

STUDY TIP

You could solve the problem in Example 1 *without* algebra by simply subtracting the bonus of $750 from the annual salary of $27,630 and dividing the result by 24 pay periods. The reason for listing this example is to allow you to practice writing algebraic versions of the problem-solving skills *you already possess.* Your goal in this section is to practice solving problems by common sense reasoning *and* to use this reasoning to write algebraic versions of the problems. Later, you will encounter more complicated problems in which algebra is a necessary part of the solution.

EXAMPLE 1   *Mathematical Modeling*

Write an algebraic equation that represents the following problem. Then solve the equation and answer the question.

You have accepted a job at an annual salary of $27,630. This salary includes a year-end bonus of $750. If you are paid twice a month, what will your gross pay be for each paycheck?

Solution

Because there are 12 months in a year and you will be paid twice a month, it follows that you will receive 24 paychecks during the year. Construct an algebraic equation for this problem as follows. Begin with a verbal model, then assign labels, and finally form an algebraic equation.

Verbal Model: | Income for year | = | 24 paychecks | + | Bonus |

Labels: Income for year = 27,630   (dollars)
Amount of each paycheck = x (dollars)
Bonus = 750   (dollars)

Equation:
$$27,630 = 24x + 750$$
$$27,630 - 750 = 24x + 750 - 750$$
$$26,880 = 24x$$
$$\frac{26,880}{24} = \frac{24x}{24}$$
$$1120 = x$$

Each paycheck will be $1120. Check this in the original statement of the problem.

Percent Problems

Numbers that describe rates, increases, decreases, and discounts are often given as percents. In applications involving percents, you need to convert the percent number to decimal (or fraction) form before performing any arithmetic operations. Some examples are listed in the following table.

Percent	10%	$12\frac{1}{2}\%$	20%	25%	$33\frac{1}{3}\%$	50%	$66\frac{2}{3}\%$	75%
Decimal	0.1	0.125	0.2	0.25	0.33...	0.5	0.66...	0.75
Fraction	$\frac{1}{10}$	$\frac{1}{8}$	$\frac{1}{5}$	$\frac{1}{4}$	$\frac{1}{3}$	$\frac{1}{2}$	$\frac{2}{3}$	$\frac{3}{4}$

The primary use of percents is to compare two numbers. For example, you can compare 3 to 6 by saying that 3 is 50% of 6. In this statement, 6 is the **base number,** and 3 is the number being compared to the base number. The following model, which is called the **percent equation,** is helpful.

Verbal Model: $\boxed{\dfrac{\text{Compared}}{\text{number}}} = \boxed{\dfrac{\text{Percent}}{\text{(decimal form)}}} \cdot \boxed{\dfrac{\text{Base}}{\text{number}}}$

Labels: Compared number $= a$

           Percent $= p$ (decimal form)

           Base number $= b$

Equation: $a = p \cdot b$ Percent equation

Remember to convert p to a decimal value before multiplying by b.

EXAMPLE 2   Solving a Percent Equation

The number 15.6 is 26% of what number?

Solution

Verbal Model: $\boxed{\dfrac{\text{Compared}}{\text{number}}} = \boxed{\dfrac{\text{Percent}}{\text{(decimal form)}}} \cdot \boxed{\dfrac{\text{Base}}{\text{number}}}$

Labels: Compared number $= 15.6$

           Percent $= 0.26$ (decimal form)

           Base number $= b$

Equation: $15.6 = 0.26b$

             $\dfrac{15.6}{0.26} = b$

               $60 = b$

Therefore, 15.6 is 26% of 60.

EXAMPLE 3 Solving a Percent Equation

The number 28 is what percent of 80?

Solution

Verbal Model:

| Compared number | = | Percent (decimal form) | · | Base number |

Labels: Compared number $= 28$
Percent $= p$ (decimal form)
Base number $= 80$

Equation: $28 = p(80)$

$$\frac{28}{80} = p$$

$$0.35 = p$$

Therefore, 28 is 35% of 80. Check this by multiplying 80 by 0.35 to see that you obtain 28.

In most real-life applications, the base number b and the number a are much more disguised than in Examples 2 and 3. It sometimes helps to think of a as a "new" amount and b as the "original" amount.

EXAMPLE 4 A Percent Application

A real estate agency receives a commission of $8092.50 for the sale of a $124,500 house. What percent commission is this?

Solution

Verbal Model:

| Commission | = | Percent (decimal form) | · | Sale price |

Labels: Commission $= 8092.50$ (dollars)
Percent $= p$ (decimal form)
Sale price $= 124,500$ (dollars)

Equation: $8092.50 = p(124,500)$

$$\frac{8092.50}{124,500} = p$$

$$0.065 = p$$

The real estate agency receives a commission of 6.5%. Use your calculator to check this solution in the original statement of the problem.

Ratios and Unit Prices

If a and b have the same units of measure, then a/b is called the **ratio** of a to b. Note the *order* implied by a ratio. The ratio of a to b means a/b, whereas the ratio of b to a means b/a.

EXAMPLE 5   Using a Ratio

Find the ratio of 4 feet to 8 inches.

Solution

You can convert 4 feet to 48 inches (by multiplying by 12) to obtain

$$\frac{4 \text{ feet}}{8 \text{ inches}} = \frac{48 \text{ inches}}{8 \text{ inches}} = \frac{48}{8} = \frac{6}{1}.$$

Or, you can convert 8 inches to $\frac{8}{12}$ feet (by dividing by 12) to obtain

$$\frac{4 \text{ feet}}{8 \text{ inches}} = \frac{4 \text{ feet}}{\frac{8}{12} \text{ feet}} = 4 \cdot \frac{12}{8} = \frac{6}{1}.$$

The **unit price** of an item is the quotient of the total price divided by the total units. That is, Unit price = total price/total units. To state unit prices, use the word "per." For instance, the unit price for a particular brand of coffee might be 4.79 dollars *per* pound.

EXAMPLE 6   Comparing Unit Prices

Which is the better buy, a 12-ounce box of breakfast cereal for $2.69 or a 16-ounce box of the same cereal for $3.49?

Solution

The unit price for the smaller box is

$$\text{Unit price} = \frac{\text{total price}}{\text{total units}} = \frac{\$2.69}{12 \text{ ounces}} \approx \$0.224 \text{ per ounce.}$$

The unit price for the larger box is

$$\text{Unit price} = \frac{\text{total price}}{\text{total units}} = \frac{\$3.49}{16 \text{ ounces}} \approx \$0.218 \text{ per ounce.}$$

The larger box has a slightly smaller unit price; therefore, it is the better buy.

Solving Proportions

A **proportion** is a statement that equates two ratios. For example, if the ratio of a to b is the same as the ratio of c to d, you can write the proportion as $a/b = c/d$. In typical problems, you know three of the values and need to find the fourth.

EXAMPLE 7   Solving a Proportion

The ratio of x to 8 is the same as the ratio of 2 to 5. What is x?

Solution

$$\frac{x}{8} = \frac{2}{5}$$ Set up proportion.

$$8 \cdot \frac{x}{8} = 8 \cdot \frac{2}{5}$$ Multiply both sides by 8.

$$x = \frac{16}{5}$$ Simplify.

Thus, $x = \frac{16}{5}$. Check this in the original statement of the problem.

EXAMPLE 8   An Application of Proportion

You are driving from Arizona to New York, a trip of 2750 miles. You begin the trip with a full tank of gas and after traveling 424 miles, you refill the tank for $22.00. How much should you plan to spend on gasoline for the entire trip?

Solution

Verbal Model:
$$\frac{\text{Dollars for trip}}{\text{Dollars for tank}} = \frac{\text{Miles for trip}}{\text{Miles for tank}}$$

Labels:

Cost of gas for entire trip $= x$	(dollars)
Cost of gas for tank $= 22$	(dollars)
Miles for entire trip $= 2750$	(miles)
Miles for tank $= 424$	(miles)

Proportion:
$$\frac{x}{22} = \frac{2750}{424}$$

$$x = 22 \cdot \left(\frac{2750}{424}\right)$$

$$x \approx 142.69$$

You should plan to spend approximately $142.69 for gasoline on the trip. Check this in the original statement of the problem.

The following list summarizes the strategy for modeling and solving real-life problems.

Strategy for Solving Word Problems

1. Ask yourself what you need to know to solve the problem. Then *write a verbal model* that will give you what you need to know.

2. *Assign labels* to each part of the verbal model—numbers to the known quantities and letters (or expressions) to the variable quantities.

3. Use the labels to *write an algebraic model* based on the verbal model.

4. *Solve* the resulting algebraic equation.

5. *Answer* the original question and *check* that your answer satisfies the original problem as stated.

In previous mathematics courses, you studied several other problem-solving strategies, such as *drawing a diagram*, *making a table*, *looking for a pattern*, and *solving a simpler problem*. Each of these strategies can also help you to solve problems in algebra.

Group Activities   Extending the Concept

Checking the Sensibility of an Answer   When you solve problems related to real-life situations, you should always ask yourself whether your answer makes sense. In your group, discuss why the following answers are suspicious and decide what a more reasonable answer might be.

a. A problem asked you to find the life expectancy of an American male born in 1970, and you use a formula to obtain a preliminary answer of 124 years.

b. A problem asked you to find the Celsius temperature equivalent to 80° Fahrenheit, and you obtain a preliminary answer of −24°C.

c. A problem asked you to find the diameter of a household electrical extension cord, and you obtain a preliminary answer of 5.0 inches.

d. A problem asked you to find the floor area of a gymnasium, and you obtain a preliminary answer of 300 square feet.

1.2 Exercises

Discussing the Concepts

1. Explain the meaning of the word *percent*.
2. Explain how to change percents to decimals and decimals to percents. Give examples.
3. Is it true that $\frac{1}{2}\% = 50\%$? Explain.
4. Define the term *ratio*. Give an example of a ratio.
5. You are told that the ratio of the number of boys to the number of girls in a class is 2 to 1. Is this enough information to determine the number of students in the class? Explain your reasoning.
6. In your own words, describe the meaning of *mathematical modeling*. Give an example.

Problem Solving

In Exercises 7 and 8, construct a verbal model and write an algebraic equation that represents the problem. Do not solve the equation.

7. *Geometry*   A "Slow Moving Vehicle" sign has the shape of an equilateral triangle. The sign has a perimeter of 129 centimeters. Find the length of each side. Include a labeled diagram with your model.

8. *Mathematical Modeling*   You have a job on an assembly line for which you are paid $10 per hour plus $0.75 per unit assembled. Find the number of units produced in an 8-hour day if your earnings for the day are $146.

In Exercises 9–14, complete the table, which shows the equivalent forms of various percents.

Percent	Parts out of 100	Decimal	Fraction
9. 30%			
10. 75%			
11.		0.075	
12.			$\frac{2}{3}$
13.			$\frac{1}{8}$
14.	100		

15. What is 35% of 250?
16. What is 300% of 16?
17. 96 is 0.8% of what number?
18. 496 is what percent of 800?
19. 1650 is what percent of 5000?
20. 2.4 is what percent of 480?

21. *Monthly Rent*   If you spend 15% of your monthly income of $2800 for rent, what is your monthly rent payment?

22. *Number of Eligible Voters*   The news media reported that 7387 votes were cast in the last election and this represented 63% of the eligible voters of a district. Assuming this is true, how many eligible voters are in the district?

23. *Salary Adjustment*   During a year of financial difficulties your company reduces your salary by 7%. What percent increase in this reduced salary is required to raise your salary to the amount it was prior to the reduction? Why isn't the percent increase the same as the percent of the reduction?

24. *Geometry*   The floor of a room is 10 feet by 12 feet and is partially covered by a circular rug with a radius of 4 feet, as shown in the figure. What percent of the floor is covered by the rug? (The area of a circle is $A = \pi r^2$, where r is the radius of the circle.)

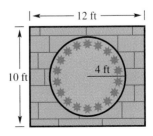

In Exercises 25 and 26, express the verbal expression as a ratio. Use the same units in both the numerator and denominator, and simplify.

25. 36 inches to 48 inches

26. 125 centimeters to 2 meters

27. *Compression Ratio* The compression ratio of a cylinder is the ratio of its expanded volume to its compressed volume (see figure). The expanded volume of one cylinder of a diesel engine is 425 cubic centimeters, and its compressed volume is 20 cubic centimeters. Find the compression ratio of this engine.

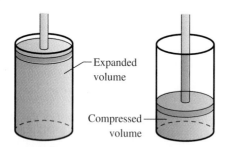

Expanded volume

Compressed volume

28. *Price-Earnings Ratio* The ratio of the price of a stock to its earnings is called the **price-earnings ratio.** Find the price-earnings ratio of a stock that sells for $56.25 per share and earns $6.25 per share.

Unit Prices In Exercises 29–32, find the unit price (in dollars per ounce) of the product.

29. A 20-ounce can of pineapple for 95¢

30. A 32-ounce bottle of cola for $1.89

31. A 1-pound, 4-ounce loaf of bread for $1.39

32. A 28-ounce box of cereal for $3.49

Comparison Shopping In Exercises 33–36, use unit prices to determine the better buy.

33. (a) A $14\frac{1}{2}$-ounce bag of chips for $2.32

(b) A 32-ounce bag of chips for $4.85

34. (a) A $10\frac{1}{2}$-ounce package of cookies for $1.79

(b) A 16-ounce package of cookies for $2.39

35. (a) An 8-ounce tube of toothpaste for $1.69

(b) A 12-ounce tube of toothpaste for $2.39

36. (a) A 2-pound package of hamburger for $3.49

(b) A 3-pound package of hamburger for $5.29

In Exercises 37–40, solve the proportion.

37. $\dfrac{t}{4} = \dfrac{3}{2}$

38. $\dfrac{5}{16} = \dfrac{x}{4}$

39. $\dfrac{y+1}{10} = \dfrac{y-1}{6}$

40. $\dfrac{a}{5} = \dfrac{a+4}{8}$

41. *Property Tax* The taxes on property with an assessed value of $75,000 are $1125. Find the taxes on property with an assessed value of $120,000.

42. *Map Scale* Use the map scale in the figure to approximate the straight-line distance from Los Angeles to San Francisco.

0 40
miles

San Francisco

CALIFORNIA

Los Angeles

Geometry In Exercises 43 and 44, solve for the length *x* of the side of the triangle by using the fact that corresponding sides of similar triangles are proportional.

43.

44.

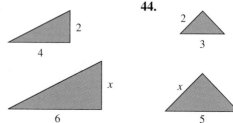

Reviewing the Major Concepts

In Exercises 45–50, solve the equation. (If it is not possible, state the reason.)

45. $x + \dfrac{x}{2} = 4$

46. $\dfrac{x}{3} + 1 = 10$

47. $8(x - 14) = 0$

48. $(1 + r)500 = 550$

49. $12(3 - x) = 5 - 7(2x + 1)$

50. $-(2x + 8) = \frac{1}{3}(6x + 5)$

Additional Problem Solving

In Exercises 51–54, construct a verbal model and write an algebraic equation that represents the problem. Do not solve the equation.

51. *Mathematical Modeling*   Find a number such that the sum of the number and 30 is 82.

52. *Mathematical Modeling*   The selling price of a jacket in a department store is $85. The cost of the jacket to the store is $62.95. What is the markup?

53. *Mathematical Modeling*   The bill for the repair of an automobile is $380. Included in this bill is a charge of $275 for parts. If the remainder of the bill is for labor, how many hours were spent to repair the car? (The charge for labor is $35 per hour.)

54. *Geometry*   The length of a rectangle is three times its width. The perimeter of the rectangle is 64 inches. Find the dimensions of the rectangle.

In Exercises 55 and 56, what percent of the entire figure is shaded? (Assume that each rectangular portion of the figure has the same area.)

55.

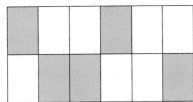

56.

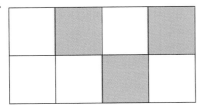

57. What is 8.5% of 816?

58. What is 68% of 800?

59. What is 0.4% of 150,000?

60. What is $33\frac{1}{3}$% of 816?

61. 84 is 24% of what number?

62. 416 is 65% of what number?

63. 42 is $10\frac{1}{2}$% of what number?

64. 168 is 350% of what number?

65. 2100 is what percent of 1200?

66. 900 is what percent of 500?

67. *Pension Fund*   Your employer withholds $6\frac{1}{2}$% of your gross monthly income for your retirement. Determine the amount withheld each month if your gross monthly income is $3800.

68. *College Enrollment*   Thirty-eight percent of the students enrolled at a college are freshmen. Determine the number of freshmen if the enrollment of the college is 3000.

69. *Company Layoff*   Because of slumping sales, a small company laid off 25 of its 160 employees. What percent of the work force was laid off?

70. *Population Growth*   In the 1970 census, the population of a city was 60,000. The population grew by 4% during the decade of the seventies and it grew by 6% during the decade of the eighties.

(a) Use the given information to approximate the population of the city at the time of the 1980 census and at the time of the 1990 census.

(b) Use the results of part (a) to approximate the percent increase in the population between 1970 and 1990. Explain why it isn't 10%.

71. *Defective Parts*   A quality control engineer reports that 1.5% of a sample of parts are defective. The engineer found three defective parts. How large was the sample?

72. *Price Inflation*   A new van costs $25,750, which is about 115% of what it was 3 years ago. What did it cost 3 years ago?

73. *Course Grade*   You missed a B by 8 points. Your point total for the course is 372. How many points were possible in the course? (Assume that you needed 80% of the course total for a B.)

74. *Real Estate Commission*   A real estate agency receives a commission of $9100 for the sale of a $130,000 house. What percent commission is this?

Graphical Estimation   In Exercises 75–78, use the bar graph to answer the questions. The graph shows the per capita food consumption of selected meats for 1970, 1980, and 1991. (Source: U.S. Department of Agriculture)

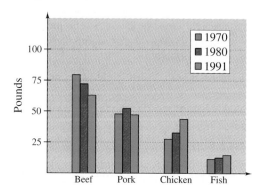

75. Approximate the decrease in the per capita consumption of beef from 1970 to 1991. Use this estimate to approximate the percent decrease.

76. Approximate the increase in the per capita consumption of chicken from 1970 to 1991. Use this estimate to approximate the percent increase.

77. Approximate the total number of pounds of pork consumed in 1991 if the population of the United States was approximately 250 million.

78. Of the four categories of meats shown in the graph, what percent of the meat diet was met by fish in 1991?

79. *Reading a Circle Graph*   The expenses for a small company for January are shown in the circle graph. What percent of the total monthly expenses is each budget item?

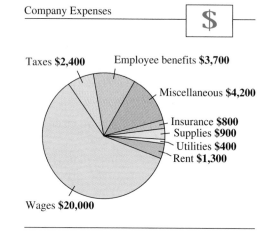

80. *Reading a Circle Graph*   The populations of six counties are shown in the circle graph. What percent of the total population is each county's population?

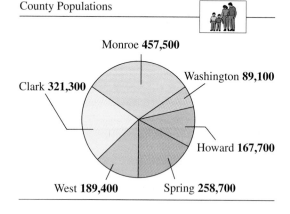

In Exercises 81–84, write the fraction as a ratio. Use the same units in both the numerator and denominator, and simplify.

81. 1 pint to 1 gallon

82. 5 pounds to 24 ounces

83. 40 milliliters to 1 liter

84. 45 minutes to 2 hours

85. *State Income Tax*   You have $12.50 of state tax withheld from your paycheck per week when your gross pay is $625. Find the ratio of tax to gross pay.

86. *Gear Ratio*   The gear ratio of two gears is the number of teeth in one gear to the number of teeth in the other gear (see figure). If two gears in a gearbox have 60 teeth and 40 teeth, find the gear ratio.

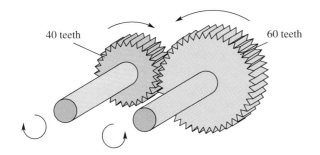

40 teeth   60 teeth

In Exercises 87–92, solve the proportion.

87. $\dfrac{x}{6} = \dfrac{2}{3}$ **88.** $\dfrac{y}{36} = \dfrac{6}{7}$

89. $\dfrac{5}{4} = \dfrac{t}{6}$ **90.** $\dfrac{7}{8} = \dfrac{x}{2}$

91. $\dfrac{y+5}{6} = \dfrac{y-2}{4}$ **92.** $\dfrac{z-3}{3} = \dfrac{z+8}{12}$

93. *Spring Length*   A force of 32 pounds stretches a spring 6 inches. Determine the number of pounds of force required to stretch it 9 inches.

94. *Fuel Mixture*   The gasoline-to-oil ratio of a two-cycle engine is 40 to 1. Determine the amount of gasoline required to produce a mixture that has $\frac{1}{2}$ pint of oil.

95. *Defective Units*   A quality control engineer for a manufacturer found one defective unit in a sample of 75. At that rate, what is the expected number of defective units in a shipment of 200,000?

96. *Public Opinion Poll*   In a public opinion poll, 870 people from a sample of 1500 indicated they would vote for Candidate A. Assuming this poll to be correct, how many votes can the candidate expect to receive if 80,000 votes are cast?

Geometry   In Exercises 97 and 98, solve for the length *x* of the side of the triangle by using the fact that corresponding sides of similar triangles are proportional.

97.

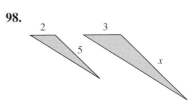

5.5   7

4   x

98.

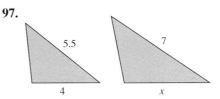

2   3

5

x

99. *Geometry*   Find the length of the shadow of a man who is 6 feet tall and is standing 15 feet from a streetlight that is 20 feet high (see figure).

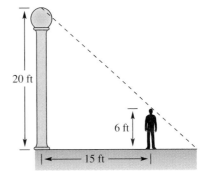

20 ft

6 ft

|← 15 ft →|

100. *Geometry*   Find the ratio of the areas of the two circles in the figure. (Note: The area of a circle is $A = \pi r^2$.)

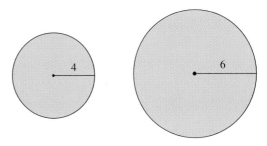

4   6

1.3 Business and Scientific Problems

Business Problems ▪ Mixture Problems ▪
Rate Problems ▪ Formulas

Business Problems

Many business problems can be represented by mathematical models involving the sum of a fixed term and a variable term. The variable term is often a *hidden product* in which one of the factors is a percent or some other type of rate. Watch for these occurrences in the discussion and examples that follow.

The **markup** on a consumer item is the difference between the **cost** a retailer pays for an item and the **price** at which the retailer sells the item. A verbal model for this relationship is as follows.

$$\boxed{\text{Selling price}} = \boxed{\text{Cost}} + \boxed{\text{Markup}}$$ Markup is a hidden product.

The markup is the product of the **markup rate** and the cost.

$$\boxed{\text{Markup}} = \boxed{\text{Markup rate}} \cdot \boxed{\text{Cost}}$$

EXAMPLE 1   *Finding the Markup Rate*

A clothing store sells a pair of jeans for $42. If the cost of the jeans is $16.80, what is the markup rate?

Solution

Verbal Model: $\boxed{\text{Selling price}} = \boxed{\text{Cost}} + \boxed{\text{Markup}}$

Labels: Selling price $= 42$ (dollars)
Cost $= 16.80$ (dollars)
Markup rate $= p$ (percent in decimal form)
Markup $= p(16.80)$ (dollars)

Equation:
$$42 = 16.8 + p(16.8)$$
$$42 - 16.8 = p(16.8)$$
$$25.2 = p(16.8)$$
$$\frac{25.2}{16.8} = p$$
$$1.5 = p$$

Because $p = 1.5$, it follows that the markup rate is 150%. Check this in the original statement of the problem.

The model for a **discount** is similar to that for a markup.

EXAMPLE 2   *Finding the Discount and the Discount Rate*

A compact disc player is marked down from its list price of $820 to a sale price of $574. What is the discount rate?

Solution

Verbal Model: $\boxed{\text{Discount}} = \boxed{\begin{array}{c}\text{Discount}\\\text{rate}\end{array}} \cdot \boxed{\begin{array}{c}\text{List}\\\text{price}\end{array}}$

Labels: Discount $= 820 - 574 = 246$ (dollars)
List price $= 820$ (dollars)
Discount rate $= p$ (percent in decimal form)

Equation: $246 = p(820)$

$$\frac{246}{820} = p$$

$$0.30 = p$$

The discount rate is 30%. Check this in the original statement of the problem.

EXAMPLE 3   *Finding the Hours of Labor*

An auto repair bill of $338 lists $170 for parts and the rest for labor. If the labor is $28 per hour, how many hours did it take to repair the auto?

Solution

Verbal Model: $\boxed{\begin{array}{c}\text{Total}\\\text{bill}\end{array}} = \boxed{\begin{array}{c}\text{Price}\\\text{of parts}\end{array}} + \boxed{\begin{array}{c}\text{Price}\\\text{of labor}\end{array}}$

Labels: Total bill $= 338$ (dollars)
Price of parts $= 170$ (dollars)
Hours of labor $= x$ (hours)
Hourly rate for labor $= 28$ (dollars per hour)
Price of labor $= 28x$ (dollars)

Equation: $338 = 170 + 28x$

$$168 = 28x$$

$$\frac{168}{28} = x$$

$$6 = x$$

It took 6 hours to repair the auto. Check this in the original statement of the problem.

STUDY TIP

Although markup and discount are similar, it is important to remember that markup is based on cost and discount is based on list price.

Mixture Problems

Many real-life problems involve combinations of two or more quantities that make up new or different quantities. Such problems are called **mixture problems.** They are usually composed of the sum of two or more "hidden products" that fit the following verbal model.

$$\boxed{\text{First rate}} \cdot \boxed{\text{Amount}} + \boxed{\text{Second rate}} \cdot \boxed{\text{Amount}} = \boxed{\text{Final rate}} \cdot \boxed{\text{Final amount}}$$

EXAMPLE 4   A Mixture Problem

A nursery wants to mix two types of lawn seed; one type sells for $10 per pound and the other type sells for $15 per pound. To obtain 20 pounds of a mixture at $12 per pound, how many pounds of each type of seed are needed?

Solution

Verbal Model:

$$\boxed{\text{Total cost of \$10 seed}} + \boxed{\text{Total cost of \$15 seed}} = \boxed{\text{Total cost of \$12 seed}}$$

Labels:
Cost of $10 seed = 10 (dollars per pound)
Pounds of $10 seed = x (pounds)
Cost of $15 seed = 15 (dollars per pound)
Pounds of $15 seed = $20 - x$ (pounds)
Cost of $12 seed = 12 (dollars per pound)
Pounds of $12 seed = 20 (pounds)

Equation:
$$10x + 15(20 - x) = 12(20)$$
$$10x + 300 - 15x = 240$$
$$-5x = -60$$
$$x = 12$$

The mixture should contain 12 pounds of the $10 seed and $20 - 12 = 8$ pounds of the $15 seed.

STUDY TIP

When you set up a verbal model, be sure to check that you are working with the *same type of units* in each part of the model. For instance, in Example 4 note that each of the three parts of the verbal model is measuring cost. (If two parts were measuring cost and the other part was measuring pounds, you would know that the model was incorrect.)

Remember that when you have found a solution, you should always go back to the original statement of the problem and check to see that the solution makes sense—both algebraically and from a common sense point of view. For instance, you can check the result of Example 4 as follows.

$$\overbrace{\left(\tfrac{\$10\text{ per}}{\text{pound}}\right)\left(\tfrac{12}{\text{pounds}}\right)}^{\$10\text{ seed}} + \overbrace{\left(\tfrac{\$15\text{ per}}{\text{pound}}\right)\left(\tfrac{8}{\text{pounds}}\right)}^{\$15\text{ seed}} = \overbrace{\left(\tfrac{\$12\text{ per}}{\text{pound}}\right)\left(\tfrac{20}{\text{pounds}}\right)}^{\$12\text{ seed}}$$
$$\$120 + \$120 = \$240$$

Rate Problems

Time-dependent problems such as those involving distance can be classified as **rate problems.** They fit the verbal model

$$\boxed{\text{Distance}} = \boxed{\text{Rate}} \cdot \boxed{\text{Time}}.$$

For instance, if you travel at a constant (or average) rate of 55 miles per hour for 45 minutes, the total distance you travel is given by

$$\left(55 \; \frac{\text{miles}}{\text{hour}}\right)\left(\frac{45}{60} \; \text{hour}\right) = 41.25 \text{ miles}.$$

As with all problems involving applications, be sure to check that the units in the verbal model make sense. For instance, in this problem the rate is given in *miles per hour*. Therefore, in order for the solution to be given in *miles*, you must convert the time (from minutes) to *hours*. In the model, you can think of the two "hours" as canceling, as follows.

$$\left(55 \; \frac{\text{miles}}{\text{hour}}\right)\left(\frac{45}{60} \; \text{hour}\right) = 41.25 \text{ miles}$$

EXAMPLE 5 *Distance-Rate-Time Problem*

Suppose that you jog at an average rate of 9 kilometers per hour. How long will it take you to jog 15 kilometers?

Solution

Verbal Model: $\boxed{\text{Distance}} = \boxed{\text{Rate}} \cdot \boxed{\text{Time}}$

Labels: Distance = 15   (kilometers)
Rate = 9   (kilometers per hour)
Time = t (hours)

Equation: $15 = 9(t)$

$$\frac{15}{9} = t$$

$$\frac{5}{3} = t$$

It would take you $1\frac{2}{3}$ hours (or 1 hour and 40 minutes). You can check this in the original statement of the problem as follows.

Check

$$\left(9 \; \frac{\text{kilometers}}{\text{hour}}\right)\left(\frac{5}{3} \; \text{hours}\right) = 15 \text{ kilometers}$$

In work problems, the **rate of work** is the *reciprocal* of the time needed to do the entire job. For instance, if it takes 5 hours to complete a job, then the per hour work rate is

$$\frac{1}{5} \text{ job per hour.}$$

The next example involves two rates of work and thus fits the model for solving *mixture* problems.

EXAMPLE 6 *Work-Rate Problem*

Consider two machines in a paper manufacturing plant. Machine 1 can produce 2000 pounds of paper in 4 hours. Machine 2 is newer and can produce 2000 pounds of paper in $2\frac{1}{2}$ hours. How long will it take the two machines working together to produce 2000 pounds of paper?

Solution

Verbal Model:

Work done	=	Portion done by Machine 1	+	Portion done by Machine 2

Labels:

Work done by both machines $= 1$ (job)

Time for each machine $= t$ (hours)

Rate for Machine 1 $= \frac{1}{4}$ (job per hour)

Rate for Machine 2 $= \frac{2}{5}$ (job per hour)

Equation:

$$1 = \left(\frac{1}{4}\right)(t) + \left(\frac{2}{5}\right)(t)$$

$$1 = \left(\frac{1}{4} + \frac{2}{5}\right)(t)$$

$$1 = \left(\frac{13}{20}\right)(t)$$

$$\frac{1}{13/20} = t$$

$$\frac{20}{13} = t$$

It would take $\frac{20}{13}$ hours (or about 1.54 hours) for both machines to complete the job. Check this solution in the original statement of the problem.

Note in Example 6 that the "2000 pounds" of paper was unnecessary information. We simply represented the 2000 pounds as "one complete job." This type of unnecessary information in an applied problem is sometimes called a *red herring*.

Formulas

Many common types of geometric, scientific, and investment problems use ready-made equations, called **formulas.** Knowing formulas such as those in the following list will help you translate and solve a wide variety of real-life problems involving perimeter, area, volume, temperature, interest, and distance.

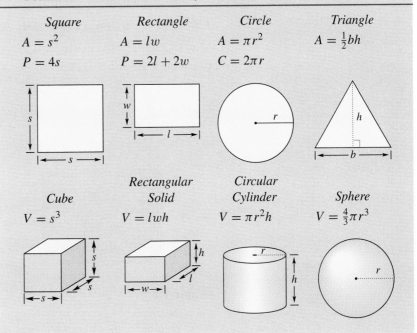

Common Formulas for Area, Perimeter, and Volume

Square: $A = s^2$, $P = 4s$
Rectangle: $A = lw$, $P = 2l + 2w$
Circle: $A = \pi r^2$, $C = 2\pi r$
Triangle: $A = \frac{1}{2}bh$

Cube: $V = s^3$
Rectangular Solid: $V = lwh$
Circular Cylinder: $V = \pi r^2 h$
Sphere: $V = \frac{4}{3}\pi r^3$

Miscellaneous Common Formulas

Temperature: $F =$ degrees Fahrenheit, $C =$ degrees Celsius

$$F = \frac{9}{5}C + 32$$

Simple Interest: $I =$ interest, $P =$ principal, $r =$ interest rate, $t =$ time

$$I = Prt$$

Distance: $d =$ distance traveled, $r =$ rate, $t =$ time

$$d = rt$$

When working with applied problems, you often need to rewrite one of the common formulas. For instance, the formula $P = 2l + 2w$ (the perimeter of a rectangle) can be rewritten or solved for w in the following manner.

$$P = 2l + 2w \qquad \text{Original formula}$$

$$P - 2l = 2w \qquad \text{Subtract } 2l \text{ from both sides.}$$

$$\frac{P - 2l}{2} = w \qquad \text{Divide both sides by 2.}$$

EXAMPLE 7 Using a Geometric Formula

A rectangular plot has an area of 120,000 square feet. The plot is 300 feet wide. How long is it?

Solution

In a problem such as this, it is helpful to begin by drawing a diagram, as shown in Figure 1.1. In this diagram, label the width of the rectangle as $w = 300$ feet, and label the unknown length as l.

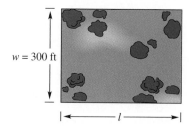

FIGURE 1.1

Now, to solve for the unknown length, use the following steps.

Verbal Model: Area = Length · Width

Labels: Area = 120,000   (square feet)
            Length = l (feet)
            Width = 300   (feet)

Equation: $120,000 = l(300)$

$$\frac{120,000}{300} = l$$

$$400 = l$$

The length of the rectangular plot is 400 feet. You can check this by multiplying 300 feet by 400 feet to obtain the area of 120,000 square feet.

EXAMPLE 8   Simple Interest

A deposit of $8000 earned $300 in interest in 6 months. What was the annual interest rate for this account?

Solution

Verbal Model: Interest = Principal · Rate · Time

Labels:
Interest = 300 (dollars)
Principal = 8000 (dollars)
Time = $\frac{1}{2}$ (year)
Annual interest rate = r (percent in decimal form)

Equation: $300 = 8000(r)\left(\dfrac{1}{2}\right)$

$$\frac{2(300)}{8000} = r$$

$$0.075 = r$$

The annual interest rate is $r = 0.075$ (or 7.5%). Check this in the original statement of the problem.

Group Activities   Communicating Mathematically

Translating a Formula   Use the information provided in the following statement to write a mathematical formula for the 10-second pulse count.

"The target heart rate is the heartbeat rate a person should have during aerobic exercise to get the full benefit of the exercise for cardiovascular conditioning. . . . Using the American College of Sports Medicine Method to calculate one's target heart rate, an individual should subtract his or her age from 220, then multiply by the desired intensity level (as a percent—sedentary persons may want to use 60% and highly fit individuals may want to use 85 to 95%) of the workout. Then divide the answer by 6 for a 10-second pulse count. (The 10-second pulse count is useful for checking whether the target heart rate is being achieved during the workout. One can easily check one's pulse—at the wrist or side of the neck—counting the number of beats in 10 seconds.)" (Source: Aerobic Fitness Association of America)

Use the formula you have found to find your own 10-second pulse count.

1.3 Exercises

Discussing the Concepts

1. Explain the difference between the markup rate and the markup.

2. Explain how to find the sale price of an item when you are given the list price and the discount rate.

3. If it takes you t hours to complete a task, what portion of the task can you complete in 1 hour?

4. If the sides of a square double, does the perimeter double? Explain.

5. If the sides of a square double, does the area double? Explain.

6. If you forgot the formula for the volume of a right circular cylinder, how could you derive it?

Problem Solving

7. *Using a Circle Graph* The circle graph shows how approximately 17 million barrels of oil per day are consumed in the United States. How many barrels are used in the transportation sector? (Source: Energy Information Administration)

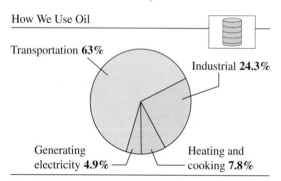

How We Use Oil

Transportation **63%**
Industrial **24.3%**
Generating electricity **4.9%**
Heating and cooking **7.8%**

8. *Using a Circle Graph* Use the figure to approximate the percent of electricity generated in the United States by hydroelectric plants. (Source: Energy Information Administration)

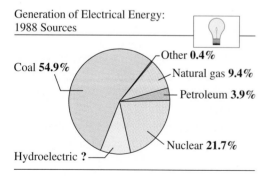

Generation of Electrical Energy: 1988 Sources

Coal **54.9%**
Other **0.4%**
Natural gas **9.4%**
Petroleum **3.9%**
Nuclear **21.7%**
Hydroelectric **?**

In Exercises 9–12, find the missing quantities. (Assume the markup rate is based on the cost.)

	Cost	Selling Price	Markup	Markup Rate
9.	$45.95	$64.33		
10.		$603.72	$184.47	
11.		$26,922.50	$4672.50	
12.	$732.00			$33\frac{1}{3}\%$

In Exercises 13–16, find the missing quantities. (Assume the discount rate is based on the list price.)

	List Price	Sale Price	Discount	Discount Rate
13.	$49.50	$25.74		
14.	$345.00		$134.55	
15.		$893.10	$251.90	
16.		$257.32	$202.18	

17. *Comparison Shopping* A department store is offering a discount of 20% on a sewing machine with a list price of $279.95. A mail-order catalog has the same machine for $228.95 plus $4.32 for shipping. Which is the better buy?

18. *Insurance Premium* The annual insurance premium for a policyholder is $862. Find the annual premium if the policyholder must pay a 20% surcharge because of an accident.

19. *Commission Rate*   Determine the commission rate for an employee who earned $450 in commissions for sales of $5000.

20. *Overtime Hours*   Last week you earned $740. If you are paid $14.50 per hour for the first 40 hours and $20 for each hour over 40, how many hours of overtime did you work?

21. *Finding a Pattern*   A rancher must purchase 500 bushels of a feed mixture for cattle and is considering oats and corn, which cost $1.70 per bushel and $3.00 per bushel, respectively.

(a) Complete the table, where x is the number of bushels of oats in the mixture.

Oats x	Corn $500 - x$	Price/Bushel of the Mixture
0		
100		
200		
300		
400		
500		

(b) How does the increase in the number of bushels of oats affect the number of bushels of corn in the mixture?

(c) How does the increase in the number of bushels of oats affect the price per bushel of the mixture?

(d) If there were an equal number of bushels of oats and corn in the mixture, how would the price of the mixture be related to the price of each component?

22. *Nut Mixture*   A grocer mixes two kinds of nuts at $3.88 per pound and $4.88 per pound, to make 100 pounds of a mixture at $4.13 per pound. How many pounds of each kind of nut were put into the mixture?

Mixtures In Exercises 23–26, find the number of units of Solutions 1 and 2 needed to obtain the desired amount and concentration of the final solution.

Concentrations

	Solution 1	Solution 2	Final Solution	Amount of Final Solution
23.	20%	60%	40%	100 gal
24.	50%	75%	60%	10 liters
25.	15%	60%	45%	24 qt
26.	60%	80%	75%	55 gal

Distance In Exercises 27–32, determine the unknown distance, rate, or time.

	Distance, d	Rate, r	Time, t
27.		650 mi/hr	$3\frac{1}{2}$ hr
28.		45 ft/sec	10 sec
29.	1000 km	110 km/hr	
30.	250 ft	32 ft/sec	
31.	1000 ft		$\frac{3}{2}$ sec
32.	385 mi		7 hr

33. *Average Speed*   Determine the time for the space shuttle to travel a distance of 5000 miles when its average speed is 17,000 miles per hour (see figure).

5000 miles

34. *Travel Time*   On the first part of a 317-mile trip, a sales representative averaged 58 miles per hour. The sales representative averaged only 52 miles per hour on the last part of the trip because of an increased volume of traffic. Find the amount of driving time at each speed if the total time was 5 hours and 45 minutes.

35. *Work Rate*   Determine the work rate for each task.

(a) A printer can print eight pages per minute.

(b) A machine shop can produce 30 units in 8 hours.

36. *Work Rate* You can complete a typing project in 5 hours, and a friend estimates that it would take him 8 hours. What fractional part of the task can be accomplished by each typist in 1 hour? If you both work on the project, in how many hours can it be completed?

In Exercises 37–40, solve for the specified variable.

37. *Ohm's Law* Solve for R in

$$E = IR.$$

38. *Markup* Solve for C in

$$S = C + rC.$$

39. *Simple Interest* Solve for r in

$$A = P + Prt.$$

40. *Compound Interest* Solve for P in

$$A = P\left(1 + \frac{r}{n}\right)^{nt}.$$

41. *Geometry* A rectangular picture frame has a perimeter of 3 feet. The width of the frame is 0.6 times its height. Find the height of the frame.

42. *Geometry* The figure shows three squares. The perimeter of square I is 12 inches and the area of square II is 36 square inches. Find the area of square III.

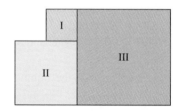

43. *Simple Interest* Find the interest on a $5000 bond that pays an annual interest rate of $9\frac{1}{2}\%$ for 6 years.

44. *Simple Interest* Find the annual interest rate on a certificate of deposit that accumulated $400 interest in 2 years on a principal of $2500.

Reviewing the Major Concepts

In Exercises 45–48, solve the equation.

45. $44 - 16x = 0$ **46.** $-4(x - 5) = 0$

47. $3[4 + 5(x - 1)] = 6x + 2$ **48.** $\frac{3}{8}x + \frac{3}{4} = 2$

In Exercises 49 and 50, write the statement using inequality notation.

49. y is no more than 45.

50. x is at least 15.

Additional Problem Solving

In Exercises 51–54, find the missing quantities. (Assume the markup rate is based on the cost.)

	Cost	Selling Price	Markup	Markup Rate
51.		$250.80	$98.80	
52.	$84.20	$113.67		
53.	$225.00			85.2%
54.		$16,440.50	$3890.50	

In Exercises 55–58, find the missing quantities. (Assume the discount rate is based on the list price.)

	List Price	Sale Price	Discount	Discount Rate
55.	$300.00		$189.00	
56.	$119.00	$79.73		
57.	$95.00			65%
58.		$15.92		20%

59. *Tire Cost*   An auto store gives the list price of a tire as $79.42. During a promotional sale, the store is selling four tires for the price of three. The store needs a markup of 10% during the sale. What is the cost to the store for each tire?

60. *Price per Pound*   The produce manager of a supermarket pays $22.60 for a 100-pound box of bananas. From past experience, the manager estimates that 10% of the bananas will spoil before they are sold. At what price per pound should the bananas be sold to give the supermarket an average markup rate on cost of 30%?

61. *Tip Rate*   A customer left a total of $10 for a meal that cost $8.45. Determine the tip rate.

62. *Tip Rate*   A customer left a total of $40 for a meal that cost $34.73. Determine the tip rate.

63. *Long Distance Rates*   The weekday rate for a telephone call is $0.75 for the first minute plus $0.55 for each additional minute. Determine the length of a call that cost $5.15. What would the cost of the call have been if it had been made during the weekend when there is a 60% discount?

64. *Labor Charges*   An appliance repair store charges $35 for the first $\frac{1}{2}$ hour of a service call. For each additional $\frac{1}{2}$ hour of labor there is a charge of $18. Find the length of a service call for which the charge is $89.

65. *Amount Financed*   A customer bought a lawn tractor that cost $4450 plus 6% sales tax. Find the amount of the sales tax and the total bill. Find the amount financed if a down payment of $1000 was made.

66. *Weekly Pay*   The weekly salary of an employee is $250 plus a 6% commission on the employee's total sales. Find the weekly pay for a week in which the sales are $5500.

67. *Number of Stamps*   You have a set of 70 stamps with a total value of $16.50. If the set includes 15¢ stamps and 30¢ stamps, find the number of each type.

68. *Number of Coins*   A person has 50 coins in dimes and quarters with a combined value of $10.25. Determine the number of coins of each type.

69. *Opinion Poll*   Fourteen hundred people were surveyed in an opinion poll. Political candidates A and B received approximately the same preference, but candidate C was preferred by twice the number of people as candidates A and B. Determine the number in the sample that preferred candidate C.

70. *Ticket Sales*   Ticket sales for a play total $2200. There are three times as many adult tickets sold as children's tickets, and the prices of the tickets for adults and children are $6 and $4, respectively. Find the number of children's tickets sold.

71. *Antifreeze Coolant*   The cooling system on a truck contains 5 gallons of coolant that is 40% antifreeze. How much must be withdrawn and replaced with 100% antifreeze to bring the coolant in the system to 50% antifreeze?

72. *Fuel Mixture*   Suppose that you mix gasoline and oil to obtain $2\frac{1}{2}$ gallons of mixture for an engine. The mixture is 40 parts gasoline and 1 part two-cycle oil. How much gasoline must be added to bring the mixture to 50 parts gasoline and 1 part oil?

73. *Flying Distance*   Two planes leave an airport at approximately the same time and fly in opposite directions. How far apart are the planes after $1\frac{1}{3}$ hours if their speeds are 480 miles per hour and 600 miles per hour?

74. *Speed of Light*   Determine the time for light to travel from the sun to earth. The distance between the sun and earth is 93,000,000 miles and the speed of light is 186,282.369 miles per second (see figure).

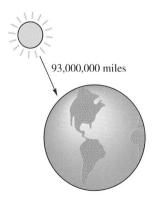

93,000,000 miles

Geometry In Exercises 75 and 76, use the closed rectangular box shown in the figure.

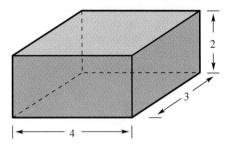

75. Find the volume of the box.

76. Find the surface area of the box. (The surface area is the combined areas of the six surfaces.)

77. *Average Wage* The average hourly wage y for public school bus drivers in the United States from 1980 to 1992 can be modeled by

$$y = 5.23 + 0.396t$$

where $t = 0$ represents 1980 (see figure). During which year was the average hourly wage $8.40? What was the average annual raise for bus drivers during this 12-year period? (Source: *National Survey of Salaries and Wages in Public Schools*)

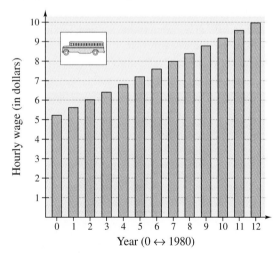

Year (0 ↔ 1980)

78. *Average Wage* The average hourly wage y for public school cafeteria workers in the United States from 1980 to 1992 can be modeled by

$$y = 3.82 + 0.301t$$

where $t = 0$ represents 1980 (see figure). During which year was the average hourly wage $6.83? What was the average annual raise for cafeteria workers during this 12-year period? (Source: *National Survey of Salaries and Wages in Public Schools*)

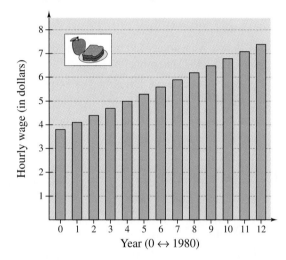

Year (0 ↔ 1980)

79. *Comparing Wage Increases* Use the information given in Exercises 77 and 78 to determine which of the two groups' average wages was increasing at a greater annual amount during the 12-year period from 1980 to 1992.

80. *New York City Marathon* Find the average speed of the woman holding the record time of 2 hours, $25\frac{1}{2}$ minutes in the New York City Marathon. The course length is 26 miles and 385 yards. (*Note:* 1 mile = 5280 feet = 1760 yards.)

81. *Average Speed* An Olympic runner completes a 5000-meter race in 13 minutes and 20 seconds. What was the average speed of the runner?

MID-CHAPTER QUIZ

Take this quiz as you would take a quiz in class. After you are done, check your work against the answers given in the back of the book.

In Exercises 1–8, solve the equation and check the result. (If it is not possible, state the reason.)

1. $4x + 3 = 11$

2. $-3(z - 2) = 0$

3. $2(y + 3) = 18 - 4y$

4. $5t + 7 = 7(t + 1) - 2t$

5. $\dfrac{1}{4}x + 6 = \dfrac{3}{2}x - 1$

6. $\dfrac{u}{4} + \dfrac{u}{3} = 1$

7. $\dfrac{4 - x}{5} + 5 = \dfrac{5}{2}$

8. $0.2x + 0.3 = 1.5$

In Exercises 9 and 10, solve the equation and round your answer to two decimal places. (A calculator may be helpful.)

9. $3x + \dfrac{11}{12} = \dfrac{5}{16}$

10. $0.42x + 6 = 5.25x - 0.80$

11. Explain how to write the decimal 0.45 as a fraction and as a percent.

12. 500 is 250% of what number?

13. Find the unit price (in dollars per ounce) of a 12-ounce box of cereal that sells for $2.35.

14. A quality control engineer for a manufacturer found one defective unit in a sample of 300. At that rate, what is the expected number of defective units in a shipment of 600,000?

15. A store is offering a discount of 25% on a computer with a list price of $1750. A mail-order catalog has the same machine for $1250 plus $24.95 for shipping. Which is the better buy?

16. Last week you earned $616. Your regular hourly wage is $12.25 for the first 40 hours, and your overtime hourly wage is $18. How many hours of overtime did you work?

17. Fifty gallons of a 30% acid solution is obtained by combining solutions that are 25% acid and 50% acid. How much of each solution is required?

18. On the first part of a 300-mile trip, a sales representative averaged 62 miles per hour. The sales representative averaged 46 miles per hour on the last part of the trip because of an increased volume of traffic. Find the amount of driving time at each speed if the total time was 6 hours.

19. You can paint a room in 6 hours and your friend can paint it in 8 hours. How long will it take both of you to paint the room?

20. The accompanying figure shows three squares. The perimeters of squares I and II are 20 inches and 32 inches, respectively. Find the area of square III.

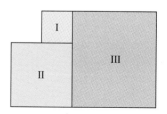

Figure for 20

1.4 Linear Inequalities

Intervals on the Real Line ▪ Properties of Inequalities ▪ Solving a Linear Inequality ▪ Applications

Intervals on the Real Line

In this section you will study **algebraic inequalities,** which are inequalities that contain one or more variable terms. Some examples are

$$x \leq 4, \quad x \geq -3, \quad x + 2 < 7, \quad \text{and} \quad 4x - 6 < 3x + 8.$$

As with an equation, you **solve** an inequality in the variable x by finding all values of x for which the inequality is true. Such values are called **solutions** and are said to **satisfy** the inequality. The set of all solutions of an inequality is the **solution set** of the inequality. The **graph** of an inequality is obtained by plotting its solution set on the real number line. Often, these graphs are intervals—either bounded or unbounded.

Bounded Intervals on the Real Number Line

Let a and b be real numbers such that $a < b$. The following intervals on the real number line are called **bounded intervals.** The numbers a and b are the **endpoints** of each interval.

Notation	Interval Type	Inequality	Graph
$[a, b]$	Closed	$a \leq x \leq b$	
(a, b)	Open	$a < x < b$	
$[a, b)$	Half-open	$a \leq x < b$	
$(a, b]$	Half-open	$a < x \leq b$	

NOTE In the list at the right, note that a closed interval contains both of its endpoints, a half-open interval contains only one of its endpoints, and an open interval does not contain either of its endpoints.

The **length** of the interval $[a, b]$ is the distance between its endpoints: $b - a$. The lengths of $[a, b]$, (a, b), $(a, b]$, and $[a, b)$ are the same. The reason that these four types of intervals are called bounded is that each has a finite length. An interval that *does not* have a finite length is **unbounded** (or **infinite**).

Unbounded Intervals on the Real Number Line

Let a and b be real numbers. The following intervals on the real number line are called **unbounded intervals.**

Notation	Interval Type	Inequality	Graph
$[a, \infty)$	Half-open	$a \leq x$	
(a, ∞)	Open	$a < x$	
$(-\infty, b]$	Half-open	$x \leq b$	
$(-\infty, b)$	Open	$x < b$	
$(-\infty, \infty)$	Entire real line		

NOTE The symbols ∞ (**positive infinity**) and $-\infty$ (**negative infinity**) do not represent real numbers. They are simply convenient symbols used to describe the unboundedness of an interval such as $(1, \infty)$.

EXAMPLE 1   Graphs of Inequalities

a. The graph of $-3 < x \leq 1$ is a bounded interval.

b. The graph of $0 < x < 2$ is a bounded interval.

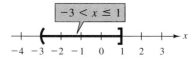

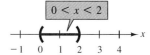

c. The graph of $-3 < x$ is an unbounded interval.

d. The graph of $x \leq 2$ is an unbounded interval.

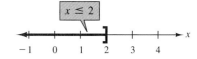

Properties of Inequalities

Solving a linear inequality is much like solving a linear equation. To isolate the variable you make use of **properties of inequalities.** These properties are similar to the properties of equality, but there are two important exceptions. When both sides of an inequality are multiplied or divided by a negative number, the direction of the inequality symbol must be reversed. Here is an example.

$-2 < 5$	Original inequality
$(-3)(-2) > (-3)(5)$	Multiply both sides by -3 and reverse the inequality.
$6 > -15$	Simplify.

Two inequalities that have the same solution set are **equivalent.** The following list describes operations that can be used to create equivalent inequalities.

NOTE These properties remain true if the symbols $<$ and $>$ are replaced by $\leq$ and $\geq$. Moreover, a, b, and c can represent real numbers, variables, or expressions. Note that you cannot multiply or divide both sides of an inequality by zero.

Properties of Inequalities

1. *Addition and Subtraction Properties*

Adding the same quantity to, or subtracting the same quantity from, both sides of an inequality produces an equivalent inequality.

If $a < b$, then $a + c < b + c$.

If $a < b$, then $a - c < b - c$.

2. *Multiplication and Division Properties: Positive Quantities*

Multiplying or dividing both sides of an inequality by a *positive* quantity produces an equivalent inequality.

If $a < b$ and c is positive, then $ac < bc$.

If $a < b$ and c is positive, then $\dfrac{a}{c} < \dfrac{b}{c}$.

3. *Multiplication and Division Properties: Negative Quantities*

Multiplying or dividing both sides of an inequality by a *negative* quantity produces an equivalent inequality in which the inequality symbol is reversed.

If $a < b$ and c is negative, then $ac > bc$.

If $a < b$ and c is negative, then $\dfrac{a}{c} > \dfrac{b}{c}$.

4. *Transitive Property*

Consider three quantities for which the first quantity is less than the second, and the second is less than the third. It follows that the first quantity must be less than the third quantity.

If $a < b$ and $b < c$, then $a < c$.

Solving a Linear Inequality

An inequality in one variable is **linear** if it can be written in one of the following forms.

$$ax + b \leq 0, \quad ax + b < 0, \quad ax + b \geq 0, \quad ax + b > 0$$

As you study the following examples, pay special attention to the steps in which the inequality symbol is reversed. Remember that when you multiply or divide an inequality by a negative number, you must reverse the inequality symbol.

STUDY TIP

Checking the solution set of an inequality is not as simple as checking the solution set of an equation. (There are usually too many x-values to substitute back into the original inequality.) You can, however, get an indication of the validity of a solution set by substituting a few convenient values of x. For instance, in Example 2 try checking that $x = 0$ satisfies the original inequality, whereas $x = 4$ does not.

EXAMPLE 2   Solving a Linear Inequality

$x + 6 < 9$	Original inequality
$x + 6 - 6 < 9 - 6$	Subtract 6 from both sides.
$x < 3$	Combine like terms.

The solution set consists of all real numbers that are less than 3. The interval notation for the solution set is $(-\infty, 3)$. The graph is shown in Figure 1.2.

FIGURE 1.2

EXAMPLE 3   Solving a Linear Inequality

$8 - 3x \leq 20$	Original inequality
$8 - 8 - 3x \leq 20 - 8$	Subtract 8 from both sides.
$-3x \leq 12$	Combine like terms.
$\dfrac{-3x}{-3} \geq \dfrac{12}{-3}$	Divide both sides by -3 and reverse the inequality symbol.
$x \geq -4$	Simplify.

The solution set consists of all real numbers that are greater than or equal to -4. The interval notation for the solution set is $[-4, \infty)$. The graph is shown in Figure 1.3.

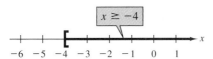

FIGURE 1.3

Technology

Most graphing utilities can sketch the graph of a linear inequality. For instance, the following steps show how to sketch the graph of $9x - 4 > 5x + 2$ on a *TI-82*.

$\boxed{Y=}$ $Y_1 = 9X - 4 > 5X + 2$

$\boxed{ZOOM}$ $\boxed{6}$

The graph produced by these steps is shown below. Notice that the graph occurs as an interval *above* the x-axis.

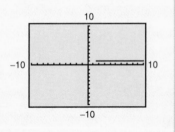

EXAMPLE 4   Solving a Linear Inequality

$9x - 4 > 5x + 2$	Original inequality
$9x - 4 + 4 > 5x + 2 + 4$	Add 4 to both sides.
$9x > 5x + 6$	Combine like terms.
$9x - 5x > 5x + 6 - 5x$	Subtract $5x$ from both sides.
$4x > 6$	Combine like terms.
$\dfrac{4x}{4} > \dfrac{6}{4}$	Divide both sides by 4.
$x > \dfrac{3}{2}$	Simplify.

The solution set consists of all real numbers that are greater than $\frac{3}{2}$. The interval notation for the solution set is $\left(\frac{3}{2}, \infty\right)$. The graph is shown in Figure 1.4.

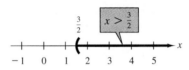

FIGURE 1.4

EXAMPLE 5   Solving a Linear Inequality

$\dfrac{2x}{3} + 12 < \dfrac{x}{6} + 18$	Original inequality
$6 \cdot \left(\dfrac{2x}{3} + 12\right) < 6 \cdot \left(\dfrac{x}{6} + 18\right)$	Multiply both sides by LCD of 6.
$4x + 72 < x + 108$	Distributive Property
$4x - x < 108 - 72$	Subtract x and 72 from both sides.
$3x < 36$	Combine like terms.
$x < 12$	Divide both sides by 3.

NOTE When one (or more) of the terms in a linear inequality involves constant denominators, it helps to multiply both sides of the equation by the least common denominator. This clears the inequality of fractions, as shown in Example 5.

The solution set consists of all real numbers that are less than 12. The interval notation for the solution set is $(-\infty, 12)$. The graph is shown in Figure 1.5.

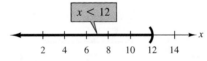

FIGURE 1.5

Sometimes it is convenient to write two inequalities as a **double inequality.** For instance, you can write the two inequalities $-4 \leq 5x - 2$ and $5x - 2 < 7$ more simply as

$$-4 \leq 5x - 2 < 7.$$

This form allows you to solve the two given inequalities together. Try solving this double inequality. You should find that the solution is $-\frac{2}{5} \leq x < \frac{9}{5}$.

EXAMPLE 6 *Solving a Double Inequality*

Write the inequalities $-7 \leq 5x - 2$ and $5x - 2 < 8$ as a double inequality. Then solve the double inequality to find the set of all real numbers that satisfy *both* inequalities.

Solution

$$-7 \leq 5x - 2 < 8 \qquad \text{Original inequality}$$
$$-7 + 2 \leq 5x - 2 + 2 < 8 + 2 \qquad \text{Add 2 to all three parts.}$$
$$-5 \leq 5x < 10 \qquad \text{Combine like terms.}$$
$$\frac{-5}{5} \leq \frac{5x}{5} < \frac{10}{5} \qquad \text{Divide each part by 5.}$$
$$-1 \leq x < 2 \qquad \text{Simplify.}$$

The solution set consists of all real numbers that are greater than or equal to -1 and less than 2. The interval notation for the solution set is $[-1, 2)$. The graph is shown in Figure 1.6.

FIGURE 1.6

The double inequality in Example 6 could have been solved in two parts as follows.

$$-7 \leq 5x - 2 \qquad \text{and} \qquad 5x - 2 < 8$$
$$-5 \leq 5x \qquad\qquad\qquad\qquad 5x < 10$$
$$-1 \leq x \qquad\qquad\qquad\qquad\quad x < 2$$

The solution set consists of all real numbers that satisfy *both* inequalities. In other words, the solution set is the set of all values of x for which $-1 \leq x < 2$.

Applications

Linear inequalities in real-life problems arise from statements that involve phrases such as "at least," "no more than," "minimum value," and so on.

EXAMPLE 7 *Translating Verbal Statements*

	Verbal Statement	*Inequality*	
a.	x is at most 3.	$x \leq 3$	"at most" means "less than or equal to."
b.	x is no more than 3.	$x \leq 3$	
c.	x is at least 3.	$x \geq 3$	"at least" means "greater than or equal to."
d.	x is more than 3.	$x > 3$	
e.	x is less than 3.	$x < 3$	

To solve real-life problems involving inequalities, you can use the same "verbal-model approach" you use with equations.

EXAMPLE 8 *Finding the Maximum Width of a Package*

An overnight delivery service will not accept any package whose combined length and girth (perimeter of a cross section) exceeds 132 inches. Suppose that you are sending a rectangular package that has square cross sections. If the length of the package is 68 inches, what is the maximum width of the sides of its square cross sections?

Solution

Begin by making a sketch. In Figure 1.7, notice that the length of the package is 68 inches, and each side is x inches wide.

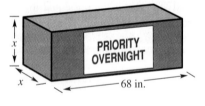

FIGURE 1.7

Verbal Model: Length + Girth ≤ 132 inches

Labels:	Width of a side $= x$	(inches)
	Length $= 68$	(inches)
	Girth $= 4x$	(inches)

Inequality: $68 + 4x \leq 132$

$$4x \leq 64$$

$$x \leq 16$$

The width of each side of the package must be less than or equal to 16 inches.

EXAMPLE 9   *Comparing Costs*

A subcompact car can be rented from Company A for $240 per week with no extra charge for mileage. A similar car can be rented from Company B for $100 per week, plus 25 cents for each mile driven. How many miles must you drive in a week so that the rental fee for Company A is less than that for Company B?

Solution

Verbal Model:

Weekly cost for Company B	>	Weekly cost for Company A

Labels: Number of miles driven in one week $= m$ (miles)
              Weekly cost for Company A $= 240$ (dollars)
              Weekly cost for Company B $= 100 + 0.25m$ (dollars)

Inequality: $100 + 0.25m > 240$
                 $0.25m > 140$
                 $m > 560$

The car from Company A is cheaper if you drive more than 560 miles in a week. A table helps confirm this conclusion.

Miles Driven	520	530	540	550	560	570
Company A	$240.00	$240.00	$240.00	$240.00	$240.00	$240.00
Company B	$230.00	$232.50	$235.00	$237.50	$240.00	$242.50

Group Activities   Communicating Mathematically

Problem Posing   Suppose your group owns a small business and must choose between two carriers for long-distance telephone service. Create realistic data for cost of the first minute of a call and cost per additional minute for each carrier, and decide what question(s) would be most helpful to ask when making such a choice. Solve the problem your group has created. Write a short memo to your company's business manager outlining the situation, explaining your mathematical solution, and summarizing your recommendations.

1.4 Exercises

Discussing the Concepts

1. Explain the meaning of $<$, $\leq$, $>$, $\geq$, and $=$.

2. Is adding -5 to both sides of an inequality the same as subtracting 5 from both sides? Explain.

3. Is dividing both sides of an inequality by 5 the same as multiplying both sides by $\frac{1}{5}$? Explain.

4. Give an example of a linear inequality that has an unbounded solution set.

5. Describe the differences between properties of equality and properties of inequality.

6. Give an example of "reversing an inequality symbol."

Problem Solving

In Exercises 7–10, determine whether the value of x satisfies the inequality.

Inequality	*Values*

7. $7x - 10 > 0$ (a) $x = 3$ (b) $x = -2$
                      (c) $x = \frac{5}{2}$ (d) $x = \frac{1}{2}$

8. $3x + 2 < \dfrac{7x}{5}$ (a) $x = 0$ (b) $x = 4$
                      (c) $x = -4$ (d) $x = -1$

9. $0 < \dfrac{x + 5}{6} < 2$ (a) $x = 10$ (b) $x = 4$
                      (c) $x = 0$ (d) $x = -6$

10. $-2 < \dfrac{3 - x}{2} \leq 2$ (a) $x = 0$ (b) $x = 3$
                      (c) $x = 9$ (d) $x = -12$

In Exercises 11–14, match the inequality with its graph. [The graphs are labeled (a), (b), (c) and (d).]

(a)
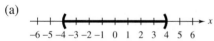

(b)

(c)

(d)

11. $x \geq 4$

12. $x < -4$ or $x \geq 4$

13. $-4 < x < 4$

14. $x < -4$

In Exercises 15–20, use a graphing utility to graph the inequality and write the interval notation for the inequality.

15. $-5 < x \leq 3$

16. $4 > x \geq 1$

17. $\frac{3}{2} \geq x > 0$

18. $-7 < x \leq -3$

19. $-\frac{15}{4} < x < -\frac{5}{2}$

20. $-\pi > x > -5$

In Exercises 21–30, solve the inequality and sketch the solution on the real number line.

21. $x + 7 \leq 9$

22. $z - 4 > 0$

23. $2x - 5 > 9$

24. $3x + 4 \leq 22$

25. $16 < 4(y + 2) - 5(2 - y)$

26. $4[z - 2(z + 1)] < 2 - 7z$

27. $-3 < \dfrac{2x - 3}{2} < 3$

28. $0 \leq \dfrac{x - 5}{2} < 4$

29. $1 > \dfrac{x - 4}{-3} > -2$

30. $-\dfrac{2}{3} < \dfrac{x - 4}{-6} \leq \dfrac{1}{3}$

In Exercises 31–34, rewrite the statement using inequality notation.

31. x is nonnegative.

32. y is more than -2.

33. n is no more than -2.

34. m is at least 4.

In Exercises 35 and 36, write a verbal description of the inequality.

35. $x \geq \frac{5}{2}$

36. $-4 \leq t \leq 4$

37. *Using a Linear Model*   A utility company has a fleet of vans. The annual operating cost per van is

$$C = 0.28m + 2900$$

where m is the number of miles traveled by a van in a year. How many miles will yield an annual operating cost that is less than $10,000?

38. *Profit*   The revenue from selling x units of a product is

$$R = 105.45x.$$

The cost of producing x units is

$$C = 78x + 25{,}850.$$

To obtain a profit, the revenue must be greater than the cost. For what values of x will this product produce a profit?

39. *Distance*   If you live 5 miles from school and your friend lives 3 miles from you, then the distance d that your friend lives from school is in what interval? Use the figure to help determine your answer.

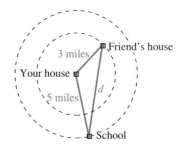

40. If $-3 \leq x \leq 10$, then $-x$ must be in what interval?

Reviewing the Major Concepts

In Exercises 41–44, place the correct inequality symbol ($<$ or $>$) between the numbers.

41. $-\frac{3}{4}$ [____] -5

42. $-\frac{1}{5}$ [____] $-\frac{1}{3}$

43. π [____] -3

44. 6 [____] $\frac{13}{2}$

In Exercises 45–48, solve the equation.

45. $-2n + 12 = 5$

46. $\dfrac{12 + x}{4} = 13$

47. $55 - 4(3 - y) = -5$

48. $8(t - 24) = 0$

Additional Problem Solving

In Exercises 49–54, match the inequality with its graph. [The graphs are labeled (a), (b), (c), (d), (e), and (f).]

(a)
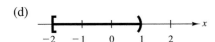

(b)
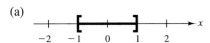

(c)

(d)

(e)
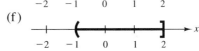

(f)

49. $-1 < x \leq 2$

50. $-1 < x \leq 1$

51. $-1 \leq x \leq 1$

52. $-1 < x < 2$

53. $-2 \leq x < 1$

54. $-2 < x < 1$

In Exercises 55–60, use a graphing utility to graph the inequality and write its interval notation.

55. $x \leq -2$

56. $x > 6$

57. $-2 < x \leq 4$

58. $-10 < x < -5$

59. $9 \geq x \geq 3$

60. $x \leq -1$ or $x > 1$

In Exercises 61–80, solve the inequality and sketch the solution on the real number line.

61. $4x < 22$

62. $2x > 5$

63. $-9x \geq 36$

64. $-6x \leq 24$

65. $-\frac{3}{4}x < -6$

66. $-\frac{1}{5}x > -2$

67. $5 - x \leq -2$

68. $1 - y \geq -5$

69. $5 - 3x < 7$

70. $12 - 5x > 5$

71. $3x - 11 > -x + 7$

72. $21x - 11 \leq 6x + 19$

73. $-3(y + 10) \geq 4(y + 10)$

74. $2(4 - z) \geq 8(1 + z)$

75. $\dfrac{x}{6} - \dfrac{x}{4} \leq 1$

76. $\dfrac{x + 3}{6} + \dfrac{x}{8} \geq 1$

77. $0 < 2x - 5 < 9$

78. $-4 \leq 2 - 3(x + 2) < 11$

79. $\dfrac{2}{5} < x + 1 < \dfrac{4}{5}$

80. $-1 < -\dfrac{x}{6} < 1$

In Exercises 81 and 82, rewrite the statement using inequality notation.

81. z is at least 2.

82. x is at least 450 but no more than 500.

In Exercises 83–86, write a verbal description of the inequality.

83. $y \geq 3$

84. $t < 4$

85. $0 < z \leq \pi$

86. $5 \geq x > -2$

87. *Travel Budget* A student group has $4500 budgeted for a field trip. The cost of transportation for the trip is $1900. To stay within the budget, all other costs C must be no more than what amount?

88. *Monthly Budget* You have budgeted $1200 a month for your total expenses. Your rent is $400 per month and you have budgeted $275 per month for food. To stay within your budget, all other costs C must be no more than what amount?

89. *Comparing Average Temperatures* The average temperature in Miami is greater than the average temperature in Washington, D.C. The average temperature in Washington, D.C. is greater than the average temperature in New York. How does the average temperature in Miami compare with that in New York?

90. *Comparing Elevations* The elevation (above sea level) of San Francisco is less than the elevation of Dallas, and the elevation of Dallas is less than the elevation of Denver. How does the elevation of San Francisco compare with the elevation of Denver?

91. *Hourly Wage* You must select one of two plans for payment when working for a company. One plan pays a straight $12.50 per hour. The second pays $8.00 per hour plus $0.75 per unit produced per hour. Write an inequality yielding the number of units that must be produced per hour so that the second plan gives the greater hourly wage. Solve the inequality.

92. *Monthly Wage* You must select one of two plans for payment when working for a company. One plan pays a straight $3000 per month. The second pays $1000 per month plus a commission of 4% of your gross sales. Write an inequality yielding the gross sales per month so that the second plan gives the greater monthly wage. Solve the inequality.

93. *Profit* The revenue and cost for selling x units of a product are given by $R = 89.95x$ and $C = 61x + 875$. To obtain a profit, the revenue must be greater than the cost. For what values of x will this product produce a profit?

94. *Operating Costs* A fuel company has a fleet of trucks. The annual operating cost per truck is $C = 0.58m + 7800$ where m is the number of miles traveled by a truck in a year. What number of miles will yield an annual operating cost that is less than $25,000?

95. *Long Distance Charges* The cost for a long distance telephone call is $0.96 for the first minute and $0.75 for each additional minute. If the total cost of the call cannot exceed $5, find the interval of time that is available for the call.

96. *Long Distance Charges* The cost for a long distance telephone call is $1.45 for the first minute and $0.95 for each additional minute. If the total cost of the call cannot exceed $15, find the interval of time that is available for the call.

Air Pollutant Emissions In Exercises 97 and 98, use the following equation, which models the amount of air pollutant emissions of lead in the continental United States from 1983 to 1991 (see figure).

$$y = 0.518 - 0.048t, \quad 3 \le t \le 11$$

In this model, y represents the amount of pollutant in micrograms per cubic meter of air and t represents the year, with $t = 0$ corresponding to 1980. (Source: U.S. Environmental Protection Agency)

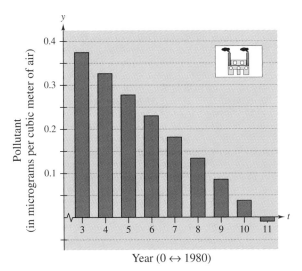

Year (0 ↔ 1980)

97. During which years between 1983 and 1991 was the air pollutant emission of lead greater than 0.23 micrograms per cubic meter?

98. During which years between 1983 and 1991 was the air pollutant emission of lead less than 0.086 micrograms per cubic meter?

99. *Geometry* If a, b, and c are the lengths of the sides of the triangle in the figure, then $a + b$ must be greater than what value?

Figure for 99

100. Four times a number n must be at least 12 and no more than 30. What interval contains this number?

101. Five times a number n must be at least 15 and no more than 45. What interval contains this number?

102. Determine all real numbers n such that $\frac{1}{3}n$ must be more than 7.

103. Determine all real numbers n such that $\frac{1}{4}n$ must be more than 8.

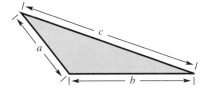

C A R E E R   I N T E R V I E W

Melanie Cansler

Shop owner

JellyBeads!

Concord, MA 01742

JellyBeads! is a per-bead bead shop where customers buy the beads and materials they need to design and create their own beaded jewelry: necklaces, bracelets, earrings, key chains, and so forth. I sell beads from all over the world, including Austrian crystal, Chinese porcelain, South American ceramic, Japanese seed, and Indian glass beads. One of my most challenging tasks is staying on top of my bead inventory and reordering stock. Because I can buy beads in bulk by the pound, strand, gross, or kilo, I need to convert from one measure to another for price comparison purposes. The unit price I ultimately pay for a particular bead varies from order to order depending on source (catalogs, suppliers, or flea markets) and quantity. When I price a particular type of bead for resale, I use a 100% markup rate on the more stable going-price for an individual bead rather the unit price that I actually paid.

1.5	**Absolute Value Equations and Inequalities**

Solving Equations Involving Absolute Value ■
Solving Inequalities Involving Absolute Value

Solving Equations Involving Absolute Value

Consider the **absolute value equation**

$$|x| = 3.$$

FIGURE 1.8

The only solutions of this equation are -3 and 3, because these are the only two real numbers whose distance from zero is 3. (See Figure 1.8.) In other words, the absolute value equation $|x| = 3$ has exactly two solutions: $x = -3$ and $x = 3$.

Solving an Absolute Value Equation

Let x be a variable or a variable expression and let a be a real number such that $a \geq 0$. The solutions of the equation $|x| = a$ are given by $x = -a$ and $x = a$. That is,

$$|x| = a \implies x = -a \quad \text{and} \quad x = a.$$

EXAMPLE 1   *Solving Absolute Value Equations*

Solve each absolute value equation.

   a. $|x| = 10$ **b.** $|x| = 0$ **c.** $|y| = -1$

Solution

a. This equation is equivalent to the two linear equations

       $x = -10$ and $x = 10.$ Equivalent linear equations

    Thus, the absolute value equation has two solutions: -10 and 10.

b. This equation is equivalent to the two linear equations

       $x = 0$ and $x = 0.$ Equivalent linear equations

    Because both equations are the same, you can conclude that the absolute value equation has only one solution: 0.

c. This absolute value equation has *no solution* because it is not possible for the absolute value of a real number to be negative.

STUDY TIP

The strategy for solving absolute value equations is to *rewrite* the equation in *equivalent forms* that can be solved by previously learned methods. This is a common strategy in mathematics. That is, when you encounter a new type of problem, you try to rewrite the problem so that it can be solved by techniques you already know.

EXAMPLE 2 *Solving Absolute Value Equations*

Solve $|3x + 4| = 10$.

Solution

$$|3x + 4| = 10 \qquad \text{Original equation}$$

$3x + 4 = -10$ or $3x + 4 = 10$		Equivalent equations
$3x + 4 - 4 = -10 - 4 \qquad 3x + 4 - 4 = 10 - 4$		Subtract 4 from both sides.
$3x = -14 \qquad\qquad 3x = 6$		Combine like terms.
$x = -\dfrac{14}{3} \qquad\qquad x = 2$		Divide both sides by 3.

The solutions are $-\frac{14}{3}$ and 2. Check these in the original equation.

NOTE When you are solving absolute value equations, remember that it is possible that they have no solution. For instance, the equation

$$|3x + 4| = -10$$

has no solution because the absolute value of a real number cannot be negative. Do not make the mistake of trying to solve such an equation by writing the "equivalent" linear equations as $3x + 4 = -10$ and $3x + 4 = 10$. These equations have solutions, but they are both extraneous.

The equation in the next example is not given in the **standard form**

$$|ax + b| = c, \qquad c \geq 0.$$

Notice that the first step in solving such an equation is to write it in standard form.

EXAMPLE 3 *An Absolute Value Equation in Nonstandard Form*

Solve $|2x - 1| + 3 = 8$.

Solution

$|2x - 1| + 3 = 8$	Original equation
$|2x - 1| = 5$	Standard form
$2x - 1 = -5$ or $2x - 1 = 5$	Equivalent equations
$2x = -4 \qquad\qquad 2x = 6$	Add 1 to both sides.
$x = -2 \qquad\qquad x = 3$	Divide both sides by 2.

The solutions are -2 and 3. Check these in the original equation.

If two algebraic expressions are equal in absolute value, they must either be *equal* to each other or be the *opposites* of each other. Thus, you can solve equations of the form

$$|ax + b| = |cx + d|$$

by forming the two linear equations

Expressions opposite

Expressions equal

$$ax + b = -(cx + d) \quad \text{and} \quad ax + b = cx + d.$$

EXAMPLE 4   Solving an Equation Involving Absolute Values

Solve $|3x - 4| = |7x - 16|$.

Solution

$$|3x - 4| = |7x - 16| \qquad \text{Original equation}$$

$$3x - 4 = -(7x - 16) \quad \text{or} \quad 3x - 4 = 7x - 16 \qquad \text{Equivalent equations}$$

$$3x - 4 = -7x + 16 \qquad\qquad 3x = 7x - 12$$

$$10x = 20 \qquad\qquad -4x = -12$$

$$x = 2 \qquad\qquad x = 3 \qquad \text{Solutions}$$

The solutions are 2 and 3. Check these in the original equation.

EXAMPLE 5   Solving an Equation Involving Absolute Values

NOTE   When solving equations of the form

$$|ax + b| = |cx + d|$$

it is possible that one of the resulting equations will not have a solution. Note this occurrence in Example 5.

Solve $|x + 5| = |x + 11|$.

Solution

By equating the expression $(x + 5)$ to the opposite of $(x + 11)$, you obtain

$$x + 5 = -(x + 11)$$

$$x + 5 = -x - 11$$

$$x = -x - 16$$

$$2x = -16$$

$$x = -8.$$

However, by setting the two expressions equal to each other, you obtain

$$x + 5 = x + 11$$

$$x = x + 6$$

$$0 = 6$$

which makes no sense. Therefore, the original equation has only one solution: -8. Check this solution in the original equation.

Solving Inequalities Involving Absolute Value

To see how to solve inequalities involving absolute value, consider the following comparisons.

$$|x| = 2 \qquad\qquad |x| < 2 \qquad\qquad |x| > 2$$

$$x = -2 \text{ and } x = 2 \qquad\qquad -2 < x < 2 \qquad\qquad x < -2 \text{ or } x > 2$$

These comparisons suggest the following rule for solving inequalities involving absolute value.

Solving an Absolute Value Inequality

Let x be a variable or an algebraic expression and let a be a real number such that $a > 0$.

1. The solutions of $|x| < a$ are all values of x that lie between $-a$ and a. That is,

$$|x| < a \quad \text{if and only if} \quad -a < x < a.$$

2. The solutions of $|x| > a$ are all values of x that are *less than* $-a$ or *greater than* a. That is,

$$|x| > a \quad \text{if and only if} \quad x < -a \text{ or } x > a.$$

These rules are also valid if $<$ is replaced by $\leq$ and $>$ is replaced by $\geq$.

Technology

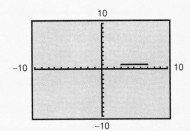

Most graphing utilities can sketch the graph of an absolute value inequality. For instance, the following steps show how to sketch the graph of

$$|x - 5| < 2$$

on a *TI-82*.

| Y= | $Y_1 =$ | ABS | (| X−5 |) | < 2 |

| ZOOM | 6 |

The graph produced by these steps is shown at the left. Notice that the graph occurs as an interval *above* the x-axis.

EXAMPLE 6   Solving an Absolute Value Inequality

Solve $|x - 5| < 2$.

Solution

$$|x - 5| < 2$$ Original inequality

$$-2 < x - 5 < 2$$ Equivalent double inequality

$$-2 + 5 < x - 5 + 5 < 2 + 5$$ Add 5 to all three parts.

$$3 < x < 7$$ Combine like terms.

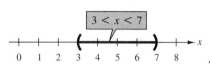

FIGURE 1.9

The solution set consists of all real numbers that are greater than 3 and less than 7. The interval notation for this solution set is $(3, 7)$. The graph of this solution set is shown in Figure 1.9.

EXAMPLE 7   Solving an Absolute Value Inequality

Solve $|3x - 4| \ge 5$.

Solution

$$|3x - 4| \ge 5$$ Original inequality

$$3x - 4 \le -5 \quad \text{or} \quad 3x - 4 \ge 5$$ Equivalent inequalities

$$3x - 4 + 4 \le -5 + 4 \qquad 3x - 4 + 4 \ge 5 + 4$$ Add 4 to both sides.

$$3x \le -1 \qquad\qquad 3x \ge 9$$ Combine like terms.

$$\frac{3x}{3} \le \frac{-1}{3} \qquad\qquad \frac{3x}{3} \ge \frac{9}{3}$$ Divide both sides by 3.

$$x \le -\frac{1}{3} \qquad\qquad x \ge 3$$ Simplify.

The solution set consists of all real numbers that are less than or equal to $-\frac{1}{3}$ or greater than or equal to 3. The interval notation for the solution set is $\left(-\infty, -\frac{1}{3}\right] \cup [3, \infty)$. The symbol $\cup$ is called a **union** symbol, and it is used to denote the combining of two sets. The graph is shown in Figure 1.10.

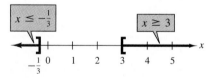

FIGURE 1.10

EXAMPLE 8 *Solving an Absolute Value Inequality*

$\left\|2 - \dfrac{x}{3}\right\| \le 0.01$	Original inequality
$-0.01 \le 2 - \dfrac{x}{3} \le 0.01$	Equivalent double inequality
$-2.01 \le -\dfrac{x}{3} \le -1.99$	Subtract 2 from all three parts.
$6.03 \ge x \ge 5.97$	Multiply all three parts by -3 and reverse both inequality symbols.
$5.97 \le x \le 6.03$	Solution set in standard form

The solution set consists of all real numbers that are greater than or equal to 5.97 and less than or equal to 6.03. The interval notation for the solution set is [5.97, 6.03]. The graph is shown in Figure 1.11.

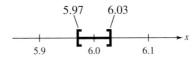

FIGURE 1.11

Group Activities   You Be the Instructor

Error Analysis   Suppose you are teaching a class in algebra and one of your students hands in the following solution.

$$|3x - 4| \ge -5$$

$$3x - 4 \le -5 \qquad \text{or} \qquad 3x - 4 \ge 5$$

$$3x - 4 + 4 \le -5 + 4 \qquad 3x - 4 + 4 \ge 5 + 4$$

$$3x \le -1 \qquad\qquad\qquad 3x \ge 9$$

$$\frac{3x}{3} \le \frac{-1}{3} \qquad\qquad\qquad \frac{3x}{3} \ge \frac{9}{3}$$

$$x \le -\frac{1}{3} \qquad\qquad\qquad x \ge 3$$

What is wrong with this solution? What could you say to help your students avoid this type of error?

1.5 Exercises

Discussing the Concepts

1. Give a graphical description of the absolute value of a real number.

2. Give an example of an absolute value equation that has only one solution.

3. In your own words, explain how to solve an absolute value equation. Illustrate your explanation with an example.

4. Describe the solution of the inequality $|x| > 3$.

5. The graph of the inequality $|x - 3| < 2$ can be described as *all real numbers that are within two units of 3*. Give a similar description of $|x - 4| < 1$.

6. The graph of the inequality $|y - 1| > 3$ can be described as *all real numbers greater than three units from 1*. Give a similar description of $|y + 2| > 4$.

Problem Solving

In Exercises 7–10, determine whether the value is a solution of the equation.

Equation	Value
7. $|4x + 5| = 10$	$x = -3$
8. $|2x - 16| = 10$	$x = 3$
9. $|6 - 2w| = 2$	$w = 4$
10. $\left|\frac{1}{2}t + 4\right| = 8$	$t = 6$

In Exercises 11–14, transform the absolute value equation into two linear equations.

11. $|x - 10| = 17$ **12.** $|7 - 2t| = 5$

13. $|4x + 1| = \frac{1}{2}$ **14.** $|22k + 6| = 9$

In Exercises 15–24, solve the equation. (Some of the equations may have no solution.)

15. $|t| = 45$ **16.** $|s| = 16$

17. $|2s + 3| = 25$ **18.** $|7a + 6| = 8$

19. $|3x + 4| = -16$ **20.** $4|x + 5| = 9$

21. $\left|\frac{2}{3}x + 4\right| = 9$ **22.** $|3.2 - 1.054x| = 2$

23. $|x + 2| = |3x - 1|$ **24.** $|x - 2| = |2x - 15|$

Think About It   In Exercises 25 and 26, write a single equation that is equivalent to the two equations.

25. $2x + 3 = 5, 2x + 3 = -5$

26. $4x - 6 = 7, 4x - 6 = -7$

In Exercises 27–30, determine whether the x-values are solutions of the inequality.

Inequality	Values	
27. $|x| < 3$	(a) $x = 2$	(b) $x = -4$
	(c) $x = 4$	(d) $x = -1$
28. $|x| \leq 5$	(a) $x = -7$	(b) $x = -4$
	(c) $x = 4$	(d) $x = 9$
29. $|x - 7| \geq 3$	(a) $x = 9$	(b) $x = -4$
	(c) $x = 11$	(d) $x = 6$
30. $|x - 3| > 5$	(a) $x = 16$	(b) $x = 3$
	(c) $x = -2$	(d) $x = -3$

In Exercises 31–34, transform the absolute value inequality into a double inequality or two separate inequalities.

31. $|y + 5| < 3$ **32.** $|6x + 7| \leq 5$

33. $|7 - 2h| \geq 9$ **34.** $|8 - x| > 25$

In Exercises 35–44, solve the inequality and sketch the graph of the solution.

35. $|y| < 4$ **36.** $|x| < 6$

37. $|y + 2| < 4$ **38.** $|x + 3| < 6$

39. $|x| > 6$ **40.** $|y| \geq 8$

41. $|y + 2| \geq 4$ **42.** $|x + 3| \geq 6$

43. $|2x + 3| > 9$ **44.** $|7r - 3| > 11$

In Exercises 45–48, solve the inequality. (If it is not possible, state the reason.)

45. $|0.2x - 3| < 4$

46. $|1.5t - 8| \leq 16$

47. $\left|\dfrac{z}{10} - 3\right| > 8$

48. $\left|\dfrac{x}{8} + 1\right| < 0$

49. *Body Temperature*   Physicians consider an adult's body temperature to be normal if it is between 97.6°F and 99.6°F. Write an absolute value inequality that describes this normal temperature range.

50. *Accuracy of Measurements*   In woodshop class, you must cut several pieces of wood within $\frac{3}{16}$ inch of the instructor's specifications. Let $(s - x)$ represent the difference between the specification s and the measured length x of a cut piece.

(a) Write an absolute value inequality that describes the values of x that are within specifications.

(b) The length of one of the pieces of wood is specified to be $s = 5\frac{1}{8}$ inches. Describe the acceptable lengths for this piece.

Reviewing the Major Concepts

51. What is $7\frac{1}{2}\%$ of 25?

52. What is 150% of 6000?

53. 225 is what percent of 150?

54. What percent of 240 is 160?

55. 0.5% of what number is 400?

56. 48% of what number is 132?

Additional Problem Solving

In Exercises 57–72, solve the equation. (Some of the equations may have no solution.)

57. $|h| = 0$

58. $|x| = -82$

59. $|x - 16| = 5$

60. $|z - 100| = 100$

61. $|32 - 3y| = 16$

62. $|3 - 5x| = 13$

63. $|3x - 2| = -5$

64. $|4 - 3x| = 0$

65. $|5x - 3| + 8 = 22$

66. $|20 - 5t| = 50$

67. $|0.32x - 2| = 4$

68. $\left|\frac{3}{2} - \frac{4}{5}x\right| = 1$

69. $|x + 8| = |2x + 1|$

70. $|10 - 3x| = |x + 7|$

71. $|5x + 4| = |3x + 8|$

72. $|45 - 4x| = |32 - 3x|$

In Exercises 73–76, determine whether the x-values are solutions of the inequality.

	Inequality		Values	
73.	$|x| \geq 3$	(a) $x = 2$	(b) $x = -4$	
		(c) $x = 4$	(d) $x = -1$	
74.	$|x| > 5$	(a) $x = -7$	(b) $x = -4$	
		(c) $x = 4$	(d) $x = 9$	
75.	$|x - 7| < 3$	(a) $x = 9$	(b) $x = -4$	
		(c) $x = 11$	(d) $x = 6$	
76.	$|x - 3| \leq 5$	(a) $x = 16$	(b) $x = 3$	
		(c) $x = -2$	(d) $x = -3$	

In Exercises 77–80, sketch a graph that represents the statement.

77. All x greater than -2 *and* less than 5

78. All x greater than or equal to 3 *and* less than 10

79. All x less than or equal to 4 *or* greater than 7

80. All x less than -6 *or* greater than or equal to 6

In Exercises 81–94, solve the inequality. Use a graphing utility to verify the solution.

81. $|y - 2| \leq 4$

82. $|x - 3| \leq 6$

83. $|y| \geq 4$

84. $|x| \geq 6$

85. $|y - 2| > 4$

86. $|x - 3| > 6$

87. $|2x| < 14$

88. $|4z| \leq 9$

89. $\left|\dfrac{y}{3}\right| \leq 3$

90. $\left|\dfrac{t}{2}\right| < 4$

91. $|3x + 2| > 4$

92. $|2x - 1| \geq 3$

93. $|x - 5| + 3 > 5$

94. $|a + 1| - 4 < 0$

Think About It   In Exercises 95 and 96, complete the statement so that the solution is $0 \leq x \leq 6$.

95. $|x - 3| \leq \rule{1.5cm}{0.4cm}$

96. $|2x - 6| \leq \rule{1.5cm}{0.4cm}$

In Exercises 97–102, solve the inequality. (If it is not possible, state the reason.)

97. $\dfrac{|x+2|}{10} \le 8$ **98.** $\dfrac{|y-16|}{4} < 30$

99. $|6t+15| \ge 30$ **100.** $|3t+1| > 5$

101. $\dfrac{|s-3|}{5} > 4$ **102.** $\dfrac{|a+6|}{2} \ge 16$

In Exercises 103–106, match the inequality with its graph. [The graphs are labeled (a), (b), (c), and (d).]

(a)

(b)

(c)

(d)

103. $|x-4| \le 4$ **104.** $|x-4| < 1$

105. $\frac{1}{2}|x-4| > 4$ **106.** $|2(x-4)| \ge 4$

In Exercises 107–112, write an absolute value inequality that represents the given interval.

107.

108.

109.

110.

111.

112.

113. *Temperature* The operating temperature for an electronic device must satisfy the inequality

$|t-72| < 10$

where t is given in degrees Fahrenheit. Sketch the graph of the solution of the inequality.

114. *Time Study* A time study was conducted to determine the length of time required to perform a particular task in a manufacturing process. The time required by approximately two-thirds of the workers in the study satisfied the inequality

$\left|\dfrac{t-15.6}{1.9}\right| < 1$

where t is time in minutes. Sketch the graph of the solution of the inequality.

In Exercises 115–118, write an absolute value inequality that represents the verbal statement.

115. The set of all real numbers x whose distance from 0 is less than 3

116. The set of all real numbers x whose distance from 0 is more than 2

117. The set of all real numbers x whose distance from 5 is more than 6

118. The set of all real numbers x whose distance from 16 is less than 5

CHAPTER PROJECT: Sounds of Animals

Some animals, such as bats, rely more heavily on their sense of sound than on their sense of sight.

Humans hear sounds with frequencies between 20 hertz and 20,000 hertz. Many animals, however, can hear sounds with frequencies far above 20,000 hertz. The frequency of a sound determines whether the sound is high or low. A high-pitched sound has a greater frequency than a low-pitched sound.

The frequency ranges in which several animals emit and receive sounds are shown in the table below.

Animal	Frequency Emittance Range	Frequency Reception Range
Human	$85 \leq f \leq 1100$	$20 \leq f \leq 20,000$
Dolphin	$7000 \leq f \leq 120,000$	$150 \leq f \leq 150,000$
Robin	$2000 \leq f \leq 13,000$	$250 \leq f \leq 21,000$
Dog	$450 \leq f \leq 1080$	$15 \leq f \leq 50,000$
Cat	$760 \leq f \leq 1520$	$60 \leq f \leq 65,000$
Bat	$10,000 \leq f \leq 120,000$	$1000 \leq f \leq 120,000$
Grasshopper	$7000 \leq f \leq 100,000$	$100 \leq f \leq 15,000$

Use this information to investigate the following questions.

1. Create a horizontal bar graph that represents the data in the table.
2. Which animals can emit sounds that humans cannot hear?
3. Which animals cannot hear the full range of human sounds?
4. Match each of the following with one of the frequency ranges in the table.
   (a) $-6700 \leq 20f - 7000 \leq 993,000$
   (b) $110,000 \leq 100(f + 1000) \leq 1,600,000$
   (c) $330 \leq \frac{1}{2}(f + 600) \leq 32,800$
   (d) $7000 + 20f \leq 21f \leq 120,000 + 20f$
   (e) $40,000 \leq 40f \leq 4,800,000$
   (f) $10 \leq \frac{1}{25}f \leq 840$
   (g) $-6500 \leq 10f - 8000 \leq 1,492,000$
5. *Research Project*   Use your school's library or some other reference source to find the frequency emittance ranges and frequency reception ranges for several other animals. Can you find any patterns between different types of animals? For instance, can you find examples of predators that cannot hear sounds made by their prey? Can you find examples of prey that cannot hear sounds made by their predators?

CHAPTER SUMMARY

After studying this chapter, you should have acquired the following skills. These skills are keyed to the Review Exercises that begin on page 117. Answers to odd-numbered Review Exercises are given in the back of the book.

- Determine whether the values of variables are solutions of equations. *(Section 1.1)* ...... **Review Exercises 1–4**

- Solve equations and check the results. *(Section 1.1)* ...... **Review Exercises 5–16**

- Solve equations with the aid of a calculator and round the results to two decimal places. *(Section 1.1)* ...... **Review Exercises 17–20**

- Complete tables showing equivalent forms of percents. *(Section 1.2)* ...... **Review Exercises 21, 22**

- Solve percent equations. *(Section 1.2)* ...... **Review Exercises 23–28**

- Solve real-life problems using percent equations. *(Section 1.2)* ...... **Review Exercises 29, 30, 38, 39**

- Express verbal expressions as ratios. *(Section 1.2)* ...... **Review Exercises 31, 32**

- Solve proportions. *(Section 1.2)* ...... **Review Exercises 33, 34**

- Translate real-life situations into proportions and solve. *(Section 1.2)* ...... **Review Exercises 35–37**

- Solve real-life problems involving markups. *(Section 1.3)* ...... **Review Exercises 40, 41**

- Solve real-life problems involving discounts. *(Section 1.3)* ...... **Review Exercises 42, 43**

- Solve real-life problems involving commissions. *(Section 1.3)* ...... **Review Exercise 44**

- Solve real-life problems involving mixtures. *(Section 1.3)* ...... **Review Exercises 45, 46**

- Solve real-life problems involving distances. *(Section 1.3)* ...... **Review Exercises 47, 48**

- Solve real-life problems involving work rates. *(Section 1.3)* ...... **Review Exercises 49, 50**

- Solve real-life problems involving simple interest. *(Section 1.3)* ...... **Review Exercises 51–56**

- Solve problems involving geometry. *(Sections 1.2, 1.3)* ...... **Review Exercises 57, 58**

- Solve equations for specified variables. *(Section 1.3)* ...... **Review Exercises 59–62**

- Solve absolute value equations. *(Section 1.5)* ...... **Review Exercises 63–66**

- Solve inequalities and sketch the solutions on the real number line. *(Sections 1.4, 1.5)* ...... **Review Exercises 67–80**

- Write inequalities that represent verbal statements. *(Section 1.4)* ...... **Review Exercises 81–84**

- Write absolute value inequalities that represent intervals. *(Section 1.4)* ...... **Review Exercises 85, 86**

REVIEW EXERCISES

In Exercises 1–4, determine whether the values of the variable are solutions of the equation.

Equation	*Values*

1. $45 - 7x = 3$ (a) $x = 3$ (b) $x = 6$

2. $3(3 - x) = -x$ (a) $x = \frac{9}{2}$ (b) $x = -\frac{2}{3}$

3. $\frac{x}{7} + \frac{x}{5} = 1$ (a) $x = \frac{35}{12}$ (b) $x = -\frac{2}{35}$

4. $\frac{x + 2}{6} = \frac{7}{2}$ (a) $x = -12$ (b) $x = 19$

In Exercises 5–16, solve the equation and check the result. (Some of the equations have no solution.)

5. $17 - 7x = 3$ 6. $3 + 6x = 51$

7. $4y - 6(y - 5) = 2$ 8. $7x + 2(7 - x) = 8$

9. $1.4t + 4.1 = 0.9t$ 10. $8(x - 2) = 3(x - 2)$

11. $\frac{3x}{4} = 4$ 12. $-\frac{5x}{14} = \frac{1}{2}$

13. $\frac{4}{5}x - \frac{1}{10} = \frac{3}{2}$ 14. $\frac{1}{4}s + \frac{3}{8} = \frac{5}{2}$

15. $\frac{v - 20}{-8} = 2v$ 16. $x + \frac{2x}{5} = 1$

In Exercises 17–20, solve the equation using a calculator. Round the result to two decimal places.

17. $382x - 575 = 715$ 18. $3.625x + 3.5 = 22.125$

19. $\frac{x}{2.33} = 14.302$ 20. $\frac{7x}{3} + 2.5 = 8.125$

In Exercises 21 and 22, complete the table.

21.

Percent	Parts out of 100	Decimal	Fraction
87%			

22.

Percent	Parts out of 100	Decimal	Fraction
			$\frac{1}{6}$

23. What is 130% of 50?

24. What is 0.4% of 7350?

25. 645 is $21\frac{1}{2}\%$ of what number?

26. 498 is 83% of what number?

27. 250 is what percent of 200?

28. 162.5 is what percent of 6500?

29. *Completing a Table*   Complete the table on instant replays in the National Football League.   (Source: National Football League)

Year	Reviewed	Reversed	Percent Reversed
1986	374	38	
1987	490	47	
1988	537	53	
1989	492	65	

30. *Interpreting a Graph*   Use the information in the figure to determine the percent increase in the per capita cost of census taking from the 1980 census to the 1990 census. Using a 1990 population of 250 million, determine the total cost of the 1990 census. (Source: U.S. Bureau of Census)

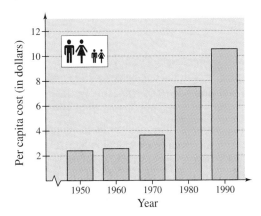

In Exercises 31 and 32, express as a ratio. (Use the same units for the numerator and denominator.)

31. 16 feet to 4 yards   **32.** 3 quarts to 5 pints

In Exercises 33 and 34, solve the proportion.

33. $\dfrac{7}{8} = \dfrac{y}{4}$ **34.** $\dfrac{x}{16} = \dfrac{5}{12}$

35. *Property Tax*   The tax on property with an assessed value of $80,000 is $1350. Find the tax on property with an assessed value of $110,000.

36. *Map Scale*   One-third inch represents 50 miles on a map. Approximate the distance between two cities that are $3\frac{1}{4}$ inches apart on the map.

37. *Geometry*   You want to measure the height of a flagpole. To do this, you measure the flagpole's shadow and find that it is 30 feet long. You also measure the height of a five-foot lamp post and find its shadow to be 3 feet long. How tall is the flagpole?

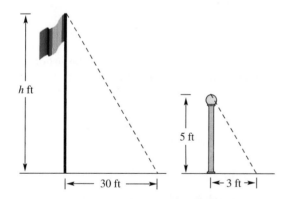

38. *Revenue Increase*   The revenues for a corporation in millions of dollars in the years 1993 and 1994 were $6521.4 and $6679.0, respectively. Determine the percentage increase in revenue from 1993 to 1994.

39. *Price Increase*   The manufacturer's suggested retail price for a certain truck model is $25,750. Estimate the price of a comparably equipped truck for the next model year if it is projected that truck prices will increase by $5\frac{1}{2}\%$.

40. *Retail Price*   A camera that costs a retailer $259.95 is marked up by 35%. Find the price to the consumer.

41. *Markup Rate*   A calculator selling for $175.00 costs the retailer $95.00. Find the markup rate.

42. *Sale Price*   The list price of a coat is $259. Find the sale price of the coat if it is reduced by 25%.

43. *Comparison Shopping*   A mail-order catalog has an attaché case with a list price of $99.97 plus $4.50 for shipping and handling. A department store has the same case for $125.95. The department store has a special 20% off sale. Which is the better buy?

44. *Sales Goal*   The weekly salary of an employee is $150 plus a 6% commission on total sales. The employee needs a minimum salary of $650 per week. How much must be sold to produce this salary?

45. *Mixture*   Determine the number of liters of a 30% saline solution and the number of liters of a 60% saline solution that are required to make 10 liters of a 50% saline solution.

46. *Mixture*   Determine the number of gallons of a 25% alcohol solution and the number of gallons of a 50% alcohol solution that are required to make 8 gallons of a 40% alcohol solution.

47. *Travel Time*   Determine the time for a bus to travel 330 miles if its average speed is 52 miles per hour.

48. *Average Speed*   An Olympic cross-country skier completed the 15-kilometer event in 41 minutes and 20 seconds. What was the average speed of the skier?

49. *Work Rate*   Find the time for two people working together to complete a task if it takes them 4.5 hours and 6 hours working individually.

50. *Work Rate*   Find the time for two people working together to complete half of a task if it takes them 8 hours and 10 hours to complete the entire task working individually.

51. *Simple Interest*   Find the total simple interest you will earn on a $1000 corporate bond that matures in 4 years and has an 8.5% interest rate.

52. *Simple Interest*   Find the annual simple interest rate on a certificate of deposit that pays $37.50 per year in interest on a principal of $500.

53. *Simple Interest*   Find the principal required to have an annual interest income of $20,000 if the annual simple interest rate on the principal is 9.5%.

54. *Simple Interest*   A corporation borrows 3.25 million dollars for 2 years to modernize one of its manufacturing facilities. If it pays an annual simple interest rate of 12%, what will be the total principal and interest that must be repaid?

55. *Simple Interest*   An inheritance of $50,000 is divided between two investments earning 8.5% and 10% simple interest. How much is in each investment if the total interest for 1 year is $4700?

56. *Simple Interest*   You invest $1000 in a certificate of deposit that has an annual simple interest rate of 7%. After 6 months the interest is computed and added to the principal. During the second 6 months the interest is computed using the original investment plus the interest earned during the first 6 months. What is the total interest earned during the first year of the investment?

57. *Geometry*   The area of the shaded region is 48 square inches. Find the dimensions of the rectangle.

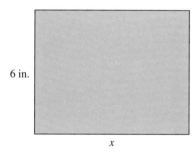

6 in.

x

58. *Geometry*   Solve for the length x by using the fact that corresponding sides of similar triangles are proportional.

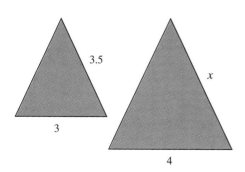

3.5

3

x

4

In Exercises 59–62, solve the equation for the specified variable.

59. Solve for x in $2x - 7y + 4 = 0$.

60. Solve for v in $\frac{2}{3}u - 4v = 2v + 3$.

61. Solve for h in $V = \pi r^2 h$.

62. Solve for h in $S = 2\pi r^2 + 2\pi r h$.

In Exercises 63–66, solve the equation.

63. $|x - 2| - 2 = 4$ **64.** $|2x + 3| = 7$

65. $|3x - 4| = |x + 2|$ **66.** $|5x + 6| = |2x - 1|$

In Exercises 67–80, solve the inequality and sketch the solution on the real number line.

67. $5x + 3 > 18$ **68.** $-11x \geq 44$

69. $\frac{1}{3} - \frac{1}{2}y < 12$ **70.** $3(2 - y) \geq 2(1 + y)$

71. $-4 < \dfrac{x}{5} \leq 4$ **72.** $-13 \leq 3 - 4x < 13$

73. $5 > \dfrac{x + 1}{-3} > 0$ **74.** $12 \geq \dfrac{x - 3}{2} > 1$

75. $|2x - 7| < 15$ **76.** $|5x - 1| < 9$

77. $|x - 4| > 3$ **78.** $|t + 3| > 2$

79. $|b + 2| - 6 > 1$ **80.** $\left|\dfrac{t}{3}\right| < 1$

In Exercises 81–84, write an inequality for the given statement.

81. z is no more than 10.

82. x is nonnegative.

83. y is at least 7 but less than 14.

84. The volume V is less than 27 cubic feet.

In Exercises 85 and 86, write an absolute value inequality that represents the interval.

85.

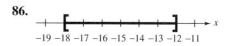

0 1 2 3 4 5 6

86.

-19 -18 -17 -16 -15 -14 -13 -12 -11

CHAPTER TEST

Take this test as you would take a test in class. After you are done, check your work against the answers given in the back of the book.

1. Determine whether (a) $x = -4$ and (b) $x = 2$ are solutions of $3(5 - 2x) - (3x - 1) = -2$.

In Exercises 2–5, solve the equation.

2. $6x - 5 = 19$

3. $15 - 7(1 - x) = 3(x + 8)$

4. $\dfrac{2x}{3} = \dfrac{x}{2} + 4$

5. $\dfrac{t - 5}{12} = \dfrac{3}{8}$

6. What is 27% of 3200?

7. 1200 is what percent of 800?

8. A store is offering a 20% discount on all items in its inventory. Find the list price on a tractor that has a sale price of $6400.

9. Which of the packages at the right is a better buy? Explain your reasoning.

Figure for 9

10. The tax on property with an assessed value of $90,000 is $1200. Estimate the tax on property with an assessed value of $110,000.

11. The bill (including parts and labor) for the repair of a home appliance was $165. The cost for parts was $85. How many hours were spent to repair the appliance if the cost of labor was $16 per half hour?

12. Two solutions—10% concentration and 40% concentration—are mixed to make 100 liters of a 30% solution. Determine the numbers of liters of the 10% solution and the 40% solution that are required.

13. Two cars start at a given time and travel in the same direction at average speeds of 40 miles per hour and 55 miles per hour. How much time must elapse before the two cars are 10 miles apart?

14. One number is four times a second number. Find the numbers if their difference is 39.

15. Find the principal required to earn $300 in simple interest in 2 years if the annual interest rate is 7.5%.

In Exercises 16–19, solve the inequality and sketch its solution.

16. $1 + 2x > 7 - x$

17. $0 \le \dfrac{1 - x}{4} < 2$

18. $|x - 3| \le 2$

19. $|x + 4| > 1$

20. Use inequality notation to denote the phrase, "t is at least 8."

Functions and Graphs

2

In the early history of the United States, most people were self-employed in agriculture and men's and women's roles were roughly equivalent. Everyone—men, women, and children—was involved in running the family farm.

As the country became less rural and more industrial, working roles changed. Men's roles became associated with working outside the home, and women's roles became associated with homemaking and caring for children. Now, since World War II, the roles of men and women in the work force are once again becoming more alike.

The table at the right shows the numbers of men and women (in thousands) that made up the civilian labor force in selected years from 1960 to 1992. The graph shows a scatter plot of the data.

Year	1960	1970	1980	1990	1991	1992
Women	21,874	29,688	42,117	53,479	53,284	53,793
Men	43,904	48,990	57,186	64,435	63,593	63,805

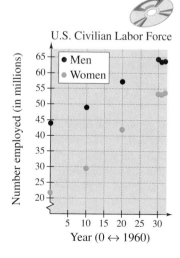

U.S. Civilian Labor Force

The chapter project related to this information is on page 191.

121

2.1	**The Rectangular Coordinate System**
	The Rectangular Coordinate System ▪ Ordered Pairs as Solutions ▪ The Distance Formula

The Rectangular Coordinate System

Just as you can represent real numbers by points on the real number line, you can represent ordered pairs of real numbers by points in a plane. This plane is called a **rectangular coordinate system** or the **Cartesian plane,** after the French mathematician René Descartes.

A rectangular coordinate system is formed by two real lines intersecting at a right angle, as shown in Figure 2.1. The horizontal number line is usually called the **x-axis,** and the vertical number line is usually called the **y-axis.** (The plural of axis is *axes*.) The point of intersection of the two axes is called the **origin,** and the axes separate the plane into four regions called **quadrants.**

René Descartes (1596–1650) was a French mathematician, philosopher, and scientist. He is sometimes called the father of modern philosophy, and his phrase "I think, therefore I am," has been quoted often. In mathematics, Descartes is known as the father of analytic geometry. Prior to Descartes's time, geometry and algebra were separate mathematical studies—it was Descartes's introduction of the rectangular coordinate system that brought the two studies together.

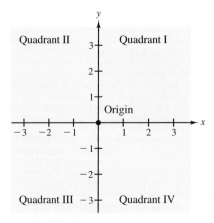

FIGURE 2.1

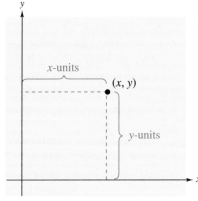

FIGURE 2.2

Each point in the plane corresponds to an **ordered pair** (x, y) of real numbers x and y, called the **coordinates** of the point. The first number (or **x-coordinate**) tells how far to the left or right the point is from the vertical axis, and the second number (or **y-coordinate**) tells how far up or down the point is from the horizontal axis, as shown in Figure 2.2.

> **NOTE** The signs of the coordinates tell you which quadrant the point lies in. For instance, if x and y are positive, the point (x, y) lies in Quadrant I.

Locating a given point in a plane is called **plotting** the point. Example 1 shows how this is done.

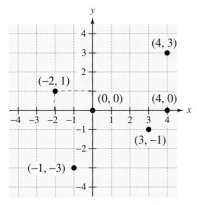

FIGURE 2.3

EXAMPLE 1   *Plotting Points on a Rectangular Coordinate System*

Plot the points $(-2, 1)$, $(4, 0)$, $(3, -1)$, $(4, 3)$, $(0, 0)$, and $(-1, -3)$ on a rectangular coordinate system.

Solution

The point $(-2, 1)$ is two units to the *left* of the vertical axis and one unit *above* the horizontal axis.

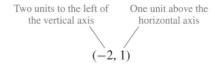

Similarly, the point $(4, 0)$ is four units to the *right* of the vertical axis and *on* the horizontal axis. (It is on the horizontal axis because its y-coordinate is 0.) The other four points can be plotted in a similar way, as shown in Figure 2.3.

In Example 1, you were given the coordinates of several points and asked to plot the points on a rectangular coordinate system. Example 2 looks at the reverse problem. That is, you are given points on a rectangular coordinate system and are asked to determine their coordinates.

EXAMPLE 2   *Finding Coordinates of Points*

Determine the coordinates of each of the points shown in Figure 2.4.

Solution

Point A lies two units to the *right* of the vertical axis and one unit *below* the horizontal axis. Therefore, point A must be given by the ordered pair $(2, -1)$. The coordinates of the other four points can be determined in a similar way; the results are summarized as follows.

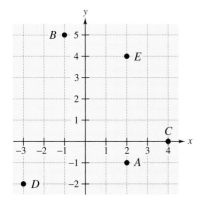

FIGURE 2.4

Point	Position	Coordinates
A	2 units *right*, 1 unit *down*	$(2, -1)$
B	1 unit *left*, 5 units *up*	$(-1, 5)$
C	4 units *right*, 0 units *up*	$(4, 0)$
D	3 units *left*, 2 units *down*	$(-3, -2)$
E	2 units *right*, 4 units *up*	$(2, 4)$

Notice that because point C lies on the x-axis, it has a y-coordinate of 0.

The primary value of a rectangular coordinate system is that it allows you to visualize relationships between two variables. Today, Descartes's ideas are commonly used in virtually every scientific and business-related field.

EXAMPLE 3 *Representing Data Graphically*

The population (in millions) of California from 1975 through 1990 is listed in the table. Plot these points on a rectangular coordinate system. (Source: U.S. Bureau of Census)

Year	1975	1976	1977	1978	1979	1980	1981	1982
Population	21.2	21.5	21.9	22.3	23.3	23.7	24.3	24.8

Year	1983	1984	1985	1986	1987	1988	1989	1990
Population	25.3	26.4	27.0	27.0	27.7	28.3	28.6	29.8

Solution

Begin by choosing which variable will be plotted on the horizontal axis and which will be plotted on the vertical axis. For these data, it seems natural to plot the years on the horizontal axis (which means that the population must be plotted on the vertical axis). Next, use the data in the table to form ordered pairs. For instance, the first three ordered pairs are (1975, 21.2), (1976, 21.5), and (1977, 21.9). All 16 points are shown in Figure 2.5. Note that the break in the x-axis indicates that the numbers between 0 and 1975 have been omitted. The break in the y-axis indicates that the numbers between 0 and 21,000 have been omitted.

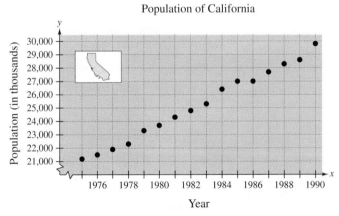

FIGURE 2.5

Ordered Pairs as Solutions

In Example 3, the relationship between the year and the population was given by a **table of values.** In mathematics, the relationship between the variables x and y is often given by an equation. From the equation, you can construct your own table of values.

EXAMPLE 4   *Constructing a Table of Values*

Construct a table of values for $y = 3x + 2$. Then plot the solution points on a rectangular coordinate system. Choose x-values of $-3, -2, -1, 0, 1, 2,$ and 3.

Solution

For each x-value, you must calculate the corresponding y-value. For example, if you choose $x = 1$, then the y-value is

$$y = 3(1) + 2 = 5.$$

The ordered pair $(x, y) = (1, 5)$ is a **solution point** (or **solution**) of the equation.

Choose x	Calculate y from $y = 3x + 2$	Solution Point
$x = -3$	$y = 3(-3) + 2 = -7$	$(-3, -7)$
$x = -2$	$y = 3(-2) + 2 = -4$	$(-2, -4)$
$x = -1$	$y = 3(-1) + 2 = -1$	$(-1, -1)$
$x = 0$	$y = 3(0) + 2 = 2$	$(0, 2)$
$x = 1$	$y = 3(1) + 2 = 5$	$(1, 5)$
$x = 2$	$y = 3(2) + 2 = 8$	$(2, 8)$
$x = 3$	$y = 3(3) + 2 = 11$	$(3, 11)$

Once you have constructed a table of values, you can get a visual idea of the relationship between the variables x and y by plotting the solution points on a rectangular coordinate system, as shown in Figure 2.6.

DISCOVERY

In the table of values in Example 4, successive x-values differ by 1. How do successive y-values differ? If successive x-values differed by 1, how would successive y-values differ for the following equations?

a. $y = x + 2$

b. $y = 2x + 2$

c. $y = 4x + 2$

d. $y = -x + 2$

Describe the pattern. *Hint*: If you have a *TI-82*, use it to form the tables, as shown on page 126.

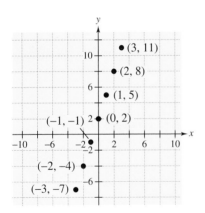

FIGURE 2.6

When making up a table of values for an equation, it is helpful to first solve the equation for y. Here is an example.

$5x + 3y = 4$	Original equation
$3y = -5x + 4$	Subtract $5x$ from both sides.
$y = -\dfrac{5}{3}x + \dfrac{4}{3}$	Divide both sides by 3.

Technology

1. Y=

$Y_1=(2/3)X^2-(5/3)$
$Y_2=$
$Y_3=$
$Y_4=$
$Y_5=$
$Y_6=$

2. TblSet

TABLE SETUP
TblMin=-3
ΔTbl=1
Indpnt: Auto Ask
Depend: Auto Ask

3. TABLE

X	Y_1	
-3	4.3333	
-2	1	
-1	-1	
0	-1.667	
1	-1	
2	1	
3	4.3333	

X=-3

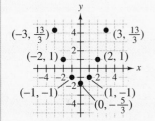

Creating a Table with a Graphing Utility

Some graphing utilities, such as the *TI-82*, have built-in programs that can create tables. Suppose, for instance, that you want to create a table of values for the equation $2x^2 - 3y = 5$. To begin, solve the equation for y.

$$2x^2 - 3y = 5 \qquad \text{Original equation}$$

$$-3y = -2x^2 + 5 \qquad \text{Subtract } 2x^2 \text{ from both sides.}$$

$$y = \frac{2}{3}x^2 - \frac{5}{3} \qquad \text{Divide both sides by } -3.$$

Now, using the equation $y = \frac{2}{3}x^2 - \frac{5}{3}$, you can use the following steps.

1. Press the Y= key and enter the equation as Y_1.

2. Use the TblSet feature and enter the values of TblMin and ΔTbl. The value of TblMin is the beginning x-value you want displayed in the table, and the value of ΔTbl is the increment for the x-values in the table. For instance, using TblMin= −3 and ΔTbl=1, the table has x-values of −3, −2, −1, and so on.

3. Use the TABLE feature to obtain the table of values. You can use the cursor keys to view x- and y-values that are not shown on the default screen.

You can confirm the values shown on the utility's screen by substituting the appropriate x-values, as shown in the table below. (Note that the utility lists decimal approximations of fractional values. For instance, $-\frac{5}{3}$ is listed as −1.667.)

Choose x	Calculate y from $y = \frac{2}{3}x^2 - \frac{5}{3}$	Solution Point
$x = -3$	$y = \frac{2}{3}(-3)^2 - \frac{5}{3} = \frac{13}{3}$	$\left(-3, \frac{13}{3}\right)$
$x = -2$	$y = \frac{2}{3}(-2)^2 - \frac{5}{3} = 1$	$(-2, 1)$
$x = -1$	$y = \frac{2}{3}(-1)^2 - \frac{5}{3} = -1$	$(-1, -1)$
$x = 0$	$y = \frac{2}{3}(0)^2 - \frac{5}{3} = -\frac{5}{3}$	$\left(0, -\frac{5}{3}\right)$
$x = 1$	$y = \frac{2}{3}(1)^2 - \frac{5}{3} = -1$	$(1, -1)$
$x = 2$	$y = \frac{2}{3}(2)^2 - \frac{5}{3} = 1$	$(2, 1)$
$x = 3$	$y = \frac{2}{3}(3)^2 - \frac{5}{3} = \frac{13}{3}$	$\left(3, \frac{13}{3}\right)$

After creating the table, you can plot the points, as shown at the left.

NOTE When a substitution of (x, y) produces a true equation, the ordered pair (x, y) is said to **satisfy** the equation.

In the next example, you are given several ordered pairs and are asked to determine whether they are solutions of the given equation. To do this, you need to substitute the values of x and y into the equation. If the substitution produces a true equation, then the ordered pair (x, y) is a solution.

EXAMPLE 5 Verifying Solutions of an Equation

Which of the ordered pairs are solutions of $x^2 - 2y = 6$?

a. $(2, 1)$ **b.** $(0, -3)$ **c.** $(-2, -5)$ **d.** $\left(1, -\frac{5}{2}\right)$

Solution

a. For the ordered pair $(2, 1)$, substitute $x = 2$ and $y = 1$ into the equation.

$$x^2 - 2y = 6 \qquad \text{Original equation}$$
$$(2)^2 - 2(1) \overset{?}{=} 6 \qquad \text{Substitute 2 for } x \text{ and 1 for } y.$$
$$2 \neq 6 \qquad \text{Is not a solution} \; ✗$$

Because the substitution does not satisfy the given equation, you can conclude that the ordered pair $(2, 1)$ *is not* a solution of the given equation.

b. For the ordered pair $(0, -3)$, substitute $x = 0$ and $y = -3$ into the equation.

$$x^2 - 2y = 6 \qquad \text{Original equation}$$
$$(0)^2 - 2(-3) \overset{?}{=} 6 \qquad \text{Substitute 0 for } x \text{ and } -3 \text{ for } y.$$
$$6 = 6 \qquad \text{Is a solution} \; ✓$$

Because the substitution satisfies the given equation, you can conclude that the ordered pair $(0, -3)$ *is* a solution of the given equation.

c. For the ordered pair $(-2, -5)$, substitute $x = -2$ and $y = -5$ into the equation.

$$x^2 - 2y = 6 \qquad \text{Original equation}$$
$$(-2)^2 - 2(-5) \overset{?}{=} 6 \qquad \text{Substitute } -2 \text{ for } x \text{ and } -5 \text{ for } y.$$
$$14 \neq 6 \qquad \text{Is not a solution} \; ✗$$

Because the substitution does not satisfy the given equation, you can conclude that the ordered pair $(-2, -5)$ *is not* a solution of the given equation.

d. For the ordered pair $\left(1, -\frac{5}{2}\right)$, substitute $x = 1$ and $y = -\frac{5}{2}$ into the equation.

$$x^2 - 2y = 6 \qquad \text{Original equation}$$
$$(1)^2 - 2\left(-\frac{5}{2}\right) \overset{?}{=} 6 \qquad \text{Substitute 1 for } x \text{ and } -\frac{5}{2} \text{ for } y.$$
$$6 = 6 \qquad \text{Is a solution} \; ✓$$

Because the substitution satisfies the given equation, you can conclude that the ordered pair $\left(1, -\frac{5}{2}\right)$ *is* a solution of the given equation.

The Distance Formula

You know from Section P.1 of the Prerequisites chapter that the distance d between two points a and b on the real number line is simply $d = |b - a|$. The same "absolute value rule" is used to find the distance between two points that lie on the same *vertical or horizontal line*, as shown in Example 6.

EXAMPLE 6   *Finding Horizontal and Vertical Distances*

a. Find the distance between the points $(2, -2)$ and $(2, 4)$.

b. Find the distance between the points $(-3, -2)$ and $(2, -2)$.

Solution

a. Because the x-coordinates are equal, you can visualize a vertical line through the points $(2, -2)$ and $(2, 4)$, as shown in Figure 2.7. The distance between these two points is the absolute value of the difference of their y-coordinates.

$$\text{Vertical distance} = |4 - (-2)| \qquad \text{Subtract } y\text{-coordinates.}$$
$$= 6 \qquad \text{Evaluate absolute value.}$$

b. Because the y-coordinates are equal, you can visualize a horizontal line through the points $(-3, -2)$ and $(2, -2)$, as shown in Figure 2.7. The distance between these two points is the absolute value of the difference of their x-coordinates.

$$\text{Horizontal distance} = |2 - (-3)| \qquad \text{Subtract } x\text{-coordinates.}$$
$$= 5 \qquad \text{Evaluate absolute value.}$$

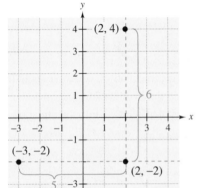

FIGURE 2.7

NOTE In Figure 2.7, note that the horizontal distance between the points $(-3, -2)$ and $(2, -2)$ is the absolute value of the difference of the x-coordinates, and the vertical distance between the points $(2, -2)$ and $(2, 4)$ is the absolute value of the difference of the y-coordinates.

The technique applied in Example 6 can be used to develop a general formula for finding the distance between two points in the plane. This general formula will work for any two points, even if they do not lie on the same vertical or horizontal line. To develop the formula, you use the **Pythagorean Theorem,** which states that for a right triangle, the hypotenuse c and sides a and b are related by the formula

$$a^2 + b^2 = c^2 \qquad \text{Pythagorean Theorem}$$

as shown in Figure 2.8. (The converse is also true. That is, if $a^2 + b^2 = c^2$, then the triangle is a right triangle.)

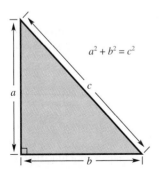

FIGURE 2.8   Pythagorean Theorem

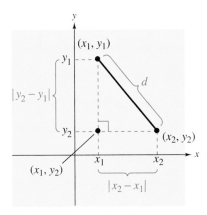

FIGURE 2.9 Distance Between Two Points

To develop a general formula for the distance between two points, let (x_1, y_1) and (x_2, y_2) represent two points in the plane (that do not lie on the same horizontal or vertical line). With these two points, a right triangle can be formed, as shown in Figure 2.9. Note that the third vertex of the triangle is (x_1, y_2). Because (x_1, y_1) and (x_1, y_2) lie on the same vertical line, the length of the vertical side of the triangle is $|y_2 - y_1|$. Similarly, the length of the horizontal side is $|x_2 - x_1|$. By the Pythagorean Theorem, the square of the distance between (x_1, y_1) and (x_2, y_2) is

$$d^2 = |x_2 - x_1|^2 + |y_2 - y_1|^2.$$

Because the distance d must be positive, you can choose the positive square root and write

$$d = \sqrt{|x_2 - x_1|^2 + |y_2 - y_1|^2}.$$

Finally, replacing $|x_2 - x_1|^2$ and $|y_2 - y_1|^2$ by the equivalent expressions $(x_2 - x_1)^2$ and $(y_2 - y_1)^2$ gives you the **Distance Formula.**

DISCOVERY

Plot the points $A(-1, -3)$ and $B(5, 2)$ and sketch the line segment from A to B. How could you verify that point $C(2, -0.5)$ is the midpoint of the segment? Why is it not sufficient to show that the distances from A to C and from C to B are equal? Find a formula for the coordinates of the midpoint of the line segment connecting (x_1, y_1) and (x_2, y_2).

The Distance Formula

The distance d between two points (x_1, y_1) and (x_2, y_2) is

$$d = \sqrt{(x_2 - x_1)^2 + (y_2 - y_1)^2}.$$

Note that for the special case in which the two points lie on the same vertical or horizontal line, the Distance Formula still works. For instance, applying the Distance Formula to the points $(2, -2)$ and $(2, 4)$ produces

$$d = \sqrt{(2 - 2)^2 + [4 - (-2)]^2} = \sqrt{6^2} = 6$$

which is the same result obtained in Example 6.

EXAMPLE 7   Finding the Distance Between Two Points

Find the distance between the points $(-1, 2)$ and $(2, 4)$, as shown in Figure 2.10.

Solution

Let $(x_1, y_1) = (-1, 2)$ and $(x_2, y_2) = (2, 4)$, and apply the Distance Formula.

$$d = \sqrt{[2 - (-1)]^2 + (4 - 2)^2} \qquad \text{Substitute coordinates of points.}$$
$$= \sqrt{3^2 + 2^2} \qquad \text{Simplify.}$$
$$= \sqrt{13} \qquad \text{Simplify.}$$
$$\approx 3.61 \qquad \text{Use a calculator.}$$

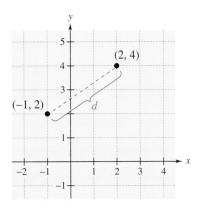

FIGURE 2.10

The Distance Formula has many applications in mathematics. For instance, the next example shows how you can use the Distance Formula and the converse of the Pythagorean Theorem to verify that three points form the vertices of a right triangle.

EXAMPLE 8   An Application of the Distance Formula

Show that the points $(1, 2)$, $(3, 1)$, and $(4, 3)$ are vertices of a right triangle.

Solution

The three points are plotted in Figure 2.11. Using the Distance Formula, you can find the lengths of the three sides of the triangle.

$$d_1 = \sqrt{(3-1)^2 + (1-2)^2} = \sqrt{4+1} = \sqrt{5}$$
$$d_2 = \sqrt{(4-3)^2 + (3-1)^2} = \sqrt{1+4} = \sqrt{5}$$
$$d_3 = \sqrt{(4-1)^2 + (3-2)^2} = \sqrt{9+1} = \sqrt{10}$$

Because

$$d_1{}^2 + d_2{}^2 = 5 + 5$$
$$= 10$$
$$= d_3{}^2$$

you can conclude from the converse of the Pythagorean Theorem that the triangle is a right triangle.

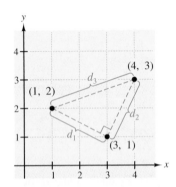

FIGURE 2.11

Group Activities   Extending the Concept

Determining Collinearity   Three or more points are **collinear** if they all lie on the same line. Use the following steps to determine if the set of points $\{A(3, 1), B(5, 4), C(9, 10)\}$ and the set of points $\{A(2, 2), B(4, 3), C(5, 4)\}$ are collinear.

a. For each set of points, use the Distance Formula to find the distances from A to B, from B to C, and from A to C. What relationship exists among these distances for each set of points?

b. Plot each set of points on a rectangular coordinate system. Do all the points of either set appear to lie on the same line?

c. Compare your conclusions from part (a) with the conclusions you made from the graphs in part (b). Make a general statement about how to use the Distance Formula to determine collinearity.

2.1 Exercises

Discussing the Concepts

1. Discuss the significance of the word *order* when referring to an ordered pair (x, y).

2. When plotting the point $(3, -2)$, what does the x-coordinate measure? What does the y-coordinate measure?

3. What is the x-coordinate of any point on the y-axis? What is the y-coordinate of any point on the x-axis?

4. Explain why the ordered pair $(-3, 4)$ is not a solution point of the equation $y = 4x + 15$.

5. When plotting points on the rectangular coordinate system, is it true that the scales on the x- and y-axes must be the same? Explain.

6. State the Pythagorean Theorem and give examples of its use.

Problem Solving

In Exercises 7–10, plot the points on a rectangular coordinate system.

7. $(4, 3), (-5, 3), (3, -5)$

8. $(-2, 5), (-2, -5), (3, 5)$

9. $\left(\frac{5}{2}, -2\right), \left(-2, \frac{1}{4}\right), \left(\frac{3}{2}, -\frac{7}{2}\right)$

10. $\left(-\frac{2}{3}, 3\right), \left(\frac{1}{4}, -\frac{5}{4}\right), \left(-5, -\frac{7}{4}\right)$

In Exercises 11 and 12, approximate the coordinates of the points.

11.

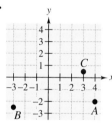

12.

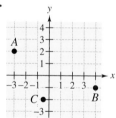

Geometry   In Exercises 13–16, plot the points and connect them with line segments to form the figure. (*Note:* A *rhombus* is a parallelogram whose sides are all the same length.)

13. *Triangle:* $(-1, 2), (2, 0), (3, 5)$

14. *Rectangle:* $(7, 0), (9, 1), (4, 6), (6, 7)$

15. *Parallelogram:* $(4, 0), (6, -2), (0, -4), (-2, -2)$

16. *Rhombus:* $(-3, -3), (-2, -1), (-1, -2), (0, 0)$

In Exercises 17–22, determine the quadrant or quadrants in which the point is located.

17. $(-3, -5)$

18. $(4, -2)$

19. $(x, 4)$

20. $(-10, y)$

21. $(x, y),\quad xy < 0$

22. $(x, y),\quad x > 0, y > 0$

In Exercises 23–26, find the coordinates of the point.

23. The point is located five units to the left of the y-axis and two units above the x-axis.

24. The point is located 10 units to the right of the y-axis and four units below the x-axis.

25. The point is on the positive x-axis 10 units from the origin.

26. The point is on the negative y-axis five units from the origin.

In Exercises 27 and 28, plot the points whose coordinates are given in the table.

27. *Exam Score*   The table gives the time x in hours invested in concentrated study for five different algebra exams and the resulting exam score y.

x	5	2	3	6.5	4
y	81	71	88	92	86

28. *Price of Stock* The year-end market prices y per share of common stock in PepsiCo, Inc. for the years 1987 through 1993 are given in the table. The time in years is given by x. (Source: PepsiCo, Inc.)

x	1987	1988	1989	1990	1991	1992	1993
y	$11\frac{1}{4}$	$13\frac{1}{8}$	$21\frac{3}{8}$	$25\frac{3}{4}$	$33\frac{3}{4}$	$42\frac{1}{4}$	$41\frac{7}{8}$

In Exercises 29 and 30, determine whether the ordered pairs are solutions of the equation.

29. $y = 3x + 8$
   (a) $(3, 17)$ (b) $(-1, 10)$
   (c) $(0, 0)$ (d) $(-2, 2)$

30. $5x - 2y + 50 = 0$
   (a) $(-10, 0)$ (b) $(-5, 5)$
   (c) $(0, 25)$ (d) $(20, -2)$

In Exercises 31 and 32, complete the table of values. Then plot the solution points on a rectangular coordinate system.

31.

x	-2	0	2	4	6
$y = 5x - 1$					

32.

x	-2	0	2	4	6
$y = \frac{3}{4}x + 2$					

In Exercises 33 and 34, use a graphing utility to complete the table of values.

33.

x	-2	0	2	4	6
$y = 4x^2 + x - 2$					

34.

x	-2	0	2	4	6
$y = |3x - 4| + 1$					

In Exercises 35–38, plot the points and find the distance between them. State whether the points lie on a horizontal or vertical line.

35. $(3, -2), (3, 5)$ **36.** $(-2, 8), (-2, 1)$
37. $\left(\frac{1}{2}, \frac{7}{8}\right), \left(\frac{11}{2}, \frac{7}{8}\right)$ **38.** $\left(\frac{3}{4}, 1\right), \left(\frac{3}{4}, -10\right)$

In Exercises 39–42, find the distance between the points.

39. $(1, 3), (5, 6)$ **40.** $(3, 10), (15, 5)$
41. $(0, 0), (12, -9)$ **42.** $(-5, 0), (3, 15)$

In Exercises 43 and 44, find the coordinates of (x, y), the lengths of the vertical and horizontal sides of the right triangle, and the length of the hypotenuse.

43. 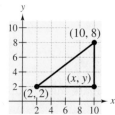   **44.**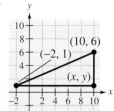

45. *Making a Conjecture* Plot the points $(2, 1), (-3, 5)$, and $(7, -3)$ on a rectangular coordinate system. Then change the sign of the x-coordinate of each point and plot the three new points on the same rectangular coordinate system. What conjecture can you make about the location of a point when the sign of the x-coordinate is changed?

46. *Making a Conjecture* Plot the points $(2, 1), (-3, 5)$, and $(7, -3)$ on a rectangular coordinate system. Then change the sign of the y-coordinate of each point and plot the three new points on the same rectangular coordinate system. What conjecture can you make about the location of a point when the sign of the y-coordinate is changed?

Reviewing the Major Concepts

In Exercises 47–50, solve the inequality.

47. $-4 < 10x + 1 < 6$ **48.** $-2 \leq 1 - 2x \leq 2$

49. $-3 \leq -\dfrac{x}{2} \leq 3$ **50.** $-5 < x - 25 < 5$

51. *Price Inflation* A new van costs $32,500, which is about 112% of what it was 3 years ago. What was its price 3 years ago?

52. *Pension Fund* You withhold $3\frac{1}{2}\%$ of your gross monthly income for your pension. Your gross monthly income is $3100. How much is withheld for pension?

Additional Problem Solving

In Exercises 53–56, plot the points on a rectangular coordinate system.

53. $(-8, -2), (6, -2), (6, 5)$

54. $(0, 4), (0, 0), (3, 0)$

55. $\left(\frac{3}{2}, 1\right), (4, -3), \left(-\frac{4}{3}, \frac{7}{3}\right)$

56. $(-3, -5), \left(\frac{9}{4}, \frac{3}{4}\right), \left(\frac{5}{2}, -2\right)$

In Exercises 57 and 58, approximate the coordinates of the points.

57.

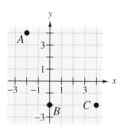

58.

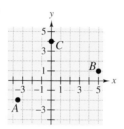

Geometry In Exercises 59–62, plot the points and connect them with line segments to form the figure.

59. *Square:* $(0, 6), (3, 3), (0, 0), (-3, 3)$

60. *Triangle:* $(1, 4), (0, -1), (5, 9)$

61. *Rhombus:* $(0, 0), (3, 2), (2, 3), (5, 5)$

62. *Parallelogram:* $(-1, 1), (0, 4), (4, -2), (5, 1)$

In Exercises 63 and 64, without plotting it, determine the quadrant in which each point is located.

63. (a) $\left(3, -\frac{5}{8}\right)$ (b) $(-6.2, 8.05)$

64. (a) $(200, 1365.6)$ (b) $\left(-\frac{5}{11}, -\frac{3}{8}\right)$

In Exercises 65–68, determine the quadrant or quadrants in which the point is located.

65. $(-3, y)$

66. $(x, 5)$

67. $(x, y), \quad xy > 0$

68. $(x, y), \quad x > 0, \ y < 0$

In Exercises 69–72, find the coordinates of the point.

69. The point is located three units to the right of the y-axis and four units below the x-axis.

70. The point is located two units to the left of the y-axis and five units above the x-axis.

71. The coordinates of the point are equal, and it is located in the third quadrant 10 units to the left of the y-axis.

72. The coordinates of the point are equal in magnitude and opposite in sign, and it is located seven units to the right of the y-axis.

Graphical Interpretation In Exercises 73 and 74, plot the points whose coordinates are given in the table. Then write a paragraph that summarizes the relationship between x and y.

73. *Fuel Efficiency* The table gives the speed x of a car in miles per hour and the approximate fuel efficiency y in miles per gallon.

x	50	55	60	65	70
y	28	26.4	24.8	23.4	22

74. *Average Temperature*   The table gives the average temperature y (degrees Fahrenheit) for Duluth, Minnesota for each month of the year, with $x = 1$ representing January.   (Source: PC USA)

x	1	2	3	4	5	6
y	6.3	12.0	22.9	38.3	50.3	59.4

x	7	8	9	10	11	12
y	65.3	63.2	54.0	44.2	28.2	13.8

In Exercises 75–78, determine whether the ordered pairs are solutions of the equation.

75. $y = \frac{7}{8}x$ (a) $\left(\frac{8}{7}, 1\right)$ (b) $\left(4, \frac{7}{2}\right)$
               (c) $(0, 0)$ (d) $(-16, 14)$

76. $y = \frac{5}{8}x - 2$ (a) $(0, 0)$ (b) $(2, 2)$
               (c) $(-4, -7)$ (d) $(32, 49)$

77. $4y - 2x = -1$ (a) $(0, 0)$ (b) $\left(\frac{1}{2}, 0\right)$
               (c) $\left(-3, -\frac{7}{4}\right)$ (d) $\left(1, -\frac{3}{4}\right)$

78. $y = 10x - 7$ (a) $(2, 10)$ (b) $(-2, -27)$
               (c) $(5, 43)$ (d) $(1, 5)$

In Exercises 79 and 80, complete the table of values. Then plot the solution points on a rectangular coordinate system.

79.

x		-4	$\frac{2}{5}$	4	8	12
$y = -\frac{5}{2}x + 4$						

80.

x		-6	-3	0	$\frac{3}{4}$	10
$y = \frac{4}{3}x - \frac{1}{3}$						

81. *Numerical Interpretation*   The cost C of producing x units is given by $C = 28x + 3000$. Use a table to help write a paragraph that describes the relationship between x and C.

82. *Numerical Interpretation*   When an employee produces x units per hour, the hourly wage y is given by $y = 0.75x + 8$. Use a table to help write a paragraph that describes the relationship between x and y.

In Exercises 83–86, plot the points and find the distance between them. State whether the points lie on a horizontal or vertical line.

83. $(3, 2), (10, 2)$ **84.** $(-120, -2), (130, -2)$

85. $\left(-3, \frac{3}{2}\right), \left(-3, \frac{9}{4}\right)$ **86.** $\left(\frac{1}{3}, -4\right), \left(\frac{1}{3}, \frac{5}{2}\right)$

In Exercises 87 and 88, find the coordinates of (x, y), the lengths of the vertical and horizontal sides of the right triangle, and the length of the hypotenuse.

87.

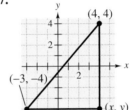

88.

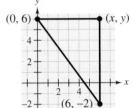

In Exercises 89–92, find the distance between the points.

89. $(-2, -3), (4, 2)$ **90.** $(-5, 4), (10, -3)$

91. $(1, 3), (3, -2)$ **92.** $\left(\frac{1}{2}, 1\right), \left(\frac{3}{2}, 2\right)$

In Exercises 93–96, use the Distance Formula to determine whether the three points lie on a line.

93. $(2, 3), (2, 6), (6, 3)$

94. $(2, 4), (-1, 6), (-3, 1)$

95. $(8, 3), (5, 2), (2, 1)$

96. $(2, 4), (1, 1), (0, -2)$

Geometry   In Exercises 97 and 98, find the perimeter of the triangle having the given vertices.

97. $(-2, 0), (0, 5), (1, 0)$

98. $(-5, -2), (-1, 4), (3, -1)$

In Exercises 99–102, plot the points and the midpoint of the line segment joining the points. The coordinates of the midpoint of the line segment joining the points (x_1, y_1) and (x_2, y_2) are

$$\text{Midpoint} = \left(\frac{x_1 + x_2}{2}, \frac{y_1 + y_2}{2} \right).$$

99. $(-2, 0), (4, 8)$ **100.** $(-3, -2), (7, 2)$

101. $(1, 6), (6, 3)$ **102.** $(2, 7), (9, -1)$

Shifting a Graph   In Exercises 103 and 104, the figure is shifted to a new location in the plane. Find the coordinates of the vertices of the figure in its new location.

103.

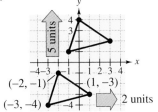

104.

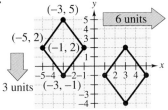

105. *Housing Construction*   A house is 30 feet wide and the ridge of the roof is 7 feet above the tops of the walls (see figure). The rafters overhang the edges of the walls by 2 feet. How long are the rafters?

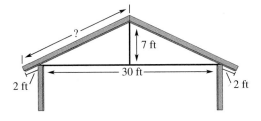

106. *Housing Construction*   Determine the length of the handrail over the stairs shown in the figure.

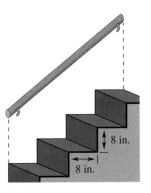

CAREER INTERVIEW

Dean R. Brookie

Architect/Planner

Brookie Architecture and Planning, Inc.

Durango, CO 81301

Mathematics is integral to the work of an architect. I use algebra and geometry in both the analytical and creative components of my work, planning and designing commercial buildings, multifamily housing, and custom-built homes. I routinely develop project budgets within certain parameters, calculate forces on various parts of buildings, and compute areas and volumes of various geometric figures. I have also found that raw geometric shapes are the building blocks for the creative process. Geometry in the form of rhythms and proportions can be used to give a building a harmonious, comforting feel or to create interesting visual tension.

2.2 Graphs of Equations

The Graph of an Equation ▪ Intercepts: An Aid to Sketching Graphs ▪ Real-Life Application of Graphs

The Graph of an Equation

In Section 2.1, you saw that the solutions of an equation in x and y can be represented by points on a rectangular coordinate system. The set of *all* solutions of an equation is called its **graph.** In this section, you will study a basic technique for sketching the graph of an equation—the **point-plotting method.**

EXAMPLE 1 *Sketching the Graph of an Equation*

Sketch the graph of $3x - y = 2$.

Solution

To begin, solve the equation for y to obtain $y = 3x - 2$. Next, create a table of values. The choice of x-values to use in the table is somewhat arbitrary. However, the more x-values you choose, the easier it will be to recognize a pattern.

NOTE The equation in Example 1 is an example of a **linear equation** in two variables—it is of first degree in both variables and its graph is a straight line. You will study this type of equation in Section 2.4.

x	-2	-1	0	1	2	3
$y = 3x - 2$	-8	-5	-2	1	4	7
Solution	$(-2, -8)$	$(-1, -5)$	$(0, -2)$	$(1, 1)$	$(2, 4)$	$(3, 7)$

Now, plot the points, as shown in Figure 2.12(a). It appears that all six points lie on a line, so complete the sketch by drawing a line through the points, as shown in Figure 2.12(b).

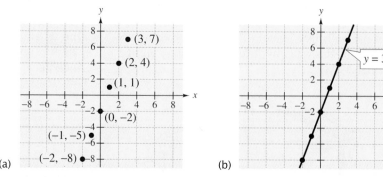

(a)   (b)

FIGURE 2.12

	The Point-Plotting Method of Sketching a Graph

NOTE   By drawing a line (curve) through the plotted points, we are implying that every point on this line (curve) is a solution point of the given equation and conversely.

1. If possible, rewrite the equation by isolating one of the variables.

2. Make up a table of values showing several solution points.

3. Plot these points on a rectangular coordinate system.

4. Connect the points with a smooth curve or line.

EXAMPLE 2   Sketching the Graph of a Nonlinear Equation

Sketch the graph of $-x^2 + 2x + y = 0$.

Solution

Begin by solving the equation for y to obtain $y = x^2 - 2x$. Next, create a table of values.

x	-2	-1	0	1	2	3	4
$y = x^2 - 2x$	8	3	0	-1	0	3	8
Solution	$(-2, 8)$	$(-1, 3)$	$(0, 0)$	$(1, -1)$	$(2, 0)$	$(3, 3)$	$(4, 8)$

Now, plot the seven solution points, as shown in Figure 2.13(a). Finally, connect the points with a smooth curve, as shown in Figure 2.13(b).

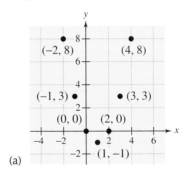
(a)

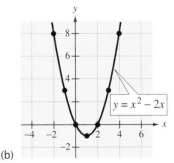
(b)

FIGURE 2.13

NOTE   The graph of the equation given in Example 2 is called a **parabola.** You will study this type of graph in detail in Section 7.3.

Example 3 looks at the graph of an equation that involves an absolute value. Remember that to find the absolute value of a number you disregard the sign of the number. For instance, $|-5| = 5$, $|2| = 2$, and $|0| = 0$.

EXAMPLE 3   *The Graph of an Absolute Value Equation*

Sketch the graph of $y = |x - 2|$.

Solution

This equation is already written in a form with y isolated on the left. So begin by creating a table of values. Be sure that you understand how the absolute value is evaluated. For instance, when $x = -2$, the value of y is

$$y = |-2 - 2| = |-4| = 4$$

and when $x = 3$, the value of y is

$$y = |3 - 2| = |1| = 1.$$

x	-2	-1	0	1	2	3	4	5
$y = |x - 2|$	4	3	2	1	0	1	2	3
Solution	$(-2, 4)$	$(-1, 3)$	$(0, 2)$	$(1, 1)$	$(2, 0)$	$(3, 1)$	$(4, 2)$	$(5, 3)$

Next, plot the points, as shown in Figure 2.14(a). It appears that the points lie in a "V-shaped" pattern, with the point $(2, 0)$ lying at the bottom of the "V." Following this pattern, connect the points to form the graph shown in Figure 2.14(b).

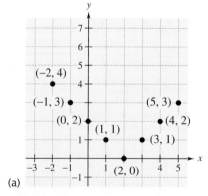

(a)

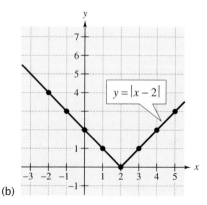

(b)

FIGURE 2.14

Intercepts: An Aid to Sketching Graphs

Two types of solution points that are especially useful are those having zero as the x-coordinate and those having zero as the y-coordinate. Such points are called **intercepts** because they are the points at which the graph intersects, respectively, the y- and x-axes.

Definition of Intercepts

The point $(a, 0)$ is called an **x-intercept** of the graph of an equation if it is a solution point of the equation. To find the x-intercepts, let $y = 0$ and solve the equation for x.

The point $(0, b)$ is called a **y-intercept** of the graph of an equation if it is a solution point of the equation. To find the y-intercepts, let $x = 0$ and solve the equation for y.

EXAMPLE 4 *Finding the Intercepts of a Graph*

Find the intercepts and sketch the graph of $y = 2x - 3$.

Solution

Find the x-intercept by letting $y = 0$ and solving for x.

$$y = 2x - 3$$
$$0 = 2x - 3$$
$$\tfrac{3}{2} = x$$

Find the y-intercept by letting $x = 0$ and solving for y.

$$y = 2x - 3$$
$$y = 2(0) - 3$$
$$y = -3$$

Therefore, the graph has one x-intercept, which occurs at the point $\left(\tfrac{3}{2}, 0\right)$, and one y-intercept, which occurs at the point $(0, -3)$. To sketch the graph of the equation, create a table of values. (Include the intercepts in the table.) Finally, using the solution points given in the table, sketch the graph of the equation, as shown in Figure 2.15.

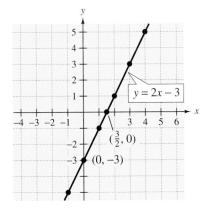

FIGURE 2.15

x	-1	0	1	$\tfrac{3}{2}$	2	3	4
$y = 2x - 3$	-5	-3	-1	0	1	3	5
Solution	$(-1, -5)$	$(0, -3)$	$(1, -1)$	$\left(\tfrac{3}{2}, 0\right)$	$(2, 1)$	$(3, 3)$	$(4, 5)$

It is possible for a graph to have no intercepts or several intercepts. For instance, consider the three graphs in Figure 2.16.

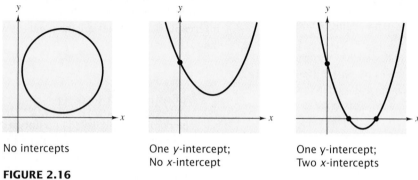

No intercepts

One y-intercept;
No x-intercept

One y-intercept;
Two x-intercepts

FIGURE 2.16

EXAMPLE 5   A Graph That Has Two x-Intercepts

Find the intercepts and sketch the graph of $y = x^2 - 5x + 4$.

Solution

Begin by making a table of values, as shown below. From the table, you can see that the graph has two x-intercepts, which occur at $(1, 0)$ and $(4, 0)$, and one y-intercept, which occurs at $(0, 4)$.

x	-1	0	1	2	3	4	5
$y = x^2 - 5x + 4$	10	4	0	-2	-2	0	4
Solution	$(-1, 10)$	$(0, 4)$	$(1, 0)$	$(2, -2)$	$(3, -2)$	$(4, 0)$	$(5, 4)$

Next, plot the points given in the table, as shown in Figure 2.17(a). Then connect them with a smooth curve, as shown in Figure 2.17(b).

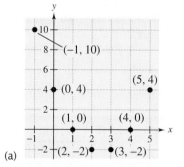

(a)

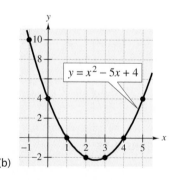

(b)

FIGURE 2.17

Real-Life Application of Graphs

Newspapers and news magazines frequently use graphs to show real-life relationships between variables. Example 6 shows how such a graph can help you visualize the concept of **straight-line depreciation.**

EXAMPLE 6 *Straight-Line Depreciation*

Your small business buys a new printing press for $65,000. For income tax purposes, you decide to depreciate the printing press over a 10-year period. At the end of the 10 years, the salvage value of the printing press is expected to be $5000. Find an equation that relates the depreciated value of the printing press to the number of years. Then sketch the graph of the equation.

Solution

The total depreciation over the 10-year period is $65,000 - 5000 = \$60,000$. Because the same amount is depreciated each year, it follows that the annual depreciation is $60,000/10 = \$6000$. Thus, after 1 year, the value of the printing press is

Value after 1 year $= 65,000 - (1)6000 = \$59,000.$

By similar reasoning, you can see that the value after 2 years is

Value after 2 years $= 65,000 - (2)6000 = \$53,000.$

Let y represent the value of the printing press after t years and follow the pattern determined for the first 2 years to obtain

$$y = 65,000 - 6000t.$$

A sketch of the graph of this equation is shown in Figure 2.18.

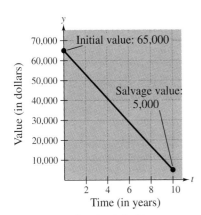

FIGURE 2.18 Straight Line Depreciation

Group Activities   Extending the Concept

Straight-Line Depreciation   In Example 6, suppose that you depreciated the printing press over 8 years instead of 10 years. Write an equation that represents the depreciated value of the printing press during the 8-year period. Then graph both depreciation models on the same rectangular coordinate system and compare the results. What are the advantages and disadvantages of each model?

2.2 Exercises

Discussing the Concepts

1. Define the graph of an equation.

2. How many solution points make up the graph of $y = 2x - 1$? Explain.

3. Explain how to find the intercepts of a graph. Give examples.

4. An equation gives the relationship between profit y and time t. Profit is decreasing at a lower rate than it has in the past. Is it possible to sketch the graph of such an equation? If so, sketch a representative graph.

Problem Solving

In Exercises 5–10, match the equation with its graph. [The graphs are labeled (a), (b), (c), (d), (e), and (f).]

(a)
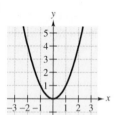

(b)
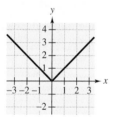

(c)
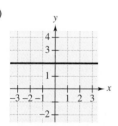

(d)
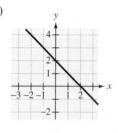

(e)
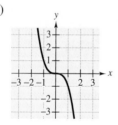

(f)
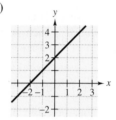

5. $y = 2$

6. $y = 2 + x$

7. $y = 2 - x$

8. $y = x^2$

9. $y = -x^3$

10. $y = |x|$

In Exercises 11–14, complete the table and use the results to sketch the graph of the equation.

11. $2x + y = 3$

x	-4			2	4
y		7	3		

12. $2x - 3y = 6$

x	-3	0		3	
y			$-\frac{2}{3}$		5

13. $y = 4 - x^2$

x		-1		2	
y	0		4		-5

14. $y = \frac{1}{2}x^3 - 4$

x	-1		1		3
y		-4		0	

In Exercises 15 and 16, use a graphing utility to create a table of values. Then sketch the graph.

15. $y = 4 - |x|$

16. $y = 4 - x^2$

In Exercises 17–24, sketch the graph of the equation.

17. $y = 3x$

18. $y = \frac{1}{3}x$

19. $y = 2x - 3$

20. $y = -x + 2$

21. $y = x^2 - 1$

22. $y = -x^2$

23. $y = |x| - 1$

24. $y = |x - 1|$

In Exercises 25–28, find the x- and y-intercepts (if any) of the graph of the equation.

25. $x + 2y = 10$

26. $3x - 2y + 12 = 0$

27. $y = (x + 5)(x - 5)$

28. $y = (x + 1)^2$

In Exercises 29–38, sketch the graph of the equation and show the coordinates of three solution points.

29. $y = 3 - x$

30. $y = x - 3$

31. $y = 4$

32. $x = -6$

33. $4x + y = 3$

34. $y - 2x = -4$

35. $y = x^2 - 4$

36. $y = 1 - x^2$

37. $y = |x + 2|$

38. $y = |x| + 2$

39. *Using a Graph*   The force F (in pounds) to stretch a spring x inches from its natural length is given by

$$F = \frac{4}{3}x, \quad 0 \le x \le 12.$$

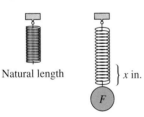

Natural length   } x in.

F

(a) Use the model to complete the following table.

x	0	3	6	9	12
F					

(b) Sketch the graph of the model.

(c) Use the graph to determine how the length of the spring changes each time the force is doubled. Explain your reasoning.

40. *Comparing Data with a Model*   The number of farms in the United States with milk cows has been decreasing. The number of farms N (in thousands) for 1984 through 1991 is given in the table.

t	4	5	6	7	8	9	10	11
N	282	269	249	228	217	204	194	182

A model for this data is

$$N = -14.5t + 337.1$$

where t is time in years, with $t = 0$ representing 1980.   (Source: U.S. Department of Agriculture)

(a) Sketch the graph of the model and plot the data in the table on the same graph.

(b) How well does the model represent the data? Explain your reasoning.

(c) Use the model to predict the number of farms with milk cows in 1994.

(d) Explain why this model may not be accurate in the future.

41. *Misleading Graphs*   Graphs can help you visualize relationships between two variables, but they can also be misused to imply results that are not correct. The two graphs below represent the *same* data points. Which graph is misleading, and why?

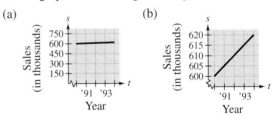

42. *Exploration*   Sketch the graphs of $y = x^2 + 1$ and $y = -(x^2 + 1)$ on the same set of coordinate axes. Explain how the graph of an equation changes when the expression for y is multiplied by -1. Justify your answer by giving additional examples.

Reviewing the Major Concepts

In Exercises 43–46, name the property of real numbers that justifies the statement.

43. $8x \cdot \dfrac{1}{8x} = 1$

44. $3x + 0 = 3x$

45. $-4(x + 10) = -4 \cdot x + (-4)(10)$

46. $5 + (-3 + x) = (5 - 3) + x$

47. *Weight of Sand*   The ratio of cement to sand in a 90-pound bag of dry mix is 1 to 4. Find the number of pounds of sand in the bag.

48. *Defective Units*   A quality control engineer for a manufacturer found one defective unit in a sample of 125. At that rate, what is the expected number of defective units in a shipment of 150,000?

Additional Problem Solving

In Exercises 49–54, sketch the graph of the equation.

49. $y = 4 - x$

50. $y = \frac{1}{2}x - 2$

51. $y = x^2 - 4$

52. $y = x^3$

53. $y = |x|$

54. $y = |x + 3|$

In Exercises 55–72, sketch the graph of the equation and show the coordinates of three solution points.

55. $y = 2x - 3$

56. $y = -4x + 8$

57. $2x - 3y = 6$

58. $3x + 4y = 12$

59. $x + 5y = 10$

60. $5x - y = 10$

61. $x - 1 = 0$

62. $y + 3 = 0$

63. $y = x^2 - 9$

64. $y = 9 - x^2$

65. $y = x(x - 2)$

66. $y = -x(x + 4)$

67. $y = x^3 - 1$

68. $y = 1 - x^4$

69. $y = |x| - 3$

70. $y = |x - 3|$

71. $y = -|x|$

72. $y = |x| + |x - 2|$

Graphical and Algebraic Solutions   In Exercises 73–76, graphically estimate the intercepts of the graph. Then check your results algebraically.

73. $y = x^2 + 3$

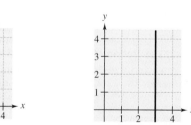

74. $x = 3$

75. $y = |x - 2|$

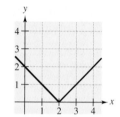

76. $y = (x - 3)(x - 4)$

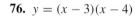
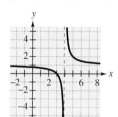

77. *Straight-Line Depreciation*   A manufacturing plant purchases a new molding machine for $225,000. The depreciated value y after t years is given by

$$y = 225,000 - 20,000t, \quad 0 \le t \le 8.$$

Sketch the graph of this model.

78. *Geometry*   A rectangle of length l and width w has a perimeter of 12 meters.

(a) Show that the width of the rectangle is $w = 6 - l$ and its area is $A = l(6 - l)$.

(b) Sketch the graph of the equation for the area.

(c) From the graph in part (b), which dimensions produce a rectangle of maximum area?

Exploration   In Exercises 79 and 80, graph the equations on the same set of coordinate axes. What conclusions can you make by comparing the graphs?

79. (a) $y = x^2$
   (b) $y = (x - 2)^2$
   (c) $y = (x + 2)^2$
   (d) $y = (x + 4)^2$

80. (a) $y = x^2$
   (b) $y = x^2 - 2$
   (c) $y = x^2 - 4$
   (d) $y = x^2 + 4$

2.3	**Graphs and Graphing Utilities**
	Introduction ▪ Using a Graphing Utility ▪ Using a Graphing Utility's Special Features

Introduction

In Section 2.2, you studied the point-plotting method of sketching the graph of an equation. One of the disadvantages of the point-plotting method is that to get a good idea about the shape of a graph you need to plot *many* points. By plotting only a few points, you can badly misrepresent the graph.

For instance, consider the equation $y = \frac{1}{30}x(x^4 - 10x^2 + 39)$. To graph this equation, suppose you calculated only the five points in the table below.

x	-3	-1	0	1	3
$y = \frac{1}{30}x(x^4 - 10x^2 + 39)$	-3	-1	0	1	3

By plotting the five points, as shown in Figure 2.19(a), you might assume that the graph of the equation is a straight line. This, however, is not correct. By plotting several more points, as shown in Figure 2.19(b), you can see that the actual graph is not straight at all.

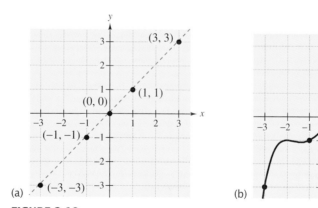

(a)   (b)

FIGURE 2.19

Thus, the point-plotting method leaves you with a dilemma. The method can be very inaccurate if only a few points are plotted, but it is very time consuming to plot a dozen (or more) points. Technology can help you solve this dilemma. Plotting several points (or even hundreds of points) on a rectangular coordinate system is something that a graphing utility can do easily.

Using a Graphing Utility

There are many different graphing utilities: some are graphing packages for computers and some are hand-held graphing calculators. In this section we describe the steps used to graph an equation with a graphing utility. We will often give keystroke sequences for illustration; however, these may not agree precisely with the steps required by *your* graphing utility.*

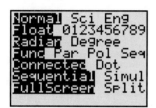

FIGURE 2.20

Graphing an Equation with a *TI-82* Graphing Utility

Before performing the following steps, set your utility so that all the standard defaults are active. For instance, all of the options at the left of the MODE screen should be highlighted (see Figure 2.20).

1. Set the viewing window for the graph. (See Example 3.) To set the standard viewing window, press ZOOM 6.

2. Rewrite the equation so that y is isolated on the left side of the equation.

3. Press the Y= key. Then enter the right side of the equation on the first line of the display. (The first line is labeled Y_1=.)

4. Press the GRAPH key.

EXAMPLE 1 *Graphing a Linear Equation*

Sketch the graph of $3y + x = 5$.

Solution

To begin, solve the given equation for y in terms of x.

$$3y + x = 5 \qquad \text{Original equation}$$
$$3y = -x + 5 \qquad \text{Subtract } x \text{ from both sides.}$$
$$y = -\frac{1}{3}x + \frac{5}{3} \qquad \text{Divide both sides by 3.}$$

Press the Y= key, and enter the following keystrokes.

(-) X,T,θ ÷ 3 + 5 ÷ 3

The top row of the display should now be as follows.

$$Y_1 = -X/3 + 5/3$$

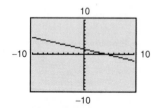

FIGURE 2.21

Press the GRAPH key, and the screen should look like that shown in Figure 2.21.

*The keystrokes given in this section correspond to the *TI-82* Graphics Calculator by Texas Instruments. For other graphing utilities, the keystrokes may differ.

In Figure 2.21, notice that the graphing utility screen does not label the tick marks on the x-axis or the y-axis. To see what the tick marks represent, you can press WINDOW . If you set your graphing utility to the standard graphing defaults before working Example 1, the screen should show the following values.

Xmin=-10   The minimum x-value is -10.
Xmax=10   The maximum x-value is 10.
Xscl=1   The x-scale is one unit per tick mark.
Ymin=-10   The minimum y-value is -10.
Ymax=10   The maximum y-value is 10.
Yscl=1   The y-scale is one unit per tick mark.

These settings are summarized visually in Figure 2.22.

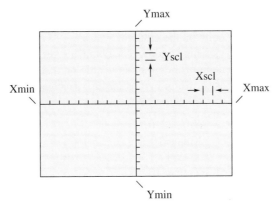

FIGURE 2.22

EXAMPLE 2   *Graphing an Equation Involving Absolute Value*

Sketch the graph of $y = |x - 4|$.

Solution

This equation is already written so that y is isolated on the left side of the equation. Press the Y= key, and enter the following keystrokes.

ABS (X,T,θ − 4)

The top row of the display should now be as follows.

$Y_1 = \text{abs}\,(X - 4)$

Press the GRAPH key, and the screen should look like that shown in Figure 2.23.

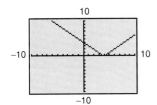

FIGURE 2.23

Using a Graphing Utility's Special Features

To use your graphing utility to its best advantage, you must learn how to set the viewing rectangle, as illustrated in the next example.

EXAMPLE 3 *Setting the Viewing Window*

Sketch the graph of $y = x^2 + 14$.

Solution

Press $\boxed{\text{Y=}}$ and enter the following keystrokes.

$$\boxed{\text{X,T,}\theta} \ \boxed{x^2} \ \boxed{+} \ 14$$

Press the $\boxed{\text{GRAPH}}$ key. If your graphing utility is set to the standard viewing window, nothing will appear on the screen. The reason for this is that the lowest point on the graph of $y = x^2 + 14$ occurs at the point $(0, 14)$. Using the standard viewing window, you obtain a screen whose largest y-value is 10. In other words, none of the graph is visible on a screen whose y-values range from -10 to 10, as shown in Figure 2.24(a). To change these settings, press $\boxed{\text{WINDOW}}$ and enter the following values.

Xmin=-10	The minimum x-value is -10.
Xmax=10	The maximum x-value is 10.
Xscl=1	The x-scale is one unit per tick mark.
Ymin=-10	The minimum y-value is -10.
Ymax=30	The maximum y-value is 30.
Yscl=5	The y-scale is five units per tick mark.

Press $\boxed{\text{GRAPH}}$, and you will obtain the graph shown in Figure 2.24(b). On this graph, note that each tick mark on the y-axis represents five units because you changed the y-scale to 5. Also note that the highest point on the y-axis is now 30 because you changed the maximum value of y to 30.

NOTE If you changed the maximum y-value and the y-scale on your utility as indicated in Example 3, you should return to the standard settings before working Example 4. To do this, press $\boxed{\text{ZOOM}}$ $\boxed{6}$.

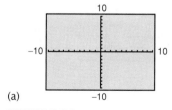

(a)

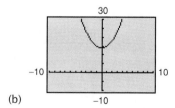

(b)

FIGURE 2.24

EXAMPLE 4   Using a Square Setting

Sketch the graph of $y = x$. The graph of this equation is a straight line that makes a 45° angle with the x-axis and y-axis. From the graph on your graphing utility, does the angle appear to be 45°?

Solution

Press Y= and enter $y = x$ on the first line.

$$Y_1 = X$$

Press the GRAPH key, and you will obtain the graph shown in Figure 2.25(a). Note that the angle the line makes with the x-axis does not appear to be 45°. The reason for this is that the screen is wider than it is tall. This makes the tick marks on the x-axis appear farther apart than the tick marks on the y-axis. To obtain the same distance between tick marks on both axes, change the graphing settings from "standard" to "square." To do this, press the following keys.

ZOOM 5   Square setting

The screen should look like that shown in Figure 2.25(b). Note in this figure that the square setting has changed the viewing window so that the x-values vary between -15 and 15.

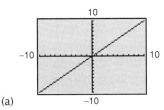

(a)

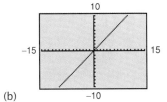

(b)

FIGURE 2.25

NOTE   There are many possible square settings on a graphing utility. To create a square setting, you need the following ratio to be $\frac{2}{3}$.

$$\frac{Y\text{max} - Y\text{min}}{X\text{max} - X\text{min}}$$

For instance, the setting in Figure 2.25(b) is square because $(Y\text{max} - Y\text{min}) = 20$ and $(X\text{max} - X\text{min}) = 30$.

EXAMPLE 5   Sketching More than One Graph on the Same Screen

Sketch the graphs of the following equations on the same screen.

$$y = -x + 5, \quad y = -x, \quad \text{and} \quad y = -x - 5$$

Solution

To begin, press $\boxed{\text{Y=}}$ and enter all three equations on the first three lines. The display should now be as follows.

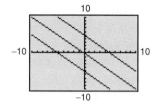

$Y_1 = \text{-X} + 5$ $\boxed{\text{(-)}}$ $\boxed{\text{X,T,}\theta}$ $\boxed{+}$ 5

$Y_2 = \text{-X}$ $\boxed{\text{(-)}}$ $\boxed{\text{X,T,}\theta}$

$Y_3 = \text{-X} - 5$ $\boxed{\text{(-)}}$ $\boxed{\text{X,T,}\theta}$ $\boxed{-}$ 5

Press the $\boxed{\text{GRAPH}}$ key and you will obtain the graph shown in Figure 2.26. Note that the graph of each equation is a straight line, and that the lines are parallel to each other.

FIGURE 2.26

Group Activities   Exploring with Technology

Graphing Equations   For each of the following equations, discuss what must first be done to the equation before you can graph it with a graphing utility. If it is possible, graph each equation with a graphing utility and approximate the x-intercept. If it is not possible to graph the equation by entering it into a graphing utility, explain why.

a. $3x + 5y = 15$ **b.** $2x - 7y = 21$

c. $x = -4$ **d.** $-x - 3y - 9 = 0$

2.3 Exercises

Discussing the Concepts

1. Suppose you correctly enter an expression for the variable y on a graphing utility. However, no graph appears in the viewing window when you graph the equation. Give a possible explanation and what steps you could take to remedy the problem. Illustrate your explanation with an example.

2. In your own words, explain the change on the display if Ymax is increased from 10 to 20.

3. In your own words, explain the change on the display if Xscl is increased from 1 to 2.

4. You graph $y = x$ and $y = x + 1$ on the same screen with $-100 \le y \le 100$. Why is this inappropriate?

Problem Solving

In Exercises 5–10, use a graphing utility to match the equation with its graph. [The graphs are labeled (a), (b), (c), (d), (e), and (f).]

(a)
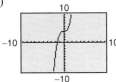

(b)
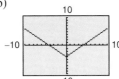

(c)
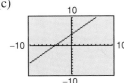

(d)
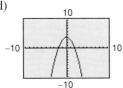

(e)
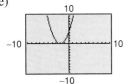

(f)
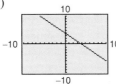

5. $y = 4 + x$

6. $y = 4 - x$

7. $y = 4 - x^2$

8. $y = x^2 + 4x + 4$

9. $y = x^3 + 4$

10. $y = |x| - 4$

In Exercises 11–18, use a graphing utility to graph the equation. Use a standard setting.

11. $y = 2x - 6$

12. $y = x - 2$

13. $y = x^2 - 3$

14. $y = 6 - x^2$

15. $y = \sqrt{x + 4}$

16. $y = 2\sqrt{2x + 3}$

17. $y = |x| - 6$

18. $y = |x - 6|$

In Exercises 19–22, use a graphing utility to graph the equation. Begin by using a standard setting. Then graph the equation a second time using the specified setting. Which setting is better? Explain.

19. $y = \frac{1}{2}x - 10$

Xmin = -1
Xmax = 20
Xscl = 1
Ymin = -15
Ymax = 5
Yscl = 1

20. $y = -2x + 25$

Xmin = -1
Xmax = 20
Xscl = 1
Ymin = -2
Ymax = 28
Yscl = 1

21. $y = 2x^3 - 5x^2$

Xmin = -5
Xmax = 5
Xscl = 1
Ymin = -5
Ymax = 5
Yscl = 1

22. $y = 6\sqrt{x + 4}$

Xmin = -5
Xmax = 5
Xscl = 1
Ymin = -1
Ymax = 20
Yscl = 2

In Exercises 23–26, find a setting on a graphing utility such that the graph of the equation agrees with the given graph.

23. $y = -3x + 15$

24. $y = 2x^2 - 16x + 26$

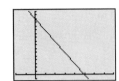

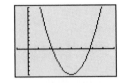

25. $y = x^3 - 3x + 2$

26. $y = (x^2 - 4)^2$

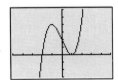

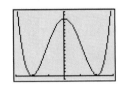

In Exercises 27–30, use the ZOOM feature of a graphing utility to estimate the lowest point on the graph.

27. $y = \frac{1}{4}(x^2 - 1)$

28. $y = x^2 + 3x - 2$

29. $y = x^3(3x + 4)$

30. $y = x\sqrt{x + 1}$

31. *Think About It*   In Exercises 27–30, you graphically estimated the lowest points on graphs. Explain how you could confirm your estimates algebraically with tables.

In Exercises 32–35, use a graphing utility to approximate the x-intercepts of the graph of the equation. (*Hint:* The ZOOM and TRACE features can help you get a better view of the intercepts.)

32. $y = 14 + 3x - 2x^2$

33. $y = 0.4x^2 - 0.2x - 3$

34. $y = x^3 - 3x^2 + 2$

35. $y = x^3 + x - 1$

36. *Think About It*   In Exercises 32–35, you graphically estimated the intercepts of graphs. Explain how you could confirm your estimates algebraically.

In Exercises 37–40, explain how to use a graphing utility to verify that $y_1 = y_2$. Then identify the rule of algebra that is illustrated.

37. $y_1 = \frac{1}{2}(x - 4)$

$y_2 = \frac{1}{2}x - 2$

38. $y_1 = 3x - x^2$

$y_2 = -x^2 + 3x$

39. $y_1 = x + (2x - 1)$

$y_2 = (x + 2x) - 1$

40. $y_1 = 2x + 0$

$y_2 = 2x$

Stopping Distance   In Exercises 41 and 42, consider the stopping distance y of an automobile (in feet), which is modeled by

$$y = 30.00 + 0.08x^2, \quad 25 \le x \le 75$$

where x is speed in miles per hour.

41. Find a graphing utility viewing rectangle that gives a good view of the graph of this model.

42. Use the graph to approximate the stopping distance when the car's speed is (a) $x = 35$ miles per hour and (b) $x = 55$ miles per hour.

Reviewing the Major Concepts

In Exercises 43–46, solve for y in terms of x.

43. $3x + y = 4$

44. $2x + 3y = 2$

45. $x^2 + 3y = 4$

46. $x^2 + y - 4 = 0$

In Exercises 47 and 48, use the point-plotting method to sketch the graph of the equation.

47. $2x - 7y + 14 = 0$

48. $y = 3 - |x - 2|$

In Exercises 49 and 50, write an algebraic expression that represents the quantity in the verbal statement, and simplify it if possible.

49. The total hourly wage for an employee when the base pay is $9.35 per hour plus 75 cents for each of q units produced per hour

50. The sum of two consecutive odd integers, the first of which is $2n - 1$

Additional Problem Solving

In Exercises 51–64, use a graphing utility to graph the equation. Use a standard setting.

51. $y = \frac{2}{3}x + 1$ **52.** $y = \frac{1}{4}x - 2$

53. $y = -\frac{3}{4}x + 4$ **54.** $y = -2x + 5$

55. $y = -x^2 + 4x - 1$ **56.** $y = \frac{1}{4}x^2 - x$

57. $y = x^3 - 2$ **58.** $y = 1 - x^3$

59. $y = 2\left(1 - \sqrt{x}\right)$ **60.** $y = x\sqrt{x + 3}$

61. $y = |x| - 4$ **62.** $y = 4 - |x|$

63. $y = |x - 4|$ **64.** $y = \frac{1}{2}|x|$

In Exercises 65–70, use a graphing utility to graph the equation. Begin by using a standard setting. Then graph the equation a second time using the specified setting. Which setting is better? Explain.

65. $y = \frac{1}{6}(2x + 3)$ **66.** $y = |x| + |x - 5|$

| Xmin = 0 |
| Xmax = 20 |
| Xscl = 1 |
| Ymin = 0 |
| Ymax = 10 |
| Yscl = 1 |

| Xmin = -3 |
| Xmax = 8 |
| Xscl = 1 |
| Ymin = 0 |
| Ymax = 10 |
| Yscl = 1 |

67. $y = 20 - \frac{1}{4}x^2$ **68.** $y = x^2 - 12x + 36$

| Xmin = -10 |
| Xmax = 10 |
| Xscl = 1 |
| Ymin = -2 |
| Ymax = 22 |
| Yscl = 1 |

| Xmin = -5 |
| Xmax = 15 |
| Xscl = 1 |
| Ymin = -2 |
| Ymax = 20 |
| Yscl = 1 |

69. $y = 0.6x^3 - 2x - 1$ **70.** $y = 0.3x\sqrt{64 - x^2}$

| Xmin = -2.5 |
| Xmax = 2.5 |
| Xscl = 1 |
| Ymin = -3.5 |
| Ymax = 1.5 |
| Yscl = 1 |

| Xmin = 0 |
| Xmax = 20 |
| Xscl = 1 |
| Ymin = 0 |
| Ymax = 20 |
| Yscl = 1 |

In Exercises 71–74, find a setting on a graphing utility such that the graph of the equation agrees with the given graph.

71. $y = 100 - 3x$ **72.** $y = \frac{2}{3}x - 8$

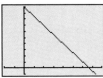   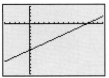

73. $y = -\frac{1}{8}x^2 + x$ **74.** $y = (x - 5)^2(x - 15)$

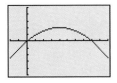   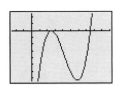

In Exercises 75–78, use a graphing utility to graph both equations on the same screen. Do the graphs intersect? If so, how many times do they intersect?

75. $y = x^2$ **76.** $y = x + 2$
  $y = 4x - x^2$ $y = 2 + 4x - x^2$

77. $y = x$ **78.** $y = 2x^2$
  $y = 3\sqrt{x}$ $y = x^4 - 2x^2$

In Exercises 79–82, use a graphing utility to estimate the x-intercepts of the graph of the equation.

79. $y = x^2 - 4x - 3$ **80.** $y = -2x^2 - x + 8$

81. $y = -x^2 + 12x - 37$ **82.** $y = x^2 + 14x + 49$

Exploration In Exercises 83 and 84, use a graphing utility to find a graph with the given intercepts. Explain your strategy.

83. $(2, 0), (0, 2)$ **84.** $(5, 0), (0, -3)$

Comparing Views   In Exercises 85 and 86, the equation gives the profit y (in thousands of dollars) for selling x units of a product. Which graphing utility viewing rectangle gives a better view of the graph? Explain your reasoning.

85. $y = 0.02x^2 + 0.75x - 10$

Xmin = 0	Xmin = 0
Xmax = 50	Xmax = 50
Xscl = 5	Xscl = 5
Ymin = -10	Ymin = -10
Ymax = 100	Ymax = 400
Yscl = 20	Yscl = 50

86. $y = \dfrac{6(4x - 1)}{x + 1}$

Xmin = 0	Xmin = 0
Xmax = 10	Xmax = 10
Xscl = 1	Xscl = 1
Ymin = -2	Ymin = -2
Ymax = 25	Ymax = 100
Yscl = 2	Yscl = 10

87. *Income*   The net income per share of common stock for McDonald's Corporation from 1984 through 1993 can be approximated by the model

$$y = (0.081x + 0.648)^2, \quad 4 \le x \le 13$$

where y is the net income per share and x represents the year, with $x = 0$ corresponding to 1980. (Source: McDonald's Corporation)

(a) Use a graphing utility to graph the model.

(b) The actual net income per share of stock in 1993 was $2.91. How closely does the model predict this value?

88. *Depreciation*   A manufacturing plant purchases a new numerically controlled machine for $500,000. The depreciated value y after x years is given by

$$y = 500,000 - 60,000x, \quad 0 \le x \le 6.$$

Use the constraints of the model to determine an appropriate viewing rectangle and graph the equation.

Strength of Copper   In Exercises 89 and 90, use the following model, which relates the percent strength of copper y in terms of its Celsius temperature x.

$$y = \sqrt{10,700 - 17.6x}, \quad 100 \le x \le 500$$

The percent strength is relative to 100% at 20°C.

89. Use a graphing utility to sketch the graph of the equation using the following range settings.

Xmin = 0
Xmax = 500
Xscl = 50
Ymin = 30
Ymax = 110
Yscl = 10

90. Use the TRACE key to find the approximate percent strength when the temperature is (a) $x = 150°C$ and (b) $x = 400°C$.

McDonald's restaurant chain owner Ray Kroc opened his first restaurant in Des Plaines, Illinois, shortly after buying business rights in 1955. At that time a regular hamburger cost $0.15 and a regular order of fries cost $0.11.

MID·CHAPTER QUIZ

Take this quiz as you would take a quiz in class. After you are done, check your work against the answers given in the back of the book.

In Exercises 1 and 2, plot the points on a rectangular coordinate system and find the distance between them.

1. $(-1, 5)$, $(3, 2)$ **2.** $(-3, -2)$, $(2, 10)$

3. Determine the quadrants in which the point $(x, 4)$ must be located if x is a real number. Explain your reasoning.

4. Find the coordinates of the point that lies 10 units to the right of the y-axis and three units below the x-axis.

5. Determine whether the following ordered pairs are solution points of the equation $4x - 3y = 10$.

    (a) $(2, 1)$ (b) $(1, -2)$ (c) $(2.5, 0)$ (d) $\left(2, -\frac{2}{3}\right)$

6. Find the x- and y-intercepts of the graph of the equation $6x - 8y + 48 = 0$.

In Exercises 7–12, sketch the graph of the equation and show the coordinates of three solution points (including intercepts).

7. $y = 2x - 3$ **8.** $y = 5$

9. $3x + y - 6 = 0$ **10.** $y = x^2 - 4$

11. $y = 6x - x^2$ **12.** $y = |x - 2| - 3$

In Exercises 13–16, use a graphing utility to graph the equation. Use a standard setting.

13. $y = 7 - 2x$ **14.** $y = x^2 + 4x$

15. $y = \sqrt{8 - x}$ **16.** $y = |x - 2| + |x + 2|$

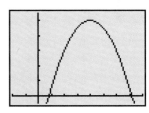

Figure for 17

In Exercises 17 and 18, find a setting on a graphing utility such that the graph of the equation is approximately the same as the given graph.

17. $y = -2(x^2 - 8x + 6)$ **18.** $y = x^3 - 5x + 5$

19. Use a graphing utility to approximate the x-intercepts of the graph of $y = x^2 - x - 2.75$.

20. Use a graphing utility to verify that $y_1 = y_2$ if $y_1 = -x + (2x - 2)$ and $y_2 = (-x + 2x) - 2$. Identify the rule of algebra that is illustrated.

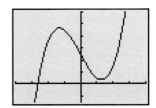

Figure for 18

2.4 Slope and Graphs of Linear Equations

The Slope of a Line ▪ Slope as a Graphing Aid ▪
Parallel and Perpendicular Lines

The Slope of a Line

The **slope** of a nonvertical line is the number of units the line rises or falls vertically for each unit of horizontal change from left to right. For example, the line in Figure 2.27 rises two units for each unit of horizontal change from left to right, and we say that this line has a slope of $m = 2$.

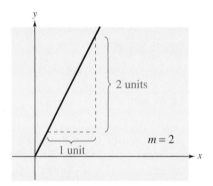

FIGURE 2.27   **FIGURE 2.28**

Definition of the Slope of a Line

The **slope** m of the nonvertical line passing through the points (x_1, y_1) and (x_2, y_2) is

$$m = \frac{y_2 - y_1}{x_2 - x_1} = \frac{\text{change in } y}{\text{change in } x}$$

where $x_1 \neq x_2$ (see Figure 2.28).

When the formula for slope is used, the *order of subtraction* is important. Given two points on a line, you are free to label either of them (x_1, y_1) and the other (x_2, y_2). However, once this is done, you must form the numerator and denominator using the same order of subtraction.

$$m = \frac{y_2 - y_1}{x_2 - x_1} \qquad m = \frac{y_1 - y_2}{x_1 - x_2} \qquad m = \frac{y_2 - y_1}{x_1 - x_2}$$

Correct   Correct   Incorrect

EXAMPLE 1   Finding the Slope of a Line Through Two Points

Find the slope of the line passing through each pair of points.

a. $(1, 2)$ and $(4, 5)$ **b.** $(1, 4)$ and $(3, 4)$ **c.** $(-1, 4)$ and $(2, 1)$

Solution

a. Let $(x_1, y_1) = (1, 2)$ and $(x_2, y_2) = (4, 5)$.

$$m = \frac{y_2 - y_1}{x_2 - x_1}$$ Difference in y-values

$$ \qquad \qquad \text{Difference in } x\text{-values}$$

$$= \frac{5 - 2}{4 - 1}$$

$$= 1$$

NOTE   You can use slope to determine if three points are collinear—i.e., all lie on the same line. Consider any three points A, B, and C. If the slope of the line through points A and B is the same as the slope of the line through points B and C, the three points are collinear.

b. The slope of the line through $(1, 4)$ and $(3, 4)$ is

$$m = \frac{4 - 4}{3 - 1} = \frac{0}{2} = 0.$$

c. The slope of the line through $(-1, 4)$ and $(2, 1)$ is

$$m = \frac{1 - 4}{2 - (-1)} = \frac{-3}{3} = -1.$$

The graphs of the three lines are shown in Figure 2.29.

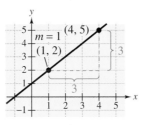

(a) Positive Slope   (b) Zero Slope   (c) Negative Slope

FIGURE 2.29

The definition of slope does not apply to vertical lines. For instance, consider the points $(3, 1)$ and $(3, 3)$ on the vertical line shown in Figure 2.30. Applying the formula for slope, you have

$$\frac{3 - 1}{3 - 3} = \frac{2}{0}.$$ Undefined

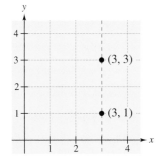

FIGURE 2.30   Slope is undefined.

Because division by zero is not defined, the slope of a vertical line is not defined.

From the slopes of the lines shown in Figures 2.29 and 2.30, you can make the following generalizations about the slope of a line.

1. A line with positive slope ($m > 0$) *rises* from left to right.
2. A line with negative slope ($m < 0$) *falls* from left to right.
3. A line with zero slope ($m = 0$) is *horizontal*.
4. A line with undefined slope is *vertical*.

EXAMPLE 2   *Using Slope to Describe Lines*

Describe the line through each pair of points.

a. $(2, -1), (2, 3)$ **b.** $(-2, 4), (3, 1)$ **c.** $(1, 3), (4, 3)$ **d.** $(-1, 1), (2, 5)$

Solution

a. Because the slope is undefined, the line is vertical.

$$m = \frac{3 - (-1)}{2 - 2} = \frac{4}{0}$$ Undefined slope (See Figure 2.31a.)

b. Because the slope is negative, the line falls from left to right.

$$m = \frac{1 - 4}{3 - (-2)} = -\frac{3}{5} < 0$$ Negative slope (See Figure 2.31b.)

c. Because the slope is zero, the line is horizontal.

$$m = \frac{3 - 3}{4 - 1} = \frac{0}{3} = 0$$ Zero slope (See Figure 2.31c.)

d. Because the slope is positive, the line rises from left to right.

$$m = \frac{5 - 1}{2 - (-1)} = \frac{4}{3} > 0$$ Positive slope (See Figure 2.31d.)

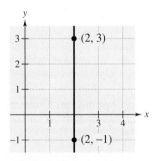

(a) Vertical line
    Undefined slope

FIGURE 2.31

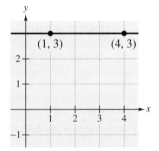

(b) Line falls
     Negative slope

(c) Horizontal line
     Zero slope

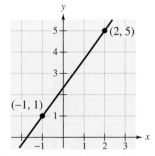

(d) Line rises
     Positive slope

Any two points on a nonvertical line can be used to calculate its slope. This is demonstrated in the next example.

EXAMPLE 3   Finding the Slope of a Line

Sketch the graph of the line given by $2x + 3y = 6$. Then find the slope of the line. (Choose two different pairs of points on the line and show that the same slope is obtained from either pair.)

Solution

Begin by solving the given equation for y.

$$y = -\frac{2}{3}x + 2 \qquad \text{y is a function of x.}$$

Then construct a table of values, as shown below.

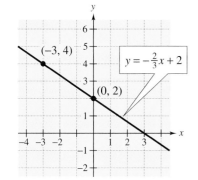

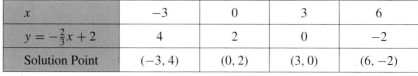

x	-3	0	3	6
$y = -\frac{2}{3}x + 2$	4	2	0	-2
Solution Point	$(-3, 4)$	$(0, 2)$	$(3, 0)$	$(6, -2)$

From the solution points shown in the table, sketch the graph of the line, as shown in Figure 2.32. To calculate the slope of the line using two different sets of points, first use the points $(-3, 4)$ and $(0, 2)$, and obtain a slope of

$$m = \frac{2 - 4}{0 - (-3)} = -\frac{2}{3}.$$

Next, use the points $(3, 0)$ and $(6, -2)$ to obtain a slope of

$$m = \frac{-2 - 0}{6 - 3} = -\frac{2}{3}.$$

Try some other pairs of points on the line to see that you obtain a slope of $m = -\frac{2}{3}$ regardless of which two points you use.

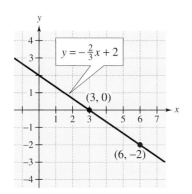

FIGURE 2.32

Technology

Setting the viewing window on a graphing utility affects the appearance of its slope. When you are using a graphing utility, remember that you cannot judge whether a slope is steep or shallow *unless* you use a square setting.

Slope as a Graphing Aid

You have seen that, before creating a table of values for an equation, you should first solve the equation for y. When you do this for a linear equation, you obtain some very useful information. Consider the results of Example 3.

$$2x + 3y = 6 \qquad \text{Original equation}$$
$$2x - 2x + 3y = -2x + 6 \qquad \text{Subtract } 2x \text{ from both sides.}$$
$$3y = -2x + 6 \qquad \text{Combine like terms.}$$
$$\frac{3y}{3} = \frac{-2x + 6}{3} \qquad \text{Divide both sides by 3.}$$
$$y = -\frac{2}{3}x + 2 \qquad \text{Simplify.}$$

Observe that the coefficient of x is the slope of the graph of this equation (see Example 3). Moreover, the constant term, 2, gives the y-intercept of the graph.

$$y = -\frac{2}{3}x + 2$$

Slope y-intercept $(0, 2)$

This form is called the **slope-intercept** form of the equation of the line.

NOTE The slope-intercept form of the equation of a line identifies y as a function of x. Hence, the term **linear function** is often used as an alternative description of this form of the equation of a line.

Slope-Intercept Form of the Equation of a Line

The graph of the equation

$$y = mx + b$$

is a line whose slope is m and whose y-intercept is $(0, b)$. (See Figure 2.33.)

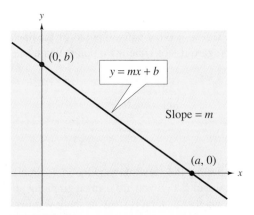

FIGURE 2.33

So far, you have been plotting several points to sketch the equation of a line. However, now that you can recognize equations of lines (linear functions), you don't have to plot as many points—two points are enough. (You might remember from geometry that *two points are all that are necessary to determine a line.*)

EXAMPLE 4 Using the Slope and y-Intercept to Sketch a Line

Use the slope and y-intercept to sketch the graph of

$$y = \frac{2}{3}x + 1.$$

Solution

The equation is already in slope-intercept form.

$$y = mx + b$$

$$y = \frac{2}{3}x + 1 \qquad\qquad \text{Slope-intercept form}$$

Thus, the slope of the line is $m = \frac{2}{3}$ and the y-intercept is $(0, b) = (0, 1)$. Now you can sketch the graph of the line as follows. First, plot the y-intercept. Then, using a slope of $\frac{2}{3}$

$$m = \frac{2}{3} = \frac{\text{change in } y}{\text{change in } x},$$

locate a second point on the line by moving three units to the right and two units up (or two units up and three units to the right), as shown in Figure 2.34(a). Finally, obtain the graph by drawing a line through the two points, as shown in Figure 2.34(b).

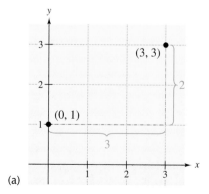
(a)

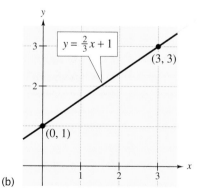
(b)

FIGURE 2.34

Parallel and Perpendicular Lines

You know from geometry that two lines in a plane are *parallel* if they do not intersect. What this means in terms of their slopes is suggested by Example 5.

EXAMPLE 5   Lines That Have the Same Slope

On the same set of axes, sketch the lines given by $y = 2x$ and $y = 2x - 3$.

Solution

For the line given by

$$y = 2x$$

the slope is $m = 2$ and the y-intercept is $(0, 0)$. For the line given by

$$y = 2x - 3$$

the slope is also $m = 2$ and the y-intercept is $(0, -3)$. The graphs of these two lines are shown in Figure 2.35.

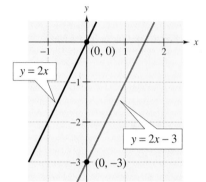

FIGURE 2.35

In Example 5, notice that the two lines have the same slope *and* appear to be parallel. The following rule states that this is always the case. That is, two (nonvertical) lines are parallel *if and only if* they have the same slope.

Parallel Lines

Two distinct nonvertical lines are parallel if and only if they have the same slope.

Another rule from geometry is that two lines in a plane are *perpendicular* if they intersect at right angles. In terms of their slopes, this means that two nonvertical lines are perpendicular if their slopes are negative reciprocals of each other.

Perpendicular Lines

Consider two nonvertical lines whose slopes are m_1 and m_2. The two lines are perpendicular if and only if their slopes are *negative reciprocals* of each other. That is,

$$m_1 = -\frac{1}{m_2}, \quad \text{or equivalently,} \quad m_1 \cdot m_2 = -1.$$

NOTE   The phrase "if and only if" in this rule is used in mathematics as a way to write two statements in one. The first statement says that *if two distinct nonvertical lines have the same slope, they must be parallel.* The second statement says that *if two distinct nonvertical lines are parallel, they must have the same slope.*

EXAMPLE 6 *Parallel or Perpendicular?*

Are the following pairs of lines parallel, perpendicular, or neither?

a. $y = -2x + 4$, $y = \frac{1}{2}x + 1$ **b.** $y = \frac{1}{3}x + 2$, $y = \frac{1}{3}x - 3$

Solution

a. The first line has a slope of $m_1 = -2$, and the second line has a slope of $m_2 = \frac{1}{2}$. Because these slopes are negative reciprocals of each other, the two lines must be perpendicular, as shown in Figure 2.36.

b. Each of these two lines has a slope of $m = \frac{1}{3}$. Therefore, the two lines must be parallel, as shown in Figure 2.37.

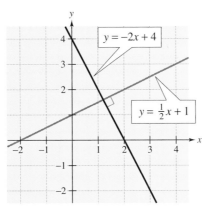

FIGURE 2.36

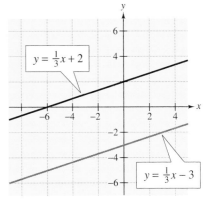

FIGURE 2.37

Group Activities Extending the Concept

Interpreting Slope   Write a function for the given verbal model. Identify the slope and then interpret the slope in the real-life setting.

1. Total pay per hour = Piecework rate · Number of pieces + Fixed hourly rate

2. Total cost = Tax rate · List price + List price

2.4 Exercises

Discussing the Concepts

1. Can any pair of points on a line be used to calculate the slope of the line? Explain.

2. In your own words, give interpretations of a negative slope, a zero slope, and a positive slope.

3. The slopes of two lines are -3 and $\frac{3}{2}$. Which is steeper? Explain.

4. In the form $y = mx + b$, what does m represent? What does b represent?

5. What is the relationship between the x-intercept of the line $y = mx + b$ and the solution to the equation $mx + b = 0$? Explain.

6. Is it possible for two lines with positive slopes to be perpendicular to each other? Explain.

Problem Solving

In Exercises 7–12, estimate the slope of the line from its graph.

7.
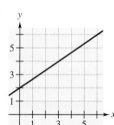

8.
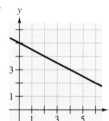

9.
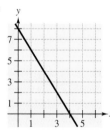

10.
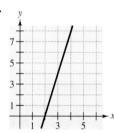

11.
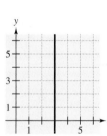

12.
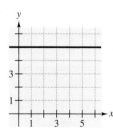

In Exercises 13–20, plot the two points and find the slope (if possible) of the line passing through them. State whether the line is rising, falling, horizontal, or vertical.

13. $(0, 0)$, $(5, -4)$
14. $(0, 0)$, $(-2, -1)$
15. $(-2, -3)$, $(6, 1)$
16. $(3, 6)$, $(5, -2)$
17. $(3, -2)$, $(3, 6)$
18. $(-4, 3)$, $(4, 3)$
19. $\left(\frac{3}{4}, 2\right)$, $\left(\frac{7}{2}, 0\right)$
20. $(-3.5, 1.6)$, $(5.1, 4.3)$

In Exercises 21–24, a point on a line and the slope of the line are given. Find two additional points on the line.

21. $(5, 2)$, $m = 0$
22. $(-4, 3)$, m is undefined.
23. $(3, -4)$, $m = 3$
24. $(-2, 6)$, $m = -3$

In Exercises 25–28, sketch the graph of a line through the point $(3, 2)$ having the given slope.

25. $m = 3$
26. $m = \frac{3}{2}$
27. $m = -\frac{1}{3}$
28. $m = 0$

In Exercises 29 and 30, which line is steeper?

29. $m_1 = 3$, $m_2 = -4$
30. $m_1 = 4$, $m_2 = 5$

In Exercises 31–34, plot the x- and y-intercepts and sketch the graph of the line.

31. $2x - y + 4 = 0$ **32.** $3x + 5y + 15 = 0$

33. $-5x + 2y - 20 = 0$ **34.** $3x - 5y - 15 = 0$

In Exercises 35–38, write the equation of the line in slope-intercept form and sketch the line. Use a graphing utility to confirm your sketch.

35. $3x - y - 2 = 0$ **36.** $x - y - 5 = 0$

37. $3x + 2y - 2 = 0$ **38.** $y + 4 = 0$

In Exercises 39–42, use a graphing utility to graph the pair of equations on the same viewing rectangle. Are the lines parallel, perpendicular, or neither? (Use the *square* setting so the slopes of the lines appear visually correct.)

39. $L_1: y = \frac{1}{2}x - 2$ **40.** $L_1: y = 3x - 2$

   $L_2: y = \frac{1}{2}x + 3$ $L_2: y = 3x + 1$

41. $L_1: y = \frac{3}{4}x - 3$ **42.** $L_1: y = -\frac{2}{3}x - 5$

   $L_2: y = -\frac{4}{3}x + 1$ $L_2: y = \frac{3}{2}x + 1$

43. *Geometry*   The length and width of a rectangular flower garden are 40 feet and 30 feet, respectively (see figure). A walkway of width x surrounds the garden.

(a) Write the outside perimeter y of the walkway in terms of x.

(b) Use a graphing utility to graph the equation for the perimeter.

(c) Determine the slope of the graph of part (b). For each additional 1-foot increase in the width of the walkway, determine the increase in its perimeter.

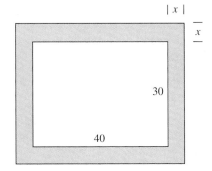

44. *Simple Interest*   An inheritance of $8000 is invested in two different mutual funds. One fund pays 6% simple interest and the other pays $7\frac{1}{2}$% simple interest.

(a) If x dollars is invested in the fund paying 6%, how much is invested in the fund paying $7\frac{1}{2}$%?

(b) Use the result of part (a) to write the annual interest y in terms of x.

(c) Use a graphing utility to graph the function in part (b).

(d) Explain why the slope of the line in part (b) is negative.

Reviewing the Major Concepts

In Exercises 45–48, evaluate the expression.

45. $|-15|$ **46.** $-|72|$

47. $-|-6|$ **48.** $|14 - 32|$

49. *Work Rate*   Two people can complete a task in t hours where t must satisfy the equation

$$\frac{t}{4} + \frac{t}{6} = 1.$$

Solve this equation and interpret the result.

50. *Maximum Height of an Object*   The velocity v of an object projected vertically with an initial velocity of 96 feet per second is given by

$$v = 96 - 32t$$

where t is time in seconds and air resistance is neglected. Find the time when the maximum height (occurs when $v = 0$) of an object is obtained.

Additional Problem Solving

In Exercises 51 and 52, identify the line that has the specified slope m.

51. (a) $m = \frac{3}{4}$ (b) $m = 0$ (c) $m = -3$

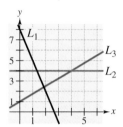

52. (a) $m = -\frac{5}{2}$ (b) m is undefined. (c) $m = 2$

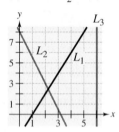

In Exercises 53–66, plot the points and, if possible, find the slope of the line passing through them. State whether the line is rising, falling, horizontal, or vertical.

53. $(0, 0), (7, 5)$

54. $(0, 0), (-3, -4)$

55. $(0, 12), (8, 0)$

56. $(0, -4), (6, 0)$

57. $(-5, -3), (-5, 4)$

58. $(0, -8), (-5, 0)$

59. $(2, -5), (7, -5)$

60. $(-2, 1), (-4, -3)$

61. $\left(\frac{3}{4}, 2\right), \left(5, -\frac{5}{2}\right)$

62. $\left(-\frac{3}{2}, -\frac{1}{2}\right), \left(\frac{5}{8}, \frac{1}{2}\right)$

63. $(4.2, -1), (-4.2, 6)$

64. $(3.4, 0), (3.4, 1)$

65. $(2.5, -2), (4.75, 5.25)$

66. $(0, 4.5), (3, 4.5)$

In Exercises 67 and 68, solve for x so that the line through the points has the given slope.

67. $(4, 5), (x, 7); m = -\frac{2}{3}$

68. $(x, -2), (5, 0); m = \frac{3}{4}$

In Exercises 69 and 70, solve for y so that the line through the points has the given slope.

69. $(-3, y), (9, 3); m = \frac{3}{2}$

70. $(-3, 20), (2, y); m = -6$

In Exercises 71–76, a point on a line and the slope of the line are given. Find two additional points on the line.

71. $(0, 3), m = -1$

72. $(-1, -5), m = 2$

73. $(-5, 0), m = \frac{4}{3}$

74. $(-1, 1), m = -\frac{3}{4}$

75. $(4, 2), m$ is undefined.

76. $(-2, -2), m = 0$

In Exercises 77–82, sketch the graph of a line through the point $(0, 1)$ having the given slope.

77. $m = 2$

78. $m = 0$

79. m is undefined.

80. $m = -1$

81. $m = -\frac{4}{3}$

82. $m = \frac{2}{3}$

In Exercises 83–90, write the equation of the line in slope-intercept form and sketch the line. Use a graphing utility to confirm your sketch.

83. $x + y = 0$

84. $x - y = 0$

85. $x - 4y + 2 = 0$

86. $x - 2y - 2 = 0$

87. $\frac{1}{3}x + \frac{1}{2}y = 1$

88. $8x + 6y - 3 = 0$

89. $y - 2 = 0$

90. $2y + 3 = 0$

In Exercises 91–94, determine whether the lines L_1 and L_2 passing through the given pairs of points are parallel, perpendicular, or neither.

91. L_1: $(1, 3), (2, 1)$; L_2: $(0, 0), (4, 2)$

92. L_1: $(-3, -3), (1, 7)$; L_2: $(0, 4), (5, -2)$

93. L_1: $(-2, 0), (4, 4)$; L_2: $(1, -2), (4, 0)$

94. L_1: $(-5, 3), (3, 0)$; L_2: $(1, 2), \left(3, \frac{22}{3}\right)$

In Exercises 95–98, use a graphing utility to graph the three equations on the same viewing rectangle. Describe the relationships among the graphs. (Use the *square* setting so the slopes of the lines appear visually correct.)

95. $y_1 = 3x$

$y_2 = -3x$

$y_3 = \frac{1}{3}x$

96. $y_1 = \frac{3}{4}x$

$y_2 = -\frac{4}{3}x$

$y_3 = \frac{4}{3}$

97. $y_1 = \frac{1}{4}x$

$y_2 = \frac{1}{4}x - 2$

$y_3 = \frac{1}{4}x + 3$

98. $y_1 = 2x$

$y_2 = 2x - 5$

$y_3 = 2x + \frac{3}{2}$

99. *Graphical Estimation* The graph shows the earnings per share of common stock for Johnson & Johnson for the years 1987 through 1993. Use the slope of each segment to determine the year when earnings (a) decreased most rapidly and (b) increased most rapidly. (Source: *Johnson & Johnson 1993 Annual Report*)

Johnson & Johnson

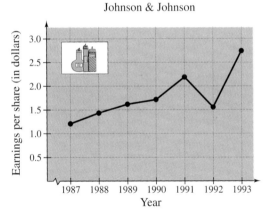

100. *Road Grade* When driving down a mountain road, you notice warning signs indicating that it is a "12% grade." This means that the slope of the road is $-\frac{12}{100}$. Over a stretch of road, your elevation drops by 2000 feet. What is the horizontal change in your position?

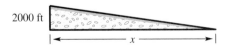

101. *Graphical Estimation* The graph gives the declared dividend per share of common stock for Emerson Electric Company for the years 1987 through 1993. Use the slope of each segment to determine the year when the dividend increased most rapidly. (Source: Emerson Electric Company)

Emerson Electric Company

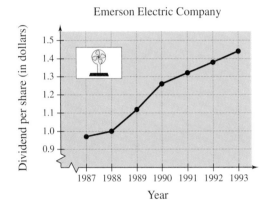

102. *Height of an Attic* The slope, or pitch, of a roof is such that it rises (or falls) 3 feet for every 4 feet of horizontal distance. Determine the maximum height in the attic of the house if the house is 30 feet wide.

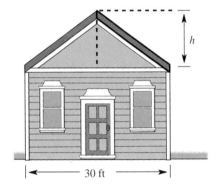

2.5 Relations and Functions

Relations ▪ Functions ▪ Function Notation ▪
Finding the Domain and Range of a Function ▪ Application

Relations

Many everyday occurrences involve two quantities that are paired or matched with each other by some rule of correspondence. The mathematical term for such a correspondence is a **relation.**

Definition of a Relation

A **relation** is any set of ordered pairs. The set of first components in the ordered pairs is the **domain** of the relation and the set of second components is the **range** of the relation.

EXAMPLE 1 Analyzing a Relation

Find the domain and range of $\{(0, 1), (1, 3), (2, 5), (3, 5), (0, 3)\}$.

Solution

The domain is the set of all first components of the relation, and the range is the set of all second components.

$$\text{Domain: } \{0, 1, 2, 3\}$$
$$\uparrow \qquad \uparrow \qquad \uparrow \qquad \uparrow \qquad \uparrow$$
$$\{(0, 1), (1, 3), (2, 5), (3, 5), (0, 3)\}$$
$$\downarrow \qquad \downarrow \qquad \downarrow \qquad \downarrow \qquad \downarrow$$
$$\text{Range: } \{1, 3, 5\}$$

A graphical representation is given in Figure 2.38.

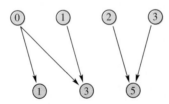

FIGURE 2.38

Functions

In modeling real-life situations, you will work with a special type of relation called a function. A **function** is a relation in which no two ordered pairs have the same first component and different second components. For instance, (2, 3) and (2, 4) could not be ordered pairs of a function.

Definition of a Function

A **function** f from a set A to a set B is a rule of correspondence that assigns to each element x in the set A exactly one element y in the set B.

The set A is called the **domain** (or set of inputs) of the function f, and the set B contains the **range** (or set of outputs) of the function.

The rule of correspondence for a function establishes a set of "input-output" ordered pairs of the form (x, y), where x is an input and y is the corresponding output. In some cases, the rule may generate only a finite set of ordered pairs, whereas in other cases the rule may generate an infinite set of ordered pairs.

EXAMPLE 2   *Input-Output Ordered Pairs for Functions*

a. For the function that pairs the year from 1991 to 1994 with the winner of the Super Bowl, each ordered pair is of the form (year, winner).

{(1991, Giants), (1992, Redskins), (1993, Cowboys), (1994, Cowboys)}

b. For the function given by $y = x - 2$, each ordered pair is of the form (x, y).

{All points on the graph of $y = x - 2$}

c. For the function that pairs the positive integers that are less than 7 with their squares, each ordered pair is of the form (n, n^2).

{(1, 1), (2, 4), (3, 9), (4, 16), (5, 25), (6, 36)}

d. For the function that pairs each real number with its square, each ordered pair is of the form (x, x^2).

{All points (x, x^2), where x is a real number}

NOTE   In Example 2, the sets in parts (a) and (c) have only finite numbers of ordered pairs, whereas the sets in parts (b) and (d) have infinite numbers of ordered pairs.

Characteristics of a Function

1. Each element in the domain *A* must be matched with an element in the range, which is contained in the set *B*.

2. Some elements in the set *B* may not be matched with any element in the domain *A*.

3. Two or more elements of the domain may be matched with the same element in the range.

4. No element of the domain is matched with two different elements in the range.

EXAMPLE 3   Test for Functions Represented by Ordered Pairs

Let $A = \{a, b, c\}$ and let $B = \{1, 2, 3, 4, 5\}$. Which represents a function from A to B?

a. $\{(a, 2), (b, 3), (c, 4)\}$ **b.** $\{(a, 4), (b, 5)\}$

c. 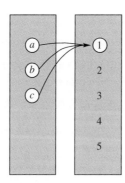   **d.**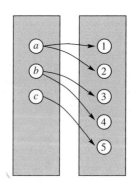

Solution

a. This set of ordered pairs *does* represent a function from *A* to *B*. Each element of *A* is matched with exactly one element of *B*.

b. This set of ordered pairs *does not* represent a function from *A* to *B*. Not all elements of *A* are matched with an element of *B*.

c. This diagram *does* represent a function from *A* to *B*. It does not matter that each element of *A* is matched with the same element in *B*.

d. This diagram *does not* represent a function from *A* to *B*. The element *a* in *A* is matched with *two* elements, 1 and 2, in *B*. This is also true of *b*.

Representing functions by sets of ordered pairs is a common practice in the study of *discrete mathematics*, which deals mainly with finite sets of data or with finite subsets of the set of real numbers. In algebra, however, it is more common to represent functions by equations or formulas involving two variables. For instance, the equation

$$y = x^2 \qquad\qquad \text{Squaring function}$$

represents the variable y as a function of the variable x. The variable x is the **independent variable** and the variable y is the **dependent variable.** In this context, the domain of the function is the set of all *allowable* real values for the independent variable x, and the range of the function is the *resulting* set of all values taken on by the dependent variable y.

EXAMPLE 4   *Testing for Functions Represented by Equations*

Which of the equations represents y as a function of x?

a. $y = x^2 + 1$ **b.** $x - y^2 = 2$ **c.** $-2x + 3y = 4$

Solution

a. From the equation

$$y = x^2 + 1$$

you can see that for each value of x there corresponds just one value of y. For instance, when $x = 1$, the value of y is $1^2 + 1 = 2$. Therefore, y *is* a function of x.

b. By writing the equation $x - y^2 = 2$ in the form

$$y^2 = x - 2$$

you can see that some values of x correspond to *two* values of y. For instance, when $x = 3$, $y^2 = 3 - 2 = 1$ and y can be 1 or -1. Hence, the solution points $(3, 1)$ and $(3, -1)$ show that y *is not* a function of x.

c. By writing the equation $-2x + 3y = 4$ in the form

$$y = \tfrac{2}{3}x + \tfrac{4}{3}$$

you can see that for each value of x there corresponds just one value of y. For instance, when $x = 2$, the value of y is $\tfrac{4}{3} + \tfrac{4}{3} = \tfrac{8}{3}$. Therefore, y *is* a function of x.

NOTE   An equation that defines y as a function of x may or may not also define x as a function of y. For instance, the equation in part (a) does not define x as a function of y, but the equation in part (c) does.

Function Notation

When an equation is used to represent a function, it is convenient to name the function so that it can be easily referenced. For example, the function $y = x^2 + 1$ in Example 4(a) can be given the name "f" and written in **function notation** as

$$f(x) = x^2 + 1.$$

Function Notation

In the notation $f(x)$:

f is the **name** of the function,

x is the **domain** (or input) value, and

$f(x)$ is a **range** (or output) value y for a given x.

The symbol $f(x)$ is read as *the value of f at x* or simply *f of x.*

The process of finding the value of $f(x)$ for a given value of x is called **evaluating a function.** This is accomplished by substituting a given x-value (input) into the equation to obtain the value of $f(x)$ (output). Here is an example.

Function	*x-Value*	*Function Value*
$f(x) = 3 - 4x$	$x = -1$	$f(-1) = 3 - 4(-1) = 3 + 4 = 7$

Although f is often used as a convenient function name and x as the independent variable, you can use other letters. For instance, the equations

$$f(x) = 2x^2 + 5, \quad f(t) = 2t^2 + 5, \quad \text{and} \quad g(s) = 2s^2 + 5$$

all define the same function. In fact, the letters used are simply "placeholders" and this same function is well described by the form

$$f(\quad) = 2(\quad)^2 + 5$$

where the parentheses are used in place of a letter. To evaluate $f(-2)$, simply place -2 in each set of parentheses, as follows.

$$f(-2) = 2(-2)^2 + 5$$
$$= 8 + 5$$
$$= 13$$

When evaluating a function, you are not restricted to substituting only numerical values into the parentheses. For instance, the value of $f(3x)$ is

$$f(3x) = 2(3x)^2 + 5$$
$$= 18x^2 + 5.$$

EXAMPLE 5   Evaluating a Function

Let $g(x) = 3x - x^2$ and find the following.

a. $g(1)$ **b.** $g(x + 1)$ **c.** $g(x) + g(1)$

Solution

a. Replacing x by 1 produces $g(1) = 3(1) - (1)^2 = 3 - 1 = 2$.

b. Replacing x with $x + 1$ produces

$$g(x + 1) = 3(x + 1) - (x + 1)^2$$
$$= 3x + 3 - (x^2 + 2x + 1)$$
$$= -x^2 + (3x - 2x) + (3 - 1)$$
$$= -x^2 + x + 2.$$

c. Using the result of part (a), we have

$$g(x) + g(1) = (3x - x^2) + 2 = -x^2 + 3x + 2.$$

Note that $g(x + 1) \neq g(x) + g(1)$. In general, $g(a + b)$ is not equal to $g(a) + g(b)$.

Sometimes a function is defined by more than one equation. An illustration of this is given in Example 6.

EXAMPLE 6   A Function Defined by Two Equations

Evaluate the function given by

$$f(x) = \begin{cases} x^2 + 1, & \text{if } x < 0 \\ x - 2, & \text{if } x \geq 0 \end{cases}$$

at (a) $x = -1$, (b) $x = 0$, and (c) $x = 1$.

Solution

a. Because $x = -1 < 0$, we use $f(x) = x^2 + 1$ to obtain

$$f(-1) = (-1)^2 + 1 = 2.$$

b. Because $x = 0 \geq 0$, we use $f(x) = x - 2$ to obtain

$$f(0) = 0 - 2 = -2.$$

c. Because $x = 1 \geq 0$, we use $f(x) = x - 2$ to obtain

$$f(1) = 1 - 2 = -1.$$

Finding the Domain and Range of a Function

The domain of a function may be explicitly described along with the function, or it may be *implied* by the expression used to define the function. The **implied domain** is the set of all real numbers (inputs) that yield real number values for the function. For instance, the function given by

$$f(x) = \frac{1}{x^2 - 9} \qquad \text{Domain: all } x \neq \pm 3$$

has an implied domain that consists of all real values of x other than $x = \pm 3$. These two values are excluded from the domain because division by zero is undefined. Another common type of implied domain is that used to avoid even roots of negative numbers. For instance, the function given by

$$f(x) = \sqrt{x} \qquad \text{Domain: all } x \geq 0$$

is defined only for $x \geq 0$. Therefore, its implied domain is the interval $[0, \infty)$. More will be said about the domains of square root functions in Chapter 5.

EXAMPLE 7   *Finding the Domain and Range of a Function*

Find the domain and range of each function.

a. $f: \{(-3, 0), (-1, 2), (0, 4), (2, 4), (4, -1)\}$

b. Area of a circle: $A = \pi r^2$

Solution

a. The domain of f consists of all first coordinates in the set of ordered pairs. The range consists of all second coordinates in the set of ordered pairs. Thus, the domain is

$$\text{Domain} = \{-3, -1, 0, 2, 4\}$$

and the range is

$$\text{Range} = \{0, 2, 4, -1\}.$$

b. For the area of a circle, you must choose nonnegative values for the radius r. Thus, the domain is the set of all real numbers r such that $r \geq 0$. The range is therefore the set of all real numbers A such that $A \geq 0$.

Note in Example 7(b) that the domain of a function can be implied by a physical context. For instance, from the equation $A = \pi r^2$, we would have no reason to restrict r to positive values. However, because we know this function represents the area of a circle, we conclude that the radius must be positive.

Application

To find mathematical models to represent functions, use the same guidelines that you use to model equations.

EXAMPLE 8   *Finding an Equation to Represent a Function*

Is the area of a square a *function* of the length of one of its sides? If so, find an equation that represents this function.

Solution

Figure 2.39 shows a square.

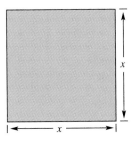

FIGURE 2.39

Verbal Model: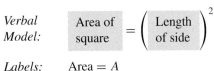

$$\text{Area of square} = \left(\text{Length of side} \right)^2$$

Labels: Area $= A$ (square units)
           Length of side $= x$ (linear units)

Function: $A = x^2$

Thus, the area of a square *is* a function of the length of one of its sides.

<hr>

Group Activities   Extending the Concept

Determining Relationships That Are Functions   Compile a list of statements describing relationships in everyday life. For each statement, identify the dependent and independent variables and discuss whether the statement *is* a function or *is not* a function and why. Here are two examples.

a. In the statement, "The number of ceramic tiles required to floor a kitchen is a function of the floor's area," the dependent variable is the required number of ceramic tiles and the independent variable is the area of the floor. This statement *is* a mathematically correct use of the word "function" because for each possible floor area there corresponds exactly one number of tiles needed to do the job.

b. In the statement, "Interest rates are a function of economic conditions," the dependent variable is interest rates and the independent variable is economic conditions. This statement *is not* a mathematically correct use of the word "function" because "economic conditions" is ambiguous; it is difficult to tell if one set of economic conditions would always result in the same interest rates.

2.5 Exercises

Discussing the Concepts

1. Explain the difference between relations and functions.

2. Is every relation a function? Explain.

3. In your own words, explain the meaning of *domain* and *range*.

4. In the equation $|y| = x$, is y a function of x? Explain your reasoning.

5. In the equation $y = |x|$, is y a function of x? Explain your reasoning.

6. Describe an advantage of function notation.

Problem Solving

In Exercises 7–10, state the domain and range of the relation. Then draw a graphic representation.

7. $\{(-2, 0), (0, 1), (1, 4), (0, -1)\}$

8. $\{(3, 10), (4, 5), (6, -2), (8, 3)\}$

9. $\{(0, 0), (4, -3), (2, 8), (5, 5), (6, 5)\}$

10. $\{(-3, 6), (-3, 2), (-3, 5)\}$

In Exercises 11–14, write a set of ordered pairs that represents the rule of correspondence.

11. In a given week, a salesperson travels a distance d in t hours at an average speed of 50 mph. The travel times for each day are 3 hours, 2 hours, 8 hours, 6 hours, and $\frac{1}{2}$ hour.

12. The cubes of all positive integers less than 8

13. The winners of the World Series from 1990 to 1993

14. The men inaugurated president of the United States in 1969, 1974, 1977, 1981, 1989, and 1993.

In Exercises 15–18, decide whether the relation is a function.

15.
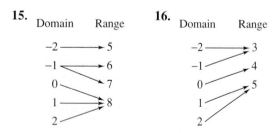

16.
(shown with Exercise 15)

17.
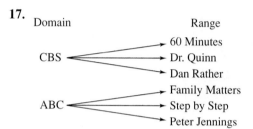

18.
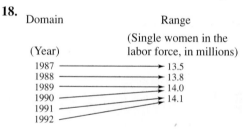

In Exercises 19 and 20, determine which sets of ordered pairs represent functions from A to B.

19. $A = \{0, 1, 2, 3\}$, $B = \{-2, -1, 0, 1, 2\}$

(a) $\{(0, 1), (1, -2), (2, 0), (3, 2)\}$

(b) $\{(0, -1), (2, 2), (1, -2), (3, 0), (1, 1)\}$

(c) $\{(0, 0), (1, 0), (2, 0), (3, 0)\}$

(d) $\{(0, 2), (3, 0), (1, 1)\}$

20. $A = \{1, 2, 3\}$, $B = \{9, 10, 11, 12\}$

(a) $\{(1, 10), (3, 11), (3, 12), (2, 12)\}$

(b) $\{(1, 10), (2, 11), (3, 12)\}$

(c) $\{(1, 10), (1, 9), (3, 11), (2, 12)\}$

(d) $\{(3, 9), (2, 9), (1, 12)\}$

In Exercises 21–24, decide whether the equation represents y as a function of x.

21. $y = 10x + 12$

22. $x - 9y + 3 = 0$

23. $|y - 2| = x$

24. $|y| = x + 2$

In Exercises 25 and 26, fill in the blank and simplify.

25. $f(x) = 3x + 5$

   (a) $f(2) = 3(\quad) + 5$

   (b) $f(-2) = 3(\quad) + 5$

   (c) $f(k) = 3(\quad) + 5$

   (d) $f(k + 1) = 3(\quad) + 5$

26. $f(x) = 3 - 2x$

   (a) $f(0) = 3 - 2(\quad)$

   (b) $f(-3) = 3 - 2(\quad)$

   (c) $f(m) = 3 - 2(\quad)$

   (d) $f(t + 2) = 3 - 2(\quad)$

In Exercises 27–32, evaluate the function as indicated, and simplify.

27. $f(x) = 12x - 7$

   (a) $f(3)$ (b) $f\left(\frac{3}{2}\right)$

   (c) $f(a) + f(1)$ (d) $f(a + 1)$

28. $h(x) = x(x - 2)$

   (a) $h(2)$ (b) $h(0)$

   (c) $h(1)$ (d) $h(t + 2)$

29. $f(x) = \begin{cases} x + 8, & \text{if } x < 0 \\ 10 - 2x, & \text{if } x \geq 0 \end{cases}$

   (a) $f(4)$ (b) $f(-10)$

   (c) $f(0)$ (d) $f(6) - f(-2)$

30. $f(x) = \begin{cases} -x, & \text{if } x \leq 0 \\ x^2 - 3x, & \text{if } x > 0 \end{cases}$

   (a) $f(0)$ (b) $f\left(-\frac{3}{2}\right)$

   (c) $f(4)$ (d) $f(-2) + f(25)$

31. $f(x) = 2x + 5$

   (a) $\dfrac{f(x + 2) - f(2)}{x}$ (b) $\dfrac{f(x - 3) - f(3)}{x}$

32. $f(x) = 2 - 3x$

   (a) $f(x + h)$ (b) $\dfrac{f(x + h) - f(x)}{h}$

In Exercises 33–36, find the domain of the function.

33. $f(x) = \dfrac{2x}{x - 3}$ **34.** $h(x) = 4x - 3$

35. $f(x) = \sqrt{2x - 1}$ **36.** $G(x) = \sqrt{x^2 - 25}$

37. *Geometry* Express the perimeter P of a square as a function of the length x of one of its sides.

38. *Geometry* Express the surface area S of a cube as a function of the length x of one of its edges.

39. *Geometry* An open box is made from a square piece of material 24 inches on a side by cutting equal squares from the corners and turning up the sides (see figure). Write the volume V of the box as a function of x.

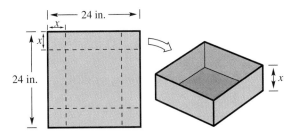

40. *Geometry* Strips of width x are cut from the four sides of a square that is 32 inches on a side (see figure). Write the area A of the remaining square as a function of x.

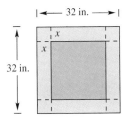

Reviewing the Major Concepts

In Exercises 41–44, simplify the expression.

41. $4 - 3(2x + 1)$

42. $24\left(\dfrac{y}{3} + \dfrac{y}{6}\right)$

43. $5(x + 2) - 4(2x - 3)$

44. $0.12x + 0.05(2000 - 2x)$

45. *Work Rate*   You can mow the lawn in 4 hours and your friend can mow it in 5 hours. What fractional part of the lawn can each of you mow in 1 hour? How long will it take both of you to mow the lawn?

46. *Average Speed*   A truck traveled at an average speed of 50 miles per hour on a 200-mile trip. On the return trip, the average speed was 42 miles per hour. Find the average speed for the round trip.

Additional Problem Solving

In Exercises 47–54, decide whether the relation is a function.

47.
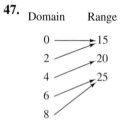

48.
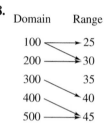

49.
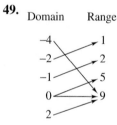

50.
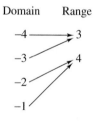

51.

Input Value	0	1	2	3	4
Output Value	0	1	4	9	16

52.

Input Value	0	1	2	1	0
Output Value	1	8	12	15	20

53.

Input Value	4	7	9	7	4
Output Value	2	4	6	8	10

54.

Input Value	0	2	4	6	8
Output Value	5	5	5	5	5

Interpreting a Graph   In Exercises 55 and 56, use the graph, which shows numbers of high school and college students in the United States. (Source: U.S. National Center for Education Statistics)

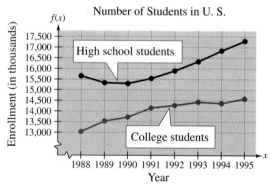

55. Is the high school enrollment a function of the year? Is the college enrollment a function of the year? Explain.

56. Let $f(x)$ represent the number of high school students in year x. Find $f(1993)$.

In Exercises 57–62, decide whether the equation represents y as a function of x.

57. $3x + 7y - 2 = 0$ **58.** $8x + y = 3$

59. $y = x(x - 10)$ **60.** $y = (x + 2)^2 + 3$

61. $|y| = 8 - x$ **62.** $x^2 + y^2 = 25$

In Exercises 63–72, evaluate the function as indicated, and simplify.

63. $g(x) = \frac{1}{2}x^2$

(a) $g(4)$ (b) $g\left(\frac{2}{3}\right)$

(c) $g(2y)$ (d) $g(4) + g(6)$

64. $f(x) = 3 - 7x$

(a) $f(-1)$ (b) $f\left(\frac{1}{2}\right)$

(c) $f(t) + f(-2)$ (d) $f(w)$

65. $f(x) = \sqrt{x + 5}$

(a) $f(-1)$ (b) $f(4)$

(c) $f\left(\frac{16}{3}\right)$ (d) $f(5z)$

66. $g(x) = 8 - |x - 4|$

(a) $g(0)$ (b) $g(8)$

(c) $g(16) - g(-1)$ (d) $g(x - 2)$

67. $f(x) = \dfrac{3x}{x - 5}$

(a) $f(0)$ (b) $f\left(\frac{5}{3}\right)$

(c) $f(2) - f(-1)$ (d) $f(x + 4)$

68. $g(x) = \dfrac{|x + 1|}{x + 1}$

(a) $g(2)$ (b) $g\left(-\frac{1}{3}\right)$

(c) $g(-4)$ (d) $g(3) + g(-5)$

69. $g(x) = 1 - x^2$

(a) $g(2.2)$ (b) $\dfrac{g(2.2) - g(2)}{0.2}$

70. $f(x) = 3x + 4$

(a) $\dfrac{f(x + 1) - f(1)}{x}$ (b) $\dfrac{f(x - 5) - f(5)}{x}$

71. $h(x) = \begin{cases} 4 - x^2, & \text{if } x \le 2 \\ x - 2, & \text{if } x > 2 \end{cases}$

(a) $h(2)$ (b) $h\left(-\frac{3}{2}\right)$

(c) $h(5)$ (d) $h(-3) + h(7)$

72. $f(x) = \begin{cases} x^2, & \text{if } x < 1 \\ x^2 - 3x + 2, & \text{if } x \ge 1 \end{cases}$

(a) $f(1)$ (b) $f(-1)$

(c) $f(2)$ (d) $f(-3) + f(3)$

In Exercises 73–76, find the domain and range of the function.

73. $f: \{(0, 0), (2, 1), (4, 8), (6, 27)\}$

74. $f: \left\{\left(-3, -\frac{17}{2}\right), \left(-1, -\frac{5}{2}\right), (4, 2), (10, 11)\right\}$

75. Circumference of a circle: $C = 2\pi r$

76. Area of a square of side s: $A = s^2$

In Exercises 77–84, find the domain of the function.

77. $h(x) = \dfrac{9}{x^2 + 1}$ **78.** $g(x) = \dfrac{x + 5}{x + 4}$

79. $f(t) = \dfrac{t + 3}{t(t + 2)}$ **80.** $g(s) = \dfrac{s - 2}{(s - 6)(s - 10)}$

81. $g(x) = \sqrt{x + 4}$ **82.** $f(x) = \sqrt{2 - x}$

83. $f(t) = t^2 + 4t - 1$ **84.** $f(x) = |x + 3|$

85. *Geometry* Express the volume V of a cube as a function of the length x of one of its edges.

86. *Distance* A plane is flying at a speed of 230 miles per hour. Express the distance d traveled by the plane as a function of time t in hours.

87. *Cost* The inventor of a new game believes that the variable cost for producing the game is \$1.95 per unit and the fixed costs are \$8000. Write the total cost C as a function of x, the number of games produced.

88. *Geometry*   Strips of width x are cut from two adjacent sides of a square that is 32 inches on a side (see figure). Write the area A of the remaining square as a function of x.

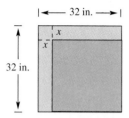

89. *Safe Load*   A solid rectangular beam has a height of 6 inches and a width of 4 inches. The safe load S of the beam with the load at the center is a function of its length L and is approximated by the model

$$S(L) = \frac{128{,}160}{L}$$

where S is measured in pounds and L is measured in feet. Find (a) $S(12)$ and (b) $S(16)$.

90. *Profit*   The marketing department of a business has determined that the profit from selling x units of a product is approximated by the model

$$P(x) = 50\sqrt{x} - 0.5x - 500.$$

Find (a) $P(1600)$ and (b) $P(2500)$.

In Exercises 91 and 92, determine whether the statements use the word *function* in ways that are *mathematically* correct.

91. (a) The sales tax on a purchased item is a function of the selling price.

     (b) Your score on the next algebra exam is a function of the number of hours you study the night before the exam.

92. (a) The amount in your savings account is a function of your salary.

     (b) The speed at which a freely falling baseball strikes the ground is a function of the height from which it was dropped.

Math Matters   Finding the Shortest Path

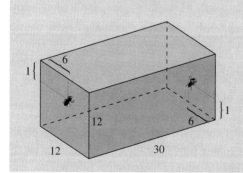

We know that the shortest distance between two points is a straight line. However, when we travel from one place to another, we often cannot travel in a straight line (we must follow the roads or sidewalks). Here is a problem involving the shortest path from one point to another. Suppose a spider and a fly are on opposite walls of a rectangular room, as shown in the figure. The spider wants to visit the fly, and assuming that the spider must travel on the surfaces of the room, what is the shortest path to the fly? (The answer is given in the back of the book.)

2.6 | **Graphs of Functions**

The Graph of a Function ▪ The Vertical Line Test ▪
Graphs of Basic Functions ▪ Transformations of Graphs of Functions

The Graph of a Function

Consider a function f whose domain and range are the set of real numbers. The
graph of f is the set of ordered pairs $(x, f(x))$, where x is in the domain of f.

$$x = x\text{-coordinate of the ordered pair}$$
$$f(x) = y\text{-coordinate of the ordered pair}$$

Figure 2.40 shows a typical graph of such a function.

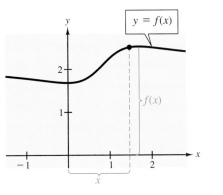

FIGURE 2.40

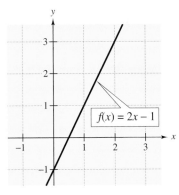

FIGURE 2.41

EXAMPLE 1 *Sketching the Graph of a Function*

Sketch the graph of $f(x) = 2x - 1$.

Solution

Another way to write this function is $y = 2x - 1$. From Section 2.2, you can
recognize the graph to be a line, as shown in Figure 2.41.

In Example 1, the (implied) domain of the function is the set of all real
numbers. When writing the equation of a function, we sometimes choose to
restrict its domain by writing a condition to the right of the equation. For instance,
the domain of the function

$$f(x) = 4x + 5, \quad x \geq 0$$

is the set of all nonnegative real numbers (all $x \geq 0$).

The Vertical Line Test

By the definition of a function, at most one y-value corresponds to a given x-value. This implies that any vertical line can intersect the graph of a function at most once.

> **Vertical Line Test for Functions**
>
> A set of points on a rectangular coordinate system is the graph of y as a function of x if and only if no vertical line intersects the graph at more than one point.

EXAMPLE 2   Using the Vertical Line Test

Decide whether the equation represents y as a function of x.

a. $y = x^2 - 3x + \frac{1}{4}$ **b.** $x = y^2 - 1$

Solution

a. From the graph of the equation in Figure 2.42(a), you can see that every vertical line intersects the graph at most once. Therefore, by the Vertical Line Test, the equation does represent y as a function of x.

b. From the graph of the equation in Figure 2.42(b), you can see that a vertical line intersects the graph twice. Therefore, by the Vertical Line Test, the equation does not represent y as a function of x.

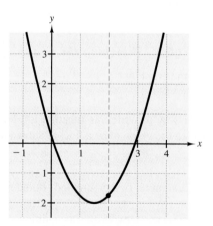
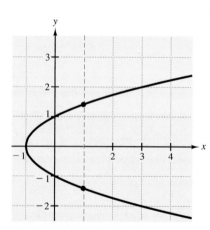

(a) Graph of a function of x.
Vertical line intersects once.

(b) Not a graph of a function of x.
Vertical line intersects twice.

FIGURE 2.42

Graphs of Basic Functions

To become good at sketching the graphs of functions, it helps to be familiar with the graphs of some basic functions. The functions shown in Figure 2.43, and variations of them, occur frequently in applications.

NOTE   Try using a graphing utility to verify the graphs at the right. The names of these functions are as follows.

(a) Constant function
(b) Identity function
(c) Absolute value function
(d) Square root function
(e) Squaring function
(f) Cubing function

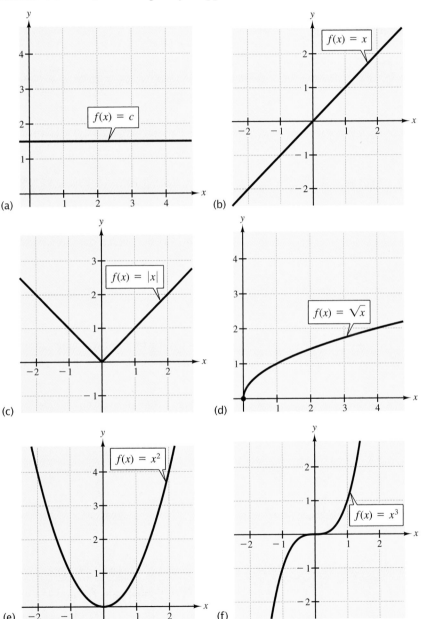

FIGURE 2.43

Transformations of Graphs of Functions

Many functions have graphs that are simple transformations of the basic graphs shown in Figure 2.43. The following list summarizes the various types of **horizontal** and **vertical shifts** of the graphs of functions.

Vertical and Horizontal Shifts

Let c be a positive real number. **Vertical** and **horizontal shifts** of the graph of the function $y = f(x)$ are represented as follows.

1. Vertical shift c units **upward:** $h(x) = f(x) + c$

2. Vertical shift c units **downward:** $h(x) = f(x) - c$

3. Horizontal shift c units to the **right:** $h(x) = f(x - c)$

4. Horizontal shift c units to the **left:** $h(x) = f(x + c)$

EXAMPLE 3   Shifts of the Graphs of Functions

Use the graph of $f(x) = x^2$ to sketch the graph of each function.

a. $g(x) = x^2 - 2$ **b.** $h(x) = (x + 3)^2$

Solution

a. Relative to the graph of $f(x) = x^2$, the graph of $g(x) = x^2 - 2$ represents a *downward shift* of two units, as shown in Figure 2.44.

b. Relative to the graph of $f(x) = x^2$, the graph of $h(x) = (x + 3)^2$ represents a *left shift* of three units, as shown in Figure 2.45.

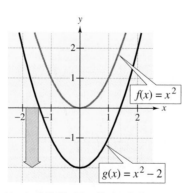

Vertical Shift: Two Units Down
FIGURE 2.44

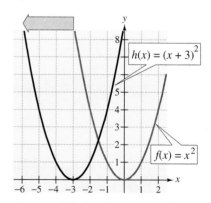

Horizontal Shift: Three Units Left
FIGURE 2.45

Some graphs can be obtained from *combinations* of vertical and horizontal shifts, as shown in part (b) of the next example.

EXAMPLE 4   *Shifts of the Graphs of Functions*

Use the graph of $f(x) = x^3$ to sketch the graph of each function.

a. $g(x) = x^3 + 2$ **b.** $h(x) = (x - 1)^3 + 2$

Solution

a.   Relative to the graph of $f(x) = x^3$, the graph of $g(x) = x^3 + 2$ represents an *upward shift* of two units, as shown in Figure 2.46.

b.   Relative to the graph of $f(x) = x^3$, the graph of $h(x) = (x - 1)^3 + 2$ represents a *right shift* of one unit, followed by an *upward shift* of two units, as shown in Figure 2.47.

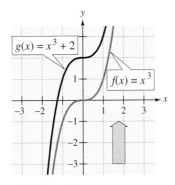

FIGURE 2.46
Vertical Shift: Two Units Up

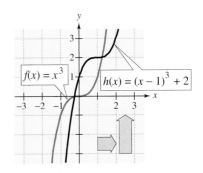

FIGURE 2.47
Horizontal Shift: One Unit Right;
Vertical Shift: Two Units Up

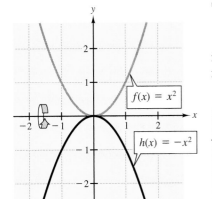

FIGURE 2.48   Reflection

The second basic type of transformation is a **reflection.** For instance, if you imagine that the x-axis represents a mirror, then the graph of $h(x) = -x^2$ is the mirror image (or reflection) of the graph of $f(x) = x^2$, as shown in Figure 2.48.

Reflections in the Coordinate Axes

Reflections of the graph of $y = f(x)$ are represented as follows.

1. Reflection in the x-axis: $h(x) = -f(x)$

2. Reflection in the y-axis: $h(x) = f(-x)$

EXAMPLE 5   *Reflections of the Graphs of Functions*

Use the graph of $f(x) = \sqrt{x}$ to sketch the graph of each function.

a. $g(x) = -\sqrt{x}$ **b.** $h(x) = \sqrt{-x}$

Solution

a. Relative to the graph of $f(x) = \sqrt{x}$, the graph of $g(x) = -\sqrt{x}$ represents a *reflection in the x-axis*, as shown in Figure 2.49.

b. Relative to the graph of $f(x) = \sqrt{x}$, the graph of $h(x) = \sqrt{-x}$ represents a *reflection in the y-axis*, as shown in Figure 2.50.

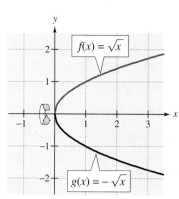

FIGURE 2.49   Reflection in *x*-Axis

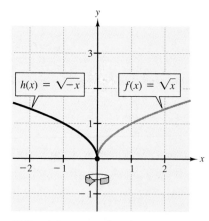

FIGURE 2.50   Reflection in *y*-Axis

Group Activities   Exploring with Technology

Constructing Transformations   Use a graphing utility to graph $f(x) = x^2 + 2$. Decide how to alter this function to produce each of the following transformation descriptions. Graph each transformation on the same screen with f; confirm that the transformation moved f as described.

a. The graph of f shifted to the left three units.

b. The graph of f shifted downward five units.

c. The graph of f shifted upward one unit.

d. The graph of f shifted to the right two units.

2.6 Exercises

Discussing the Concepts

1. Explain the change in the range of $f(x) = 2x$ if the domain is changed from $[0, 2]$ to $[0, 4]$.

2. In your own words, explain how to use the Vertical Line Test.

3. Describe the four types of shifts of the graph of a function.

4. Describe the relationship between the graphs of $f(x)$ and $g(x) = -f(x)$.

5. Describe the relationship between the graphs of $f(x)$ and $g(x) = f(-x)$.

6. Describe the relationship between the graphs of $f(x)$ and $g(x) = f(x - 2)$.

Problem Solving

In Exercises 7–10, use a graphing utility to graph the function and find its domain and range.

7. $g(x) = 1 - x^2$

8. $f(x) = |x + 1|$

9. $f(x) = \sqrt{x - 2}$

10. $h(t) = \sqrt{4 - t^2}$

In Exercises 11–14, use the Vertical Line Test to determine whether y is a function of x.

11. $y = \frac{1}{3}x^3$

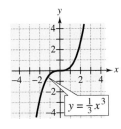

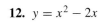

12. $y = x^2 - 2x$

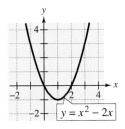

13. $y^2 = x$

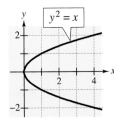

14. $|y| = x$

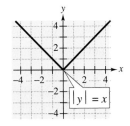

15. *Think About It* Does the graph in Exercise 11 represent x as a function of y? Explain your reasoning.

16. *Think About It* Does the graph in Exercise 12 represent x as a function of y? Explain your reasoning.

In Exercises 17–20, match the function with its graph. [The graphs are labeled (a), (b), (c), and (d).]

(a)

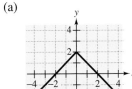

(b)

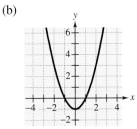

(c)

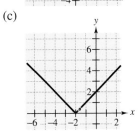

(d)

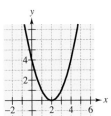

17. $f(x) = x^2 - 1$

18. $f(x) = (x - 2)^2$

19. $f(x) = 2 - |x|$

20. $f(x) = |x + 2|$

In Exercises 21–28, sketch the graph of the function. Then determine its domain and range.

21. $g(x) = \frac{1}{2}x^2$

22. $h(x) = \frac{1}{4}x^2 - 1$

23. $f(x) = -(x - 1)^2$

24. $g(x) = (x + 2)^2 + 3$

25. $K(s) = |s - 4| + 1$

26. $Q(t) = 1 - |t + 1|$

27. $h(x) = \begin{cases} 2x + 3, & \text{if } x < 0 \\ 3 - x, & \text{if } x \geq 0 \end{cases}$

28. $f(t) = \begin{cases} \sqrt{4 + t}, & \text{if } t < 0 \\ \sqrt{4 - t}, & \text{if } t \geq 0 \end{cases}$

In Exercises 29 and 30, select the viewing rectangle that shows the most complete graph of the function.

29. $f(x) = -(x^2 - 20x + 50)$

(a)
Xmin = 0
Xmax = 10
Xscl = 2
Ymin = 0
Ymax = 30
Yscl = 2

(b)
Xmin = 0
Xmax = 20
Xscl = 2
Ymin = -10
Ymax = 60
Yscl = 6

(c)
Xmin = 15
Xmax = 30
Xscl = 2
Ymin = -10
Ymax = 60
Yscl = 5

30. $f(x) = x^4 - 10x^3$

(a)
Xmin = -10
Xmax = 10
Xscl = 1
Ymin = -10
Ymax = 10
Yscl = 1

(b)
Xmin = -5
Xmax = 5
Xscl = 1
Ymin = -10
Ymax = 20
Yscl = 2

(c)
Xmin = -3
Xmax = 12
Xscl = 1
Ymin = -1200
Ymax = 400
Yscl = 100

In Exercises 31–36, identify the transformation of the graph of $f(x) = x^2$ and sketch the graph of h.

31. $h(x) = x^2 + 2$ **32.** $h(x) = x^2 - 4$

33. $h(x) = (x + 2)^2$ **34.** $h(x) = (x - 4)^2$

35. $h(x) = -x^2$ **36.** $h(x) = -x^2 + 4$

In Exercises 37–42, use the graph of $f(x) = \sqrt{x}$ to write a function that represents the graph.

37.
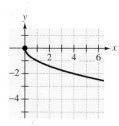

38.
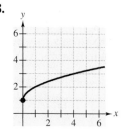

39.
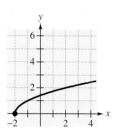

40.
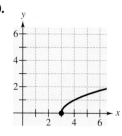

41.
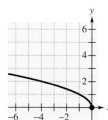

42.
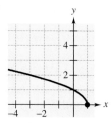

43. *Geometry* The perimeter of a rectangle is 200 meters (see figure).

(a) Show that the area of the rectangle is given by $A = x(100 - x)$, where x is its length.

(b) Use a graphing utility to graph the area function.

(c) Graphically, what value of x yields the largest value of A? Interpret the results.

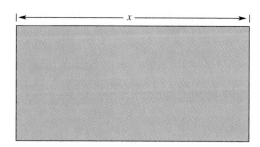

44. *Profit* The profit P when x units of a product are sold is given by $P(x) = 0.47x - 100$ for x in the interval $0 \leq x \leq 1000$.

(a) Use a graphing utility to graph the profit function over the specified domain.

(b) Approximately how many units must be sold for the company to break even $(P = 0)$?

(c) Approximately how many units must be sold for the company to make a profit of $300?

Reviewing the Major Concepts

In Exercises 45–48, solve the equation.

45. $4 - \frac{1}{2}x = 6$

46. $500 - 0.75x = 235$

47. $4(x - 3) = 2x$

48. $12(3 - x) = 5 - 7x$

49. *Phone Charges* The cost for a phone call is $1.10 for the first minute and $0.45 for each additional minute. The total cost of a call cannot exceed $11. Find the interval of time that is available for the call.

50. *Operating Cost* A fuel company has a fleet of trucks. The annual operating cost per truck is

$$C = 0.65m + 4500$$

where m is the number of miles traveled by a truck in a year. What number of miles will yield an annual operating cost that is less than $20,000?

Additional Problem Solving

In Exercises 51–54, use the Vertical Line Test to determine whether y is a function of x.

51. $y = (x + 2)^2$

52. $x - 2y^2 = 0$

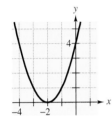
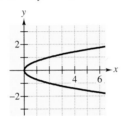

53. $x^2 + y^2 = 16$

54. $y = x^3 - 3$

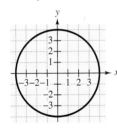
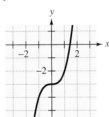

In Exercises 55–58, sketch the graph of the equation. Does the graph represent y as a function of x?

55. $3x - 5y = 15$

56. $y = x^2 + 2$

57. $y^2 = x + 1$

58. $x = y^4$

In Exercises 59–76, sketch the graph of the function. Then determine its domain and range.

59. $f(x) = 2x - 7$

60. $f(x) = 3 - 2x$

61. $h(x) = x^2 - 6x + 8$

62. $f(x) = -x^2 - 2x + 1$

63. $f(t) = \sqrt{t - 2}$

64. $h(x) = \sqrt{4 - x}$

65. $g(s) = \frac{1}{2}s^3$

66. $f(x) = x^3 - 4$

67. $f(x) = |x + 3|$

68. $g(x) = 2 - |x - 1|$

69. $f(x) = 6 - 3x, \quad 0 \le x \le 2$

70. $f(x) = \frac{1}{3}x - 2, \quad 6 \le x \le 12$

71. $h(x) = x^3, \quad -2 \le x \le 2$

72. $h(x) = x(6 - x), \quad 0 \le x \le 6$

73. $f(x) = \begin{cases} x + 6, & \text{if } x < 0 \\ 6 - 2x, & \text{if } x \ge 0 \end{cases}$

74. $f(x) = \begin{cases} -x, & \text{if } x \le 0 \\ x^2 - 4x, & \text{if } x > 0 \end{cases}$

75. $h(x) = \begin{cases} 4 - x^2, & \text{if } x \le 2 \\ x - 2, & \text{if } x > 2 \end{cases}$

76. $f(x) = \begin{cases} x^2, & \text{if } x < 1 \\ x^2 - 3x + 2, & \text{if } x \ge 1 \end{cases}$

In Exercises 77–82, identify the transformation of the graph of $f(x) = x^3$ and sketch the graph of h.

77. $h(x) = x^3 + 3$

78. $h(x) = x^3 - 5$

79. $h(x) = (x - 3)^3$

80. $h(x) = -x^3$

81. $h(x) = 2 - (x - 1)^3$

82. $h(x) = (x + 2)^3 - 3$

In Exercises 83–88, identify the transformation of the graph of $f(x) = |x|$ and use a graphing utility to sketch the graph of h.

83. $h(x) = |x - 5|$

84. $h(x) = |x + 3|$

85. $h(x) = |x| - 5$

86. $h(x) = |-x|$

87. $h(x) = -|x|$

88. $h(x) = 5 - |x|$

In Exercises 89–96, use the graph of $f(x) = x^2$ to write an equation that represents the transformation of f given by the graph.

89.

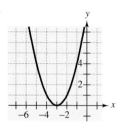

90.

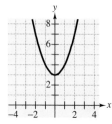

91.

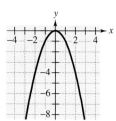

92.

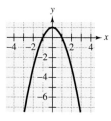

93.

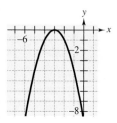

94.

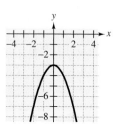

95.

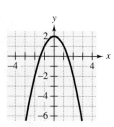

96.

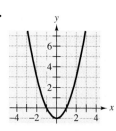

97. *Civilian Population of the United States* For 1950 through 1990, the civilian population P (in thousands) of the United States can be modeled by

$$P_1(t) = -12t^2 + 2900t + 149{,}300$$

where $t = 0$ represents 1950. (Source: U. S. Bureau of Census)

(a) Use a graphing utility to graph the function over the appropriate domain.

(b) In the transformation of the population function

$$P_2(t) = -12(t+20)^2 + 2900(t+20) + 149{,}300$$

$t = 0$ corresponds to what calendar year? Explain.

(c) Use a graphing utility to graph P_2 over the appropriate domain.

98. *Aircraft Orders* The following table gives the number N of new civil jet aircraft orders for U.S. aircraft manufacturers for the years 1986 through 1992. (Source: Aerospace Industries Association of America)

Year	1986	1987	1988	1989	1990	1991	1992
N	332	519	956	1015	670	280	231

A model for this data is given by

$$N(t) = \frac{3.13t + 617}{0.20t^2 + 0.58t + 1}$$

where $t = 0$ represents 1990.

(a) What is the domain of the function?

(b) Use a graphing utility to graph the data and the model on the same set of coordinate axes.

(c) For which year does the model most accurately estimate the actual data? During which year is it least accurate?

CHAPTER PROJECT: Working Men and Women

Data is frequently represented in three ways: *numerically* by a table, *graphically*, or *algebraically* by an equation, formula, or algebraic model. Because each type of representation has its own advantages, the best way to present data is to use a mix of all three representations.

In gathering data, it may be easiest to first organize the data in a table. Finding patterns or trends in the data may be easier to recognize with a graph. And, finally, making future predictions may be easier with an algebraic model.

For instance, the table below shows the numbers of men and women (in thousands) who were practicing physicians from 1970 through 1990. The data is represented graphically by the scatter plot at the left. While investigating the questions below, you will be asked to find algebraic models to represent the data, and then to use the models to interpret the data.   (Source: American Medical Association)

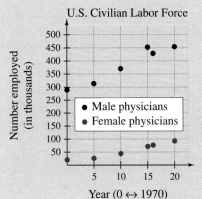

U.S. Civilian Labor Force

Number employed (in thousands)

● Male physicians
● Female physicians

Year (0 ↔ 1970)

Year	1970	1975	1980	1985	1986	1990
Women Physicians	21.4	27.2	44.7	71.9	76.8	93.3
Men Physicians	289.5	313.1	370.2	452.3	429.0	454.0

1. *Graphical Reasoning*   Use the scatter plot to write a verbal description of the data. Discuss any trend or pattern that is evident from the scatter plot.

2. *Linear Modeling*   Use graph paper to redraw the scatter plot. Approximate each of the data sets with a line. Then find an equation for each line.

3. *Linear Modeling*   Use the linear regression program on a graphing utility to find a linear model for each data set. Compare the results with those obtained in Question 2.

4. *How Well Does It Fit?*   When the regression program in Question 3 is run, it will display a correlation coefficient r that measures how well the linear model fits the data. The closer r is to 1, the better the model fits the data. Which of the two models fits its data better? Does your answer seem reasonable from the graphical point of view?

5. *Prediction*   Predict the numbers of women and men physicians in 1995. Discuss different ways that you could obtain the prediction. Which method do you prefer? Why?

6. *Prediction*   Extend the lines you drew in Question 2 until they intersect. What interpretation can you make? Is the interpretation realistic in the context of the data?

7. *Research Project*   Use your school's library or another reference source to find data for the numbers of men and women in an occupation. Organize the data numerically, graphically, and algebraically. What can you conclude?

CHAPTER SUMMARY

After studying this chapter, you should have acquired the following skills. These skills are keyed to the Review Exercises that begin on page 193. Answers to odd-numbered Review Exercises are given in the back of the book.

- Plot points on a rectangular coordinate system. *(Section 2.1)* **Review Exercises 1, 2**

- Plot points that form polygons. *(Section 2.1)* **Review Exercises 3, 4**

- Determine the quadrants in which points are located. *(Section 2.1)* **Review Exercises 5–8**

- Plot pairs of points and find the distances between them. *(Section 2.1)* **Review Exercises 9–12**

- Determine whether ordered pairs are solutions of equations. *(Section 2.1)* **Review Exercises 13, 14**

- Sketch graphs of equations and label the *x*- and *y*-intercepts. *(Section 2.2)* **Review Exercises 15–20**

- Find the slopes of the lines through pairs of points. *(Section 2.4)* **Review Exercises 21–26**

- Find missing values so that sets of points are collinear. *(Section 2.4)* **Review Exercises 27, 28**

- Find additional points on lines given a point on each line and the slope. *(Section 2.4)* **Review Exercises 29–34**

- Write equations of lines in slope-intercept form and sketch the lines. *(Section 2.4)* **Review Exercises 35–38**

- Use a graphing utility to graph pairs of equations to determine whether the pairs of lines are parallel, perpendicular, or neither. *(Sections 2.3, 2.4)* **Review Exercises 39–42**

- Decide whether relations are functions. *(Section 2.5)* **Review Exercises 43–46**

- Estimate the intercepts of graphs, check them algebraically, and determine whether the graphs represent *y* as a function of *x*. *(Sections 2.2, 2.6)* **Review Exercises 47–50**

- Evaluate functions and simplify. *(Section 2.5)* **Review Exercises 51–58**

- Find the domains of functions. *(Section 2.5)* **Review Exercises 59–62**

- Select viewing rectangles on a graphing utility that show the most complete graphs of functions. *(Sections 2.3, 2.6)* **Review Exercises 63, 64**

- Sketch graphs of functions and check results using a graphing utility. *(Sections 2.3, 2.6)* **Review Exercises 65–76**

- Identify and sketch transformations of graphs. *(Section 2.6)* **Review Exercises 77–80**

- Solve real-life problems modeled by functions. *(Sections 2.5, 2.6)* **Review Exercises 81–84**

REVIEW EXERCISES

In Exercises 1 and 2, plot the points on a rectangular coordinate system.

1. $(0, -3)$, $\left(\frac{5}{2}, 5\right)$, $(-2, -4)$

2. $\left(1, -\frac{3}{2}\right)$, $\left(-2, 2\frac{3}{4}\right)$, $(5, 10)$

Geometry In Exercises 3 and 4, plot the points and verify that the points form the indicated polygon.

3. *Right Triangle:* $(1, 1)$, $(12, 9)$, $(4, 20)$

4. *Parallelogram:* $(0, 0)$, $(7, 1)$, $(8, 4)$, $(1, 3)$

In Exercises 5–8, determine the quadrant(s) in which the point is located.

5. $(2, -6)$

6. $(-4.8, -2)$

7. $(4, y)$

8. (x, y), $xy > 0$

In Exercises 9–12, plot the points and find the distance between them.

9. $(4, 3)$, $(4, 8)$

10. $(2, -5)$, $(6, -5)$

11. $(-5, -1)$, $(1, 2)$

12. $(-2, 10)$, $(3, -2)$

In Exercises 13 and 14, determine whether the ordered pairs are solutions of the equation.

13. $y = 4 - \frac{1}{2}x$ (a) $(4, 2)$ (b) $(-1, 5)$
                                  (c) $(-4, 0)$ (d) $(8, 0)$

14. $3x - 2y + 18 = 0$ (a) $(3, 10)$ (b) $(0, 9)$
                               (c) $(-4, 3)$ (d) $(-8, 0)$

In Exercises 15–20, sketch the graph of the equation. Label the x- and y-intercepts.

15. $y = 6 - \frac{1}{3}x$

16. $y = \frac{3}{4}x - 2$

17. $3y - 2x - 3 = 0$

18. $3x + 4y + 12 = 0$

19. $x = |y - 3|$

20. $y = 1 - x^3$

In Exercises 21–26, find the slope of the line through the points.

21. $(-1, 1)$, $(6, 3)$

22. $(-2, 5)$, $(3, -8)$

23. $(-1, 3)$, $(4, 3)$

24. $(7, 2)$, $(7, 8)$

25. $(0, 6)$, $(8, 0)$

26. $(0, 0)$, $\left(\frac{7}{2}, 6\right)$

In Exercises 27 and 28, find t so that the three points are collinear. (*Note:* Collinear means that the points lie on the same straight line.)

27. $(-3, -3)$, $(0, t)$, $(1, 3)$

28. $(2, 1)$, $(1, t)$, $(8, 3)$

In Exercises 29–34, a point on a line and the slope of the line are given. Find two additional points on the line.

29. $(2, -4)$, $m = -3$

30. $\left(-4, \frac{1}{2}\right)$, $m = 2$

31. $(3, 1)$, $m = \frac{5}{4}$

32. $(7, -2)$, $m = 0$

33. $(3, 7)$, m is undefined.

34. $\left(-3, -\frac{3}{2}\right)$, $m = -\frac{1}{3}$

In Exercises 35–38, write the equation of the line in slope-intercept form and sketch the line.

35. $5x - 2y - 4 = 0$

36. $x - 3y - 6 = 0$

37. $x + 2y - 2 = 0$

38. $y - 6 = 0$

In Exercises 39–42, use a graphing utility to graph the equations on the same viewing rectangle. Are the lines parallel, perpendicular, or neither?

39. $L_1: y = \frac{3}{2}x + 1$
        $L_2: y = \frac{2}{3}x - 1$

40. $L_1: y = 2x - 5$
        $L_2: y = 2x + 3$

41. $L_1: y = \frac{3}{2}x - 2$
        $L_2: y = -\frac{2}{3}x + 1$

42. $L_1: y = -0.3x - 2$
        $L_2: y = 0.3x + 1$

In Exercises 43–46, decide whether the relation is a function.

43.

Domain	Range
6	0
7	1
8	2
9	

44.

Domain	Range
-4	-2
-2	-1
0	0
2	1
4	2

45.

Domain	Range
1	
2	
3	10
4	
5	

46.

Domain	Range
15	3
25	5
35	7
45	9
55	11

In Exercises 47–50, graphically estimate the intercepts. Then check your estimate algebraically. Does the graph represent y as a function of x?

47. $9y^2 = 4x^3$

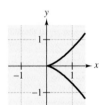

48. $y = 4x^3 - x^4$

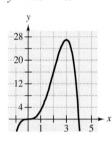

49. $y = x^2(x - 3)$

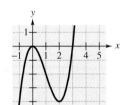

50. $x^3 + y^3 - 6xy = 0$

In Exercises 51–58, evaluate the function for the specified values of the independent variable and simplify when possible.

51. $f(x) = 4 - \frac{5}{2}x$

(a) $f(-10)$ (b) $f\left(\frac{2}{5}\right)$

(c) $f(t) + f(-4)$ (d) $f(x + h)$

52. $h(x) = x(x - 8)$

(a) $h(8)$ (b) $h(10)$

(c) $h(-3)$ (d) $h(t + 4)$

53. $f(t) = \sqrt{5 - t}$

(a) $f(-3)$ (b) $f(5)$

(c) $f\left(\frac{11}{3}\right)$ (d) $f(5z)$

54. $g(x) = \dfrac{|x + 4|}{4}$

(a) $g(0)$ (b) $g(-8)$

(c) $g(2) - g(-5)$ (d) $g(x - 2)$

55. $f(x) = \begin{cases} -3x, & \text{if } x \le 0 \\ 1 - x^2, & \text{if } x > 0 \end{cases}$

(a) $f(2)$ (b) $f\left(-\frac{2}{3}\right)$

(c) $f(1)$ (d) $f(4) - f(3)$

56. $h(x) = \begin{cases} x^3, & \text{if } x \le 1 \\ (x - 1)^2 + 1, & \text{if } x > 1 \end{cases}$

(a) $h(2)$ (b) $h\left(-\frac{1}{2}\right)$

(c) $h(0)$ (d) $h(4) - h(3)$

57. $f(x) = 3 - 2x$

(a) $\dfrac{f(x + 2) - f(2)}{x}$ (b) $\dfrac{f(x - 3) - f(3)}{x}$

58. $f(x) = 7x + 10$

(a) $\dfrac{f(x + 1) - f(1)}{x}$ (b) $\dfrac{f(x - 5) - f(5)}{x}$

In Exercises 59–62, find the domain of the function.

59. $h(x) = 4x^2 - 7$ **60.** $g(s) = \dfrac{s+1}{(s-1)(s+5)}$

61. $f(x) = \sqrt{5 - 2x}$ **62.** $f(x) = |x - 6| + 10$

In Exercises 63 and 64, select the viewing rectangle on a graphing utility that shows the most complete graph of the function.

63. $f(x) = x^4 - 2x^3$

(a)

```
Xmin = 0
Xmax = 20
Xscl = 1
Ymin = -10
Ymax = 10
Yscl = 1
```

(b)

```
Xmin = -5
Xmax = 5
Xscl = 1
Ymin = -100
Ymax = 100
Yscl = 10
```

(c)

```
Xmin = -3
Xmax = 3
Xscl = 1
Ymin = -3
Ymax = 5
Yscl = 1
```

64. $f(x) = 5x\sqrt{16 - x^2}$

(a)

```
Xmin = -6
Xmax = 6
Xscl = 1
Ymin = -50
Ymax = 50
Yscl = 10
```

(b)

```
Xmin = -20
Xmax = 20
Xscl = 4
Ymin = -600
Ymax = 600
Yscl = 100
```

(c)

```
Xmin = -10
Xmax = 10
Xscl = 2
Ymin = -20
Ymax = 20
Yscl = 4
```

In Exercises 65–76, sketch the graph of the function. Use a graphing utility to confirm your graph.

65. $g(x) = \frac{1}{8}x^2$ **66.** $y = 4 - (x - 3)^2$

67. $y = (x - 2)^2$ **68.** $h(x) = 9 - (x - 2)^2$

69. $y = \dfrac{1}{2}x(2 - x)$ **70.** $f(t) = \sqrt{\dfrac{t}{2}}$

71. $y = 8 - 2|x|$ **72.** $f(x) = |x + 1| - 2$

73. $g(x) = \frac{1}{4}x^3, \quad -2 \le x \le 2$

74. $h(x) = x(4 - x), \quad 0 \le x \le 4$

75. $f(x) = \begin{cases} 2 - (x-1)^2, & \text{if } x < 1 \\ 2 + (x-1)^2, & \text{if } x \ge 1 \end{cases}$

76. $f(x) = \begin{cases} 2x, & \text{if } x \le 0 \\ x^2 + 1, & \text{if } x > 0 \end{cases}$

In Exercises 77–80, identify the transformation of the graph of $f(x) = x^4$ and sketch the graph of h.

77. $h(x) = -x^4$ **78.** $h(x) = x^4 + 2$

79. $h(x) = (x - 1)^4$ **80.** $h(x) = 1 - x^4$

81. *Path of a Projectile*   The height y (in feet) of a projectile is given by

$$y = -\frac{1}{16}x^2 + 5x$$

where x is the horizontal distance (in feet) from where the projectile was launched.

(a) Sketch the path of the projectile.

(b) How high is the projectile when it is at its maximum height?

(c) How far from the launch point does the projectile strike the ground?

82. *Velocity of a Ball*   The velocity of a ball thrown upward from ground level is given by

$$v = -32t + 80$$

where t is time in seconds and v is velocity in feet per second.

(a) Find the velocity when $t = 2$.

(b) Find the time when the ball reaches its maximum height. (*Hint:* Find the time when $v = 0$.)

(c) Find the velocity when $t = 3$.

83. *Power Generation*   The power generated by a wind turbine is given by the function

$$P = kw^3$$

where P is the number of kilowatts produced at a wind speed of w miles per hour and k is the constant of proportionality.

(a) Find k if $P = 1000$ when $w = 20$.

(b) Find the output for a wind speed of 25 miles per hour.

84. *Geometry*   A wire 100 inches long is to be cut into four pieces to form a rectangle whose shortest side has a length of x. Express the area A of the rectangle as a function of x. Use a graphing utility to graph the function.

CHAPTER TEST

Take this test as you would take a test in class. After you are done, check your work against the answers given in the back of the book.

1. Determine the quadrant in which the point (x, y) lies if $x > 0$ and $y < 0$.

2. Plot the points $(0, 5)$ and $(3, 1)$. Then find the distance between them.

3. Find the x- and y-intercepts of the graph of $y = -3(x + 1)$.

4. Sketch the graph of $y = |x - 2|$.

5. Find the slope of the line through $(-4, 7)$ and $(2, 3)$.

6. Find the slope of the line through $(3, -2)$ and $(3, 6)$.

7. Sketch the line passing through $(0, -6)$ with a slope of $m = \frac{4}{3}$.

8. Plot the x- and y-intercepts of the graph of $2x + 5y = 10$. Use the result to sketch the graph.

9. Write the equation $5x + 3y - 9 = 0$ in slope-intercept form. Find the slope of a line that is perpendicular to this line.

10. The graph of $y^2(4 - x) = x^3$ is shown at the right. Does the graph represent y as a function of x? Explain your reasoning.

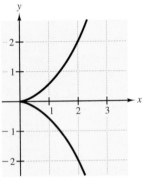

Figure for 10

11. Does $3x^2 - y^2 = 9$ represent y as a function of x? Explain.

12. Does $3x - y = 0$ represent y as a function of x? Explain.

In Exercises 13–15, evaluate $g(x) = x/(x - 3)$ for the indicated value.

13. $g(2)$

14. $g\left(\frac{7}{2}\right)$

15. $g(x + 2)$

16. Find the domain of $h(t) = \sqrt{9 - t}$.

17. Find the domain of $f(x) = \dfrac{x + 1}{x - 4}$.

18. Sketch the graph of $g(x) = \sqrt{2 - x}$.

19. Use a graphing utility to graph $y = -0.3x^2 + 2x + 5$. Then use the graph to estimate the intercepts of the graph.

20. Use the graph of $y = |x|$ to write an equation for each graph.

(a)

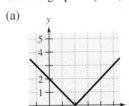

(b)

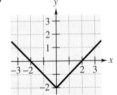

(c)

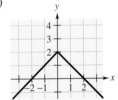

Polynomials and Factoring

3

- Adding and Subtracting Polynomials
- Exponents and Multiplying Polynomials
- Factoring Polynomials: An Introduction
- Factoring Trinomials
- Solving Polynomial Equations

Consider four cubes that have side lengths of 1 inch, 2 inches, 3 inches, and 4 inches. The surface area S of each cube is given by

$$S = 6x^2 \qquad \text{Surface area}$$

where x is the length of each side. The volume V of each cube is given by

$$V = x^3. \qquad \text{Volume}$$

The table at the right gives the surface areas and volumes of the four cubes.

From the graphs of S and V at the right, notice that surface area and volume are increasing at different rates. For $0 < x < 4$, the surface area is increasing at a greater rate and for $x > 4$, the volume is increasing at a greater rate. For which value of x do the two graphs intersect?

Cube	x	S	V
Cube 1	1 inch	6 square inches	1 cubic inch
Cube 2	2 inches	24 square inches	8 cubic inches
Cube 3	3 inches	54 square inches	27 cubic inches
Cube 4	4 inches	96 square inches	64 cubic inches

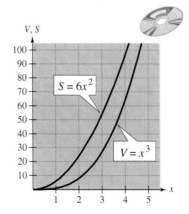

The chapter project related to this information is on page 249.

197

3.1 Adding and Subtracting Polynomials

Basic Definitions ▪ Adding and Subtracting Polynomials ▪ Applications

Basic Definitions

A **polynomial in x** is an algebraic expression whose terms are all of the form ax^k, where a is any real number and k is a nonnegative integer.

Definition of a Polynomial in x

Let $a_n, \ldots, a_2, a_1, a_0$ be real numbers and let n be a *nonnegative integer.* A **polynomial in x** is an expression of the form

$$a_n x^n + a_{n-1} x^{n-1} + \cdots + a_2 x^2 + a_1 x + a_0$$

where $a_n \neq 0$. The polynomial is of **degree n,** and the number a_n is the **leading coefficient.** The number a_0 is the **constant term.**

NOTE The following *are not* polynomials for the reasons stated.

- The expression $2x^{-1} + 5$ is not a polynomial because the exponent in $2x^{-1}$ is not nonnegative.

- The expression $x^3 + 3x^{1/2}$ is not a polynomial because the exponent in $3x^{1/2}$ is not an integer.

In the term ax^k, a is the **coefficient** and k is the **degree** of the term. Note that the degree of the term ax is 1, and the degree of the constant term is 0. Because a polynomial is an algebraic *sum,* the coefficients take on the signs between the terms. For instance,

$$x^3 - 4x^2 + 3 = (1)x^3 + (-4)x^2 + (0)x + 3$$

has coefficients $1, -4, 0$, and 3. A polynomial that is written in order of descending powers of the variable is said to be in **standard form.** A polynomial with only one term is a **monomial.** Polynomials with two unlike terms are called **binomials,** and those with three unlike terms are called **trinomials.**

EXAMPLE 1 Identifying Leading Coefficients and Degrees

	Polynomial	Standard Form	Degree	Leading Coefficient
a.	$5x^2 - 2x^7 + 4 - 2x$	$-2x^7 + 5x^2 - 2x + 4$	7	-2
b.	$16 - 8x^3$	$-8x^3 + 16$	3	-8
c.	10	10	0	10
d.	$5 + x^4 - 6x^3$	$x^4 - 6x^3 + 5$	4	1

Adding and Subtracting Polynomials

To add two polynomials, combine like terms. This can be done in either a horizontal or vertical format, as shown in Examples 2 and 3.

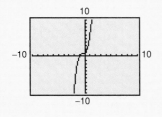

EXAMPLE 2   *Adding Polynomials Horizontally*

a. $(2x^3 + x^2 - 5) + (x^2 + x + 6)$ Given polynomials

$\qquad = (2x^3) + (x^2 + x^2) + (x) + (-5 + 6)$ Group like terms.

$\qquad = 2x^3 + 2x^2 + x + 1$ Combine like terms.

b. $(3x^2 + 2x + 4) + (3x^2 - 6x + 3) + (-x^2 + 2x - 4)$

$\qquad = (3x^2 + 3x^2 - x^2) + (2x - 6x + 2x) + (4 + 3 - 4)$

$\qquad = 5x^2 - 2x + 3$

EXAMPLE 3   *Using a Vertical Format to Add Polynomials*

Use a vertical format to perform the operations.

$$(5x^3 + 2x^2 - x + 7) + (3x^2 - 4x + 7) + (-x^3 + 4x^2 - 8)$$

Solution

To use a vertical format, align the terms of the polynomials by their degrees.

$$
\begin{array}{r}
5x^3 + 2x^2 - x + 7 \\
3x^2 - 4x + 7 \\
-x^3 + 4x^2 - 8 \\
\hline
4x^3 + 9x^2 - 5x + 6
\end{array}
$$

To subtract one polynomial from another, *add the opposite*. You can do this by changing the sign of each of the terms of the polynomial that is being subtracted and then adding the resulting like terms.

EXAMPLE 4   *Subtracting Polynomials Horizontally*

$(3x^3 - 5x^2 + 3) - (x^3 + 2x^2 - x - 4)$ Given polynomials

$\quad = (3x^3 - 5x^2 + 3) + (-x^3 - 2x^2 + x + 4)$ Add the opposite.

$\quad = (3x^3 - x^3) + (-5x^2 - 2x^2) + (x) + (3 + 4)$ Group like terms.

$\quad = 2x^3 - 7x^2 + x + 7$ Combine like terms.

Be especially careful to use the correct signs when subtracting one polynomial from another. One of the most common mistakes in algebra is to forget to change signs correctly when subtracting one expression from another. Here is an example.

Wrong sign

$\downarrow$

$$(x^2 - 2x + 3) - (x^2 + 2x - 2) \neq x^2 - 2x + 3 - x^2 + 2x - 2 \quad \text{Common error}$$

$\uparrow$

Wrong sign

The error illustrated above is forgetting to change two of the signs in the polynomial that is being subtracted. Remember to add the *opposite* of *every* term of the subtracted polynomial.

EXAMPLE 5   Using a Vertical Format to Subtract Polynomials

Use a vertical format to perform the operations.

$$(4x^4 - 2x^3 + 5x^2 - x + 8) - (3x^4 - 2x^3 + 3x - 4)$$

Solution

$$
\begin{array}{l}
(4x^4 - 2x^3 + 5x^2 - x + 8) \\
-(3x^4 - 2x^3 \qquad\quad + 3x - 4)
\end{array}
\quad\Longrightarrow\quad
\begin{array}{l}
4x^4 - 2x^3 + 5x^2 - x + 8 \\
-3x^4 + 2x^3 \qquad\quad - 3x + 4 \\
\hline
x^4 \qquad\quad + 5x^2 - 4x + 12
\end{array}
$$

EXAMPLE 6   Combining Polynomials

a. $(2x^2 - 7x + 2) - (4x^2 + 5x - 1) + (-x^2 + 4x + 4)$

$$= 2x^2 - 7x + 2 - 4x^2 - 5x + 1 - x^2 + 4x + 4$$

$$= (2x^2 - 4x^2 - x^2) + (-7x - 5x + 4x) + (2 + 1 + 4)$$

$$= -3x^2 - 8x + 7$$

b. $(-x^2 + 4x - 3) - [(4x^2 - 3x + 8) - (-x^2 + x + 7)]$

$$= (-x^2 + 4x - 3) - [4x^2 - 3x + 8 + x^2 - x - 7]$$

$$= (-x^2 + 4x - 3) - [(4x^2 + x^2) + (-3x - x) + (8 - 7)]$$

$$= (-x^2 + 4x - 3) - [5x^2 - 4x + 1]$$

$$= -x^2 + 4x - 3 - 5x^2 + 4x - 1$$

$$= (-x^2 - 5x^2) + (4x + 4x) + (-3 - 1)$$

$$= -6x^2 + 8x - 4$$

Applications

There are many applications that involve polynomials. One commonly used second-degree polynomial is called a **position polynomial.** This polynomial has the form

$$h = -16t^2 + v_0 t + s_0 \qquad \text{Position polynomial}$$

where the height h is measured in feet and the time t is measured in seconds.

This position polynomial gives the height (above ground) of a free-falling object. The coefficient of t, v_0, is the **initial velocity** of the object, and the constant term s_0 is the **initial height** of the object. If the initial velocity is positive, the object was projected upward (at $t = 0$), and if the initial velocity is negative, the object was projected downward.

Galileo Galilei (1564–1642) believed that all objects, regardless of their weight (or mass), accelerate at the same rate. This law of falling bodies was not proven until the invention of the vacuum tube in the 1650s.

EXAMPLE 7   *Finding the Height of a Free-Falling Object*

An object is thrown downward from the top of a 200-foot building. The initial velocity is -10 feet per second. Use the position polynomial

$$h = -16t^2 - 10t + 200$$

to find the height of the object when $t = 1$, $t = 2$, and $t = 3$. (See Figure 3.1.)

Solution

When $t = 1$, the height of the object is

$$
\begin{aligned}
h &= -16(1)^2 - 10(1) + 200 & \text{Substitute 1 for } t. \\
&= -16 - 10 + 200 & \text{Simplify.} \\
&= 174 \text{ feet.}
\end{aligned}
$$

When $t = 2$, the height of the object is

$$
\begin{aligned}
h &= -16(2)^2 - 10(2) + 200 & \text{Substitute 2 for } t. \\
&= -64 - 20 + 200 & \text{Simplify.} \\
&= 116 \text{ feet.}
\end{aligned}
$$

When $t = 3$, the height of the object is

$$
\begin{aligned}
h &= -16(3)^2 - 10(3) + 200 & \text{Substitute 3 for } t. \\
&= -144 - 30 + 200 & \text{Simplify.} \\
&= 26 \text{ feet.}
\end{aligned}
$$

Use your calculator to determine the height of the object when $t = 3.2368$. What can you conclude?

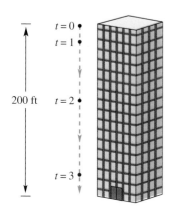

FIGURE 3.1

A second commonly used mathematical model that often involves polynomials is called a **profit equation.**

In 1990, there were about 15 million sole proprietorships (businesses owned by a single person or family) in the United States. Most of these were small businesses with 15 or fewer employees.

EXAMPLE 8   *Finding a Profit*

A small manufacturing company can produce and sell x flashlights per week. The total cost (in dollars) for producing x flashlights is given by $C = 2x + 900$, and the total revenue from selling x flashlights is given by $R = 6x$. Find the profit obtained by selling 700 flashlights per week.

Solution

Verbal Model:

| Profit | = | Revenue | − | Cost |

Labels: Weekly profit $= P$ (dollars)
             Weekly revenue $R = 6x$ (dollars)
             Weekly cost $C = 2x + 900$ (dollars)
             Number of flashlights $= 700$ (flashlights)

Equation: $P = R - C$

$= 6x - (2x + 900)$

$= 6(700) - [2(700) + 900]$

$= 4200 - (1400 + 900)$

$= \$1900$

Therefore, the total weekly profit from selling 700 flashlights is $1900.

Group Activities   Extending the Concept

Problem Posing   Work as a group to create a problem involving cost and revenue equations and requiring a profit equation for the production and sale of a product of your choice. Be sure to describe your cost and revenue situations such that the profit is negative when no units are sold and positive when 200 units are sold. Solve your problem and interpret the solution. Exchange your group's problem for that of another group, and solve each other's problems.

3.1 Exercises

Discussing the Concepts

1. Explain the difference between the degree of a term in a polynomial and the degree of a polynomial.

2. What operation separates the terms of a polynomial? What operation separates the factors of a term?

3. Give an example of combining like terms.

4. Can two third-degree polynomials be added to produce a second-degree polynomial? If so, give an example.

5. Is every trinomial a second-degree polynomial? Explain.

6. Describe the method for subtracting polynomials.

Problem Solving

In Exercises 7–10, write the polynomial in standard form, and find its degree and leading coefficient.

7. $3x^2 + 2 - x$

8. $5t + 3$

9. $5 - 3y^4$

10. $8z - 16z^2$

In Exercises 11–16, determine whether the polynomial is a monomial, binomial, or trinomial.

11. $12 - 5y^2$

12. $-6y + 3$

13. $x^3 + 2x^2 - 4$

14. t^3

15. 5

16. $25 - 2u^2$

In Exercises 17–20, use a horizontal format to find the sum.

17. $(2x^2 - 3) + (5x^2 + 6)$

18. $(3x^3 - 2x + 8) + (3x - 5)$

19. $(2 - 8y) + (-2y^4 + 3y + 2)$

20. $(20s - 12s^2 - 32) + (15s^2 + 6s)$

In Exercises 21–24, use a vertical format to find the sum.

21. $5x^2 - 3x + 4$
    $-3x^2 \qquad - 4$

22. $4x^3 - 2x^2 + 8x$
    $\qquad 4x^2 + \ x - 6$

23. $(2b - 3) + (b^2 - 2b) + (7 - b^2)$

24. $(v^2 + v - 3) + (4v + 1) + (2v^2 - 3v)$

In Exercises 25–28, use a horizontal format to find the difference.

25. $(3x^2 - 2x + 1) - (2x^2 + x - 1)$

26. $(5y^4 - 2) - (3y^4 + 2)$

27. $(8x^3 - 4x^2 + 3x) - [(x^3 - 4x^2 + 5) + (x - 5)]$

28. $(y^4 - y^2) - (y^2 + 3y^4)$

In Exercises 29 and 30, use a vertical format to find the difference.

29. $x^2 - \ x + 3$
    $- \qquad (x - 2)$

30. $0.2t^4 - 0.5t^2$
    $- \quad (-t^4 + 0.3t^2 - 1.4)$

In Exercises 31–34, perform the operations.

31. $-(2x^3 - 3) + (4x^3 - 2x)$

32. $(p^3 + 4) - [(p^2 + 4) + (3p - 9)]$

33. $15v - 3(3v - v^2) + 9(8v + 3)$

34. $9(7x^2 - 3x + 3) - 4(15x + 2) - (3x^2 - 7x)$

Graphical Reasoning   In Exercises 35 and 36, use a graphing utility to graph the expressions for y_1 and y_2. What conclusion can you make?

35. $y_1 = (x^3 - 3x^2 - 2) - (x^2 + 1)$
    $y_2 = x^3 - 4x^2 - 3$

36. $y_1 = \left(\frac{1}{2}x^3 + 2x\right) + (x^3 - x^2 - x + 1)$
    $y_2 = \frac{3}{2}x^3 - x^2 + x + 1$

In Exercises 37 and 38, evaluate the polynomial function for the given values of the variable. Then use a graphing utility to graph the function and graphically confirm your results.

37. $f(x) = x^3 - 12x$ (a) $x = -2$ (b) $x = 0$
                                (c) $x = 2$ (d) $x = 4$

38. $f(x) = \frac{1}{4}x^4 - 2x^2$ (a) $x = -2$ (b) $x = 0$
                                 (c) $x = 2$ (d) $x = 3$

Free-Falling Object In Exercises 39 and 40, find the height of a free-falling object at the specified times using the position polynomial. Then write a paragraph that describes the path of the object.

39. $h = -16t^2 + 32$ (a) $t = 0$ (b) $t = \frac{1}{2}$
                               (c) $t = 1$ (d) $t = \frac{3}{2}$

40. $h = -16t^2 + 96t$ (a) $t = 0$ (b) $t = 2$
                               (c) $t = 3$ (d) $t = 6$

41. *Profit* A manufacturer can produce and sell x radios per week. The total cost (in dollars) of producing the radios is given by

$C = 8x + 15{,}000$

and the total revenue is given by

$R = 14x.$

Find the profit obtained by selling 5000 radios per week.

42. *Profit* A manufacturer can produce and sell x golf clubs per week. The total cost (in dollars) of producing the golf clubs is given by

$C = 12x + 8000$

and the total revenue is given by

$R = 17x.$

Find the profit obtained by selling 10,000 golf clubs per week.

Reviewing the Major Concepts

In Exercises 43 and 44, evaluate the expression.

43. $5 - |-2|$ **44.** $-15 - 3(4 - 18)$

45. Use a graphing utility to graph each of the functions and identify the transformation of $f(x) = x^5$.
   (a) $g(x) = x^5 - 2$ (b) $g(x) = (x - 2)^5$
   (c) $h(x) = -x^5$ (d) $h(x) = (-x)^5$

46. *Interpreting a Model* A manufacturing plant purchases a new computer system for $175,000. The depreciated value y after t years is given by

$y = 175{,}000 - 25{,}000t.$

Write a verbal description of the depreciation.

Additional Problem Solving

In Exercises 47–54, write the polynomial in standard form, and find its degree and leading coefficient.

47. $10x - 4$ **48.** $3x^2 + 8$

49. $-3x^3 - 2x^2 - 3$ **50.** $6t + 4t^5 - t^2 + 3$

51. -4 **52.** 28

53. $v_0 t - 16t^2$ (v_0 is constant)

54. $48 - \frac{1}{2}at^2$ (a is constant)

In Exercises 55–58, give an example of a polynomial in x that satisfies the given conditions. (*Note:* Each problem has many correct answers.)

55. A monomial of degree 3

56. A trinomial of degree 4 and leading coefficient of -2

57. A binomial of degree 2 and leading coefficient of 8

58. A polynomial of degree 4 that has three terms and a leading coefficient of -8

In Exercises 59–66, use a horizontal format to perform the addition.

59. $(8 - t^4) + (5 + t^4)$

60. $(z^3 + 6z - 2) + (3z^2 - 6z)$

61. $x^2 + [3 + (x - 4x^2)]$

62. $(3a^2 + 5a) + (7 - a^2 - 5a) + (2a^2 + 8)$

63. $(x^2 - 3x + 8) + (2x^2 - 4x) + 3x^2$

64. $(5y + 6) + (4y^2 - 6y - 3)$

65. Add $3x^2 + 8$ to $7 - 5x^2$.

66. Add $3t^3 - 4t$ to $t^2 + 2t$.

In Exercises 67–70, use a vertical format to find the sum.

67. $\begin{array}{r} 3x^4 - 2x^2 - 9 \\ -5x^4 + \ x^2 \quad\quad \\ \hline \end{array}$

68. $\begin{array}{r} 4x^3 + 8x^2 - 5x + 3 \\ x^3 - 3x^2 \quad\quad - 7 \\ \hline \end{array}$

69. $(5p^2 - 4p + 2) + (-3p^2 + 2p - 7)$

70. $(16 - 32t) + (64 + 48t - 16t^2)$

In Exercises 71–78, use a horizontal format to find the difference.

71. $(4 - y^3) - (4 + y^3)$

72. $(6t^3 - 12) - (-t^3 + t - 2)$

73. $(5q^2 - 3q + 5) - (4q^2 - 3q - 10)$

74. $(-10s^2 - 5) - (2s^2 + 6s)$

75. $(x^3 - 3x) - [3x^3 - (x^2 + 5x)]$

76. $(5y^2 - 2y) - [(y^2 + y) - (3y^2 - 6y + 2)]$

77. Subtract $6x^3 - x + 11$ from $10x^3 + 15$.

78. Subtract $2y^5 + 3y^2$ from $2y^3 + y^2$.

In Exercises 79 and 80, use a vertical format to find the difference.

79. $(25 - 15x - 2x^3) - (12 - 13x + 2x^3)$

80. $(z^2 - 5) - (z^2 - 5)$

In Exercises 81–88, perform the operations.

81. $(4x^2 + 5x - 6) - (2x^2 - 4x + 5)$

82. $(13x^3 - 9x^2 + 4x - 5) - (5x^3 + 7x + 3)$

83. $(10x^2 - 11) - (-7x^3 - 12x^2 - 15)$

84. $(15y^4 - 18y - 18) - (-11y^4 - 8y - 8)$

85. $5s - [6s - (30s + 8)]$

86. $3x^2 - 2[3x + (9 - x^2)]$

87. $2(t^2 + 12) - 5(t^2 + 5) + 6(t^2 + 5)$

88. $-10(v + 2) + 8(v - 1) - 3(v - 9)$

In Exercises 89 and 90, evaluate the polynomial for the values of the variable.

89. $-4x^3 + 16x - 16$ (a) $x = -1$ (b) $x = 0$
                               (c) $x = 2$ (d) $x = \frac{5}{2}$

90. $3t^4 + 4t^3$ (a) $t = -1$ (b) $t = -\frac{2}{3}$
                     (c) $t = 0$ (d) $t = 1$

Free-Falling Object In Exercises 91 and 92, use the position polynomial to find the height of the free-falling object at the specified times. Then write a paragraph that describes the path of the object.

91. $h = -16t^2 + 80t + 50$ (a) $t = 0$ (b) $t = 2$
                                (c) $t = 4$ (d) $t = 5$

92. $h = -16t^2 + 128$ (a) $t = \frac{1}{2}$ (b) $t = 1$
                            (c) $t = \frac{3}{2}$ (d) $t = 2$

Free-Falling Object In Exercises 93–96, use the position polynomial to determine whether the free-falling object was dropped, was thrown upward, or was thrown downward. Also determine the height of the object at time $t = 0$.

93. $h = -16t^2 + 100$

94. $h = -16t^2 + 50t$

95. $h = -16t^2 - 24t + 50$

96. $h = -16t^2 + 32t + 300$

97. *Free-Falling Object*   An object is thrown upward from the top of a 200-foot building (see figure). The initial velocity is 40 feet per second. Use the position polynomial

$$h = -16t^2 + 40t + 200$$

to find the height of the object when $t = 1$, $t = 2$, and $t = 3$.

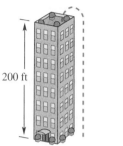

200 ft

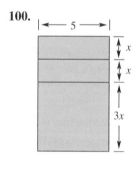

200 ft

Figure for 97   **Figure for 98**

98. *Free-Falling Object*   An object is dropped from a hot air balloon that is 200 feet above the ground (see figure). Use the position polynomial

$$h = -16t^2 + 200$$

to find the height of the object when $t = 1$, $t = 2$, and $t = 3$.

Geometry   In Exercises 99 and 100, find the area of the entire region.

99.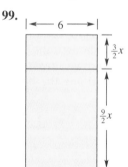

100.

$\boxed{101.}$ *Consumption of Milk*   The per capita consumption (average consumption per person) of whole milk and lowfat milk in the United States from 1970 to 1991 can be approximated by the following two polynomial models.

$$y = 25.23 - 0.84t + 0.006t^2 \qquad \text{Whole milk}$$

$$y = 4.52 + 0.54t - 0.007t^2 \qquad \text{Lowfat milk}$$

In these models, y represents per capita consumption in gallons and t represents the year, with $t = 0$ corresponding to 1970.   (Source: U.S. Department of Agriculture)

(a) Find a polynomial model that represents the per capita consumption of milk (both types) during this time period. Use this model to find the per capita consumption of milk in 1975 and in 1985.

(b) During the given period, the per capita consumption of whole milk was decreasing and the per capita consumption of lowfat milk was increasing (see figure). Use a graphing utility to graph the model for the per capita consumption of milk (both types). Is this consumption increasing or decreasing?

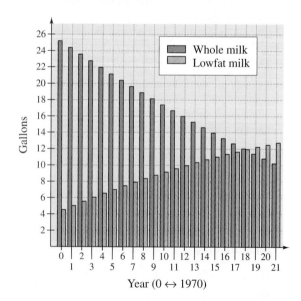

3.2   # Exponents and Multiplying Polynomials

Properties of Exponents ▪ Multiplying Polynomials ▪
Special Products ▪ Application

Properties of Exponents

In the Prerequisites chapter, you reviewed two properties of exponents:

$$a^m \cdot a^n = a^{m+n} \quad \text{and} \quad (ab)^m = a^m \cdot b^m.$$

The following summary lists these and three additional properties.

Properties of Exponents

Let m and n be positive integers, and let a and b represent real numbers, variables, or algebraic expressions.

Property	*Example*
1. $a^m \cdot a^n = a^{m+n}$	$x^3(x^2) = x^{3+2} = x^5$
2. $(ab)^m = a^m \cdot b^m$	$(2x)^3 = 2^3(x^3) = 8x^3$
3. $(a^m)^n = a^{mn}$	$(x^2)^3 = x^{2 \cdot 3} = x^6$
4. $\dfrac{a^m}{a^n} = a^{m-n}, \quad m > n, \quad a \neq 0$	$\dfrac{x^4}{x^2} = x^{4-2} = x^2$
5. $\left(\dfrac{a}{b}\right)^m = \dfrac{a^m}{b^m}, \quad b \neq 0$	$\left(\dfrac{x}{2}\right)^3 = \dfrac{x^3}{2^3} = \dfrac{x^3}{8}$

The validity of each property is illustrated in Examples 1 and 2.

EXAMPLE 1   *Illustrating the Properties of Exponents*

a. To multiply exponential expressions that have the *same base*, add exponents.

$$x^2 \cdot x^3 = \underbrace{x \cdot x}_{\text{2 factors}} \cdot \underbrace{x \cdot x \cdot x}_{\text{3 factors}} = \underbrace{x \cdot x \cdot x \cdot x \cdot x}_{\text{5 factors}} = x^{2+3} = x^5$$

b. To raise the product of two factors to the *same power*, raise each factor to the power and multiply.

$$(3x)^3 = \underbrace{3x \cdot 3x \cdot 3x}_{\text{3 factors}} = \underbrace{3 \cdot 3 \cdot 3}_{\text{3 factors}} \cdot \underbrace{x \cdot x \cdot x}_{\text{3 factors}} \cdot = 3^3 \cdot x^3 = 27x^3$$

EXAMPLE 2 Illustrating the Properties of Exponents

a. To raise an exponential expression to a power, multiply the powers.

$$(x^3)^2 = \underbrace{(x \cdot x \cdot x)}_{3 \text{ factors}} \cdot \underbrace{(x \cdot x \cdot x)}_{3 \text{ factors}} = \underbrace{(x \cdot x \cdot x \cdot x \cdot x \cdot x)}_{6 \text{ factors}} = x^{3 \cdot 2} = x^6$$

b. To divide exponential expressions that have the same base, *subtract* exponents.

$$\frac{x^4}{x^2} = \frac{\overbrace{x \cdot x \cdot x \cdot x}^{4 \text{ factors}}}{\underbrace{x \cdot x}_{2 \text{ factors}}} = x^{4-2} = x^2$$

c. To raise the quotient of two expressions to a power, raise each expression to the power and divide.

$$\left(\frac{x}{3}\right)^3 = \frac{x}{3} \cdot \frac{x}{3} \cdot \frac{x}{3} = \frac{\overbrace{x \cdot x \cdot x}^{3 \text{ factors}}}{\underbrace{3 \cdot 3 \cdot 3}_{3 \text{ factors}}} = \frac{x^3}{3^3} = \frac{x^3}{27}$$

The next example shows how the properties of exponents can be used to simplify expressions that involve products, powers, and quotients.

EXAMPLE 3 Applying Properties of Exponents

Simplify each expression.

a. $(x^2 y^4)(3x)$ **b.** $(-2y^2)^3$ **c.** $\dfrac{14a^5 b^3}{7a^2 b^2}$ **d.** $\left(\dfrac{x^2}{2y}\right)^3$ **e.** $\dfrac{x^{n+2} y^{3n}}{x^2 y^n}$

Solution

a. $(x^2 y^4)(3x) = 3(x^2 \cdot x)(y^4) = 3x^3 y^4$

b. $(-2y^2)^3 = (-2)^3 (y^2)^3 = -8y^6$

c. $\dfrac{14a^5 b^3}{7a^2 b^2} = 2(a^{5-2})(b^{3-2}) = 2a^3 b$

d. $\left(\dfrac{x^2}{2y}\right)^3 = \dfrac{(x^2)^3}{(2y)^3} = \dfrac{x^6}{2^3 y^3} = \dfrac{x^6}{8y^3}$

e. $\dfrac{x^{n+2} y^{3n}}{x^2 y^n} = x^{(n+2)-2} y^{3n-n} = x^n y^{2n}$

Multiplying Polynomials

The simplest type of polynomial multiplication involves a monomial multiplier. The product is obtained by direct application of the Distributive Property. For instance, to multiply the monomial $3x$ by the polynomial $(2x^2 - 5x + 3)$, multiply *each* term of the polynomial by $3x$.

$$(3x)(2x^2 - 5x + 3) = (3x)(2x^2) - (3x)(5x) + (3x)(3)$$
$$= 6x^3 - 15x^2 + 9x$$

EXAMPLE 4   *Finding Products with Monomial Multipliers*

a. $(2x - 7)(-3x) = 2x(-3x) - 7(-3x)$ Distributive Property

$$= -6x^2 + 21x$$ Properties of exponents

b. $4x^2(3x - 2x^3 + 1)$

$$= 4x^2(3x) - 4x^2(2x^3) + 4x^2(1)$$ Distributive Property

$$= 12x^3 - 8x^5 + 4x^2$$ Properties of exponents

$$= -8x^5 + 12x^3 + 4x^2$$ Standard form

c. $(-x)(5x^2 - x) = (-x)(5x^2) - (-x)(x)$ Distributive Property

$$= -5x^3 + x^2$$ Properties of exponents

To multiply two binomials, you can use both (left and right) forms of the Distributive Property. For example, if you treat the binomial $(2x + 7)$ as a single quantity, you can multiply $(3x - 2)$ by $(2x + 7)$ as follows.

$$(3x - 2)(2x + 7) = 3x(2x + 7) - 2(2x + 7)$$
$$= (3x)(2x) + (3x)(7) - (2)(2x) - 2(7)$$
$$= 6x^2 + 21x - 4x - 14$$

Product of First terms	Product of Outer terms	Product of Inner terms	Product of Last terms

$$= 6x^2 + 17x - 14$$

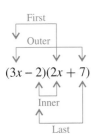

First
Outer
$(3x - 2)(2x + 7)$
Inner
Last

The FOIL Method

With practice you should be able to multiply two binomials without writing out all of the above steps. In fact, the four products in the boxes above suggest that the product of two binomials can be written in just one step, which is called the **FOIL Method.** Note that the words *first, outer, inner,* and *last* refer to the positions of the terms in the original product, as shown at the left.

EXAMPLE 5 *Multiplying Binomials (FOIL Method)*

Use the FOIL Method to multiply the following.

a. $(3x + 4)(2x + 1)$ **b.** $(x - 3)(x + 3)$

Solution

a. $(3x + 4)(2x + 1) = 6x^2 + 3x + 8x + 4$
$$= 6x^2 + 11x + 4 \qquad \text{Combine like terms.}$$

b. $(x - 3)(x + 3) = x^2 + 3x - 3x - 9$
$$= x^2 - 9 \qquad \text{Combine like terms.}$$

When multiplying polynomials that have three or more terms, use the same basic principle as for multiplying monomials and binomials. That is, *each term of one polynomial must be multiplied by each term of the other polynomial.* This can be done using either a horizontal or vertical format.

EXAMPLE 6 *Multiplying Polynomials (Horizontal Format)*

$(4x^2 - 3x - 1)(2x - 5)$

$= (4x^2 - 3x - 1)(2x) - (4x^2 - 3x - 1)(5)$ Distributive Property

$= 8x^3 - 6x^2 - 2x - (20x^2 - 15x - 5)$ Distributive Property

$= 8x^3 - 6x^2 - 2x - 20x^2 + 15x + 5$ Subtract (change signs).

$= 8x^3 - 26x^2 + 13x + 5$ Combine like terms.

STUDY TIP

When multiplying two polynomials, it is best to write each in standard form before using either the horizontal or the vertical format.

EXAMPLE 7 *Multiplying Polynomials (Vertical Format)*

$$
\begin{array}{r}
4x^2 + x - 2 \\
\times \quad -x^2 + 3x + 5 \\
\hline
20x^2 + 5x - 10 \\
12x^3 + 3x^2 - 6x \\
-4x^4 - x^3 + 2x^2 \\
\hline
-4x^4 + 11x^3 + 25x^2 - x - 10
\end{array}
$$

Standard form
Standard form

$5(4x^2 + x - 2)$
$3x(4x^2 + x - 2)$
$-x^2(4x^2 + x - 2)$

Special Products

Some binomial products, such as that in Example 5(b), have special forms that occur frequently in algebra.

Special Products

Let u and v be real numbers, variables, or algebraic expressions. Then the following formulas are true.

| *Special Product* | *Example* |

Sum and Difference of Two Terms

$$(u + v)(u - v) = u^2 - v^2 \qquad (3x - 4)(3x + 4) = 9x^2 - 16$$

NOTE When squaring a binomial, the resulting middle term is always *twice* the product of the two terms.

Square of a Binomial

$$(u + v)^2 = u^2 + 2uv + v^2 \qquad (2x + 5)^2 = 4x^2 + 2(2x)(5) + 25$$
$$= 4x^2 + 20x + 25$$

$$(u - v)^2 = u^2 - 2uv + v^2 \qquad (x - 6)^2 = x^2 - 2(x)(6) + 36$$
$$= x^2 - 12x + 36$$

EXAMPLE 8   *Product of the Sum and Difference of Two Terms*

a. $(3x - 2)(3x + 2) = (3x)^2 - 4^2 = 9x^2 - 16$

b. $(6 + 5x)(6 - 5x) = 6^2 - (5x)^2 = 36 - 25x^2$

EXAMPLE 9   *Squaring a Binomial*

$$(2x - 7)^2 = (2x)^2 - 2(2x)(7) + 7^2 = 4x^2 - 28x + 49$$

EXAMPLE 10   *Cubing a Binomial*

$$(x - 4)^3 = (x - 4)^2(x - 4)$$
$$= (x^2 - 8x + 16)(x - 4)$$
$$= x^2(x - 4) - 8x(x - 4) + 16(x - 4)$$
$$= x^3 - 4x^2 - 8x^2 + 32x + 16x - 64$$
$$= x^3 - 12x^2 + 48x - 64$$

Application

EXAMPLE 11 Finding Area and Volume

The closed box shown in Figure 3.2 has sides whose lengths (in inches) are consecutive integers. Find a polynomial that describes the volume of the box. What is the volume if the length of the shortest side is 4 inches?

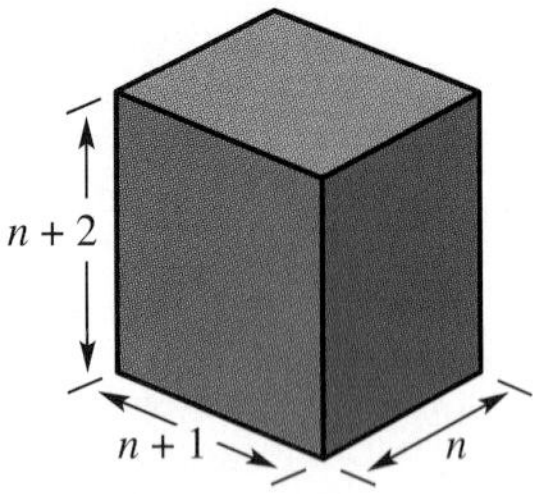

FIGURE 3.2

Solution

Verbal Model:

| Volume | = | Length | · | Width | · | Height |

Labels:
Length $= n$ (inches)
Width $= n + 1$ (inches)
Height $= n + 2$ (inches)
Volume $= V$ (cubic inches)

Equation:
$$V = n(n + 1)(n + 2)$$
$$= n(n^2 + 3n + 2)$$
$$= n^3 + 3n^2 + 2n$$

If $n = 4$, the volume of the box is

$$V = (4)^3 + 3(4)^2 + 2(4) = 64 + 48 + 8 = 120 \text{ cubic inches.}$$

Group Activities Extending the Concept

Investigating a Demand Function A company determines that the number of units of a product that it can sell depends on the price of the product. A model for this relationship is

$$p = 24 - 0.001x$$

where p is the price (in dollars) and x is the number of units.

a. Sketch the graph of this model and use the result to describe the relationship between the price and the number of units sold.

b. The revenue obtained from selling x units is given by $R = xp$. Sketch the graph of the revenue function and describe the relationship between the number of units sold and the revenue.

3.2 Exercises

Discussing the Concepts

1. Write, from memory, the rules of exponents.

2. Discuss the difference between $(2x)^3$ and $2x^3$.

3. Give an example of how to use the Distributive Property to multiply two binomials.

4. Explain the meaning of each letter of FOIL as it relates to multiplying two binomials.

5. What is the degree of the product of two polynomials of degrees m and n?

6. *True or False?*   Decide whether the statement is true or false. Explain your reasoning.

   (a) The product of two monomials is a monomial.

   (b) The product of two binomials is a binomial.

Problem Solving

In Exercises 7–14, use the expression to illustrate a property of exponents. Use Examples 1 and 2 as models.

7. $t^3 \cdot t^4$

8. $(2y)^3$

9. $\left(\dfrac{y}{5}\right)^4$

10. $\left(\dfrac{3}{t}\right)^5$

11. $\dfrac{x^6}{x^4}$

12. $\dfrac{2^4 y^5}{2^2 y^3}$

13. $\dfrac{3^7 x^5}{3^3 x^3}$

14. $\dfrac{4^3(x+2)^4}{4(x+2)^2}$

In Exercises 15–22, simplify the expressions.

15. (a) $-3x^3 \cdot x^5$ (b) $(-3x)^2 \cdot x^5$

16. (a) $5^2 y^4 \cdot y^2$ (b) $(5y)^2 \cdot y^4$

17. (a) $5x^2 y(-2y)^2(3)$ (b) $5x^2 y(-2y)^3(3)$

18. (a) $-(m^3 n^2)(mn^3)$ (b) $-(m^3 n^2)^2(-mn^3)$

19. (a) $\dfrac{12m^5 n^6}{9mn^3}$ (b) $\dfrac{-18m^3 n^6}{-6mn^3}$

20. (a) $-\dfrac{(-3x^2 y)^3}{9x^2 y^2}$ (b) $-\dfrac{(-2xy^3)^2}{6y^2}$

21. (a) $\dfrac{(-5u^3 v)^2}{10u^2 v}$ (b) $\dfrac{-5(u^3 v)^2}{10u^2 v}$

22. (a) $\left[\dfrac{(3x^2)(2x)^2}{(-2x)(6x)}\right]^2$ (b) $\left[\dfrac{(3x^2)(2x)^4}{(-2x)^2(6x)}\right]^2$

In Exercises 23–28, simplify the expression.

23. $(-2a^2)(-8a)$

24. $(-6n)(3n^2)$

25. $-2x(5+3x^2-7x^3)$

26. $-3a^2(11a-3)$

27. $5a(a+2)-3a(2a-3)$

28. $(2t-1)(t+1)+3(2t-5)$

In Exercises 29–32, use a horizontal format to perform the multiplication.

29. $(u+5)(2u^2+3u-4)$

30. $(x^2+4)(x^2-2x-4)$

31. $(x-1)(x^2-4x+6)$

32. $(2x^2-3)(2x^2-2x+3)$

In Exercises 33–36, use a vertical format to perform the multiplication.

33.
$$7x^2 - 14x + 9$$
$$\underline{\times \qquad\quad x + 3}$$

34.
$$4x^4 - 6x^2 + 9$$
$$\underline{\times \qquad\quad 2x^2 + 3}$$

35. $(u-2)(u^2+u+3)$

36. $(x+5)(x^2-x+6)$

In Exercises 37–42, expand the product.

37. $(x+2)(x-2)$

38. $(5u+12v)(5u-12v)$

39. $(x+5)^2$

40. $(3x-8)^2$

41. $[(x+2)+y]^2$

42. $[(x-4)-y]^2$

Geometry   In Exercises 43 and 44, write an expression that represents the area of the shaded portion of the figure. Then simplify the expression.

43.

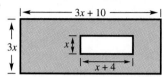

44.

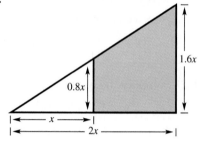

Geometrical Modeling   In Exercises 45 and 46, use the area model to write two different expressions for the total area. Then equate the two expressions and name the algebraic property that is illustrated.

45.

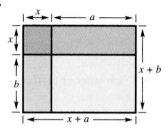

46.

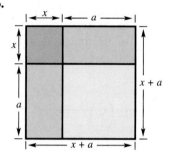

Reviewing the Major Concepts

In Exercises 47–50, plot the points on a rectangular coordinate system and find the slope of the line passing through the points.

47. $(-3, 2)$, $(5, -4)$ **48.** $(2, 8)$, $(7, -3)$

49. $\left(\frac{5}{2}, \frac{7}{2}\right)$, $\left(\frac{7}{3}, -2\right)$ **50.** $\left(-\frac{9}{4}, -\frac{1}{4}\right)$, $\left(-3, \frac{9}{2}\right)$

51. *Distance*   An automobile is traveling at an average speed of 48 miles per hour. Express the distance d traveled as a function of time t in hours.

52. *Geometry*   A square has side lengths of x centimeters. Express its area A as a function of x.

Additional Problem Solving

In Exercises 53–56, use the expression to illustrate a property of exponents. Model after Example 1.

53. $(-5x)^5$ **54.** $(x^2)^3$

55. $\dfrac{3^4 x^5}{3x^2}$ **56.** $\dfrac{2^3(u-3)^2}{2^2(u-3)}$

In Exercises 57–64, simplify the expressions.

57. (a) $(-5z)^3$ (b) $(-5z)^2$

58. (a) $-5z^3$ (b) $-z^2 \cdot z^2$

59. (a) $(-5z)^4$ (b) $-5z^4 \cdot (-5z)^4$

60. (a) $(3y)^3(2y^2)$ (b) $3y^3 \cdot 2y^2$

61. (a) $\dfrac{28x^2 y^3}{21xy^2}$ (b) $\dfrac{21xy^2}{28y}$

62. (a) $\dfrac{(-3xy)^3}{9xy^2}$ (b) $\dfrac{(-3xy)^4}{-3(xy)^2}$

63. (a) $\left(\dfrac{3x}{4y}\right)^2$ (b) $\left(\dfrac{5u}{3v}\right)^3$

64. (a) $\left(\dfrac{2a}{3y}\right)^5$ (b) $-\left(\dfrac{2a}{3y}\right)^2$

In Exercises 65–78, simplify the expression.

65. $5x(2x)^2$

66. $(3y)^3 \left(\frac{y}{2}\right)^2$

67. $2y(5-y)$

68. $5z(2z-7)$

69. $4x(2x^2-3x+5)$

70. $3y^2(-3y^2+7y-3)$

71. $-3x(-5x)(5x+2)$

72. $4t(-3t)(t^2-1)$

73. $\left(4y-\frac{1}{3}\right)(12y+9)$

74. $\left(5t-\frac{3}{4}\right)(2t-16)$

75. $(2x+y)(3x+2y)$

76. $(2x-y)(3x-2y)$

77. $(s-3t)(s+t)-(s-3t)(s-t)$

78. $2x(x-5y)+3y(x+5y)$

In Exercises 79–82, use a horizontal format to perform the multiplication.

79. $(x^3-3x+2)(x-2)$

80. $(t+3)(t^2-5t+1)$

81. $(t^2+t-2)(t^2-t+2)$

82. $(y^2+3y+5)(2y^2-3y-1)$

In Exercises 83–86, use a vertical format to perform the multiplication.

83.
$$\begin{array}{r} -x^2+2x-1 \\ \times \qquad 2x+1 \\ \hline \end{array}$$

84.
$$\begin{array}{r} 2s^2-5s+6 \\ \times \qquad 3s-4 \\ \hline \end{array}$$

85. $(x^3+x)(x^2-4)$

86. $(z-2)(z^2+z+1)$

In Exercises 87–98, use a special product formula to perform the multiplication.

87. $(2+7y)(2-7y)$

88. $(4+3z)(4-3z)$

89. $\left(2x-\frac{1}{4}\right)\left(2x+\frac{1}{4}\right)$

90. $\left(\frac{2}{3}x+7\right)\left(\frac{2}{3}x-7\right)$

91. $(0.2t+0.5)(0.2t-0.5)$

92. $(4a-0.1b)(4a+0.1b)$

93. $(x+10)^2$

94. $(x+9)^2$

95. $(2x-7y)^2$

96. $(5-3z)^2$

97. $[u-(v-3)]^2$

98. $[2z+(y+1)]^2$

In Exercises 99–104, simplify the expression.

99. $(t+3)^2-(t-3)^2$

100. $(a+6)^2+(a-6)^2$

101. $(x+3)^3$

102. $(y-2)^3$

103. $(u+v)^3$

104. $(u-v)^3$

Graphical and Algebraic Reasoning   In Exercises 105–108, use a graphing utility to graph the expressions for y_1 and y_2. What conclusion can you make? Verify your conclusion algebraically.

105. $y_1=(x+1)(x^2-x+2)$

$y_2=x^3+x+2$

106. $y_1=(x-3)^2$

$y_2=x^2-6x+9$

107. $y_1=(2x-3)(x+2)$

$y_2=2x^2+x-6$

108. $y_1=\left(x+\frac{1}{2}\right)\left(x-\frac{1}{2}\right)$

$y_2=x^2-\frac{1}{4}$

Geometry   In Exercises 109 and 110, find the area of the shaded region.

109.

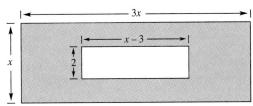

110.

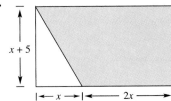

111. Given the function $f(x) = x^2 - 2x$, find and simplify each of the following.

(a) $f(t-3)$ (b) $f(2+h) - f(2)$

112. Given the function $f(x) = 2x^2 - 5x + 4$, find and simplify each of the following.

(a) $f(y+2)$ (b) $f(1+h) - f(1)$

113. *Geometry*   The length of a rectangle is $1\frac{1}{2}$ times its width w. Find (a) the perimeter and (b) the area of the rectangle.

114. *Geometry*   The base of a triangle is $3x$ and its height is $x + 5$. Find the area A of the triangle.

115. *Compound Interest*   After 2 years, an investment of $1000 compounded annually at an interest rate r will yield an amount $1000(1+r)^2$. Find this product.

116. *Compound Interest*   After 2 years, an investment of $1000 compounded annually at an interest rate of 9.5% will yield an amount $1000(1 + 0.095)^2$. Find this product.

117. *Finding a Pattern*   Perform the multiplications.

(a) $(x-1)(x+1)$

(b) $(x-1)(x^2+x+1)$

(c) $(x-1)(x^3+x^2+x+1)$

From the pattern formed by these products, can you predict the result of $(x-1)(x^4+x^3+x^2+x+1)$?

118. *Evaluating Expressions*   Verify that $(x+y)^2$ is not equal to $x^2 + y^2$ by letting $x = 3$ and $y = 4$ and evaluating both expressions.

Math Matters   The Rolling Coin Problem

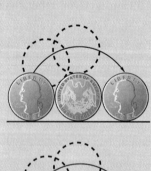

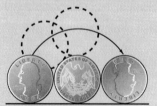

The history of mathematics contains many paradoxes—problems that appear to be contradictory or to have absurd solutions. Here is one problem that concerns two coins.

Suppose one coin is fixed to the edge of a table. A second coin is placed to the left of the fixed coin, with the head upright, as shown in the figure. If the second coin is rolled along the fixed coin with *no slippage*, will it arrive on the right side of the fixed coin with its head upright (upper figure) or turned down (lower figure)?

To many people, the correct answer appears to be the second possibility because, they reason, the coin rolls around half of its circumference, which should result in a rotation of 180°, and should therefore turn the head of the coin down. Which of these two possibilities seems most likely to you? Try the experiment using two quarters. What is the result? (The answer is given in the back of the book.)

3.3	**Factoring Polynomials: An Introduction**

Common Monomial Factors ▪ Factoring by Grouping ▪
Factoring Special Products ▪ Factoring Completely

Common Monomial Factors

In Section 3.2, you studied ways to multiply polynomials. In this section, you will study the reverse process—**factoring polynomials.** Here is an example.

Use Distributive Property to multiply.   *Use Distributive Property to factor.*
$$3x(4 - 5x) = 12x - 15x^2 \qquad 12x - 15x^2 = 3x(4 - 5x)$$

Notice that factoring changes a *sum of terms* into a *product of factors.*

To be efficient in factoring expressions, you need to understand the concept of the *greatest common factor* of two (or more) integers or terms. Recall from arithmetic that every integer can be factored into a product of prime numbers. The **greatest common factor** of two or more integers is the greatest integer that is a factor of each number.

EXAMPLE 1 *Finding the Greatest Common Factor*

Find the greatest common factor of $6x^5$, $30x^4$, and $12x^3$.

Solution

From the factorizations

$$6x^5 = 2 \cdot 3 \cdot x \cdot x \cdot x \cdot x \cdot x = (6x^3)(x^2)$$
$$30x^4 = 2 \cdot 3 \cdot 5 \cdot x \cdot x \cdot x \cdot x = (6x^3)(5x)$$
$$12x^3 = 2 \cdot 2 \cdot 3 \cdot x \cdot x \cdot x = (6x^3)(2)$$

you conclude that the greatest common factor is $6x^3$.

Consider the three terms given in Example 1 as terms of the polynomial

$$6x^5 + 30x^4 + 12x^3.$$

The common factor, $6x^3$, of these terms is the **greatest common monomial factor** of the polynomial. When you use the Distributive Property to remove this factor from each term of the polynomial, you are **factoring out** the greatest common monomial factor.

$$6x^5 + 30x^4 + 12x^3 = 6x^3(x^2) + 6x^3(5x) + 6x^3(2) \qquad \text{Factor each term.}$$
$$= 6x^3(x^2 + 5x + 2) \qquad \text{Factor out common monomial factor.}$$

EXAMPLE 2   *Factoring Out the Greatest Common Monomial Factor*

Factor the polynomial $24x^3 - 32x^2$.

Solution

The greatest common factor of $24x^3$ and $32x^2$ is $8x^2$.

$$24x^3 - 32x^2 = (8x^2)(3x) - (8x^2)(4)$$
$$= 8x^2(3x - 4)$$

If a polynomial in x with integer coefficients has a greatest common monomial factor of the form ax^n, the following statements must be true.

1. The coefficient, a, of the greatest common monomial factor must be the greatest integer that *divides* each of the coefficients in the polynomial.

2. The variable factor, x^n, of the greatest common monomial factor has the same power as the lowest power of x of all terms of the polynomial.

The greatest common monomial factor of a polynomial is usually considered to have a positive coefficient. However, sometimes it is convenient to factor a negative number out of a polynomial. You can see how this is done in the next example.

STUDY TIP

Whenever you are factoring a polynomial, remember that you can check your results by multiplying. That is, if you multiply the factors, you should obtain the original polynomial.

EXAMPLE 3   *A Negative Common Monomial Factor*

Factor the polynomial $-3x^2 + 12x - 18$ in two ways.

a. Factor out 3.   **b.** Factor out -3.

Solution

a. To factor out the common monomial factor of 3, write the following.

$$-3x^2 + 12x - 18 = 3(-x^2) + 3(4x) + 3(-6)$$
$$= 3(-x^2 + 4x - 6)$$

Check this result by multiplying.

b. To factor -3 out of the polynomial, write the following.

$$-3x^2 + 12x - 18 = -3(x^2) + (-3)(-4x) + (-3)(6)$$
$$= -3(x^2 - 4x + 6)$$

Check this result by multiplying.

Factoring by Grouping

Some polynomials have common factors that are not simple monomials. For instance, the polynomial

$$x^2(2x - 3) + 4(2x - 3)$$

has the common *binomial* factor $(2x - 3)$. Factoring out this common factor produces

$$x^2(2x - 3) + 4(2x - 3) = (2x - 3)(x^2 + 4).$$

This type of factoring is part of a procedure called **factoring by grouping.**

EXAMPLE 4   *Common Binomial Factors*

Factor $5x^2(6x - 5) - 2(6x - 5)$.

Solution

Each of the terms of this polynomial has a binomial factor of $(6x - 5)$. Factoring this binomial out of each term produces the following.

$$5x^2(6x - 5) - 2(6x - 5) = (6x - 5)(5x^2 - 2)$$

In Example 4, the given polynomial was already grouped so that it was easy to determine the common binomial factor. In practice, you will have to do the grouping as well as the factoring.

EXAMPLE 5   *Factoring By Grouping*

Factor $x^3 - 5x^2 + x - 5$.

Solution

By grouping the first two terms together and the third and fourth terms together, you obtain the following.

$$\begin{aligned}
x^3 - 5x^2 + x - 5 &= (x^3 - 5x^2) + (x - 5) \\
&= x^2(x - 5) + (x - 5) \\
&= (x - 5)(x^2 + 1)
\end{aligned}$$

You can *check* to see that you have factored the expression correctly by multiplying out the factors and comparing the result with the original expression.

DISCOVERY

Use your calculator to verify the special polynomial form called the "difference of two squares." To do so, evaluate the equation when $u = 8$ and $v = 4$. Try more values, including negative values. What can you conclude?

Factoring Special Products

Some polynomials have special forms that you should learn to recognize so that you can factor them easily. One of the easiest special polynomial forms to recognize and to factor is the form $u^2 - v^2$, called a **difference of two squares.**

Difference of Two Squares

Let u and v be real numbers, variables, or algebraic expressions. Then the expression $u^2 - v^2$ can be factored as follows.

$$u^2 - v^2 = (u + v)(u - v)$$

Difference   Opposite signs

To recognize perfect squares, look for coefficients that are squares of integers and for variables raised to *even* powers.

EXAMPLE 6   *Factoring the Difference of Two Squares*

a. $x^2 - 64 = x^2 - 8^2$ Write as difference of two squares.

$= (x + 8)(x - 8)$ Factored form

b. $49x^2 - 81 = (7x)^2 - 9^2$ Write as difference of two squares.

$= (7x + 9)(7x - 9)$ Factored form

Remember that the rule $u^2 - v^2 = (u + v)(u - v)$ applies to polynomials or expressions in which u and v are themselves expressions.

EXAMPLE 7   *Factoring the Difference of Two Squares*

Factor $(x + 2)^2 - 9$.

Solution

$(x + 2)^2 - 9 = (x + 2)^2 - 3^2$ Write as difference of two squares.

$= [(x + 2) + 3][(x + 2) - 3]$ Factored form

$= (x + 5)(x - 1)$ Simplify.

To check this result, write the original polynomial in standard form. Then multiply the factored form to see that you obtain the standard form.

Sum and Difference of Two Cubes

Let u and v be real numbers, variables, or algebraic expressions. Then the expressions $u^3 + v^3$ and $u^3 - v^3$ can be factored as follows.

$$1.\ u^3 + v^3 = (u + v)(u^2 - uv + v^2)$$

Like signs / Unlike signs

$$2.\ u^3 - v^3 = (u - v)(u^2 + uv + v^2)$$

Like signs / Unlike signs

EXAMPLE 8   Factoring Sums and Differences of Cubes

Factor each polynomial.

a. $x^3 - 125$ **b.** $8y^3 + 1$

Solution

a. This polynomial is the difference of two cubes because x^3 is the cube of x and 125 is the cube of 5.

$$x^3 - 125 = x^3 - 5^3 \qquad \text{Difference of two cubes}$$
$$= (x - 5)(x^2 + 5x + 5^2) \qquad \text{Factored form}$$
$$= (x - 5)(x^2 + 5x + 25) \qquad \text{Simplify.}$$

b. This polynomial is the sum of two cubes because $8y^3$ is the cube of $2y$ and 1 is the cube of 1.

$$8y^3 + 1 = (2y)^3 + 1^3 \qquad \text{Sum of two cubes}$$
$$= (2y + 1)[(2y)^2 - (2y)(1) + 1^2] \qquad \text{Factored form}$$
$$= (2y + 1)(4y^2 - 2y + 1) \qquad \text{Simplify.}$$

You can check the results of Example 8 by multiplying, as follows.

$$
\begin{array}{r}
x^2 + 5x + 25 \\
\times \qquad\quad x - 5 \\
\hline
-5x^2 - 25x - 125 \\
x^3 + 5x^2 + 25x \qquad\quad \\
\hline
x^3 \qquad\qquad\qquad - 125
\end{array}
$$

$$
\begin{array}{r}
4y^2 - 2y + 1 \\
\times \qquad\quad 2y + 1 \\
\hline
4y^2 - 2y + 1 \\
8y^3 - 4y^2 + 2y \qquad\quad \\
\hline
8y^3 \qquad\qquad\qquad + 1
\end{array}
$$

Factoring Completely

Sometimes the difference of two squares can be hidden by the presence of a common monomial factor. Remember that with *all* factoring techniques, you should first remove any common monomial factors.

EXAMPLE 9   Removing a Common Monomial Factor First

Factor $125x^2 - 80$.

Solution

$$125x^2 - 80 = 5(25x^2 - 16)$$ Remove common monomial factor.

$$= 5[(5x)^2 - 4^2]$$ Write as difference of two squares.

$$= 5(5x + 4)(5x - 4)$$ Factored form

The polynomial in Example 9 is said to be **completely factored** because none of its factors can be further factored using integer coefficients.

EXAMPLE 10   Factoring Completely

NOTE   Note in Example 10 that the *sum of two squares* cannot be factored further using integer coefficients. Such polynomials are called **prime** with respect to the integers. For instance, $x^2 + 4$ and $4x^2 + 9$ are both prime.

a. $x^4 - y^4 = (x^2 + y^2)(x^2 - y^2)$ Factor as difference of two squares.

$$= (x^2 + y^2)(x^2 - y^2)$$ Find second difference of two squares.

$$= (x^2 + y^2)(x + y)(x - y)$$ Factored completely

b. $81m^4 - 1 = (9m^2 + 1)(9m^2 - 1)$ Factor as difference of two squares.

$$= (9m^2 + 1)(9m^2 - 1)$$ Find second difference of two squares.

$$= (9m^2 + 1)(3m + 1)(3m - 1)$$ Factored completely

Group Activities   Problem Solving

Position Polynomials   Write a position polynomial for the height of an object thrown upward from ground level at initial velocities (in feet per second) of (a) 80, (b) 15, and (c) 44. Completely factor the polynomials and demonstrate their equivalence to the original polynomials.

3.3 Exercises

Discussing the Concepts

1. Explain what is meant by saying that a polynomial is in factored form?

2. How do you check your result after factoring a polynomial?

3. Describe a method of finding the greatest common factor of two or more integers.

4. Explain how the word *factor* can be used as a noun and as a verb.

5. Give an example of using the Distributive Property to factor a polynomial.

6. Give an example of a polynomial that is prime with respect to the integers.

Problem Solving

In Exercises 7–10, find the greatest common factor of the expressions.

7. $48, 90$

8. $36, 150, 100$

9. $28b^2, 14b^3, 42b^5$

10. $16x^2y, 84xy^2, 36x^2y^2$

In Exercises 11–16, factor out the greatest common monomial factor. (Some of the polynomials may have no common monomial factor other than 1 and −1.)

11. $8z - 8$

12. $5x + 5$

13. $21u^2 - 14u$

14. $36y^4 + 24y^2$

15. $15xy^2 - 3x^2y + 9xy$

16. $4x^2 - 2xy + 3y^2$

In Exercises 17–20, factor a negative real number out of the polynomial and then write the polynomial factor in standard form.

17. $10 - x$

18. $32 - x^4$

19. $y - 3y - 2y^2$

20. $-2t^3 + 4t^2 + 7$

In Exercises 21–24, factor the expression.

21. $\frac{3}{2}x + \frac{5}{4} = \frac{1}{4}()$

22. $\frac{1}{3}x - \frac{5}{6} = \frac{1}{6}()$

23. $\frac{5}{8}x + \frac{5}{16}y = \frac{5}{16}()$

24. $\frac{7}{12}u - \frac{21}{8}v = \frac{7}{24}()$

In Exercises 25–28, factor the polynomial by factoring out the greatest common factor.

25. $2y(y - 3) + 5(y - 3)$

26. $7t(s + 9) - 6(s + 9)$

27. $a(a + 6) - a^2(a + 6)$

28. $(5x + y)(x - y) - 5x(x - y)$

In Exercises 29–32, factor the expression by grouping.

29. $x^2 + 25x + x + 25$

30. $x^2 - 7x + x - 7$

31. $a^3 - 4a^2 + 2a - 8$

32. $3s^3 + 6s^2 + 5s + 10$

In Exercises 33–38, factor the difference of two squares.

33. $x^2 - 64$

34. $y^2 - 144$

35. $4z^2 - y^2$

36. $9u^2 - \frac{1}{4}v^2$

37. $(x - 1)^2 - 16$

38. $36 - (y - 6)^2$

In Exercises 39–42, factor the sum or difference of cubes.

39. $x^3 - 8$

40. $t^3 - 27$

41. $x^3 + 64y^3$

42. $u^3 + 125v^3$

In Exercises 43–46, factor the polynomial completely.

43. $3x^4 - 300x^2$

44. $8x^3 - 64$

45. $2a^4 - 32$

46. $625 - u^4$

47. *Revenue* The revenue from selling x units of a product at a price of p dollars per unit is given by $R = xp$. For a particular commodity, the revenue is

$R = 800x - 0.25x^2$.

Factor this expression and determine an expression that gives the price in terms of x.

48. *Geometry* The surface area of a right circular cylinder is $S = \pi r^2 + 2\pi rh$ (see figure). Factor the expression for the surface area.

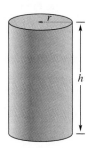

Figure for 48

Reviewing the Major Concepts

In Exercises 49–52, simplify the expression by combining like terms.

49. $\frac{3}{8}x - \frac{1}{12}x + 8$

50. $6y - 3x + 3x - 10y$

51. $3x^2 - 5x + 3 + 28x - 33x^2$

52. $4x^3 - 3x^2y + 4xy^2 + 15x^2y + y^3$

In Exercises 53 and 54, write an absolute value inequality that represents the verbal statement.

53. The set of all real numbers x whose distance from 0 is less than 5

54. The set of all real numbers x whose distance from 6 is more than 3

Additional Problem Solving

In Exercises 55–60, find the greatest common factor of the expressions.

55. $3x^2, 12x$

56. $27x^4, 18x^3$

57. $30z^2, -12z^3$

58. $-45y, 150y^3$

59. $42(x+8)^2, 63(x+8)^3$

60. $66(3-y), 44(3-y)^2$

In Exercises 61–74, factor out the greatest common monomial factor. (Some of the polynomials may have no common monomial factor other than 1 and −1.)

61. $4u + 10$

62. $-15t - 10$

63. $24x^2 - 18$

64. $14z^3 + 21$

65. $2x^2 + x$

66. $-a^3 - 4a$

67. $11u^2 + 9$

68. $16x^2 - 3y^3$

69. $3x^2y^2 - 15y$

70. $4uv + 6u^2v^2$

71. $28x^2 + 16x - 8$

72. $9 - 27y - 15y^2$

73. $14x^4 + 21x^3 + 9x^2$

74. $17x^5y^3 - xy^2 + 34y^2$

In Exercises 75–78, factor a negative real number out of the polynomial and then write the polynomial factor in standard form.

75. $7 - 14x$

76. $15 - 5x$

77. $5 + 4x - x^2$

78. $12x - 6x^2 - 18$

In Exercises 79–82, factor the polynomial.

79. $\frac{2}{3}t + \frac{3}{4} = \frac{1}{12}(\quad)$

80. $\frac{3}{4}x - \frac{2}{5} = \frac{1}{20}(\quad)$

81. $2y - \frac{3}{5} = \frac{1}{5}(\quad)$

82. $3z + \frac{3}{8} = \frac{1}{8}(\quad)$

In Exercises 83–86, factor the polynomial by factoring out the greatest common factor.

83. $5x(x+2) - 3(x+2)$

84. $6(t-3) - 5t(t-3)$

85. $(10+b)(c+7) - (10+b)(c+7)^2$

86. $y^4(y-12)^2 + y^2(y-12)^3$

In Exercises 87–92, factor by grouping.

87. $y^2 - 6y + 2y - 12$ **88.** $y^2 + 3y + 4y + 12$

89. $x^3 + 2x^2 + x + 2$ **90.** $t^3 - 11t^2 + t - 11$

91. $z^4 + 3z^3 - 2z - 6$ **92.** $4u^4 - 2u^3 - 6u + 3$

In Exercises 93–100, factor the difference of two squares.

93. $16y^2 - 9$ **94.** $9z^2 - 25$

95. $225x^2 - 9y^2$ **96.** $625a^2 - 49b^2$

97. $u^2 - \frac{1}{16}$ **98.** $v^2 - \frac{9}{25}$

99. $81 - (z + 5)^2$ **100.** $(x - 3)^2 - 4$

In Exercises 101–106, factor the sum or difference of cubes.

101. $a^3 + b^3$ **102.** $m^3 - 8n^3$

103. $8t^3 - 27$ **104.** $27s^3 + 64$

105. $27u^3 + 1$ **106.** $64v^3 - 125$

In Exercises 107–112, factor completely.

107. $8 - 50x^2$ **108.** $a^3 - 16a$

109. $y^4 - 81$ **110.** $u^4 - 256$

111. $2x^3 - 54$ **112.** $5y^3 - 625$

Graphical Reasoning   In Exercises 113–116, use a graphing utility to compare the two graphs. What can you conclude?

113. $y_1 = 3x - 6$, $y_2 = 3(x - 2)$

114. $y_1 = x^3 - 2x^2$, $y_2 = x^2(x - 2)$

115. $y_1 = x^2 - 4$, $y_2 = (x + 2)(x - 2)$

116. $y_1 = x(x + 1) - 4(x + 1)$, $y_2 = (x + 1)(x - 4)$

Mental Math   In Exercises 117 and 118, evaluate the quantity mentally using the sample as a model.

$48 \cdot 52 = (50 - 2)(50 + 2) = 50^2 - 2^2 = 2496$

117. $79 \cdot 81$ **118.** $18 \cdot 22$

Think About It   In Exercises 119 and 120, show all the different groupings that can be used to completely factor the polynomial. Carry out the various factorizations to show that they yield the same result.

119. $3x^3 + 4x^2 - 3x - 4$ **120.** $6x^3 - 8x^2 + 9x - 12$

121. *Simple Interest*   The total amount of money from a principal of P invested at $r\%$ simple interest for t years is given by $P + Prt$. Factor this expression.

122. *Revenue*   The revenue from selling x units of a product at a price of p dollars per unit is given by $R = xp$. For a particular commodity, the revenue is

$$R = 1000x - 0.4x^2.$$

Factor the expression for the revenue and determine an expression that gives the price in terms of x.

123. *Geometry*   The area of a rectangle of length l is given by $45l - l^2$. Factor this expression to determine the width of the rectangle.

124. *Geometry*   The area of a rectangle of width w is given by $32w - w^2$. Factor this expression to determine the length of the rectangle.

125. *Product Design*   A washer on the drive train of a car has an inside radius of r centimeters and an outside radius of R centimeters (see figure). Find the area of one of the flat surfaces of the washer and express the area in factored form.

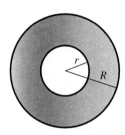

126. *Chemical Reaction*   The rate of change of a chemical reaction is given by $kQx - kx^2$, where Q is the amount of the original substance, x is the amount of substance formed, and k is a constant of proportionality. Factor the expression for this rate of change.

MID-CHAPTER QUIZ

Take this quiz as you would take a quiz in class. After you are done, check your work against the answers given in the back of the book.

1. Determine the degree and leading coefficient of the polynomial

    $3 - 2x + 4x^3 - 2x^4$.

2. Explain why $2x - 3x^{1/2} + 5$ is not a polynomial.

In Exercises 3–14, perform the indicated operations and simplify.

3. Add $2t^3 + 3t^2 - 2$ to $t^3 + 9$

4. $(3 - 7y) + (7y^2 + 2y - 3)$

5. $(7x^3 - 3x^2 + 1) - (x^2 - 2x^3)$

6. $(5 - u) - 2[3 - (u^2 + 1)]$

7. $(-5n^2)(-2n^3)$

8. $\dfrac{6x^7}{(-2x^2)^3}$

9. $7y(4 - 3y)$

10. $2z(z + 5) - 7(z + 5)$

11. $(6r + 5)(6r - 5)$

12. $(2x - 3)^2$

13. $(x + 1)(x^2 - x + 1)$

14. $(v - 3)^2 - (v + 3)^2$

In Exercises 15–18, factor the expression completely.

15. $28a^2 - 21a$

16. $25 - 4x^2$

17. $z^3 + 3z^2 - 9z - 27$

18. $4y^3 - 32$

19. Find all possible products of the form $(5x + m)(2x + n)$ such that $mn = 10$.

20. Find the area of the shaded portion of the figure.

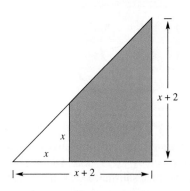

3.4	**Factoring Trinomials**

Perfect Square Trinomials ▪ Factoring Trinomials ▪
Factoring Trinomials by Grouping (Optional) ▪ Summary of Factoring

Perfect Square Trinomials

A **perfect square trinomial** is the square of a binomial. For instance,

$$x^2 + 6x + 9 = (x + 3)^2$$

is the square of the binomial $(x + 3)$. Perfect square trinomials come in two forms, one in which the middle term is positive and the other in which it is negative.

Perfect Square Trinomials

Let u and v represent real numbers, variables, or algebraic expressions.

1. $u^2 + 2uv + v^2 = (u + v)^2$ **2.** $u^2 - 2uv + v^2 = (u - v)^2$

Same sign   Same sign

To recognize a perfect square trinomial, remember that the first and last terms must be perfect squares and positive, and the middle term must be twice the product of u and v. (The middle term can be positive or negative.)

EXAMPLE 1   Factoring Perfect Square Trinomials

a. $x^2 - 4x + 4 = x^2 - 2(2x) + 2^2 = (x - 2)^2$
b. $16y^2 + 24y + 9 = (4y)^2 + 2(4y)(3) + 3^2 = (4y + 3)^2$
c. $9x^2 - 30xy + 25y^2 = (3x)^2 - 2(3x)(5y) + (5y)^2 = (3x - 5y)^2$

EXAMPLE 2   Removing a Common Monomial Factor First

a. $3x^2 - 30x + 75 = 3(x^2 - 10x + 25)$ Remove common monomial factor.
$= 3(x - 5)^2$ Factor.
b. $16y^3 + 80y^2 + 100y = 4y(4y^2 + 20y + 25)$ Remove common monomial factor.
$= 4y(2y + 5)^2$ Factor.

Factoring Trinomials

To factor a trinomial of the form $x^2 + bx + c$, consider the following.

$$(x + m)(x + n) = x^2 + nx + mx + mn$$
$$= x^2 + \underbrace{(m + n)}x + \underbrace{mn}$$

Sum of   Product
terms   of terms

$$= x^2 + \boxed{b}\,x + \boxed{c}$$

From this, you can see that to factor a trinomial $x^2 + bx + c$ into a product of two binomials, you must find *factors of c whose sum is b.* There are many different techniques for factoring trinomials. The most common is to use "*Guess, Check, and Revise*" with mental math.

EXAMPLE 3   Factoring a Trinomial

Factor $x^2 + 3x - 4$.

Solution

You need to find two numbers whose product is -4 and whose sum is 3. Using mental math, you can determine that the numbers are 4 and -1.

$$x^2 + 3x - 4 = (x + 4)(x - 1)$$

EXAMPLE 4   Factoring Trinomials

Factor each trinomial.

a. $x^2 - 2x - 8$ **b.** $x^2 - 5x + 6$

Solution

a. You need to find two numbers whose product is -8 and whose sum is -2.

The product of -4 and 2 is -8.

$$x^2 - 2x - 8 = (x - 4)(x + 2)$$

The sum of -4 and 2 is -2.

b. You need to find two numbers whose product is 6 and whose sum is -5.

The product of -3 and -2 is 6.

$$x^2 - 5x + 6 = (x - 3)(x - 2)$$

The sum of -3 and -2 is -5.

STUDY TIP

With *any* factoring problem, remember that you can check your result by multiplying. For instance, in Example 3, you can check the result by multiplying $(x + 4)$ by $(x - 1)$ to see that you obtain $x^2 + 3x - 4$.

Remember that not all trinomials are factorable using integers. For instance, $x^2 - 2x - 4$ is not factorable using integers because there is no pair of factors of -4 whose sum is -2.

When factoring a trinomial of the form $x^2 + bx + c$, if you have trouble finding two factors of c whose sum is b, try making a list of all the distinct pairs of factors. For instance, consider the trinomial

$$x^2 - 5x - 24.$$

For this trinomial, $c = -24$ and $b = -5$. Thus, you need two factors of -24 whose sum is -5. Here is the complete list.

Factors of -24	Sum of Factors	
$(1)(-24)$	$1 - 24 = -23$	
$(-1)(24)$	$-1 + 24 = 23$	
$(2)(-12)$	$2 - 12 = -10$	
$(-2)(12)$	$-2 + 12 = 10$	
$(3)(-8)$	$3 - 8 = -5$	⬅ Correct choice
$(-3)(8)$	$-3 + 8 = 5$	
$(4)(-6)$	$4 - 6 = -2$	
$(-4)(6)$	$-4 + 6 = 2$	

With experience, you will be able to *mentally* narrow this list down to only two or three possibilities whose sums can then be tested to determine the correct factorization. Here are some suggestions for narrowing down the list.

Guidelines for Factoring $x^2 + bx + c$

1. If c is *positive*, its factors have like signs that match the sign of b.

2. If c is *negative*, its factors have different signs.

3. If $|b|$ is small relative to $|c|$, first try those factors of c that are closest to each other in absolute value.

4. If $|b|$ is near $|c|$, first try those factors of c that are farthest from each other in absolute value.

EXAMPLE 5   Factoring a Trinomial

Factor $x^2 - 17x - 18$.

Solution

You need to find two numbers whose product is -18 and whose sum is -17.

The product of -18 and 1 is -18.

$$x^2 - 17x - 18 = (x - 18)(x + 1)$$

The sum of -18 and 1 is -17.

To factor a trinomial whose leading coefficient is not 1, use the following pattern.

$$ax^2 + bx + c = (\boxed{}x + \boxed{})(\boxed{}x + \boxed{})$$

Factors of a (top), Factors of c (bottom)

The goal is to find a combination of factors of a and c such that the outer and inner products add up to the middle term bx.

EXAMPLE 6 *Factoring a Trinomial of the Form ax² + bx + c*

Factor $4x^2 + 5x - 6$.

Solution

First, observe that $4x^2 + 5x - 6$ has no common monomial factor. The leading coefficient 4 factors as $(1)(4)$ or as $(2)(2)$. The constant term -6 factors as $(-1)(6)$, $(1)(-6)$, $(-2)(3)$, or $(2)(-3)$. A test of the many possibilities is shown below.

<table>
<tr><td>$(x + 1)(4x - 6) = 4x^2 - 2x - 6$</td><td>$(4x - 6)$ has a common factor of 2.</td></tr>
<tr><td>$(x - 1)(4x + 6) = 4x^2 + 2x - 6$</td><td>$(4x + 6)$ has a common factor of 2.</td></tr>
<tr><td>$(x + 6)(4x - 1) = 4x^2 + 23x - 6$</td><td></td></tr>
<tr><td>$(x - 6)(4x + 1) = 4x^2 - 23x - 6$</td><td></td></tr>
<tr><td>$(x - 2)(4x + 3) = 4x^2 - 5x - 6$</td><td>Middle term has incorrect sign.</td></tr>
<tr><td>$(x + 2)(4x - 3) = 4x^2 + 5x - 6$</td><td>⇐ Correct factorization</td></tr>
<tr><td>$(2x + 1)(2x - 6) = 4x^2 - 10x - 6$</td><td>$(2x - 6)$ has a common factor of 2.</td></tr>
<tr><td>$(2x - 1)(2x + 6) = 4x^2 + 10x - 6$</td><td>$(2x + 6)$ has a common factor of 2.</td></tr>
<tr><td>$(2x + 2)(2x - 3) = 4x^2 - 2x - 6$</td><td>$(2x + 2)$ has a common factor of 2.</td></tr>
<tr><td>$(2x - 2)(2x + 3) = 4x^2 + 2x - 6$</td><td>$(2x - 2)$ has a common factor of 2.</td></tr>
<tr><td>$(x + 3)(4x - 2) = 4x^2 + 10x - 6$</td><td>$(4x - 2)$ has a common factor of 2.</td></tr>
<tr><td>$(x - 3)(4x + 2) = 4x^2 - 10x - 6$</td><td>$(4x + 2)$ has a common factor of 2.</td></tr>
</table>

STUDY TIP

If the original trinomial has no common monomial factor, its binomial factors cannot have common monomial factors. Thus, in Example 6, you do not have to test factors, such as $(4x - 6)$, that have a common factor of 2.

Thus, the correct factorization is

$$4x^2 + 5x - 6 = (x + 2)(4x - 3).$$

Check this result by multiplying.

Guidelines for Factoring $ax^2 + bx + c$

---

1. First, factor out any common monomial factors.

2. Exclude any binomial factors that have a common factor.

3. If the middle term is the opposite of b, switch signs for the factors of c.

EXAMPLE 7 Factoring a Trinomial of the Form $ax^2 + bx + c$

Factor $2x^2 - x - 21$.

Solution

For this trinomial, $a = 2$, which factors as $(1)(2)$, and $c = -21$, which factors as $(1)(-21)$, $(-1)(21)$, $(3)(-7)$, or $(-3, 7)$. Because b is small, avoid the large factors of -21, and test the smaller ones.

$$(2x + 3)(x - 7) = 2x^2 - 11x - 21$$
$$(2x + 7)(x - 3) = 2x^2 + x - 21 \qquad \text{Middle term has incorrect sign.}$$
$$(2x - 7)(x + 3) = 2x^2 - x - 21 \qquad \Longleftarrow \quad \text{Correct factorization}$$

Therefore, the correct factorization is

$$2x^2 - x - 21 = (2x - 7)(x + 3).$$

Check this result by multiplying.

EXAMPLE 8 Factoring a Trinomial of the Form $ax^2 + bx + c$

Factor $3x^2 + 11x + 10$.

Solution

For this trinomial, $a = 3$, which factors as $(1)(3)$, and $c = 10$, which factors as $(1)(10)$ or $(2)(5)$.

$$(x + 10)(3x + 1) = 3x^2 + 31x + 10$$
$$(x + 1)(3x + 10) = 3x^2 + 13x + 10$$
$$(x + 5)(3x + 2) = 3x^2 + 17x + 10$$
$$(x + 2)(3x + 5) = 3x^2 + 11x + 10 \qquad \Longleftarrow \quad \text{Correct factorization}$$

Therefore, the correct factorization is

$$3x^2 + 11x + 10 = (x + 2)(3x + 5).$$

EXAMPLE 9   Factoring Completely

Factor $8x^3 - 60x^2 + 28x$.

Solution

Begin by factoring out the common monomial factor $4x$.

$$8x^3 - 60x^2 + 28x = 4x(2x^2 - 15x + 7)$$

Now, for the new trinomial $2x^2 - 15x + 7$, $a = 2$ and $c = 7$. The possible factorizations of this trinomial are as follows.

$$(2x - 7)(x - 1) = 2x^2 - 9x + 7$$
$$(2x - 1)(x - 7) = 2x^2 - 15x + 7 \qquad \Longleftarrow \quad \text{Correct factorization}$$

Therefore, the complete factorization of the original trinomial is

$$8x^3 - 60x^2 + 28x = 4x(2x^2 - 15x + 7) = 4x(2x - 1)(x - 7).$$

Check this result by multiplying.

When you are factoring a trinomial with a negative leading coefficient, we suggest that you first factor -1 out of the trinomial.

EXAMPLE 10   A Trinomial with a Negative Leading Coefficient

Factor $-3x^2 + 16x + 35$.

Solution

Begin by factoring out -1.

$$-3x^2 + 16x + 35 = (-1)(3x^2 - 16x - 35)$$

For the new trinomial $3x^2 - 16x - 35$, $a = 3$ and $c = -35$. Some possible factorizations of this trinomial are as follows.

$$(3x - 1)(x + 35) = 3x^2 + 104x - 35$$
$$(3x - 35)(x + 1) = 3x^2 - 32x - 35$$
$$(3x - 7)(x + 5) = 3x^2 + 8x - 35$$
$$(3x - 5)(x + 7) = 3x^2 + 16x - 35 \qquad \text{Middle term has incorrect sign.}$$
$$(3x + 5)(x - 7) = 3x^2 - 16x - 35 \qquad \Longleftarrow \quad \text{Correct factorization}$$

Thus, the correct factorization is

$$-3x^2 + 16x + 35 = (-1)(3x + 5)(x - 7) = (3x + 5)(-x + 7).$$

Factoring Trinomials by Grouping (Optional)

So far in this section, you have been using "Guess, Check, and Revise" to factor trinomials. An alternative technique is to use factoring by grouping to factor a trinomial. For instance, suppose you rewrote the trinomial $2x^2 + 7x - 15$ as

$$2x^2 + 7x - 15 = 2x^2 + 10x - 3x - 15.$$

Then, by grouping the first two terms and the third and fourth terms, you could factor the polynomial as follows.

$$
\begin{aligned}
2x^2 + 7x - 15 &= 2x^2 + (10x - 3x) - 15 & \text{Rewrite middle term.}\\
&= (2x^2 + 10x) - (3x + 15) & \text{Group terms.}\\
&= 2x(x + 5) - 3(x + 5) & \text{Factor groups.}\\
&= (2x - 3)(x + 5) & \text{Distributive Property}
\end{aligned}
$$

The key to this method of factoring is knowing how to rewrite the middle term. In general, *to factor a trinomial $ax^2 + bx + c$ by grouping, choose factors of the product ac that add up to b and use these factors to rewrite the middle term.*

EXAMPLE 11   Factoring a Trinomial by Grouping

Use factoring by grouping to factor the trinomial.

$$3x^2 + 5x - 2$$

Solution

For the trinomial $3x^2 + 5x - 2$, $ac = 3(-2) = -6$, which has factors 6 and -1 that add up to 5. Therefore, rewrite the middle term as $5x = 6x - x$. This produces the following.

$$
\begin{aligned}
3x^2 + 5x - 2 &= 3x^2 + (6x - x) - 2 & \text{Rewrite middle term.}\\
&= (3x^2 + 6x) - (x + 2) & \text{Group terms.}\\
&= 3x(x + 2) - (x + 2) & \text{Factor groups.}\\
&= (x + 2)(3x - 1) & \text{Distributive Property}
\end{aligned}
$$

Therefore, the trinomial factors as

$$3x^2 + 5x - 2 = (x + 2)(3x - 1).$$

Check this result by multiplying.

What do you think of this optional technique? Some people think that it is more efficient than the trial-and-error process, especially when the coefficients a and c have many factors.

Summary of Factoring

Guidelines for Factoring Polynomials

1. Factor out any common factors.

2. Factor according to one of the special polynomial forms: difference of squares, sum or difference of cubes, or perfect square trinomials.

3. Factor trinomials using the methods for $a = 1$ and $a \neq 1$.

4. Factor by grouping—for polynomials with four terms.

5. Check to see if the factors themselves can be factored further.

6. Check the results by multiplying the factors.

EXAMPLE 12   Factoring Polynomials

a. $\quad 3x^2 - 108 = 3(x^2 - 36)$ $\qquad\qquad\qquad$ Remove common factor.

$\qquad\qquad\quad = 3(x + 6)(x - 6)$ $\qquad\qquad$ Difference of two squares

b. $\quad 4x^3 - 32x^2 + 64x = 4x(x^2 - 8x + 16)$ $\qquad$ Remove common factor.

$\qquad\qquad\qquad\quad = 4x(x - 4)^2$ $\qquad\qquad$ Factor.

c. $\quad 6x^3 + 27x^2 - 15x = 3x(2x^2 + 9x - 5)$ $\qquad$ Remove common factor.

$\qquad\qquad\qquad\quad = 3x(2x - 1)(x + 5)$ $\qquad$ Factor.

d. $\quad x^3 - 3x^2 - 4x + 12 = (x^3 - 3x^2) + (-4x + 12)$ $\qquad$ Group terms.

$\qquad\qquad\qquad\quad = x^2(x - 3) - 4(x - 3)$ $\qquad$ Remove common factors.

$\qquad\qquad\qquad\quad = (x - 3)(x^2 - 4)$ $\qquad\qquad$ Distributive Property

$\qquad\qquad\qquad\quad = (x - 3)(x + 2)(x - 2)$ $\qquad$ Factor.

Group Activities   You Be the Instructor

Creating a Test   Create five factoring problems that you think represent a fair test of a person's factoring skills. Discuss how it is possible to *create* polynomials that are factorable. Exchange problems with another person in your class. Do each other's problems, then check each other's work.

3.4 Exercises

Discussing the Concepts

1. In your own words, explain how you would factor $x^2 - 5x + 6$.

2. Give an example of a prime trinomial.

3. Explain how you can check the factors of a trinomial. Give an example.

4. *Error Analysis*   Identify the error.

$$9x^2 - 9x - 54 = (3x + 6)(3x - 9)$$
$$= 3(x + 2)(x - 3)$$

5. Is $x(x + 2) - 2(x + 2)$ in factored form? Explain.

6. Is $(2x - 4)(x + 1)$ completely factored? Explain.

Problem Solving

In Exercises 7–10, factor the perfect square trinomial.

7. $x^2 + 4x + 4$

8. $z^2 + 6z + 9$

9. $25y^2 - 10y + 1$

10. $4z^2 + 28z + 49$

In Exercises 11–14, find all values of b for which the polynomial is a perfect square trinomial.

11. $x^2 + bx + 81$

12. $x^2 + bx + \frac{9}{16}$

13. $4x^2 + bx + 9$

14. $16x^2 + bxy + 25y^2$

In Exercises 15–18, find a real number c such that the polynomial is a perfect square trinomial.

15. $x^2 + 8x + c$

16. $x^2 + 12x + c$

17. $y^2 - 6y + c$

18. $z^2 - 20z + c$

In Exercises 19–22, find the missing factor.

19. $x^2 + 5x + 4 = (x + 4)(\quad)$

20. $a^2 + 2a - 8 = (a + 4)(\quad)$

21. $y^2 - y - 20 = (y + 4)(\quad)$

22. $y^2 + 6y + 8 = (y + 4)(\quad)$

In Exercises 23–28, factor the trinomial.

23. $x^2 + 4x + 3$

24. $x^2 + 7x + 10$

25. $t^2 - 4t - 21$

26. $x^2 + 4x - 12$

27. $x^2 - 2xy - 35y^2$

28. $a^2 - 21ab + 110b^2$

In Exercises 29 and 30, find all values of b for which the trinomial can be factored.

29. $x^2 + bx + 35$

30. $x^2 + bx - 38$

In Exercises 31 and 32, find two values of c for which the trinomial can be factored.

31. $x^2 + 6x + c$

32. $x^2 - 12x + c$

In Exercises 33 and 34, find the missing factor.

33. $5x^2 + 18x + 9 = (x + 3)(\quad)$

34. $3y^2 - y - 30 = (y + 3)(\quad)$

In Exercises 35–46, factor the trinomial, if possible. (Some of the trinomials may be prime.)

35. $5x^2 + 7x + 2$

36. $3x^2 - 2x - 5$

37. $6x^2 - 11x + 3$

38. $4y^2 - 5y - 9$

39. $3t^2 - 4t - 10$

40. $2z^2 + 3z + 8$

41. $6u^2 - 5uv - 4v^2$

42. $10a^2 + 23ab + 6b^2$

43. $-2x^2 - x + 6$

44. $-6x^2 + 5x - 6$

45. $1 - 11x - 60x^2$

46. $2 + 5x - 12x^2$

In Exercises 47–50, factor the trinomial by grouping.

47. $3x^2 + 10x + 8$

48. $6x^2 - x - 15$

49. $15x^2 - 11x + 2$

50. $12x^2 - 14x + 1$

Reviewing the Major Concepts

In Exercises 51–54, find the missing coordinate of the solution point.

51. $y = \frac{3}{5}x + 4$

  (15, ▮)

52. $y = 3 - \frac{5}{9}x$

  (12, ▮)

53. $y = 5.5 - 0.95x$

  (▮ , −1)

54. $y = 3 + 0.2x$

  (▮ , 4.4)

55. *Recipe Proportions* Two and one-half cups of flour are required to make one batch of cookies. How many cups are required to make $3\frac{1}{2}$ batches?

56. *Fuel Mixture* The gasoline-to-oil ratio for a two-cycle engine is 32 to 1. Determine the amount of gasoline required to produce a mixture that has $\frac{1}{2}$ pint of oil.

Additional Problem Solving

In Exercises 57–62, factor the perfect square trinomial.

57. $a^2 - 12a + 36$

58. $y^2 - 14y + 49$

59. $b^2 + 4b + 4$

60. $4x^2 - 4x + 1$

61. $u^2 + 8uv + 16v^2$

62. $4y^2 + 20yz + 25z^2$

In Exercises 63 and 64, fill in the missing number.

63. $x^2 + 12x + 50 = (x + 6)^2 + $ ▮

64. $x^2 + 10x + 22 = (x + 5)^2 + $ ▮

In Exercises 65–68, find the missing factor.

65. $x^2 - 2x - 24 = (x + 4)($ ▮ $)$

66. $x^2 + 7x + 12 = (x + 4)($ ▮ $)$

67. $z^2 - 6z + 8 = (z - 4)($ ▮ $)$

68. $z^2 + 2z - 24 = (z - 4)($ ▮ $)$

In Exercises 69–76, factor the trinomial.

69. $x^2 - 5x + 6$

70. $x^2 - 10x + 24$

71. $y^2 + 7y - 30$

72. $m^2 - 3m - 10$

73. $x^2 - 20x + 96$

74. $y^2 - 35y + 300$

75. $x^2 + 30x + 216$

76. $u^2 + 5uv + 6v^2$

In Exercises 77–80, find all values of b for which the trinomial can be factored.

77. $x^2 + bx + 18$

78. $x^2 + bx + 14$

79. $x^2 + bx - 21$

80. $x^2 + bx - 7$

In Exercises 81 and 82, find two values of c for which the trinomial can be factored.

81. $x^2 - 3x + c$

82. $x^2 + 9x + c$

In Exercises 83–86, find the missing factor.

83. $5a^2 + 12a - 9 = (a + 3)($ ▮ $)$

84. $5x^2 + 19x + 12 = (x + 3)($ ▮ $)$

85. $2y^2 - 3y - 27 = (y + 3)($ ▮ $)$

86. $5c^2 + 11c - 12 = (c + 3)($ ▮ $)$

In Exercises 87–96, factor the trinomial, if possible. (Some of the trinomials may be prime.)

87. $3x^2 + 4x + 1$

88. $5x^2 + 7x + 2$

89. $2x^2 - 9x + 9$

90. $2t^2 - 13t + 20$

91. $6b^2 + 19b - 7$

92. $10x^2 - 24x - 18$

93. $24x^2 - 14xy - 3y^2$

94. $20x^2 + x - 12$

95. $3y^2 + 4y + 12$

96. $10x^2 + 9xy - 9y^2$

In Exercises 97 and 98, factor by grouping.

97. $6x^2 + x - 2$

98. $2x^2 + 9x + 9$

In Exercises 99–106, factor completely.

99. $3x^4 - 12x^3$

100. $20y^2 - 45$

101. $10t^3 + 2t^2 - 36t$

102. $16z^2 - 56z + 49$

103. $36 - (z + 3)^2$

104. $3t^3 - 24$

105. $54x^3 - 2$

106. $v^3 + 3v^2 + 5v$

In Exercises 107–110, use a graphing utility to graph the two equations on the same screen. What can you conclude?

107. $y_1 = x^2 + 6x + 9$, $y_2 = (x + 3)^2$

108. $y_1 = 4x^2 - 4x + 1$, $y_2 = (2x - 1)^2$

109. $y_1 = x^2 + 2x - 3$, $y_2 = (x - 1)(x + 3)$

110. $y_1 = 3x^2 - 8x - 16$, $y_2 = (3x + 4)(x - 4)$

Mental Math   In Exercises 111 and 112, use mental math to evaluate the expression.

Sample: $29^2 = (30 - 1)^2$

$$= 30^2 - 2(30)(1) + 1^2$$

$$= 900 - 60 + 1$$

$$= 841$$

111. 52^2 **112.** 39^2

Geometric Factoring Models   In Exercises 113 and 114, factor the trinomial and represent the result with a geometric factoring model. The sample shows the factoring of $x^2 + 3x + 2$ as $(x + 1)(x + 2)$.

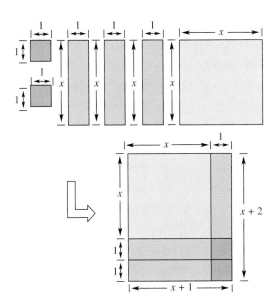

113. $x^2 + 4x + 3$ **114.** $x^2 + 5x + 4$

Geometric Factoring Models   In Exercises 115–118, match the geometric factoring model with the correct factoring formula. [The models are labeled (a), (b), (c), and (d).]

(a)

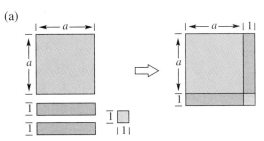

(b)

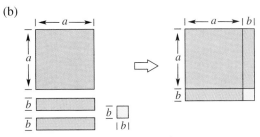

(c)

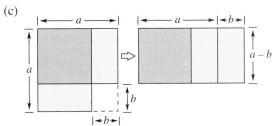

(d)

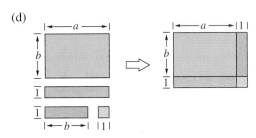

115. $a^2 - b^2 = (a + b)(a - b)$

116. $a^2 + 2a + 1 = (a + 1)^2$

117. $a^2 + 2ab + b^2 = (a + b)^2$

118. $ab + a + b + 1 = (a + 1)(b + 1)$

| **3.5** | **Solving Polynomial Equations** |

The Zero-Factor Property ▪ Solving Quadratic Equations by Factoring ▪
Solving Polynomial Equations by Factoring ▪ Applications

The Zero-Factor Property

You have spent the first part of this chapter developing skills for simplifying and
factoring polynomials. In this section, you will use these skills with the following
Zero-Factor Property to solve equations.

NOTE The Zero-Factor
Property is just another way of
saying that the only way the
product of two or more factors
can be zero is if one or more of
the factors is zero.

Zero-Factor Property

Let a and b be real numbers, variables, or algebraic expressions. If a and
b are factors such that

$$ab = 0$$

then $a = 0$ or $b = 0$. This property also applies to three or more factors.

The Zero-Factor Property is the primary property for solving equations in
algebra. For instance, to solve the equation

$$(x - 2)(x + 3) = 0 \qquad \text{Original equation}$$

you can use the Zero-Factor Property to conclude that either $(x - 2)$ or $(x + 3)$
must be zero. Setting the first factor equal to zero implies that $x = 2$ is a solution.

$$x - 2 = 0 \quad \Longrightarrow \quad x = 2 \qquad \text{First solution}$$

Similarly, setting the second factor equal to zero implies that $x = -3$ is a solution.

$$x + 3 = 0 \quad \Longrightarrow \quad x = -3 \qquad \text{Second solution}$$

Thus, the equation $(x - 2)(x + 3) = 0$ has exactly two solutions: 2 and -3. You
can check these solutions by substituting into the original equation.

Check

$$(x - 2)(x + 3) = 0 \qquad \text{Original equation}$$

$$(2 - 2)(2 + 3) \overset{?}{=} 0 \qquad \text{Substitute 2 for } x.$$

$$(0)(5) = 0 \qquad \text{First solution checks. } ✓$$

$$(-3 - 2)(-3 + 3) \overset{?}{=} 0 \qquad \text{Substitute } -3 \text{ for } x.$$

$$(-5)(0) = 0 \qquad \text{Second solution checks. } ✓$$

Solving Quadratic Equations by Factoring

A **quadratic equation** is an equation of the form $ax^2 + bx + c = 0$. Here are some examples.

$$x^2 - x - 6 = 0, \quad 3x^2 + 2x - 1 = 0, \quad \text{and} \quad x^2 + 3x = 0$$

In the next four examples, note how you can combine your factoring skills with the Zero-Factor Property to solve quadratic equations.

EXAMPLE 1   Using Factoring to Solve a Quadratic Equation

Solve $x^2 - x - 12 = 0$.

Solution

First, check to see that the right side of the equation is zero. Next, factor the left side of the equation. Finally, apply the Zero-Factor Property to find the solutions.

$x^2 - x - 12 = 0$	Original equation
$(x + 3)(x - 4) = 0$	Factor left side of equation.
$x + 3 = 0 \quad \Longrightarrow \quad x = -3$	Set 1st factor equal to 0.
$x - 4 = 0 \quad \Longrightarrow \quad x = 4$	Set 2nd factor equal to 0.

The equation has two solutions: -3 and 4.

Check

$x^2 - x - 12 = 0$	Original equation
$(-3)^2 - (-3) - 12 \overset{?}{=} 0$	Substitute -3 for x.
$9 + 3 - 12 \overset{?}{=} 0$	Simplify.
$0 = 0$	Solution checks. ✓

Check

$x^2 - x - 12 = 0$	Original equation
$(4)^2 - 4 - 12 \overset{?}{=} 0$	Substitute 4 for x.
$16 - 4 - 12 \overset{?}{=} 0$	Simplify.
$0 = 0$	Solution checks. ✓

Factoring and the Zero-Factor Property allow you to solve a quadratic equation by converting it into two *linear* equations, which you already know how to solve. This is a common strategy in algebra—breaking down a given problem into simpler parts, each of which can be solved by previously learned methods.

Technology

You can use a graphing utility to estimate the solutions of an equation. For instance, to estimate the solutions of the equation in Example 1, sketch the graph of

$$y = x^2 - x - 12$$

as shown below. The two solutions correspond to the x-intercepts of the graph: $x = -3$ and $x = 4$.

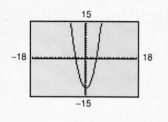

If the Zero-Factor Property is to be used, a polynomial equation *must* be written in **standard form.** That is, the polynomial must be on one side of the equation and zero must be the only term on the other side. For instance, to write $x^2 - 3x = 18$ in standard form, subtract 18 from both sides of the equation.

$$x^2 - 3x = 18 \qquad\qquad \text{Original equation}$$

$$x^2 - 3x - 18 = 18 - 18 \qquad\qquad \text{Subtract 18 from both sides.}$$

$$x^2 - 3x - 18 = 0 \qquad\qquad \text{Standard form}$$

To solve this equation, factor the left side as $(x + 3)(x - 6)$ and then form the linear equations $x + 3 = 0$ and $x - 6 = 0$. The solutions of these two linear equations are -3 and 6, respectively. The general strategy for solving a quadratic equation by factoring is summarized in the following diagram.

Write in standard form.	$\rightarrow$	Factor left side of equation.	$\rightarrow$	Set factors equal to zero.	$\rightarrow$	Solve linear equations.	$\rightarrow$	Check in original equation.

EXAMPLE 2   *Solving a Quadratic Equation by Factoring*

Solve $3x^2 + 5x = 12$.

Solution

$$3x^2 + 5x = 12 \qquad\qquad \text{Original equation}$$

$$3x^2 + 5x - 12 = 0 \qquad\qquad \text{Write in standard form.}$$

$$(3x - 4)(x + 3) = 0 \qquad\qquad \text{Factor left side of equation.}$$

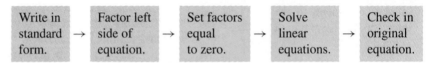

$$3x - 4 = 0 \quad\Longrightarrow\quad x = \frac{4}{3} \qquad \text{Set 1st factor equal to 0.}$$

$$x + 3 = 0 \quad\Longrightarrow\quad x = -3 \qquad \text{Set 2nd factor equal to 0.}$$

The solutions are $\frac{4}{3}$ and -3. Check these solutions in the original equation.

Be sure you see that the Zero-Factor Property can be applied only to a product that is equal to *zero*. For instance, you cannot conclude from the equation

$$x(x - 3) = 10$$

that $x = 10$ and $x - 3 = 10$ yield solutions. Instead, you must first write the equation in standard form and then factor the left side, as follows.

$$x^2 - 3x - 10 = 0 \qquad\Longrightarrow\qquad (x - 5)(x + 2) = 0$$

Now, from the factored form you can see that the solutions are 5 and -2.

In Examples 1 and 2, the original equations each involved a second-degree (quadratic) polynomial and each had *two different* solutions. You will sometimes encounter second-degree polynomial equations that have only one (repeated) solution. This occurs when the left side of the equation is a perfect square trinomial, as shown in Example 3.

EXAMPLE 3 *A Quadratic Equation with a Repeated Solution*

Solve $x^2 - 6x + 11 = 2$.

Solution

$x^2 - 6x + 11 = 2$	Original equation
$x^2 - 6x + 9 = 0$	Write in standard form.
$(x - 3)^2 = 0$	Factor left side of equation.
$x - 3 = 0 \quad \Longrightarrow \quad x = 3$	Set factor equal to 0.

Note that even though the left side of this equation has two factors, the factors are the same. Thus, the only solution of the equation is 3.

Check

$x^2 - 6x + 11 = 2$	Original equation
$(3)^2 - 6(3) + 11 \stackrel{?}{=} 2$	Substitute 3 for x.
$9 - 18 + 11 \stackrel{?}{=} 2$	Simplify.
$2 = 2$	Solution checks. ✓

EXAMPLE 4 *Solving a Polynomial Equation in Nonstandard Form*

Solve $(x + 3)(x + 6) = 4$.

Solution

Begin by multiplying the factors on the left side.

$(x + 3)(x + 6) = 4$	Original equation
$x^2 + 9x + 18 = 4$	Multiply factors.
$x^2 + 9x + 14 = 0$	Write in standard form.
$(x + 2)(x + 7) = 0$	Factor left side of equation.
$x + 2 = 0 \quad \Longrightarrow \quad x = -2$	Set 1st factor equal to 0.
$x + 7 = 0 \quad \Longrightarrow \quad x = -7$	Set 2nd factor equal to 0.

The equation has two solutions: -2 and -7. Check these in the original equation.

DISCOVERY

In Example 3, use a graphing utility to graph the equations

$$y = x^2 - 6x + 11$$

and

$$y = 2$$

on the same screen. From the graph, determine the number of solutions of the equation. Explain how to use a graphing utility to solve

$$2x^3 - 3x^2 - 5x + 1 = 0.$$

How many solutions does the equation have? How does the number of solutions relate to the degree of the equation?

Solving Polynomial Equations by Factoring

EXAMPLE 5   Solving a Polynomial Equation with Three Factors

Solve $3x^3 = 15x^2 + 18x$.

Solution

$$3x^3 = 15x^2 + 18x$$ Original equation

$$3x^3 - 15x^2 - 18x = 0$$ Write in standard form.

$$3x(x^2 - 5x - 6) = 0$$ Remove common factor.

$$3x(x - 6)(x + 1) = 0$$ Factor.

$$3x = 0 \implies x = 0$$ Set 1st factor equal to 0.

$$x - 6 = 0 \implies x = 6$$ Set 2nd factor equal to 0.

$$x + 1 = 0 \implies x = -1$$ Set 3rd factor equal to 0.

There are three solutions: 0, 6, and -1. Check these in the original equation.

Notice that the equation in Example 5 is a third-degree equation and has three solutions. This is not a coincidence. In general, a polynomial equation can have *at most* as many solutions as its degree. For instance, a second-degree equation can have zero, one, or two solutions, but it cannot have three or more solutions.

EXAMPLE 6   Solving a Polynomial Equation with Four Factors

Solve $x^4 + x^3 - 4x^2 - 4x = 0$.

Solution

$$x^4 + x^3 - 4x^2 - 4x = 0$$ Original equation

$$x(x^3 + x^2 - 4x - 4) = 0$$ Remove common factor.

$$x[x^2(x + 1) - 4(x + 1)] = 0$$ Remove common factors.

$$x[(x + 1)(x^2 - 4)] = 0$$ Distributive Property

$$x(x + 1)(x + 2)(x - 2) = 0$$ Factor.

$$x = 0 \implies x = 0$$

$$x + 1 = 0 \implies x = -1$$

$$x + 2 = 0 \implies x = -2$$

$$x - 2 = 0 \implies x = 2$$

There are four solutions: 0, -1, -2, and 2. Check these in the original equation.

Applications

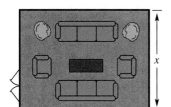

FIGURE 3.3

EXAMPLE 7   Geometry

A rectangular room has an area of 192 square feet. The length of the room is 4 feet more than its width, as shown in Figure 3.3. Find the dimensions of the room.

Solution

Verbal Model: Length · Width = Area

Labels: Width = x (feet)
Length = $x + 4$ (feet)
Area = 192 (square feet)

Equation: $x(x + 4) = 192$

$$x^2 + 4x - 192 = 0$$

$$(x + 16)(x - 12) = 0$$

$$x = -16 \text{ or } 12$$

Because the negative solution does not make sense, choose the positive solution. Thus, the room is 12 feet by 16 feet. Check this in the original statement of the problem.

EXAMPLE 8   A Falling-Body Problem

A rock is dropped into a well that is 64 feet deep, as shown in Figure 3.4. The height of the rock (above the water level) is given by $h = -16t^2 + 64$, where the height is measured in feet and the time t is measured in seconds. How long will it take the rock to hit the water at the bottom of the well?

Solution

In Figure 3.4, note that the bottom of the well corresponds to a height of 0 feet. Thus, substitute a height of 0 into the equation and solve for t.

$$0 = -16t^2 + 64 \qquad \text{Substitute 0 for } h.$$

$$16t^2 - 64 = 0 \qquad \text{Write in standard form.}$$

$$16(t^2 - 4) = 0 \qquad \text{Remove common factor.}$$

$$16(t + 2)(t - 2) = 0 \qquad \text{Factor.}$$

$$t = -2 \text{ or } 2 \qquad \text{Solutions}$$

Because a time of -2 seconds does not make sense in this problem, choose the positive solution and conclude that the rock hits the bottom of the well 2 seconds after it is dropped. Check this solution in the original statement of the problem.

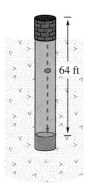

FIGURE 3.4

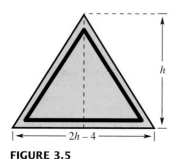

FIGURE 3.5

EXAMPLE 9 *Geometry*

A triangular roadside sign has a base that is to be 4 feet less than twice its height, as shown in Figure 3.5. A local zoning ordinance restricts the area of signs to a maximum of 24 square feet. Find the base and height of the largest sign that meets the zoning ordinance.

Solution

Verbal Model:

$$\frac{1}{2} \cdot \boxed{\text{Base}} \cdot \boxed{\text{Height}} = \boxed{\text{Area}}$$

Labels: Height $= h$ (feet)
Base $= 2h - 4$ (feet)
Area $= 24$ (square feet)

Equation:

$$\frac{1}{2}(2h - 4)(h) = 24$$
$$h^2 - 2h = 24$$
$$h^2 - 2h - 24 = 0$$
$$(h + 4)(h - 6) = 0$$
$$h = -4 \text{ or } 6$$

Because the height must be positive, discard -4 as a solution and choose $h = 6$. Thus, the height of the sign is 6 feet, and the base is $2(6) - 4 = 8$ feet. Check this solution in the original statement of the problem.

Group Activities   Exploring with Technology

Approximating a Solution   The polynomial equation $x^3 - x - 3 = 0$ cannot be solved *algebraically* using any of the techniques described in this book. It does, however, have one solution that is a real number.

a. *Graphical Solution:* Use a graphing utility to estimate the solution.

b. *Numerical Solution:* Use a graphing utility to create a table such as the one below and estimate the solution.

x	1.0	1.1	1.2	1.3	1.4	1.5	1.6	1.7	1.8	1.9	2.0
$x^3 - x - 3$	-3										3

3.5 Exercises

Discussing the Concepts

1. Give an example of how the Zero-Factor Property can be used to solve a quadratic equation.

2. *True or False?* If

$$(2x - 5)(x + 4) = 1,$$

then $2x - 5 = 1$ or $x + 4 = 1$. Explain.

3. Is it possible for a quadratic equation to have only one solution? Explain.

4. What is the maximum number of solutions of an nth-degree polynomial equation? Give an example of a third-degree equation that has only one real number solution.

Problem Solving

In Exercises 5–10, use the Zero-Factor Property to solve the equation.

5. $2x(x - 8) = 0$

6. $z(z + 6) = 0$

7. $25(a + 4)(a - 2) = 0$

8. $17(t - 3)(t + 8) = 0$

9. $(x - 3)(2x + 1)(x + 4) = 0$

10. $\frac{1}{5}x(x - 2)(3x + 4) = 0$

In Exercises 11–30, solve the equation by factoring.

11. $5y - y^2 = 0$

12. $3x^2 + 9x = 0$

13. $x^2 - 25 = 0$

14. $x^2 - 121 = 0$

15. $(x + 2)^2 - 9 = 0$

16. $1 - (y + 3)^2 = 0$

17. $m^2 - 8m + 16 = 0$

18. $a^2 + 4a + 4 = 0$

19. $x(x - 3) = 10$

20. $s(s + 4) = 96$

21. $x(x + 2) - 10(x + 2) = 0$

22. $x(x - 15) + 3(x - 15) = 0$

23. $7 + 13x - 2x^2 = 0$

24. $11 + 32y - 3y^2 = 0$

25. $x^3 - 19x^2 + 84x = 0$

26. $x^3 + 18x^2 + 45x = 0$

27. $z^2(z + 2) - 4(z + 2) = 0$

28. $16(3 - u) - u^2(3 - u) = 0$

29. $c^3 - 3c^2 - 9c + 27 = 0$

30. $v^3 + 4v^2 - 4v - 16 = 0$

Graphical Reasoning In Exercises 31–34, explain how the x-intercepts of the graph correspond to the solutions of the polynomial equation when $y = 0$.

31. $y = x^2 - 9$

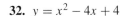

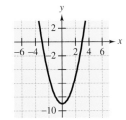

32. $y = x^2 - 4x + 4$

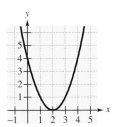

33. $y = x^2 - 2x - 3$

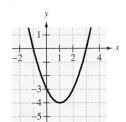

34. $y = x^3 - 3x^2 - x + 3$

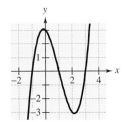

In Exercises 35–38, use a graphing utility to solve the equation graphically.

35. $x^2 - 6x = 0$

36. $x^2 - 11x + 28 = 0$

37. $2x^2 + 5x - 12 = 0$

38. $2 + x - 2x^2 - x^3 = 0$

39. *Geometry*   An open box is to be made from a piece of material that is 5 meters long and 4 meters wide. The box is to be made by cutting squares of dimension x from the corners and turning up the sides, as shown at the right. The volume V of a rectangular solid is the product of its length, width, and height.

(a) Show that the volume is given by

$$V = (5 - 2x)(4 - 2x)x.$$

(b) Determine the values of x for which $V = 0$. Determine an appropriate domain for the function V in the context of this problem.

(c) Complete the table.

x	0.25	0.50	0.75	1.00	1.25	1.50	1.75
V							

(d) Use the table to determine x when $V = 3$. Verify the result algebraically.

(e) Use a graphing utility to graph the volume function. Use the graph to approximate the value of x that yields the box of greatest volume.

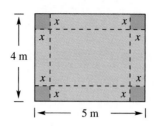

Figure for 39

40. *Free-Falling Object*   An object is thrown upward from a height of 64 feet with an initial velocity of 48 feet per second. Find the time t for the object to reach the ground by solving the following equation.

$$-16t^2 + 48t + 64 = 0$$

41. Find two consecutive positive integers whose product is 132.

42. *Geometry*   The length of a rectangle is $2\frac{1}{4}$ times the width. Find the dimensions of the rectangle if its area is 900 square inches.

Reviewing the Major Concepts

In Exercises 43–46, simplify the expression.

43. $2(x + 5) - 3 - (2x - 3)$

44. $3(y + 4) + 5 - (3y + 5)$

45. $4 - 2[3 + 4(x + 1)]$

46. $5x + x[3 - 2(x - 3)]$

47. *Simple Interest*   Evaluate the expression Prt when $P = \$2000$, $r = 0.07$, and $t = 10$.

48. *Geometry*   Evaluate the expression $\pi r^2 h$ for the volume of a right cylinder when $r = 3$ inches and $h = 5$ inches.

Additional Problem Solving

In Exercises 49–54, use the Zero-Factor Property to solve the equation.

49. $(y - 3)(y + 10) = 0$

50. $(s - 16)(s + 15) = 0$

51. $(2t + 5)(3t + 1) = 0$

52. $(5x - 3)(x - 8) = 0$

53. $4x(2x - 3)(2x + 25) = 0$

54. $(y - 39)(2y + 7)(y + 12) = 0$

In Exercises 55–84, solve the equation by factoring.

55. $3y^2 - 48 = 0$

56. $25z^2 - 100 = 0$

57. $(t - 2)^2 - 16 = 0$

58. $(s + 4)^2 - 49 = 0$

59. $9x^2 + 15x = 0$

60. $4x^2 - 6x = 0$

61. $x^2 - 3x - 10 = 0$

62. $x^2 - x - 12 = 0$

63. $x^2 + 16x + 64 = 0$

64. $x^2 - 12x + 36 = 0$

65. $4z^2 - 12z + 9 = 0$

66. $16t^2 + 48t + 36 = 0$

67. $14x^2 + 9x + 1 = 0$ **68.** $4x^2 + 15x - 25 = 0$

69. $y(y + 6) = 72$ **70.** $x(x - 3) = 10$

71. $t(2t - 3) = 35$ **72.** $3u(3u + 1) = 20$

73. $(a + 2)(a + 5) = 10$

74. $(x - 8)(x - 7) = 20$

75. $(x - 4)(x + 5) = 10$

76. $(u - 8)(u + 10) = 63$

77. $x(x + 10) - 2(x + 10) = 0$

78. $u(u - 3) + 3(u - 3) = 0$

79. $6t^3 - t^2 - t = 0$

80. $3u^3 - 5u^2 - 2u = 0$

81. $x^2(x - 25) - 16(x - 25) = 0$

82. $y^2(y + 250) - (y + 250) = 0$

83. $a^3 + 2a^2 - 9a - 18 = 0$

84. $x^3 - 2x^2 - 4x + 8 = 0$

In Exercises 85–88, use a graphing utility to solve the equation graphically.

85. $x^2 - 8x + 12 = 0$ **86.** $(x - 2)^2 - 9 = 0$

87. $x^3 - 4x = 0$ **88.** $2x^3 - 5x^2 - 12x = 0$

89. The sum of a positive number and its square is 240. Find the number.

90. Find two consecutive positive even integers whose product is 168.

91. *Geometry* The rectangular floor of a storage shed has an area of 330 square feet. The length of the floor is 7 feet more than its width (see figure). Find the dimensions of the floor.

92. *Geometry* The outside dimensions of a picture frame are 28 centimeters and 23 centimeters (see figure). The area of the exposed part of the picture is 414 square centimeters. Find the width w of the frame.

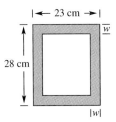

93. *Geometry* The triangular cross section of a machined part must have an area of 48 square inches (see figure). Find the base and height of the triangle if the height is $1\frac{1}{2}$ times the base.

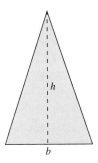

94. *Geometry* The height of a triangle is 4 inches less than its base. Find the base and height of the triangle if its area is 70 square inches.

95. *Geometry* An open box with a square base is to be constructed from 880 square inches of material (see figure). What should the dimensions of the base be if the height of the box is to be 6 inches? (*Hint:* The surface area is given by $S = x^2 + 4xh$.)

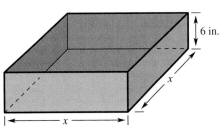

96. *Free-Falling Object*   An object is dropped from a weather balloon 6400 feet above the ground (see figure). Find the time t for the object to reach the ground by solving the equation $-16t^2 + 6400 = 0$.

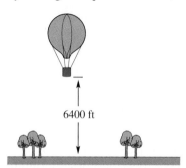

6400 ft

97. *Break-Even Analysis*   The revenue R from the sale of x units of a product is given by $R = 90x - x^2$. The cost of producing x units of the product is given by $C = 200 + 60x$. How many units must the company produce and sell in order to break even?

98. *United States Exports*   For 1980 through 1992, the total U.S. exports E (in billions of dollars) can be modeled by $E = -0.23t^3 + 7.02t^2 - 33.03t + 240.58$, where $t = 0$ represents 1980.

(a) Use a graphing utility to graph the function for $0 \le t \le 15$.

(b) Use the graph to approximate the value of t when $E = 522.96$ billion dollars.

99. *Numerical Reasoning*   For the following items, use the product $P = (x + 5)(x - 4)$.

(a) Complete the table.

x	3	4	5	6	7	8
P						

(b) Use the table to determine how P changes for each one-unit increase in x.

(c) Use the values of P in the table to solve the equation $(x + 5)(x - 4) = 70$.

100. Let a and b be real numbers such that $a \ne 0$. Find the solutions of $ax^2 + bx = 0$.

101. Let a be a nonzero real number. Find the solutions of $ax^2 - ax = 0$.

102. *Exploration*   Solve the equation $2(x + 3)^2 + (x + 3) - 15 = 0$ in two ways.

(a) Let $u = x + 3$, and solve the resulting equation. Find values of x that satisfy the given equation.

(b) Expand and collect like terms in the given equation, and solve the resulting equation for x.

103. *Exploration*   Solve the equations using either method (or both methods) described in Exercise 102.

(a) $3(x + 6)^2 - 10(x + 6) - 8 = 0$

(b) $8(x + 2)^2 - 18(x + 2) + 9 = 0$

CAREER INTERVIEW

Laura Balaoro
Real Estate Broker
Laura Balaoro and Associates Real Estate Services
San Jose, CA 95110

I own and operate an independent real estate brokerage. I deal primarily in residential sales.

One of the more important decisions to be made in the actual purchase of a house is the size of the buyer's down payment. The larger the down payment a buyer can make, the less mortgage interest he or she will pay in the long run. A low down payment can also result in higher closing costs as a result of additional mortgage insurance. However, it is often difficult for a first-time buyer to make a substantial down payment. We normally figure the size of the down payment as a percent of the price of the house, so we use this formula: down payment = house price × percent. It is standard in my area for first-time buyers to use 10% as a minimum down payment. With an average home price of $210,000 in my region, a first-time buyer might expect to need a down payment of at least $210,000 × 10%, or $21,000.

CHAPTER PROJECT: Volume of a Box

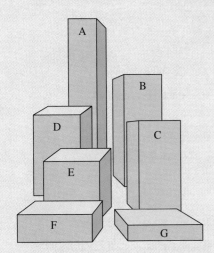

Each of the seven boxes shown at the left has a surface area S of 216 square inches. If, for each box, x is the side length of each square base and h is the height, then

$$S = 2x^2 + 4xh \qquad \text{Formula for surface area of closed box}$$

$$216 = 2x^2 + 4xh. \qquad \text{Substitute 216 for } S.$$

By solving this equation for h, you can write the volume as a function of x.

$$V = x^2 h \qquad \text{Formula for volume}$$

$$= x^2 \left(\frac{54}{x} - \frac{x}{2} \right) \qquad \text{Substitute } \left(\frac{54}{x} \right) - \left(\frac{x}{2} \right) \text{ for } h.$$

$$= 54x - \frac{1}{2}x^3 \qquad \text{Simplify.}$$

Use this information to investigate the questions below.

	Side, x	Height, h	Volume, V
Box A	3.0 inches		
Box B	4.0 inches		
Box C	4.5 inches		
Box D	5.0 inches		
Box E	6.0 inches		
Box F	8.0 inches		
Box G	9.0 inches		

1. Complete the table above, assuming that each closed box has a surface area of 216 square inches. Which box has the greatest volume?

2. Use a graphing utility to graph the volume function. Explain how to use the graph to find the box with the greatest volume.

3. Suppose the seven boxes are open (without tops). In this case, if the surface area remains 216 square inches, the surface area formula for each box is $S = x^2 + 4xh = 216$. Solve this equation for h in terms of x and use the result to write the volume as a function of x.

4. Using the formulas you found in Question 3, complete the table above for the seven open boxes. Which box has the greatest volume?

5. Use a graphing utility to graph the volume function for the open boxes. Explain how to use the graph to find the box with the greatest volume. Does the result agree with that obtained in Question 4? Explain your reasoning.

CHAPTER SUMMARY

After studying this chapter, you should have acquired the following skills.
These skills are keyed to the Review Exercises that begin on page 251.
Answers to odd-numbered Review Exercises are given in the back of the
book.

- State why certain algebraic expressions are not polynomials.
  (Section 3.1)   **Review Exercises 1–4**

- Find and correct errors involving operations with polynomials.
  (Sections 3.1, 3.2)   **Review Exercises 5–8**

- Simplify expressions by performing arithmetic operations.
  (Sections 3.1, 3.2)   **Review Exercises 9–32**

- Multiply polynomials. *(Section 3.2)*   **Review Exercises 33–40**

- Factor out the greatest common factors from polynomials. *(Section 3.3)*   **Review Exercises 41–44, 66**

- Factor polynomials by grouping. *(Section 3.3)*   **Review Exercises 45–48, 67, 68**

- Factor polynomials that are differences of two squares. *(Section 3.3)*   **Review Exercises 49–52, 65**

- Factor polynomials that are sums or differences of two cubes.
  (Section 3.3)   **Review Exercises 53–56, 69, 70**

- Factor perfect square trinomials. *(Section 3.4)*   **Review Exercises 57–60, 71, 72**

- Factor trinomials of the form $x^2 + bx + c$ and $ax^2 + bx + c$. *(Section 3.4)*   **Review Exercises 61–64, 73–76**

- Graph pairs of equations using a graphing utility, and draw conclusions.
  (Sections 3.1, 3.2, 3.3, 3.4)   **Review Exercises 77, 78**

- Find values of variables for which trinomials are factorable.
  (Section 3.4)   **Review Exercises 79–82**

- Solve polynomial equations. *(Section 3.5)*   **Review Exercises 83–92**

- Solve polynomial equations using a graphing utility. *(Section 3.5)*   **Review Exercises 93, 94**

- Solve real-life problems represented by polynomials. *(Sections 3.2, 3.5)*   **Review Exercises 95, 98**

- Solve problems involving geometry. *(Sections 3.2, 3.4, 3.5)*   **Review Exercises 96, 97, 100, 101**

- Translate a verbal statement into a polynomial equation and solve.
  (Section 3.5)   **Review Exercise 99**

REVIEW EXERCISES

In Exercises 1–4, state why the algebraic expression is not a polynomial.

1. $y^2 - \dfrac{2}{y}$

2. $x^2 + 2 + 3x^{1/2}$

3. $4 + 2|x^3|$

4. $z^2 - 2 + 4z^{-2}$

In Exercises 5–8, correct the error.

5. $-4(x - 2) = -4x - 8$

6. $6x - (x + 7) = 5x + 7$

7. $(x + 3)^2 = x^2 + 9$

8. $(x + 4)(x - 4) = x^2 + 8x - 16$

In Exercises 9–32, simplify the expression.

9. $5x + 3x^2 + x - 4x^2$

10. $\frac{1}{2}x + \frac{2}{3} + 4x + \frac{1}{3}$

11. $3t - 2(t^2 - t - 5)$

12. $10y^2 - 5(y^2 - 9)$

13. $(-x^3 - 3x) - 4(2x^3 - 3x + 1)$

14. $(7z^2 + 6z) - 3(5z^2 + 2z)$

15. $3y^2 - [2y - 3(y^2 + 5)]$

16. $(16a^3 + 5a) - 5[a + (2a^3 - 1)]$

17. $(-6z)^3$

18. $4x^3y^2(-3y)^2(2)$

19. $-(u^2v)^2(-4u^3v)$

20. $(12x^2y)(3x^2y^4)$

21. $\dfrac{120u^5v^3}{15u^3v}$

22. $-\dfrac{(-2x^2y^3)^2}{-3xy^2}$

23. $\dfrac{72x^4}{6x^2}$

24. $\left(-\dfrac{1}{2}y^2\right)^2$

25. $(-2x)^3(x + 4)$

26. $3y(-4y)(y - 2)$

27. $(5x + 3)(3x - 4)$

28. $(3y^2 + 2)(4y^2 - 5)$

29. $(2x^2 - 3x + 2)(2x + 3)$

30. $(5s^3 + 4s - 3)(4s - 5)$

31. $2u(u - 7) - (u + 1)(u - 7)$

32. $(3v + 2)(-5v) + 5v(3v + 2)$

In Exercises 33–40, find the product.

33. $(4x - 7)^2$

34. $(8 - 3x)^2$

35. $(5u - 8)(5u + 8)$

36. $(7a + 4)(7a - 4)$

37. $(2z + 3)(3z - 5)$

38. $(6t + 1)(t - 11)$

39. $[(u - 3) + v][(u - 3) - v]$

40. $[(m - 5) + n]^2$

In Exercises 41–44, factor out the greatest common factor.

41. $6x^2 + 15x^3$

42. $8y - 12y^4$

43. $28(x + 5) - 70(x + 5)^2$

44. $(u - 9v)(u - v) + v(u - 9v)$

In Exercises 45–48, factor by grouping.

45. $v^3 - 2v^2 - v + 2$

46. $y^3 + 4y^2 - y - 4$

47. $t^3 + 3t^2 - 9t - 9$

48. $x^3 + 7x^2 + 3x + 21$

In Exercises 49–52, factor the difference of two squares.

49. $9a^2 - 100$

50. $b^2 - 900$

51. $(u + 6)^2 - 81$

52. $(y - 3)^2 - 16$

In Exercises 53–56, factor the sum or difference of two cubes.

53. $u^3 - 1$

54. $t^3 - 125$

55. $8x^3 + 27$

56. $27x^3 + 64$

In Exercises 57–60, factor the perfect square trinomial.

57. $x^2 - 40x + 400$

58. $y^2 + 26y + 169$

59. $4s^2 + 40s + 100$

60. $u^2 - 10uv + 25v^2$

In Exercises 61–64, factor the trinomial.

61. $x^2 + 2x - 35$

62. $x^2 - 12x + 32$

63. $18x^2 + 27x + 10$

64. $5x^2 + 11x - 12$

In Exercises 65–76, factor the expression completely.

65. $4a - 64a^3$

66. $3b + 27b^3$

67. $8x(2x - 3) - 4(2x - 3)$

68. $x^3 + 3x^2 - 4x - 12$

69. $8x^3 + 1$

70. $t^3 - 216$

71. $4u^2 - 28u + 49$

72. $x^2 - \frac{2}{3}x + \frac{1}{9}$

73. $\frac{1}{4}x^2 + x - 1$

74. $3x^2 + 23x - 8$

75. $4t^2 - t + 13$

76. $6h^2 - h - \frac{1}{3}$

In Exercises 77 and 78, use a graphing utility to graph the equations on the same screen. What can you conclude?

77. $y_1 = 1 - (x - 2)^2,\ y_2 = -(x - 1)(x - 3)$

78. $y_1 = 2x^2 + 3x - 27,\ y_2 = (2x + 9)(x - 3)$

In Exercises 79 and 80, find all values of b for which the trinomial is factorable.

79. $x^2 + bx + 5$

80. $x^2 + bx + 6$

In Exercises 81 and 82, find two values of c for which the trinomial is factorable.

81. $x^2 + 7x + c$

82. $3x^2 - 5x + c$

In Exercises 83–92, solve the equation.

83. $10x(x - 3) = 0$

84. $3x(4x + 7) = 0$

85. $v^2 - 100 = 0$

86. $(x + 3)^2 - 25 = 0$

87. $x^2 - 25x = -150$

88. $4t^2 - 12t = -9$

89. $3s^2 - 2s - 8 = 0$

90. $2y^3 + 2y^2 - 24y = 0$

91. $z(5 - z) + 36 = 0$

92. $b^3 - 6b^2 - b + 6 = 0$

In Exercises 93 and 94, use a graphing utility to solve the equation. Explain your strategy.

93. $y = x^2 - 10x + 21$

94. $y = x^3 - 6x^2$

95. *Probability* The probability of three successes in five trials of an experiment is given by $10p^3(1 - p)^2$. Find this product.

96. *Geometry* A rectangle has a length of l units and a width of $l - 5$ units. Find (a) the perimeter and (b) the area of the rectangle.

97. *Geometry* The figure shows a square with sides of length x, within which is a smaller square with sides of length y.

(a) Remove the smaller square from the larger square. What is the area of the remaining figure?

(b) After removing the small square, slide and rotate the remaining top rectangle so that it fits against the right side of the figure. What are the dimensions of the resulting rectangle and what special product formula have you demonstrated geometrically?

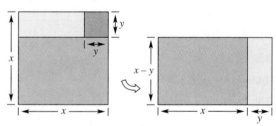

98. *Free-Falling Object* An object is thrown upward from a height of 48 feet with an initial velocity of 32 feet per second. Find the time t for the object to reach the ground by solving the following equation.

$$-16t^2 + 32t + 48 = 0$$

99. Find two consecutive positive odd integers whose product is 195.

100. *Geometry* The width of a rectangle is $\frac{3}{4}$ of its length. Find the dimensions of the rectangle if its area is 432 square inches.

101. *Geometry* The perimeter of a rectangular storage lot at a car dealership is 800 feet. The lot is surrounded by fencing that costs $15 per foot for the front and $10 per foot for the remaining three sides. Find the dimensions of the parking lot if the total cost of the fencing is $9500.

CHAPTER TEST

Take this test as you would take a test in class. After you are done, check your work against the answers given in the back of the book.

1. Determine the degree and leading coefficient of $-5.2x^3 + 3x^2 - 8$.

2. Explain why the following expression is not a polynomial.

$$\frac{4}{x^2 + 2}$$

In Exercises 3–9, perform the operations and simplify.

3. (a) $(5a^2 - 3a + 4) + (a^2 - 4)$ (b) $(16 - y^2) - (16 + 2y + y^2)$

4. (a) $-2(2x^4 - 5) + 4x(x^3 + 2x - 1)$ (b) $4t - [3t - (10t + 7)]$

5. (a) $(-2u^2v)^3(3v^2)$ (b) $3(5x)(2xy)^2$

6. (a) $2y\left(\dfrac{y}{4}\right)^2$ (b) $\dfrac{(-3x^2y)^4}{6x^2}$

7. (a) $-3x(x - 4)$ (b) $(2x - 3y)(x + 5y)$

8. (a) $(x - 1)[2x + (x - 3)]$ (b) $(2s - 3)(3s^2 - 4s + 7)$

9. (a) $(4x - 3)^2$ (b) $[4 - (a + b)][4 + (a + b)]$

In Exercises 10–15, factor the expression completely.

10. $18y^2 - 12y$

11. $v^2 - \dfrac{16}{9}$

12. $x^3 - 3x^2 - 4x + 12$

13. $9u^2 - 6u + 1$

14. $6x^2 - 26x - 20$

15. $x^3 + 27$

In Exercises 16 and 17, solve the equation.

16. $(y + 2)^2 - 9 = 0$

17. $12 + 5y - 3y^2 = 0$

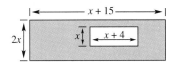

Figure for 18

18. Find the area of the shaded region in the figure at the right.

19. The length of a rectangle is $1\frac{1}{2}$ times its width. Find the dimensions of the rectangle if its area is 54 square centimeters.

20. The height of a free-falling object is given by

$$h = -16t^2 - 40t + 144$$

where h is measured in feet and t is measured in seconds. The object is projected downward when $t = 0$. How long does it take to hit the ground?

CUMULATIVE TEST: CHAPTERS P–3

Take this test as you would take a test in class. After you are done, check your work against the answers given in the back of the book.

1. Evaluate: $-\dfrac{8}{45} \div \dfrac{12}{25}$

2. Write an algebraic expression for the statement, "The number n is tripled and the product is decreased by 8."

In Exercises 3 and 4, perform the operations and simplify.

3. (a) $(2a^2b)^3(-ab^2)^2$ (b) $3x(x^2 - 2) - x(x^2 + 5)$

4. (a) $t(3t - 1) - 2t(t + 4)$ (b) $[2 + (x - y)]^2$

In Exercises 5 and 6, solve the equations.

5. (a) $12 - 5(3 - x) = x + 3$ (b) $1 - \dfrac{x + 2}{4} = \dfrac{7}{8}$

6. (a) $y^2 - 64 = 0$ (b) $2t^2 - 5t - 3 = 0$

7. Your annual automobile insurance premium is $1225. Because of a driving violation, your premium is increased 15%. What is your new premium?

8. The triangles at the right are similar. Solve for x by using the fact that corresponding sides of similar triangles are proportional.

9. Solve $|x - 2| \geq 3$ and sketch its solution.

10. The revenue from selling x units of a product is $R = 12.90x$. The cost of producing x units is $C = 8.50x + 450$. To obtain a profit, the revenue must be greater than the cost. For what values of x will this product produce a profit? Explain your reasoning.

11. Determine whether the equation $x - y^3 = 0$ represents y as a function of x.

12. Find the domain of the function $f(x) = \sqrt{x - 2}$.

13. Given $f(x) = x^2 - 3x$, find (a) $f(4)$ and (b) $f(c + 3)$.

14. Find the slope of the line passing through $(-4, 0)$ and $(4, 6)$. Then, find the distance between the points.

15. Factor $y^3 - 3y^2 - 9y + 27$ by grouping.

16. Factor $3x^2 - 8x - 35$.

In Exercises 17–20, graph the equation.

17. $4x + 3y - 12 = 0$ **18.** $y = 1 - (x - 2)^2$

19. $y = |x + 3|$ **20.** $y = x^3 - 1$

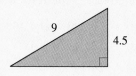

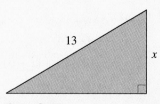

Figure for 8

Rational Expressions and Equations

4

- Simplifying Rational Expressions
- Multiplying and Dividing Rational Expressions
- Adding and Subtracting Rational Expressions
- Dividing Polynomials
- Graphing Rational Functions
- Solving Rational Equations

When air resistance is not considered, the velocity v of an object that falls from rest can be modeled by

$$v = -32t \qquad \text{Velocity function}$$

where v is measured in feet per second and t is the time in seconds. With this model, notice that the longer the object falls, the faster it falls. This model works well for compact objects that fall for only a few seconds.

For objects with large surface areas or objects that fall for several seconds, however, air resistance hampers the object's fall so that it reaches a *terminal velocity*. The table and graph at the right compare the velocity of a sky diver with no air resistance (v_1), with air resistance but no parachute (v_2), and with air resistance and a parachute (v_3).

t	0 sec	1 sec	2 sec	3 sec	4 sec
v_1	0 ft/sec	−32 ft/sec	−64 ft/sec	−96 ft/sec	−128 ft/sec
v_2	0 ft/sec	−24 ft/sec	−38 ft/sec	−52 ft/sec	−60 ft/sec
v_3	0 ft/sec	−18 ft/sec	−20 ft/sec	−20 ft/sec	−20 ft/sec

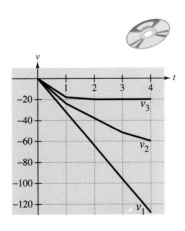

The chapter project related to this information is on page 315.

255

4.1	**Simplifying Rational Expressions**
	The Domain of a Rational Expression ▪ Simplifying Rational Expressions

The Domain of a Rational Expression

Algebra may seem simpler when you realize that it consists primarily of a basic set of operations, including addition, subtraction, multiplication, division, factoring, and simplifying, that are applied to different types of algebraic expressions. For instance, in Chapter 3 you applied these operations to *polynomials*. In this chapter, you will apply these operations to *rational expressions*.

NOTE Like polynomials, rational expressions can be used to describe functions. Such functions are called **rational functions.**

> **Definition of a Rational Expression**
>
> Let u and v be polynomials. The algebraic expression
>
> $$\frac{u}{v}$$
>
> is a **rational expression.** The **domain** of this rational expression is the set of all real numbers for which $v \neq 0$.

EXAMPLE 1 Finding the Domain of a Rational Function

Find the domains of the following rational functions.

a. $f(x) = \dfrac{4}{x-2}$ **b.** $g(x) = \dfrac{2x+5}{8}$ **c.** $h(x) = 3x^2 + 2x - 5$

Solution

a. The denominator is zero when $x - 2 = 0$ or $x = 2$. Therefore, the domain is all real values of x such that $x \neq 2$. In interval notation, you can write the domain as

Domain $= (-\infty, 2) \cup (2, \infty)$.

b. The denominator, 8, is never zero, hence, the domain is the set of *all* real numbers. In interval notation, you can write the domain as

Domain $= (-\infty, \infty)$.

c. Note that any polynomial is also a rational expression, because you can consider its denominator to be 1. The domain of this function is the set of all real numbers.

DISCOVERY

Use a graphing utility to graph the equation

$$y = \frac{4}{x-2}.$$

Then use the TRACE feature of the utility to determine the behavior of the graph near $x = 2$. Try graphing equations that correspond to parts (b) and (c) of Example 1. How does each of these graphs differ from the graph of $y = 4/(x - 2)$?

In applications involving rational functions, it is often necessary to further restrict the domain. To indicate such a restriction, write the domain to the right of the fraction. For instance, the domain of the rational function

$$f(x) = \frac{x^2 + 20}{x + 4}, \qquad x > 0$$

is the set of positive real numbers, as indicated by the inequality $x > 0$. (Note that the normal domain of this function would be all real values of x such that $x \neq -4$. However, because "$x > 0$" is listed to the right of the function, the domain is restricted by this inequality.)

EXAMPLE 2   An Application Involving a Restricted Domain

You have started a small manufacturing business. The initial investment for the business is $120,000. The cost of each unit that you manufacture is $15. Thus, your total cost of producing x units is

$$C = 15x + 120,000. \qquad \text{Cost function}$$

Your average cost per unit depends on the number of units produced. For instance, the average cost per unit $\overline{C}$ for producing 100 units is

$$\overline{C} = \frac{15(100) + 120,000}{100} = \$1215. \qquad \text{Average cost per unit for 100 units}$$

The average cost per unit decreases as the number of units increases. For instance, the average cost per unit $\overline{C}$ for producing 1000 units is

$$\overline{C} = \frac{15(1000) + 120,000}{1000} = \$135. \qquad \text{Average cost per unit for 1000 units}$$

In general, the average cost of producing x units is

$$\overline{C} = \frac{15x + 120,000}{x}. \qquad \text{Average cost per unit for } x \text{ units}$$

What is the domain of this rational function?

Solution

If you were considering this function from only a mathematical point of view, you would say that the domain is all real values of x such that $x \neq 0$. However, because this fraction is a mathematical model representing a real-life situation, you must consider which values of x make sense in real life. For this model, the variable x represents the number of units that you produce. Assuming that you cannot produce a fractional number of units, you conclude that the domain is the set of positive integers. That is,

Domain = $\{1, 2, 3, 4, \ldots\}$.

Simplifying Rational Expressions

As with numerical fractions, a rational expression is said to be **simplified** or **in reduced form** if its numerator and denominator have no factors in common (other than ±1). To reduce fractions, you can apply the following rule.

In the 19th century, mathematicians were expanding their knowledge of calculus and laying the foundation for complex numbers. At that time, the properties of real numbers had not been finalized. Teaching at the University of Berlin, Karl Weierstrass (1815–1897) recognized the need for a logical foundation for the real number system. His work contributed much to the formal real number system that forms the foundation of modern algebra.

Cancellation Rule for Fractions

Let u, v, and w represent numbers, variables, or algebraic expressions such that $v \neq 0$ and $w \neq 0$. Then the following Cancellation Rule is valid.

$$\frac{u\cancel{w}}{v\cancel{w}} = \frac{u}{v}.$$

Be sure you see that this Cancellation Rule allows us to cancel only factors, not terms. For instance, consider the following.

$$\frac{\cancel{2} \cdot 2}{\cancel{2}(x + 5)} \qquad \text{You can cancel common factor 2.}$$

$$\frac{3 + x}{3 + 2x} \qquad \text{You cannot cancel common term 3.}$$

Using the Cancellation Rule to simplify a rational expression requires two steps: (1) completely factor the numerator and denominator and (2) apply the Cancellation Rule to cancel any *factors* that are common to both the numerator and denominator. Thus, your success in simplifying rational expressions actually lies in your ability to *completely factor* the polynomials in both the numerator and denominator.

EXAMPLE 3   *Simplifying a Rational Expression*

Simplify $\dfrac{2x^3 - 6x}{6x^2}$.

Solution

To begin, completely factor both the numerator and denominator.

$$\frac{2x^3 - 6x}{6x^2} = \frac{2x(x^2 - 3)}{2x(3x)} \qquad \text{Factor numerator and denominator.}$$

$$= \frac{\cancel{2x}(x^2 - 3)}{\cancel{2x}(3x)} \qquad \text{Cancel common factor } 2x.$$

$$= \frac{x^2 - 3}{3x} \qquad \text{Simplified form}$$

EXAMPLE 4   *Adjusting the Domain After Simplifying*

Simplify $\dfrac{x^2 + 2x - 15}{3x - 9}$.

Solution

$$\frac{x^2 + 2x - 15}{3x - 9} = \frac{(x + 5)(x - 3)}{3(x - 3)} \qquad \text{Factor numerator and denominator.}$$

$$= \frac{(x + 5)(x - 3)}{3(x - 3)} \qquad \text{Cancel common factor } (x - 3).$$

$$= \frac{x + 5}{3}, \quad x \neq 3 \qquad \text{Simplified form}$$

Canceling common factors from the numerator and denominator of a rational expression can change its domain. For instance, in Example 4 the domain of the original expression is all real values of x such that $x \neq 3$. Thus, the original expression is equal to the simplified expression for all real numbers *except* 3.

EXAMPLE 5   *Simplifying a Rational Expression*

Simplify $\dfrac{x^3 - 16x}{x^2 - 2x - 8}$.

Solution

$$\frac{x^3 - 16x}{x^2 - 2x - 8} = \frac{x(x^2 - 16)}{(x + 2)(x - 4)} \qquad \text{Partially factor.}$$

$$= \frac{x(x + 4)(x - 4)}{(x + 2)(x - 4)} \qquad \text{Factor completely.}$$

$$= \frac{x(x + 4)(x - 4)}{(x + 2)(x - 4)} \qquad \text{Cancel common factor } (x - 4).$$

$$= \frac{x(x + 4)}{x + 2}, \quad x \neq 4 \qquad \text{Simplified form}$$

In this text, when simplifying a rational expression, we follow the convention of listing *by the simplified expression* all values of x that must be specifically excluded from the domain in order to make the domains of the simplified and original expressions agree. For instance, in Example 5 the restriction $x \neq 4$ must be listed with the simplified expression in order to make the two domains agree. (Note that the value of -2 is excluded from both domains, so it is not necessary to list this value.)

EXAMPLE 6 *Simplification Involving a Change of Sign*

Simplify $\dfrac{2x^2 - 9x + 4}{12 + x - x^2}$.

Solution

$$\frac{2x^2 - 9x + 4}{12 + x - x^2} = \frac{(2x - 1)(x - 4)}{(4 - x)(3 + x)} \qquad \text{Factor numerator and denominator.}$$

$$= \frac{(2x - 1)(x - 4)}{-(x - 4)(3 + x)} \qquad (4 - x) = -(x - 4)$$

$$= \frac{(2x - 1)(x - 4)}{-(x - 4)(3 + x)} \qquad \text{Cancel common factor } (x - 4).$$

$$= -\frac{2x - 1}{3 + x}, \qquad x \neq 4 \qquad \text{Simplified form}$$

The simplified form is equivalent to the original expression for all values of x except 4. (Note that -3 is excluded from the domains of both the original and simplified expressions.)

In Example 6, be sure you see that when dividing the numerator and denominator by the common factor of $(x - 4)$, you keep the minus sign. In the simplified form of the fraction, we usually like to move the minus sign out in front of the fraction. However, this is a personal preference. All of the following forms are legitimate.

$$-\frac{2x - 1}{3 + x} = \frac{-(2x - 1)}{3 + x} = \frac{2x - 1}{-3 - x} = \frac{2x - 1}{-(3 + x)}$$

In the next two examples, the Cancellation Rule is used to simplify rational expressions that involve more than one variable.

EXAMPLE 7 *A Rational Expression Involving Two Variables*

Simplify $\dfrac{3xy + y^2}{2y}$.

Solution

$$\frac{3xy + y^2}{2y} = \frac{y(3x + y)}{2y} \qquad \text{Factor numerator and denominator.}$$

$$= \frac{y(3x + y)}{2y} \qquad \text{Cancel common factor } y.$$

$$= \frac{3x + y}{2}, \qquad y \neq 0 \qquad \text{Simplified form}$$

EXAMPLE 8   A Rational Expression Involving Two Variables

Simplify $\dfrac{2x^2 + 2xy - 4y^2}{5x^3 - 5xy^2}$.

Solution

$$\frac{2x^2 + 2xy - 4y^2}{5x^3 - 5xy^2} = \frac{2(x - y)(x + 2y)}{5x(x - y)(x + y)} \qquad \text{Factor numerator and denominator.}$$

$$= \frac{2(x - y)(x + 2y)}{5x(x - y)(x + y)} \qquad \text{Cancel common factor } (x - y).$$

$$= \frac{2(x + 2y)}{5x(x + y)}, \quad x \neq y \qquad \text{Simplified form}$$

As you study the examples and work the exercises in this and the following three sections, keep in mind that you are *rewriting expressions in simpler forms.* You are not solving equations. Equal signs are used in the steps of the simplification process only to indicate that the new form of the expression is *equivalent* to the previous one.

Group Activities   You Be the Instructor

Error Analysis   Suppose you are the instructor of an algebra course. One of your students turns in the following incorrect solutions. Find the errors, discuss the student's misconceptions, and construct correct solutions.

a. $\dfrac{3x^2 + 5x - 4}{x} = 3x + 5 - 4 = 3x + 1$

b. $\dfrac{x^2 + 7x}{x + 7} = \dfrac{x^2}{x} + \dfrac{7x}{7} = x + x = 2x$

4.1 Exercises

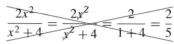

Discussing the Concepts

1. Define the term *rational expression*.

2. Give an example of a rational function whose domain is the set of all real numbers.

3. How do you determine whether a rational expression is in reduced form?

4. Can you cancel common terms from the numerator and denominator of a rational expression? Explain.

5. *Error Analysis*   Describe the error.

$$\frac{2x^2}{x^2+4} = \frac{2x^2}{x^2+4} = \frac{2}{1+4} = \frac{2}{5}$$

6. Is the following statement true? Explain.

$$\frac{6x-5}{5-6x} = -1$$

Problem Solving

In Exercises 7–12, find the domain of the expression.

7. $\dfrac{5}{x-8}$

8. $\dfrac{9}{x-13}$

9. $\dfrac{x}{x^2+4}$

10. $\dfrac{x}{x^2-4}$

11. $\dfrac{y+5}{y^2-3y}$

12. $\dfrac{3t}{t^2-2t-3}$

In Exercises 13–16, evaluate the function as indicated. If it is not possible, state the reason.

13. $f(x) = \dfrac{4x}{x+3}$

    (a) $f(1)$ (b) $f(-2)$

    (c) $f(-3)$ (d) $f(0)$

14. $g(t) = \dfrac{t-2}{2t-5}$

    (a) $g(2)$ (b) $g\left(\frac{5}{2}\right)$

    (c) $g(-2)$ (d) $g(0)$

15. $h(s) = \dfrac{s^2}{s^2-s-2}$

    (a) $h(10)$ (b) $h(0)$

    (c) $h(-1)$ (d) $h(2)$

16. $f(x) = \dfrac{x^3+1}{x^2-6x+9}$

    (a) $f(-1)$ (b) $f(3)$

    (c) $f(-2)$ (d) $f(2)$

In Exercises 17 and 18, describe the domain.

17. *Inventory Cost*   The inventory cost I when x units of a product are ordered from a supplier is given by $I = (0.25x + 2000)/x$.

18. *Average Cost*   The average cost $\overline{C}$ for a manufacturer to produce x units of a product is given by $\overline{C} = (1.35x + 4570)/x$.

In Exercises 19–22, complete the statement.

19. $\dfrac{5}{6} = \dfrac{5(\rule{1cm}{0.4pt})}{6(x+3)}$, $x \neq -3$

20. $\dfrac{5x}{12} = \dfrac{25x^2(x-10)}{12(\rule{1cm}{0.4pt})}$, $x \neq 10$

21. $\dfrac{8x}{x-5} = \dfrac{8x(\rule{1cm}{0.4pt})}{x^2-3x-10}$, $x \neq -2$

22. $\dfrac{3-z}{z^2} = \dfrac{(3-z)(\rule{1cm}{0.4pt})}{z^3+2z^2}$, $z \neq -2$

In Exercises 23–34, simplify the expression.

23. $\dfrac{5x}{25}$

24. $\dfrac{32y}{24}$

25. $\dfrac{18x^2y}{15xy^4}$

26. $\dfrac{16y^2z^2}{60y^5z}$

27. $\dfrac{3xy^2}{xy^2+x}$

28. $\dfrac{x+3x^2y}{3xy+1}$

29. $\dfrac{y^2 - 64}{5(3y + 24)}$

30. $\dfrac{x^2 - 25z^2}{x + 5z}$

31. $\dfrac{3 - x}{2x^2 - 3x - 9}$

32. $\dfrac{2y^2 + 13y + 20}{2y^2 + 17y + 30}$

33. $\dfrac{3m^2 - 12n^2}{m^2 + 4mn + 4n^2}$

34. $\dfrac{x^2 + xy - 2y^2}{x^2 + 3xy + 2y^2}$

Using a Table In Exercises 35 and 36, complete the table. What can you conclude?

35.

x	-2	-1	0	1	2	3	4
$\dfrac{x^2 - x - 2}{x - 2}$							
$x + 1$							

36.

x	-2	-1	0	1	2	3	4
$\dfrac{x^2 + 5x}{x}$							
$x + 5$							

Geometry In Exercises 37 and 38, find the ratio of the area of the shaded portion to the total area of the figure.

37.

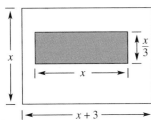

38.

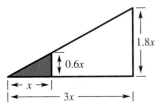

Creating and Using a Model In Exercises 39 and 40, use the following polynomial models, which give the cost of Medicare and the U.S. population aged 65 and older, for the years 1990 to 1995 (see figures).

$$C = 107.1 + 12.64t + 0.54t^2 \qquad \text{Cost of Medicare}$$

$$P = 31.6 + 0.51t - 0.14t^2 \qquad \text{Population}$$

In these models, C represents the total annual cost of Medicare (in billions of dollars), P represents the U.S. population aged 65 and older, and t represents the year, with $t = 0$ corresponding to 1990. (Source: Congressional Budget Office and U.S. Bureau of Census)

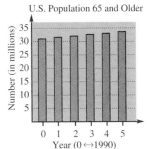

39. Find a rational model that represents the average cost of Medicare *per person aged 65 and older* during the years 1990 to 1995.

40. Use the model found in Exercise 39 to complete the following table, which shows the average cost of Medicare per person aged 65 and older.

Year	Average Cost
1990	
1991	
1992	
1993	
1994	
1995	

Reviewing the Major Concepts

In Exercises 41–44, simplify the expression.

41. $-6x(10 - 7x)$ **42.** $t(t^2 + 1) - t(t^2 - 1)$

43. $(11 - x)(11 + x)$ **44.** $(x - 2)(x^2 + 2x + 4)$

In Exercises 45 and 46, write the quantity as a repeated multiplication.

45. $\left(\dfrac{y}{3}\right)^4$

46. $(-2x)^5$

Additional Problem Solving

In Exercises 47–54, find the domain of the expression.

47. $\dfrac{7x}{x + 4}$ **48.** $x^4 + 2x^2 - 5$

49. $\dfrac{4}{x^2 + 9}$ **50.** $\dfrac{y^2 - 3}{7}$

51. $\dfrac{5t}{t^2 - 16}$ **52.** $\dfrac{z + 2}{z(z - 4)}$

53. $\dfrac{u^2}{u^2 - 2u - 5}$ **54.** $\dfrac{y + 5}{4y^2 - 5y - 6}$

In Exercises 55 and 56, describe the domain.

55. *Geometry*   A rectangle of length x inches has an area of 500 square inches. The perimeter of the rectangle is given by

$$P = 2\left(x + \frac{500}{x}\right).$$

56. *Cost*   The cost in millions of dollars for the government to seize $p\%$ of a certain illegal drug as it enters the country is given by

$$C = \frac{528p}{100 - p}.$$

In Exercises 57–60, complete the statement.

57. $\dfrac{x}{2} = \dfrac{3x(x + 16)^2}{2(\quad)}, \quad x \neq -16$

58. $\dfrac{7}{15} = \dfrac{7(\quad)}{45(x - 10)^2}, \quad x \neq 10$

59. $\dfrac{x + 5}{3x} = \dfrac{(x + 5)(\quad)}{3x^2(x - 2)}, \quad x \neq 2$

60. $\dfrac{3y - 7}{y + 2} = \dfrac{(3y - 7)(\quad)}{y^2 - 4}, \quad y \neq 2$

In Exercises 61–80, simplify the expression.

61. $\dfrac{12y^2}{2y}$ **62.** $\dfrac{15z^3}{15z^3}$

63. $\dfrac{x^2(x - 8)}{x(x - 8)}$ **64.** $\dfrac{a^2b(b - 3)}{b^3(b - 3)^2}$

65. $\dfrac{2x - 3}{4x - 6}$ **66.** $\dfrac{y^2 - 81}{2y - 18}$

67. $\dfrac{5 - x}{3x - 15}$ **68.** $\dfrac{x^2 - 36}{6 - x}$

69. $\dfrac{a + 3}{a^2 + 6a + 9}$ **70.** $\dfrac{u^2 - 12u + 36}{u - 6}$

71. $\dfrac{x^2 - 7x}{x^2 - 14x + 49}$ **72.** $\dfrac{z^2 + 22z + 121}{3z + 33}$

73. $\dfrac{y^3 - 4y}{y^2 + 4y - 12}$ **74.** $\dfrac{x^2 - 7x}{x^2 - 4x - 21}$

75. $\dfrac{15x^2 + 7x - 4}{15x^2 + x - 2}$ **76.** $\dfrac{56z^2 - 3z - 20}{49z^2 - 16}$

77. $\dfrac{5xy + 3x^2y^2}{xy^3}$ **78.** $\dfrac{4u^2v - 12uv^2}{18uv}$

79. $\dfrac{u^2 - 4v^2}{u^2 + uv - 2v^2}$ **80.** $\dfrac{x^2 + 4xy}{x^2 - 16y^2}$

Think About It   In Exercises 81 and 82, explain how you can show that the two expressions are not equivalent.

81. $\dfrac{x - 4}{4} \neq x - 1$ **82.** $\dfrac{x - 4}{x} \neq -4$

In Exercises 83 and 84, write the rational expression in reduced form. (Assume n is a positive integer.)

83. $\dfrac{x^{2n} - 4}{x^n + 2}$ **84.** $\dfrac{x^{2n} + x^n - 12}{x^{n+1} + 4x}$

85. *Average Cost*   A machine shop has a setup cost of $2500 for the production of a new product. The cost of labor and material to produce each unit is $9.25.

   (a) Write a rational expression that gives the average cost per unit when *x* units are produced.

   (b) Determine the domain of the expression in part (a).

   (c) Find the average cost per unit when $x = 100$ units.

86. *Pollution Removal*   The cost in dollars of removing *p*% of the air pollutants in the stack emission of a utility company is given by the rational function

$$C = \frac{80,000p}{100 - p}.$$

Determine the domain of the rational function.

87. *Geometry*   One swimming pool is circular and another is rectangular. The rectangular pool's width is three times its depth, and its length is 6 feet more than its width. The circular pool has a diameter that is twice the width of the rectangular pool, and it is 2 feet deeper. Find the ratio of the volume of the circular pool to the volume of the rectangular pool.

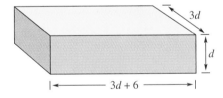

88. *Distance Traveled*   A van starts on a trip and travels at an average speed of 45 miles per hour. Three hours later, a car starts on the same trip and travels at an average speed of 60 miles per hour (see figure).

   (a) Find the distance each vehicle has traveled when the car has been on the road for *t* hours.

   (b) Use the result of part (a) to determine the ratio of the distance the car has traveled to the distance the van has traveled.

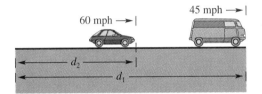

CAREER INTERVIEW

Lisa M. Deitemeyer

Civil Engineer

Johnson-Brittain & Associates, Inc.

Tucson, AZ 85701

Johnson-Brittain does highway design work, primarily for the Arizona Department of Transportation. I am responsible for drainage design of roadways and intersections. It is important that water properly drain off the road surface to avoid flooding problems. One strategy for removing excess water is to use a pipe drainage system that empties into a retention pond. When designing a pipe system and choosing pipe size, I use the equation $V = Q/A$ to find the velocity V of water moving at flow rate Q (volume per unit time) through a given pipe of cross-sectional area A. Finding the water velocity is very important. If it is too fast, erosion can occur in the retention pond. If it is too slow, sedimentation can clog the pipe. As you can see, algebra is very important to my work. I am always solving for different variables that are needed for drainage design.

4.2	**Multiplying and Dividing Rational Expressions**
	Multiplying Rational Expressions ▪ Dividing Rational Expressions ▪ Complex Fractions

Multiplying Rational Expressions

The rule for multiplying rational expressions is the same as the rule for multiplying numerical fractions.

$$\frac{3}{4} \cdot \frac{7}{6} = \frac{21}{24} = \frac{\cancel{3} \cdot 7}{\cancel{3} \cdot 8} = \frac{7}{8}$$

That is, you *multiply numerators, multiply denominators, and write the new fraction in reduced form.*

Multiplying Rational Expressions

Let u, v, w, and z be real numbers, variables, or algebraic expressions such that $v \neq 0$ and $z \neq 0$. Then the product of u/v and w/z is given by

$$\frac{u}{v} \cdot \frac{w}{z} = \frac{uw}{vz}.$$

In order to recognize common factors, write the numerators and denominators in factored form, as demonstrated in Example 1.

EXAMPLE 1 *Multiplying Rational Expressions*

NOTE The result in Example 1 can be read as

$$\frac{4x^3 y}{3xy^4} \cdot \frac{-6x^2 y^2}{10x^4}$$

is equal to

$$-\frac{4}{5y}$$

for all values of x except 0.

Multiply the rational expressions.

$$\frac{4x^3 y}{3xy^4} \cdot \frac{-6x^2 y^2}{10x^4}$$

Solution

$$\frac{4x^3 y}{3xy^4} \cdot \frac{-6x^2 y^2}{10x^4} = \frac{(4x^3 y) \cdot (-6x^2 y^2)}{(3xy^4) \cdot (10x^4)} \qquad \text{Multiply numerators and denominators.}$$

$$= \frac{-24x^5 y^3}{30x^5 y^4} \qquad \text{Simplify.}$$

$$= \frac{-4\cancel{(6)}\cancel{(x^5)}\cancel{(y^3)}}{5\cancel{(6)}\cancel{(x^5)}\cancel{(y^3)}(y)} \qquad \text{Factor and cancel.}$$

$$= -\frac{4}{5y}, \quad x \neq 0 \qquad \text{Simplified form}$$

EXAMPLE 2   *Multiplying Rational Expressions*

Multiply the rational expressions.

$$\frac{x}{5x^2 - 20x} \cdot \frac{x - 4}{2x^2 + x - 3}$$

Solution

$$\frac{x}{5x^2 - 20x} \cdot \frac{x - 4}{2x^2 + x - 3}$$

$$= \frac{x \cdot (x - 4)}{(5x^2 - 20x) \cdot (2x^2 + x - 3)} \qquad \text{Multiply numerators and denominators.}$$

$$= \frac{x(x - 4)}{5x(x - 4)(x - 1)(2x + 3)} \qquad \text{Factor.}$$

$$= \frac{\cancel{x}(\cancel{x - 4})}{5\cancel{x}(\cancel{x - 4})(x - 1)(2x + 3)} \qquad \text{Cancel common factors.}$$

$$= \frac{1}{5(x - 1)(2x + 3)}, \quad x \ne 0, \ x \ne 4 \qquad \text{Simplified form}$$

EXAMPLE 3   *Multiplying Rational Expressions*

Multiply the rational expressions.

$$\frac{4x^2 - 4x}{x^2 + 2x - 3} \cdot \frac{x^2 + x - 6}{4x}$$

Solution

$$\frac{4x^2 - 4x}{x^2 + 2x - 3} \cdot \frac{x^2 + x - 6}{4x}$$

$$= \frac{4x(x - 1)(x + 3)(x - 2)}{(x - 1)(x + 3)(4x)} \qquad \text{Multiply and factor.}$$

$$= \frac{\cancel{4x}(\cancel{x - 1})(\cancel{x + 3})(x - 2)}{(\cancel{x - 1})(\cancel{x + 3})(\cancel{4x})} \qquad \text{Cancel common factors.}$$

$$= x - 2, \quad x \ne 0, \ x \ne 1, \ x \ne -3 \qquad \text{Simplified form}$$

Technology

You can use a graphing utility to check your results when multiplying rational expressions. For instance, in Example 3, try graphing the equations

$$y_1 = \frac{4x^2 - 4x}{x^2 + 2x - 3} \cdot \frac{x^2 + x - 6}{4x}$$

and

$$y_2 = x - 2$$

on the same screen. If the two graphs coincide, as shown below, you can conclude that the two functions are equivalent.

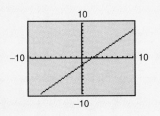

The rule for multiplying fractions can be extended to cover products involving expressions that are not in fractional form. To do this, rewrite the nonfractional expression as a fraction whose denominator is 1. Here is a simple example.

$$\frac{x + 3}{x - 2} \cdot (5x) = \frac{x + 3}{x - 2} \cdot \frac{5x}{1} = \frac{(x + 3)(5x)}{x - 2} = \frac{5x(x + 3)}{x - 2}$$

In the next example, note how to divide out a factor that differs only in sign. Note the step in which $(y - x)$ is rewritten as $(-1)(x - y)$.

EXAMPLE 4 *Multiplying Rational Expressions*

Multiply the rational expressions.

$$\frac{x - y}{y^2 - x^2} \cdot \frac{x^2 - xy - 2y^2}{3x - 6y}$$

Solution

$$\frac{x - y}{y^2 - x^2} \cdot \frac{x^2 - xy - 2y^2}{3x - 6y} = \frac{x - y}{(y + x)(y - x)} \cdot \frac{(x - 2y)(x + y)}{3(x - 2y)}$$

$$= \frac{x - y}{(y + x)(-1)(x - y)} \cdot \frac{(x - 2y)(x + y)}{3(x - 2y)}$$

$$= \frac{(x - y)(x - 2y)(x + y)}{(y + x)(-1)(x - y)(3)(x - 2y)}$$

$$= \frac{\cancel{(x - y)}\cancel{(x - 2y)}\cancel{(x + y)}}{\cancel{(x + y)}(-1)\cancel{(x - y)}(3)\cancel{(x - 2y)}}$$

$$= -\frac{1}{3}, \quad x \neq y,\ x \neq -y,\ x \neq 2y$$

The rule for multiplying rational expressions can be extended to cover the product of three or more fractions, as shown in Example 5.

EXAMPLE 5 *Multiplying Three Rational Expressions*

Multiply the rational expressions.

$$\frac{x^2 - 3x + 2}{x + 2} \cdot \frac{3x}{x - 2} \cdot \frac{2x + 4}{x^2 - 5x}$$

Solution

$$\frac{x^2 - 3x + 2}{x + 2} \cdot \frac{3x}{x - 2} \cdot \frac{2x + 4}{x^2 - 5x}$$

$$= \frac{(x - 1)(x - 2)(3)(x)(2)(x + 2)}{(x + 2)(x - 2)(x)(x - 5)} \qquad \text{Multiply and factor.}$$

$$= \frac{(x - 1)\cancel{(x - 2)}(3)\cancel{(x)}(2)\cancel{(x + 2)}}{\cancel{(x + 2)}\cancel{(x - 2)}\cancel{(x)}(x - 5)} \qquad \text{Cancel common factors.}$$

$$= \frac{6(x - 1)}{x - 5}, \quad x \neq 0,\ x \neq 2,\ x \neq -2 \qquad \text{Simplified form}$$

Dividing Rational Expressions

To divide two rational expressions, multiply the first fraction by the *reciprocal* of the second. That is, simply *invert the divisor and multiply*. For instance, to perform the following division

$$\frac{x}{x+3} \div \frac{4}{x-1}$$

invert the fraction $4/(x-1)$ and multiply, as follows.

$$\frac{x}{x+3} \div \frac{4}{x-1} = \frac{x}{x+3} \cdot \frac{x-1}{4}$$

$$= \frac{x(x-1)}{(x+3)(4)}$$

$$= \frac{x(x-1)}{4(x+3)}$$

Dividing Rational Expressions

Let u, v, w, and z be real numbers, variables, or algebraic expressions such that $v \neq 0$, $w \neq 0$, and $z \neq 0$. The quotient of u/v and w/z is

$$\frac{u}{v} \div \frac{w}{z} = \frac{u}{v} \cdot \frac{z}{w} = \frac{uz}{vw}.$$

EXAMPLE 6   *Dividing Rational Expressions*

Perform the division.

$$\frac{2x}{3x-12} \div \frac{x^2-2x}{x^2-6x+8}$$

Solution

$$\frac{2x}{3x-12} \div \frac{x^2-2x}{x^2-6x+8}$$

$$= \frac{2x}{3x-12} \cdot \frac{x^2-6x+8}{x^2-2x} \qquad \text{Invert divisor and multiply.}$$

$$= \frac{(2x)(x-2)(x-4)}{(3)(x-4)(x)(x-2)} \qquad \text{Factor.}$$

$$= \frac{(2x)(x-2)(x-4)}{(3)(x-4)(x)(x-2)} \qquad \text{Cancel common factors.}$$

$$= \frac{2}{3}, \quad x \neq 0, \ x \neq 2, \ x \neq 4 \qquad \text{Simplified form}$$

Complex Fractions

Problems involving division of two rational expressions are sometimes written as **complex fractions.** The rules for dividing fractions still apply in such cases.

EXAMPLE 7   *Simplifying a Complex Fraction*

$$\frac{\left(\dfrac{x^2 + 2x - 3}{x - 3}\right)}{4x + 12} = \frac{\left(\dfrac{x^2 + 2x - 3}{x - 3}\right)}{\left(\dfrac{4x + 12}{1}\right)} \qquad \text{Rewrite denominator.}$$

$$= \frac{x^2 + 2x - 3}{x - 3} \cdot \frac{1}{4x + 12} \qquad \text{Invert divisor and multiply.}$$

$$= \frac{(x - 1)(x + 3)}{(x - 3)(4)(x + 3)} \qquad \text{Factor.}$$

$$= \frac{(x - 1)\cancel{(x + 3)}}{(x - 3)(4)\cancel{(x + 3)}} \qquad \text{Cancel common factors.}$$

$$= \frac{x - 1}{4(x - 3)}, \quad x \ne -3 \qquad \text{Simplified form}$$

Group Activities   Extending the Concept

Using a Table   Complete the following table with the given values of x.

x	60	100	1000	10,000	100,000	1,000,000
$\dfrac{x - 10}{x + 10}$						
$\dfrac{x + 50}{x - 50}$						
$\dfrac{x - 10}{x + 10} \cdot \dfrac{x + 50}{x - 50}$						

What kind of pattern do you see? Try to explain what is going on. Can you see why?

4.2 Exercises

Discussing the Concepts

1. In your own words, explain how to divide rational expressions.

2. Explain how to divide a rational expression by a polynomial. Give an example.

3. Define the term *complex fraction.* Give an example and show how to simplify the fraction.

4. *Error Analysis* Describe the error.

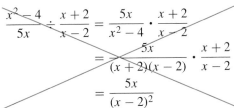

$$\frac{x^2-4}{5x} \div \frac{x+2}{x-2} = \frac{5x}{x^2-4} \cdot \frac{x+2}{x-2}$$

$$= \frac{5x}{(x+2)(x-2)} \cdot \frac{x+2}{x-2}$$

$$= \frac{5x}{(x-2)^2}$$

Problem Solving

In Exercises 5 and 6, evaluate the expression for each value of x. If it is not possible, state the reason.

Expression   *Values*

5. $\dfrac{x-10}{4x}$ (a) $x = 10$ (b) $x = 0$
              (c) $x = -2$ (d) $x = 12$

6. $\dfrac{x^2-4x}{x^2-9}$ (a) $x = 0$ (b) $x = 4$
              (c) $x = 3$ (d) $x = -3$

In Exercises 7–10, complete the statement.

7. $\dfrac{7}{3y} = \dfrac{7x^2}{3y(\quad)}$, $\quad x \ne 0$

8. $\dfrac{2x}{x-3} = \dfrac{14x(x-3)^2}{(x-3)(\quad)}$, $\quad x \ne 0$

9. $\dfrac{13x}{x-2} = \dfrac{13x(\quad)}{4-x^2}$, $\quad x \ne -2$

10. $\dfrac{x^2}{10-x} = \dfrac{x^2(\quad)}{x^2-10x}$, $\quad x \ne 0$

In Exercises 11–18, multiply and simplify.

11. $\dfrac{7x^2}{3} \cdot \dfrac{9}{14x}$

12. $\dfrac{6}{5a} \cdot (25a)$

13. $\dfrac{8}{3+4x} \cdot (9+12x)$

14. $\dfrac{1-3x}{4} \cdot \dfrac{46}{15-45x}$

15. $\dfrac{(2x-3)(x+8)}{x^3} \cdot \dfrac{x}{3-2x}$

16. $\dfrac{x+14}{x^3(10-x)} \cdot \dfrac{x(x-10)}{5}$

17. $\dfrac{xu-yu+xv-yv}{xu+yu-xv-yv} \cdot \dfrac{xu+yu+xv+yv}{xu-yu-xv+yv}$

18. $\dfrac{t^2+4t+3}{2t^2-t-10} \cdot \dfrac{t}{t^2+3t+2} \cdot \dfrac{2t^2+4t^3}{t^2+3t}$

In Exercises 19–26, divide and simplify.

19. $\dfrac{3x}{4} \div \dfrac{x^2}{2}$

20. $\dfrac{u}{10} \div u^2$

21. $\dfrac{3(a+b)}{4} \div \dfrac{(a+b)^2}{2}$

22. $\dfrac{x^2+9}{5(x+2)} \div \dfrac{x+3}{5(x^2-4)}$

23. $\dfrac{16x^2+8x+1}{3x^2+8x-3} \div \dfrac{4x^2-3x-1}{x^2+6x+9}$

24. $\dfrac{x^2-25}{x} \div \dfrac{x^3-5x^2}{x^2+x}$

25. $\dfrac{\left(\dfrac{x^2}{12}\right)}{\left(\dfrac{5x}{18}\right)}$

26. $\dfrac{\left[\dfrac{(3u^2v)^2}{6v^3}\right]}{\left[\dfrac{(uv^3)^2}{3uv}\right]}$

In Exercises 27–30, perform the operations and simplify.

27. $\left[\dfrac{x^2}{9} \cdot \dfrac{3(x+4)}{x^2+2x}\right] \div \dfrac{x}{x+2}$

28. $\left(\dfrac{x^2+6x+9}{x^2} \cdot \dfrac{2x+1}{x^2-9}\right) \div \dfrac{4x^2+4x+1}{x^2-3x}$

29. $\left[\dfrac{xy+y}{4x} \div (3x+3)\right] \div \dfrac{y}{3x}$

30. $\left(\dfrac{3u^2-u-4}{u^2}\right)^2 \div \dfrac{3u^2+12u+4}{u^4-3u^3}$

🖩 *Graphical Reasoning* In Exercises 31–34, use a graphing utility to graph the two equations in the same viewing rectangle. Use the graphs to verify that the expressions are equivalent.

31. $y_1 = \dfrac{3x+2}{x} \cdot \dfrac{x^2}{9x^2-4}$, $y_2 = \dfrac{x}{3x-2}$

32. $y_1 = \dfrac{x^2-10x+25}{x^2-25} \cdot \dfrac{x+5}{2}$, $y_2 = \dfrac{x-5}{2}$

33. $y_1 = \dfrac{3x+15}{x^4} \div \dfrac{x+5}{x^2}$, $y_2 = \dfrac{3}{x^2}$

34. $y_1 = (x^2+6x+9) \div \dfrac{2x(x+3)}{3}$, $y_2 = \dfrac{3(x+3)}{2x}$

Geometry In Exercises 35 and 36, write an expression for the area of the shaded region. Then simplify the expression.

35.

36.

Reviewing the Major Concepts

In Exercises 37–40, factor the expression completely.

37. $3x^2 - 21x$

38. $-x^3 + 3x^2 - x + 3$

39. $4t^2 - 169$

40. $y^3 - 64$

41. *Simple Interest* You borrow $12,000 for 6 months. At simple interest of 12%, how much will you owe?

42. *Geometry* A rectangle is 2 inches longer than it is wide. Write an expression that represents its area.

Additional Problem Solving

In Exercises 43–46, complete the statement.

43. $\dfrac{3x}{x-4} = \dfrac{3x(x+2)^2}{(x-4)()}$, $x \neq -2$

44. $\dfrac{x+1}{x} = \dfrac{(x+1)^3}{x()}$, $x \neq -1$

45. $\dfrac{3u}{7v} = \dfrac{3u()}{7v(u+1)}$, $u \neq -1$

46. $\dfrac{3t+5}{t} = \dfrac{(3t+5)()}{5t^2(3t-5)}$, $t \neq \dfrac{5}{3}$

In Exercises 47–62, multiply and simplify.

47. $\dfrac{8s^3}{9s} \cdot \dfrac{6s^2}{32s}$

48. $\dfrac{3x^4}{7x} \cdot \dfrac{8x^2}{9}$

49. $16u^4 \cdot \dfrac{12}{8u^2}$

50. $\dfrac{25}{8x} \cdot \dfrac{8x}{35}$

51. $\dfrac{8u^2v}{3u+v} \cdot \dfrac{u+v}{12u}$

52. $\dfrac{x+25}{8} \cdot \dfrac{8}{x+25}$

53. $\dfrac{12-r}{3} \cdot \dfrac{3}{r-12}$

54. $\dfrac{8-z}{8+z} \cdot \dfrac{z+8}{z-8}$

55. $\dfrac{6r}{r-2} \cdot \dfrac{r^2-4}{33r^2}$

56. $\dfrac{5y-20}{5y+15} \cdot \dfrac{2y+6}{y-4}$

57. $(u-2v)^2 \cdot \dfrac{u+2v}{u-2v}$

58. $\dfrac{x-2y}{x+2y} \cdot \dfrac{x^2+4y^2}{x^2-4y^2}$

59. $\dfrac{2t^2-t-15}{t+2} \cdot \dfrac{t^2-t-6}{t^2-6t+9}$

60. $\dfrac{y^2-16}{2y^3} \cdot \dfrac{4y}{y^2-6y+8}$

61. $\dfrac{x+5}{x-5} \cdot \dfrac{2x^2-9x-5}{3x^2+x-2} \cdot \dfrac{x^2-1}{x^2+7x+10}$

62. $\dfrac{x^3+3x^2-4x-12}{x^3-3x^2-4x+12} \cdot \dfrac{x^2-9}{x}$

In Exercises 63–70, divide and simplify.

63. $\dfrac{7xy^2}{10u^2v} \div \dfrac{21x^3}{45uv}$

64. $\dfrac{25x^2y}{60x^3y^2} \div \dfrac{5x^4y^3}{16x^2y}$

65. $\dfrac{(x^3y)^2}{(x+2y)^2} \div \dfrac{x^2y}{(x+2y)^3}$

66. $\dfrac{x^2-y^2}{2x^2-8x} \div \dfrac{(x-y)^2}{2xy}$

67. $\dfrac{\left(\dfrac{25x^2}{x-5}\right)}{\left(\dfrac{10x}{5-x}\right)}$

68. $\dfrac{\left(\dfrac{5x}{x+7}\right)}{\left(\dfrac{10}{x^2+8x+7}\right)}$

69. $\dfrac{x(x+3)-2(x+3)}{x^2-4} \div \dfrac{x}{x^2+4x+4}$

70. $\dfrac{t^3+t^2-9t-9}{t^2-5t+6} \div \dfrac{t^2+6t+9}{t-2}$

In Exercises 71 and 72, perform the operations and simplify your answer.

71. $\dfrac{2x^2+5x-25}{3x^2+5x+2} \cdot \dfrac{3x^2+2x}{x+5} \div \left(\dfrac{x}{x+1}\right)^2$

72. $\dfrac{t^2-100}{4t^2} \cdot \dfrac{t^3-5t^2-50t}{t^4+10t^3} \div \dfrac{(t-10)^2}{5t}$

73. *Photocopy Rate*   A photocopier produces copies at a rate of 20 pages per minute.

(a) Determine the time required to copy one page.

(b) Determine the time required to copy x pages.

(c) Determine the time required to copy 35 pages.

74. *Pumping Rate*   The rate for a pump is 15 gallons per minute.

(a) Determine the time required to pump 1 gallon.

(b) Determine the time required to pump x gallons.

(c) Determine the time required to pump 130 gallons.

Probability   In Exercises 75 and 76, consider an experiment in which a marble is tossed into a rectangular box whose dimensions are $3x-2$ by x inches. The probability that the marble will come to rest in the unshaded portion of the box is equal to the ratio of the unshaded area to the total area of the figure. Find the probability.

75.

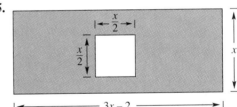

76.

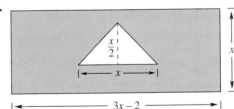

4.3	**Adding and Subtracting Rational Expressions**
	Adding or Subtracting with Like Denominators ■
	Adding or Subtracting with Unlike Denominators ■ Complex Fractions

Adding or Subtracting with Like Denominators

As with numerical fractions, the procedure used to add or subtract two rational expressions depends upon whether the expressions have *like* or *unlike* denominators. To add or subtract two rational expressions with *like* denominators, simply combine their numerators and place the result over the common denominator.

Adding or Subtracting with Like Denominators

If u, v, and w are real numbers, variables, or variable expressions, and $w \neq 0$, the following rules are valid.

1. $\dfrac{u}{w} + \dfrac{v}{w} = \dfrac{u+v}{w}$ Add fractions with like denominators.

2. $\dfrac{u}{w} - \dfrac{v}{w} = \dfrac{u-v}{w}$ Subtract fractions with like denominators.

EXAMPLE 1   *Adding and Subtracting with Like Denominators*

a. $\dfrac{x}{4} + \dfrac{5-x}{4} = \dfrac{x+(5-x)}{4} = \dfrac{5}{4}$

b. $\dfrac{7}{2x-3} - \dfrac{3x}{2x-3} = \dfrac{7-3x}{2x-3}$

STUDY TIP

After adding or subtracting two (or more) rational expressions, check the resulting fraction to see if it can be simplified, as illustrated in Example 2.

EXAMPLE 2   *Subtracting Rational Expressions and Simplifying*

$$\frac{x}{x^2-2x-3} - \frac{3}{x^2-2x-3} = \frac{x-3}{x^2-2x-3} \qquad \text{Subtract.}$$

$$= \frac{x-3}{(x-3)(x+1)} \qquad \text{Factor.}$$

$$= \frac{(x-3)(1)}{(x-3)(x+1)} \qquad \text{Cancel common factor.}$$

$$= \frac{1}{x+1}, \quad x \neq 3 \qquad \text{Simplified form}$$

The rules for adding and subtracting rational expressions with like denominators can be extended to cover sums and differences involving three or more rational expressions, as illustrated in Example 3.

EXAMPLE 3 Combining Three Rational Expressions

$$\frac{x^2 - 26}{x - 5} - \frac{2x + 4}{x - 5} + \frac{10 + x}{x - 5} = \frac{(x^2 - 26) - (2x + 4) + (10 + x)}{x - 5}$$

$$= \frac{x^2 - 26 - 2x - 4 + 10 + x}{x - 5}$$

$$= \frac{x^2 - x - 20}{x - 5}$$

$$= \frac{(x - 5)(x + 4)}{x - 5}$$

$$= x + 4, \quad x \neq 5$$

Adding or Subtracting with Unlike Denominators

To add or subtract rational expressions with *unlike* denominators, you must first rewrite each expression using the **least common multiple** of the denominators of the individual expressions. The least common multiple of two (or more) polynomials is the simplest polynomial that is a multiple of each of the original polynomials.

EXAMPLE 4 Finding Least Common Multiples

a. The least common multiple of

$$6x = 2 \cdot 3 \cdot x \quad \text{and} \quad 2x^2 = 2 \cdot x^2$$

is $2 \cdot 3 \cdot x^2 = 6x^2$.

b. The least common multiple of

$$x^2 - x = x(x - 1) \quad \text{and} \quad 2x - 2 = 2(x - 1)$$

is $2x(x - 1)$.

c. The least common multiple of

$$3x^2 + 6x = 3x(x + 2) \quad \text{and} \quad x^2 + 4x + 4 = (x + 2)^2$$

is $3x(x + 2)^2$.

To add or subtract rational expressions with *unlike* denominators, you must first rewrite the rational expressions so that they have *like* denominators. The like denominator that you use is the least common multiple of the original denominators and is called the **least common denominator** of the original rational expressions. Once the rational expressions have been written with like denominators, you can simply add or subtract the rational expressions using the rules given at the beginning of this section.

EXAMPLE 5 *Adding with Unlike Denominators*

Add the rational expressions: $\dfrac{7}{6x} + \dfrac{5}{8x}$.

Solution

The least common denominator of $6x$ and $8x$ is $24x$, so the first step is to rewrite each fraction with this denominator.

$$\frac{7}{6x} + \frac{5}{8x} = \frac{7(4)}{6x(4)} + \frac{5(3)}{8x(3)} \qquad \text{Rewrite fractions using least common denominator.}$$

$$= \frac{28}{24x} + \frac{15}{24x} \qquad \text{Like denominators}$$

$$= \frac{28 + 15}{24x} \qquad \text{Add fractions.}$$

$$= \frac{43}{24x} \qquad \text{Simplified form}$$

EXAMPLE 6 *Subtracting with Unlike Denominators*

Subtract the rational expressions: $\dfrac{3}{x-3} - \dfrac{5}{x+2}$.

Solution

The least common denominator is $(x-3)(x+2)$.

$$\frac{3}{x-3} - \frac{5}{x+2} = \frac{3(x+2)}{(x-3)(x+2)} - \frac{5(x-3)}{(x-3)(x+2)}$$

$$= \frac{3x+6}{(x-3)(x+2)} - \frac{5x-15}{(x-3)(x+2)}$$

$$= \frac{(3x+6)-(5x-15)}{(x-3)(x+2)}$$

$$= \frac{3x+6-5x+15}{(x-3)(x+2)}$$

$$= \frac{-2x+21}{(x-3)(x+2)}$$

Technology

You can use a graphing utility to check your results when adding or subtracting rational expressions. For instance, in Example 5, try graphing the equations

$$y_1 = \frac{7}{6x} + \frac{5}{8x}$$

and

$$y_2 = \frac{43}{24x}$$

on the same screen. If the two graphs coincide, as shown below, you can conclude that the two functions are equivalent.

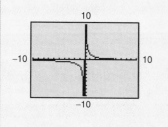

EXAMPLE 7 Adding with Unlike Denominators

NOTE In Example 7, note that the factors in the denominator are $x^2 - 4 = (x + 2)(x - 2)$ and $2 - x$. Because

$$(2 - x) = (-1)(x - 2)$$

the original addition problem can be written as a subtraction problem.

Add the rational expressions: $\dfrac{6x}{x^2 - 4} + \dfrac{3}{2 - x}$.

Solution

$$\frac{6x}{x^2 - 4} + \frac{3}{(-1)(x - 2)} = \frac{6x}{(x + 2)(x - 2)} - \frac{3}{x - 2}$$

$$= \frac{6x}{(x + 2)(x - 2)} - \frac{3(x + 2)}{(x + 2)(x - 2)}$$

$$= \frac{6x}{(x + 2)(x - 2)} - \frac{3x + 6}{(x + 2)(x - 2)}$$

$$= \frac{6x - (3x + 6)}{(x + 2)(x - 2)}$$

$$= \frac{6x - 3x - 6}{(x + 2)(x - 2)}$$

$$= \frac{3x - 6}{(x + 2)(x - 2)}$$

$$= \frac{3(x - 2)}{(x + 2)(x - 2)}$$

$$= \frac{3}{x + 2}, \quad x \neq 2$$

EXAMPLE 8 Combining Three Rational Expressions

$$\frac{2x - 5}{6x + 9} - \frac{4}{2x^2 + 3x} + \frac{1}{x} = \frac{(2x - 5)(x)}{3(2x + 3)(x)} - \frac{(4)(3)}{x(2x + 3)(3)} + \frac{3(2x + 3)}{(x)(3)(2x + 3)}$$

$$= \frac{2x^2 - 5x}{3x(2x + 3)} - \frac{12}{3x(2x + 3)} + \frac{6x + 9}{3x(2x + 3)}$$

$$= \frac{2x^2 - 5x - 12 + 6x + 9}{3x(2x + 3)}$$

$$= \frac{2x^2 + x - 3}{3x(2x + 3)}$$

$$= \frac{(x - 1)(2x + 3)}{3x(2x + 3)}$$

$$= \frac{(x - 1)(2x + 3)}{3x(2x + 3)}$$

$$= \frac{x - 1}{3x}, \quad x \neq -\frac{3}{2}$$

Complex Fractions

Complex fractions can have numerators or denominators that are the sums or differences of fractions. To simplify a complex fraction, first combine its numerator and its denominator into single fractions. Then divide by inverting the denominator and multiplying.

EXAMPLE 9 Simplifying a Complex Fraction

$$\frac{\left(\dfrac{x}{4}+\dfrac{3}{2}\right)}{\left(2-\dfrac{3}{x}\right)}=\frac{\left(\dfrac{x}{4}+\dfrac{6}{4}\right)}{\left(\dfrac{2x}{x}-\dfrac{3}{x}\right)} \qquad \text{Find least common denominators.}$$

$$=\frac{\left(\dfrac{x+6}{4}\right)}{\left(\dfrac{2x-3}{x}\right)} \qquad \begin{array}{l}\text{Add fractions in numerator and de-}\\ \text{nominator.}\end{array}$$

$$=\frac{x+6}{4}\cdot\frac{x}{2x-3} \qquad \text{Invert divisor and multiply.}$$

$$=\frac{x(x+6)}{4(2x-3)}, \qquad x\neq 0 \qquad \text{Simplified form}$$

Another way to simplify the complex fraction given in Example 9 is to multiply the numerator and denominator by the least common denominator of *every* fraction in the numerator and denominator. For this fraction, notice what happens when we multiply the numerator and denominator by $4x$.

$$\frac{\left(\dfrac{x}{4}+\dfrac{3}{2}\right)}{\left(2-\dfrac{3}{x}\right)}=\frac{\left(\dfrac{x}{4}+\dfrac{3}{2}\right)}{\left(2-\dfrac{3}{x}\right)}\cdot\frac{4x}{4x}$$

$$=\frac{\dfrac{x}{4}(4x)+\dfrac{3}{2}(4x)}{2(4x)-\dfrac{3}{x}(4x)}$$

$$=\frac{x^2+6x}{8x-12}$$

$$=\frac{x(x+6)}{4(2x-3)}, \qquad x\neq 0$$

Which of these two methods do you prefer?

EXAMPLE 10   Complex Fractions

Simplify $\dfrac{\left(\dfrac{2}{x+2}\right)}{\left(\dfrac{1}{x+2}+\dfrac{2}{x}\right)}$.

Solution

$$\frac{\left(\dfrac{2}{x+2}\right)}{\left(\dfrac{1}{x+2}+\dfrac{2}{x}\right)}=\frac{\left(\dfrac{2}{x+2}\right)(x)(x+2)}{\dfrac{1}{x+2}(x)(x+2)+\dfrac{2}{x}(x)(x+2)}$$

$$=\frac{2x}{x+2(x+2)}$$

$$=\frac{2x}{3x+4},\quad x\neq -2, x\neq 0$$

Group Activities   Extending the Concept

Comparing Two Methods   Each person in your group should evaluate each of the following expressions at the given variable in two different ways: (1) combine and simplify the rational expressions first and then evaluate the simplified expression at the given variable value, and (2) substitute the given value for the variable first and then simplify the resulting expression. Did you get the same result with each method? Discuss which method you prefer and why. List any advantages and/or disadvantages of each method.

a. $\dfrac{1}{m-4}-\dfrac{1}{m+4}+\dfrac{3m}{m^2-16}, m=2$

b. $\dfrac{x-2}{x^2-9}+\dfrac{3x+2}{x^2-5x+6}, x=4$

c. $\dfrac{3y^2+16y-8}{y^2+2y-8}-\dfrac{y-1}{y-2}+\dfrac{y}{y+4}, y=3$

4.3 Exercises

Discussing the Concepts

1. In your own words, describe how to add or subtract rational expressions with like denominators.

2. In your own words, describe how to add or subtract rational expressions with unlike denominators.

3. Is it possible for the least common denominator of two fractions to be the same as one of the fraction's denominators? If so, give an example.

4. *Error Analysis*   Describe the error.

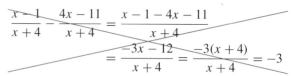

$$\frac{x-1}{x+4} - \frac{4x-11}{x+4} = \frac{x-1-4x-11}{x+4}$$
$$= \frac{-3x-12}{x+4} = \frac{-3(x+4)}{x+4} = -3$$

Problem Solving

In Exercises 5–10, combine and simplify.

5. $\dfrac{5}{8} + \dfrac{7}{8}$

6. $\dfrac{7}{12} - \dfrac{5}{12}$

7. $\dfrac{x}{9} - \dfrac{x+2}{9}$

8. $\dfrac{z^2}{3} + \dfrac{z^2-2}{3}$

9. $\dfrac{5x-1}{x+4} + \dfrac{5-4x}{x+4}$

10. $\dfrac{2x-1}{x(x-3)} + \dfrac{1-x}{x(x-3)}$

In Exercises 11–16, find the least common multiple of the expressions.

11. $5x^2, 20x^3$

12. $14t^2, 42t^5$

13. $15x^2, 3(x+5)$

14. $18y^3, 27y(y-3)^2$

15. $6(x^2-4), 2x(x+2)$

16. $t^3 + 3t^2 + 9t, 2t^2(t^2-9)$

In Exercises 17–20, find the least common denominator of the two fractions and rewrite each fraction using the least common denominator.

17. $\dfrac{n+8}{3n-12}, \dfrac{10}{6n^2}$

18. $\dfrac{8s}{(s+2)^2}, \dfrac{3}{s^3+s^2-2s}$

19. $\dfrac{x-8}{x^2-25}, \dfrac{9x}{x^2-10x+25}$

20. $\dfrac{3y}{y^2-y-12}, \dfrac{y-4}{y^2+3y}$

In Exercises 21–32, perform the operation and simplify.

21. $\dfrac{5}{4x} - \dfrac{3}{5}$

22. $\dfrac{10}{b} + \dfrac{1}{10b^2}$

23. $\dfrac{20}{x-4} + \dfrac{20}{4-x}$

24. $\dfrac{15}{2-t} - \dfrac{7}{t-2}$

25. $25 + \dfrac{10}{x+4}$

26. $\dfrac{100}{x-10} - 8$

27. $\dfrac{x}{x^2-9} + \dfrac{3}{x(x-3)}$

28. $\dfrac{x}{x^2-x-30} - \dfrac{1}{x+5}$

29. $\dfrac{3u}{u^2-2uv+v^2} + \dfrac{2}{u-v}$

30. $\dfrac{1}{x} - \dfrac{3}{y} + \dfrac{3x-y}{xy}$

31. $\dfrac{x+2}{x-1} - \dfrac{2}{x+6} - \dfrac{14}{x^2+5x-6}$

32. $\dfrac{x}{x^2+15x+50} + \dfrac{7}{2(x+10)} - \dfrac{3}{2(x+5)}$

In Exercises 33 and 34, use a graphing utility to graph the two equations on the same screen. Use the graphs to verify that the expressions are equivalent. Verify the results algebraically.

33. $y_1 = \dfrac{2}{x} + \dfrac{4}{x(x-2)}, y_2 = \dfrac{2x}{x(x-2)}$

34. $y_1 = 3 - \dfrac{1}{x-1}, y_2 = \dfrac{3x-4}{x-1}$

In Exercises 35–40, simplify the complex fraction.

35. $\dfrac{\dfrac{1}{2}}{\left(3+\dfrac{1}{x}\right)}$

36. $\dfrac{\left(\dfrac{1}{t}-1\right)}{\left(\dfrac{1}{t}+1\right)}$

37. $\dfrac{\left(3+\dfrac{9}{x-3}\right)}{\left(4+\dfrac{12}{x-3}\right)}$

38. $\dfrac{\left(x+\dfrac{2}{x-3}\right)}{\left(x+\dfrac{6}{x-3}\right)}$

39. $\dfrac{\left(\dfrac{y}{x}-\dfrac{x}{y}\right)}{\left(\dfrac{x+y}{xy}\right)}$

40. $\dfrac{\left(x-\dfrac{2y^2}{x-y}\right)}{x-2y}$

41. *Parallel Resistance*   When two resistors are connected in parallel (see figure), the total resistance is

$$\dfrac{1}{\left(\dfrac{1}{R_1}+\dfrac{1}{R_2}\right)}.$$

Simplify this complex fraction.

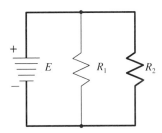

42. *Average of Two Numbers*   Determine the average of the two real numbers $x/3$ and $x/5$. Copy the real number line shown below and plot the average. What can you conclude?

Interpreting a Table   In Exercises 43 and 44, use a graphing utility to complete the table. Comment on the domain and equivalence of the expressions.

43.

x	-3	-2	-1	0	1	2	3
$\dfrac{\left(1-\dfrac{1}{x}\right)}{\left(1-\dfrac{1}{x^2}\right)}$							
$\dfrac{x}{x+1}$							

44.

x	-3	-2	-1	0	1	2	3
$\dfrac{\left(1+\dfrac{4}{x}+\dfrac{4}{x^2}\right)}{\left(1-\dfrac{4}{x^2}\right)}$							
$\dfrac{x+2}{x-2}$							

Reviewing the Major Concepts

In Exercises 45–50, simplify the expression.

45. $-(-3x^2)^3(2x^4)$

46. $(4x^3y^2)(-2xy^3)$

47. $(a^2+b^2)^0$

48. $[(2x^2y)^2]^3$

49. $\left(\dfrac{5}{x^2}\right)^2$

50. $-\dfrac{(2u^2v)^2}{-3uv^2}$

51. *Monthly Wage*   A company offers two wage plans. One plan pays a straight $2500 per month. The second pays $1500 per month plus a commission of 4% on gross sales. Let x represent the gross sales and write an inequality that represents gross sales in which the second plan gives the greater monthly wage. Solve the inequality.

52. Three times a number n must be at least 15 and no more than 60. What interval contains this number?

Additional Problem Solving

In Exercises 53–62, combine and simplify.

53. $\dfrac{4-y}{4} + \dfrac{3y}{4}$

54. $\dfrac{10x^2+1}{3} - \dfrac{10x^2}{3}$

55. $\dfrac{2}{3a} - \dfrac{11}{3a}$

56. $\dfrac{16+z}{5z} - \dfrac{11-z}{5z}$

57. $\dfrac{2x+5}{3} + \dfrac{1-x}{3}$

58. $\dfrac{6x}{13} - \dfrac{7x}{17}$

59. $\dfrac{3y}{3} - \dfrac{3y-3}{3} - \dfrac{7}{3}$

60. $\dfrac{3y-22}{y-6} - \dfrac{2y+16}{y-6}$

61. $\dfrac{-16u}{9} - \dfrac{27-16u}{9} + \dfrac{2}{9}$

62. $\dfrac{7s-5}{2s+5} + \dfrac{3(s+10)}{2s+5}$

In Exercises 63–66, find the least common multiple of the expressions.

63. $9y^3,\ 12y$

64. $8t(t+2),\ 14(t^2-4)$

65. $6x^2,\ 15x(x-1)$

66. $2y^2+y-1,\ 4y^2-2y$

In Exercises 67–70, find the least common denominator of the two fractions and rewrite each fraction using the least common denominator.

67. $\dfrac{v}{2v^2+2v},\ \dfrac{4}{3v^2}$

68. $\dfrac{2}{x^2(x-3)},\ \dfrac{5}{x(x+3)}$

69. $\dfrac{4x}{(x+5)^2},\ \dfrac{x-2}{x^2-25}$

70. $\dfrac{5t}{2t(t-3)^2},\ \dfrac{4}{t(t-3)}$

In Exercises 71–92, perform the operation and simplify.

71. $\dfrac{7}{a} + \dfrac{14}{a^2}$

72. $\dfrac{1}{6u^2} - \dfrac{2}{9u}$

73. $\dfrac{3x}{x-8} - \dfrac{6}{8-x}$

74. $\dfrac{1}{y-6} + \dfrac{y}{6-y}$

75. $\dfrac{3x}{3x-2} + \dfrac{2}{2-3x}$

76. $\dfrac{y}{5y-3} - \dfrac{3}{3-5y}$

77. $-\dfrac{1}{6x} + \dfrac{1}{6(x-3)}$

78. $\dfrac{1}{x} - \dfrac{1}{x+2}$

79. $\dfrac{x}{x+3} - \dfrac{5}{x-2}$

80. $\dfrac{3}{t(t+1)} + \dfrac{4}{t}$

81. $\dfrac{3}{x+1} - \dfrac{2}{x}$

82. $\dfrac{5}{x-4} - \dfrac{3}{x}$

83. $\dfrac{3}{x-5} + \dfrac{2}{x+5}$

84. $\dfrac{7}{2x-3} + \dfrac{3}{2x+3}$

85. $\dfrac{4}{x^2} - \dfrac{4}{x^2+1}$

86. $\dfrac{2}{y} + \dfrac{1}{2y^2}$

87. $\dfrac{4}{x-4} + \dfrac{16}{(x-4)^2}$

88. $\dfrac{3}{x-2} - \dfrac{1}{(x-2)^2}$

89. $\dfrac{y}{x^2+xy} - \dfrac{x}{xy+y^2}$

90. $\dfrac{5}{x+y} + \dfrac{5}{x-y}$

91. $\dfrac{4}{x} - \dfrac{2}{x^2} + \dfrac{4}{x+3}$

92. $\dfrac{5}{2(x+1)} - \dfrac{1}{2x} - \dfrac{3}{2(x+1)^2}$

In Exercises 93–100, simplify the complex fraction.

93. $\dfrac{\left(16x - \dfrac{1}{x}\right)}{\left(\dfrac{1}{x} - 4\right)}$

94. $\dfrac{\left(\dfrac{36}{y} - y\right)}{6+y}$

95. $\dfrac{\left(\dfrac{3}{x^2} + \dfrac{1}{x}\right)}{\left(2 - \dfrac{4}{5x}\right)}$

96. $\dfrac{\left(16 - \dfrac{1}{x^2}\right)}{\left(\dfrac{1}{4x^2} - 4\right)}$

97. $\dfrac{\left(1 - \dfrac{1}{y^2}\right)}{\left(1 - \dfrac{4}{y} + \dfrac{3}{y^2}\right)}$

98. $\dfrac{\left(\dfrac{x+1}{x+2} - \dfrac{1}{x}\right)}{\left(\dfrac{2}{x+2}\right)}$

99. $\dfrac{\left(\dfrac{x}{x-3} - \dfrac{2}{3}\right)}{\left(\dfrac{10}{3x} + \dfrac{x^2}{x-3}\right)}$

100. $\dfrac{\left(\dfrac{1}{2x} - \dfrac{6}{x+5}\right)}{\left(\dfrac{x}{x-5} + \dfrac{1}{x}\right)}$

Difference Quotient In Exercises 101 and 102, use the function to find and simplify the expression for

$$\frac{f(2+h) - f(2)}{h}.$$

This expression is called a difference quotient and is used in calculus.

101. $f(x) = \dfrac{1}{x}$ **102.** $f(x) = \dfrac{x}{x-1}$

103. *Work Rate* After two workers work together for t hours on a common task, the fractional parts of the job done by the two workers are $t/4$ and $t/6$. What fractional part of the task has been completed?

104. *Work Rate* After two workers work together for t hours on a common task, the fractional parts of the job done by the two workers are $t/3$ and $t/5$. What fractional part of the task has been completed?

105. *Average of Two Numbers* Determine the average of the two real numbers $x/4$ and $x/6$.

106. *Average of Three Numbers* Determine the average of the three real numbers x, $x/2$, and $x/3$.

107. *Equal Parts* Find three real numbers that divide the real number line between $x/6$ and $x/2$ into four equal parts (see figure).

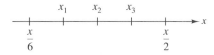

108. *Equal Parts* Find two real numbers that divide the real number line between $x/5$ and $x/3$ into three equal parts (see figure).

109. *Monthly Payment* The approximate annual interest rate r of a monthly installment loan is given by

$$r = \frac{\left[\dfrac{24(NM - P)}{N}\right]}{\left(P + \dfrac{MN}{12}\right)}$$

where N is the total number of payments, M is the monthly payment, and P is the amount financed.

(a) Approximate the annual interest rate for a 4-year car loan of \$10,000 that has monthly payments of \$300.

(b) Simplify the expression for the annual interest rate r, and then rework part (a).

110. *Error Analysis* Determine whether the following is correct. If it is not, find and correct any errors.

$$\frac{2}{x} - \frac{3}{x+1} + \frac{x+1}{x^2} = \frac{2x(x+1) - 3x^2 + (x+1)^2}{x^2(x+1)}$$

$$= \frac{2x^2 + x - 3x^2 + x^2 + 1}{x^2(x+1)}$$

$$= \frac{x+1}{x^2(x+1)}$$

$$= \frac{1}{x^2}$$

MID-CHAPTER QUIZ

Take this quiz as you would take a quiz in class. After you are done, check your work against the answers given in the back of the book.

1. Determine the domain of $\dfrac{y+2}{y(y-4)}$.

2. Evaluate $h(x) = (x^2 - 9)/(x^2 - x - 2)$ as indicated. If it is not possible, state the reason.

    (a) $h(-3)$ (b) $h(0)$ (c) $h(-1)$ (d) $h(5)$

In Exercises 3–8, write the expression in reduced form.

3. $\dfrac{9y^2}{6y}$

4. $\dfrac{8u^3 v^2}{36uv^3}$

5. $\dfrac{4x^2 - 1}{x - 2x^2}$

6. $\dfrac{(z+3)^2}{2z^2 + 5z - 3}$

7. $\dfrac{7ab + 3a^2 b^2}{a^2 b}$

8. $\dfrac{2mn^2 - n^3}{2m^2 + mn - n^2}$

In Exercises 9–18, perform the operations and simplify your answer.

9. $\dfrac{11t^2}{6} \cdot \dfrac{9}{33t}$

10. $(x^2 + 2x) \cdot \dfrac{5}{x^2 - 4}$

11. $\dfrac{4}{3(x-1)} \cdot \dfrac{12x}{6(x^2 + 2x - 3)}$

12. $\dfrac{5u}{3(u+v)} \cdot \dfrac{2(u^2 - v^2)}{3v} \div \dfrac{25u^2}{18(u - v)}$

13. $\dfrac{\left(\dfrac{9t^2}{3-t}\right)}{\left(\dfrac{6t}{t-3}\right)}$

14. $\dfrac{\left(\dfrac{10}{x^2 + 2x}\right)}{\left(\dfrac{15}{x^2 + 3x + 2}\right)}$

15. $\dfrac{4x}{x+5} - \dfrac{3x}{4}$

16. $4 + \dfrac{x}{x^2 - 4} - \dfrac{2}{x^2}$

17. $\dfrac{\left(1 - \dfrac{2}{x}\right)}{\left(\dfrac{3}{x} - \dfrac{4}{5}\right)}$

18. $\dfrac{\left(\dfrac{3}{x} + \dfrac{x}{3}\right)}{\left(\dfrac{x+3}{6x}\right)}$

19. You start a business with a setup cost of $6000. The cost of material for producing each unit of your product is $10.50.

    (a) Write an algebraic fraction that gives the average cost per unit when x units are produced. Explain your reasoning.

    (b) Find the average cost per unit when $x = 500$ units are produced.

20. Find the ratio of the shaded portion of the figure to the total area of the figure.

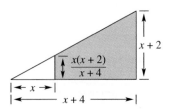

Figure for 20

4.4	**Dividing Polynomials**
	Dividing a Polynomial by a Monomial ▪ Long Division ▪ Synthetic Division ▪ Factoring and Division

Dividing a Polynomial by a Monomial

To divide a polynomial by a monomial, reverse the procedure used to add (or subtract) two rational expressions. Here is an example.

$$2 + \frac{1}{x} = \frac{2x}{x} + \frac{1}{x} = \frac{2x+1}{x} \qquad \text{Add fractions.}$$

$$\frac{2x+1}{x} = \frac{2x}{x} + \frac{1}{x} = 2 + \frac{1}{x} \qquad \text{Divide by monomial.}$$

Dividing a Polynomial by a Monomial

Let u, v, and w be real numbers, variables, or algebraic expressions such that $w \neq 0$.

1. $\dfrac{u+v}{w} = \dfrac{u}{w} + \dfrac{v}{w}$ **2.** $\dfrac{u-v}{w} = \dfrac{u}{w} - \dfrac{v}{w}$

When dividing a polynomial by a monomial, remember to reduce the resulting expressions to simplest form, as illustrated in Example 1.

EXAMPLE 1 *Dividing a Polynomial by a Monomial*

Perform the division and simplify.

$$\frac{12x^2 - 20x + 8}{4x}$$

Solution

$$\frac{12x^2 - 20x + 8}{4x} = \frac{12x^2}{4x} - \frac{20x}{4x} + \frac{8}{4x} \qquad \text{Divide each term by } 4x.$$

$$= \frac{3(4x)(x)}{4x} - \frac{5(4x)}{4x} + \frac{2(4)}{4x} \qquad \text{Cancel common factors.}$$

$$= 3x - 5 + \frac{2}{x} \qquad \text{Simplified form}$$

Long Division

In Section 4.2, you learned how to divide one polynomial by another by factoring and canceling common factors. For instance, you can divide $(x^2 - 2x - 3)$ by $(x - 3)$ as follows.

$$(x^2 - 2x - 3) \div (x - 3) = \frac{x^2 - 2x - 3}{x - 3} \qquad \text{Write as fraction.}$$

$$= \frac{(x + 1)(x - 3)}{x - 3} \qquad \text{Factor.}$$

$$= \frac{(x + 1)(x\!\!-\!\!3)}{x\!\!-\!\!3} \qquad \text{Cancel common factor.}$$

$$= x + 1, \quad x \neq 3 \qquad \text{Simplified form}$$

This procedure works well for polynomials that factor easily. For those that do not, you can use a more general procedure that follows a "long division algorithm" similar to the algorithm used for dividing positive integers. We review that procedure in Example 2.

EXAMPLE 2   Long Division Algorithm for Positive Integers

Use the long division algorithm to divide 6584 by 28.

Solution

Think $\frac{65}{28} \approx 2.$

Think $\frac{98}{28} \approx 3.$

Think $\frac{144}{28} \approx 5.$

$$
\begin{array}{r}
235 \\
28\overline{)6584} \\
\underline{56} \\
98 \\
\underline{84} \\
144 \\
\underline{140} \\
4
\end{array}
$$

Multiply $2 \cdot 28.$
Subtract and bring down 8.
Multiply $3 \cdot 28.$
Subtract and bring down 4.
Multiply $5 \cdot 28.$
Remainder

Thus, you have

$$6584 \div 28 = 235 + \frac{4}{28} = 235 + \frac{1}{7}.$$

In Example 2, the number 6584 is the **dividend,** 28 is the **divisor,** 235 is the **quotient,** and 4 is the **remainder.**

In the next several examples, you will see how the long division algorithm can be extended to cover division of one polynomial by another.

EXAMPLE 3 *Long Division Algorithm for Polynomials*

Use the long division algorithm to perform the division.

$$(x^2 + 2x + 4) \div (x - 1)$$

Solution

$$
\begin{array}{r}
x + 3 \\
x - 1 \overline{)\; x^2 + 2x + 4} \\
\underline{x^2 - x} \\
3x + 4 \\
\underline{3x - 3} \\
7
\end{array}
$$

Multiply $x(x - 1)$.

Subtract and bring down 4.

Multiply $3(x - 1)$.

Subtract.

Considering the remainder as a fractional part of the divisor, you can write the result as

$$
\underbrace{\frac{\overbrace{x^2 + 2x + 4}^{\text{Dividend}}}{\underbrace{x - 1}_{\text{Divisor}}}}_{} = \overbrace{x + 3}^{\text{Quotient}} + \frac{\overset{\text{Remainder}}{7}}{\underbrace{x - 1}_{\text{Divisor}}}.
$$

You can check a long division problem by multiplying. For instance, you can check the results of Example 3 as follows.

$$\frac{x^2 + 2x + 4}{x - 1} \overset{?}{=} x + 3 + \frac{7}{x - 1}$$

$$(x - 1)\left(\frac{x^2 + 2x + 4}{x - 1}\right) \overset{?}{=} (x - 1)\left(x + 3 + \frac{7}{x - 1}\right)$$

$$x^2 + 2x + 4 \overset{?}{=} (x + 3)(x - 1) + 7$$

$$x^2 + 2x + 4 \overset{?}{=} (x^2 + 2x - 3) + 7$$

$$x^2 + 2x + 4 = x^2 + 2x + 4 \qquad \text{Result checks. } \checkmark$$

In a long division problem, if the remainder is 0, the divisor is said to **divide evenly** into the dividend. For instance, $x + 2$ divides evenly into $3x^3 + 10x^2 + 6x - 4$:

$$\frac{3x^3 + 10x^2 + 6x - 4}{x + 2} = 3x^2 + 4x - 2.$$

When using the long division algorithm for polynomials, be sure that both the divisor and dividend are written in standard form before beginning the division process.

EXAMPLE 4 *Writing in Standard Form Before Dividing*

Divide $-13x^3 + 10x^4 + 8x - 7x^2 + 4$ by $3 - 2x$.

Solution

First write the divisor and dividend in standard polynomial form.

$$
\begin{array}{r}
-5x^3 - x^2 + 2x - 1 \\
-2x + 3 \overline{\smash{)}\ 10x^4 - 13x^3 - 7x^2 + 8x + 4} \\
\underline{10x^4 - 15x^3} \\
2x^3 - 7x^2 \\
\underline{2x^3 - 3x^2} \\
-4x^2 + 8x \\
\underline{-4x^2 + 6x} \\
2x + 4 \\
\underline{2x - 3} \\
7
\end{array}
$$

Multiply $-5x^3(-2x + 3)$.

Subtract and bring down $-7x^2$.
Multiply $-x^2(-2x + 3)$.

Subtract and bring down $8x$.
Multiply $2x(-2x + 3)$.

Subtract and bring down 4.
Multiply $(-1)(-2x + 3)$.

This shows that

$$
\frac{10x^4 - 13x^3 - 7x^2 + 8x + 4}{-2x + 3} = -5x^3 - x^2 + 2x - 1 + \frac{7}{-2x + 3}.
$$

Technology

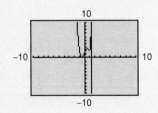

You can check the result of a division problem *algebraically*, as shown at the bottom of page 287. You can check the result *graphically* with a graphing utility by comparing the graphs of the original quotient and the simplified form. For example, the figure at the left shows the graphs of

$$
y_1 = \frac{10x^4 - 13x^3 - 7x^2 + 8x + 4}{-2x + 3}
$$

and

$$
y_2 = -5x^3 - x^2 + 2x - 1 + \frac{7}{-2x + 3}.
$$

Because the graphs coincide, it follows that the expressions are equivalent.

When the dividend is missing some powers of x, the long division algorithm requires that you account for the missing powers, as shown in Example 5.

EXAMPLE 5   Accounting for Missing Powers of x

Divide $(x^3 - 2)$ by $(x - 1)$.

Solution

Note how the missing x^2- and x-terms are accounted for.

$$
\begin{array}{r}
x^2 + x + 1 \\
x - 1 \overline{\smash{)}\ x^3 + 0x^2 + 0x - 2} \\
\underline{x^3 - x^2} \\
x^2 + 0x \\
\underline{x^2 - x} \\
x - 2 \\
\underline{x - 1} \\
-1
\end{array}
$$

Multiply $x^2(x-1)$.

Subtract and bring down $0x$.
Multiply $x(x-1)$.

Subtract and bring down -2.
Multiply $(1)(x-1)$.

Subtract.

Thus, you have

$$\frac{x^3 - 2}{x - 1} = x^2 + x + 1 - \frac{1}{x - 1}.$$

EXAMPLE 6   A Second-Degree Divisor

Divide $x^4 + 6x^3 + 6x^2 - 10x - 3$ by $x^2 + 2x - 3$.

Solution

NOTE In each of the long division examples so far, the divisor has been a first-degree polynomial. The long division algorithm works just as well with polynomial divisors of degree 2 or more, as shown in Example 6.

$$
\begin{array}{r}
x^2 + 4x + 1 \\
x^2 + 2x - 3 \overline{\smash{)}\ x^4 + 6x^3 + 6x^2 - 10x - 3} \\
\underline{x^4 + 2x^3 - 3x^2} \\
4x^3 + 9x^2 - 10x \\
\underline{4x^3 + 8x^2 - 12x} \\
x^2 + 2x - 3 \\
\underline{x^2 + 2x - 3} \\
0
\end{array}
$$

Multiply $x^2(x^2 + 2x - 3)$.

Subtract and bring down $-10x$.
Multiply $4x(x^2 + 2x - 3)$.

Subtract and bring down -3.
Multiply $(1)(x^2 + 2x - 3)$.

Subtract.

Thus, $x^2 + 2x - 3$ divides evenly into $x^4 + 6x^3 + 6x^2 - 10x - 3$:

$$\frac{x^4 + 6x^3 + 6x^2 - 10x - 3}{x^2 + 2x - 3} = x^2 + 4x + 1.$$

Synthetic Division

There is a nice shortcut for division by polynomials of the form $x - k$. It is called **synthetic division,** and is outlined for a third-degree polynomial as follows.

Synthetic Division for a Third-Degree Polynomial

Use synthetic division to divide $ax^3 + bx^2 + cx + d$ by $x - k$ as follows.

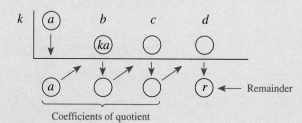

Vertical Pattern: Add terms.
Diagonal Pattern: Multiply by k.

NOTE Be sure you see that synthetic division works *only* for divisors of the form $x - k$. Remember that $x + k = x - (-k)$. Moreover, the degree of the quotient is always one less than the degree of the dividend.

EXAMPLE 7 Using Synthetic Division

Use synthetic division to divide $x^3 + 3x^2 - 4x - 10$ by $x - 2$.

Solution

The coefficients of the dividend form the top row of the synthetic division tableau. Because you are dividing by $(x - 2)$, write 2 at the top left of the tableau. To begin the algorithm, bring down the first coefficient. Then multiply this coefficient by 2, write the result in the second row of the tableau, and add the two numbers in the second column. By continuing this pattern, you obtain the following tableau.

$$
\begin{array}{r|rrrr}
2 & 1 & 3 & -4 & -10 \\
  &   & 2 & 10 & 12 \\
\hline
  & 1 & 5 & 6 & \boxed{2} \leftarrow \text{Remainder}
\end{array}
$$

The bottom row of the tableau shows that the quotient is

$$(1)x^2 + (5)x + (6)$$

and the remainder is 2. Thus, the result of the division problem is

$$\frac{x^3 + 3x^2 - 4x - 10}{x - 2} = x^2 + 5x + 6 + \frac{2}{x - 2}.$$

Factoring and Division

If the remainder in a synthetic division problem turns out to be zero, you can conclude that the divisor divides *evenly* into the dividend.

EXAMPLE 8   Factoring a Polynomial

Completely factor $x^3 - 7x + 6$. Use the fact that $x - 1$ is one of the factors.

Solution

Because you are given one of the factors, divide this factor into the given polynomial to obtain

$$\frac{x^3 - 7x + 6}{x - 1} = x^2 + x - 6.$$

From this result, you can factor the original polynomial as follows.

$$x^3 - 7x + 6 = (x - 1)(x^2 + x - 6)$$
$$= (x - 1)(x + 3)(x - 2)$$

Group Activities   Exploring with Technology

Investigating Polynomials and Their Factors   Use a graphing utility to graph the following polynomials on the same viewing rectangle using the standard setting. Use the TRACE feature to find the x-intercepts. What can you conclude about the polynomials? Verify your conclusion algebraically.

a. $y = (x - 4)(x - 2)(x + 1)$

b. $y = (x^2 - 6x + 8)(x + 1)$

c. $y = x^3 - 5x^2 + 2x + 8$

Now use your graphing utility to graph the function

$$f(x) = \frac{x^3 - 5x^2 + 2x + 8}{x - 2}.$$

Use the TRACE feature to find the x-intercepts. Why does it have only two x-intercepts? To what other function does the graph of $f(x)$ appear to be equivalent? What is the difference between the two graphs? (*Hint:* Zoom in at $x = 2$.)

4.4 Exercises

Discussing the Concepts

1. *Error Analysis*   Describe the error.

$$\frac{6x + 5y}{x} = \frac{6\cancel{x} + 5y}{\cancel{x}} = 6 + 5y$$

2. Create a polynomial division problem and identify the dividend, divisor, quotient, and remainder.

3. What does it mean for a divisor to *divide evenly* into a dividend?

4. Explain how you can check a polynomial division problem. Give an example.

5. *True or False?*   If the divisor divides evenly into the dividend, the divisor and quotient are factors of the dividend. Explain.

6. For synthetic division, what form must the divisor have?

Problem Solving

In Exercises 7–10, perform the division of a polynomial by a monomial and check your result.

7. $\dfrac{50z^3 + 30z}{-5z}$

8. $\dfrac{18c^4 - 24c^2}{-6c}$

9. $(5x^2y - 8xy + 7xy^2) \div 2xy$

10. $(-14s^4t^2 + 7s^2t^2 - 18t) \div 2s^2t$

In Exercises 11–24, perform the division and check your result.

11. $\dfrac{x^2 - 8x + 15}{x - 3}$

12. $\dfrac{t^2 - 18t + 72}{t - 6}$

13. Divide $21 - 4x - x^2$ by $3 - x$.

14. Divide $10t^2 - 7t - 12$ by $2t - 3$.

15. $\dfrac{x^3 - 2x^2 + 4x - 8}{x - 2}$

16. $\dfrac{x^3 - 28x - 48}{x + 4}$

17. $\dfrac{x^2 + 16}{x + 4}$

18. $\dfrac{y^2 + 8}{y + 2}$

19. $\dfrac{6z^2 + 7z}{5z - 1}$

20. $\dfrac{8y^2 - 2y}{3y + 5}$

21. $x^5 \div (x^2 + 1)$

22. $x^4 \div (x - 2)$

23. $(x^3 + 4x^2 + 7x + 6) \div (x^2 + 2x + 3)$

24. $(2x^3 + 2x^2 - 2x - 15) \div (2x^2 + 4x + 5)$

In Exercises 25–28, use synthetic division to perform the division.

25. $\dfrac{x^3 + 3x^2 - 1}{x + 4}$

26. $\dfrac{2x^5 - 3x^3 + x}{x - 3}$

27. $\dfrac{0.1x^2 + 0.8x + 1}{x - 0.2}$

28. $\dfrac{x^3 - 0.8x + 2.4}{x + 1}$

In Exercises 29–32, use synthetic division to perform the division. Use the result to factor the dividend.

29. $\dfrac{x^2 - 15x + 56}{x - 8}$

30. $\dfrac{24 + 13t - 2t^2}{8 - t}$

31. $\dfrac{x^4 - x^3 - 3x^2 + 4x - 1}{x - 1}$

32. $\dfrac{x^4 - 16}{x - 2}$

In Exercises 33 and 34, use a graphing utility to graph the two equations on the same screen. Use the graphs to verify that the expressions are equivalent. Verify the results algebraically.

33. $y_1 = \dfrac{x + 4}{2x}, \ y_2 = \dfrac{1}{2} + \dfrac{2}{x}$

34. $y_1 = \dfrac{x^2 + 2}{x + 1}, \ y_2 = x - 1 + \dfrac{3}{x + 1}$

Finding a Pattern In Exercises 35 and 36, complete the table for the given polynomial. The first row is completed for Exercise 35. What conclusion can you draw as you compare the polynomial values with the remainders? (Use synthetic division to find the remainders.)

x	Polynomial Value	Divisor	Remainder
-2	-8	$x + 2$	-8
-1			
0			
$\frac{1}{2}$			
1			
2			

35. $x^3 - x^2 - 2x$ **36.** $2x^3 - x^2 - 2x + 1$

Geometry In Exercises 37 and 38, you are given the expression for the volume of the solid shown. Find an expression for the missing dimension.

37. $V = x^3 + 18x^2 + 80x + 96$

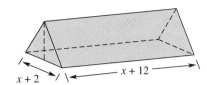

38. $V = h^4 + 3h^3 + 2h^2$

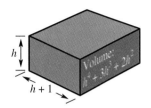

Reviewing the Major Concepts

In Exercises 39–42, multiply.

39. $(x + 1)^2$

40. $(2 - y)(3 + 2y)$

41. $(4 - 5z)(4 + 5z)$

42. $t(t - 4)(2t + 3)$

43. *Geometry* The base of a triangle is $5x$ and its height is $2x + 9$. Find the area A of the triangle.

44. Given the function $f(x) = 3x - x^2$, find and simplify $f(2 + t) - f(2)$.

Additional Problem Solving

In Exercises 45–54, perform the division of a polynomial by a monomial and check your result.

45. $\dfrac{6z + 10}{2}$

46. $\dfrac{9x + 12}{3}$

47. $\dfrac{10z^2 + 4z - 12}{4}$

48. $\dfrac{4u^2 + 8u - 24}{16}$

49. $(7x^3 - 2x^2) \div x$

50. $(6a^2 + 7a) \div a$

51. $\dfrac{8z^3 + 3z^2 - 2z}{2z}$

52. $\dfrac{6x^4 + 8x^3 - 18x^2}{3x^2}$

53. $\dfrac{m^4 + 2m^2 - 7}{m}$

54. $\dfrac{l^2 - 8}{-l}$

In Exercises 55–76, perform the division and check your result.

55. $\dfrac{4(x + 5)^2 + 8(x + 5)}{x + 5}$

56. $\dfrac{12(r - 9)^3 - 18(r - 9)^2}{4(r - 9)^2}$

57. $(x^2 + 15x + 50) \div (x + 5)$

58. $(y^2 - 6y - 16) \div (y + 2)$

59. Divide $2y^2 + 7y + 3$ by $2y + 1$.

60. Divide $5 + 4x - x^2$ by $1 + x$.

61. $\dfrac{12t^2 - 40t + 25}{2t - 5}$

62. $\dfrac{15 - 14u - 8u^2}{5 + 2u}$

63. $\dfrac{16x^2 - 1}{4x + 1}$

64. $\dfrac{81y^2 - 25}{9y - 5}$

65. $\dfrac{x^3 + 125}{x + 5}$

66. $\dfrac{x^3 - 27}{x - 3}$

67. $\dfrac{2x + 9}{x + 2}$

68. $\dfrac{12x - 5}{2x + 3}$

69. $\dfrac{5x^2 + 2x + 3}{x + 2}$

70. $\dfrac{2x^2 + 5x + 2}{x + 4}$

71. $\dfrac{12x^2 - 17x - 5}{3x + 2}$

72. $\dfrac{8x^2 + 2x + 3}{4x - 1}$

73. $\dfrac{2x^3 - 5x^2 + x - 6}{x - 3}$

74. $\dfrac{5x^3 + 3x^2 + 12x + 20}{x + 1}$

75. $\dfrac{x^6 - 1}{x - 1}$

76. $\dfrac{x^3}{x - 1}$

Think About It   In Exercises 77 and 78, perform the division assuming that n is a positive integer.

77. $\dfrac{x^{3n} + 3x^{2n} + 6x^n + 8}{x^n + 2}$

78. $\dfrac{x^{3n} - x^{2n} + 5x^n - 5}{x^n - 1}$

In Exercises 79–82, use synthetic division to perform the division.

79. $\dfrac{x^4 - 4x^3 + x + 10}{x - 2}$

80. $\dfrac{x^4}{x + 2}$

81. $\dfrac{5x^3 + 12}{x + 5}$

82. $\dfrac{8x + 35}{x - 10}$

In Exercises 83–88, use synthetic division to perform the division. Use the result to factor the dividend.

83. $\dfrac{2a^2 + 14a + 45}{a + 9}$

84. $\dfrac{x^2 + 3x - 154}{x + 14}$

85. $\dfrac{15x^2 - 2x - 8}{x - \frac{4}{5}}$

86. $\dfrac{18x^2 - 9x - 20}{x + \frac{5}{6}}$

87. $\dfrac{2t^3 + 15t^2 + 19t - 30}{t + 5}$

88. $\dfrac{5t^3 - 27t^2 - 14t - 24}{t - 6}$

Think About It   In Exercises 89 and 90, find the constant c so that the denominator will divide evenly into the numerator.

89. $\dfrac{x^3 + 2x^2 - 4x + c}{x - 2}$

90. $\dfrac{x^4 - 3x^2 + c}{x + 6}$

91. *Geometry*   The rectangle's area is $2x^3 + 3x^2 - 6x - 9$. Find its width if its length is $2x + 3$.

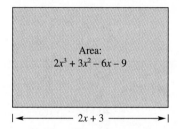

Area:
$2x^3 + 3x^2 - 6x - 9$

$\longleftarrow \quad 2x + 3 \quad \longrightarrow$

92. *Geometry*   A rectangular house has a volume of

$$x^3 + 55x^2 + 650x + 2000$$

cubic feet (the space in the attic is not included). The height of the house is $x + 5$ (see figure). Find the number of square feet of floor space *on the first floor* of the house.

$x + 5$

4.5 Graphing Rational Functions

Introduction ▪ Horizontal and Vertical Asymptotes ▪
Graphing Rational Functions ▪ Application

Introduction

Recall that the domain of a rational function consists of all values of x for which the denominator is not zero. For instance, the domain of

$$f(x) = \frac{x+2}{x-1}$$

is all real numbers except $x = 1$. When graphing a rational function, pay special attention to the shape of the graph near x-values that are not in the domain.

EXAMPLE 1 Sketching the Graph of a Rational Function

Sketch the graph of $f(x) = \dfrac{x+2}{x-1}$.

Solution

Begin by noticing that the domain is all real numbers except $x = 1$. Next, construct a table of values, including x-values that are close to 1 on the left *and* the right.

x-Values to the Left of 1

x	-3	-2	-1	0	0.5	0.9
$f(x)$	0.25	0	-0.5	-2	-5	-29

x-Values to the Right of 1

x	1.1	1.5	2	3	4	5
$f(x)$	31	7	4	2.5	2	1.75

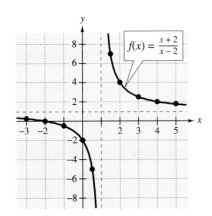

FIGURE 4.1

Plot the points to the left of 1 and connect them with a smooth curve, as shown in Figure 4.1. Do the same for the points to the right of 1. *Do not* connect the two portions of the graph, which are called its **branches.**

NOTE In Figure 4.1, as x approaches 1 from the left, the values of $f(x)$ approach negative infinity, and as x approaches 1 from the right, the values of $f(x)$ approach positive infinity.

Horizontal and Vertical Asymptotes

An **asymptote** of a graph is a line to which the graph becomes arbitrarily close as $|x|$ or $|y|$ increases without bound. In other words, if a graph has an asymptote, it is possible to move far enough out on the graph so that there is almost no difference between the graph and the asymptote.

The graph in Example 1 has two asymptotes: the line $x = 1$ is a **vertical asymptote,** and the line $y = 1$ is a **horizontal asymptote.** Other examples of asymptotes are shown in Figure 4.2.

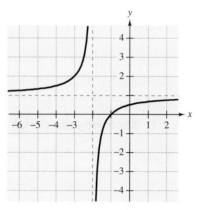

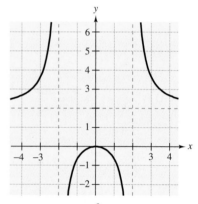

Graph of $y = \dfrac{x+1}{x+2}$
Horizontal asymptote: $y = 1$
Vertical asymptote: $x = -2$

Graph of $y = \dfrac{2x^2}{x^2-4}$
Horizontal asymptote: $y = 2$
Vertical asymptotes: $x = \pm 2$

FIGURE 4.2

The graph of a rational function may have no horizontal or vertical asymptotes, or it may have several.

Guidelines for Finding Asymptotes

Let $f(x) = p(x)/q(x)$, where $p(x)$ and $q(x)$ have no common factors.

1. The graph of f has a vertical asymptote at each x-value for which the denominator is zero.

2. The graph of f has at most one horizontal asymptote. If the degree of $p(x)$ is less than the degree of $q(x)$, the line $y = 0$ is a horizontal asymptote. If the degree of $p(x)$ is equal to the degree of $q(x)$, the line $y = a/b$ is a horizontal asymptote, where a is the leading coefficient of $p(x)$ and b is the leading coefficient of $q(x)$. If the degree of $p(x)$ is greater than the degree of $q(x)$, the graph has no horizontal asymptote.

EXAMPLE 2 *Finding Horizontal and Vertical Asymptotes*

Find all horizontal and vertical asymptotes of the graph of

$$f(x) = \frac{2x}{3x^2 + 1}.$$

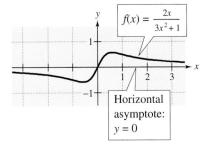

$f(x) = \dfrac{2x}{3x^2 + 1}$

Horizontal asymptote: $y = 0$

FIGURE 4.3

Solution

For this rational function, the degree of the numerator is less than the degree of the denominator. This implies that the graph has the line

$$y = 0 \qquad \text{Horizontal asymptote}$$

as a horizontal asymptote, as shown in Figure 4.3. To find any vertical asymptotes, set the denominator equal to zero and solve the resulting equation for x.

$$3x^2 + 1 = 0 \qquad \text{Set denominator equal to zero.}$$

Because this equation has no real solution, you can conclude that the graph has no vertical asymptote.

Remember that the graph of a rational function can have at most one horizontal asymptote, but it can have several vertical asymptotes. For instance, the graph in Example 3 has two vertical asymptotes.

EXAMPLE 3 *Finding Horizontal and Vertical Asymptotes*

Find all horizontal and vertical asymptotes of the graph of

$$f(x) = \frac{2x^2}{x^2 - 1}.$$

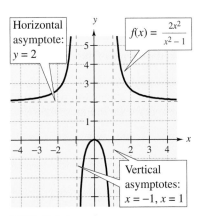

Horizontal asymptote: $y = 2$

$f(x) = \dfrac{2x^2}{x^2 - 1}$

Vertical asymptotes: $x = -1, x = 1$

FIGURE 4.4

Solution

For this rational function, the degree of the numerator is equal to the degree of the denominator. The leading coefficient of the numerator is 2, and the leading coefficient of the denominator is 1. Thus, the graph has the line

$$y = \frac{2}{1} = 2 \qquad \text{Horizontal asymptote}$$

as a horizontal asymptote, as shown in Figure 4.4. To find any vertical asymptotes, set the denominator equal to zero and solve the resulting equation for x.

$$x^2 - 1 = 0 \qquad \text{Set denominator equal to zero.}$$

This equation has two real solutions: -1 and 1. Thus, the graph has two vertical asymptotes: the lines $x = -1$ and $x = 1$.

Graphing Rational Functions

To sketch the graph of a rational function, we suggest the following guidelines.

Guidelines for Graphing Rational Functions

Let $f(x) = p(x)/q(x)$, where $p(x)$ and $q(x)$ have no common factors.

1. Find and plot the y-intercept (if any) by evaluating $f(0)$.

2. Set the numerator equal to zero and solve the equation for x. The real solutions represent the x-intercepts of the graph. Plot these intercepts.

3. Find and sketch the horizontal and vertical asymptotes of the graph.

4. Plot at least one point both between and beyond each x-intercept and vertical asymptote.

5. Use smooth curves to complete the graph between and beyond the vertical asymptotes.

EXAMPLE 4   Sketching the Graph of a Rational Function

Sketch the graph of $f(x) = \dfrac{2}{x - 3}$.

Solution

Begin by noting that the numerator and denominator have no common factors. Following the above guidelines produces the following.

- Because $f(0) = -\frac{2}{3}$, the y-intercept is $\left(0, -\frac{2}{3}\right)$.
- Because the numerator is never zero, there are no x-intercepts.
- Because the denominator is zero when $x = 3$, the line $x = 3$ is a vertical asymptote.
- Because the degree of the numerator is less than the degree of the denominator, the line $y = 0$ is a horizontal asymptote.

Plot the intercepts, asymptotes, and the additional points from the following table. Then complete the graph by drawing two branches, as shown in Figure 4.5. Note that the two branches are not connected.

x	-2	1	2	4	5
$f(x)$	$-\frac{2}{5}$	-1	-2	2	1

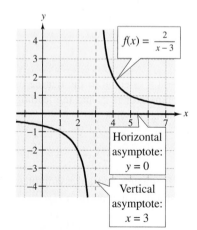

FIGURE 4.5

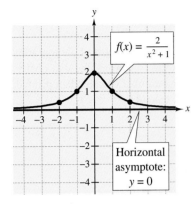

FIGURE 4.6

EXAMPLE 5 *Sketching the Graph of a Rational Function*

Sketch the graph of $f(x) = \dfrac{2}{x^2 + 1}$.

Solution

Begin by noting that the numerator and denominator have no common factors.

- Because $f(0) = 2$, the y-intercept is $(0, 2)$.
- Because the numerator is never zero, there are no x-intercepts.
- Because the denominator is never zero, there are no vertical asymptotes.
- Because the degree of the numerator is less than the degree of the denominator, the line $y = 0$ is a horizontal asymptote.

By plotting the intercepts, asymptotes, and the additional points from the following table, you can obtain the graph shown in Figure 4.6.

x	-2	-1	1	2
$f(x)$	$\frac{2}{5}$	1	1	$\frac{2}{5}$

Technology

Graphing a Rational Function

A graphing utility can help you sketch the graph of a rational function. With most graphing utilities, however, there are problems with graphs of rational functions. If you use the *connected mode*, the graphing utility will try to connect any branches of the graph. If you use the *dot mode*, the graphing utility will draw a dotted (rather than a solid) graph. Both of these options are shown below for the graph of $y = (x - 1)/(x - 3)$.

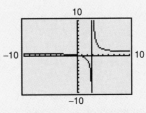

Connected Mode

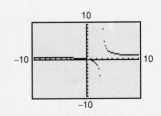

Dot Mode

Application

EXAMPLE 6 *Finding the Average Cost*

As a fund-raising project, a club is publishing a calendar. The cost of photography and typesetting is $850. In addition to these "one-time" charges, the unit cost of printing each calendar is $3.25. Let x represent the number of calendars printed. Write a model that represents the average cost per calendar.

Solution

The total cost C of printing x calendars is

$$C = 3.25x + 850.$$ Total cost function

The average cost per calendar $\bar{A}$ for printing x calendars is

$$\bar{A} = \frac{3.25x + 850}{x}.$$ Average cost function

From the graph shown in Figure 4.7, notice that the average cost decreases as the number of calendars increases.

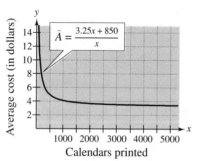

$$\bar{A} = \frac{3.25x + 850}{x}$$

Average cost (in dollars)

Calendars printed

FIGURE 4.7

In 1995, about 14,000 individuals were members of the Professional Photographers of America.

Group Activities   Extending the Concept

More About the Average Cost   In Example 6, what is the horizontal asymptote of the graph of the average cost function? What is the significance of this asymptote in the problem? Is it possible to sell enough calendars to obtain an average cost of $3.00 per calendar? Explain your reasoning.

4.5 Exercises

Discussing the Concepts

1. In your own words, describe how to determine the domain of a rational function. Give an example of a rational function whose domain is all real numbers except 2.

2. In your own words, describe what is meant by an *asymptote* of a graph.

3. *True or False?*   If the graph of rational function *f* has a vertical asymptote at $x = 3$, it is possible to sketch the graph without lifting your pencil from the paper. Explain.

4. Does every rational function have a vertical asymptote? Explain.

Problem Solving

In Exercises 5 and 6, (a) complete each table, (b) determine the vertical and horizontal asymptotes of the graph, and (c) find the domain of the function.

x	0	0.5	0.9	0.99	0.999
y					

x	2	1.5	1.1	1.01	1.001
y					

x	2	5	10	100	1000
y					

5. $f(x) = \dfrac{4}{x - 1}$

6. $f(x) = \dfrac{2x}{x - 1}$

In Exercises 7–12, find the domain of the function and identify any horizontal and vertical asymptotes.

7. $f(x) = \dfrac{5}{x^2}$

8. $g(x) = \dfrac{3}{x - 5}$

9. $g(t) = \dfrac{3}{t^2 + 1}$

10. $h(s) = \dfrac{2s^2}{s + 3}$

11. $y = \dfrac{5x^2}{x^2 - 1}$

12. $y = \dfrac{3x + 2}{2x - 1}$

In Exercises 13–16, match the function with its graph.

(a)
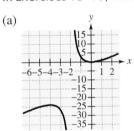

(b)
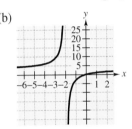

(c)
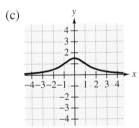

(d)
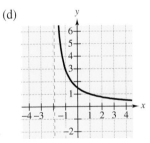

13. $f(x) = \dfrac{3}{x + 2}$

14. $f(x) = \dfrac{3x}{x + 2}$

15. $f(x) = \dfrac{3x^2}{x + 2}$

16. $f(x) = \dfrac{3}{x^2 + 2}$

In Exercises 17–24, sketch the graph of the rational function. As sketching aids, check for intercepts, vertical asymptotes, and horizontal asymptotes.

17. $f(x) = \dfrac{1}{x - 2}$

18. $f(x) = \dfrac{3}{x + 1}$

19. $g(x) = \dfrac{1}{2 - x}$

20. $g(x) = \dfrac{-3}{x + 1}$

21. $y = \dfrac{2x+4}{x}$

22. $y = \dfrac{2x}{x+4}$

23. $y = \dfrac{2x^2}{x^2+1}$

24. $y = \dfrac{10}{(x-2)^2}$

In Exercises 25–28, use a graphing utility to graph the function. Give the domain of the function and identify any horizontal or vertical asymptotes.

25. $h(x) = \dfrac{x-3}{x-1}$

26. $h(x) = \dfrac{x^2}{x-2}$

27. $f(t) = \dfrac{6}{t^2+1}$

28. $g(t) = 2 + \dfrac{3}{t+1}$

In Exercises 29–32, use the graph of $f(x) = 1/x$ to sketch the graph of g.

29. $g(x) = -\dfrac{1}{x}$

30. $g(x) = \dfrac{1}{x} + 2$

31. $g(x) = \dfrac{1}{x-2}$

32. $g(x) = \dfrac{1}{x+3}$

33. *Average Cost*   The cost of producing x units is $C = 2500 + 0.50x$, $0 < x$.

(a) Write the average cost $\bar{A}$ as a function of x.

(b) Find the average costs of producing $x = 1000$ and $x = 10{,}000$ units.

(c) Use a graphing utility to graph the average cost function. Determine the horizontal asymptote of the graph.

34. *Concentration of a Mixture*   A 25-liter container contains 5 liters of a 25% brine solution. You add x liters of a 75% brine solution to the container. The concentration C of the resulting mixture is

$$C = \dfrac{3x+5}{4(x+5)}.$$

(a) Determine the domain of the rational function within the physical constraints of the problem.

(b) Use a graphing utility to graph the function. As the container is filled, what percent does the concentration of the brine appear to approach?

Reviewing the Major Concepts

In Exercises 35–38, solve the inequality and sketch the graph of the solution on the real number line.

35. $2x - 12 \geq 0$

36. $7 - 3x < 4 - x$

37. $|x-3| < 2$

38. $|x-5| > 3$

39. Determine all real numbers n such that $\frac{1}{3}n$ must be at least 10 and no more than 50.

40. *Operating Costs*   The annual operating cost for a truck is

$$C = 0.45m + 6200$$

where m is the number of miles traveled by the truck in a year. What number of miles will yield an annual operating cost that is less than \$15,000?

Additional Problem Solving

In Exercises 41–50, find the domain of the function and identify any horizontal and vertical asymptotes.

41. $f(x) = 2 + \dfrac{1}{x-3}$

42. $f(x) = \dfrac{2}{x-3}$

43. $f(x) = \dfrac{3x}{x^2-9}$

44. $f(x) = \dfrac{5x^2}{x^2-9}$

45. $f(x) = \dfrac{x}{x+8}$

46. $f(u) = \dfrac{u^2}{u-10}$

47. $g(t) = \dfrac{3}{t(t-1)}$

48. $h(x) = 4 - \dfrac{3}{x}$

49. $y = \dfrac{2x^2}{x^2+1}$

50. $y = \dfrac{3-5x}{1-3x}$

In Exercises 51–62, sketch the graph of the rational function. As sketching aids, check for intercepts, vertical asymptotes, and horizontal asymptotes. Use a graphing utility to verify your graph.

51. $g(x) = \dfrac{5}{x}$

52. $g(x) = \dfrac{5}{x-4}$

53. $f(x) = \dfrac{5}{x^2}$

54. $f(x) = \dfrac{5}{(x-4)^2}$

55. $g(t) = 3 - \dfrac{2}{t}$

56. $g(v) = \dfrac{2v}{v+1}$

57. $y = \dfrac{3x}{x+4}$

58. $y = \dfrac{x-2}{x}$

59. $y = \dfrac{4}{x^2+1}$

60. $y = \dfrac{4x^2}{x^2+1}$

61. $y = -\dfrac{x}{x^2-4}$

62. $y = \dfrac{3x^2}{x^2-x-2}$

In Exercises 63–66, use a graphing utility to graph the function. Give its domain.

63. $y = \dfrac{2(x^2+1)}{x^2}$

64. $y = \dfrac{2(x^2-1)}{x^2}$

65. $y = \dfrac{3}{x} + \dfrac{1}{x-2}$

66. $y = \dfrac{x}{2} - \dfrac{2}{x}$

In Exercises 67–70, use the graph of $f(x) = 4/x^2$ to sketch the graph of g.

67. $g(x) = 2 + \dfrac{4}{x^2}$

68. $g(x) = -\dfrac{4}{x^2}$

69. $g(x) = -\dfrac{4}{(x-2)^2}$

70. $g(x) = 5 - \dfrac{4}{x^2}$

Think About It In Exercises 71 and 72, use a graphing utility to graph the function. Explain why there is no vertical asymptote when a superficial examination of the function may indicate that there should be one.

71. $g(x) = \dfrac{4-2x}{x-2}$

72. $h(x) = \dfrac{x^2-9}{x+3}$

73. *Medicine* The concentration of a certain chemical in the bloodstream t hours after injection into the muscle tissue is given by

$$C = \dfrac{2t}{4t^2+25}, \quad 0 \le t.$$

(a) Determine the horizontal asymptote of the function and interpret its meaning in the context of the problem.

(b) Graph the function on a graphing utility. Approximate the time when the concentration is greatest.

74. *Average Cost* The cost of producing x units is $C = 30{,}000 + 1.25x$, $0 < x$.

(a) Write the average cost $\bar{A}$ as a function of x.

(b) Find the average costs of producing $x = 10{,}000$ and $x = 100{,}000$ units.

(c) Use a graphing utility to graph the average cost function. Determine the horizontal asymptote of the graph.

75. *Geometry* A rectangular region of length x and width y has an area of 400 square meters.

(a) Verify that the perimeter P is given by

$$P = 2\left(x + \dfrac{400}{x}\right).$$

(b) Determine the domain of the function within the physical constraints of the problem.

(c) Sketch a graph of the function and approximate the dimensions of the rectangle that has a minimum perimeter.

76. *Sales* The cumulative number N (in thousands) of units of a product sold over a period of t years on the market is modeled by

$$N = \dfrac{150t(1+4t)}{1+0.15t^2}, \quad 0 \le t.$$

(a) Estimate cumulative sales when $t = 1$, $t = 2$, and $t = 4$.

(b) Use a graphing utility to graph the function. Determine the horizontal asymptote of the graph.

(c) Explain the meaning of the horizontal asymptote in the context of the problem.

4.6	**Solving Rational Equations**
	Equations Containing Constant Denominators ▪
	Equations Containing Variable Denominators ▪ Applications

Equations Containing Constant Denominators

In Section 1.1, you studied a strategy for solving equations that contain fractions with *constant* denominators. We review that procedure here because it is the basis for solving more general equations involving fractions. Recall from Section 1.1 that you can "clear an equation of fractions" by multiplying both sides of the equation by the least common denominator of the fractions in the equation. Note how this is done in Example 1.

EXAMPLE 1 An Equation Containing Constant Denominators

Solve $\dfrac{3}{5} = \dfrac{x}{2}$.

Solution

The least common denominator of the two fractions is 10, so begin by multiplying both sides of the equation by 10.

$$\frac{3}{5} = \frac{x}{2} \qquad \text{Original equation}$$

$$10\left(\frac{3}{5}\right) = 10\left(\frac{x}{2}\right) \qquad \text{Multiply both sides by 10.}$$

$$6 = 5x \qquad \text{Simplify.}$$

$$\frac{6}{5} = x \qquad \text{Divide both sides by 5.}$$

The solution is $\frac{6}{5}$. You can check this as follows.

Check

$$\frac{3}{5} = \frac{x}{2} \qquad \text{Original equation}$$

$$\frac{3}{5} \stackrel{?}{=} \frac{6/5}{2} \qquad \text{Substitute } \tfrac{6}{5} \text{ for } x.$$

$$\frac{3}{5} \stackrel{?}{=} \frac{6}{5} \cdot \frac{1}{2} \qquad \text{Invert and multiply.}$$

$$\frac{3}{5} = \frac{3}{5} \qquad \text{Solution checks.} ✔$$

EXAMPLE 2   An Equation Containing Constant Denominators

Solve $\dfrac{x}{6} = 7 - \dfrac{x}{12}$.

Solution

$$\dfrac{x}{6} = 7 - \dfrac{x}{12}$$ Original equation

$$12\left(\dfrac{x}{6}\right) = 12\left(7 - \dfrac{x}{12}\right)$$ Multiply both sides by 12.

$$2x = 84 - x$$ Distribute and simplify.

$$3x = 84$$ Add x to both sides.

$$x = 28$$ Divide both sides by 3.

The solution is 28. Check this in the original equation.

EXAMPLE 3   An Equation Containing Constant Denominators

Solve $\dfrac{x+2}{6} - \dfrac{x-4}{8} = \dfrac{2}{3}$.

Solution

$$\dfrac{x+2}{6} - \dfrac{x-4}{8} = \dfrac{2}{3}$$ Original equation

$$24\left(\dfrac{x+2}{6} - \dfrac{x-4}{8}\right) = 24\left(\dfrac{2}{3}\right)$$ Multiply both sides by 24.

$$4(x+2) - 3(x-4) = 8(2)$$ Distribute and simplify.

$$4x + 8 - 3x + 12 = 16$$ Distributive Property

$$x + 20 = 16$$ Combine like terms.

$$x = -4$$ Subtract 20 from both sides.

The solution is -4. You can check this as follows.

Check

$$\dfrac{x+2}{6} - \dfrac{x-4}{8} = \dfrac{2}{3}$$ Original equation

$$\dfrac{-4+2}{6} - \dfrac{-4-4}{8} \stackrel{?}{=} \dfrac{2}{3}$$ Substitute -4 for x.

$$-\dfrac{1}{3} + 1 \stackrel{?}{=} \dfrac{2}{3}$$ Simplify.

$$\dfrac{2}{3} = \dfrac{2}{3}$$ Solution checks. ✔

Equations Containing Variable Denominators

Remember that you always *exclude* those values of a variable that make the denominator of a rational expression zero. This is especially critical for solving equations that contain variable denominators. You will see why in the examples that follow.

EXAMPLE 4 An Equation Containing Variable Denominators

Solve the equation.

$$\frac{7}{x} - \frac{1}{3x} = \frac{8}{3}$$

Solution

For this equation the least common denominator is $3x$. Therefore, begin by multiplying both sides of the equation by $3x$.

$$\frac{7}{x} - \frac{1}{3x} = \frac{8}{3} \qquad \text{Original equation}$$

$$3x\left(\frac{7}{x} - \frac{1}{3x}\right) = 3x\left(\frac{8}{3}\right) \qquad \text{Multiply both sides by } 3x.$$

$$\frac{21x}{x} - \frac{3x}{3x} = \frac{24x}{3} \qquad \text{Distributive Property}$$

$$21 - 1 = 8x \qquad \text{Simplify.}$$

$$\frac{20}{8} = x \qquad \text{Combine like terms and divide both sides by 8.}$$

$$x = \frac{5}{2} \qquad \text{Simplify.}$$

The solution is $\frac{5}{2}$. You can check this as follows.

Check

$$\frac{7}{x} - \frac{1}{3x} = \frac{8}{3} \qquad \text{Original equation}$$

$$\frac{7}{5/2} - \frac{1}{3(5/2)} \overset{?}{=} \frac{8}{3} \qquad \text{Substitute } \frac{5}{2} \text{ for } x.$$

$$7\left(\frac{2}{5}\right) - \frac{2}{15} \overset{?}{=} \frac{8}{3} \qquad \text{Invert and multiply.}$$

$$\frac{14}{5} - \frac{2}{15} \overset{?}{=} \frac{8}{3} \qquad \text{Simplify.}$$

$$\frac{40}{15} \overset{?}{=} \frac{8}{3} \qquad \text{Combine like terms.}$$

$$\frac{8}{3} = \frac{8}{3} \qquad \text{Solution checks. } \checkmark$$

Technology

You can use a graphing utility to estimate the solution of the equation in Example 4. To do this, graph the left side of the equation and the right side of the equation on the same screen.

$$y = \frac{7}{x} - \frac{1}{3x} \text{ and } y = \frac{8}{3}$$

The solution of the equation is the x-coordinate of the point at which the two graphs intersect, as shown below.

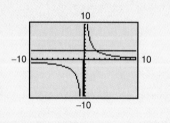

Throughout the text, we have emphasized the importance of checking solutions. Up to this point, the main reason for checking has been to make sure that you did not make errors in the solution process. In the next example you will see that there is another reason for checking solutions in the *original* equation. That is, even with no mistakes in the solution process, it can happen that a "trial solution" does not satisfy the original equation. This type of "solution" is called **extraneous.** An extraneous solution of an equation does not, by definition, satisfy the original equation, and therefore *must not* be listed as an actual solution.

EXAMPLE 5   *An Equation with No Solution*

Solve the equation.

$$\frac{5x}{x-2} = 7 + \frac{10}{x-2}$$

Solution

The least common denominator for this equation is $x - 2$. Therefore, begin by multiplying both sides of the equation by $x - 2$.

$$\frac{5x}{x-2} = 7 + \frac{10}{x-2} \qquad \text{Original equation}$$

$$(x-2)\left(\frac{5x}{x-2}\right) = (x-2)\left(7 + \frac{10}{x-2}\right) \qquad \text{Multiply both sides by } x - 2.$$

$$5x = 7(x-2) + 10, \quad x \neq 2 \qquad \text{Distribute and simplify.}$$

$$5x = 7x - 14 + 10 \qquad \text{Distributive Property}$$

$$5x = 7x - 4 \qquad \text{Combine like terms.}$$

$$-2x = -4 \qquad \text{Subtract } 7x \text{ from both sides.}$$

$$x = 2 \qquad \text{Divide both sides by } -2.$$

NOTE   In Example 5, can you see why $x = 2$ is extraneous? By looking back at the original equation you can see that 2 is excluded from the domain of two of the fractions that occur in the equation.

At this point, the solution appears to be 2. However, by performing the following check, you will see that this "trial solution" is extraneous.

Check

$$\frac{5x}{x-2} = 7 + \frac{10}{x-2} \qquad \text{Original equation}$$

$$\frac{5(2)}{2-2} \stackrel{?}{=} 7 + \frac{10}{2-2} \qquad \text{Substitute 2 for } x.$$

$$\frac{10}{0} \stackrel{?}{=} 7 + \frac{10}{0} \qquad \text{Solution does not check. } \times$$

Because the check results in *division by zero*, 2 is extraneous. Therefore, the original equation has no solution.

EXAMPLE 6   An Equation Containing Variable Denominators

Solve $\dfrac{4}{x-2} + \dfrac{3x}{x+1} = 3$.

Solution

The least common denominator is $(x-2)(x+1)$.

$$\frac{4}{x-2} + \frac{3x}{x+1} = 3$$

$$(x-2)(x+1)\left(\frac{4}{x-2} + \frac{3x}{x+1}\right) = 3(x-2)(x+1)$$

$$4(x+1) + 3x(x-2) = 3(x^2 - x - 2), \quad x \neq 2, x \neq -1$$

$$4x + 4 + 3x^2 - 6x = 3x^2 - 3x - 6$$

$$x = -10$$

The solution is -10. Check this in the original equation.

So far in this section, each of the equations has had one solution or no solution. The equation in the next example has two solutions.

DISCOVERY

Use a graphing utility to graph the equation

$$y = \frac{3x}{x+1} - \frac{12}{x^2-1} - 2.$$

Then use the ZOOM and TRACE features of the utility to determine the x-intercepts. How do the x-intercepts compare with the solutions to Example 7? What can you conclude?

EXAMPLE 7   An Equation That Has Two Solutions

Solve $\dfrac{3x}{x+1} = \dfrac{12}{x^2-1} + 2$.

Solution

The least common denominator is $(x+1)(x-1) = x^2 - 1$.

$$\frac{3x}{x+1} = \frac{12}{x^2-1} + 2$$

$$(x^2-1)\left(\frac{3x}{x+1}\right) = (x^2-1)\left(\frac{12}{x^2-1} + 2\right)$$

$$(x-1)(3x) = 12 + 2(x^2-1), \quad x \neq \pm 1$$

$$3x^2 - 3x = 12 + 2x^2 - 2$$

$$x^2 - 3x - 10 = 0$$

$$(x+2)(x-5) = 0$$

$$x + 2 = 0 \implies x = -2$$

$$x - 5 = 0 \implies x = 5$$

The solutions are -2 and 5. Check these in the original equation.

Applications

EXAMPLE 8   Average Cost

A manufacturing plant can produce x units of a certain item for $26 per unit *plus* an initial investment of $80,000. How many units must be produced to have an average cost of $30 per unit?

Solution

Verbal Model:

$$\boxed{\text{Average cost per unit}} = \boxed{\text{Total cost}} \div \boxed{\text{Number of units}}$$

Labels: Number of units $= x$ (units)
Average cost per unit $= 30$ (dollars per unit)
Total cost $= 26x + 80,000$ (dollars)

Equation:

$$30 = \frac{26x + 80,000}{x}$$

$$30x = 26x + 80,000, \quad x \neq 0$$

$$4x = 80,000$$

$$x = 20,000$$

The plant should produce 20,000 units. Check this in the original statement of the problem.

EXAMPLE 9   A Work-Rate Problem

With only the cold water valve open, it takes 8 minutes to fill the tub of a washer. With both the hot and cold water valves open, it takes only 5 minutes. How long will it take the tub to fill with only the hot water valve open?

Solution

Verbal Model:

$$\boxed{\text{Rate for cold water}} + \boxed{\text{Rate for hot water}} = \boxed{\text{Rate for warm water}}$$

Labels: Rate for cold water $= \frac{1}{8}$ (tub per minute)
Rate for hot water $= 1/t$ (tub per minute)
Rate for warm water $= \frac{1}{5}$ (tub per minute)

Equation:

$$\frac{1}{8} + \frac{1}{t} = \frac{1}{5}$$

Try solving this equation. You should discover that it takes $13\frac{1}{3}$ minutes to fill the tub with hot water. Check this in the original statement of the problem.

EXAMPLE 10   Batting Average

In this year's playing season, a baseball player has been at bat 140 times and has hit the ball safely 35 times. Thus, the "batting average" for the player is 35/140 = .250. How many consecutive times must the player hit safely to obtain a batting average of .300?

Solution

Verbal Model:

Batting average	$=$	Total hits	$\div$	Total times at bat

Labels: Current times at bat $= 140$
Current hits $= 35$
Additional consecutive hits $= x$

Equation:

$$.300 = \frac{x + 35}{x + 140}$$

$$.300(x + 140) = x + 35$$

$$.3x + 42 = x + 35$$

$$7 = 0.7x$$

$$10 = x$$

The player must hit safely the next 10 times at bat. After that, the batting average will be 45/150 = .300.

Group Activities   Problem Solving

Interpreting Average Cost   You buy sand in bulk for a construction project. You find that the total cost of your order depends on the weight of the order. Suppose the total cost C in dollars is given by $C = 100 + 50x - 0.2x^2$, $1 \leq x \leq 50$, where x is the weight in thousands of pounds. Construct a rational function representing the average cost per thousand pounds. Because of cost constraints, you may proceed with the project only if the average cost of the sand is less than $50 per thousand pounds. What is the smallest order you can place and still proceed with the project?

4.6 Exercises

Discussing the Concepts

1. Explain the difference between the following.

$$\frac{5}{x+3} + \frac{5}{3} = 3, \quad \frac{5}{x+3} + \frac{5}{3} + 3$$

2. Describe how to solve a rational equation.

3. Define the term *extraneous solution*. How do you identify an extraneous solution?

4. Describe the steps that can be used to transform an equation into an equivalent equation.

5. Explain how you can use a graphing utility to estimate the solution of a rational equation.

6. When can you use cross-multiplication to solve a rational equation? Explain.

Problem Solving

In Exercises 7–10, decide whether the values of x are solutions of the equation.

Equation   *Values*

7. $\dfrac{x}{3} - \dfrac{x}{5} = \dfrac{4}{3}$ (a) $x = 0$ (b) $x = -1$
                 (c) $x = \frac{1}{8}$ (d) $x = 10$

8. $x = 4 + \dfrac{21}{x}$ (a) $x = 0$ (b) $x = -3$
                 (c) $x = 7$ (d) $x = -1$

9. $\dfrac{x}{4} + \dfrac{3}{4x} = 1$ (a) $x = -1$ (b) $x = 1$
                 (c) $x = 3$ (d) $x = \frac{1}{2}$

10. $5 - \dfrac{1}{x-3} = 2$ (a) $x = \frac{10}{3}$ (b) $x = -\frac{1}{3}$
                  (c) $x = 0$ (d) $x = 1$

In Exercises 11–28, solve the equation.

11. $\dfrac{x}{4} = \dfrac{3}{8}$ 12. $\dfrac{x}{10} = \dfrac{12}{5}$

13. $\dfrac{h}{5} - \dfrac{h+2}{9} = \dfrac{2}{3}$ 14. $\dfrac{u}{6} + \dfrac{u+6}{15} = 3$

15. $\dfrac{7}{x} = 21$ 16. $\dfrac{9}{t} = -\dfrac{4}{3}$

17. $\dfrac{12}{y+5} + \dfrac{1}{2} = 2$ 18. $\dfrac{7}{8} - \dfrac{16}{t-2} = \dfrac{3}{4}$

19. $\dfrac{4}{2x+3} + \dfrac{17}{5x-3} = 3$

20. $\dfrac{2}{6q+5} - \dfrac{3}{4(6q+5)} = \dfrac{1}{28}$

21. $\dfrac{x}{x+4} + \dfrac{4}{x+4} + 2 = 0$

22. $\dfrac{2}{(x-4)(x-2)} = \dfrac{1}{x-4} + \dfrac{2}{x-2}$

23. $\dfrac{32}{t} = 2t$ 24. $\dfrac{45}{u} = \dfrac{u}{5}$

25. $\dfrac{1}{x-1} + \dfrac{3}{x+1} = 2$ 26. $\dfrac{x+42}{x} = x$

27. $\dfrac{x}{2} = \dfrac{1 - \dfrac{3}{x}}{1 - \dfrac{1}{x}}$ 28. $\dfrac{2x}{3} = \dfrac{1 - \dfrac{1}{x}}{1 + \dfrac{1}{x}}$

In Exercises 29–32, (a) use the graph to determine any x-intercepts of the equation, and (b) set $y = 0$ and solve the resulting equation to confirm your result.

29. $y = \dfrac{x+2}{x-2}$ 30. $y = \dfrac{2x}{x+4}$

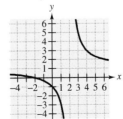

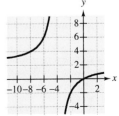

31. $y = x - \dfrac{1}{x}$

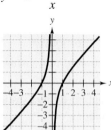

32. $y = x - \dfrac{2}{x} - 1$

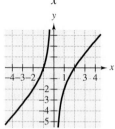

 Graphical Reasoning In Exercises 33–36, (a) use a graphing utility to graph the equation and determine any x-intercepts of the equation, and (b) set $y = 0$ and solve the resulting rational equation to confirm the result of part (a).

33. $y = \dfrac{1}{x} + \dfrac{4}{x-5}$

34. $y = 20\left(\dfrac{2}{x} - \dfrac{3}{x-1}\right)$

35. $y = (x+1) - \dfrac{6}{x}$

36. $y = \dfrac{x^2 - 4}{x}$

37. *Wind Speed* A plane with a speed of 300 miles per hour in still air travels 680 miles with a tail wind in the same time it could travel 520 miles with a head wind of equal speed. Find the speed of the wind.

38. *Average Speed* During the first part of a 6-hour, 320-mile trip, you travel at an average speed of r miles per hour. During the last part of the trip, you increase your average speed by 10 miles per hour. What were your two average speeds?

39. *Partnership Costs* Some partners buy a piece of property for $78,000 by sharing the cost equally. To ease the financial burden, they look for three additional partners to reduce the cost per person by $1300. How many partners are presently in the group?

40. *Population Model* A biologist introduces 100 insects into a culture. The population P in the culture is approximated by the model

$$P = \dfrac{500(1 + 3t)}{5 + t}$$

where t is the time in hours. Find the time required for the population to increase to 1000 insects.

41. *Swimming Pool* The flow rate of one pipe is $1\frac{1}{4}$ times that of a second pipe. A swimming pool can be filled in 5 hours using both pipes. Find the time required to fill the pool using only the pipe with the lower flow rate.

42. *Swimming Pool* Assume that the pipe with the higher flow rate in Exercise 41 is shut off after 1 hour and that it takes an additional 10 hours to fill the pool. Find the flow rate of each pipe.

Using a Model In Exercises 43 and 44, use the following model, which approximates the total revenue y (in billions of dollars) for the car and truck rental industry in the United States from 1985 to 1991.

$$y = 43.31 - \dfrac{275.25}{t} + \dfrac{654.53}{t^2}, \quad 5 \le t \le 11$$

In this model, $t = 0$ represents 1980. (Source: *Current Business Reports*)

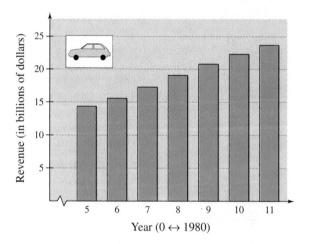

Year ($0 \leftrightarrow 1980$)

43. (a) Use the bar graph to *graphically* determine the year the total revenue first exceeded $20 billion.

(b) Use the model to *algebraically* confirm your answer to part (a).

(c) What would you estimate the revenue to be in 1996? Explain your reasoning.

44. Use a graphing utility to graph the model over the specified domain.

Reviewing the Major Concepts

In Exercises 45–48, solve the equation. Show how to use a graphing utility to check your solution.

45. $125 - 50x = 0$

46. $t^2 - 8t = 0$

47. $x^2 + x - 42 = 0$

48. $x(10 - x) = 25$

49. Find two consecutive positive even integers whose product is 624.

50. *Free-Falling Object* An object is dropped from a construction project 576 feet above the ground. Find the time t for the object to reach the ground by solving the equation $-16t^2 + 576 = 0$.

Additional Problem Solving

In Exercises 51–84, solve the equation.

51. $\dfrac{t}{2} = \dfrac{1}{8}$

52. $\dfrac{y}{5} = \dfrac{3}{2}$

53. $\dfrac{z+2}{3} = \dfrac{z}{12}$

54. $5 + \dfrac{y}{3} = y + 2$

55. $\dfrac{4t}{3} = 15 - \dfrac{t}{6}$

56. $\dfrac{x}{3} + \dfrac{x}{6} = 10$

57. $\dfrac{9}{25 - y} = -\dfrac{1}{4}$

58. $\dfrac{2}{u+4} = \dfrac{5}{8}$

59. $5 - \dfrac{12}{a} = \dfrac{5}{3}$

60. $\dfrac{6}{b} + 22 = 24$

61. $\dfrac{5}{x} = \dfrac{25}{3(x+2)}$

62. $\dfrac{10}{x+4} = \dfrac{15}{4(x+1)}$

63. $\dfrac{8}{3x+5} = \dfrac{1}{x+2}$

64. $\dfrac{500}{3x+5} = \dfrac{50}{x-3}$

65. $\dfrac{3}{x+2} - \dfrac{1}{x} = \dfrac{1}{5x}$

66. $\dfrac{12}{x+5} + \dfrac{5}{x} = \dfrac{20}{x}$

67. $\dfrac{10}{x(x-2)} + \dfrac{4}{x} = \dfrac{5}{x-2}$

68. $\dfrac{x}{x-2} + \dfrac{1}{x-4} = \dfrac{2}{x^2 - 6x + 8}$

69. $\dfrac{10}{x+3} + \dfrac{10}{3} = 6$

70. $3\left(\dfrac{1}{x} + 4\right) = 2 + \dfrac{4}{3x}$

71. $\dfrac{1}{x-5} + \dfrac{1}{x+5} = \dfrac{x+3}{x^2 - 25}$

72. $\dfrac{2}{x-10} - \dfrac{3}{x-2} = \dfrac{6}{x^2 - 12x + 20}$

73. $\dfrac{1}{2} = \dfrac{18}{x^2}$

74. $\dfrac{1}{6} = \dfrac{150}{z^2}$

75. $x + 1 = \dfrac{72}{x}$

76. $t + 4\left(\dfrac{2t+5}{t+4}\right) = 0$

77. $1 = \dfrac{16}{y} - \dfrac{39}{y^2}$

78. $\dfrac{2x}{3x-10} - \dfrac{5}{x} = 0$

79. $\dfrac{2x}{5} = \dfrac{x^2 - 5x}{5x}$

80. $\dfrac{8(x-1)}{x^2 - 4} = \dfrac{4}{x-2}$

81. $\dfrac{2(x+7)}{x+4} - 2 = \dfrac{2x+20}{2x+8}$

82. $\dfrac{10}{x^2 - 2x} + \dfrac{4}{x} = \dfrac{5}{x-2}$

83. $x - \dfrac{24}{x} = 5$

84. $\dfrac{3x}{2} + \dfrac{4}{x} = 5$

Graphical Reasoning In Exercises 85–88, (a) use a graphing utility to determine any x-intercepts of the equation, and (b) set $y = 0$ and solve the resulting rational equation to confirm the results of part (a).

85. $y = \dfrac{x-4}{x+5}$

86. $y = \dfrac{1}{x} - \dfrac{3}{x+4}$

87. $y = (x-1) - \dfrac{12}{x}$

88. $y = \dfrac{x}{2} - \dfrac{4}{x} - 1$

89. What number can be added to its reciprocal to obtain $\dfrac{65}{8}$?

90. The sum of twice a number and three times its reciprocal is $\dfrac{97}{4}$. What is the number?

91. *Pollution Removal* The cost C in dollars of removing $p\%$ of the air pollution in the stack emission of a utility company is modeled by

$$C = \frac{120,000p}{100 - p}.$$

(a) Use a graphing utility to graph the model. Use the result to graphically estimate the percent of stack emission that can be removed for $680,000.

(b) Use the model to algebraically determine the percent of stack emission that can be removed for $680,000.

92. *Average Cost* The average cost for producing x units of a product is given by

$$\text{Average cost} = 1.50 + \frac{4200}{x}.$$

Determine the number of units that must be produced to have an average cost of $2.90.

93. *Comparing Two Speeds* One person runs 2 miles per hour faster than a second person. The first person runs 5 miles in the same time the second runs 4 miles. Find the speed of each.

94. *Comparing Two Speeds* The speed of a commuter plane is 150 miles per hour slower than a passenger jet. The commuter plane travels 450 miles in the same time the jet travels 1150 miles. Find the speed of each plane.

95. *Speed* A boat travels at a speed of 20 miles per hour in still water. It travels 48 miles upstream, and then returns to the starting point, in a total of 5 hours. Find the speed of the current.

96. *Speed* You travel 72 miles in a certain time. If you had traveled 6 miles per hour faster, the trip would have taken 10 minutes less time. What was your speed?

97. *Partnership Costs* A group plans to start a new business that will require $240,000 for start-up capital. The partners in the group will share the cost equally. If two additional people join the group, the cost per person will decrease by $4000. How many partners are currently in the group?

98. *Partnership Costs* A group of people agree to share equally in the cost of a $150,000 endowment to a college. If they could find four more people to join the group, each person's share of the cost would decrease by $6250. How many people are presently in the group?

Using a Table In Exercises 99 and 100, complete the table by finding the time for two individuals to complete a task. The first two columns in the table give the times required for two individuals to complete the task working alone. (Assume that when they work together their individual rates do not change.)

99.

Person #1	Person #2	Together
6 hours	6 hours	
3 minutes	5 minutes	
5 hours	$2\frac{1}{2}$ hours	

100.

Person #1	Person #2	Together
4 days	4 days	
$5\frac{1}{2}$ hours	3 hours	
a days	b days	

101. *Work Rate* One landscaper works $1\frac{1}{2}$ times as fast as a second landscaper. Find their individual rates if it takes them 9 hours working together to complete a certain job.

102. *Work Rate* The slower of the two landscapers in Exercise 101 is given another job after 4 hours. The faster of the two must work an additional 10 hours to complete the task. Find their individual rates.

CHAPTER PROJECT: Air Resistance

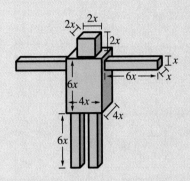

Consider a human, a mouse, and an ant, each of which falls from a height of 30 feet. The mouse and the ant will hit the ground without injury, and simply walk away. The human, on the other hand, is likely to be injured. Why?

The answer is that the mouse and ant encounter such great air resistance that they never fall fast enough to get hurt. The human encounters less air resistance and can fall fast enough to become injured. The amount of air resistance that an object encounters depends on the ratio of its surface area to its volume. A falling object that has a large surface area relative to its volume will encounter great air resistance, but a falling object that has a small surface area relative to its volume will encounter little air resistance.

To model the surface areas and volumes of different animals, consider the geometric model at the left. At first glance, it may appear that the ratio of the geometric model's surface area to its volume is constant. As you investigate the questions below, however, you will discover that the ratio is a function of x. In other words, small animals have different ratios from large animals.

1. Find algebraic models for the total surface area and total volume of the geometric model at the left.

Each Arm and Leg	Head	Trunk
$S = x^2 + 4(6x^2)$	$S = 5(4x^2)$	$S = 2(16x^2) + 4(24x^2) - 4x^2 - 4x^2$
$V = 6x^3$	$V = 8x^3$	$V = 96x^3$

2. Find a rational expression that gives the ratio R of the surface area of the geometric model to its volume.

$$R = \frac{S}{V}$$

Simplify the result.

3. Complete the following table by calculating the ratios of surface area to volume for five animals. What can you conclude?

Animal	Mouse	Squirrel	Cat	Human	Elephant
x	$\frac{1}{24}$	$\frac{1}{12}$	$\frac{1}{6}$	$\frac{2}{5}$	2
Ratio					

4. The geometric model at the left is a better fit for a mouse, cat, human, and elephant than for a squirrel. Why?

5. Approximate the ratio for an ant. Explain your reasoning.

6. Approximate the ratio for a human with a parachute. Explain your reasoning.

CHAPTER SUMMARY

After studying this chapter, you should have acquired the following skills. These skills are keyed to the Review Exercises that begin on page 317. Answers to odd-numbered Review Exercises are given in the back of the book.

- Find the domains of rational expressions. *(Section 4.1)*   **Review Exercises 1–4**

- Simplify rational expressions. *(Section 4.1)*   **Review Exercises 5–12**

- Match functions with their graphs. *(Section 4.5)*   **Review Exercises 13–16**

- Multiply and divide rational expressions. *(Section 4.2)*   **Review Exercises 17–28**

- Add and subtract rational expressions. *(Section 4.3)*   **Review Exercises 29–40**

- Simplify compound fractions. *(Sections 4.2, 4.3)*   **Review Exercises 41–46**

- Divide polynomials. *(Section 4.4)*   **Review Exercises 47–52**

- Divide polynomials using synthetic division. *(Section 4.4)*   **Review Exercises 53–56**

- Graph pairs of equations using a graphing utility, and verify their equivalence. *(Sections 4.2, 4.3, 4.4)*   **Review Exercises 57–60**

- Graph rational functions using a graphing utility. *(Section 4.5)*   **Review Exercises 61–74**

- Solve rational equations. *(Section 4.6)*   **Review Exercises 75–88**

- Graph rational functions using a graphing utility to determine any x-intercepts, and confirm them algebraically. *(Section 4.6)*   **Review Exercises 89, 90**

- Translate real-life situations into rational equations and solve. *(Section 4.6)*   **Review Exercises 91–93**

REVIEW EXERCISES

In Exercises 1–4, find the domain of the expression.

1. $\dfrac{3y}{y-8}$

2. $\dfrac{t+4}{t+12}$

3. $\dfrac{u}{u^2-7u+6}$

4. $\dfrac{x-12}{x(x^2-16)}$

In Exercises 5–12, simplify the expression.

5. $\dfrac{6x^4y^2}{15xy^2}$

6. $\dfrac{2(y^3z)^2}{28(yz^2)^2}$

7. $\dfrac{5b-15}{30b-120}$

8. $\dfrac{4a}{10a^2+26a}$

9. $\dfrac{9x-9y}{y-x}$

10. $\dfrac{x+3}{x^2-x-12}$

11. $\dfrac{x^2-5x}{2x^2-50}$

12. $\dfrac{x^2+3x+9}{x^3-27}$

In Exercises 13–16, match the function with its graph.

(a)

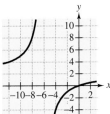

(b)

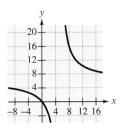

(c)

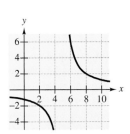

(d)

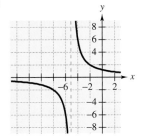

13. $f(x)=\dfrac{5}{x-6}$

14. $f(x)=\dfrac{6}{x+5}$

15. $f(x)=\dfrac{6x}{x-5}$

16. $f(x)=\dfrac{2x}{x+6}$

In Exercises 17–52, perform the operation(s) and simplify your answer.

17. $\dfrac{7}{8}\cdot\dfrac{2x}{y}\cdot\dfrac{y^2}{14x^2}$

18. $\dfrac{15(x^2y)^3}{3y^3}\cdot\dfrac{12y}{x}$

19. $\dfrac{60z}{z+6}\cdot\dfrac{z^2-36}{5}$

20. $\dfrac{1}{6}(x^2-16)\cdot\dfrac{3}{x^2-8x+16}$

21. $\dfrac{u}{u-3}\cdot\dfrac{3u-u^2}{4u^2}$

22. $x^2\cdot\dfrac{x+1}{x^2-x}\cdot\dfrac{(5x-5)^2}{x^2+6x+5}$

23. $\dfrac{6/x}{2/x^3}$

24. $\dfrac{0}{5x^2/2y}$

25. $25y^2\div\dfrac{xy}{5}$

26. $\dfrac{6}{z^2}\div4z^2$

27. $\dfrac{x^2-7x}{x+1}\div\dfrac{x^2-14x+49}{x^2-1}$

28. $\left(\dfrac{6x}{y^2}\right)^2\div\left(\dfrac{3x}{y}\right)^3$

29. $\dfrac{4}{9}-\dfrac{11}{9}$

30. $\dfrac{2(3y+4)}{2y+1}+\dfrac{3-y}{2y+1}$

31. $\dfrac{15}{16}-\dfrac{5}{24}-1$

32. $-\dfrac{3}{8}+\dfrac{7}{6}-\dfrac{1}{12}$

33. $\dfrac{1}{x+5}+\dfrac{3}{x-12}$

34. $\dfrac{2}{x-10}+\dfrac{3}{4-x}$

35. $5x+\dfrac{2}{x-3}-\dfrac{3}{x+2}$

36. $4-\dfrac{4x}{x+6}+\dfrac{7}{x-5}$

37. $\dfrac{6}{x}-\dfrac{6x-1}{x^2+4}$

38. $\dfrac{5}{x+2}+\dfrac{25-x}{x^2-3x-10}$

39. $\dfrac{5}{x+3}-\dfrac{4x}{(x+3)^2}-\dfrac{1}{x-3}$

40. $\dfrac{8}{y}-\dfrac{3}{y+5}+\dfrac{4}{y-2}$

41. $\dfrac{\left(\dfrac{6x^2}{x^2+2x-35}\right)}{\left(\dfrac{x^3}{x^2-25}\right)}$

42. $\dfrac{\left[\dfrac{24-18x}{(2-x)^2}\right]}{\left(\dfrac{60-45x}{x^2-4x-4}\right)}$

43. $\dfrac{3t}{\left(5 - \dfrac{2}{t}\right)}$

44. $\dfrac{\left(x - 3 + \dfrac{2}{x}\right)}{\left(1 - \dfrac{2}{x}\right)}$

45. $\dfrac{\left(\dfrac{1}{a^2 - 16} - \dfrac{1}{a}\right)}{\left(\dfrac{1}{a^2 + 4a} + 4\right)}$

46. $\dfrac{\left(\dfrac{1}{x^2} - \dfrac{1}{y^2}\right)}{\left(\dfrac{1}{x} + \dfrac{1}{y}\right)}$

47. $(4x^3 - x) \div 2x$

48. $(10x + 15) \div (5x - 2)$

49. $\dfrac{6x^3 + 2x^2 - 4x + 2}{3x - 1}$

50. $\dfrac{4x^4 - x^3 - 7x^2 + 18x}{x - 2}$

51. $\dfrac{x^4 - 3x^2 + 2}{x^2 - 1}$

52. $\dfrac{3x^6}{x^2 - 1}$

In Exercises 53–56, use synthetic division to perform the division.

53. $\dfrac{x^3 + 7x^2 + 3x - 14}{x + 2}$

54. $\dfrac{x^4 - 2x^3 - 15x^2 - 2x + 10}{x - 5}$

55. $(x^4 - 3x^2 - 25) \div (x - 3)$

56. $(2x^3 + 5x - 2) \div \left(x + \tfrac{1}{2}\right)$

In Exercises 57–60, use a graphing utility to graph the equations on the same screen. Use the graphs to verify that the expressions are equivalent. Verify the results algebraically.

57. $y_1 = \dfrac{x^2 + 6x + 9}{x^2} \cdot \dfrac{x^2 - 3x}{x + 3}, \; y_2 = \dfrac{x^2 - 9}{x}$

58. $y_1 = \dfrac{1}{x} - \dfrac{3}{x(x + 3)}, \; y_2 = \dfrac{1}{x + 3}$

59. $y_1 = \dfrac{\left(\dfrac{1}{x} - \dfrac{1}{2}\right)}{2x}, \; y_2 = \dfrac{2 - x}{4x^2}$

60. $y_1 = \dfrac{x^3 - 2x^2 - 7}{x - 2}, \; y_2 = x^2 - \dfrac{7}{x - 2}$

In Exercises 61–74, use a graphing utility to graph the function.

61. $g(x) = \dfrac{2 + x}{1 - x}$

62. $h(x) = \dfrac{x - 3}{x - 2}$

63. $f(x) = \dfrac{x}{x^2 + 1}$

64. $f(x) = \dfrac{2x}{x^2 + 4}$

65. $P(x) = \dfrac{3x + 6}{x - 2}$

66. $s(x) = \dfrac{2x - 6}{x + 4}$

67. $h(x) = \dfrac{4}{(x - 1)^2}$

68. $g(x) = \dfrac{-2}{(x + 3)^2}$

69. $f(x) = -\dfrac{5}{x^2}$

70. $f(x) = \dfrac{4}{x}$

71. $y = \dfrac{x}{x^2 - 1}$

72. $y = \dfrac{2x}{x^2 - 4}$

73. $y = \dfrac{2x^2}{x^2 - 4}$

74. $y = \dfrac{2}{x + 3}$

In Exercises 75–88, solve the equation.

75. $\dfrac{3x}{8} = -15$

76. $\dfrac{t + 1}{8} = \dfrac{1}{2}$

77. $3\left(8 - \dfrac{12}{t}\right) = 0$

78. $\dfrac{1}{3y - 4} = \dfrac{6}{4(y + 1)}$

79. $\dfrac{2}{y} - \dfrac{1}{3y} = \dfrac{1}{3}$

80. $8\left(\dfrac{6}{x} - \dfrac{1}{x + 5}\right) = 15$

81. $r = 2 + \dfrac{24}{r}$

82. $\dfrac{3}{y + 1} - \dfrac{8}{y} = 1$

83. $\dfrac{2}{x} - \dfrac{x}{6} = \dfrac{2}{3}$

84. $\dfrac{2x}{x - 3} - \dfrac{3}{x} = 0$

85. $\dfrac{12}{x^2 + x - 12} - \dfrac{1}{x - 3} = -1$

86. $\dfrac{3}{x - 1} + \dfrac{6}{x^2 - 3x + 2} = 2$

87. $\dfrac{5}{x^2 - 4} - \dfrac{6}{x - 2} = -5$

88. $\dfrac{3}{x^2 - 9} + \dfrac{4}{x + 3} = 1$

In Exercises 89 and 90, (a) use a graphing utility to determine any x-intercepts of the graph of the equation, and (b) set $y = 0$ and solve the resulting rational equation to confirm the results of part (a).

89. $y = \dfrac{1}{x} - \dfrac{1}{2x + 3}$ **90.** $y = \dfrac{x}{4} - \dfrac{2}{x} - \dfrac{1}{2}$

91. *Average Speed*   You drive 56 miles on a service call for your company. On the return trip, which takes 10 minutes less than the original trip, your average speed is 8 miles an hour faster. What is your average speed on the return trip?

92. *Batting Average*   In this year's playing season, a baseball player has been at bat 150 times and has hit the ball safely 45 times. Thus, the batting average for the player is 45/150 = .300. How many consecutive times must the player hit safely to obtain a batting average of .400?

93. *Forming a Partnership*   A group of people agree to share equally in the cost of a $60,000 piece of machinery. If they could find two more people to join the group, each person's share of the cost would decrease by $5000. How many people are presently in the group?

Math Matters   Numbers in Other Languages

The numeral systems for several languages are shown below. One numeral system that is not shown is the Roman numeral system. Use a reference book to find how the Roman numeral system works. Then write the standard numerals for the following—each represents a famous date in American history.

(a) MDCCLXXVI   (b) MCDXCII   (c) MDCXX

Modern	1	2	3	4	5	6	7	8	9	10	100
Egyptian											
Babylonian											
Early Roman	I	II	III	IIII	V	VI	VII	VIII	IX	X	C
Chinese											
Hindu											
Mayan											

CHAPTER TEST

Take this test as you would take a test in class. After you are done, check your work against the answers given in the back of the book.

1. Find the domain of $\dfrac{3y}{y^2 - 25}$.

2. Find the least common denominator of $\dfrac{3}{x^2}$, $\dfrac{x}{x - 3}$, $\dfrac{2x}{x^3(x + 3)}$, and $\dfrac{10}{x^2 + 6x + 9}$.

3. Simplify the rational expression. (a) $\dfrac{2 - x}{3x - 6}$ (b) $\dfrac{2a^2 - 5a - 12}{5a - 20}$

In Exercises 4–15, perform the operation and simplify.

4. $\dfrac{4z^3}{5} \cdot \dfrac{25}{12z^2}$

5. $\dfrac{y^2 + 8y + 16}{2(y - 2)} \cdot \dfrac{8y - 16}{(y + 4)^3}$

6. $(4x^2 - 9) \cdot \dfrac{2x + 3}{2x^2 - x - 3}$

7. $\dfrac{(2xy^2)^3}{15} \div \dfrac{12x^3}{21}$

8. $\dfrac{\left(\dfrac{3x}{x + 2}\right)}{\left(\dfrac{12}{x^3 + 2x^2}\right)}$

9. $\dfrac{\left(9x - \dfrac{1}{x}\right)}{\left(\dfrac{1}{x} - 3\right)}$

10. $2x + \dfrac{1 - 4x^2}{x + 1}$

11. $\dfrac{5x}{x + 2} - \dfrac{2}{x^2 - x - 6}$

12. $\dfrac{3}{x} - \dfrac{5}{x^2} + \dfrac{2x}{x^2 + 2x + 1}$

13. $\dfrac{4}{x + 1} + \dfrac{4x}{x + 1}$

14. $\dfrac{t^4 + t^2 - 6t}{t^2 - 2}$

15. $\dfrac{2x^4 - 15x^2 - 7}{x - 3}$

16. Sketch the graph of the function. Describe how to find the asymptotes of the function.

(a) $f(x) = \dfrac{3}{x - 3}$ (b) $g(x) = \dfrac{3x}{x - 3}$

In Exercises 17–19, solve the equation.

17. $\dfrac{3}{h + 2} = \dfrac{1}{8}$

18. $\dfrac{2}{x + 5} - \dfrac{3}{x + 3} = \dfrac{1}{x}$

19. $\dfrac{1}{x + 1} + \dfrac{1}{x - 1} = \dfrac{2}{x^2 - 1}$

20. One painter works $1\frac{1}{2}$ times as fast as another. Find their individual rates for painting a room if it takes them 4 hours working together.

Radicals and Complex Numbers

- Integer Exponents and Scientific Notation
- Rational Exponents and Radicals
- Simplifying and Combining Radicals
- Multiplying and Dividing Radical Expressions
- Solving Radical Equations
- Complex Numbers

Complex numbers such as -1, i, and $1 + i$ can be plotted in a complex plane, as shown in the graph at the right. The vertical axis represents the imaginary part of the complex number and the horizontal axis represents the real part.

In the graph at the right, the black region is a *fractal* called the *Mandelbrot Set*, after Benoit Mandelbrot. Fractals can be used to model objects in nature, such as coastlines, ferns, or moonscapes. For example, the moonscape at the right is a fractal that was generated by a computer.

A complex number c is in the Mandelbrot Set if the sequence

$$c, \ c^2 + c, \ (c^2 + c)^2 + c, \ \ldots,$$

is bounded. The table at the right indicates that $c = -1$ and $c = i$ are in the Mandelbrot Set, but $c = 1 + i$ is not.

c	$c^2 + c$	$(c^2 + c)^2 + c$	$((c^2 + c)^2 + c)^2 + c$
-1	0	-1	0
i	$-1 + i$	$-i$	$-1 + i$
$1 + i$	$1 + 3i$	$-7 + 7i$	$1 - 97i$

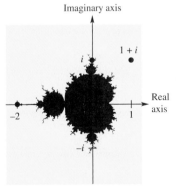

The chapter project related to this information is on page 376.

5.1	**Integer Exponents and Scientific Notation**
	Integer Exponents ▪ Scientific Notation

Integer Exponents

So far in the text, all exponents have been positive integers. In this section, the definition of an exponent is extended to include zero and negative integers. If a is a real number such that $a \neq 0$, then a^0 is defined as 1. Moreover, if m is an integer, then a^{-m} is defined as the reciprocal of a^m.

Definitions of Zero Exponents and Negative Exponents

Let a be a real number such that $a \neq 0$, and let m be an integer.

1. $a^0 = 1, \quad a \neq 0$ **2.** $a^{-m} = \dfrac{1}{a^m}, \quad a \neq 0$

These definitions are consistent with the properties of exponents given in Section 3.2. For instance, consider the following.

$$x^0 \cdot x^m = x^{0+m} = x^m = 1 \cdot x^m$$

(x^0 is the same as 1)

EXAMPLE 1   Zero Exponents and Negative Exponents

Rewrite each expression without using zero exponents or negative exponents.

a. 3^0 **b.** 0^0 **c.** 2^{-1} **d.** 3^{-2}

Solution

a. $3^0 = 1$

b. 0^0 is undefined. Remember when raising a number to the zero power, you must be sure the number is not zero.

c. $2^{-1} = \dfrac{1}{2^1} = \dfrac{1}{2}$

d. $3^{-2} = \dfrac{1}{3^2} = \dfrac{1}{9}$

The following properties of exponents are valid for all integer exponents, including integer exponents that are zero or negative. (The first five properties were listed in Section 3.2.)

Properties of Exponents

Let m and n be integers, and let a and b represent real numbers, variables, or algebraic expressions.

Property	*Example*
1. $a^m \cdot a^n = a^{m+n}$	$x^4(x^3) = x^{4+3} = x^7$
2. $(ab)^m = a^m \cdot b^m$	$(3x)^2 = 3^2(x^2) = 9x^2$
3. $(a^m)^n = a^{mn}$	$(x^3)^3 = x^{3 \cdot 3} = x^9$
4. $\dfrac{a^m}{a^n} = a^{m-n}, \quad a \neq 0$	$\dfrac{x^3}{x} = x^{3-1} = x^2, \quad x \neq 0$
5. $\left(\dfrac{a}{b}\right)^m = \dfrac{a^m}{b^m}, \quad b \neq 0$	$\left(\dfrac{x}{3}\right)^2 = \dfrac{x^2}{3^2} = \dfrac{x^2}{9}$
6. $\left(\dfrac{a}{b}\right)^{-m} = \left(\dfrac{b}{a}\right)^m, \quad a \neq 0, \; b \neq 0$	$\left(\dfrac{x}{3}\right)^{-2} = \left(\dfrac{3}{x}\right)^2 = \dfrac{3^2}{x^2} = \dfrac{9}{x^2}$
7. $a^{-m} = \dfrac{1}{a^m}, \quad a \neq 0$	$x^{-2} = \dfrac{1}{x^2}, \quad x \neq 0$
8. $a^0 = 1, \quad a \neq 0$	$(x^2 + 1)^0 = 1$

EXAMPLE 2   Using Properties of Exponents

a. $2x^{-1} = 2(x^{-1}) = 2\left(\dfrac{1}{x}\right) = \dfrac{2}{x}$

b. $(2x)^{-1} = \dfrac{1}{(2x)^1} = \dfrac{1}{2x}$

c. $\dfrac{3}{x^{-2}} = \dfrac{3}{\left(\dfrac{1}{x^2}\right)} = 3\left(\dfrac{x^2}{1}\right) = 3x^2$

d. $\dfrac{1}{(3x)^{-2}} = \dfrac{1}{\left[\dfrac{1}{(3x)^2}\right]} = \dfrac{1}{\left(\dfrac{1}{3^2x^2}\right)} = \dfrac{1}{\left(\dfrac{1}{9x^2}\right)} = (1)\left(\dfrac{9x^2}{1}\right) = 9x^2$

EXAMPLE 3   Using Properties of Exponents

Rewrite each expression using only positive exponents. (For each expression, assume that $x \neq 0$ and $y \neq 0$.)

a. $(-5x^{-3})^2$ **b.** $-\left(\dfrac{7x}{y^2}\right)^{-2}$

Solution

a. $(-5x^{-3})^2 = (-5)^2(x^{-3})^2$ Property 2

$\qquad\qquad\quad = 25x^{-6}$ Property 3

$\qquad\qquad\quad = \dfrac{25}{x^6}$ Property 7

b. $-\left(\dfrac{7x}{y^2}\right)^{-2} = -\left(\dfrac{y^2}{7x}\right)^{2}$ Property 6

$\qquad\qquad\quad = -\dfrac{(y^2)^2}{(7x)^2}$ Property 5

$\qquad\qquad\quad = -\dfrac{y^4}{49x^2}$ Property 3

EXAMPLE 4   Using Properties of Exponents

Rewrite each expression using only positive exponents. (For each expression, assume that $x \neq 0$ and $y \neq 0$.)

a. $\left(\dfrac{8x^{-1}y^4}{4x^3y^2}\right)^{-3}$ **b.** $\dfrac{3xy^0}{x^2(5y)^0}$

Solution

a. $\left(\dfrac{8x^{-1}y^4}{4x^3y^2}\right)^{-3} = \left(\dfrac{2y^2}{x^4}\right)^{-3}$ Simplify.

$\qquad\qquad\quad = \left(\dfrac{x^4}{2y^2}\right)^{3}$ Property 6

$\qquad\qquad\quad = \dfrac{x^{12}}{2^3y^6}$ Property 5

$\qquad\qquad\quad = \dfrac{x^{12}}{8y^6}$ Simplify.

b. $\dfrac{3xy^0}{x^2(5y)^0} = \dfrac{3x(1)}{x^2(1)} = \dfrac{3}{x}$ Property 8

Scientific Notation

Exponents provide an efficient way of writing and computing with very large (or very small) numbers. For instance, a drop of water contains more than 33 billion billion molecules—that is, 33 followed by 18 zeros.

$$33,000,000,000,000,000,000$$

It is convenient to write such numbers in **scientific notation.** This notation has the form $c \times 10^n$, where $1 \leq c < 10$ and n is an integer. Thus, the number of molecules in a drop of water can be written in scientific notation as

$$3.3 \times 10,000,000,000,000,000,000 = 3.3 \times 10^{19}.$$

The *positive* exponent 19 indicates that the number being written in scientific notation is *large* (10 or more) and that the decimal point has been moved 19 places. A *negative* exponent in scientific notation indicates that the number is *small* (less than 1).

EXAMPLE 5 Writing Scientific Notation

Write each real number in scientific notation.

a. 0.0000684   **b.** 937,200,000

Solution

a. $0.0000684 = 6.84 \times 10^{-5}$

Five places

b. $937,200,000.0 = 9.372 \times 10^8$

Eight places

EXAMPLE 6 Writing Decimal Notation

Convert each number from scientific notation to decimal notation.

a. 2.486×10^2 **b.** 1.81×10^{-6}

Solution

a. $2.486 \times 10^2 = 248.6$

Two places

b. $1.81 \times 10^{-6} = 0.00000181$

Six places

Technology

Using Scientific Notation

Most scientific and graphing calculators automatically switch to scientific notation when they are showing large (or small) numbers that exceed the display range. Try multiplying $86,500,000 \times 6000$. If your calculator follows standard conventions, its display should be

$$\boxed{5.19 \quad 11} \quad \text{or} \quad \boxed{5.19 \quad \text{E} \quad 11}.$$

This means that $c = 5.19$ and the exponent of 10 is $n = 11$, which implies that the number is 5.19×10^{11}.

To *enter* numbers in scientific notation, your calculator should have an exponential entry key labeled $\boxed{\text{EE}}$ or $\boxed{\text{EXP}}$. If you were to perform the preceding multiplication using scientific notation, you could begin by writing

$$86,500,000 \times 6000 = (8.65 \times 10^7)(6.0 \times 10^3)$$

and then entering the following.

8.65 $\boxed{\text{EXP}}$ 7 $\boxed{\times}$ 6 $\boxed{\text{EXP}}$ 3 $\boxed{=}$ Scientific

8.65 $\boxed{\text{EE}}$ 7 $\boxed{\times}$ 6 $\boxed{\text{EE}}$ 3 $\boxed{\text{ENTER}}$ Graphing

---

EXAMPLE 7   *Using Scientific Notation with a Calculator*

Use a scientific or graphing calculator to find the following.

a. $65,000 \times 3,400,000,000$ **b.** $0.000000348 \div 870$

Solution

a. 6.5 $\boxed{\text{EXP}}$ 4 $\boxed{\times}$ 3.4 $\boxed{\text{EXP}}$ 9 $\boxed{=}$ Scientific

     6.5 $\boxed{\text{EE}}$ 4 $\boxed{\times}$ 3.4 $\boxed{\text{EE}}$ 9 $\boxed{\text{ENTER}}$ Graphing

     The calculator display should read $\boxed{2.21 \quad 14}$, which implies that

$$(6.5 \times 10^4)(3.4 \times 10^9) = 2.21 \times 10^{14} = 221,000,000,000,000.$$

b. 3.48 $\boxed{\text{EXP}}$ 7 $\boxed{+/-}$ $\boxed{\div}$ 8.7 $\boxed{\text{EXP}}$ 2 $\boxed{=}$ Scientific

     3.48 $\boxed{\text{EE}}$ $\boxed{(-)}$ 7 $\boxed{\div}$ 8.7 $\boxed{\text{EE}}$ 2 $\boxed{\text{ENTER}}$ Graphing

     The calculator display should read $\boxed{4.0 \quad -10}$, which implies that

$$\frac{3.48 \times 10^{-7}}{8.7 \times 10^2} = 4.0 \times 10^{-10} = 0.0000000004.$$

EXAMPLE 8 *Using Scientific Notation*

Evaluate $\dfrac{(2,400,000,000)(0.00000345)}{(0.00007)(3800)}$.

Solution

Begin by rewriting each number in scientific notation and simplifying.

$$
\begin{aligned}
\frac{(2,400,000,000)(0.00000345)}{(0.00007)(3800)} &= \frac{(2.4 \times 10^9)(3.45 \times 10^{-6})}{(7.0 \times 10^{-5})(3.8 \times 10^3)} \\
&= \frac{(2.4)(3.45)(10^3)}{(7)(3.8)(10^{-2})} \\
&= \frac{(8.28)(10^5)}{26.6} \\
&\approx 0.3112782(10^5) \\
&= 31,127.82
\end{aligned}
$$

Group Activities   Communicating Mathematically

Developing a Mathematical Method   Discuss why scientific notation is used and give examples of its usefulness. Develop an easy-to-use saying, description, or method for converting numbers to and from scientific notation. Each member of your group should demonstrate the method to the rest of the group, using one of the following as an example.

a. 0.0000042

b. 293,600,000,000

c. 3.1×10^{-6}

d. 5.12×10^{11}

Discuss how to tell which of two numbers written in scientific notation is larger. Describe a comparison method and test it by deciding which is larger: 7×10^{51} or 8×10^{50}.

5.1 Exercises

Discussing the Concepts

1. In $(3x)^4$, what is $3x$ called? What is 4 called?

2. Discuss any differences between $(-2x)^4$ and $-2x^4$.

3. The expressions $4x$ and x^4 each represent repeated operations. What are the operations? Write the expressions showing the repeated operations.

4. In your own words, describe how you can "move" a factor from the numerator to the denominator or vice versa.

5. Is 32.5×10^5 in scientific notation? Explain.

6. When is scientific notation an efficient way of writing and computing real numbers?

Problem Solving

In Exercises 7–20, evaluate the expression.

7. 5^{-2}

8. 2^{-4}

9. $\dfrac{1}{(-2)^{-5}}$

10. $-\dfrac{1}{6^2}$

11. $\left(\frac{2}{3}\right)^{-1}$

12. $\left(\frac{4}{5}\right)^{-3}$

13. $27 \cdot 3^{-3}$

14. $4^2 \cdot 4^{-3}$

15. $\dfrac{10^3}{10^{-2}}$

16. $\dfrac{10^{-5}}{10^{-6}}$

17. $(4^2 \cdot 4^{-1})^{-2}$

18. $(5^3 \cdot 5^{-4})^{-3}$

19. $(5^0 - 4^{-2})^{-1}$

20. $(32 + 4^{-3})^0$

In Exercises 21–34, rewrite the expression using only positive exponents, and simplify. (Assume any variables in the expression are nonzero.)

21. $y^4 \cdot y^{-2}$

22. $x^{-2} \cdot x^{-5}$

23. $\dfrac{(4t)^0}{t^{-2}}$

24. $\dfrac{(5u)^4}{(5u)^4}$

25. $(-3x^{-3}y^2)(4x^2y^{-5})$

26. $(5s^5t^{-5})\left(\dfrac{3s^{-2}}{50t^{-1}}\right)$

27. $(3x^2y^{-2})^{-2}$

28. $(-4y^{-3}z)^{-3}$

29. $\dfrac{6^2x^3y^{-3}}{12x^{-2}y}$

30. $\dfrac{2^{-4}y^{-1}z^{-3}}{4^{-2}yz^{-3}}$

31. $\left(\dfrac{3u^2v^{-1}}{3^3u^{-1}v^3}\right)^{-2}$

32. $\left(\dfrac{5^2x^3y^{-3}}{125xy}\right)\left(\dfrac{5x}{3y}\right)^{-1}$

33. $\dfrac{a+b}{ba^{-1} - ab^{-1}}$

34. $\dfrac{u^{-1} - v^{-1}}{u^{-1} + v^{-1}}$

In Exercises 35–40, write in scientific notation.

35. *Land Area of Earth:* 57,500,000 square miles

36. *Ocean Area of Earth:* 139,400,000 square miles

37. *Light Year:* 9,461,000,000,000,000 kilometers

38. *Thickness of Soap Bubble:* 0.0000001 meter

39. *Relative Density of Hydrogen:* 0.0000899

40. *One Micron (Millionth of Meter):* 0.00003937 inch

In Exercises 41–46, write in decimal notation.

41. *1993 PepsiCo Beverage Sales:* $\$2.82 \times 10^{10}$

42. *Number of Air Sacs in Lungs:* 3.5×10^8

43. *Temperature of Sun:* 1.3×10^7 degrees Celsius

44. *Width of Air Molecule:* 9.0×10^{-9} meter

45. *Charge of Electron:* 4.8×10^{-10} electrostatic unit

46. *Width of Human Hair:* 9.0×10^{-4} meter

In Exercises 47–50, evaluate without a calculator. Write your answer in scientific notation.

47. $(6.5 \times 10^6)(2 \times 10^4)$

48. $5 \times (10^8)^3$

49. $\dfrac{3.6 \times 10^{12}}{6 \times 10^5}$

50. $\dfrac{72,000,000,000}{0.00012}$

In Exercises 51 and 52, evaluate with a calculator.

51. $\dfrac{(3.82 \times 10^5)^2}{(8.5 \times 10^4)(5.2 \times 10^{-3})}$

52. $\dfrac{(6,200,000)(0.005)^3}{(0.00035)^5}$

53. *Masses of Earth and Sun*   The masses of the earth and sun are approximately 5.975×10^{24} and 1.99×10^{30} kilograms, respectively. The mass of the sun is approximately how many times that of the earth?

54. *Kepler's Third Law*   In 1619, Johannes Kepler, a German astronomer, discovered that the period T (in years) of each planet in our solar system is related to the planet's mean distance R (in astronomical units) from the sun by the equation

$$\frac{T^2}{R^3} = k.$$

Test Kepler's equation for the nine planets in our solar system, using the table at the right. Do you get approximately the same value of k for each planet?

Planet	T	R
Mercury	0.241	0.387
Venus	0.615	0.723
Earth	1.000	1.000
Mars	1.881	1.523
Jupiter	11.861	5.203
Saturn	21.457	9.541
Uranus	84.008	19.190
Neptune	164.784	30.086
Pluto	248.350	39.508

Reviewing the Major Concepts

In Exercises 55–58, factor the expression.

55. $x^2 - 3x + 2$

56. $2x^2 + 5x - 7$

57. $11x^2 + 6x - 5$

58. $4x^2 - 28x + 49$

59. *Comparing Memberships*   The current membership of a public television station is 8415, which is 110% of what it was a year ago. How many members did the station have last year?

60. *Buy Now or Wait?*   A sales representative indicates that if a customer waits another month for a new car that currently costs $23,500, the price will increase by 4%. However, the customer will pay an interest penalty of $725 for the early withdrawal of a certificate of deposit if the car is purchased now. Determine whether the customer should buy now or wait another month.

Additional Problem Solving

In Exercises 61–76, evaluate the expression.

61. -10^{-3}

62. -20^{-2}

63. $(-3)^{-5}$

64. 25^0

65. $\dfrac{1}{4^{-3}}$

66. $\dfrac{1}{-8^{-2}}$

67. $\left(\dfrac{3}{16}\right)^0$

68. $\left(-\dfrac{5}{8}\right)^{-2}$

69. $\dfrac{3^4}{3^{-2}}$

70. $\dfrac{5^{-1}}{5^2}$

71. $(2^{-3})^2$

72. $(-4^{-1})^{-2}$

73. $2^{-3} + 2^{-4}$

74. $4 - 3^{-2}$

75. $\left(\dfrac{3}{4} + \dfrac{5}{8}\right)^{-2}$

76. $\left(\dfrac{1}{2} - \dfrac{2}{3}\right)^{-1}$

In Exercises 77–94, rewrite the expression using only positive exponents, and simplify. (Assume any variables in the expression are nonzero.)

77. $z^5 \cdot z^{-3}$

78. $t^{-1} \cdot t^{-6}$

79. $\dfrac{1}{x^{-6}}$

80. $\dfrac{x^{-3}}{y^{-1}}$

81. $\dfrac{a^{-6}}{a^{-7}}$

82. $\dfrac{6u^{-2}}{15u^{-1}}$

83. $(2x^2)^{-2}$

84. $(4a^{-2}b^3)^{-3}$

85. $\left(\dfrac{x}{10}\right)^{-1}$

86. $\left(\dfrac{4}{z}\right)^{-2}$

87. $\left(\dfrac{a^{-2}}{b^{-2}}\right)\left(\dfrac{b}{a}\right)^3$

88. $\left(\dfrac{a^{-3}}{b^{-3}}\right)\left(\dfrac{b}{a}\right)^3$

89. $\left[(x^{-4}y^{-6})^{-1}\right]^2$

90. $(ab)^{-2}(a^2b^2)^{-1}$

91. $(u + v^{-2})^{-1}$

92. $x^{-2}(x^2 + y^2)$

93. $\left[(2x^{-3}y^{-2})^2\right]^{-2}$

94. $\left[\left(\dfrac{2x^2}{4y}\right)^{-3}\right]^2$

In Exercises 95–98, write in scientific notation.

95. 3,600,000

96. 98,100,000

97. 0.0000000381

98. 0.0007384

In Exercises 99–102, write in decimal notation.

99. 6×10^7

100. 5.05×10^{12}

101. 1.359×10^{-7}

102. 8.6×10^{-9}

In Exercises 103–108, evaluate without a calculator. Write your answer in scientific notation.

103. $(2 \times 10^9)(3.4 \times 10^{-4})$

104. $(5 \times 10^4)^3$

105. $\dfrac{3.6 \times 10^9}{9 \times 10^5}$

106. $\dfrac{2.5 \times 10^{-3}}{5 \times 10^2}$

107. $(4,500,000)(2,000,000,000)$

108. $\dfrac{64,000,000}{0.00004}$

In Exercises 109–112, evaluate with a calculator. Round your answer to two decimal places.

109. $\dfrac{1.357 \times 10^{12}}{(4.2 \times 10^2)(6.87 \times 10^{-3})}$

110. $(8.67 \times 10^4)^7$

111. $\dfrac{(0.0000565)(2,850,000,000,000)}{0.00465}$

112. $\dfrac{(5,000,000)^3(0.000037)^2}{(0.005)^4}$

113. *Distance to the Sun*   The distance from the earth to the sun is approximately 93 million miles. Write this distance in scientific notation.

114. *Copper Electrons*   A cube of copper with an edge of 1 centimeter has approximately 8×10^{22} free electrons. Write this real number in decimal form.

115. *Light-Year*   One light-year (the distance light can travel in 1 year) is approximately 9.45×10^{15} meters. Approximate the time for light to travel from the sun to the earth if that distance is approximately 1.49×10^{11} meters.

116. *Distance to a Star*   The star Alpha Andromedae is 90 light years from the earth (see figure). Use the definition of a light year in Exercise 115 to write this distance in meters.

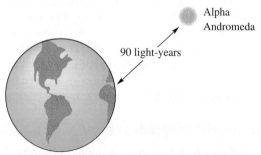

Alpha
Andromeda

90 light-years

117. *Federal Debt*   In 1993, the population of the United States was 257 million, and the federal debt was 4410 billion dollars. Use these two numbers to determine the amount each person would have had to pay (per capita debt) to remove the debt.   (Source: U.S. Bureau of Census)

| | 5.2 | **Rational Exponents and Radicals** |

Roots and Radicals ▪ Rational Exponents ▪
Radicals and Calculators ▪ Radical Functions

Roots and Radicals

The **square root** of a number is defined as one of its two equal factors. For example, 5 is a square root of 25 because 5 is one of the two equal factors of 25. In a similar way, a **cube root** of a number is one of its three equal factors.

Number	Equal Factors	Root	Type
$9 = 3^2$	$3 \cdot 3$	3	Square root
$25 = (-5)^2$	$(-5)(-5)$	-5	Square root
$-27 = (-3)^3$	$(-3)(-3)(-3)$	-3	Cube root
$16 = 2^4$	$2 \cdot 2 \cdot 2 \cdot 2$	2	Fourth root

Definition of nth Root of a Number

Let a and b be real numbers and let n be an integer such that $n \geq 2$. If

$$a = b^n$$

then b is an **nth root of a.** If $n = 2$, the root is a **square root,** and if $n = 3$, the root is a **cube root.**

Some numbers have more than one nth root. For example, both 5 and -5 are square roots of 25 because $25 = 5^2$ and $25 = (-5)^2$. To avoid ambiguity about which root of a number you are talking about, the **principal nth root** of a number is defined in terms of a radical symbol $\sqrt[n]{}$.

Principal nth Root of a Number

NOTE "Having the same sign" means that the principal nth root of a is positive if a is positive and negative if a is negative. For example, $\sqrt{4} = 2$ and $\sqrt[3]{-8} = -2$.

Let a be a real number that has at least one (real number) nth root. The **principal nth root of a** is the nth root that has the same sign as a, and it is denoted by the **radical symbol**

$$\sqrt[n]{a}. \qquad \text{Principal } n\text{th root}$$

The positive integer n is the **index** of the radical, and the number a is the **radicand.** If $n = 2$, omit the index and write $\sqrt{a}$ rather than $\sqrt[2]{a}$.

EXAMPLE 1 Finding Roots of a Number

Find the roots.

a. $\sqrt{36}$ **b.** $-\sqrt{36}$ **c.** $\sqrt{-4}$ **d.** $\sqrt[3]{8}$ **e.** $\sqrt[3]{-8}$

Solution

a. $\sqrt{36} = 6$ because $6 \cdot 6 = 6^2 = 36$.

b. $-\sqrt{36} = -6$ because $6 \cdot 6 = 6^2 = 36$.

c. $\sqrt{-4}$ is not real because there is no real number that when multiplied by itself yields -4.

d. $\sqrt[3]{8} = 2$ because $2 \cdot 2 \cdot 2 = 2^3 = 8$.

e. $\sqrt[3]{-8} = -2$ because $(-2)(-2)(-2) = (-2)^3 = -8$.

Properties of nth Roots

1. If a is a positive real number and n is *even*, then a has exactly two (real) nth roots, which are denoted by $\sqrt[n]{a}$ and $-\sqrt[n]{a}$.

2. If a is any real number and n is *odd*, then a has only one (real) nth root, which is denoted by $\sqrt[n]{a}$.

3. If a is a negative real number and n is *even*, then a has no (real) nth root.

Integers such as 1, 4, 9, 16, 49, and 81 are called **perfect squares** because they have integer square roots. Similarly, integers such as 1, 8, 27, 64, and 125 are called **perfect cubes** because they have integer cube roots.

EXAMPLE 2 Classifying Perfect Roots

State whether the number is a perfect square root, a perfect cube root, both, or neither.

a. 81   **b.** 64   **c.** 32

Solution

a. 81 is a perfect square root because $9^2 = 81$. It is not a perfect cube root.

b. 64 is a perfect square root because $8^2 = 64$, and it is also a perfect cube root because $4^3 = 64$.

c. 32 is not a perfect square root or a perfect cube root. (It is a perfect 5th root because $2^5 = 32$.)

Raising a number to the nth power and taking the principal nth root of a number can be thought of as *inverse* operations. Here are two examples.

$$\left(\sqrt{4}\right)^2 = 2^2 = 4 \quad \text{and} \quad \sqrt{2^2} = \sqrt{4} = 2$$

$$\left(\sqrt[3]{27}\right)^3 = 3^3 = 27 \quad \text{and} \quad \sqrt[3]{3^3} = \sqrt[3]{27} = 3$$

Inverse Properties of nth Powers and nth Roots

Let a be a real number, and let n be an integer such that $n \geq 2$.

1. If a has a principal nth root, then

$$\left(\sqrt[n]{a}\right)^n = a.$$

2. If n is *odd*, then

$$\sqrt[n]{a^n} = a.$$

If n is *even*, then

$$\sqrt[n]{a^n} = |a|.$$

EXAMPLE 3   *Evaluating Radical Expressions*

Evaluate each radical expression.

a. $\sqrt[3]{5^3}$ **b.** $\sqrt[3]{(-2)^3}$ **c.** $\left(\sqrt{7}\right)^2$ **d.** $\sqrt{(-3)^2}$ **e.** $\sqrt{-3^2}$

Solution

a. Because the index of the radical is odd, you can write

$$\sqrt[3]{5^3} = 5.$$

b. Because the index of the radical is odd, you can write

$$\sqrt[3]{(-2)^3} = -2.$$

c. Using the inverse property of powers and roots, you can write

$$\left(\sqrt{7}\right)^2 = 7.$$

d. Because the index of the radical is even, you must include absolute value signs, and write

$$\sqrt{(-3)^2} = |-3| = 3.$$

e. Because $\sqrt{-3^2} = \sqrt{-9}$ is an even root of a negative number, its value is not a real number.

Rational Exponents

NOTE The numerator of a rational exponent denotes the *power* to which the base is raised, and the denominator denotes the *root* to be taken.

Definition of Rational Exponents

Let a be a real number, and let n be an integer such that $n \geq 2$. If the principal nth root of a exists, we define $a^{1/n}$ to be

$$a^{1/n} = \sqrt[n]{a}.$$

If m is a positive integer that has no common factor with n, then

$$a^{m/n} = (a^{1/n})^m = \left(\sqrt[n]{a}\right)^m \quad \text{and} \quad a^{m/n} = (a^m)^{1/n} = \sqrt[n]{a^m}.$$

It does not matter in which order the two operations are performed, provided the nth root exists. Here is an example.

$$8^{2/3} = \left(\sqrt[3]{8}\right)^2 = 2^2 = 4 \qquad \text{Cube root, then second power}$$
$$8^{2/3} = \sqrt[3]{8^2} = \sqrt[3]{64} = 4 \qquad \text{Second power, then cube root}$$

The properties of exponents that we listed in Section 5.1 also apply to rational exponents (provided the roots indicated by the denominators exist). We relist the first seven of those properties here, with different examples.

D I S C O V E R Y

Use a calculator to evaluate the expressions below.

$$\frac{3.4^{4.6}}{3.4^{3.1}} \quad \text{and} \quad 3.4^{1.5}$$

How are these two expressions related? Use your calculator to verify some of the other properties of exponents.

Properties of Exponents

Let r and s be rational numbers, and let a and b be real numbers, variables, or algebraic expressions.

Property	*Example*
1. $a^r \cdot a^s = a^{r+s}$	$4^{1/2}(4^{1/3}) = 4^{5/6}$
2. $(ab)^r = a^r \cdot b^r$	$(2x)^{1/2} = 2^{1/2}(x^{1/2})$
3. $(a^r)^s = a^{rs}$	$(x^3)^{1/3} = x$
4. $\dfrac{a^r}{a^s} = a^{r-s}, \quad a \neq 0$	$\dfrac{x^2}{x^{1/2}} = x^{2-(1/2)} = x^{3/2}$
5. $\left(\dfrac{a}{b}\right)^r = \dfrac{a^r}{b^r}, \quad b \neq 0$	$\left(\dfrac{x}{3}\right)^{1/3} = \dfrac{x^{1/3}}{3^{1/3}}$
6. $\left(\dfrac{a}{b}\right)^{-r} = \left(\dfrac{b}{a}\right)^r, \quad a \neq 0, \quad b \neq 0$	$\left(\dfrac{x}{4}\right)^{-1/2} = \left(\dfrac{4}{x}\right)^{1/2} = \dfrac{2}{x^{1/2}}$
7. $a^{-r} = \dfrac{1}{a^r}, \quad a \neq 0$	$4^{-1/2} = \dfrac{1}{4^{1/2}} = \dfrac{1}{2}$

EXAMPLE 4   *Evaluating Expressions with Rational Exponents*

a. $8^{4/3} = \left(\sqrt[3]{8}\right)^4 = 2^4 = 16$

b. $(4^2)^{3/2} = \left(\sqrt{4^2}\right)^3 = 4^3 = 64$

c. $25^{-3/2} = \dfrac{1}{25^{3/2}} = \dfrac{1}{\left(\sqrt{25}\right)^3} = \dfrac{1}{5^3} = \dfrac{1}{125}$

d. $\left(\dfrac{64}{125}\right)^{2/3} = \left(\sqrt[3]{\dfrac{64}{125}}\right)^2 = \left(\dfrac{\sqrt[3]{64}}{\sqrt[3]{125}}\right)^2 = \left(\dfrac{4}{5}\right)^2 = \dfrac{16}{25}$

e. $-9^{1/2} = -\sqrt{9} = -3$

f. $(-9)^{1/2} = \sqrt{-9}$ is not a real number.

In parts (e) and (f) of Example 4, be sure that you see the distinction between the expressions $-9^{1/2}$ and $(-9)^{1/2}$.

EXAMPLE 5   *Using Properties of Exponents*

Rewrite each expression using rational exponents.

a. $x\sqrt[4]{x^3}$ **b.** $\dfrac{\sqrt[3]{x^2}}{\sqrt{x^3}}$

Solution

a. $x\sqrt[4]{x^3} = x(x^{3/4}) = x^{1+(3/4)} = x^{7/4}$

b. $\dfrac{\sqrt[3]{x^2}}{\sqrt{x^3}} = \dfrac{x^{2/3}}{x^{3/2}} = x^{(2/3)-(3/2)} = x^{-5/6} = \dfrac{1}{x^{5/6}}$

EXAMPLE 6   *Using Properties of Exponents*

Use properties of exponents to rewrite each expression in simpler form.

a. $\sqrt{\sqrt[3]{x}}$ **b.** $\dfrac{(2x-1)^{4/3}}{\sqrt[3]{2x-1}}$

Solution

a. $\sqrt{\sqrt[3]{x}} = \sqrt{x^{1/3}} = (x^{1/3})^{1/2} = x^{1/6}$

b. $\dfrac{(2x-1)^{4/3}}{\sqrt[3]{2x-1}} = \dfrac{(2x-1)^{4/3}}{(2x-1)^{1/3}} = (2x-1)^{(4/3)-(1/3)} = 2x-1$

Radicals and Calculators

There are two methods of evaluating radicals on most calculators. For square roots, you can use the *square root key* $\boxed{\sqrt{\ }}$. For other roots, you should first convert the radical to exponential form and then use the *exponential key* $\boxed{y^x}$ or $\boxed{\wedge}$.

EXAMPLE 7 *Evaluating Roots with a Calculator*

Evaluate the following. Round the result to three decimal places.

a. $\sqrt{5}$ **b.** $\sqrt[5]{25}$ **c.** $\sqrt[3]{-4}$ **d.** $(1.4)^{-2/5}$

Solution

a. 5 $\boxed{\sqrt{\ }}$ Scientific

    $\boxed{\sqrt{\ }}$ 5 $\boxed{\text{ENTER}}$ Graphing

The display is 2.236068. Rounded to three decimal places, $\sqrt{5} \approx 2.236$.

b. First rewrite the expression as $\sqrt[5]{25} = 25^{1/5}$. Then use one of the following keystroke sequences.

    25 $\boxed{y^x}$ $\boxed{(}$ 1 $\boxed{\div}$ 5 $\boxed{)}$ $\boxed{=}$ Scientific

    25 $\boxed{\wedge}$ $\boxed{(}$ 1 $\boxed{\div}$ 5 $\boxed{)}$ $\boxed{\text{ENTER}}$ Graphing

The display is 1.9036539. Rounded to three decimal places, $\sqrt[5]{25} \approx 1.904$.

c. If your calculator does not have a cube root key, use the fact that

$$\sqrt[3]{-4} = \sqrt[3]{(-1)(4)} = \sqrt[3]{-1}\sqrt[3]{4} = -\sqrt[3]{4}$$

and attach the negative sign of the radicand as the last keystroke.

    4 $\boxed{y^x}$ $\boxed{(}$ 1 $\boxed{\div}$ 3 $\boxed{)}$ $\boxed{=}$ $\boxed{+/-}$ Scientific

    $\boxed{\sqrt[3]{\ }}$ $\boxed{(-)}$ 4 $\boxed{\text{ENTER}}$ Graphing

The display is -1.5874011. Rounded to three decimal places, $\sqrt[3]{-4} \approx -1.587$.

d. 1.4 $\boxed{y^x}$ $\boxed{(}$ 2 $\boxed{\div}$ 5 $\boxed{+/-}$ $\boxed{)}$ $\boxed{=}$ Scientific

    1.4 $\boxed{\wedge}$ $\boxed{(}$ $\boxed{(-)}$ 2 $\boxed{\div}$ 5 $\boxed{)}$ $\boxed{\text{ENTER}}$ Graphing

The display is 0.8740752. Rounded to three decimal places, $(1.4)^{-2/5} \approx 0.874$.

NOTE Some calculators have a cube root key or submenu command. If your calculator does, try using it to evaluate the expression in Example 7(c).

Radical Functions

The **domain** of the radical $\sqrt[n]{x}$ is the set of all real numbers such that x has a principal nth root.

Domain of a Radical

Let n be an integer that is greater than or equal to 2.

1. If n is odd, the domain of $\sqrt[n]{x}$ is the set of all real numbers.

2. If n is even, the domain of $\sqrt[n]{x}$ is the set of all nonnegative real numbers.

EXAMPLE 8   *Finding the Domain of a Radical Function*

Describe the domain of each function.

a. $f(x) = \sqrt{x}$ b. $f(x) = \sqrt[3]{x}$ c. $f(x) = \sqrt{x^2}$ d. $f(x) = \sqrt{x^3}$

Solution

a. The domain of $f(x) = \sqrt{x}$ is the set of all nonnegative real numbers. For instance, 2 is in the domain, but -2 is not because $\sqrt{-2}$ is not a real number.

b. The domain of $f(x) = \sqrt[3]{x}$ is the set of all real numbers because for any real number x, the expression $\sqrt[3]{x}$ is a real number.

c. The domain of $f(x) = \sqrt{x^2}$ is the set of all real numbers because for any real number x, the expression x^2 is a nonnegative real number.

d. The domain of $f(x) = \sqrt{x^3}$ is the set of all nonnegative real numbers. For instance, 1 is in the domain, but -1 is not because $\sqrt{(-1)^3} = \sqrt{-1}$ is not a real number.

Group Activities   Exploring with Technology

Describing Domains and Ranges   In your group, discuss the domain and range of each of the following functions. Use a graphing utility to verify your conclusions.

a. $y = x^{3/2}$ b. $y = x^2$ c. $y = x^{1/3}$ d. $y = \left(\sqrt{x}\right)^2$ e. $y = x^{-4/5}$

5.2 Exercises

Discussing the Concepts

1. In your own words, define an nth root of a number.

2. Define the *radicand* of a radical and the *index* of a radical.

3. If n is even, what must be true about the radicand for the nth root to be a real number? Explain.

4. Determine the values of x for which $\sqrt{x^2} \neq x$. Explain your answer.

5. Is it true that $\sqrt{2} = 1.414$? Explain.

6. Given a real number x, state the conditions of n for each of the following.

    (a) $\sqrt[n]{x^n} = x$ (b) $\sqrt[n]{x^n} = |x|$

Problem Solving

In Exercises 7–12, complete the statement.

7. Because $7^2 = 49$, ____ is a square root of 49.

8. Because $24.5^2 = 600.25$, 24.5 is a ____.

9. Because $4.2^3 = 74.088$, 4.2 is a ____.

10. Because $6^4 = 1296$, ____ is a fourth root of 1296.

11. Because $45^2 = 2025$, 45 is called the ____ of 2025.

12. Because $12^3 = 1728$, 12 is called the ____ of 1728.

In Exercises 13–26, evaluate the root without a calculator. If it is not possible, state the reason.

13. $\sqrt{81}$

14. $\sqrt{144}$

15. $\sqrt{\frac{9}{16}}$

16. $\sqrt{0.09}$

17. $\sqrt{-64}$

18. $\sqrt{0.36}$

19. $-\sqrt{\frac{4}{9}}$

20. $\sqrt{-\frac{9}{25}}$

21. $\sqrt[3]{125}$

22. $\sqrt[3]{-8}$

23. $\sqrt[4]{81}$

24. $\sqrt[5]{32}$

25. $\sqrt[5]{-0.00243}$

26. $\sqrt[6]{64}$

In Exercises 27–30, evaluate without a calculator.

27. $25^{1/2}$

28. $81^{-3/4}$

29. $32^{-2/5}$

30. $-\left(\frac{1}{125}\right)^{2/3}$

In Exercises 31–36, fill in the missing description.

Radical Form	Rational Exponent Form
31. $\sqrt{16} = 4$	
32. $\sqrt[4]{81} = 3$	
33. $\sqrt[3]{27^2} = 9$	
34.	$125^{1/3} = 5$
35.	$256^{3/4} = 64$
36.	$64^{2/3} = 16$

In Exercises 37–40, use a calculator to evaluate the expression. Round the result to four decimal places.

37. $\dfrac{8 - \sqrt{35}}{2}$

38. $962^{2/3}$

39. $\sqrt[4]{342}$

40. $\sqrt[5]{-35^3}$

In Exercises 41–50, simplify the expression.

41. $\sqrt[3]{t^6}$

42. $\sqrt[4]{z^4}$

43. $3^{1/4} \cdot 3^{3/4}$

44. $(2^{1/2})^{2/3}$

45. $\dfrac{2^{1/5}}{2^{6/5}}$

46. $\dfrac{5^{-3/4}}{5}$

47. $x^{2/3} \cdot x^{7/3}$

48. $z^{3/5} \cdot z^{-2/5}$

49. $\left(\dfrac{x^{1/4}}{x^{1/6}}\right)^3$

50. $\left(\dfrac{3m^{1/6}n^{1/3}}{4n^{-2/3}}\right)^2$

Mathematical Modeling   In Exercises 51 and 52, use the formula for the *declining balances method*

$$r = 1 - \left(\frac{S}{C}\right)^{1/n}$$

to find the depreciation rate r. In the formula, n is the useful life of the item (in years), S is the salvage value (in dollars), and C is the original cost (in dollars).

51. A truck whose original cost is $75,000 is depreciated over an 8-year period, as shown in the graph.

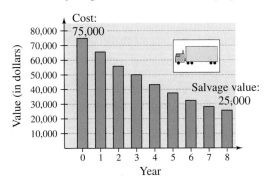

52. A printing press whose original cost is $125,000 is depreciated over a 10-year period, as shown in the graph.

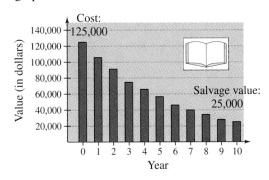

In Exercises 53 and 54, determine the domain of the function.

53. $f(x) = 3\sqrt{x}$

54. $g(x) = \dfrac{10}{\sqrt[3]{x}}$

Reviewing the Major Concepts

In Exercises 55–58, evaluate the quantity.

55. $-\dfrac{13}{35} \cdot \dfrac{-25}{104}$

56. $\dfrac{5}{12} \cdot \dfrac{9}{75}$

57. $\dfrac{14}{3} \div \dfrac{42}{45}$

58. $\dfrac{-8}{105} \div \dfrac{2}{3}$

In Exercises 59 and 60, use the function to find and simplify the expression for

$$\frac{f(2+h) - f(2)}{h}.$$

59. $f(x) = x^2 - 3$

60. $f(x) = \dfrac{3}{x+5}$

Additional Problem Solving

In Exercises 61–74, evaluate without a calculator. If it is not possible, state the reason.

61. $\sqrt{64}$

62. $-\sqrt{100}$

63. $\sqrt{-100}$

64. $\sqrt{169}$

65. $\sqrt{0.16}$

66. $-\sqrt{0.0009}$

67. $\sqrt{49 - 4(2)(-15)}$

68. $\sqrt{\dfrac{75}{3}}$

69. $\sqrt[3]{1000}$

70. $\sqrt[3]{64}$

71. $\sqrt[3]{-\dfrac{1}{64}}$

72. $-\sqrt[3]{0.008}$

73. $-\sqrt[4]{-625}$

74. $-\sqrt[4]{\dfrac{1}{625}}$

In Exercises 75–78, determine whether the square root is a rational or irrational number.

75. $\sqrt{6}$

76. $\sqrt{\dfrac{9}{16}}$

77. $\sqrt{900}$

78. $\sqrt{72}$

In Exercises 79–86, evaluate without a calculator.

79. $49^{1/2}$

80. $-121^{1/2}$

81. $16^{3/4}$

82. $243^{-3/5}$

83. $\left(\dfrac{8}{27}\right)^{2/3}$

84. $\left(\dfrac{256}{625}\right)^{1/4}$

85. $\left(\dfrac{121}{9}\right)^{-1/2}$

86. $\left(\dfrac{27}{1000}\right)^{-4/3}$

In Exercises 87–94, use a calculator to evaluate the expression. (If it is not possible, state the reason.) Round the result to four decimal places.

87. $\sqrt{73}$

88. $\sqrt{-532}$

89. $\dfrac{3-\sqrt{17}}{9}$

90. $\dfrac{-5+\sqrt{3215}}{10}$

91. $1698^{-3/4}$

92. $315^{2/5}$

93. $\sqrt[3]{545^2}$

94. $\sqrt[3]{159}$

In Exercises 95–108, simplify the expression.

95. $\sqrt{t^2}$

96. $\sqrt[3]{z^3}$

97. $\sqrt[3]{y^9}$

98. $\sqrt[4]{a^8}$

99. $\left(\frac{2}{3}\right)^{5/3}\cdot\left(\frac{2}{3}\right)^{1/3}$

100. $(4^{1/3})^{9/4}$

101. $(3x^{-1/3}y^{3/4})^2$

102. $(-2u^{3/5}v^{-1/5})^3$

103. $\dfrac{18y^{4/3}z^{-1/3}}{24y^{-2/3}z}$

104. $\dfrac{a^{3/4}\cdot a^{1/2}}{a^{5/2}}$

105. $(c^{3/2})^{1/3}$

106. $(k^{-1/3})^{3/2}$

107. $\sqrt{\sqrt[4]{y}}$

108. $\sqrt[3]{\sqrt{2x}}$

In Exercises 109–112, multiply and simplify.

109. $x^{1/2}(2x-3)$

110. $x^{4/3}(3x^2-4x+5)$

111. $y^{-1/3}(y^{1/3}+5y^{4/3})$

112. $(x^{1/2}-3)(x^{1/2}+3)$

In Exercises 113 and 114, determine the domain of the function.

113. $g(x)=\dfrac{2}{\sqrt[4]{x}}$

114. $h(x)=\sqrt[4]{x}$

In Exercises 115–118, use a graphing utility to graph the function. Check the domain of the function algebraically. Did the graphing utility skip part of the domain? If so, complete the graph by hand.

115. $y=\dfrac{5}{\sqrt[4]{x^3}}$

116. $y=4\sqrt[3]{x}$

117. $g(x)=2x^{3/5}$

118. $h(x)=5x^{2/3}$

119. *Perfect Squares* Find all possible "last digits" of perfect squares. (For instance, the last digit of 81 is 1 and the last digit of 64 is 4.) Is it possible that 4,322,788,987 is a perfect square?

120. *Geometry* The usable space in a particular microwave oven is in the form of a cube (see figure). The sales brochure indicates that the interior space of the oven is 2197 cubic inches. Find the inside dimensions of the oven.

121. *Velocity of a Stream* A stream of water moving at the rate of v feet per second can carry particles of size $0.03\sqrt{v}$ inches. Find the particle size that can be carried by a stream flowing at the rate of $\frac{3}{4}$ foot per second.

5.3	**Simplifying and Combining Radicals**

Simplifying Radicals ▪ Rationalization Techniques ▪
Adding and Subtracting Radicals ▪ Application of Radicals

Simplifying Radicals

In this section, you will study ways to simplify and combine radicals. For instance, the expression $\sqrt{12}$ can be simplified as

$$\sqrt{12} = \sqrt{4 \cdot 3} = \sqrt{4}\sqrt{3} = 2\sqrt{3}.$$

This rewriting is based on the following rules for multiplying and dividing radicals.

Multiplying and Dividing Radicals

Let u and v be real numbers, variables, or algebraic expressions. If the nth roots of u and v are real, the following properties are true.

1. $\sqrt[n]{u}\sqrt[n]{v} = \sqrt[n]{uv}$ Multiplication Property

2. $\dfrac{\sqrt[n]{u}}{\sqrt[n]{v}} = \sqrt[n]{\dfrac{u}{v}}, \quad v \neq 0$ Division Property

---

You can use these properties of radicals to *simplify* radical expressions as follows.

$$\sqrt{48} = \sqrt{16 \cdot 3} = \sqrt{16}\sqrt{3} = 4\sqrt{3}$$

This simplification process is called **removing perfect square factors from the radical.**

EXAMPLE 1   *Removing Constant Factors from Radicals*

Simplify each radical by removing as many factors as possible.

a. $\sqrt{75}$ **b.** $\sqrt{72}$ **c.** $\sqrt{162}$

Solution

a. $\sqrt{75} = \sqrt{25 \cdot 3} = \sqrt{25}\sqrt{3} = 5\sqrt{3}$

b. $\sqrt{72} = \sqrt{36 \cdot 2} = \sqrt{36}\sqrt{2} = 6\sqrt{2}$

c. $\sqrt{162} = \sqrt{81 \cdot 2} = \sqrt{81}\sqrt{2} = 9\sqrt{2}$

When removing *variable* factors from a square root radical, remember that it is not valid to write $\sqrt{x^2} = x$ *unless* you happen to know that x is nonnegative. Without knowing anything about x, the only way you can simplify $\sqrt{x^2}$ is to include absolute value signs when you remove x from the radical.

$$\sqrt{x^2} = |x| \qquad\qquad \text{Restricted by absolute value signs}$$

When simplifying the expression $\sqrt{x^3}$, it is not necessary to include absolute value signs because the domain of this expression does not include negative numbers.

$$\sqrt{x^3} = \sqrt{x^2(x)} = x\sqrt{x} \qquad\qquad \text{Restricted by domain of radical}$$

EXAMPLE 2 Removing Variable Factors from Radicals

Simplify the radical expression.

a. $\sqrt{25x^2}$ **b.** $\sqrt{12x^3}, \quad x \geq 0$ **c.** $\sqrt{144x^4}$

Solution

a. $\sqrt{25x^2} = \sqrt{5^2 x^2} = \sqrt{5^2}\sqrt{x^2} = 5|x| \qquad \sqrt{x^2} = |x|$

b. $\sqrt{12x^3} = \sqrt{2^2 x^2 (3x)} = 2x\sqrt{3x} \qquad \sqrt{2^2}\sqrt{x^2} = 2x, \ x \geq 0$

c. $\sqrt{144x^4} = \sqrt{12^2 (x^2)^2} = 12x^2 \qquad \sqrt{12^2}\sqrt{(x^2)^2} = 12|x^2| = 12x^2$

In the same way that perfect squares can be removed from square root radicals, perfect nth powers can be removed from nth root radicals.

EXAMPLE 3 Removing Factors from Radicals

Simplify the radical expressions.

a. $\sqrt[3]{40}$ **b.** $\sqrt[4]{x^5}, \quad x \geq 0$ **c.** $\sqrt[3]{54x^3 y^5}$

Solution

a. $\sqrt[3]{40} = \sqrt[3]{8(5)} = \sqrt[3]{2^3 (5)} = 2\sqrt[3]{5} \qquad \sqrt[3]{2^3} = 2$

b. $\sqrt[4]{x^5} = \sqrt[4]{x^4(x)} = x\sqrt[4]{x} \qquad \sqrt[4]{x^4} = x, \ x \geq 0$

c. $\sqrt[3]{54x^3 y^5} = \sqrt[3]{27x^3 y^3 (2y^2)} \qquad \sqrt[3]{3^3}\sqrt[3]{x^3}\sqrt[3]{y^3} = 3xy$

$\qquad\qquad = \sqrt[3]{3^3 x^3 y^3 (2y^2)}$

$\qquad\qquad = 3xy\sqrt[3]{2y^2}$

Rationalization Techniques

Removing factors from radicals is only one of three techniques that we use to simplify radicals. We summarize all three techniques as follows.

Simplifying Radical Expressions

A radical expression is said to be in simplest form if all three of the following are true.

1. All possible factors have been removed from each radical.

2. No radical contains a fraction.

3. No denominator of a fraction contains a radical.

To meet the last two conditions, you can use a technique called **rationalizing the denominator.** This involves multiplying both the numerator and denominator by a factor that creates a perfect nth power in the denominator.

STUDY TIP

When rationalizing a denominator, remember that for square roots you want a perfect square in the denominator, for cube roots you want a perfect cube, and so on.

EXAMPLE 4   Rationalizing the Denominator

a. $\sqrt{\dfrac{3}{5}} = \dfrac{\sqrt{3}}{\sqrt{5}} = \dfrac{\sqrt{3}}{\sqrt{5}} \cdot \dfrac{\sqrt{5}}{\sqrt{5}} = \dfrac{\sqrt{15}}{\sqrt{5^2}} = \dfrac{\sqrt{15}}{5}$ Multiply by $\sqrt{5}/\sqrt{5}$ to create a perfect square in the denominator.

b. $\dfrac{4}{\sqrt[3]{9}} = \dfrac{4}{\sqrt[3]{9}} \cdot \dfrac{\sqrt[3]{3}}{\sqrt[3]{3}} = \dfrac{4\sqrt[3]{3}}{\sqrt[3]{3^3}} = \dfrac{4\sqrt[3]{3}}{3}$ Multiply by $\sqrt[3]{3}/\sqrt[3]{3}$ to create a perfect cube in the denominator.

c. $\dfrac{8}{3\sqrt{18}} = \dfrac{8}{3\sqrt{18}} \cdot \dfrac{\sqrt{2}}{\sqrt{2}} = \dfrac{8\sqrt{2}}{3\sqrt{36}} = \dfrac{8\sqrt{2}}{3(6)} = \dfrac{4\sqrt{2}}{9}$

EXAMPLE 5   Rationalizing the Denominator

Simplify the expression.

a. $\sqrt{\dfrac{8x}{12y^5}}$ **b.** $\sqrt[3]{\dfrac{54x^6y^3}{5z^2}}$

Solution

a. $\sqrt{\dfrac{8x}{12y^5}} = \sqrt{\dfrac{2x}{3y^5}} = \dfrac{\sqrt{2x}}{\sqrt{3y^5}} \cdot \dfrac{\sqrt{3y}}{\sqrt{3y}} = \dfrac{\sqrt{6xy}}{\sqrt{9y^6}} = \dfrac{\sqrt{6xy}}{3y^3}$

b. $\sqrt[3]{\dfrac{54x^6y^3}{5z^2}} = \dfrac{\sqrt[3]{(3^3)(2)(x^6)(y^3)}}{\sqrt[3]{5z^2}} \cdot \dfrac{\sqrt[3]{25z}}{\sqrt[3]{25z}} = \dfrac{3x^2y\sqrt[3]{50z}}{\sqrt[3]{5^3z^3}} = \dfrac{3x^2y\sqrt[3]{50z}}{5z}$

Adding and Subtracting Radicals

Two or more radical expressions are *alike* if they have the same radicand and the same index. For instance, $\sqrt{2}$ and $3\sqrt{2}$ are alike, but $\sqrt{3}$ and $\sqrt[3]{3}$ are not alike. Two radical expressions that are alike can be added or subtracted by adding or subtracting their coefficients.

EXAMPLE 6 *Combining Radicals*

NOTE Notice in Example 6(c) that *before* concluding that two radicals cannot be combined, you should check to see that they are written in simplest form.

a. $\sqrt{7} + 5\sqrt{7} - 2\sqrt{7} = (1 + 5 - 2)\sqrt{7} = 4\sqrt{7}$

b. $6\sqrt{x} - \sqrt[3]{4} - 5\sqrt{x} + 2\sqrt[3]{4} = 6\sqrt{x} - 5\sqrt{x} - \sqrt[3]{4} + 2\sqrt[3]{4}$
$$= (6 - 5)\sqrt{x} + (-1 + 2)\sqrt[3]{4}$$
$$= \sqrt{x} + \sqrt[3]{4}$$

c. $3\sqrt[3]{x} + \sqrt[3]{8x} = 3\sqrt[3]{x} + 2\sqrt[3]{x} = (3 + 2)\sqrt[3]{x} = 5\sqrt[3]{x}$

EXAMPLE 7 *Simplifying Radical Expressions*

a. $\sqrt{45x} + 3\sqrt{20x} = 3\sqrt{5x} + 6\sqrt{5x} = 9\sqrt{5x}$

b. $5\sqrt{x^3} - x\sqrt{4x} = 5x\sqrt{x} - 2x\sqrt{x} = 3x\sqrt{x}$

c. $\sqrt[3]{54y^5} + 4\sqrt[3]{2y^2} = 3y\sqrt[3]{2y^2} + 4\sqrt[3]{2y^2} = (3y + 4)\sqrt[3]{2y^2}$

In some instances, it may be necessary to rationalize denominators before combining radicals.

EXAMPLE 8 *Rationalizing Denominators Before Simplifying*

Simplify the expression $\sqrt{7} - \dfrac{5}{\sqrt{7}}$.

Solution

$$\sqrt{7} - \frac{5}{\sqrt{7}} = \sqrt{7} - \left(\frac{5}{\sqrt{7}} \cdot \frac{\sqrt{7}}{\sqrt{7}} \right)$$
$$= \sqrt{7} - \frac{5\sqrt{7}}{7}$$
$$= \left(1 - \frac{5}{7} \right) \sqrt{7}$$
$$= \frac{2}{7}\sqrt{7}$$

Application of Radicals

A common use of radicals occurs in applications involving right triangles. Recall that a right triangle is one that contains a right (or 90°) angle, as shown in Figure 5.1. The relationship among the three sides of a right triangle is described by the **Pythagorean Theorem,** which says that if a and b are the lengths of the legs and c is the length of the hypotenuse, then

$$c = \sqrt{a^2 + b^2}. \qquad \text{Pythagorean Theorem}$$

For instance, if $a = 6$ and $b = 9$, then

$$c = \sqrt{6^2 + 9^2} = \sqrt{117} = \sqrt{9}\sqrt{13} = 3\sqrt{13}.$$

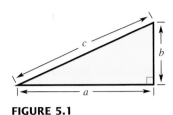

FIGURE 5.1

EXAMPLE 9   An Application of the Pythagorean Theorem

A softball diamond has the shape of a square with 60-foot sides (see Figure 5.2). The catcher is 5 feet behind home plate. How far does the catcher have to throw to reach second base?

Solution

In Figure 5.2, let x be the hypotenuse of a right triangle with 60-foot legs. Thus, by the Pythagorean Theorem, you have the following.

$$x = \sqrt{60^2 + 60^2} \qquad \text{Pythagorean Theorem}$$
$$x = \sqrt{7200}$$
$$x \approx 84.9 \text{ feet}$$

Thus, the distance from home plate to second base is approximately 84.9 feet. Because the catcher is 5 feet behind home plate, the catcher must make a throw of

$$x + 5 \approx 84.9 + 5 = 89.9 \text{ feet}.$$

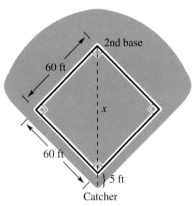

FIGURE 5.2

Group Activities   You Be the Instructor

Error Analysis   Suppose you are an algebra instructor and one of your students hands in the following work. Find and correct the errors, and discuss how you can help your student avoid such errors in the future.

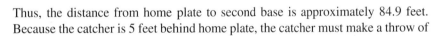

a. $7\sqrt{3} + 4\sqrt{2} = 11\sqrt{5}$ b. $3\sqrt[3]{k} - 6\sqrt{k} = -3\sqrt{k}$

5.3 Exercises

Discussing the Concepts

1. Give an example of multiplying two radicals.

2. Describe the three conditions that characterize a simplified radical expression.

3. Is $\sqrt{2} + \sqrt{18}$ in simplest form? Explain.

4. Describe the steps you would use to simplify $\dfrac{1}{\sqrt{3}}$.

5. For what values of x is $\sqrt{x^2} \neq x$? Explain.

6. Explain what it means for two radical expressions to be alike.

Problem Solving

In Exercises 7–10, write the expression as a single radical.

7. $\sqrt[3]{11} \cdot \sqrt[3]{10}$

8. $\sqrt[4]{35} \cdot \sqrt[4]{3}$

9. $\dfrac{\sqrt{15}}{\sqrt{31}}$

10. $\dfrac{\sqrt[3]{84}}{\sqrt[3]{9}}$

In Exercises 11–14, write the expression as a product or quotient of radicals and simplify.

11. $\sqrt{9 \cdot 35}$

12. $\sqrt[3]{27 \cdot 4}$

13. $\sqrt[3]{\dfrac{1000}{11}}$

14. $\sqrt[5]{\dfrac{243}{2}}$

In Exercises 15–22, simplify the radical.

15. $\sqrt{20}$

16. $\sqrt{50}$

17. $\sqrt[3]{24}$

18. $\sqrt[3]{54}$

19. $\sqrt{\dfrac{15}{4}}$

20. $\sqrt{\dfrac{5}{36}}$

21. $\sqrt[5]{\dfrac{15}{243}}$

22. $\sqrt[3]{\dfrac{1}{1000}}$

In Exercises 23–30, simplify the expression.

23. $\sqrt{9x^5}$

24. $\sqrt{64x^3}$

25. $\sqrt[4]{3x^4y^2}$

26. $\sqrt[3]{16x^4y^5}$

27. $\sqrt[5]{\dfrac{32x^2}{y^5}}$

28. $\sqrt[3]{\dfrac{16z^3}{y^6}}$

29. $\sqrt{\dfrac{32a^4}{b^2}}$

30. $\sqrt{\dfrac{18x^2}{z^6}}$

In Exercises 31–38, rationalize the denominator and simplify further, if possible.

31. $\sqrt{\dfrac{1}{3}}$

32. $\sqrt{\dfrac{1}{5}}$

33. $\sqrt[4]{\dfrac{5}{4}}$

34. $\sqrt[3]{\dfrac{9}{25}}$

35. $\dfrac{1}{\sqrt{y}}$

36. $\sqrt{\dfrac{5}{c}}$

37. $\sqrt[3]{\dfrac{2x}{3y}}$

38. $\sqrt[3]{\dfrac{20x^2}{9y^2}}$

In Exercises 39–44, combine the radical expressions, if possible.

39. $3\sqrt{2} - \sqrt{2}$

40. $\dfrac{2}{5}\sqrt{5} - \dfrac{6}{5}\sqrt{5}$

41. $2\sqrt[3]{54} + 12\sqrt[3]{16}$

42. $4\sqrt[4]{48} - \sqrt[4]{243}$

43. $\sqrt{25y} + \sqrt{64y}$

44. $\sqrt[3]{16t^4} - \sqrt[3]{54t^4}$

In Exercises 45 and 46, use a graphing utility to graph the equations on the same screen. Use the graphs to verify that the expressions are equivalent. Verify the results algebraically.

45. $y_1 = \sqrt{\dfrac{3}{x}}, \ y_2 = \dfrac{\sqrt{3x}}{x}$

46. $y_1 = \sqrt[3]{8x^4} - \sqrt[3]{x}, \ y_2 = x\sqrt[3]{x}$

Geometry In Exercises 47 and 48, find the length of the hypotenuse of the right triangle.

47.

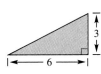

48.

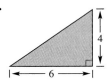

49. *Vibrating String* The frequency f in cycles per second of a vibrating string is given by

$$f = \frac{1}{100}\sqrt{\frac{400 \times 10^6}{5}}.$$

Use a calculator to approximate this number. (Round your answer to two decimal places.)

50. *Geometry* The foundation of a house is 40 feet long and 30 feet wide. The height of the attic is 5 feet (see figure).

(a) Use the Pythagorean Theorem to find the length of the hypotenuse of the right triangle formed by the roof line.

(b) Use the result of part (a) to determine the total area of the roof.

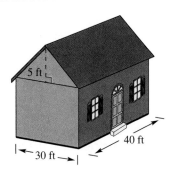

Reviewing the Major Concepts

In Exercises 51–54, simplify the expression.

51. $(x^2 - 3xy)^0$

52. $(-3x^2y^3)^2(4xy^2)$

53. $\dfrac{64r^2s^4}{16rs^2}$

54. $\left(\dfrac{3x}{4y^3}\right)^2$

55. *Geometry* The height of a triangle is 12 inches less than its base. The area is 110 square inches. Find the height and base.

56. *Geometry* An open box with a square base is to be constructed from 825 square inches of material. If the length of an edge of the square base is x inches, the surface area is given by

$$S = x^2 + 4xh.$$

What should be the dimensions of the base if the height of the box is to be 10 inches?

Additional Problem Solving

In Exercises 57–60, write as a single radical.

57. $\sqrt{3} \cdot \sqrt{10}$

58. $\sqrt[5]{9} \cdot \sqrt[5]{19}$

59. $\dfrac{\sqrt[5]{152}}{\sqrt[5]{3}}$

60. $\dfrac{\sqrt[4]{633}}{\sqrt[4]{5}}$

In Exercises 61–64, write the expression as a product or quotient of radicals and simplify.

61. $\sqrt[4]{81 \cdot 11}$

62. $\sqrt[5]{100,000 \cdot 3}$

63. $\sqrt{\dfrac{35}{9}}$

64. $\sqrt[4]{\dfrac{165}{16}}$

In Exercises 65–72, simplify the radical.

65. $\sqrt{27}$

66. $\sqrt{125}$

67. $\sqrt[4]{30,000}$

68. $\sqrt[5]{96}$

69. $\sqrt{\dfrac{15}{49}}$

70. $\sqrt{\dfrac{5}{9}}$

71. $\sqrt[3]{\dfrac{35}{64}}$

72. $\sqrt[4]{\dfrac{5}{16}}$

In Exercises 73–76, simplify the expression.

73. $\sqrt{4 \times 10^{-4}}$

74. $\sqrt{8.5 \times 10^2}$

75. $\sqrt[3]{2.4 \times 10^6}$

76. $\sqrt[4]{4.4 \times 10^{-4}}$

In Exercises 77–86, simplify the expression.

77. $\sqrt{48y^4}$

78. $\sqrt[4]{32x^6}$

79. $\sqrt[3]{x^4 y^3}$

80. $\sqrt[3]{a^5 b^6}$

81. $\sqrt[5]{32x^5 y^6}$

82. $\sqrt[4]{128u^4 v^7}$

83. $\sqrt{\dfrac{13}{25}}$

84. $\sqrt{\dfrac{15}{36}}$

85. $\sqrt[3]{\dfrac{54a^4}{b^9}}$

86. $\sqrt[4]{\dfrac{3u^2}{16v^8}}$

In Exercises 87–96, rationalize the denominator and simplify further, if possible.

87. $\dfrac{12}{\sqrt{3}}$

88. $\dfrac{5}{\sqrt{10}}$

89. $\dfrac{6}{\sqrt[3]{32}}$

90. $\dfrac{10}{\sqrt[5]{16}}$

91. $\sqrt{\dfrac{4}{x}}$

92. $\dfrac{1}{\sqrt{2x}}$

93. $\dfrac{6}{\sqrt{3b^3}}$

94. $\dfrac{1}{\sqrt{xy}}$

95. $\dfrac{a^3}{\sqrt[3]{ab^2}}$

96. $\dfrac{3u^2}{\sqrt[4]{8u^3}}$

In Exercises 97–104, combine the radical expressions, if possible.

97. $\sqrt[4]{3} - 5\sqrt[4]{7} - 12\sqrt[4]{3}$

98. $9\sqrt[3]{17} + 7\sqrt[3]{2} - 4\sqrt[3]{17} + \sqrt{2}$

99. $12\sqrt{8} - 3\sqrt[3]{8}$

100. $4\sqrt{32} + 7\sqrt{32}$

101. $5\sqrt{9x} - 3\sqrt{x}$

102. $3\sqrt{x+1} + 10\sqrt{x+1}$

103. $10\sqrt[3]{z} - \sqrt[3]{z^4}$

104. $5\sqrt[3]{24u^2} + 2\sqrt[3]{81u^5}$

In Exercises 105–108, perform the addition or subtraction and simplify your answer.

105. $\sqrt{5} - \dfrac{3}{\sqrt{5}}$

106. $\sqrt{10} + \dfrac{5}{\sqrt{10}}$

107. $\sqrt{20} - \sqrt{\dfrac{1}{5}}$

108. $\dfrac{x}{\sqrt{3x}} + \sqrt{27x}$

In Exercises 109–112, place the correct symbol (<, >, or =) between the numbers.

109. $\sqrt{7} + \sqrt{18}$ ___ $\sqrt{7 + 18}$

110. $\sqrt{10} - \sqrt{6}$ ___ $\sqrt{10 - 6}$

111. 5 ___ $\sqrt{3^2 + 2^2}$

112. 5 ___ $\sqrt{3^2 + 4^2}$

Geometry In Exercises 113 and 114, find the length of the hypotenuse of the right triangle.

113.

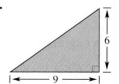

114.

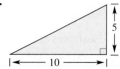

115. *Period of a Pendulum* The period T in seconds of a pendulum (see figure) is given by

$$T = 2\pi \sqrt{\dfrac{L}{32}}$$

where L is the length of the pendulum in feet. Find the period of a pendulum whose length is 4 feet. (Round your answer to two decimal places.)

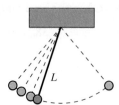

116. *Geometry*   All four corners are cut from a 4-foot-by-8-foot sheet of plywood as shown in the figure. Find the perimeter of the remaining piece of plywood.

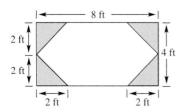

117. *Calculator Experiment*   Enter any positive real number in your calculator and find its square root. Then repeatedly take the square root of the result.

$$\sqrt{x}, \sqrt{\sqrt{x}}, \sqrt{\sqrt{\sqrt{x}}}, \dots$$

What real number does the display appear to be approaching?

118. *Think About It*   Square the real number $5/\sqrt{3}$ and note that the radical is eliminated from the denominator. Is this equivalent to rationalizing the denominator? Why or why not?

Math Matters   Standard Metric Prefixes

In the metric system, a *kilo*gram is equal to 1000 grams, a *centi*meter is equal to $\frac{1}{100}$ of a meter, and a *milli*liter is equal to $\frac{1}{1000}$ of a liter. "Kilo," "centi," and "milli" are the most well-known prefixes in the metric system. However, there are several other prefixes that can be used to denote powers of 10 in the metric system.

How many of the prefixes listed have you heard used? Use the list of prefixes to answer the following questions. (The answers are given in the back of the book.)

Factor by Which Unit Is Multiplied	Prefix
10^{12}	tera
10^{9}	giga
10^{6}	mega
10^{3}	kilo
10^{2}	hecto
10^{1}	deca
10^{-1}	deci
10^{-2}	centi
10^{-3}	milli
10^{-6}	micro
10^{-9}	nano
10^{-12}	pico
10^{-15}	femto
10^{-18}	atto

How many nanoseconds are in a second?
How many millimeters are in a meter?
How many watts are in a kilowatt?

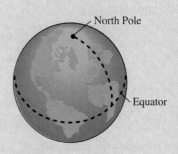

In 1793 the French government adopted the metric system, in which the basic unit of length was defined as one ten-millionth of the distance between the North Pole and the equator. For working standards, a platinum bar was marked with lines a meter apart. Later, when refinements in measuring the earth were discovered, the meter was redefined in terms of the platinum bar that had been constructed.

MID-CHAPTER QUIZ

Take this quiz as you would take a quiz in class. After you are done, check your work against the answers given in the back of the book.

In Exercises 1–4, evaluate the expression.

1. -12^{-2}

2. $\left(\frac{3}{4}\right)^{-3}$

3. $\sqrt{\frac{25}{9}}$

4. $(-64)^{1/3}$

In Exercises 5–8, rewrite the expression using only positive exponents, and simplify. (Assume any variables in the expression are nonzero.)

5. $(t^3)^{-1/2}(3t^3)$

6. $(3x^2y^{-1})(4x^{-2}y)^{-2}$

7. $\dfrac{10u^{-2}}{15u}$

8. $\dfrac{(10x)^0}{(x^2+4)^{-1}}$

9. Write each number in scientific notation: (a) 13,400,000; (b) 0.00075.

10. Evaluate the expression without using a calculator.

(a) $(3 \times 10^3)^4$

(b) $\dfrac{3.2 \times 10^4}{16 \times 10^7}$

In Exercises 11–14, simplify the expression.

11. (a) $\sqrt{150}$ (b) $\sqrt[3]{54}$

12. (a) $\sqrt{27x^2}$ (b) $\sqrt[4]{81x^6}$

13. (a) $\sqrt[4]{\dfrac{5}{16}}$ (b) $\sqrt{\dfrac{24}{49}}$

14. (a) $\sqrt{\dfrac{40u^3}{9}}$ (b) $\sqrt[3]{\dfrac{16}{u^6}}$

In Exercises 15 and 16, rationalize the denominator and simplify.

15. (a) $\sqrt{\dfrac{2}{3}}$ (b) $\dfrac{24}{\sqrt{12}}$

16. (a) $\dfrac{10}{\sqrt{5x}}$ (b) $\sqrt[3]{\dfrac{3}{2a}}$

In Exercises 17 and 18, combine the radical expressions, if possible.

17. $\sqrt{200y} - 3\sqrt{8y}$

18. $6x\sqrt[3]{5x^2} + 2\sqrt[3]{40x^4}$

19. Explain why $\sqrt{5^2 + 12^2} \neq 17$. Determine the correct value of the radical.

20. The four corners are cut from an $8\frac{1}{2}$-inch-by-11-inch sheet of paper as shown in the figure at the right. Find the perimeter of the remaining piece of paper.

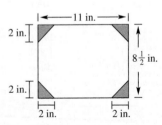

Figure for 20

5.4	**Multiplying and Dividing Radical Expressions**
	Multiplying Radical Expressions ▪ Dividing Radical Expressions

Multiplying Radical Expressions

You can multiply radical expressions by using the Distributive Property or the FOIL Method. In both procedures, you also make use of the Multiplication Property of Radicals. Recall from Section 5.3 that the product of two radicals is given by

$$\sqrt[n]{a}\,\sqrt[n]{b} = \sqrt[n]{ab}$$

where a and b are real numbers whose nth roots are also real numbers.

EXAMPLE 1   Multiplying Radical Expressions

$$\sqrt{3}\left(2 + \sqrt{5}\right) = 2\sqrt{3} + \sqrt{3}\sqrt{5} \qquad \text{Distributive Property}$$

$$= 2\sqrt{3} + \sqrt{15} \qquad \text{Multiplication Property of Radicals}$$

In Example 1, the product of $\sqrt{3}$ and $\sqrt{5}$ is best left as $\sqrt{15}$. In some cases, however, the product of two radicals can be simplified, as shown in Example 2.

EXAMPLE 2   Multiplying Radical Expressions

Find the products and simplify.

a. $\sqrt{2}\left(4 - \sqrt{8}\right)$ **b.** $\sqrt{6}\left(\sqrt{12} - \sqrt{3}\right)$

Solution

a. $\sqrt{2}\left(4 - \sqrt{8}\right) = 4\sqrt{2} - \sqrt{2}\sqrt{8}$ Distributive Property

$\qquad\qquad\qquad = 4\sqrt{2} - \sqrt{16}$ Multiplication Property of Radicals

$\qquad\qquad\qquad = 4\sqrt{2} - 4$ Simplify.

b. $\sqrt{6}\left(\sqrt{12} - \sqrt{3}\right) = \sqrt{6}\sqrt{12} - \sqrt{6}\sqrt{3}$ Distributive Property

$\qquad\qquad\qquad\quad = \sqrt{72} - \sqrt{18}$ Multiplication Property of Radicals

$\qquad\qquad\qquad\quad = 6\sqrt{2} - 3\sqrt{2}$ Find perfect square factors.

$\qquad\qquad\qquad\quad = 3\sqrt{2}$ Simplify.

In Examples 1 and 2, the Distributive Property was used to multiply radical expressions. In Example 3, note how the FOIL Method can be used to multiply binomial radical expressions.

EXAMPLE 3 Using the FOIL Method

$$\begin{matrix} & \overset{F}{\overbrace{}} & \overset{O}{\overbrace{}} & \overset{I}{\overbrace{}} & \overset{L}{\mid} \end{matrix}$$

a. $\left(2\sqrt{7} - 4\right)\left(\sqrt{7} + 1\right) = 2\left(\sqrt{7}\right)^2 + 2\sqrt{7} - 4\sqrt{7} - 4$ FOIL Method

$$= 2(7) + (2 - 4)\sqrt{7} - 4 \qquad \text{Combine like radicals.}$$

$$= 10 - 2\sqrt{7} \qquad \text{Combine like terms.}$$

b. $\left(3 - \sqrt{x}\right)\left(1 + \sqrt{x}\right) = 3 + 3\sqrt{x} - \sqrt{x} - \left(\sqrt{x}\right)^2$

$$= 3 + 2\sqrt{x} - x, \quad x \geq 0 \qquad \text{Combine like radicals.}$$

In addition to the FOIL Method, you can also use special product formulas to multiply binomial radical expressions.

EXAMPLE 4 Using a Special Product Formula

a. $\left(2 - \sqrt{5}\right)\left(2 + \sqrt{5}\right) = 2^2 - \left(\sqrt{5}\right)^2$ Special product formula

$$= 4 - 5$$

$$= -1$$

b. $\left(\sqrt{3} + \sqrt{x}\right)\left(\sqrt{3} - \sqrt{x}\right) = \left(\sqrt{3}\right)^2 - \left(\sqrt{x}\right)^2$ Special product formula

$$= 3 - x, \quad x \geq 0$$

In Example 4(a), the expressions $\left(2 - \sqrt{5}\right)$ and $\left(2 + \sqrt{5}\right)$ are called **conjugates** of each other. The product of two conjugates is the difference of two squares, which is given by the special product formula $(a + b)(a - b) = a^2 - b^2$. Here are some other examples.

Expression	Conjugate	Product
$\left(1 - \sqrt{3}\right)$	$\left(1 + \sqrt{3}\right)$	$(1)^2 - \left(\sqrt{3}\right)^2 = 1 - 3 = -2$
$\left(\sqrt{5} + \sqrt{2}\right)$	$\left(\sqrt{5} - \sqrt{2}\right)$	$\left(\sqrt{5}\right)^2 - \left(\sqrt{2}\right)^2 = 5 - 2 = 3$
$\left(\sqrt{10} - 3\right)$	$\left(\sqrt{10} + 3\right)$	$\left(\sqrt{10}\right)^2 - (3)^2 = 10 - 9 = 1$
$\sqrt{x} + 2$	$\sqrt{x} - 2$	$\left(\sqrt{x}\right)^2 - (2)^2 = x - 4, \quad x \geq 0$

Dividing Radical Expressions

To simplify a *quotient* involving radicals, we rationalize the denominator. For single-term denominators, you can use the rationalizing process described in Section 5.3. To rationalize a denominator involving two terms, multiply both the numerator and denominator by the *conjugate of the denominator*.

EXAMPLE 5   *Simplifying Quotients Involving Radicals*

a.

$$\frac{\sqrt{3}}{1 - \sqrt{5}} = \frac{\sqrt{3}}{1 - \sqrt{5}} \cdot \frac{1 + \sqrt{5}}{1 + \sqrt{5}}$$ Multiply by conjugate of denominator.

$$= \frac{\sqrt{3}\left(1 + \sqrt{5}\right)}{1^2 - \left(\sqrt{5}\right)^2}$$ Special product formula

$$= \frac{\sqrt{3} + \sqrt{15}}{1 - 5}$$ Simplify.

$$= \frac{\sqrt{3} + \sqrt{15}}{-4}$$ Simplify.

b.

$$\frac{4}{2 - \sqrt{3}} = \frac{4}{2 - \sqrt{3}} \cdot \frac{2 + \sqrt{3}}{2 + \sqrt{3}}$$ Multiply by conjugate of denominator.

$$= \frac{4\left(2 + \sqrt{3}\right)}{2^2 - \left(\sqrt{3}\right)^2}$$ Special product formula

$$= \frac{8 + 4\sqrt{3}}{4 - 3}$$ Simplify.

$$= 8 + 4\sqrt{3}$$ Simplify.

EXAMPLE 6   *Simplifying Quotients Involving Radicals*

$$\frac{5\sqrt{2}}{\sqrt{7} + \sqrt{2}} = \frac{5\sqrt{2}}{\sqrt{7} + \sqrt{2}} \cdot \frac{\sqrt{7} - \sqrt{2}}{\sqrt{7} - \sqrt{2}}$$ Multiply by conjugate of denominator.

$$= \frac{5\left(\sqrt{14} - \sqrt{4}\right)}{\left(\sqrt{7}\right)^2 - \left(\sqrt{2}\right)^2}$$ Special product formula

$$= \frac{5\left(\sqrt{14} - 2\right)}{7 - 2}$$ Simplify.

$$= \frac{5\left(\sqrt{14} - 2\right)}{5}$$ Cancel common factor.

$$= \sqrt{14} - 2$$ Simplest form

EXAMPLE 7   *Dividing Radical Expressions*

Perform the division and simplify.

a. $6 \div \left(\sqrt{x} - 2 \right)$ **b.** $\left(2 - \sqrt{3} \right) \div \left(\sqrt{6} + \sqrt{2} \right)$

Solution

a. $\dfrac{6}{\sqrt{x} - 2} = \dfrac{6}{\sqrt{x} - 2} \cdot \dfrac{\sqrt{x} + 2}{\sqrt{x} + 2}$ Multiply by conjugate of denominator.

$= \dfrac{6 \left(\sqrt{x} + 2 \right)}{\left(\sqrt{x} \right)^2 - 2^2}$ Special product formula

$= \dfrac{6\sqrt{x} + 12}{x - 4}, \quad x \geq 0$ Simplify.

b. $\dfrac{2 - \sqrt{3}}{\sqrt{6} + \sqrt{2}} = \dfrac{2 - \sqrt{3}}{\sqrt{6} + \sqrt{2}} \cdot \dfrac{\sqrt{6} - \sqrt{2}}{\sqrt{6} - \sqrt{2}}$ Multiply by conjugate of denominator.

$= \dfrac{2\sqrt{6} - 2\sqrt{2} - \sqrt{18} + \sqrt{6}}{\left(\sqrt{6} \right)^2 - \left(\sqrt{2} \right)^2}$ Special product formula

$= \dfrac{3\sqrt{6} - 2\sqrt{2} - 3\sqrt{2}}{6 - 2}$ Simplify.

$= \dfrac{3\sqrt{6} - 5\sqrt{2}}{4}$ Simplify.

Group Activities   You Be the Instructor

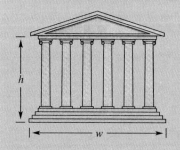

The Golden Section   The ratio of the width of the Temple of Hephaestus to its height is approximately

$$\frac{w}{h} \approx \frac{2}{\sqrt{5} - 1}.$$

This number is called the **golden section.** Early Greeks believed that the most aesthetically pleasing rectangles were those whose sides have this ratio.

a. Rationalize the denominator for this number. Approximate your answer, rounded to two decimal places.

b. Use the Pythagorean Theorem, a straightedge, and a compass to construct a rectangle whose sides have the golden section as their ratio.

5.4 Exercises

Discussing the Concepts

1. Multiply $\sqrt{3}\left(1-\sqrt{6}\right)$. State an algebraic property to justify each step.

2. Explain the meaning of each letter of the word FOIL in the FOIL Method.

3. Multiply $3-\sqrt{2}$ by its conjugate. Explain why the result has no radicals.

4. Is the number $3/\left(1+\sqrt{5}\right)$ in simplest form? If not, explain the steps for writing it in simplest form.

Problem Solving

In Exercises 5–18, multiply and simplify.

5. $\sqrt{3}\cdot\sqrt{6}$

6. $\sqrt{5}\cdot\sqrt{10}$

7. $\sqrt{5}\left(2-\sqrt{3}\right)$

8. $\sqrt{11}\left(\sqrt{5}-3\right)$

9. $\left(\sqrt{3}+2\right)\left(\sqrt{3}-2\right)$

10. $\left(\sqrt{15}+3\right)\left(\sqrt{15}-3\right)$

11. $\left(\sqrt{20}+2\right)^2$

12. $\left(4-\sqrt{20}\right)^2$

13. $\sqrt{y}\left(\sqrt{y}+4\right)$

14. $\left(5-\sqrt{3v}\right)^2$

15. $\left(9\sqrt{x}+2\right)\left(5\sqrt{x}-3\right)$

16. $\left(16\sqrt{u}-3\right)\left(\sqrt{u}-1\right)$

17. $\sqrt[3]{4}\left(\sqrt[3]{2}-7\right)$

18. $\left(\sqrt[3]{9}+5\right)\left(\sqrt[3]{5}-5\right)$

In Exercises 19–24, complete the statement.

19. $5\sqrt{3}+15\sqrt{3}=5(\quad)$

20. $x\sqrt{7}-x^2\sqrt{7}=x(\quad)$

21. $4\sqrt{12}-2x\sqrt{27}=2\sqrt{3}(\quad)$

22. $5\sqrt{50}+10y\sqrt{8}=5\sqrt{2}(\quad)$

23. $6u^2+\sqrt{18u^3}=3u(\quad)$

24. $12s^3-\sqrt{32s^4}=4s^2\ (\quad)$

In Exercises 25–28, find the conjugate of the number. Then multiply the number by its conjugate.

25. $\sqrt{11}-\sqrt{3}$

26. $\sqrt{10}+\sqrt{7}$

27. $\sqrt{x}-3$

28. $\sqrt{5a}+\sqrt{2}$

In Exercises 29–36, rationalize the denominator of the expression.

29. $\dfrac{6}{\sqrt{22}-2}$

30. $\dfrac{3}{2\sqrt{10}-5}$

31. $\dfrac{8}{\sqrt{7}+3}$

32. $\dfrac{10}{\sqrt{9}+\sqrt{5}}$

33. $\dfrac{2t^2}{\sqrt{5t}-\sqrt{t}}$

34. $\dfrac{5x}{\sqrt{x}-\sqrt{2}}$

35. $\dfrac{\sqrt{u+v}}{\sqrt{u-v}-\sqrt{u}}$

36. $\dfrac{z}{\sqrt{u+z}-\sqrt{u}}$

In Exercises 37 and 38, evaluate the function.

37. $f(x)=x^2-2x-1$

(a) $f\left(1+\sqrt{2}\right)$ (b) $f\left(\sqrt{4}\right)$

38. $g(x)=x^2-4x+1$

(a) $g\left(1+\sqrt{5}\right)$ (b) $g\left(2-\sqrt{3}\right)$

In Exercises 39 and 40, use a graphing utility to graph the two equations on the same screen. Use the graphs to verify that the expressions are equivalent. Verify the results algebraically.

39. $y_1=\dfrac{10}{\sqrt{x}+1}, y_2=\dfrac{10\left(\sqrt{x}-1\right)}{x-1}$

40. $y_1=\dfrac{2\sqrt{x}}{2-\sqrt{x}}, y_2=\dfrac{2\left(2\sqrt{x}+x\right)}{4-x}$

41. *Strength of a Wooden Beam* The rectangular cross section of a wooden beam cut from a log of diameter 24 inches (see figure) will have maximum strength if its width w and height h are given by

$$w = 8\sqrt{3} \quad \text{and} \quad h = \sqrt{24^2 - \left(8\sqrt{3}\right)^2}.$$

Find the area of the rectangular cross section and express the area in simplest form.

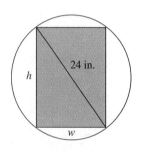

42. *Force* The force required to slide a 500-pound block across a milling machine is

$$\frac{500k}{\dfrac{1}{\sqrt{k^2+1}} + \dfrac{k^2}{\sqrt{k^2+1}}}$$

where k is the friction constant (see figure). Simplify this expression.

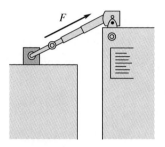

Reviewing the Major Concepts

In Exercises 43–46, solve the equation.

43. $\dfrac{4}{x} - \dfrac{2}{3} = 0$

44. $2x = 3[1 + (4 - x)]$

45. $3x^2 - 13x - 10 = 0$

46. $x(x - 3) = 40$

In Exercises 47–50, use a graphing utility to identify the transformation of the graph of $f(x) = \sqrt{x}$.

47. $h(x) = 4 + \sqrt{x}$

48. $h(x) = \sqrt{4 + x}$

49. $h(x) = -\sqrt{x}$

50. $h(x) = \sqrt{-x}$

Additional Problem Solving

In Exercises 51–68, multiply and simplify.

51. $\sqrt{2} \cdot \sqrt{8}$

52. $\sqrt{6} \cdot \sqrt{18}$

53. $\sqrt{2}\left(\sqrt{20} + 8\right)$

54. $\sqrt{7}\left(\sqrt{14} + 3\right)$

55. $\left(3 - \sqrt{5}\right)\left(3 + \sqrt{5}\right)$

56. $\left(\sqrt{11} + 3\right)\left(\sqrt{11} - 3\right)$

57. $\left(2\sqrt{2} + \sqrt{4}\right)\left(2\sqrt{2} - \sqrt{4}\right)$

58. $\left(4\sqrt{3} + \sqrt{2}\right)\left(4\sqrt{3} - \sqrt{2}\right)$

59. $\left(\sqrt{5} + 3\right)\left(\sqrt{3} - 5\right)$

60. $\left(\sqrt{30} + 6\right)\left(\sqrt{2} + 6\right)$

61. $\left(10 + \sqrt{2x}\right)^2$

62. $\sqrt{x}\left(5 - \sqrt{x}\right)$

63. $\left(3\sqrt{x} - 5\right)\left(3\sqrt{x} + 5\right)$

64. $\left(7 - 3\sqrt{3t}\right)\left(7 + 3\sqrt{3t}\right)$

65. $\left(\sqrt{x} + \sqrt{y}\right)\left(\sqrt{x} - \sqrt{y}\right)$

66. $\left(3\sqrt{u} + \sqrt{3v}\right)\left(3\sqrt{u} - \sqrt{3v}\right)$

67. $\left(\sqrt[3]{2x} + 5\right)^2$

68. $\left(\sqrt[3]{y} + 2\right)\left(\sqrt[3]{y^2} - 5\right)$

In Exercises 69–72, simplify the expression.

69. $\dfrac{4 - 8\sqrt{x}}{12}$

70. $\dfrac{-3 + 27\sqrt{2y}}{18}$

71. $\dfrac{-2y + \sqrt{12y^3}}{8y}$

72. $\dfrac{-t^2 - \sqrt{2t^3}}{3t}$

In Exercises 73–76, find the conjugate of the number. Then multiply the number by its conjugate.

73. $2 + \sqrt{5}$

74. $\sqrt{2} - 9$

75. $\sqrt{2u} - \sqrt{3}$

76. $\sqrt{t} + 7$

In Exercises 77–84, rationalize the denominator of the expression.

77. $\dfrac{2}{6+\sqrt{2}}$

78. $\dfrac{44\sqrt{5}}{3\sqrt{5}-1}$

79. $\left(\sqrt{7}+2\right)\div\left(\sqrt{7}-2\right)$

80. $\left(5-3\sqrt{3}\right)\div\left(3+\sqrt{3}\right)$

81. $\dfrac{3x}{\sqrt{15}-\sqrt{3}}$

82. $\dfrac{6(y+1)}{y^2+\sqrt{y}}$

83. $\left(\sqrt{x}-5\right)\div\left(2\sqrt{x}-1\right)$

84. $\left(2\sqrt{t}+1\right)\div\left(2\sqrt{t}-1\right)$

In Exercises 85 and 86, evaluate the function as indicated.

85. $f(x)=x^2-6x+1$

 (a) $f\left(2-\sqrt{3}\right)$ (b) $f\left(3-2\sqrt{2}\right)$

86. $g(x)=x^2+8x+11$

 (a) $g\left(-4+\sqrt{5}\right)$ (b) $g\left(-4\sqrt{2}\right)$

In Exercises 87 and 88, use a graphing utility to graph the two equations on the same screen. Use the graphs to verify that the expressions are equivalent. Verify the results algebraically.

87. $y_1=\dfrac{4x}{\sqrt{x}+4},\; y_2=\dfrac{4x\left(\sqrt{x}-4\right)}{x-16}$

88. $y_1=\dfrac{\sqrt{2x}+6}{\sqrt{2x}-2},\; y_2=\dfrac{x+6+4\sqrt{2x}}{x-2}$

89. *Geometry* The areas of the circles in the figure are 15 square centimeters and 20 square centimeters. Find the ratio of the radius of the small circle to the radius of the large circle.

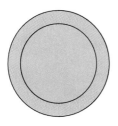

90. *Radical Spiral* Copy the spiral shown below and find the lengths of the sides labeled $a, b, c, d, e,$ and f. Explain your reasoning.

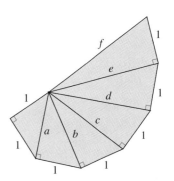

C A R E E R   I N T E R V I E W

Lt. Robert Orr

Civil Service Lieutenant for Boston Police Department

Boston, MA 02201

I am a member of the Boston Police Dive Team, serving as an underwater investigator for the inner Boston Harbor. Safe execution of a scuba dive depends on making correct computations when preparing for the dive. For instance, it is vital to know how long the compressed air in my tank will last. Pressure under water increases with depth. If I start with a tank containing 70 cubic feet of compressed air at the surface, and dive to a depth at which the pressure is twice that at the surface, the volume of air available for breathing is halved to 35 cubic feet. In general, this relationship is $y=c^{-1}x$, where y is the available underwater air volume, x is the surface air volume, and c is the multiple of surface pressure. Because the body demands the same volume of air while diving as at the surface, the tank of air will last roughly $1/c$ times as long. We use *U.S. Navy Diving Manual* tables for most calculations, but it is useful to understand this relationship to make estimates.

5.5 Solving Radical Equations

Solving Radical Equations ▪ Applications

Solving Radical Equations

Solving equations involving radicals is somewhat like solving equations that contain fractions—you try to get rid of the radicals and obtain a polynomial equation. Then you solve the polynomial equation using the standard procedures. The following property plays a key role.

Raising Both Sides of an Equation to the nth Power

Let u and v be real numbers, variables, or algebraic expressions, and let n be a positive integer. If $u = v$, then it follows that

$$u^n = v^n.$$

This is called **raising both sides of an equation to the nth power.**

To use this property to solve an equation, first try to isolate one of the radicals on one side of the equation.

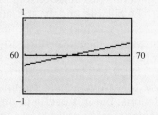

Technology

To use a graphing utility to check the solution in Example 1, sketch the graph of

$$y = \sqrt{x} - 8$$

as shown below. Notice that the graph crosses the x-axis when $x = 64$, which confirms the solution that was obtained algebraically.

EXAMPLE 1   Solving an Equation Having One Radical

Solve $\sqrt{x} - 8 = 0$.

Solution

$\sqrt{x} - 8 = 0$	Original equation
$\sqrt{x} = 8$	Isolate radical.
$\left(\sqrt{x}\right)^2 = 8^2$	Square both sides.
$x = 64$	Simplify.

Check

$\sqrt{x} - 8 = 0$	Original equation
$\sqrt{64} - 8 \overset{?}{=} 0$	Substitute 64 for x.
$8 - 8 = 0$	Solution checks. ✓

Therefore, the equation has one solution: $x = 64$.

EXAMPLE 2   Solving an Equation Having One Radical

Solve $\sqrt[3]{2x+1} - 2 = 3$.

Solution

$$\sqrt[3]{2x+1} - 2 = 3 \qquad \text{Original equation}$$
$$\sqrt[3]{2x+1} = 5 \qquad \text{Isolate radical.}$$
$$\left(\sqrt[3]{2x+1}\right)^3 = 5^3 \qquad \text{Cube both sides.}$$
$$2x + 1 = 125 \qquad \text{Simplify.}$$
$$2x = 124 \qquad \text{Subtract 1 from both sides.}$$
$$x = 62 \qquad \text{Divide both sides by 2.}$$

Check

$$\sqrt[3]{2x+1} - 2 = 3 \qquad \text{Original equation}$$
$$\sqrt[3]{2(62)+1} - 2 \stackrel{?}{=} 3 \qquad \text{Substitute 62 for } x.$$
$$\sqrt[3]{125} - 2 \stackrel{?}{=} 3 \qquad \text{Simplify.}$$
$$5 - 2 = 3 \qquad \text{Solution checks.} \checkmark$$

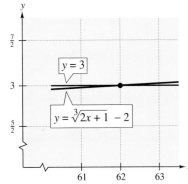

FIGURE 5.3

Therefore, the equation has one solution: $x = 62$. You can also check the solution graphically, as shown in Figure 5.3.

EXAMPLE 3   Solving an Equation Having One Radical

Solve $\sqrt{3x} + 6 = 0$.

Solution

$$\sqrt{3x} + 6 = 0 \qquad \text{Original equation}$$
$$\sqrt{3x} = -6 \qquad \text{Isolate radical.}$$
$$\left(\sqrt{3x}\right)^2 = (-6)^2 \qquad \text{Square both sides.}$$
$$3x = 36 \qquad \text{Simplify.}$$
$$x = 12 \qquad \text{Divide both sides by 3.}$$

Check

$$\sqrt{3x} + 6 = 0 \qquad \text{Original equation}$$
$$\sqrt{3(12)} + 6 \stackrel{?}{=} 0 \qquad \text{Substitute 12 for } x.$$
$$6 + 6 \neq 0 \qquad \text{Solution does not check.} \; ✗$$

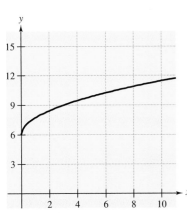

FIGURE 5.4

Therefore, the equation has no solution. You can also check this graphically, as shown in Figure 5.4.

Technology

In Example 4, you can graphically check the solution of the equation by "graphing the left side and right side on the same screen." That is, by graphing the equations

$$y = \sqrt{5x + 3}$$

and

$$y = \sqrt{x + 11}$$

on the same screen, as shown below, you can see that the two graphs intersect when $x = 2$.

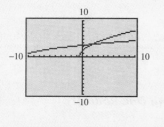

EXAMPLE 4 Solving an Equation Having Two Radicals

Solve $\sqrt{5x + 3} = \sqrt{x + 11}$.

Solution

$\sqrt{5x + 3} = \sqrt{x + 11}$	Original equation
$\left(\sqrt{5x + 3}\right)^2 = \left(\sqrt{x + 11}\right)^2$	Square both sides.
$5x + 3 = x + 11$	Simplify.
$5x = x + 8$	Subtract 3 from both sides.
$4x = 8$	Subtract x from both sides.
$x = 2$	Divide both sides by 4.

Check

$\sqrt{5x + 3} = \sqrt{x + 11}$	Original equation
$\sqrt{5(2) + 3} \stackrel{?}{=} \sqrt{2 + 11}$	Substitute 2 for x.
$\sqrt{13} = \sqrt{13}$	Solution checks. ✓

Therefore, the equation has one solution: $x = 2$.

EXAMPLE 5 Solving an Equation Having Two Radicals

Solve $\sqrt[4]{3x} + \sqrt[4]{2x - 5} = 0$.

Solution

$\sqrt[4]{3x} + \sqrt[4]{2x - 5} = 0$	Original equation
$\sqrt[4]{3x} = -\sqrt[4]{2x - 5}$	Isolate radicals.
$\left(\sqrt[4]{3x}\right)^4 = \left(-\sqrt[4]{2x - 5}\right)^4$	Raise both sides to 4th power.
$3x = 2x - 5$	Simplify.
$x = -5$	Subtract $2x$ from both sides.

Check

$\sqrt[4]{3x} + \sqrt[4]{2x - 5} = 0$	Original equation
$\sqrt[4]{3(-5)} + \sqrt[4]{2(-5) - 5} \stackrel{?}{=} 0$	Substitute -5 for x.
$\sqrt[4]{-15} + \sqrt[4]{-15} \neq 0$	Solution does not check. ✗

The solution does not check because it yields fourth roots of negative radicands. Therefore, this equation has no solution. Try checking this graphically. If you graph both sides of the equation, you will discover that they do not intersect.

EXAMPLE 6   An Equation That Converts to a Quadratic Equation

Solve $\sqrt{x} + 2 = x$.

Solution

$$\sqrt{x} + 2 = x \qquad \text{Original equation}$$

$$\sqrt{x} = x - 2 \qquad \text{Isolate radical.}$$

$$\left(\sqrt{x}\right)^2 = (x - 2)^2 \qquad \text{Square both sides.}$$

$$x = x^2 - 4x + 4 \qquad \text{Simplify.}$$

$$-x^2 + 5x - 4 = 0 \qquad \text{Standard form}$$

$$(-1)(x - 4)(x - 1) = 0 \qquad \text{Factor.}$$

$$x - 4 = 0 \quad \Longrightarrow \quad x = 4 \qquad \text{Set 1st factor equal to 0.}$$

$$x - 1 = 0 \quad \Longrightarrow \quad x = 1 \qquad \text{Set 2nd factor equal to 0.}$$

Try checking each of these solutions. When you do, you will find that $x = 4$ is a valid solution, but that $x = 1$ is extraneous.

When an equation contains two radicals, it may not be possible to isolate both. In such cases, you may have to raise both sides of the equation to a power at *two* different stages in the solution.

EXAMPLE 7   Repeatedly Squaring Both Sides of an Equation

Solve $\sqrt{3t + 1} = 2 - \sqrt{3t}$.

Solution

$$\sqrt{3t + 1} = 2 - \sqrt{3t} \qquad \text{Original equation}$$

$$\left(\sqrt{3t + 1}\right)^2 = \left(2 - \sqrt{3t}\right)^2 \qquad \text{Square both sides (1st time).}$$

$$3t + 1 = 4 - 4\sqrt{3t} + 3t \qquad \text{Simplify.}$$

$$-3 = -4\sqrt{3t} \qquad \text{Isolate radical.}$$

$$(-3)^2 = \left(-4\sqrt{3t}\right)^2 \qquad \text{Square both sides (2nd time).}$$

$$9 = 16(3t) \qquad \text{Simplify.}$$

$$\frac{3}{16} = t \qquad \text{Divide both sides by 48.}$$

The solution is $t = \frac{3}{16}$. Check this in the original equation.

Applications

EXAMPLE 8 An Application Involving Electricity

The amount of power consumed by an electrical appliance is given by

$$I = \sqrt{\frac{P}{R}}$$

where I is the current measured in amps, R is the resistance measured in ohms, and P is the power measured in watts. Find the power used by an electric heater, for which $I = 10$ amps and $R = 16$ ohms.

Solution

$$I = \sqrt{\frac{P}{R}} \qquad\qquad \text{Original equation}$$

$$10 = \sqrt{\frac{P}{16}} \qquad\qquad \text{Substitute for } I \text{ and } R.$$

$$(10)^2 = \left(\sqrt{\frac{P}{16}}\right)^2 \qquad\qquad \text{Square both sides.}$$

$$100 = \frac{P}{16} \qquad\qquad \text{Simplify.}$$

$$1600 = P \qquad\qquad \text{Multiply both sides by 16.}$$

Therefore, the electric heater uses 1600 watts of power. Check this solution in the original equation.

An alternative way to solve the problem in Example 8 would be to first solve the equation for P.

$$I = \sqrt{\frac{P}{R}} \qquad\qquad \text{Original equation}$$

$$I^2 = \left(\sqrt{\frac{P}{R}}\right)^2 \qquad\qquad \text{Square both sides.}$$

$$I^2 = \frac{P}{R} \qquad\qquad \text{Simplify.}$$

$$I^2 R = P \qquad\qquad \text{Multiply both sides by } R.$$

At this stage, you can substitute the known values of I and R to obtain

$$P = (10)^2 16 = 1600.$$

EXAMPLE 9 *The Velocity of a Falling Object*

The velocity of a free-falling object can be determined from the equation

$$v = \sqrt{2gh}$$

where v is the velocity measured in feet per second, $g = 32$ feet per second per second, and h is the distance (in feet) the object has fallen. Find the height from which a rock has been dropped if it strikes the ground with a velocity of 50 feet per second.

Solution

$v = \sqrt{2gh}$	Original equation
$50 = \sqrt{2(32)h}$	Substitute for v and g.
$(50)^2 = \left(\sqrt{64h}\right)^2$	Square both sides.
$2500 = 64h$	Simplify.
$39 \approx h$	Divide both sides by 64.

Thus, the rock has fallen approximately 39 feet when it hits the ground. Check this solution in the original equation.

Group Activities   Extending the Concept

An Experiment   Without using a stopwatch, you can find the length of time an object has been falling by using the following equation from physics

$$t = \sqrt{\frac{h}{384}}$$

where t is the length of time in seconds and h is the height in inches the object has fallen. How far does an object fall in 0.25 second? in 0.10 second?

Use this equation to test how long it takes members of your group to catch a falling ruler. Hold the ruler vertically while another group member holds his or her hands near the lower end of the ruler ready to catch it. Before releasing the ruler, record the mark on the ruler closest to the top of the catcher's hands. Release the ruler. After it has been caught, again note the mark closest to the top of the catcher's hands. (The difference between these two measurements is h.) Which member of your group reacts most quickly?

5.5 Exercises

Discussing the Concepts

1. In your own words, describe the steps that can be used to solve a radical equation.

2. Does raising both sides of an equation to the nth power always yield an equivalent equation? Explain.

3. One reason for checking a solution in the original equation is to discover errors that were made when solving the equation. Describe another reason.

4. *Error Analysis*   Describe the error.

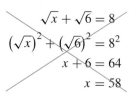

Problem Solving

In Exercises 5–8, determine whether each value of x is a solution of the equation.

Equation		Values of x	

5. $\sqrt{x} - 10 = 0$ (a) $x = -4$ (b) $x = -100$
   (c) $x = \sqrt{10}$ (d) $x = 100$

6. $\sqrt{3x} - 6 = 0$ (a) $x = \frac{2}{3}$ (b) $x = 2$
   (c) $x = 12$ (d) $x = -\frac{1}{3}\sqrt{6}$

7. $\sqrt[3]{x} - 4 = 4$ (a) $x = -60$ (b) $x = 68$
   (c) $x = 20$ (d) $x = 0$

8. $\sqrt[4]{2x} + 2 = 6$ (a) $x = 128$ (b) $x = 2$
   (c) $x = -2$ (d) $x = 0$

In Exercises 9–22, solve the equation. (Some of the equations have no solution.)

9. $\sqrt{x} = 20$

10. $\sqrt{x} = 5$

11. $\sqrt{u} + 13 = 0$

12. $\sqrt{y} + 15 = 0$

13. $\sqrt{3y + 5} - 3 = 4$

14. $\sqrt{5z - 2} + 7 = 10$

15. $\sqrt{x^2 + 5} = x + 3$

16. $\sqrt{x^2 - 4} = x - 2$

17. $\sqrt{3y - 5} - 3\sqrt{y} = 0$

18. $\sqrt{2u + 10} - 2\sqrt{u} = 0$

19. $\sqrt[3]{3x - 4} = \sqrt[3]{x + 10}$

20. $2\sqrt[3]{10 - 3x} = \sqrt[3]{2 - x}$

21. $\sqrt{8x + 1} = x + 2$

22. $\sqrt{3x + 7} = x + 3$

Graphical Reasoning   In Exercises 23–26, use a graphing utility to graph both sides of the equation on the same screen. Then use the graphs to approximate the solution. Check your approximations algebraically.

23. $\sqrt{x} = 2(2 - x)$

24. $\sqrt{2x + 3} = 4x - 3$

25. $\sqrt{x + 3} = 5 - \sqrt{x}$

26. $\sqrt[3]{5x - 8} = 4 - \sqrt[3]{x}$

Geometry   In Exercises 27–30, find the length of the side labeled x. Round to two decimal places.

27.

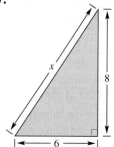

28.

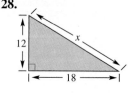

29.

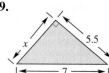

30.

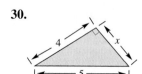

31. *Drawing a Diagram*   A basketball court is 50 feet wide and 94 feet long. Draw a diagram and find the length of the diagonal of the court.

32. *Drawing a Diagram*   For a square "25-inch" television screen, it is the diagonal of the screen that is 25 inches. Draw a diagram that shows a television screen with a 25-inch diagonal. What are its dimensions?

33. *Drawing a Diagram*   A house has a basement floor with dimensions 26 feet by 32 feet. The gas hot water heater and furnace are diagonally across the basement from where the natural gas line enters the house. Draw a diagram showing the gas line and find the length of the gas line across the basement.

34. *Drawing a Diagram*   An 8-foot plank is used to brace a basement wall during the construction of a home. The plank is nailed to the wall 5 feet above the floor. Draw a diagram showing the plank, and find its slope.

Height of an Object   In Exercises 35 and 36, use the following formula, which gives the time t in seconds for a free-falling object to fall d feet.

$$t = \sqrt{\frac{d}{16}}$$

35. A construction worker drops a nail and observes it strike a water puddle after approximately 2 seconds. Estimate the height of the worker.

36. A construction worker drops a nail and observes it strike a water puddle after approximately 3 seconds. Estimate the height of the worker.

Length of a Pendulum   In Exercises 37 and 38, use the following formula, which gives the time t in seconds for a pendulum of length L feet to go through one complete cycle (its period).

$$t = 2\pi \sqrt{\frac{L}{32}}$$

37. How long is the pendulum of a grandfather clock with a period of 1.5 seconds (see figure)?

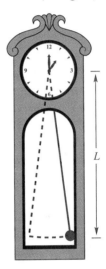

38. How long is the pendulum of a mantle clock with a period of 0.75 seconds?

Reviewing the Major Concepts

In Exercises 39–44, simplify the fraction.

39. $\dfrac{-16x^2}{12x}$

40. $\dfrac{5t^4}{45t^{-2}}$

41. $\dfrac{6u^2v^{-3}}{27uv^3}$

42. $\dfrac{-14r^4s^2}{-98rs^2}$

43. $\left(\dfrac{3x^2}{2y^{-1}}\right)^{-2}$

44. $\left(\dfrac{15a^{-3}b}{21ab^{-2}}\right)^{0}$

45. *Comparing Prices*   A department store is offering a discount of 20% on a sewing machine with a list price of $239.95. A mail-order catalog has the same machine for $188.95 plus $4.32 for shipping. Which is the better bargain?

46. *Insurance Premium*   The annual insurance premium for a policyholder is normally $739. However, after having an automobile accident, the policyholder is charged an additional 30%. What is the new annual premium?

Additional Problem Solving

In Exercises 47–64, solve the equation. (Some of the equations have no solution.)

47. $\sqrt{y} - 7 = 0$

48. $\sqrt{t} - 13 = 0$

49. $\sqrt{a + 100} = 25$

50. $\sqrt{b + 12} = 13$

51. $\sqrt{10x} = 30$

52. $\sqrt{8x} = 6$

53. $5\sqrt{x + 2} = 8$

54. $2\sqrt{x + 4} = 7$

55. $\sqrt{3x + 2} + 5 = 0$

56. $\sqrt{1 - x} + 10 = 4$

57. $\sqrt{x + 3} = \sqrt{2x - 1}$

58. $\sqrt{3t + 1} = \sqrt{t + 15}$

59. $\sqrt[3]{2x + 15} - \sqrt[3]{x} = 0$

60. $\sqrt[4]{2x} + \sqrt[4]{x + 3} = 0$

61. $(x + 4)^{2/3} = 4$

62. $2x^{3/4} = 54$

63. $\sqrt{2x} = x - 4$

64. $\sqrt{x} = x - 6$

Graphical Reasoning In Exercises 65–68, use a graphing utility to graph both sides of the equation on the same screen. Then use the graphs to approximate the solution. Check your approximations algebraically.

65. $\sqrt{x^2 + 1} = 5 - 2x$

66. $\sqrt{8 - 3x} = x$

67. $4\sqrt[3]{x} = 7 - x$

68. $\sqrt[3]{x + 4} = \sqrt{6 - x}$

Geometry In Exercises 69–72, find the length of the side labeled x. Round to two decimal places.

69.

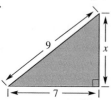

70.

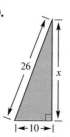

71.

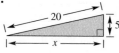

72.

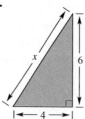

In Exercises 73–78, match the function with its graph. [The graphs are labeled (a), (b), (c), (d), (e), and (f).]

(a)

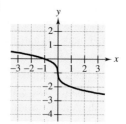

(b)

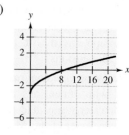

(c)

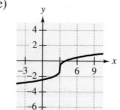

(d)

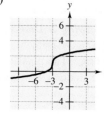

(e)

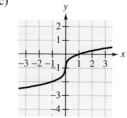

(f)

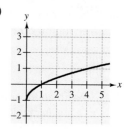

73. $f(x) = \sqrt[3]{x} - 1$

74. $f(x) = \sqrt[3]{x - 3} + 1$

75. $f(x) = \sqrt[3]{x + 3} + 1$

76. $f(x) = -\sqrt[3]{x} - 1$

77. $f(x) = \sqrt{x} - 1$

78. $f(x) = \sqrt{x} - 3$

79. *Length of a Ramp*   A ramp is 20 feet long and rests on a porch that is 4 feet high (see figure). Find the distance between the porch and the base of the ramp.

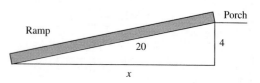

80. *Drawing a Diagram*   A guy wire on a radio tower is attached to the top of a 100-foot tower and to an anchor 50 feet from the base of the tower. Draw a diagram showing the guy wire, and find its length.

Free-Falling Object In Exercises 81–84, use the equation for the velocity of a free-falling object, $v = \sqrt{2gh}$, as described in Example 9.

81. An object is dropped from a height of 50 feet. Find the velocity of the object when it strikes the ground.

82. An object is dropped from a height of 200 feet. Find the velocity of the object when it strikes the ground.

83. An object that was dropped strikes the ground with a velocity of 60 feet per second. Find the height from which the object was dropped.

84. An object that was dropped strikes the ground with a velocity of 120 feet per second. Find the height from which the object was dropped.

85. *Demand for a Product*   The demand equation for a certain product is

$$p = 50 - \sqrt{0.8(x - 1)}$$

where x is the number of units demanded per day and p is the price per unit. Find the demand if the price is $30.02.

86. *Geometry*   The surface area of a cone is given by

$$S = \pi r \sqrt{r^2 + h^2}$$

as shown in the figure. Solve this equation for h.

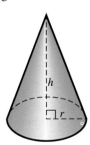

87. *Geometry*   Write a function that gives the radius r of a circle in terms of the circle's area A. Use a graphing utility to graph this function.

88. *Geometry*   Determine the length and width of a rectangle with a perimeter of 68 inches and a diagonal of length 26 inches.

89. *Reading a Graph*   An airline offers daily flights between Chicago and Denver. The total monthly cost of the flights is

$$C = \sqrt{0.2x + 1}, \quad 0 \le x$$

where C is measured in millions of dollars and x is measured in thousands of passengers (see figure). The total cost of the flights for a certain month is 2.5 million dollars. Approximately how many passengers flew that month?

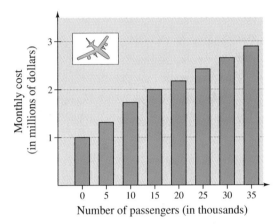

Number of passengers (in thousands)

90. *Congressional Aides*   Each member of the United States Congress has a staff of congressional aides. From 1930 through 1990, the number of aides A assigned to the members of the House of Representatives can be modeled by the function

$$A = 1220\sqrt[3]{t - 42} + 4900$$

where $t = 0$ represents 1930.

(a) Use a graphing utility to graph the function.

(b) In what year were approximately 7340 aides assigned to members of the House of Representatives?

(c) The House of Representatives has 435 members. Write a function that gives the average number of congressional aides per representative. Use a graphing utility to graph this function.

5.6	**Complex Numbers**
	The Imaginary Unit i ▪ Complex Numbers ▪ Operations with Complex Numbers ▪ Complex Conjugates

The Imaginary Unit i

In Section 5.2, you learned that a negative number has no *real* square root. For instance, $\sqrt{-1}$ is not real because there is no real number x such that $x^2 = -1$. Thus, as long as you are dealing only with real numbers, the equation

$$x^2 = -1$$

has no solution. To overcome this deficiency, mathematicians have expanded the set of numbers, using the **imaginary unit i,** defined as

$$i = \sqrt{-1}.$$ Imaginary unit

This number has the property that $i^2 = -1$. Thus, the imaginary unit i is a solution of the equation $x^2 = -1$.

The Square Root of a Negative Number

Let c be a positive real number. Then the square root of $-c$ is given by

$$\sqrt{-c} = \sqrt{c(-1)} = \sqrt{c}\sqrt{-1} = \sqrt{c}\,i.$$

When writing $\sqrt{-c}$ in the **i-form,** $\sqrt{c}\,i$, note that i is outside the radical.

DISCOVERY

Use a calculator to evaluate the following radicals. Does one result in an error message? Explain why.

a. $\sqrt{121}$

b. $\sqrt{-121}$

c. $-\sqrt{121}$

EXAMPLE 1 *Writing Numbers in i-Form*

a. $\sqrt{-36} = \sqrt{36(-1)} = \sqrt{36}\sqrt{-1} = 6i$

b. $\sqrt{-\dfrac{16}{25}} = \sqrt{\dfrac{16}{25}(-1)} = \sqrt{\dfrac{16}{25}}\sqrt{-1} = \dfrac{4}{5}i$

c. $\sqrt{-5} = \sqrt{5(-1)} = \sqrt{5}\sqrt{-1} = \sqrt{5}\,i$

d. $\sqrt{-54} = \sqrt{54(-1)} = \sqrt{54}\sqrt{-1} = 3\sqrt{6}\,i$

e. $\dfrac{\sqrt{-48}}{\sqrt{-3}} = \dfrac{\sqrt{48}\sqrt{-1}}{\sqrt{3}\sqrt{-1}} = \dfrac{\sqrt{48}\,i}{\sqrt{3}\,i} = \sqrt{\dfrac{48}{3}} = \sqrt{16} = 4$

f. $\dfrac{\sqrt{-18}}{\sqrt{2}} = \dfrac{\sqrt{18}\sqrt{-1}}{\sqrt{2}} = \dfrac{\sqrt{18}\,i}{\sqrt{2}} = \sqrt{\dfrac{18}{2}}\,i = \sqrt{9}\,i = 3i$

To perform operations with square roots of negative numbers, you must *first* write the numbers in *i*-form. Once the numbers are written in *i*-form, you add, subtract, and multiply as follows.

$$ai + bi = (a + b)i \qquad\qquad \text{Addition}$$

$$ai - bi = (a - b)i \qquad\qquad \text{Subtraction}$$

$$(ai)(bi) = ab(i^2) = ab(-1) = -ab \qquad\qquad \text{Multiplication}$$

EXAMPLE 2   Adding Square Roots of Negative Numbers

Perform the addition: $\sqrt{-9} + \sqrt{-49}$.

Solution

$$\sqrt{-9} + \sqrt{-49} = \sqrt{9}\sqrt{-1} + \sqrt{49}\sqrt{-1} \qquad \text{Property of radicals}$$

$$= 3i + 7i \qquad\qquad\qquad \text{Write in } i\text{-form.}$$

$$= 10i \qquad\qquad\qquad\qquad \text{Simplify.}$$

In the next example, notice how you can multiply radicals that involve square roots of negative numbers.

EXAMPLE 3   Multiplying Square Roots of Negative Numbers

a.
$$\sqrt{-15}\sqrt{-15} = \left(\sqrt{15}\,i\right)\left(\sqrt{15}\,i\right) \qquad \text{Write in } i\text{-form.}$$

$$= \left(\sqrt{15}\right)^2 i^2 \qquad\qquad \text{Multiply.}$$

$$= 15(-1) \qquad\qquad\quad \text{Definition of } i$$

$$= -15 \qquad\qquad\qquad \text{Simplify.}$$

b.
$$\sqrt{-5}\left(\sqrt{-45} - \sqrt{-4}\right) = \sqrt{5}\,i\left(3\sqrt{5}\,i - 2i\right) \qquad \text{Write in } i\text{-form.}$$

$$= \left(\sqrt{5}\,i\right)\left(3\sqrt{5}\,i\right) - \left(\sqrt{5}\,i\right)(2i) \qquad \text{Distributive Property}$$

$$= 3(5)(-1) - 2\sqrt{5}(-1) \qquad\qquad \text{Multiply.}$$

$$= -15 + 2\sqrt{5} \qquad\qquad\qquad\qquad \text{Simplify.}$$

STUDY TIP

When performing operations with numbers in *i*-form, you sometimes need to be able to evaluate powers of the imaginary unit *i*. The first several powers of *i* are as follows.

$$i^1 = i$$

$$i^2 = -1$$

$$i^3 = i(i^2) = i(-1) = -i$$

$$i^4 = (i^2)(i^2) = (-1)(-1) = 1$$

$$i^5 = i(i^4) = i(1) = i$$

$$i^6 = (i^2)(i^4) = (-1)(1) = -1$$

$$i^7 = (i^3)(i^4) = (-i)(1) = -i$$

$$i^8 = (i^4)(i^4) = (1)(1) = 1$$

Note how the pattern of values i, -1, $-i$, and 1 repeats itself for powers greater than 4.

When multiplying square roots of negative numbers, be sure to write them in *i*-form *before multiplying*. If you do not you can obtain incorrect answers. For instance, in Example 3(a) be sure you see that

$$\sqrt{-15}\sqrt{-15} \neq \sqrt{(-15)(-15)} = \sqrt{225} = 15.$$

Complex Numbers

A number of the form $a + bi$, where a and b are real numbers, is called a **complex number.**

Definition of a Complex Number

If a and b are real numbers, the number $a + bi$ is a **complex number,** and it is said to be written in **standard form.** If $b = 0$, the number $a + bi = a$ is a real number. If $b \neq 0$, the number $a + bi$ is called an **imaginary number.**

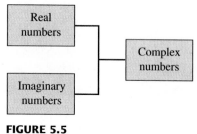

FIGURE 5.5

NOTE Two complex numbers $a + bi$ and $c + di$, in standard form, are equal if and only if $a = c$ and $b = d$.

A number cannot be both real and imaginary. For instance, the numbers -2, 0, 1, $\frac{1}{2}$, and $\sqrt{2}$ are real numbers (but they are *not* imaginary numbers), and the numbers $-3i$, $2 + 4i$, and $-1 + i$ are imaginary numbers (but they are *not* real numbers). The diagram shown in Figure 5.5 further illustrates the relationship among real, complex, and imaginary numbers.

EXAMPLE 4 Equality of Two Complex Numbers

a. Are the complex numbers $\sqrt{9} + \sqrt{-48}$ and $3 - 4\sqrt{3}\,i$ equal?

b. Find x and y so that the equation is valid.
$$3x - \sqrt{-25} = -6 + 3yi$$

Solution

a. Begin by writing the first number in standard form.
$$\sqrt{9} + \sqrt{-48} = \sqrt{3^2} + \sqrt{4^2(3)(-1)} = 3 + 4\sqrt{3}\,i$$

From this form, you can see that the two numbers are not equal because they have imaginary parts that differ in sign.

b. Begin by writing the left side of the equation in standard form.

$$3x - \sqrt{-25} = -6 + 3yi \qquad \text{Original equation}$$
$$3x - 5i = -6 + 3yi \qquad \text{Both sides in standard form}$$

For these two numbers to be equal, their real parts must be equal to each other and their imaginary parts must be equal to each other. Thus, $3x = -6$, which implies that $x = -2$, and $-5 = 3y$, which implies that $y = -\frac{5}{3}$.

Operations with Complex Numbers

The real number a is called the **real part** of the complex number $a + bi$, and the real number b is called the **imaginary part** of the complex number. To add or subtract two complex numbers, we add (or subtract) the real and imaginary parts separately. This is similar to combining like terms of a polynomial.

$$(a + bi) + (c + di) = (a + c) + (b + d)i \qquad \text{Addition of complex numbers}$$
$$(a + bi) - (c + di) = (a - c) + (b - d)i \qquad \text{Subtraction of complex numbers}$$

EXAMPLE 5   Adding and Subtracting Complex Numbers

a. $(3 - i) + (-2 + 4i) = (3 - 2) + (-1 + 4)i = 1 + 3i$

b. $3i + (5 - 3i) = 5 + (3 - 3)i = 5$

c. $4 - (-1 + 5i) + (7 + 2i) = [4 - (-1) + 7] + (-5 + 2)i = 12 - 3i$

Note in part (b) that the sum of two complex numbers can be a real number.

The Commutative, Associative, and Distributive Properties of real numbers are also valid for complex numbers.

EXAMPLE 6   Multiplying Complex Numbers

a. $(1 - i)(\sqrt{-9}) = (1 - i)(3i)$ Write in standard form.

$\qquad\qquad\quad = (1)(3i) - (i)(3i)$ Distributive Property

$\qquad\qquad\quad = 3i - 3(i^2)$ Simplify.

$\qquad\qquad\quad = 3i - 3(-1)$ Definition of i

$\qquad\qquad\quad = 3 + 3i$ Simplify.

b. $(2 - i)(4 + 3i) = 8 + 6i - 4i - 3i^2$ FOIL Method

$\qquad\qquad\qquad = 8 + 6i - 4i - 3(-1)$ Definition of i

$\qquad\qquad\qquad = 11 + 2i$ Combine like terms.

c. $(3 + 2i)(3 - 2i) = 3^2 - (2i)^2$ Special product formula

$\qquad\qquad\qquad = 9 - 2^2 i^2$ Simplify.

$\qquad\qquad\qquad = 9 - 4(-1)$ Definition of i

$\qquad\qquad\qquad = 9 + 4$ Simplify.

$\qquad\qquad\qquad = 13$ Simplify.

Until very recently it was thought that shapes in nature, such as clouds, coastlines, and mountain ranges, could not be described in mathematical terms. In the 1970s Benoit Mandelbrot discovered that many of these shapes do have a pattern in their irregularity—they are made up of smaller parts that are scaled-down versions of the shape itself. Computers using mathematical terms with complex numbers are able to generate the larger image. Mandelbrot coined the term *fractals* for these shapes and for the geometry used to describe them.

Complex Conjugates

In Example 6(c), note that the product of two complex numbers can be a real number. This occurs with pairs of complex numbers of the form $a+bi$ and $a-bi$, called **complex conjugates.** In general, the product of complex conjugates has the following form.

$$(a+bi)(a-bi) = a^2 - (bi)^2$$
$$= a^2 - b^2i^2$$
$$= a^2 - b^2(-1)$$
$$= a^2 + b^2$$

Here are some examples.

Complex Number	Complex Conjugate	Product
$4-5i$	$4+5i$	$4^2+5^2 = 41$
$3+2i$	$3-2i$	$3^2+2^2 = 13$
$-2 = -2+0i$	$-2 = -2-0i$	$(-2)^2+0^2 = 4$
$i = 0+i$	$-i = 0-i$	$0^2+1^2 = 1$

Complex conjugates are used to divide one complex number by another. To do this, multiply the numerator and denominator by the *complex conjugate of the denominator,* as shown in Example 7.

EXAMPLE 7 Division of Complex Numbers

Perform the division and write your answer in standard form.

$$5 \div (3 - 2i)$$

Solution

$$\frac{5}{3-2i} = \frac{5}{3-2i} \cdot \frac{3+2i}{3+2i}$$ Multiply by complex conjugate of denominator.

$$= \frac{5(3+2i)}{(3-2i)(3+2i)}$$ Multiply fractions.

$$= \frac{5(3+2i)}{3^2+2^2}$$ Product of complex conjugates

$$= \frac{15+10i}{13}$$ Simplify.

$$= \frac{15}{13} + \frac{10}{13}i$$ Standard form

EXAMPLE 8   Division of Complex Numbers

Divide $2 + 3i$ by $4 - 2i$.

Solution

$$\frac{2 + 3i}{4 - 2i} = \frac{2 + 3i}{4 - 2i} \cdot \frac{4 + 2i}{4 + 2i}$$ 　　Multiply by complex conjugate of denominator.

$$= \frac{8 + 16i + 6i^2}{4^2 + 2^2}$$ 　　Multiply fractions.

$$= \frac{8 + 16i + 6(-1)}{20}$$ 　　Definition of i

$$= \frac{2 + 16i}{20}$$ 　　Combine like terms.

$$= \frac{1}{10} + \frac{4}{5}i$$ 　　Standard form

EXAMPLE 9   Verifying a Complex Solution of an Equation

Show that $x = 2 + i$ is a solution of the equation $x^2 - 4x + 5 = 0$.

Solution

$$x^2 - 4x + 5 = 0$$ 　　Original equation

$$(2 + i)^2 - 4(2 + i) + 5 \overset{?}{=} 0$$ 　　Substitute $2 + i$ for x.

$$4 + 4i + i^2 - 8 - 4i + 5 \overset{?}{=} 0$$ 　　Expand.

$$i^2 + 1 \overset{?}{=} 0$$ 　　Combine like terms.

$$(-1) + 1 \overset{?}{=} 0$$ 　　Definition of i

$$0 = 0$$ 　　Solution checks. ✓

Therefore, $2 + i$ is a solution of the original equation.

Group Activities   Extending the Concept

Prime Polynomials   The polynomial $x^2 + 1$ is prime *with respect to the integers.* It is not, however, prime *with respect to the complex numbers.* Show how $x^2 + 1$ can be factored using complex numbers.

5.6 Exercises

Discussing the Concepts

1. Define the imaginary unit i.

2. Explain why the equation $x^2 = -1$ does not have real number solutions.

3. *Error Analysis* Describe the error.

$$\sqrt{-3}\sqrt{-3} = \sqrt{(-3)(-3)} = \sqrt{9} = 3$$

4. *True or False?* Some numbers are both real and imaginary. Explain.

5. Find the product of $3 - 2i$ and its complex conjugate.

6. Describe the methods for adding, subtracting, multiplying, and dividing complex numbers.

Problem Solving

In Exercises 7–12, write the number in i-form.

7. $\sqrt{-4}$

8. $\sqrt{-9}$

9. $\sqrt{-27}$

10. $\sqrt{-\frac{8}{25}}$

11. $\sqrt{-7}$

12. $\sqrt{-15}$

In Exercises 13–20, perform the operations and write your answer in standard form.

13. $\sqrt{-16} + 6i$

14. $\sqrt{-\frac{1}{4}} - \frac{3}{2}i$

15. $\sqrt{-50} - \sqrt{-8}$

16. $\sqrt{-500} + \sqrt{-45}$

17. $\sqrt{-8}\sqrt{-2}$

18. $\sqrt{-25}\sqrt{-6}$

19. $\sqrt{-5}\left(\sqrt{-3} - \sqrt{-2}\right)$

20. $\sqrt{-4}\left(\sqrt{-9} + \sqrt{-4}\right)$

In Exercises 21–32, perform the operations and write your answer in standard form.

21. $(4 - 3i) + (6 + 7i)$

22. $22 + (-5 + 8i) + 10i$

23. $13i - (14 - 7i)$

24. $(3 - 2i)^3$

25. $15i - (3 - 25i) + \sqrt{-81}$

26. $(-1 + i) - \sqrt{2} - \sqrt{-2}$

27. $(3i)(12i)$

28. $(8i)^2$

29. $(4 + 3i)(-7 + 4i)$

30. $(7 + i)^2$

31. $\left(-3 - \sqrt{-12}\right)\left(4 - \sqrt{-12}\right)$

32. $(0.05 + 2.50i) - (6.2 + 11.8i)$

In Exercises 33–36, multiply the number by its complex conjugate.

33. $-2 - 8i$

34. $-4 + \sqrt{2}i$

35. $1 + \sqrt{-3}$

36. $-3 - \sqrt{-5}$

In Exercises 37–40, perform the division and write your answer in standard form.

37. $\dfrac{-12}{2 + 7i}$

38. $\dfrac{17i}{5 + 3i}$

39. $\dfrac{20}{2i}$

40. $\dfrac{4 - 5i}{4 + 5i}$

Reviewing the Major Concepts

In Exercises 41–44, simplify the expression.

41. $\sqrt{128} + 3\sqrt{50}$

42. $3\sqrt{5}\sqrt{500}$

43. $\dfrac{8}{\sqrt{10}}$

44. $\dfrac{5}{\sqrt{12} - 2}$

45. *Real Estate Taxes* The tax on a property with an assessed value of $145,000 is $2400. Find the tax on a property with an assessed value of $90,000.

46. *Gasoline Cost* A car uses 7 gallons of gas for a trip of 200 miles. How many gallons would be used on a trip of 325 miles?

Additional Problem Solving

In Exercises 47–54, write the number in i-form.

47. $-\sqrt{-144}$ **48.** $\sqrt{-49}$

49. $\sqrt{-\frac{4}{25}}$ **50.** $-\sqrt{-\frac{36}{121}}$

51. $\sqrt{-0.09}$ **52.** $\sqrt{-0.0004}$

53. $\sqrt{-8}$ **54.** $\sqrt{-75}$

In Exercises 55–58, perform the operations and write your answer in standard form.

55. $\sqrt{-18}\sqrt{-3}$ **56.** $\sqrt{-0.16}\sqrt{-1.21}$

57. $\sqrt{-3}\left(\sqrt{-3}+\sqrt{-4}\right)$ **58.** $\sqrt{-12}\left(\sqrt{-3}-\sqrt{-12}\right)$

In Exercises 59–62, determine a and b.

59. $5-4i=(a+3)+(b-1)i$

60. $-10+12i=2a+(5b-3)i$

61. $-4-\sqrt{-8}=a+bi$

62. $\sqrt{-36}-3=a+bi$

In Exercises 63–84, perform the operations and write your answer in standard form.

63. $(-4-7i)+(-10-33i)$

64. $(-10+2i)+(4-7i)$

65. $(15+10i)-(2+10i)$

66. $(-21-50i)+(21-20i)$

67. $(30-i)-(18+6i)+3i^2$

68. $(4+6i)+(15+24i)-(1-i)$

69. $(-2i)(-10i)$ **70.** $(-5i)(4i)$

71. $(-6i)(-i)(6i)$ **72.** $\frac{1}{2}(10i)(12i)(-3i)$

73. $(-3i)^3$ **74.** $(2i)^4$

75. $-5(13+2i)$ **76.** $10(8-6i)$

77. $4i(-3-5i)$ **78.** $-3i(10-15i)$

79. $(-7+7i)(4-2i)$ **80.** $(3+5i)(2+15i)$

81. $(3-4i)^2$ **82.** $(2+i)^3$

83. $\left(-2+\sqrt{-5}\right)\left(-2-\sqrt{-5}\right)$

84. $\sqrt{-9}\left(1+\sqrt{-16}\right)$

In Exercises 85–90, find the conjugate of the number. Then find the product of the number and its conjugate.

85. $2+i$ **86.** $3+2i$

87. $5-\sqrt{6}\,i$ **88.** $10-3i$

89. $10i$ **90.** 20

In Exercises 91–96, perform the division and write your answer in standard form.

91. $\frac{4}{1-i}$ **92.** $\frac{20}{3+i}$

93. $\frac{4i}{1-3i}$ **94.** $\frac{15}{2(1-i)}$

95. $\frac{2+3i}{1+2i}$ **96.** $\frac{1+i}{3i}$

In Exercises 97–100, find the sum or difference and write your answer in standard form.

97. $\frac{1}{1-2i}+\frac{4}{1+2i}$ **98.** $\frac{3i}{1+i}+\frac{2}{2+3i}$

99. $\frac{i}{4-3i}-\frac{5}{2+i}$ **100.** $\frac{1+i}{i}-\frac{3}{5-2i}$

In Exercises 101–104, decide whether each number is a solution of the equation.

101. $x^2+2x+5=0$
  (a) $x=-1+2i$ (b) $x=-1-2i$

102. $x^2-4x+13=0$
  (a) $x=2-3i$ (b) $x=2+3i$

103. $x^3+4x^2+9x+36=0$
  (a) $x=-4$ (b) $x=-3i$

104. $x^3-8x^2+25x-26=0$
  (a) $x=2$ (b) $x=3-2i$

In Exercises 105–108, perform the operation.

105. $(a+bi)+(a-bi)$ **106.** $(a+bi)(a-bi)$

107. $(a+bi)-(a-bi)$ **108.** $(a+bi)^2+(a-bi)^2$

CHAPTER PROJECT: Fractals

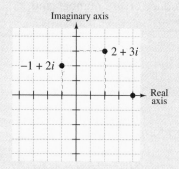

FIGURE A The Complex Plane

A complex number $a + bi$ can be represented by a point in the *complex plane,* as shown in Figure A. In the complex plane, the horizontal axis represents the real part of the complex number and the vertical axis represents the imaginary part. For instance, points representing the complex numbers $2 + 3i$ and $-1 + 2i$ are shown in Figure A.

In the hands of a person who understands "fractal geometry," the complex plane can become an easel on which stunning pictures, called *fractals,* can be drawn. The most famous such picture is the *Mandelbrot Set,* named after the Polish-born mathematician Benoit Mandelbrot. To decide whether a complex number is in the Mandelbrot Set, consider the sequence

$$c, \quad c^2 + c, \quad (c^2 + c)^2 + c, \quad ((c^2 + c)^2 + c)^2 + c, \dots$$

Note that each successive term is obtained by adding c to the square of the previous term. The behavior of this sequence depends on the value of c. Mandelbrot discovered that for some values of c, the sequence remains bounded—the terms of the sequence remain small. It is these values that are members of the Mandelbrot Set, which is represented by the black region in Figure B.

Use this information to investigate the following questions.

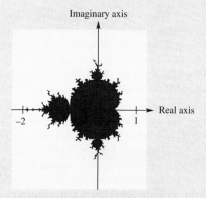

FIGURE B The Mandelbrot Set

1. The table on page 321 shows that the Mandelbrot sequences for $c = -1$ and $c = i$ are bounded and the sequence for $c = 1 + i$ is not bounded. Hence, -1 and i are in the Mandelbrot Set, but $1 + i$ is not. Decide whether the following numbers are in the Mandelbrot Set.

   (a) $c = 0$ (b) $c = 2$ (c) $c = \frac{1}{2}i$ (d) $c = -i$ (e) $c = 1$ (f) $c = -2$

2. The absolute value of the complex number $a + bi$ is defined as $|a + bi| = \sqrt{a^2 + b^2}$. If the absolute value of any term in the Mandelbrot sequence exceeds 2, the terms will eventually become unbounded. To add interest to the graph of the Mandelbrot Set, computer scientists discovered that the points that are not in the set can be assigned a variety of colors, as shown in Figure C. A color is assigned to a number c depending on how quickly its Mandelbrot sequence reaches a term whose absolute value is greater than 2. For each item below, find a complex number c such that its Mandelbrot sequence first exceeds an absolute value of 2 in the indicated term.

   (a) First term, c

   (b) Second term, $c^2 + c$

   (c) Third term, $(c^2 + c)^2 + c$

   (d) Fourth term, $((c^2 + c)^2 + c)^2 + c$

   (e) Fifth term, $(((c^2 + c)^2 + c)^2 + c)^2 + c$

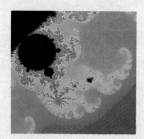

FIGURE C

3. *Research Project* Use your school's library or some other reference source to find other information about the Mandelbrot Set or about other fractals. Write a paper that describes your findings. Include illustrations in your paper.

CHAPTER SUMMARY

After studying this chapter, you should have acquired the following skills. These skills are keyed to the Review Exercises that begin on page 378. Answers to odd-numbered Review Exercises are given in the back of the book.

- Evaluate numerical expressions involving integer exponents and scientific notation. *(Section 5.1)*

  Review Exercises 1–8

- Simplify algebraic expressions involving integer exponents. *(Section 5.1)*

  Review Exercises 9–16

- Evaluate numerical expressions involving integer exponents or radicals. *(Section 5.2)*

  Review Exercises 17–26

- Translate expressions from radical form to rational exponent form and vice versa. *(Section 5.2)*

  Review Exercises 27–30

- Evaluate numerical expressions involving rational exponents. *(Section 5.2)*

  Review Exercises 31–36

- Simplify expressions involving rational exponents or radicals. *(Section 5.2)*

  Review Exercises 37–46

- Simplify expressions involving radicals by rationalizing the denominator. *(Section 5.3)*

  Review Exercises 47–54

- Add, subtract, multiply, and divide expressions involving radicals. *(Sections 5.3, 5.4)*

  Review Exercises 55–60

- Graph functions using a graphing utility. *(Sections 5.2, 5.3, 5.4)*

  Review Exercises 61–64

- Solve radical equations. *(Section 5.5)*

  Review Exercises 65–74

- Approximate solutions of equations using a graphing utility. *(Section 5.5)*

  Review Exercises 75, 76

- Write complex numbers in standard form. *(Section 5.6)*

  Review Exercises 77–82

- Add, subtract, multiply, and divide complex numbers. *(Section 5.6)*

  Review Exercises 83–92

- Solve problems using geometry. *(Sections 5.3, 5.5)*

  Review Exercises 93, 94

- Solve real-life problems modeled by radical equations. *(Section 5.5)*

  Review Exercises 95, 96

REVIEW EXERCISES

In Exercises 1–8, evaluate the expression.

1. $(2^3 \cdot 3^2)^{-1}$

2. $(2^{-2} \cdot 5^2)^{-2}$

3. $\left(\dfrac{2}{5}\right)^{-3}$

4. $\left(\dfrac{1}{3^{-2}}\right)^2$

5. $(6 \times 10^3)^2$

6. $(3 \times 10^{-3})(8 \times 10^7)$

7. $\dfrac{3.5 \times 10^7}{7 \times 10^4}$

8. $\dfrac{1}{(6 \times 10^{-3})^2}$

In Exercises 9–16, simplify the expression.

9. $\dfrac{4x^2}{2x}$

10. $4(-3x)^3$

11. $(x^3 y^{-4})^2$

12. $5yx^0$

13. $\dfrac{t^{-5}}{t^{-2}}$

14. $\dfrac{a^5 \cdot a^{-3}}{a^{-2}}$

15. $\left(\dfrac{y}{3}\right)^{-3}$

16. $(2x^2 y^4)^4 (2x^2 y^4)^{-4}$

In Exercises 17–22, evaluate the expression.

17. $\sqrt{1.44}$

18. $\sqrt{0.16}$

19. $\sqrt{\dfrac{25}{36}}$

20. $-\sqrt{\dfrac{64}{225}}$

21. $\sqrt{169 - 25}$

22. $\sqrt{16 + 9}$

In Exercises 23–26, evaluate the expression. Round the result to two decimal places.

23. $1800(1 + 0.08)^{24}$

24. $0.0024(7,658,400)$

25. $\sqrt{13^2 - 4(2)(7)}$

26. $\dfrac{-3.7 + \sqrt{15.8}}{2(2.3)}$

In Exercises 27–30, fill in the missing description.

Radical Form	Rational Exponent Form
27. $\sqrt{49} = 7$	
28. $\sqrt[3]{0.125} = 0.5$	
29.	$216^{1/3} = 6$
30.	$16^{1/4} = 2$

In Exercises 31–34, evaluate the expression.

31. $27^{4/3}$

32. $16^{3/4}$

33. $25^{3/2}$

34. $243^{-2/5}$

In Exercises 35 and 36, evaluate the expression. Round the result to two decimal places.

35. $75^{-3/4}$

36. $510^{5/3}$

In Exercises 37–46, simplify the expression.

37. $x^{3/4} \cdot x^{-1/6}$

38. $(2y^2)^{3/2}(2y^{-4})^{1/2}$

39. $\dfrac{15x^{1/4} y^{3/5}}{5x^{1/2} y}$

40. $\dfrac{48a^2 b^{5/2}}{14a^{-3} b^{-1/2}}$

41. $\sqrt{360}$

42. $\sqrt{\dfrac{50}{9}}$

43. $\sqrt{0.25x^4 y}$

44. $\sqrt{0.16s^6 t^3}$

45. $\sqrt[3]{48a^3 b^4}$

46. $\sqrt[4]{32u^4 v^5}$

In Exercises 47–54, rationalize the denominator and simplify further when possible.

47. $\sqrt{\dfrac{5}{6}}$

48. $\sqrt{\dfrac{3}{20}}$

49. $\dfrac{3}{\sqrt{12x}}$

50. $\dfrac{4y}{\sqrt{10z}}$

51. $\dfrac{2}{\sqrt[3]{2x}}$

52. $\sqrt[3]{\dfrac{16t}{s^2}}$

53. $\dfrac{6}{7 - \sqrt{7}}$

54. $\dfrac{x}{\sqrt{x} + 1}$

In Exercises 55–60, perform the operations and simplify.

55. $3\sqrt{40} - 10\sqrt{90}$

56. $9\sqrt{50} - 5\sqrt{8} + \sqrt{48}$

57. $\sqrt{25x} + \sqrt{49x} - \sqrt{x}$

58. $\left(\sqrt{5} + 6\right)^2$

59. $\left(3 - \sqrt{x}\right)\left(3 + \sqrt{x}\right)$

60. $15 \div \left(\sqrt{x} + 3\right)$

In Exercises 61–64, use a graphing utility to graph the function.

61. $y = 3\sqrt[3]{2x}$

62. $y = \dfrac{10}{\sqrt[4]{x^2 + 1}}$

63. $g(x) = 4x^{3/4}$

64. $h(x) = \frac{1}{2}x^{4/3}$

In Exercises 65–74, solve the equation.

65. $\sqrt{y} = 15$

66. $\sqrt{3x} - 9 = 0$

67. $\sqrt{2(a - 7)} = 14$

68. $\sqrt{3(2x + 3)} = \sqrt{x + 15}$

69. $\sqrt{2(x + 5)} = x + 5$

70. $\sqrt{5t} = 1 + \sqrt{5(t - 1)}$

71. $\sqrt[3]{5x + 2} - \sqrt[3]{7x - 8} = 0$

72. $\sqrt[4]{9x - 2} - \sqrt[4]{8x} = 0$

73. $\sqrt{1 + 6x} = 2 - \sqrt{6x}$

74. $\sqrt{2 + 9b} - 1 = 3\sqrt{b}$

In Exercises 75 and 76, use a graphing utility to approximate the solution of the equation.

75. $4\sqrt[3]{x} = 7 - x$

76. $2\sqrt{x} - 1 = \sqrt{10 - x}$

In Exercises 77–82, write the complex number in standard form.

77. $\sqrt{-48}$

78. $\sqrt{-0.16}$

79. $10 - 3\sqrt{-27}$

80. $3 + 2\sqrt{-500}$

81. $\frac{3}{4} - 5\sqrt{-\frac{3}{25}}$

82. $-0.5 + 3\sqrt{-1.21}$

In Exercises 83–92, perform the operation and write the answer in standard form.

83. $(-4 + 5i) - (-12 + 8i)$

84. $5(3 - 8i) + (5 + 12i)$

85. $(-2)(15i)(-3i)$

86. $-10i(4 - 7i)$

87. $(4 - 3i)(4 + 3i)$

88. $(6 - 5i)^2$

89. $(12 - 5i)(2 + 7i)$

90. $\dfrac{4}{5i}$

91. $\dfrac{5i}{2 + 9i}$

92. $\dfrac{2 + i}{1 - 9i}$

93. *Geometry*   The four corners are cut from an $8\frac{1}{2}$-inch-by-11-inch sheet of paper (see figure). Find the perimeter of the remaining piece of paper.

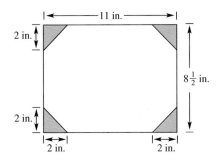

94. *Geometry*   Determine the length and width of a rectangle with a perimeter of 84 inches and a diagonal of 30 inches.

95. *Length of a Pendulum*   The time t in seconds for a pendulum of length L in feet to go through one complete cycle (its period) is given by

$$t = 2\pi \sqrt{\dfrac{L}{32}}.$$

How long is the pendulum of a grandfather clock with a period of 1.3 seconds?

96. *Height of a Bridge*   The time t in seconds for a free-falling object to fall d feet is given by

$$t = \sqrt{\dfrac{d}{16}}.$$

A child drops a pebble from a bridge and observes it strike the water after approximately 4 seconds. Estimate the height of the bridge.

CHAPTER TEST

Take this test as you would take a test in class. After you are done, check your work against the answers given in the back of the book.

In Exercises 1 and 2, evaluate the expressions without using a calculator.

1. (a) $2^{-2} + 2^{-3}$

   (b) $\dfrac{6.3 \times 10^{-3}}{2.1 \times 10^{2}}$

2. (a) $27^{-2/3}$

   (b) $\sqrt{2}\sqrt{18}$

3. Write 0.000032 in scientific notation.

4. Write 3.04×10^{7} in decimal notation.

In Exercises 5–7, simplify the expressions.

5. (a) $\dfrac{12t^{-2}}{20t^{-1}}$

   (b) $(x + y^{-2})^{-1}$

6. (a) $\left(\dfrac{x^{1/2}}{x^{1/3}}\right)^{2}$

   (b) $5^{1/4} \cdot 5^{7/4}$

7. (a) $\sqrt{\dfrac{32}{9}}$

   (b) $\sqrt[3]{24}$

8. In your own words, explain the meaning of "rationalize" and demonstrate by rationalizing the denominator: $\dfrac{3}{\sqrt{6}}$

9. Combine: $5\sqrt{3x} - 3\sqrt{75x}$

10. Multiply and simplify: $\sqrt{5}\left(\sqrt{15x} + 3\right)$ **11.** Expand: $\left(4 - \sqrt{2x}\right)^{2}$

12. Factor: $7\sqrt{27} + 14y\sqrt{12} = 7\sqrt{3}(\ \ \ \ \ \ )$

In Exercises 13 and 14, solve the equation.

13. $\sqrt{x^{2} - 1} = x - 2$ **14.** $\sqrt{x} - x + 6 = 0$

In Exercises 15–18, perform the operation and simplify.

15. $(2 + 3i) - \sqrt{-25}$ **16.** $(2 - 3i)^{2}$

17. $\sqrt{-16}\left(1 + \sqrt{-4}\right)$ **18.** $(3 - 2i)(1 + 5i)$

19. Divide $5 - 2i$ by i. Write the result in standard form.

20. The velocity v (in feet per second) of an object is given by $v = \sqrt{2gh}$, where $g = 32$ feet per second per second and h is the distance (in feet) the object has fallen. Find the height from which a rock has been dropped if it strikes the ground with a velocity of 80 feet per second.

Quadratic Equations and Inequalities

6

- Factoring and Extracting Roots
- Completing the Square
- The Quadratic Formula and the Discriminant
- Applications of Quadratic Equations
- Quadratic and Rational Inequalities

The height s (in feet) of a falling object can be modeled by

$$s = \tfrac{1}{2}gt^2 + v_0 t + s_0 \qquad \text{Position function}$$

where g is the acceleration due to gravity (in feet per second per second), t is the time (in seconds), v_0 is the initial velocity (in feet per second), and s_0 is the initial height (in feet). On earth, $g \approx -32$ feet per second per second. On other planets and moons, the value of g varies significantly. For instance, on earth's moon, $g \approx -5.3$.

The position function for an object that is propelled at a velocity of 32 feet per second straight up from a height of 48 feet is

$$s = -16t^2 + 32t + 48.$$

The height of this object is shown in the table and graph at the right.

t	0.0	0.5	1.0	1.5	2.0	2.5	3.0
s	48	60	64	60	48	28	0

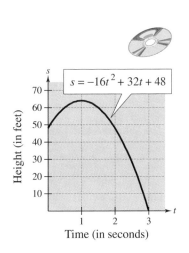

$s = -16t^2 + 32t + 48$

Height (in feet)

Time (in seconds)

The chapter project related to this information is on page 428.

381

<table>
<tr><td>

6.1

</td><td>

Factoring and Extracting Roots

</td></tr>
</table>

Solving Equations by Factoring ▪ Extracting Square Roots ▪
Equations with Complex Solutions ▪ Equations of Quadratic Form

Fermat's Last Theorem states that the equation $x^n + y^n = z^n$ has no solution when x, y, and z are nonzero integers and $n > 2$. In 1637, Pierre de Fermat wrote in the margin of a book that he had discovered a proof of this theorem; however, his proof has never been found. On June 23, 1993, 356 years later, a 200-page proof was presented at a gathering of mathematicians at Cambridge University in England by an American mathematician, Andrew Wiles.

NOTE   When the two solutions of a quadratic equation are identical, they are called a **double** or **repeated solution.** This occurs in Example 1(c).

Solving Equations by Factoring

In this chapter, you will study methods for solving quadratic equations and equations of quadratic form. To begin, let's review the method of factoring that you studied in Section 3.5.

Remember that the first step in solving a quadratic equation by factoring is to write the equation in standard form. Next, factor the left side. Finally, set each factor equal to zero and solve for x.

EXAMPLE 1   Solving Quadratic Equations by Factoring

a.

$x^2 + 5x = 24$	Original equation
$x^2 + 5x - 24 = 0$	Standard form
$(x + 8)(x - 3) = 0$	Factor.
$x + 8 = 0 \implies x = -8$	Set 1st factor equal to 0.
$x - 3 = 0 \implies x = 3$	Set 2nd factor equal to 0.

The solutions are -8 and 3. Check these in the original equation.

b.

$3x^2 = 4 - 11x$	Original equation
$3x^2 + 11x - 4 = 0$	Standard form
$(3x - 1)(x + 4) = 0$	Factor.
$3x - 1 = 0 \implies x = \frac{1}{3}$	Set 1st factor equal to 0.
$x + 4 = 0 \implies x = -4$	Set 2nd factor equal to 0.

The solutions are $\frac{1}{3}$ and -4. Check these in the original equation.

c.

$9x^2 + 12 = 3 + 12x + 5x^2$	Original equation
$4x^2 - 12x + 9 = 0$	Standard form
$(2x - 3)(2x - 3) = 0$	Factor.
$2x - 3 = 0$	Set factor equal to 0.
$x = \dfrac{3}{2}$	Repeated solution

The only solution is $\frac{3}{2}$. Check this in the original equation.

Extracting Square Roots

Consider the following equation, where $d > 0$ and u is an algebraic expression.

$$u^2 = d$$ Original equation

$$u^2 - d = 0$$ Standard form

$$\left(u + \sqrt{d}\right)\left(u - \sqrt{d}\right) = 0$$ Factor.

$$u + \sqrt{d} = 0 \quad \Longrightarrow \quad u = -\sqrt{d}$$ Set 1st factor equal to 0.

$$u - \sqrt{d} = 0 \quad \Longrightarrow \quad u = \sqrt{d}$$ Set 2nd factor equal to 0.

Because the solutions differ only in sign, they can be written together using a "plus or minus sign"

$$u = \pm\sqrt{d}.$$

This form of the solution is read as "u is equal to plus or minus the square root of d." Solving an equation of the form $u^2 = d$ *without* going through the steps of factoring is called **extracting square roots.**

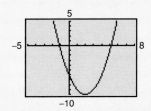

Technology

To graphically check solutions of an equation that is written in standard form, graph the left side of the equation and look at its x-intercepts. For instance, in Example 2(b), you can write the equation as

$$(x - 2)^2 - 10 = 0$$

and then sketch the graph of

$$y = (x - 2)^2 - 10$$

as shown below. From the graph, you can see that the x-intercepts are approximately 5.16 and -1.16.

Extracting Square Roots

The equation $u^2 = d$, where $d > 0$, has exactly two solutions:

$$u = \sqrt{d} \qquad \text{and} \qquad u = -\sqrt{d}.$$

These solutions can also be written as $u = \pm\sqrt{d}$.

EXAMPLE 2   *Extracting Square Roots*

a. $3x^2 = 15$ Original equation

$$x^2 = 5$$ Divide both sides by 3.

$$x = \pm\sqrt{5}$$ Extract square roots.

The solutions are $\sqrt{5}$ and $-\sqrt{5}$. Check these in the original equation.

b. $(x - 2)^2 = 10$ Original equation

$$x - 2 = \pm\sqrt{10}$$ Extract square roots.

$$x = 2 \pm \sqrt{10}$$ Add 2 to both sides.

The solutions are $2 + \sqrt{10} \approx 5.16$ and $2 - \sqrt{10} \approx -1.16$. Check these in the original equation.

Equations with Complex Solutions

Prior to Section 5.6, the only solutions you have been finding have been real numbers. But now that you have studied complex numbers, it makes sense to look for other types of solutions. For instance, although the quadratic equation $x^2 + 1 = 0$ has no solutions that are real numbers, it does have two solutions that are complex numbers: i and $-i$. To check this, substitute i and $-i$ for x.

$$(i)^2 + 1 = -1 + 1 = 0 \qquad \text{Solution checks} \checkmark$$
$$(-i)^2 + 1 = -1 + 1 = 0 \qquad \text{Solution checks} \checkmark$$

One way to find complex solutions of a quadratic equation is to extend the *extraction of square roots* technique to cover the case where d is a negative number.

Extracting Complex Square Roots

The equation $u^2 = d$, where $d < 0$, has exactly two solutions:

$$u = \sqrt{d}\, i \qquad \text{and} \qquad u = -\sqrt{d}\, i.$$

These solutions can also be written as $u = \pm\sqrt{d}\, i$.

Technology

When graphically checking solutions, it is important to realize that only the real solutions appear as x-intercepts—the complex solutions cannot be estimated from the graph. For instance, in Example 3(a), the graph of

$$y = x^2 + 8$$

has no x-intercepts. This agrees with the fact that both of its solutions are complex numbers.

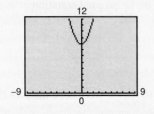

EXAMPLE 3   *Extracting Complex Square Roots*

a. $x^2 + 8 = 0$ Original equation

$\qquad x^2 = -8$ Subtract 8 from both sides.

$\qquad x = \pm\sqrt{8}\, i$ Extract complex square roots.

$\qquad x = \pm 2\sqrt{2}\, i$ Simplify.

The solutions are $2\sqrt{2}\, i$ and $-2\sqrt{2}\, i$. Check these in the original equation.

b. $2(3x - 5)^2 + 32 = 0$ Original equation

$\qquad 2(3x - 5)^2 = -32$ Subtract 32 from both sides.

$\qquad (3x - 5)^2 = -16$ Divide both sides by 2.

$\qquad 3x - 5 = \pm 4i$ Extract complex square roots.

$\qquad 3x = 5 \pm 4i$ Add 5 to both sides.

$\qquad x = \dfrac{5 \pm 4i}{3}$ Divide both sides by 3.

The solutions are $(5 + 4i)/3$ and $(5 - 4i)/3$. Check these in the original equation.

Equations of Quadratic Form

Both the factoring and extraction of square roots methods can be applied to non-quadratic equations that are of **quadratic form.** An equation is said to be of quadratic form if it has the form

$$au^2 + bu + c = 0$$

where u is an algebraic expression. Here are two examples.

Equation	*Written in Quadratic Form*
$x^4 + 5x^2 + 4 = 0$	$\left(x^2\right)^2 + 5\left(x^2\right) + 4 = 0$
$x - 5\sqrt{x} + 6 = 0$	$\left(\sqrt{x}\right)^2 - 5\left(\sqrt{x}\right) + 6 = 0$

To solve an equation of quadratic form, it helps to make a substitution and rewrite the equation in terms of u, as demonstrated in Examples 4 and 5.

EXAMPLE 4   Solving an Equation of Quadratic Form

Solve $x^4 - 13x^2 + 36 = 0$.

Solution

Begin by using a graphing utility to graph $y = x^4 - 13x^2 + 36$, as shown in Figure 6.1. The graph indicates that there are four real solutions near $x = \pm 2$ and $x = \pm 3$.

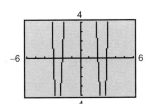

FIGURE 6.1

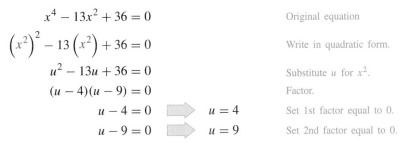

$x^4 - 13x^2 + 36 = 0$	Original equation
$\left(x^2\right)^2 - 13\left(x^2\right) + 36 = 0$	Write in quadratic form.
$u^2 - 13u + 36 = 0$	Substitute u for x^2.
$(u - 4)(u - 9) = 0$	Factor.
$u - 4 = 0 \quad\Longrightarrow\quad u = 4$	Set 1st factor equal to 0.
$u - 9 = 0 \quad\Longrightarrow\quad u = 9$	Set 2nd factor equal to 0.

At this point you have found the "u-solutions." To find the "x-solutions," replace u by x^2, as follows.

$$u = 4 \quad\Longrightarrow\quad x^2 = 4 \quad\Longrightarrow\quad x = \pm 2$$
$$u = 9 \quad\Longrightarrow\quad x^2 = 9 \quad\Longrightarrow\quad x = \pm 3$$

The solutions are 2, -2, 3, and -3. Check these in the original equation.

NOTE   Be sure you see in Example 4 that the u-solutions of 4 and 9 represent only a temporary step. They are not solutions of the original equation.

EXAMPLE 5 Solving an Equation of Quadratic Form

Solve $x - 5\sqrt{x} + 6 = 0$ for x.

Solution

This equation is of quadratic form with $u = \sqrt{x}$.

$x - 5\sqrt{x} + 6 = 0$	Original equation
$\left(\sqrt{x}\right)^2 - 5\left(\sqrt{x}\right) + 6 = 0$	Write in quadratic form.
$u^2 - 5u + 6 = 0$	Substitute u for $\sqrt{x}$.
$(u - 2)(u - 3) = 0$	Factor.
$u - 2 = 0 \implies u = 2$	Set 1st factor equal to 0.
$u - 3 = 0 \implies u = 3$	Set 2nd factor equal to 0.

Now, using the u-solutions of 2 and 3, you obtain the following x-solutions.

$u = 2 \implies \sqrt{x} = 2 \implies x = 4$

$u = 3 \implies \sqrt{x} = 3 \implies x = 9$

The solutions are 4 and 9. Check these in the original equation.

NOTE In Example 5, remember that checking the solutions of a radical equation is especially important because the trial solutions often turn out to be extraneous.

Group Activities Exploring with Technology

Analyzing Solutions of Quadratic Equations Use a graphing utility to graph each of the following equations. How many times does the graph of each equation cross the x-axis?

a. $y = 2x^2 + x - 15$
b. $y = -4(x - 2)^2 - 3$
c. $y = 9x^2 + 24x + 16$
d. $y = (3x - 1)^2 + 3$

Now set each equation equal to zero and use the techniques of this section to solve the resulting equations. How many of each type of solution (real or complex) does each equation have? Summarize the relationship between the number of x-intercepts in the graph of a quadratic equation and the number and type of roots found algebraically.

6.1 Exercises

Discussing the Concepts

1. For a quadratic equation $ax^2 + bx + c = 0$ where a, b, and c are real numbers with $a \neq 0$, explain why b and c can equal 0 but a cannot.

2. Explain the Zero-Factor Property and how it can be used to solve a quadratic equation.

3. Is it possible for a quadratic equation to have only one solution? If so, give an example.

4. *True or False?*   The only solution of the equation $x^2 = 25$ is $x = 5$. Explain.

5. Describe the steps in solving a quadratic equation by extracting square roots.

6. Describe the procedure for solving an equation of quadratic form. Give an example.

Problem Solving

In Exercises 7–16, solve the equation by factoring.

7. $4x^2 - 12x = 0$

8. $25y^2 - 75y = 0$

9. $x^2 - 12x + 36 = 0$

10. $9x^2 + 24x + 16 = 0$

11. $(y - 4)(y - 3) = 6$

12. $(6 + u)(1 - u) = 10$

13. $3x(x - 6) - 5(x - 6) = 0$

14. $3(4 - x) - 2x(4 - x) = 0$

15. $6x^2 = 54$

16. $\dfrac{x^2}{6} = 24$

In Exercises 17–22, solve the quadratic equation by extracting square roots.

17. $x^2 = 64$

18. $9z^2 = 121$

19. $(x + 4)^2 = 169$

20. $(y - 20)^2 = 625$

21. $(2x + 1)^2 = 50$

22. $(3x - 5)^2 = 48$

In Exercises 23–28, solve the equation by extracting complex square roots.

23. $x^2 + 4 = 0$

24. $x^2 = -9$

25. $9(x + 6)^2 = -121$

26. $4(x - 4)^2 = -169$

27. $(x - 1)^2 = -27$

28. $\left(y - \frac{5}{6}\right)^2 = -\frac{4}{5}$

In Exercises 29–34, find all real and complex solutions.

29. $2x^2 - 5x = 0$

30. $3x^2 + 8x - 16 = 0$

31. $x^2 - 100 = 0$

32. $(y + 12)^2 + 400 = 0$

33. $(x - 5)^2 + 100 = 0$

34. $(y + 12)^2 - 400 = 0$

In Exercises 35–42, solve the equation. List all real and complex solutions.

35. $x^6 + 7x^3 - 8 = 0$

36. $x^4 - 13x^2 + 36 = 0$

37. $2x - 9\sqrt{x} + 10 = 0$

38. $3x^{2/3} + 8x^{1/3} - 3 = 0$

39. $(x^2 - 2)^2 - 36 = 0$

40. $(u^2 + 4)^2 - 64 = 0$

41. $(x^2 - 5)^2 - 100 = 0$

42. $(y^2 + 12)^2 - 400 = 0$

In Exercises 43–46, use a graphing utility to graph the function. Use the graph to approximate any x-intercepts of the graph. Set $y = 0$ and solve the resulting equation. Compare the result with the x-intercepts of the graph.

43. $y = x^2 - 9$

44. $y = x^2 - 2x - 15$

45. $y = 4 - (x - 3)^2$

46. $y = 4(x + 1)^2 - 9$

In Exercises 47–50, use a graphing utility to graph the function and observe that the graph has no x-intercepts. Set $y = 0$ and solve the resulting equation. What type of roots does the equation have?

47. $y = (x - 1)^2 + 1$

48. $y = (x + 2)^2 + 3$

49. $y = -(x + 3)^2 - 2$

50. $y = -(x - 4)^2 - 4$

Graphical Reasoning In Exercises 51 and 52, use the model

$$y = (44.17 + 2.82t)^2, \quad 0 \le t \le 22$$

which approximates the amount of fire loss to private property in the United States from 1970 to 1992. In this model, y represents the value of private property lost to fire (in millions of dollars) and t represents the year, with $t = 0$ corresponding to 1970 (see figure).   (Source: Insurance Information Institute)

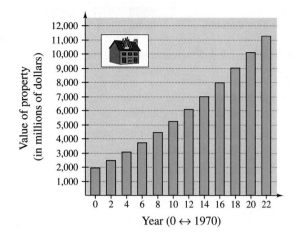

Year (0 ↔ 1970)

51. Graphically estimate the year in which loss of private property due to fire was approximately $4500 million. Verify algebraically.

52. Graphically estimate the year in which loss of private property due to fire was approximately $9500 million. Verify algebraically.

Reviewing the Major Concepts

In Exercises 53–56, factor the expression completely.

53. $16x^2 - 121$

54. $9t^2 - 24t + 16$

55. $x(x - 10) - 4(x - 10)$

56. $4x^3 - 12x^2 + 16x$

57. *Speed*   A boat's still-water speed is 18 miles per hour. It travels 35 miles upstream, and then returns to its starting point, in a total of 4 hours. Find the speed of the current.

58. *Partnership Costs*   A group of people agree to share equally in the cost of a $250,000 endowment to a college. If they could find two more people to join the group, each person's share of the cost would decrease by $6250. How many people are presently in the group?

Additional Problem Solving

In Exercises 59–68, solve the equation by factoring.

59. $u(u - 9) - 12(u - 9) = 0$

60. $16x(x - 8) - 12(x - 8) = 0$

61. $4x^2 - 25 = 0$

62. $16y^2 - 121 = 0$

63. $x^2 + 10x + 600 = 0$

64. $8x^2 - 10x + 3 = 0$

65. $2x(3x + 2) = 5 - 6x^2$

66. $(2z + 1)(2z - 1) = -4z^2 - 5z + 2$

67. $\frac{1}{2}y^2 = 32$

68. $5t^2 = 125$

In Exercises 69–78, solve the quadratic equation by extracting square roots.

69. $25x^2 = 16$

70. $z^2 = 169$

71. $4u^2 - 225 = 0$

72. $16x^2 - 1 = 0$

73. $(x - 3)^2 = 0.25$

74. $(x + 2)^2 = 0.81$

75. $(x - 2)^2 = 7$

76. $(x + 8)^2 = 28$

77. $(4x - 3)^2 - 98 = 0$

78. $(5x + 11)^2 - 300 = 0$

In Exercises 79–88, solve the equation by extracting complex square roots.

79. $u^2 + 17 = 0$

80. $4v^2 + 9 = 0$

81. $(t - 3)^2 = -25$

82. $(x + 5)^2 + 81 = 0$

83. $(2y - 3)^2 + 25 = 0$

84. $(3z + 4)^2 + 144 = 0$

85. $\left(c - \frac{2}{3}\right)^2 + \frac{1}{9} = 0$

86. $\left(u + \frac{5}{8}\right)^2 + \frac{49}{16} = 0$

87. $\left(x + \frac{7}{3}\right)^2 = -\frac{38}{9}$

88. $(2x + 3)^2 = -54$

In Exercises 89–98, find all the real and complex solutions.

89. $x^2 - 900 = 0$

90. $y^2 - 225 = 0$

91. $x^2 + 900 = 0$

92. $y^2 + 225 = 0$

93. $\frac{2}{3}x^2 = 6$

94. $\frac{1}{3}x^2 = 4$

95. $(x - 5)^2 - 100 = 0$

96. $(y + 12)^2 - 400 = 0$

97. $(x - 5)^2 + 100 = 0$

98. $(y + 12)^2 + 400 = 0$

In Exercises 99–114, solve the equation. List all real and complex solutions.

99. $x^4 - 5x^2 + 4 = 0$

100. $4x^4 - 101x^2 + 25 = 0$

101. $x^4 - 5x^2 + 6 = 0$

102. $x^4 - 11x^2 + 30 = 0$

103. $x^4 - 3x^2 - 4 = 0$

104. $x^4 - x^2 - 6 = 0$

105. $(x^2 - 4)^2 + 2(x^2 - 4) - 3 = 0$

106. $(x^2 - 1)^2 + (x^2 - 1) - 6 = 0$

107. $\left(\sqrt{x} - 1\right)^2 + 3\left(\sqrt{x} - 1\right) - 4 = 0$

108. $\left(2 + \sqrt{x}\right)^2 + 4\left(2 + \sqrt{x}\right) - 21 = 0$

109. $\frac{1}{x^2} - \frac{3}{x} + 2 = 0$

110. $\frac{2}{x^2} + \frac{3}{x} - 2 = 0$

111. $3\left(\frac{x}{x + 1}\right)^2 + 7\left(\frac{x}{x + 1}\right) - 6 = 0$

112. $4\left(\frac{x + 1}{x - 1}\right)^2 + 19\left(\frac{x + 1}{x - 1}\right) - 5 = 0$

113. $x^{2/3} - x^{1/3} - 6 = 0$

114. $2x^{2/3} - 7x^{1/3} + 5 = 0$

In Exercises 115 and 116, use a graphing utility to graph the function. Use the graph to approximate any x-intercepts of the graph. Set $y = 0$ and solve the resulting equation. Compare the result with the x-intercepts of the graph.

115. $y = 5x - x^2$

116. $y = 9 - 4(x - 3)^2$

In Exercises 117 and 118, use a graphing utility to graph the function and observe that the graph has no x-intercepts. Set $y = 0$ and solve the resulting equation. What type of roots does the equation have?

117. $y = (x - 2)^2 + 3$

118. $y = (x + 3)^2 + 5$

Think About It In Exercises 119–122, find a quadratic equation having the given solutions.

119. $5, -2$

120. $-2, \frac{1}{3}$

121. $1 + \sqrt{2}, 1 - \sqrt{2}$

122. $1 + \sqrt{2}\,i, 1 - \sqrt{2}\,i$

Free-Falling Object In Exercises 123 and 124, find the time required for an object to reach the ground when it is dropped from a height of s_0 feet. The height h (in feet) is given by

$$h = -16t^2 + s_0$$

where t measures time in seconds from the time the object is released.

123. $s_0 = 256$

124. $s_0 = 48$

125. *Free-Falling Object* The height h (in feet) of an object thrown upward from a tower 144 feet high (see figure) is given by

$$h = 144 + 128t - 16t^2$$

where t measures the time in seconds from the time the object is released. How long does it take for the object to reach the ground?

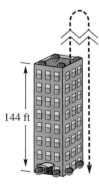

144 ft

126. *Revenue*   The revenue R (in dollars) when x units of a product are sold is given by

$$R = x\left(120 - \frac{1}{2}x\right).$$

Determine the number of units that must be sold to produce a revenue of $7000.

Reading a Graph   In Exercises 127 and 128, use the following model, which gives the federal funding for health research in the United States from 1985 to 1992.

$$y = (52.9 + 3.9t)^2, \quad 5 \le t \le 12$$

In this model, y represents funding (in millions of dollars) and $t = 0$ represents 1980 (see figure). (Source: U.S. National Science Foundation)

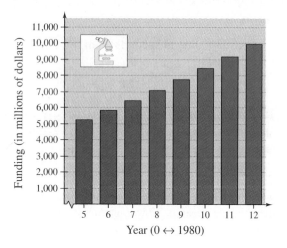

Year (0 ↔ 1980)

127. Use the graph to find the year in which federal funding for health was approximately $6500 million. Verify algebraically.

128. Use the graph to find the year in which federal funding for health was approximately $9200 million. Verify algebraically.

129. *Compound Interest*   A principal of $1500 is deposited in an account at an annual interest rate r compounded annually. If the amount after 2 years is $1685.40, the annual interest rate is the solution to the equation $1685.40 = 1500(1 + r)^2$. Find r.

Free-Falling Object   In Exercises 130 and 131, find the time required for an object to reach the ground when it is dropped from a height of s_0 feet. The height h (in feet) is given by

$$h = -16t^2 + s_0$$

where t measures time in seconds from the time the object is released.

130. $s_0 = 128$

131. $s_0 = 500$

CAREER INTERVIEW

Peggy Murray

Freelance Artist

HomeArt

Mission Hills, KS 66208

I am self-employed as a freelance artist and work out of my studio in my home on a wide variety of projects including greeting card and announcement designs; pen-and-ink house portraits; finished art, such as flyers and posters; and hand-painted Christmas tree ornaments. Recently I designed and produced a local historical society cookbook. All of my work depends heavily on careful measurement, and I frequently use proportions to change the size of type, pieces of art, or photos to fit a fixed amount of space.

Math is important not only in my creation process but also in my billing process. For most of the work I do, I charge a fixed price for a set size of a particular item, such as a house portrait or a painted ornament. However, when I take on unusual projects or do work for an ad agency, I use a different system of billing that considers my time, supplies, and any extensive driving that was involved. For these cases, I figure the amount I charge as (hourly rate × number of hours worked) + (cost of all supplies) + (rate per mile × number of miles driven).

| 6.2 | **Completing the Square** |

Constructing Perfect Square Trinomials ▪
Solving Equations by Completing the Square

Constructing Perfect Square Trinomials

STUDY TIP

In Section 3.4, a *perfect square trinomial* was defined as the square of a binomial. For instance, the perfect square trinomial $x^2 + 10x + 25 = (x + 5)^2$ is the square of the binomial $(x + 5)$.

Consider the quadratic equation

$$(x - 2)^2 = 10.$$ Completed square form

You know from Example 2(b) in the previous section that this equation has two solutions: $2 + \sqrt{10}$ and $2 - \sqrt{10}$. Suppose you had been given the equation in its standard form

$$x^2 - 4x - 6 = 0.$$ Standard form

How would you solve this equation if you were given only the standard form? You could try factoring, but after attempting to do so you would find that the left side of the equation is not factorable (using integer coefficients).

In this section, you will study a technique for rewriting an equation in a completed square form. This technique is called **completing the square.**

Completing the Square

To **complete the square** for the expression $x^2 + bx$, add $(b/2)^2$, which is the square of half the coefficient of x. Consequently,

$$x^2 + bx + \left(\frac{b}{2}\right)^2 = \left(x + \frac{b}{2}\right)^2.$$

EXAMPLE 1   *Creating a Perfect Square Trinomial*

What term should be added to $x^2 - 8x$ so that it becomes a perfect square trinomial?

Solution

For this expression, the coefficient of the x-term is -8. Divide this term by 2, and square the result to obtain $(-4)^2 = 16$. This is the term that should be added to the expression to make it a perfect square trinomial.

$$x^2 - 8x + 16 = x^2 - 8x + (-4)^2$$ Add 16 to the expression.
$$= (x - 4)^2$$ Completed square form

Solving Equations by Completing the Square

When completing the square to solve an equation, remember that it is essential to *preserve the equality*. Thus, when you add a constant term to one side of the equation, you must be sure to add the same constant to the other side of the equation.

EXAMPLE 2   Completing the Square: Leading Coefficient Is 1

NOTE   In Example 2, completing the square is used for the sake of illustration. This particular equation would be easier to solve by factoring. Try reworking the problem by factoring to see that you obtain the same two solutions. In Example 3, the equation cannot be solved by factoring (using integer coefficients).

Solve $x^2 + 12x = 0$.

Solution

$$x^2 + 12x = 0 \qquad\qquad\qquad \text{Original equation}$$

$$x^2 + 12x + \underbrace{(6)^2}_{(\text{half})^2} = 36 \qquad\qquad \text{Add } \left(\tfrac{12}{2}\right)^2 = 36 \text{ to both sides.}$$

$$(x + 6)^2 = 36 \qquad\qquad \text{Completed square form}$$

$$x + 6 = \pm\sqrt{36} \qquad\qquad \text{Extract square roots.}$$

$$x = -6 \pm 6 \qquad\qquad \text{Subtract 6 from both sides.}$$

$$x = 0 \ \text{ or } \ x = -12 \qquad\qquad \text{Solutions}$$

The solutions are 0 and -12. Check these in the original equation.

EXAMPLE 3   Completing the Square: Leading Coefficient Is 1

Solve $x^2 - 6x + 7 = 0$.

Solution

$$x^2 - 6x + 7 = 0 \qquad\qquad\qquad \text{Original equation}$$

$$x^2 - 6x = -7 \qquad\qquad\qquad \text{Subtract 7 from both sides.}$$

$$x^2 - 6x + \underbrace{(-3)^2}_{(\text{half})^2} = -7 + 9 \qquad\qquad \text{Add } (-3)^2 = 9 \text{ to both sides.}$$

$$(x - 3)^2 = 2 \qquad\qquad \text{Completed square form}$$

$$x - 3 = \pm\sqrt{2} \qquad\qquad \text{Extract square roots.}$$

$$x = 3 \pm \sqrt{2} \qquad\qquad \text{Solutions}$$

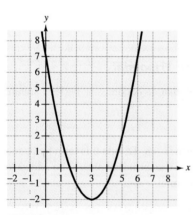

FIGURE 6.2

The solutions are $3 + \sqrt{2}$ and $3 - \sqrt{2}$. Check these in the original equation. Also try graphically checking the solutions, as shown in Figure 6.2.

EXAMPLE 4 A Leading Coefficient That Is Not 1

Solve $3x^2 + 5x = 2$.

Solution

STUDY TIP

If the leading coefficient of a quadratic expression is not 1, you must divide both sides of the equation by this coefficient *before* completing the square. This process is demonstrated in Examples 4 and 5.

$$3x^2 + 5x = 2 \qquad \text{Original equation}$$

$$x^2 + \frac{5}{3}x = \frac{2}{3} \qquad \text{Divide both sides by 3.}$$

$$x^2 + \frac{5}{3}x + \left(\frac{5}{6}\right)^2 = \frac{2}{3} + \frac{25}{36} \qquad \text{Add } \left(\frac{5}{6}\right)^2 = \frac{25}{36} \text{ to both sides.}$$

$$\left(x + \frac{5}{6}\right)^2 = \frac{49}{36} \qquad \text{Completed square form}$$

$$x + \frac{5}{6} = \pm\frac{7}{6} \qquad \text{Extract square roots.}$$

$$x = -\frac{5}{6} \pm \frac{7}{6} \qquad \text{Subtract } \frac{5}{6} \text{ from both sides.}$$

$$x = \frac{1}{3} \quad \text{or} \quad x = -2 \qquad \text{Solutions}$$

The solutions are $\frac{1}{3}$ and -2. Check these in the original equation.

EXAMPLE 5 A Leading Coefficient That Is Not 1

Solve $2x^2 - x - 2 = 0$.

Solution

$$2x^2 - x - 2 = 0 \qquad \text{Original equation}$$

$$2x^2 - x = 2 \qquad \text{Add 2 to both sides.}$$

$$x^2 - \frac{1}{2}x = 1 \qquad \text{Divide both sides by 2.}$$

$$x^2 - \frac{1}{2}x + \left(-\frac{1}{4}\right)^2 = 1 + \frac{1}{16} \qquad \text{Add } \left(-\frac{1}{4}\right)^2 = \frac{1}{16} \text{ to both sides.}$$

$$\left(x - \frac{1}{4}\right)^2 = \frac{17}{16} \qquad \text{Completed square form}$$

$$x - \frac{1}{4} = \pm\frac{\sqrt{17}}{4} \qquad \text{Extract square roots.}$$

$$x = \frac{1}{4} \pm \frac{\sqrt{17}}{4} \qquad \text{Add } \frac{1}{4} \text{ to both sides.}$$

The solutions are $\frac{1}{4}\left(1 \pm \sqrt{17}\right)$. Check these in the original equation.

EXAMPLE 6 A Quadratic Equation with Complex Solutions

Solve $x^2 - 4x + 8 = 0$.

Solution

$x^2 - 4x + 8 = 0$	Original equation
$x^2 - 4x = -8$	Subtract 8 from both sides.
$x^2 - 4x + (-2)^2 = -8 + 4$	Add $(-2)^2 = 4$ to both sides.
$(x - 2)^2 = -4$	Completed square form
$x - 2 = \pm 2i$	Extract complex square roots.
$x = 2 \pm 2i$	Add 2 to both sides.

The solutions are $2 + 2i$ and $2 - 2i$. The first of these is checked as follows. Try checking the other.

Check

$x^2 - 4x + 8 = 0$	Original equation
$(2 + 2i)^2 - 4(2 + 2i) + 8 \stackrel{?}{=} 0$	Substitute $2 + 2i$ for x.
$4 + 8i - 4 - 8 - 8i + 8 \stackrel{?}{=} 0$	Simplify.
$0 = 0$	Solution checks. ✓

Group Activities   You Be the Instructor

Error Analysis   Suppose you teach an algebra class and one of your students hands in the following solution. Find and correct the error(s). Discuss how to explain the error(s) to your student.

1. Solve $x^2 + 6x - 13 = 0$ by completing the square.

$$x^2 + 6x = 13$$
$$x^2 + 6x + \left(\frac{6}{2}\right)^2 = 13$$
$$(x + 3)^2 = 13$$
$$x + 3 = \pm\sqrt{13}$$
$$x = -3 \pm \sqrt{13}$$

6.2 Exercises

Discussing the Concepts

1. What is a perfect square trinomial?

2. What term must be added to $x^2 + 5x$ to complete the square? Explain how you found the term.

3. Explain the use of extracting square roots when solving a quadratic equation by the method of completing the square.

4. Is it possible for a quadratic equation to have no real number solution? If so, give an example.

5. When the method of completing the square is used to solve a quadratic equation, what is the first step if the leading coefficient is not 1? Is the resulting equation equivalent to the given equation? Explain.

6. *True or False?*   If you solve a quadratic equation by completing the square and obtain solutions that are rational numbers, you could have solved the equation by factoring. Explain.

Problem Solving

In Exercises 7–10, find the term that must be added to the expression so that it becomes a perfect square trinomial.

7. $x^2 + 8x +$ ▨

8. $y^2 - 2y +$ ▨

9. $t^2 + 5t +$ ▨

10. $a^2 - \frac{1}{3}a +$ ▨

In Exercises 11–14, solve the quadratic equation (a) by completing the square and (b) by factoring.

11. $x^2 - 6x = 0$

12. $t^2 + 9t = 0$

13. $x^2 + 7x + 12 = 0$

14. $y^2 - 8y + 12 = 0$

In Exercises 15–26, solve the quadratic equation by completing the square. Give the solutions in exact form and in decimal form rounded to two decimal places. (The solutions may be complex numbers.)

15. $x^2 - 4x - 3 = 0$

16. $x^2 - 6x + 7 = 0$

17. $x^2 + 2x + 3 = 0$

18. $x^2 - 6x + 12 = 0$

19. $x^2 - \frac{2}{3}x - 3 = 0$

20. $x^2 + \frac{4}{5}x - 1 = 0$

21. $t^2 + 5t + 3 = 0$

22. $u^2 - 9u - 1 = 0$

23. $2x^2 + 8x + 3 = 0$

24. $3x^2 - 24x - 5 = 0$

25. $0.1x^2 + 0.5x + 0.2 = 0$

26. $0.02x^2 + 0.10x - 0.05 = 0$

In Exercises 27–30, find the real solutions.

27. $\dfrac{x}{2} - \dfrac{1}{x} = 1$

28. $\dfrac{x}{2} + \dfrac{5}{x} = 4$

29. $\sqrt{2x + 1} = x - 3$

30. $\sqrt{3x - 2} = x - 2$

In Exercises 31–34, use a graphing utility to approximate any x-intercepts of the graph. Set $y = 0$ and solve the resulting equation. Compare the result with the x-intercepts of the graph.

31. $y = x^2 + 4x - 1$

32. $y = x^2 + 6x - 4$

33. $y = x^2 - 2x - 5$

34. $y = 2x^2 - 6x - 5$

In Exercises 35 and 36, consider a windlass that is used to pull a boat to the dock (see figure).

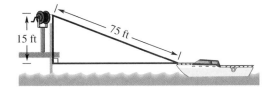

35. Find the distance from the boat to the dock when the rope is 75 feet long.

36. Find the distance from the boat to the dock when the rope is 50 feet long.

Reviewing the Major Concepts

In Exercises 37–40, simplify the expression.

37. $\sqrt[3]{16x^4y^3}$

38. $\sqrt{72x^2y^3}$

39. $\sqrt{3x^2} - 2\sqrt{12}$

40. $\dfrac{14}{\sqrt{7}}$

41. *Time* A jogger starts running at 6 miles per hour. Five minutes later, another jogger starts on the same trail, running at 8 miles per hour. When will the second jogger overtake the first?

42. Find three consecutive even integers whose sum is 102.

Additional Problem Solving

In Exercises 43–50, find the term that must be added to the expression so that it becomes a perfect square trinomial.

43. $y^2 - 20y +$ ▢

44. $x^2 + 12x +$ ▢

45. $x^2 - \frac{6}{5}x +$ ▢

46. $y^2 + \frac{4}{3}y +$ ▢

47. $y^2 - \frac{3}{5}y +$ ▢

48. $u^2 + 7u +$ ▢

49. $r^2 - 0.4r +$ ▢

50. $s^2 + 4.5s +$ ▢

In Exercises 51–60, solve the quadratic equation (a) by completing the square and (b) by factoring.

51. $x^2 - 25x = 0$

52. $x^2 + 32x = 0$

53. $t^2 - 8t + 7 = 0$

54. $x^2 + 12x + 27 = 0$

55. $x^2 + 2x - 24 = 0$

56. $z^2 + 3z - 10 = 0$

57. $x^2 - 3x - 18 = 0$

58. $t^2 - 5t - 36 = 0$

59. $2x^2 - 11x + 12 = 0$

60. $3x^2 - 5x - 2 = 0$

In Exercises 61–80, solve the quadratic equation by completing the square. Give the solutions in exact form and in decimal form rounded to two decimal places. (The solutions may be complex numbers.)

61. $x^2 + 4x - 3 = 0$

62. $x^2 + 6x + 7 = 0$

63. $u^2 - 4u + 1 = 0$

64. $a^2 - 10a - 15 = 0$

65. $x^2 - 10x - 2 = 0$

66. $x^2 + 8x - 4 = 0$

67. $y^2 + 20y + 10 = 0$

68. $y^2 + 6y - 24 = 0$

69. $v^2 + 3v - 2 = 0$

70. $z^2 - 7z + 9 = 0$

71. $-x^2 + x - 1 = 0$

72. $1 - x - x^2 = 0$

73. $3x^2 + 9x + 5 = 0$

74. $5x^2 - 15x + 7 = 0$

75. $4y^2 + 4y - 9 = 0$

76. $4z^2 - 3z + 2 = 0$

77. $0.1x^2 + 0.2x + 0.5 = 0$

78. $\frac{1}{2}t^2 + t + 2 = 0$

79. $x(x - 7) = 2$

80. $2x\left(x + \frac{4}{3}\right) = 5$

In Exercises 81–84, use a graphing utility to approximate any x-intercepts of the graph. Set $y = 0$ and solve the resulting equation. Compare the result with the x-intercepts of the graph.

81. $\frac{1}{3}x^2 + 2x - 6 = 0$

82. $\frac{1}{2}x^2 - 3x + 1 = 0$

83. $\dfrac{3}{x} - x - 1 = 0$

84. $\sqrt{x} - x + 2 = 0$

85. *Geometric Modeling*

(a) Find the area of the two adjoining rectangles and the large square in the figure.

(b) Find the area of the small square in the lower right-hand corner of the figure and add it to the area found in part (a).

(c) Find the dimensions and the area of the entire figure after adjoining the small square in the lower right-hand corner. Note that you have shown completing the square geometrically.

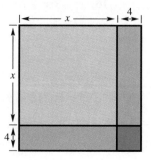

86. *Geometric Modeling*   Use the model in Exercise 85 to geometrically represent completing the square for $x^2 + 6x$.

87. *Geometry*   The area of the rectangle in the figure is 160 square feet. Find the rectangle's dimensions.

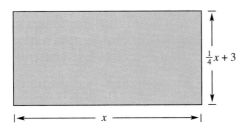

$\frac{1}{4}x + 3$

x

88. *Cutting Across the Lawn*   On the sidewalk, the distance from the dormitory to the cafeteria is 400 meters (see figure). By cutting across the lawn, the walking distance is shortened to 300 meters. How long is each part of the L-shaped sidewalk?

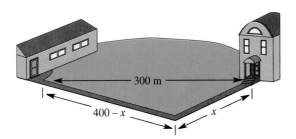

300 m

$400 - x$

x

89. *Revenue*   The revenue R from selling x units of a certain product is given by

$$R = x\left(50 - \frac{1}{2}x\right).$$

Find the number of units that must be sold to produce a revenue of $1218.

90. *Revenue*   The revenue R from selling x units of a certain product is given by

$$R = x\left(100 - \frac{1}{10}x\right).$$

Find the number of units that must be sold to produce a revenue of $12,000.

91. *Fencing in a Corral*   You have 200 feet of fencing to enclose two adjacent rectangular corrals (see figure). The total area of the enclosed region is 1400 square feet. What are the dimensions of each corral? (The corrals are the same size.)

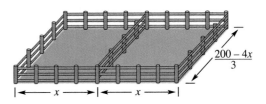

$\frac{200 - 4x}{3}$

x x

92. *Geometry*   An open box with a rectangular base of x inches by $x + 4$ inches has a height of 6 inches (see figure). Find the dimensions of the box if its volume is 840 cubic inches.

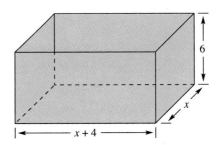

6

x

$x + 4$

93. *Geometry*   A closed box has a square base with edge x feet and a height of 3 feet (see figure). The material for constructing the base costs $1.50 per square foot and the material for the top and sides costs $1 per square foot. Find the dimensions of the box if its volume is 128 cubic feet.

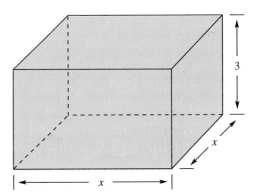

3

x

x

6.3	**The Quadratic Formula and the Discriminant**

The Quadratic Formula ▪ Solving Equations by the Quadratic Formula ▪ The Discriminant

The Quadratic Formula

A fourth technique for solving a quadratic equation involves the **Quadratic Formula.** This formula is obtained by completing the square for a general quadratic equation.

$$ax^2 + bx + c = 0 \qquad \text{Standard form, } a \neq 0$$

$$ax^2 + bx = -c \qquad \text{Subtract } c \text{ from both sides.}$$

$$x^2 + \frac{b}{a}x = -\frac{c}{a} \qquad \text{Divide both sides by } a.$$

$$x^2 + \frac{b}{a}x + \left(\frac{b}{2a}\right)^2 = -\frac{c}{a} + \left(\frac{b}{2a}\right)^2 \qquad \text{Add } \left(\frac{b}{2a}\right)^2 \text{ to both sides.}$$

$$\left(x + \frac{b}{2a}\right)^2 = \frac{b^2 - 4ac}{4a^2} \qquad \text{Simplify.}$$

$$x + \frac{b}{2a} = \pm\sqrt{\frac{b^2 - 4ac}{4a^2}} \qquad \text{Extract square roots.}$$

$$x = -\frac{b}{2a} \pm \frac{\sqrt{b^2 - 4ac}}{2|a|} \qquad \text{Subtract } \frac{b}{2a} \text{ from both sides.}$$

$$x = \frac{-b \pm \sqrt{b^2 - 4ac}}{2a} \qquad \text{Simplify.}$$

STUDY TIP

The Quadratic Formula is one of the most important formulas in algebra, and you should memorize it. We have found that it helps to try to memorize a verbal statement of the rule. For instance, you might try to remember the following verbal statement of the Quadratic Formula: "Minus b, plus or minus the square root of b squared minus $4ac$, all divided by $2a$."

The Quadratic Formula

The solutions of $ax^2 + bx + c = 0$, $a \neq 0$, are given by the **Quadratic Formula**

$$x = \frac{-b \pm \sqrt{b^2 - 4ac}}{2a}.$$

The expression inside the radical, $b^2 - 4ac$, is called the **discriminant.**

1. If $b^2 - 4ac > 0$, the equation has two real solutions.

2. If $b^2 - 4ac = 0$, the equation has one (repeated) real solution.

3. If $b^2 - 4ac < 0$, the equation has no real solutions.

Solving Equations by the Quadratic Formula

When using the Quadratic Formula, remember that *before* the formula can be applied, you must first write the quadratic equation in standard form.

EXAMPLE 1   *The Quadratic Formula: Two Distinct Solutions*

$x^2 + 6x = 16$	Original equation
$x^2 + 6x - 16 = 0$	Write in standard form.
$x = \dfrac{-b \pm \sqrt{b^2 - 4ac}}{2a}$	Quadratic Formula
$x = \dfrac{-6 \pm \sqrt{6^2 - 4(1)(-16)}}{2(1)}$	Substitute: $a = 1$, $b = 6$, $c = -16$.
$x = \dfrac{-6 \pm \sqrt{100}}{2}$	Simplify.
$x = \dfrac{-6 \pm 10}{2}$	Simplify.
$x = 2$ or $x = -8$	Solutions

The solutions are 2 and -8. Check these in the original equation.

NOTE   In Example 1, the solutions are rational numbers, which means that the equation could have been solved by factoring. Try solving the equation by factoring.

EXAMPLE 2   *The Quadratic Formula: Two Distinct Solutions*

$-x^2 - 4x + 8 = 0$	Leading coefficient is negative.
$x^2 + 4x - 8 = 0$	Multiply both sides by -1.
$x = \dfrac{-b \pm \sqrt{b^2 - 4ac}}{2a}$	Quadratic Formula
$x = \dfrac{-4 \pm \sqrt{4^2 - 4(1)(-8)}}{2(1)}$	Substitute: $a = 1$, $b = 4$, $c = -8$.
$x = \dfrac{-4 \pm \sqrt{48}}{2}$	Simplify.
$x = \dfrac{-4 \pm 4\sqrt{3}}{2}$	Simplify.
$x = \dfrac{\cancel{2}(-2 \pm 2\sqrt{3})}{\cancel{2}}$	Cancel common factor.
$x = -2 \pm 2\sqrt{3}$	Solutions

The solutions are $-2 + 2\sqrt{3}$ and $-2 - 2\sqrt{3}$. Check these in the original equation.

STUDY TIP

If the leading coefficient of a quadratic equation is negative, we suggest that you begin by multiplying both sides of the equation by -1, as shown in Example 2. This will produce a positive leading coefficient, which is less cumbersome to work with.

EXAMPLE 3 The Quadratic Formula: One Repeated Solution

$$18x^2 - 24x + 8 = 0 \qquad\qquad \text{Original equation}$$

$$9x^2 - 12x + 4 = 0 \qquad\qquad \text{Divide both sides by 2.}$$

$$x = \frac{-b \pm \sqrt{b^2 - 4ac}}{2a} \qquad\qquad \text{Quadratic Formula}$$

$$x = \frac{-(-12) \pm \sqrt{(-12)^2 - 4(9)(4)}}{2(9)}$$

$$x = \frac{12 \pm \sqrt{144 - 144}}{18} \qquad\qquad \text{Simplify.}$$

$$x = \frac{12 \pm \sqrt{0}}{18} \qquad\qquad \text{Simplify.}$$

$$x = \frac{2}{3} \qquad\qquad \text{Solution}$$

The only solution is $\frac{2}{3}$. Check this in the original equation.

Note in the next example how the Quadratic Formula can be used to solve a quadratic equation that has complex solutions.

EXAMPLE 4 The Quadratic Formula: Complex Solutions

$$2x^2 - 4x + 5 = 0 \qquad\qquad \text{Original equation}$$

$$x = \frac{-b \pm \sqrt{b^2 - 4ac}}{2a} \qquad\qquad \text{Quadratic Formula}$$

$$x = \frac{-(-4) \pm \sqrt{(-4)^2 - 4(2)(5)}}{2(2)}$$

$$x = \frac{4 \pm \sqrt{-24}}{4} \qquad\qquad \text{Simplify.}$$

$$x = \frac{4 \pm 2\sqrt{6}\,i}{4} \qquad\qquad \text{Write in } i\text{-form.}$$

$$x = \frac{2(2 \pm \sqrt{6}\,i)}{2 \cdot 2} \qquad\qquad \text{Cancel common factor.}$$

$$x = \frac{2 \pm \sqrt{6}\,i}{2} \qquad\qquad \text{Solutions}$$

The solutions are $\frac{1}{2}(2 \pm \sqrt{6}\,i)$. Check these in the original equation.

The Discriminant

The radicand in the Quadratic Formula, $b^2 - 4ac$, is called the **discriminant** because it allows you to "discriminate" among different types of solutions.

Using the Discriminant

Let a, b, and c be rational numbers such that $a \neq 0$. The **discriminant** of the quadratic equation $ax^2 + bx + c = 0$ is given by $b^2 - 4ac$, and can be used to classify the solutions of the equation as follows.

Discriminant	Solution Types
1. Perfect square	Two distinct rational solutions (Example 1)
2. Positive nonperfect square	Two distinct irrational solutions (Example 2)
3. Zero	One repeated rational solution (Example 3)
4. Negative number	Two distinct complex solutions (Example 4)

EXAMPLE 5   Using the Discriminant

Equation	Discriminant	Solution Types
a. $x^2 - x + 2 = 0$	$b^2 - 4ac = (-1)^2 - 4(1)(2)$ $= 1 - 8$ $= -7$	Two distinct complex solutions
b. $2x^2 - 3x - 2 = 0$	$b^2 - 4ac = (-3)^2 - 4(2)(-2)$ $= 9 + 16$ $= 25$	Two distinct rational solutions
c. $x^2 - 2x + 1 = 0$	$b^2 - 4ac = (-2)^2 - 4(1)(1)$ $= 4 - 4$ $= 0$	One repeated rational solution
d. $x^2 - 2x - 1 = 9$	$b^2 - 4ac = (-2)^2 - 4(1)(-10)$ $= 4 + 40$ $= 44$	Two distinct irrational solutions

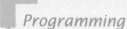

Programming

The *TI-82* program below can be used to solve equations with the Quadratic Formula. (See the Appendix for other calculator models.) Enter a, b, and c when prompted.

```
PROGRAM: QUDRATIC
: Prompt A
: Prompt B
: Prompt C
: B^2 - 4AC → D
: If D < 0
: Then
: Disp ''NO SOLUTION''
: Stop
: End
: (-B + √D)/(2A) → S
: Disp S
: (-B - √D)/(2A) → S
: Disp S
: Stop
```

You have now studied four ways to solve quadratic equations: (1) factoring, (2) extracting square roots, (3) completing the square, and (4) the Quadratic Formula.

EXAMPLE 6   Using a Calculator with the Quadratic Formula

Solve $1.2x^2 - 17.8x + 8.05 = 0$.

Solution

Using the Quadratic Formula, you can write

$$x = \frac{-(-17.8) \pm \sqrt{(-17.8)^2 - 4(1.2)(8.05)}}{2(1.2)}.$$

To evaluate these solutions, begin by calculating the square root.

17.8 $\boxed{+/-}$ $\boxed{x^2}$ $\boxed{-}$ 4 $\boxed{\times}$ 1.2 $\boxed{\times}$ 8.05 $\boxed{=}$ $\boxed{\sqrt{}}$ Scientific

$\boxed{\sqrt{}}$ $\boxed{(}$ $\boxed{(}$ $\boxed{(-)}$ 17.8 $\boxed{)}$ $\boxed{x^2}$ $\boxed{-}$ 4 $\boxed{\times}$ 1.2   Graphing
$\boxed{\times}$ 8.05 $\boxed{)}$ $\boxed{ENTER}$

The display for either of these keystroke sequences should be 16.67932852. Storing this result and using the recall key, we find the following two solutions.

$$x \approx \frac{17.8 + 16.67932852}{2.4} \approx 14.366$$ Add stored value.

$$x \approx \frac{17.8 - 16.67932852}{2.4} \approx 0.467$$ Subtract stored value.

Group Activities   You Be the Instructor

Problem Posing   Suppose you are writing a quiz that covers quadratic equations. Write four quadratic equations, including one with solutions $x = \frac{5}{3}$ and $x = -2$ and one with solutions $x = 4 \pm \sqrt{3}$, and instruct students to use any of the four solution methods: factoring, extracting square roots, completing the square, and using the Quadratic Formula. Trade quizzes with another member of your group and check one another's work.

6.3 Exercises

Discussing the Concepts

1. State the quadratic formula *in words*.
2. What is the discriminant of $ax^2 + bx + c = 0$? How is the discriminant related to the number of solutions of the equation?
3. Explain how completing the square can be used to develop the Quadratic Formula.
4. Summarize the four methods for solving a quadratic equation.

Problem Solving

In Exercises 5–8, write in standard form.

5. $2x^2 = 7 - 2x$

6. $7x^2 + 15x = 5$

7. $x(10 - x) = 5$

8. $x(3x + 8) = 15$

In Exercises 9–12, solve (a) by the Quadratic Formula and (b) by factoring.

9. $x^2 - 11x + 28 = 0$

10. $x^2 + 9x + 14 = 0$

11. $4x^2 + 12x + 9 = 0$

12. $10x^2 - 11x + 3 = 0$

In Exercises 13–16, use the discriminant to determine the type of solutions of the equation.

13. $2x^2 - 5x - 4 = 0$

14. $10x^2 + 5x + 1 = 0$

15. $3x^2 - x + 2 = 0$

16. $9x^2 - 24x + 16 = 0$

In Exercises 17–26, use the Quadratic Formula to find all real or complex solutions.

17. $x^2 - 2x - 4 = 0$

18. $x^2 - 2x - 6 = 0$

19. $t^2 + 4t + 1 = 0$

20. $y^2 + 6y + 4 = 0$

21. $x^2 + 3x + 3 = 0$

22. $2x^2 - x + 1 = 0$

23. $9z^2 + 6z - 4 = 0$

24. $8a^2 - 8a - 1 = 0$

25. $2.5x^2 + x - 0.9 = 0$

26. $0.09x^2 - 0.12x - 0.26 = 0$

In Exercises 27–30, solve by the most convenient method. Find all real or complex solutions.

27. $y^2 + 15y = 0$

28. $t^2 = 150$

29. $x^2 + 8x + 25 = 0$

30. $2x^2 + 8x + 4.5 = 0$

▦ *Graphical Reasoning* In Exercises 31–34, use a graphing utility to graph the function. Use the graph to approximate any x-intercepts of the graph. Set $y = 0$ and solve the resulting equation. Compare the result with the x-intercepts of the graph.

31. $y = 3x^2 - 6x + 1$

32. $y = x^2 + x + 1$

33. $y = -(4x^2 - 20x + 25)$

34. $y = x^2 - 4x + 3$

In Exercises 35–38, solve the equation.

35. $\dfrac{2x^2}{5} - \dfrac{x}{2} = 1$

36. $\dfrac{x}{3} + \dfrac{1}{x} = 4$

37. $\sqrt{x + 3} = x - 1$

38. $\sqrt{2x - 3} = x - 2$

Exploration In Exercises 39–42, determine the values of c such that the equation has (a) two real number solutions, (b) one real number solution, and (c) two imaginary number solutions.

39. $x^2 - 6x + c = 0$

40. $x^2 - 12x + c = 0$

41. $x^2 + 8x + c = 0$

42. $x^2 + 2x + c = 0$

43. *Geometry* A rectangle has a width of x inches, a length of $x + 6.3$ inches, and an area of 58.14 square inches. Find its dimensions.

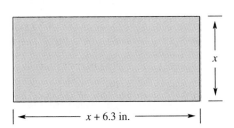

$\longleftarrow$ $x + 6.3$ in. $\longrightarrow$

44. *Analyzing Data* The numbers of cellular phone subscribers (in millions) in the United States for the years 1987 to 1992 are given in the following table. (Source: Cellular Telecommunications Industry Association)

Year	1987	1988	1989	1990	1991	1992
Subscribers	1.23	2.07	3.51	5.28	7.56	11.03

(a) Create a bar graph for the data.

(b) The data can be approximated by the model

$$s = 0.30t^2 + 2.22t + 5.29, \quad -3 \le t \le 2$$

where $t = 0$ corresponds to 1990. Use a graphing utility to graph this model.

(c) Use the model to determine the year in which the cellular phone companies had 2.1 million subscribers.

Reviewing the Major Concepts

In Exercises 45–48, perform the operations and simplify.

45. $\left(\sqrt{x} + 3\right)\left(\sqrt{x} - 3\right)$

46. $\sqrt{u}\left(\sqrt{20} - \sqrt{5}\right)$

47. $\left(2\sqrt{t} + 3\right)^2$

48. $\dfrac{50x}{\sqrt{2}}$

49. *Mixture* Determine the number of gallons of a 30% solution that must be mixed with a 60% solution to obtain 20 gallons of a 40% solution.

50. *Original Price* A suit sells for $375 during a 25% store-wide clearance sale. What was the original price of the suit?

Additional Problem Solving

In Exercises 51–56, solve the equation (a) by the Quadratic Formula and (b) by factoring.

51. $x^2 + 6x + 8 = 0$

52. $x^2 - 12x + 27 = 0$

53. $x^2 - \frac{4}{3}x + \frac{4}{9} = 0$

54. $x^2 + x + \frac{1}{4} = 0$

55. $6x^2 - x - 2 = 0$

56. $9x^2 - 30x + 25 = 0$

In Exercises 57–60, use the discriminant to determine the type of solutions of the equation.

57. $x^2 + 7x + 15 = 0$

58. $3x^2 - 2x - 5 = 0$

59. $4x^2 - 12x + 9 = 0$

60. $2x^2 + 10x + 6 = 0$

In Exercises 61–74, use the Quadratic Formula to find all real or complex solutions.

61. $x^2 + 6x - 3 = 0$

62. $x^2 + 8x - 4 = 0$

63. $x^2 - 10x + 23 = 0$

64. $u^2 - 12u + 29 = 0$

65. $2v^2 - 2v - 1 = 0$

66. $4x^2 + 6x + 1 = 0$

67. $2x^2 + 4x - 3 = 0$

68. $2x^2 + 3x + 3 = 0$

69. $x^2 - 0.4x - 0.16 = 0$

70. $x^2 + 0.6x - 0.41 = 0$

71. $4x^2 - 6x + 3 = 0$

72. $-5x^2 - 15x + 10 = 0$

73. $9x^2 = 1 + 9x$

74. $x - x^2 = 1 - 6x^2$

In Exercises 75–80, solve by the most convenient method. Find all real or complex solutions.

75. $z^2 - 169 = 0$

76. $4u^2 + 49 = 0$

77. $25(x - 3)^2 - 36 = 0$

78. $2y(y - 18) + 3(y - 18) = 0$

79. $x^2 - 24x + 128 = 0$

80. $1.2x^2 - 0.8x - 5.5 = 0$

In Exercises 81–84, use a calculator to solve the equation. Round to three decimal places.

81. $5x^2 - 18x + 6 = 0$

82. $15x^2 + 3x - 105 = 0$

83. $-0.04x^2 + 4x - 0.8 = 0$

84. $3.7x^2 - 10.2x + 3.2 = 0$

85. *Free-Falling Object*   A ball is thrown upward at a velocity of 40 feet per second from a bridge that is 50 feet above the level of the water (see figure). The height h (in feet) of the ball above the water at time t seconds after it is thrown is given by

$$h = -16t^2 + 40t + 50.$$

(a) Find the time when the ball is again at the 50-foot level above the water.

(b) Find the time when the ball strikes the water.

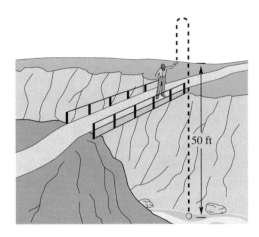

50 ft

86. *Geometry*   Two circular regions are tangent to each other (see figure). The distance between centers is 10 feet.

(a) Find the radius of each circle if their combined area is 52π square feet.

(b) Suppose the distance between the centers remained the same but radii were made equal. Would the combined area of the circles increase or decrease?

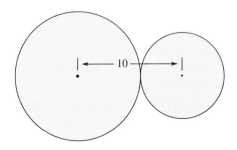

10

Aerospace Employment   In Exercises 87–89, use the following model, which approximates the number of people employed in the aerospace industry in the United States from 1987 to 1992.

$$y = 803.49 - 33.23t - 10.77t^2, \quad -3 \le t \le 2$$

In this model, y represents the number employed in the aerospace industry (in thousands) and t represents the year, with $t = 0$ corresponding to 1990. (Source: U.S. Department of Commerce)

87. Use a graphing utility to graph the model.

88. Use the TRACE feature of a graphing utility to find the year in which there were approximately 750,000 employed in the aerospace industry in the United States.

89. Use the model to estimate the number employed in the aerospace industry in 1993.

In 1991, the aerospace industry's revenue in the United States was about 135 billion dollars.

MID-CHAPTER QUIZ

Take this quiz as you would take a quiz in class. After you are done, check your work against the answers given in the back of the book.

In Exercises 1–8, solve the quadratic equation by the specified method.

1. Factoring:

$2x^2 - 72 = 0$

2. Factoring:

$2x^2 + 3x - 20 = 0$

3. Extracting square roots:

$t^2 = 12$

4. Extracting square roots:

$(u - 3)^2 - 16 = 0$

5. Completing the square:

$s^2 + 10s + 1 = 0$

6. Completing the square:

$2y^2 + 6y - 5 = 0$

7. Quadratic Formula:

$x^2 + 4x - 6 = 0$

8. Quadratic Formula:

$6v^2 - 3v - 4 = 0$

In Exercises 9–16, solve the equation by the most convenient method. (Find all the real *and* complex solutions.)

9. $x^2 + 5x + 7 = 0$

10. $36 - (t - 4)^2 = 0$

11. $x(x - 10) + 3(x - 10) = 0$

12. $x(x - 3) = 10$

13. $4b^2 - 12b + 9 = 0$

14. $3m^2 + 10m + 5 = 0$

15. $\dfrac{4}{u} - u = 3$

16. $\sqrt{2x + 5} = x + 1$

In Exercises 17 and 18, use a graphing utility to graph the function. Use the graph to approximate any *x*-intercepts of the graph. Set $y = 0$ and solve the resulting equation. Write a paragraph comparing the results of your algebraic and graphical solutions.

17. $y = \frac{1}{2}x^2 - 3x - 1$

18. $y = x^2 + 0.45x - 4$

19. The revenue R from selling x units of a certain product is given by

$R = x(20 - 0.2x).$

Find the number of units that must be sold to produce a revenue of $500.

20. The perimeter of a rectangle with sides x and $100 - x$ is 200 meters. Its area A is given by $A = x(100 - x)$. Determine the dimensions of the rectangle if its area is 2275 square meters.

6.4 Applications of Quadratic Equations

Applications of Quadratic Equations

Applications of Quadratic Equations

EXAMPLE 1   An Investment Problem

A car dealer bought a fleet of cars from a car rental agency for a total of $90,000. By the time the dealer had sold all but six of the cars, at an average profit of $2500 each, the original investment of $90,000 had been regained. How many cars did the dealer sell, and what was the average price per car?

Solution

Although this problem is stated in terms of average price and average profit per car, we can use a model that assumes that each car sold for the same price.

Verbal Model:	$\boxed{\text{Selling price per car}} = \boxed{\text{Cost per car}} + \boxed{\text{Profit per car}}$	

Labels: Number of cars sold $= x$ (cars)

Number of cars bought $= x + 6$ (cars)

Selling price per car $= \dfrac{90{,}000}{x}$ (dollars per car)

Cost per car $= \dfrac{90{,}000}{x + 6}$ (dollars per car)

Profit per car $= 2500$ (dollars per car)

Equation:

$$\frac{90{,}000}{x} = \frac{90{,}000}{x + 6} + 2500$$

$$90{,}000(x + 6) = 90{,}000x + 2500x(x + 6), \quad x \neq 0, \; x \neq -6$$

$$90{,}000x + 540{,}000 = 90{,}000x + 2500x^2 + 15{,}000x$$

$$0 = 2500x^2 + 15{,}000x - 540{,}000$$

$$0 = x^2 + 6x - 216$$

$$0 = (x - 12)(x + 18)$$

$$x - 12 = 0 \implies x = 12$$

$$x + 18 = 0 \implies x = -18$$

Choosing the positive value, it follows that the dealer sold 12 cars at an average price of $\frac{1}{12}(90{,}000) = \$7500$ per car. Check this result in the original statement of the problem.

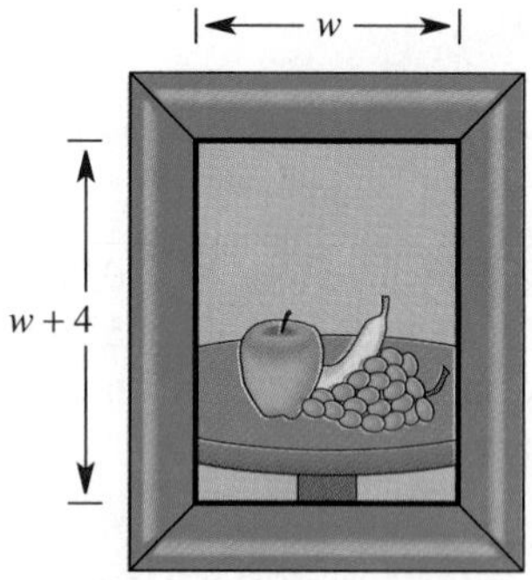

EXAMPLE 2   Geometry

A picture is 4 inches taller than it is wide and has an area of 192 square inches. What are the dimensions of the picture?

Solution

Begin by drawing a diagram, as shown in Figure 6.3.

Verbal Model: | Area of picture | = | Width | · | Height |

Labels: Picture width $= w$ (inches)
Picture height $= w + 4$ (inches)
Area $= 192$ (square inches)

Equation: $192 = w(w + 4)$

$0 = w^2 + 4w - 192$

$0 = (w + 16)(w - 12)$

$w + 16 = 0 \implies w = -16$

$w - 12 = 0 \implies w = 12$

FIGURE 6.3

Of the two possible solutions, choose the positive value of w and conclude that the picture is $w = 12$ inches wide and $w + 4$ or 16 inches tall. Check these dimensions in the original statement of the problem.

EXAMPLE 3   An Interest Problem

The formula

$$A = P(1 + r)^2$$

represents the amount of money A in an account in an which P dollars is deposited for 2 years at an annual interest rate of r (in decimal form). Find the interest rate if a deposit of $6000 increases to $6933.75 over a 2-year period.

Solution

$A = P(1 + r)^2$ Given formula

$6933.75 = 6000(1 + r)^2$ Substitute for A and P.

$1.155625 = (1 + r)^2$ Divide both sides by 6000.

$\pm 1.075 = 1 + r$ Extract square roots.

$0.075 = r$ Choose positive solution.

The annual interest rate is $r = 0.075 = 7.5\%$. Check this result in the original statement of the problem.

EXAMPLE 4   Reduced Rates

A ski club chartered a bus for a ski trip at a cost of $520. In an attempt to lower the bus fare per skier, the club invited nonmembers to go along. When five nonmembers joined the trip, the fare per skier decreased by $5.20. How many club members are going on the trip?

Solution

Verbal Model:
$$\boxed{\text{Cost per skier}} \cdot \boxed{\text{Number of skiers}} = \boxed{\$520}$$

Labels:

Number of ski club members $= x$	(people)
Number of skiers $= x + 5$	(people)
Original cost per skier $= \dfrac{520}{x}$	(dollars)
New cost per skier $= \dfrac{520}{x} - 5.20$	(dollars)

Equation:

$$\left(\frac{520}{x} - 5.20 \right)(x + 5) = 520$$

$$\left(\frac{520 - 5.2x}{x} \right)(x + 5) = 520$$

$$(520 - 5.2x)(x + 5) = 520x, \quad x \neq 0$$

$$520x - 5.2x^2 - 26x + 2600 = 520x$$

$$-5.2x^2 - 26x + 2600 = 0$$

$$x^2 + 5x - 500 = 0$$

$$(x + 25)(x - 20) = 0$$

$$x + 25 = 0 \quad \Longrightarrow \quad x = -25$$

$$x - 20 = 0 \quad \Longrightarrow \quad x = 20$$

Choosing the positive value of x implies that there are 20 ski club members. Check this solution in the original equation, as follows.

Check

Number of Skiers	Cost per Skier
20	$\dfrac{520}{20} = \$26.00$
25	$\dfrac{520}{25} = \$20.80$

From these two calculations, you can see that the difference in cost per skier is $\$26.00 - \$20.80 = \$5.20$.

EXAMPLE 5   An Application Involving the Pythagorean Theorem

An L-shaped sidewalk from building A to building B on a college campus is 200 meters long, as shown in Figure 6.4. By cutting diagonally across the grass, students shorten the walking distance to 150 meters. What are the lengths of the two legs of the existing sidewalk?

Solution

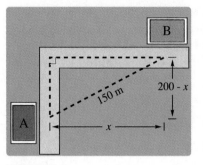

FIGURE 6.4

Verbal Model: $a^2 + b^2 = c^2$ Pythagorean Theorem

Labels: Length of one leg $= x$ (meters)
            Length of other leg $= 200 - x$ (meters)
            Length of diagonal $= 150$ (meters)

Equation: $x^2 + (200 - x)^2 = (150)^2$

$2x^2 - 400x + 40{,}000 = 22{,}500$

$2x^2 - 400x + 17{,}500 = 0$

$x^2 - 200x + 8750 = 0$

By the Quadratic Formula, you can find the solutions as follows.

$$x = \frac{200 \pm \sqrt{(-200)^2 - 4(1)(8750)}}{2(1)}$$

$$= \frac{200 \pm \sqrt{5000}}{2}$$

$$= \frac{200 \pm 50\sqrt{2}}{2}$$

$$= 100 \pm 25\sqrt{2}$$

Both solutions are positive, and it does not matter which one you choose. If you let

$$x = 100 + 25\sqrt{2} \approx 135.4 \text{ meters,}$$

the length of the other leg is

$$200 - x \approx 200 - 135.4 \approx 64.6 \text{ meters.}$$

NOTE   In Example 5, notice that you obtain the same dimensions if you choose the other value of x. That is, if the length of one leg is

$$x = 100 - 25\sqrt{2} \approx 64.6 \text{ meters,}$$

the length of the other leg is

$$200 - x \approx 200 - 64.6 \approx 135.4 \text{ meters.}$$

EXAMPLE 6 *Work Problem*

An office contains two copy machines. Machine B is known to take 12 minutes longer than machine A to copy the company's monthly report. Using both machines together, it takes 8 minutes to reproduce the report. How long would it take each machine alone to reproduce the report?

Solution

Verbal Model:

$$\boxed{\text{Work done by machine A}} + \boxed{\text{Work done by machine B}} = \boxed{\text{1 complete job}}$$

$$\boxed{\text{Rate for A}} \cdot \boxed{\text{Time for both}} + \boxed{\text{Rate for B}} \cdot \boxed{\text{Time for both}} = \boxed{1}$$

Labels:

Time for machine A $= t$	(minutes)
Rate for machine A $= 1/t$	(job per minute)
Time for machine B $= t + 12$	(minutes)
Rate for machine B $= 1/(t + 12)$	(job per minute)
Time for both machines $= 8$	(minutes)
Rate for both machines $= \frac{1}{8}$	(job per minute)

Equation:

$$\frac{1}{t}(8) + \frac{1}{t + 12}(8) = 1$$

$$8\left(\frac{1}{t} + \frac{1}{t + 12}\right) = 1$$

$$8\left[\frac{t + 12 + t}{t(t + 12)}\right] = 1$$

$$8t(t + 12)\left[\frac{2t + 12}{t(t + 12)}\right] = t(t + 12)$$

$$8(2t + 12) = t^2 + 12t$$

$$16t + 96 = t^2 + 12t$$

$$0 = t^2 - 4t - 96$$

$$0 = (t - 12)(t + 8)$$

$$t - 12 = 0 \implies t = 12$$

$$t + 8 = 0 \implies t = -8$$

Choose the positive value for t and find that

Time for machine A $= t = 12$ minutes

Time for machine B $= t + 12 = 24$ minutes.

Check these solutions in the original equation.

EXAMPLE 7   The Height of a Model Rocket

A model rocket is projected straight upward from ground level according to the height equation $h = -16t^2 + 192t, t \geq 0$, where h is the height in feet and t is the time in seconds. (a) After how many seconds will the height be 432 feet? (b) When will the rocket hit the ground?

Solution

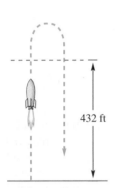

a.

$$h = -16t^2 + 192t \qquad \text{Original equation}$$

$$432 = -16t^2 + 192t \qquad \text{Substitute 432 for } h.$$

$$16t^2 - 192t + 432 = 0 \qquad \text{Standard form}$$

$$t^2 - 12t + 27 = 0 \qquad \text{Divide both sides by 16.}$$

$$(t - 3)(t - 9) = 0 \qquad \text{Factor.}$$

$$t - 3 = 0 \quad \Longrightarrow \quad t = 3 \qquad \text{Set 1st factor equal to 0.}$$

$$t - 9 = 0 \quad \Longrightarrow \quad t = 9 \qquad \text{Set 2nd factor equal to 0.}$$

432 ft

FIGURE 6.5

The rocket obtains a height of 432 feet at two different times—once (going up) after 3 seconds, and again (coming down) after 9 seconds. (See Figure 6.5.)

b.   To find the time it takes for the rocket to hit the ground, let the height be 0.

$$0 = -16t^2 + 192t$$

$$0 = t^2 - 12t$$

$$0 = t(t - 12)$$

$$t = 0 \quad \text{or} \quad t = 12$$

The rocket will hit the ground after 12 seconds. (Note that the time of $t = 0$ seconds corresponds to the time of lift-off.)

Group Activities   Exploring with Technology

Analyzing Quadratic Functions   Use a graphing utility to graph $y_1 = 3x^2 + 2x - 1$ and $y_2 = -x^2 + 5x + 4$. For each function, use the ZOOM and TRACE features to find either the maximum or minimum function value. Discuss other methods that you could use to find these values.

6.4 Exercises

Discussing the Concepts

1. In your own words, describe guidelines for solving a word problem.

2. Describe the strategies that can be used to solve a quadratic equation.

3. *Unit Analysis* Describe the units of the product.

$$\frac{9 \text{ dollars}}{\text{hour}} \cdot (20 \text{ hours})$$

4. *Unit Analysis* Describe the units of the product.

$$\frac{20 \text{ feet}}{\text{minute}} \cdot \frac{1 \text{ minute}}{60 \text{ seconds}} \cdot (45 \text{ seconds})$$

5. Give an example of a quadratic equation that has only one repeated solution.

6. Give an example of a quadratic equation that has two imaginary solutions.

Problem Solving

In Exercises 7–10, find two positive integers that satisfy the requirement.

7. The product of two consecutive integers is 240.

8. The product of two consecutive integers is 1122.

9. The product of two consecutive *even* integers is 224.

10. The product of two consecutive *odd* integers is 255.

In Exercises 11–14, complete the table of widths, lengths, perimeters, and areas of rectangles.

Width	Length	Perimeter	Area
11. $0.75l$	l	42 in.	
12. $l - 6$	l	108 ft	
13. $l - 20$	l		12,000 m²
14. w	$1.5w$		216 cm²

Compound Interest In Exercises 15–18, find the interest rate r. Use the formula $A = P(1 + r)^2$, where A is the amount after 2 years in an account earning r percent compounded annually and P is the original investment.

15. $P = \$3000.00$
    $A = \$3499.20$

16. $P = \$10,000.00$
    $A = \$11,990.25$

17. $P = \$8000.00$
    $A = \$8420.20$

18. $P = \$6500.00$
    $A = \$7372.46$

19. *Geometry* A television station claims that it covers a circular region of approximately 25,000 square miles.

    (a) Assume that the station is located at the center of the circular region. How far is the station from its farthest listener?

    (b) Assume that the station is located on the edge of the circular region. How far is the station from its farthest listener?

20. *Geometry* The height of a triangle is twice its base. The area of the triangle is 625 square inches. Find the dimensions of the triangle.

21. *Geometry* A retail lumber business plans to build a rectangular storage region adjoining the sales office (see figure). The region will be fenced on three sides, and the fourth side will be bounded by the existing building. Find the dimensions of the region if 350 feet of fencing is used and the area of the region is 12,500 square feet.

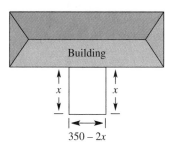
Building

$350 - 2x$

22. *Geometry*   Your home is built on a square lot. To add more space to your yard, you purchase an additional 20 feet along the side of the property (see figure). The area of the lot is now 25,500 square feet. What are the dimensions of the new lot?

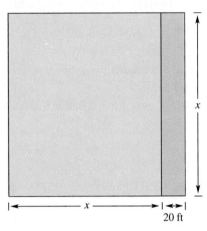

23. *Selling Price*   A store owner bought a case of grade A large eggs for $21.60. By the time all but 6 dozen of the eggs had been sold at a profit of $0.30 per dozen, the original investment of $21.60 had been regained. How many dozen eggs did the owner sell, and what was the selling price per dozen?

24. *Selling Price*   A manager of a computer store bought several computers of the same model for $27,000. When all but three of the computers had been sold at a profit of $750 per computer, the original investment of $27,000 had been regained. How many computers were sold, and what was the selling price of each computer?

25. *Reduced Ticket Price*   A service organization obtains a block of tickets to a ball game for $240. When eight more people decide to go to the game, the price per ticket is decreased by $1. How many people are going to the game?

26. *Reduced Fare*   A science club charters a bus to attend a science fair at a cost of $480. In an attempt to lower the bus fare per person, the club invites nonmembers to go along. When two nonmembers join the trip, the fare per person is decreased by $1. How many people are going on the excursion?

27. *Solving Graphically and Algebraically*   An adjustable rectangular form has minimum dimensions of 3 meters by 4 meters. The length and width can be expanded by equal amounts x (see figure).

(a) Write the length d of the diagonal as a function of x. Use a graphing utility to graph the function. Use the graph to approximate the value of x when $d = 10$ meters.

(b) Find x algebraically when $d = 10$.

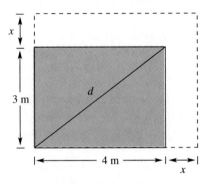

28. *Solving Graphically and Numerically*   A meteorologist is positioned 100 feet from the point where a weather balloon is launched (see figure).

(a) Write the distance d between the balloon and the meteorologist as a function of the height h of the balloon. Use a graphing utility to graph the function. Use the graph to approximate the value of h when $d = 200$ feet.

(b) Complete the table. Use the table to approximate the value of h when $d = 200$ feet.

h	160	165	170	175	180	185
d						

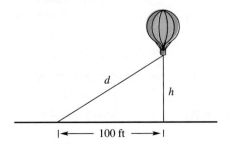

29. *Airspeed*   An airline runs a commuter flight between two cities that are 720 miles apart. If the average speed of the planes could be increased by 40 miles per hour, the travel time would be decreased by 12 minutes. What airspeed is required to obtain this decrease in travel time?

30. *Average Speed*   A truck traveled the first 100 miles of a trip at one speed and the last 135 miles at an average speed of 5 miles per hour less. If the entire trip took 5 hours, what was the average speed for the first part of the trip?

31. *Work Rate*   Working together, two people can complete a task in 5 hours. Working alone, how long would it take each to do the task if one person took 2 hours longer than the other?

32. *Work Rate*   An office contains two printers. Machine B is known to take 3 minutes longer than machine A to produce the company's monthly financial report. Using both machines together, it takes 6 minutes to produce the report. How long would it take each machine to produce the report?

Free-Falling Object   In Exercises 33–36, find the time necessary for an object to fall to ground level from an initial height of h_0 feet if its height h at any time t (in seconds) is given by $h = h_0 - 16t^2$.

33. $h_0 = 144$

34. $h_0 = 625$

35. $h_0 = 1454$ (height of the Sears Tower)

36. $h_0 = 984$ (height of the Eiffel Tower)

Cost, Revenue, and Profit   In Exercises 37 and 38, you are given the cost C of producing x units, the revenue R from selling x units, and the profit P. Find the value of x that will produce the profit P.

37. $C = 100 + 30x$, $R = x(90 - x)$, $P = \$800$

38. $C = 4000 - 40x + 0.02x^2$, $R = x(50 - 0.01x)$, $P = \$63,500$

In Exercises 39 and 40, solve for the specified variable.

39. Solve for r in $A = P(1 + r)^2$.

40. Solve for r in $A = \pi r^2$.

Reviewing the Major Concepts

In Exercises 41–44, find the product.

41. $-2x^5(5x^{-3})$

42. $(3x + 2)(7x - 10)$

43. $(2x - 15)^2$

44. $(x + 3)(x^2 - 3x + 9)$

45. *List Price*   A computer is discounted 15% from its list price to a sale price of $1955. Find the list price.

46. *Investment*   A combined total of $24,000 is invested in two bonds that pay 7.5% and 9% simple interest. The total annual interest is $1935. How much is invested in each bond?

Additional Problem Solving

In Exercises 47–50, find two positive integers that satisfy the requirement.

47. The product of two consecutive *odd* integers is 483.

48. The product of two consecutive *even* integers is 528.

49. The sum of the squares of two consecutive integers is 313.

50. The sum of the squares of two consecutive integers is 421.

In Exercises 51–54, complete the table of widths, lengths, perimeters, and areas of rectangles.

	Width	Length	Perimeter	Area
51.	w	$w + 3$	54 km	
52.	w	$1.5w$	40 m	
53.	$\frac{3}{4}l$	l		2700 in.^{2}
54.	$\frac{1}{3}l$	l		192 in.^{2}

55. *Geometry*   An open-top rectangular conduit for carrying water in a manufacturing process is made by folding up the edges of a sheet of aluminum 48 inches wide (see figure). A cross section of the conduit must have an area of 288 square inches. Find the width and height of the conduit.

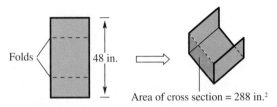

Folds   48 in.   ⇒   Area of cross section = 288 in.²

56. *Geometry*   A friend built a fence around three sides of his property (see figure). In total, he used 550 feet of fencing. By his calculations, the area of the lot is 1 acre (43,560 square feet). Is this correct? Explain your answer.

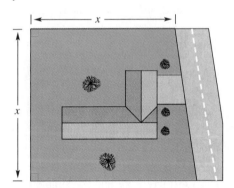

Compound Interest   In Exercises 57–60, find the interest rate r. Use the formula $A = P(1 + r)^2$, where A is the amount after 2 years in an account earning r percent compounded annually and P is the original investment.

57. $P = \$250.00$, $A = \$280.90$

58. $P = \$500.00$, $A = \$572.45$

59. $P = \$10,000.00$, $A = \$11,556.25$

60. $P = \$2000.00$, $A = \$2354.45$

61. *Venture Capital*   Eighty thousand dollars is needed to begin a small business. The cost will be divided equally among investors. Some have made a commitment to invest. If three more investors are found, the amount required from each would decrease by $6000. How many have made a commitment to invest in the business?

62. *Ticket Prices*   A service organization paid $210 for a block of tickets to a ball game. The block contained three more tickets than the organization needed for its members. By inviting three more people to attend (and share in the cost), the organization lowered the price per ticket by $3.50. How many people are going to the game?

63. *Delivery Route*   You are asked to deliver pizza to offices B and C in your city (see figure), and you are required to keep a log of all the mileages between stops. You forget to look at the odometer at stop B, but after getting to stop C you record the total distance traveled from the pizza shop as 18 miles. The return distance from C to A is 16 miles. If the route approximates a right triangle, estimate the distance from A to B.

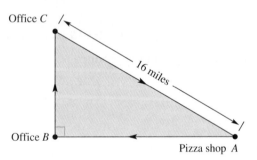

Office *C*

16 miles

Office *B*   Pizza shop *A*

64. *Geometry*   The perimeter of a rectangle is 102 inches and the length of the diagonal is 39 inches. Find the dimensions of the rectangle.

65. *Work Rate*   A builder works with two plumbing companies. Company A is known to take 3 days longer than Company B to do the plumbing in a particular style of house. Using both companies it takes 4 days. How long would it take to do the plumbing using each company individually?

66. *Work Rate* Working together, two people can complete a task in 6 hours. Working alone, one person takes 2 hours longer than the other. How long would it take each person to do the task alone?

67. *Height of a Baseball* The height h in feet of a baseball hit 3 feet above the ground is given by

$$h = 3 + 75t - 16t^2$$

where t is time in seconds. Find the time when the ball hits the ground in the outfield.

68. *Hitting Baseballs* You are hitting baseballs. When you toss the ball into the air, your hand is 5 feet above the ground (see figure). You hit the ball when it falls back to a height of 4 feet. If you toss the ball with an initial velocity of 25 feet per second, the height h of the ball t seconds after leaving your hand is given by

$$h = 5 + 25t - 16t^2.$$

How much time will pass before you hit the ball?

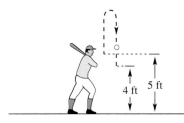

4 ft   5 ft

In Exercises 69–72, solve for the specified variable.

69. Solve for r in $S = 4\pi r^2$.

70. Solve for s in $A = \frac{1}{4}s^2\sqrt{3}$.

71. Solve for b in $I = \frac{1}{12}(M)(a^2 + b^2)$.

72. Solve for h in $S = \pi r\sqrt{r^2 + h^2}$.

73. A small business uses a minivan to make deliveries. The cost per hour for fuel for the van is

$$C = \frac{v^2}{600}$$

where v is the speed in miles per hour. The driver is paid $5 per hour. Find the speed if the cost for wages and fuel for a 110-mile trip is $20.39.

74. *Reading a Graph* For the years 1983 to 1990, the number of mountain bike owners m (in millions) in the United States can be approximated by the model

$$m = 0.337t^2 - 2.265t + 3.962, \quad 3 \le t \le 10$$

where $t = 3$ represents 1983 (see figure). In which year did 2.5 million people own mountain bikes? (Source: Bicycle Institute of America)

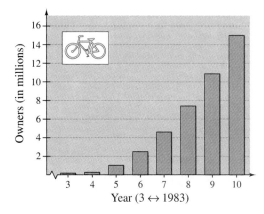

75. *Geometry* The area A of an ellipse is given by $A = \pi ab$ (see figure). For a certain ellipse it is required that $a + b = 20$.

(a) Show that $A = \pi a(20 - a)$.

(b) Complete the following table.

a	4	7	10	13	16
A					

(c) Find two values of a such that $A = 300$.

(d) Use a graphing utility to graph the area function.

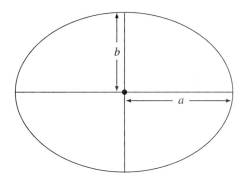

Math Matters   Olympic Games

No.	Year	City	Participants
1	1896	Athens, Greece	13
2	1900	Paris, France	22
3	1904	St. Louis, USA	12
4	1908	London, England	23
5	1912	Stockholm, Sweden	28
6	1916	Berlin, Germany	—
7	1920	Antwerp, Belgium	29
8	1924	Paris, France	44
9	1928	Amsterdam, Holland	46
10	1932	Los Angeles, USA	37
11	1936	Berlin, Germany	49
12	1940	Tokyo, Japan	—
13	1944	London, England	—
14	1948	London, England	59
15	1952	Helsinki, Finland	69
16	1956	Melbourne, Australia	67
17	1960	Rome, Italy	83
18	1964	Tokyo, Japan	93
19	1968	Mexico City, Mexico	112
20	1972	Munich, West Germany	122
21	1976	Montreal, Canada	92
22	1980	Moscow, USSR	81
23	1984	Los Angeles, USA	140
24	1988	Seoul, South Korea	160
25	1992	Barcelona, Spain	172
26	1996	Atlanta, USA	

As this book is being written, the 26th Summer Olympic Games are scheduled to be held in Atlanta, Georgia, USA, in the summer of 1996. The original Olympic Games were held once every four years in ancient Greece. The earliest recorded games took place in 776 B.C., and the games were abandoned in 393 A.D.

In 1896 the games were revived. Since then, they have taken place every four years, except during the two World Wars. Note that the three games that were not held had already been planned, and therefore kept their planned numbers.

The number of participating countries has tended to increase each year. Do you know why the number of participants was down in 1976 and 1980?

6.5   Quadratic and Rational Inequalities

Finding Test Intervals ▪ Quadratic Inequalities ▪
Rational Inequalities ▪ Application

Finding Test Intervals

When you are working with polynomial inequalities, it is important to realize that
the value of a polynomial can change signs only at its **zeros.** That is, a polynomial
can change signs only at the x-values that make the polynomial zero. For instance,
the first-degree polynomial $x + 2$ has a zero at -2, and it changes sign at that zero.
You can picture this result on the real number line, as shown in Figure 6.6.

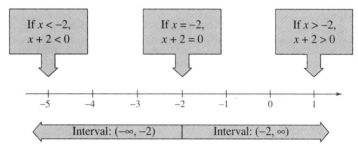

FIGURE 6.6

Note in Figure 6.6 that the zero of the polynomial partitions the real number
line into two **test intervals.** The polynomial is negative for every x-value in the
first test interval $(-\infty, -2)$, and it is positive for every x-value in the second
test interval $(-2, \infty)$. You can use the same basic approach to determine the test
intervals for any polynomial.

Finding Test Intervals for a Polynomial

1. Find all real zeros of the polynomial, and arrange the zeros in increasing order. The zeros of a polynomial are called its **critical numbers.**

2. Use the critical numbers of the polynomial to determine its test
intervals.

3. Choose a representative x-value in each test interval and evaluate the
polynomial at that value. If the value of the polynomial is negative,
the polynomial will have negative values for *every* x-value in the
interval. If the value of the polynomial is positive, the polynomial
will have positive values for *every* x-value in the interval.

Quadratic Inequalities

The concepts of critical numbers and test intervals can be used to solve nonlinear inequalities, as demonstrated in Examples 1, 2, and 4.

EXAMPLE 1 Solving a Quadratic Inequality

Solve the inequality $x^2 - 5x < 0$.

Solution

Begin by finding the critical numbers of $x^2 - 5x$.

$$x^2 - 5x < 0 \qquad \text{Original inequality}$$

$$x(x - 5) < 0 \qquad \text{Factor.}$$

From this factorization, you can see that the critical numbers of $x^2 - 5x$ are 0 and 5. This implies that the test intervals are

$$(-\infty, 0), \quad (0, 5), \quad \text{and} \quad (5, \infty).$$

To test an interval, choose a convenient number in the interval and compute the sign of $x(x - 5)$. The results are shown in Figure 6.7.

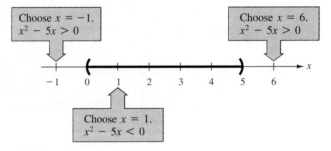

FIGURE 6.7

Because the inequality $x^2 - 5x < 0$ is satisfied only by the middle test interval, you can conclude that the solution set of the inequality is the interval $(0, 5)$.

In Example 1, note that you would have used the same basic procedure if the inequality symbol had been $\le$, $>$, or $\ge$. For instance, from Figure 6.7, you can see that the solution set of the inequality

$$x^2 - 5x \ge 0$$

consists of the union of the half-open intervals $(-\infty, 0]$ and $[5, \infty)$, which is written as $(-\infty, 0] \cup [5, \infty)$.

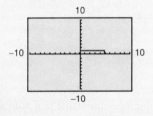

Just as with solving quadratic *equations*, the first step in solving a quadratic *inequality* is to write the inequality in **standard form,** with the polynomial on the left and zero on the right, as demonstrated in Example 2.

EXAMPLE 2 Solving a Quadratic Inequality

Solve the inequality $2x^2 + 5x > 12$.

Solution

Begin by writing the inequality in standard form.

$$2x^2 + 5x > 12 \qquad \text{Original inequality}$$

$$2x^2 + 5x - 12 > 0 \qquad \text{Write in standard form.}$$

$$(x + 4)(2x - 3) > 0 \qquad \text{Factor.}$$

From this factorization, you can see that the critical numbers of $2x^2 + 5x - 12$ are -4 and $\frac{3}{2}$. This implies that the test intervals are

$$(-\infty, -4), \quad \left(-4, \tfrac{3}{2}\right), \quad \text{and} \quad \left(\tfrac{3}{2}, \infty\right).$$

To test an interval, choose a convenient number in the interval and compute the sign of $(x + 4)(2x - 3)$. The results are shown in Figure 6.8. From the figure, you can see that the polynomial $2x^2 + 5x - 12$ is positive in the open intervals $(-\infty, -4)$ and $\left(\tfrac{3}{2}, \infty\right)$. Thus, the solution set is $(-\infty, -4) \cup \left(\tfrac{3}{2}, \infty\right)$.

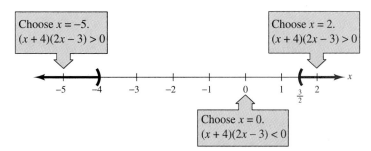

FIGURE 6.8

In Examples 1 and 2, the critical numbers were found by factoring. With quadratic polynomials that do not factor, you can use the Quadratic Formula to find the critical numbers. For instance, to solve the inequality

$$x^2 - 2x - 1 \le 0$$

you can use the Quadratic Formula to determine that the critical numbers are $1 - \sqrt{2} \approx -0.414$ and $1 + \sqrt{2} \approx 2.414$.

The solutions of the quadratic inequalities in Examples 1 and 2 consist, respectively, of a single interval and the union of two intervals. When solving the exercises for this section, you should be on the watch for some unusual solution sets, as illustrated in Example 3.

EXAMPLE 3   *Unusual Solution Sets*

Solve each inequality to verify that the given solution set is correct.

a. The solution set of the quadratic inequality

$$x^2 + 2x + 4 > 0$$

consists of the entire set of real numbers, $(-\infty, \infty)$. In other words, the quadratic $x^2 + 2x + 4$ is positive for every real value of x. (Note that this quadratic inequality has no critical numbers. In such a case, there is only one test interval—the entire real line.)

b. The solution set of the quadratic inequality

$$x^2 + 2x + 1 \leq 0$$

consists of the single real number $\{-1\}$.

c. The solution set of the quadratic inequality

$$x^2 + 3x + 5 < 0$$

is empty. In other words, the quadratic $x^2 + 3x + 5$ is not less than zero for any value of x.

d. The solution set of the quadratic inequality

$$x^2 - 4x + 4 > 0$$

consists of all real numbers *except* the number 2. In interval notation, this solution set can be written as $(-\infty, 2) \cup (2, \infty)$.

Remember that checking the solution set of an inequality is not as straightforward as checking the solutions of an equation, because inequalities tend to have infinitely many solutions. Even so, we suggest that you check several x-values in your solution set to confirm that they satisfy the inequality. Also try checking x-values that are not in the solution set to confirm that they do not satisfy the inequality.

For instance, the solution of $x^2 - 5x < 0$ is $(0, 5)$. Try checking some numbers in this interval to verify that they satisfy the inequality. Then check some numbers outside the interval to verify that they do not satisfy the inequality.

Rational Inequalities

The concepts of critical numbers and test intervals can be extended to inequalities involving rational expressions. To do this, use the fact that the value of a rational expression can change sign only at its *zeros* (the x-values for which its numerator is zero) and its *undefined values* (the x-values for which its denominator is zero). These two types of numbers make up the **critical numbers** of a rational inequality. For instance, the critical numbers of the inequality

$$\frac{x - 2}{(x - 1)(x + 3)} < 0$$

are $x = 2$ (the numerator is zero), and $x = 1$ and $x = -3$ (the denominator is zero). From these three critical numbers you can see that the inequality has *four* test intervals.

$$(-\infty, -3), \quad (-3, 1), \quad (1, 2), \quad \text{and} \quad (2, \infty)$$

STUDY TIP

When solving a rational inequality, you should begin by writing the inequality in standard form, with the rational expression (as a single fraction) on the left and zero on the right. For instance, the first step in solving

$$\frac{2x}{x + 3} < 4$$

is to write it as

$$\frac{2x}{x + 3} < 4$$

$$\frac{2x}{x + 3} - 4 < 0$$

$$\frac{2x - 4(x + 3)}{x + 3} < 0$$

$$\frac{-2x - 12}{x + 3} < 0.$$

Try solving this inequality. You should find that the solution set is $(-\infty, -6) \cup (-3, \infty)$.

EXAMPLE 4   Solving a Rational Inequality

Solve the inequality $\dfrac{x}{x - 2} > 0$.

Solution

The numerator is zero when $x = 0$ and the denominator is zero when $x = 2$. Thus, the two critical numbers are 0 and 2, which implies that the test intervals are

$$(-\infty, 0), \quad (0, 2), \quad \text{and} \quad (2, \infty).$$

A test of these intervals, as shown in Figure 6.9, reveals that the rational expression $x/(x - 2)$ is positive in the open intervals $(-\infty, 0)$ and $(2, \infty)$. Therefore, the solution set of the inequality is $(-\infty, 0) \cup (2, \infty)$.

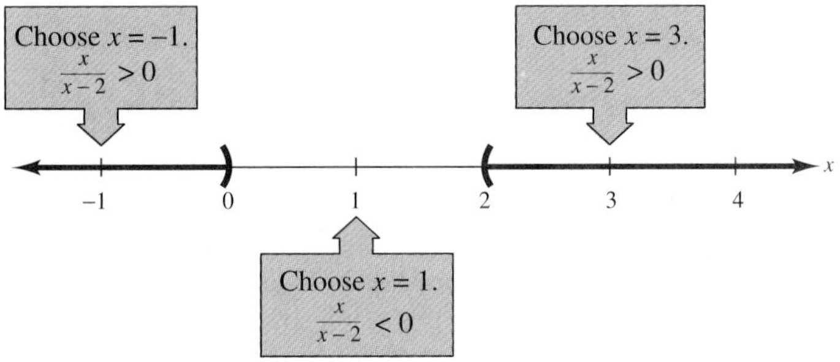

Choose $x = -1$.
$\dfrac{x}{x-2} > 0$

Choose $x = 3$.
$\dfrac{x}{x-2} > 0$

Choose $x = 1$.
$\dfrac{x}{x-2} < 0$

FIGURE 6.9

Application

EXAMPLE 5   The Height of a Projectile

A projectile is fired straight up from ground level with an initial velocity of 256 feet per second, as shown in Figure 6.10, so that its height at any time t is given by

$$h = -16t^2 + 256t$$

where the height h is measured in feet and the time t is measured in seconds. During what interval of time will the height of the projectile exceed 960 feet?

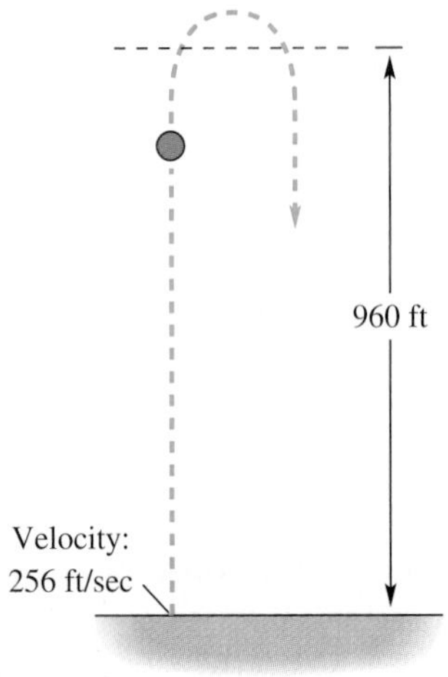

Solution

To solve this problem, find the values of t for which h is greater than 960.

$-16t^2 + 256t > 960$	Original inequality
$-16t^2 + 256t - 960 > 0$	Subtract 960 from both sides.
$t^2 - 16t + 60 < 0$	Divide both sides by -16 and reverse the inequality.
$(t - 6)(t - 10) < 0$	Factor.

Thus, the critical numbers are $t = 6$ and $t = 10$. A test of the intervals $(-\infty, 6)$, $(6, 10)$, and $(10, \infty)$ shows that the solution interval is $(6, 10)$. Therefore, the height of the object will exceed 960 feet for values of t that are greater than 6 seconds and less than 10 seconds.

**Velocity:
256 ft/sec**

960 ft

FIGURE 6.10

Group Activities   Exploring with Technology

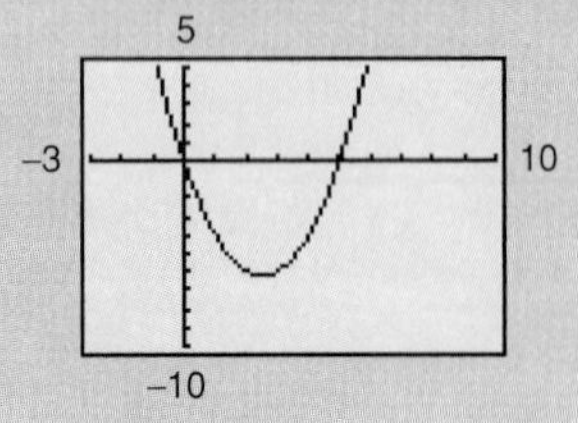

Graphing an Inequality   You can use a graph on a rectangular coordinate system as an alternative method for solving an inequality. For instance, to solve the inequality in Example 1

$$x^2 - 5x < 0$$

you can sketch the graph of $y = x^2 - 5x$. Using a graphing utility, you can obtain the graph shown at the left. From the graph, you can see that the only part of the curve that lies below the x-axis is the portion for which $0 < x < 5$. Thus, the solution of $x^2 - 5x < 0$ is $0 < x < 5$. Try using this graphing approach to solve some of the other examples in this section.

6.5 Exercises

Discussing the Concepts

1. Explain the change in an inequality that occurs when both sides are multiplied by a negative real number.

2. Give a verbal description of the union of intervals $(-\infty, 5] \cup (10, \infty)$.

3. Define the term *critical number* and explain its use in solving quadratic inequalities.

4. In your own words, describe the procedure for solving quadratic inequalities.

5. Give an example of a quadratic inequality that has no real solution.

6. Explain the distinction between the critical numbers of quadratic inequalities and rational inequalities.

Problem Solving

In Exercises 7–10, find the critical numbers.

7. $x(2x - 5)$

8. $5x(x - 3)$

9. $x^2 - 4x + 3$

10. $3x^2 - 2x - 8$

In Exercises 11–14, determine whether the x-values are solutions of the inequality.

11. $2x^2 - 7x - 4 > 0$
    (a) $x = 0$
    (b) $x = 6$
    (c) $x = -\frac{1}{2}$
    (d) $x = -\frac{3}{2}$

12. $4x^2 + 3x - 10 \leq 0$
    (a) $x = 0$
    (b) $x = 3$
    (c) $x = -2$
    (d) $x = \frac{1}{2}$

13. $\dfrac{2}{3 - x} \leq 0$
    (a) $x = 3$
    (b) $x = 4$
    (c) $x = -4$
    (d) $x = -\frac{1}{3}$

14. $\dfrac{x - 2}{x + 1} > 0$
    (a) $x = 0$
    (b) $x = -1$
    (c) $x = -4$
    (d) $x = -\frac{3}{2}$

In Exercises 15–18, determine the intervals for which the polynomial is entirely negative and entirely positive.

15. $x - 4$

16. $7x(3 - x)$

17. $x^2 - 4x - 5$

18. $2x^2 - 4x - 3$

In Exercises 19–30, solve the inequality and graph the solution on the real number line. (Some of the inequalities have no solution.)

19. $2x + 6 \geq 0$

20. $5x - 20 < 0$

21. $3x(2 - x) < 0$

22. $2x(6 - x) > 0$

23. $x^2 + 3x \leq 10$

24. $t^2 - 15t + 50 < 0$

25. $x^2 + 4x + 5 < 0$

26. $x^2 + 6x + 10 > 0$

27. $x^2 + 2x + 1 \geq 0$

28. $25 \geq (x - 3)^2$

29. $x(x - 2)(x + 2) > 0$

30. $x^2(x - 2) \leq 0$

In Exercises 31–34, determine the critical numbers of the rational expression and plot them on the real number line.

31. $\dfrac{5}{x - 3}$

32. $\dfrac{-6}{x + 2}$

33. $\dfrac{2x}{x + 5}$

34. $\dfrac{x - 2}{x - 10}$

In Exercises 35–38, solve the rational inequality. As part of your solution, include a graph that shows the test intervals.

35. $\dfrac{5}{x - 3} > 0$

36. $\dfrac{x + 5}{3x + 2} \geq 0$

37. $\dfrac{x}{x - 3} \leq 2$

38. $\dfrac{x + 4}{x - 5} \geq 10$

In Exercises 39–42, use a graphing utility to solve the inequality. Use the strategies described on page 420 and on page 423.

39. $0.5x^2 + 1.25x - 3 > 0$

40. $\frac{1}{3}x^2 - 3x < 0$

41. $\dfrac{6x}{x+5} < 2$

42. $x + \dfrac{1}{x} > 3$

43. *Height of a Projectile*   A projectile is fired straight up from ground level with an initial velocity of 128 feet per second, so that its height at any time t is given by $h = -16t^2 + 128t$, where the height h is measured in feet and the time t is measured in seconds. During what interval of time will the height of the projectile exceed 240 feet?

44. *Annual Interest Rate*   You are investing $1000 in a certificate of deposit for 2 years and you want the interest to exceed $150. The interest is compounded annually. What interest rate should you have?

Reviewing the Major Concepts

In Exercises 45–48, solve the inequality.

45. $7 - 3x > 4 - x$ **46.** $2(x + 6) - 20 < 2$

47. $|x - 3| < 2$ **48.** $|x - 5| > 3$

49. *Geometry*   Determine the length and width of a rectangle with a perimeter of 50 inches and a diagonal of length $5\sqrt{13}$ inches.

50. *Demand for a Product*   The demand equation for a certain product is

$$p = 75 - \sqrt{1.2(x - 10)}$$

where x is the number of units demanded per day and p is the price per unit. Find the demand if the price is $59.90.

Additional Problem Solving

In Exercises 51–54, find the critical numbers.

51. $4x^2 - 81$ **52.** $y(y - 4) - 3(y - 4)$

53. $4x^2 - 20x + 25$ **54.** $4x^2 - 4x - 3$

In Exercises 55–58, determine the intervals for which the polynomial is entirely negative and entirely positive.

55. $\frac{2}{3}x - 8$ **56.** $3 - x$

57. $2x(x - 4)$ **58.** $x^2 - 9$

In Exercises 59–74, solve the inequality and graph the solution on the real number line. (Some of the inequalities have no solution.)

59. $-\frac{3}{4}x + 6 < 0$ **60.** $3x - 2 \ge 0$

61. $3x(x - 2) < 0$ **62.** $2x(x - 6) > 0$

63. $x^2 - 4 \ge 0$ **64.** $z^2 \le 9$

65. $-2u^2 + 7u + 4 < 0$ **66.** $-3x^2 - 4x + 4 \le 0$

67. $x^2 - 4x + 2 > 0$ **68.** $-x^2 + 8x - 11 \le 0$

69. $(x - 5)^2 < 0$ **70.** $(y + 3)^2 \ge 0$

71. $6 - (x - 5)^2 < 0$ **72.** $(y + 3)^2 - 6 \ge 0$

73. $x^2 - 6x + 9 \ge 0$ **74.** $x^2 + 8x + 16 < 0$

In Exercises 75–84, solve the rational inequality. As part of your solution, include a graph that shows the test intervals.

75. $\dfrac{4}{x - 3} < 0$ **76.** $\dfrac{3}{4 - x} > 0$

77. $\dfrac{-5}{x - 3} > 0$ **78.** $\dfrac{-3}{4 - x} > 0$

79. $\dfrac{x}{x - 4} \le 0$ **80.** $\dfrac{z - 1}{z + 3} < 0$

81. $\dfrac{y - 3}{2y - 11} \ge 0$ **82.** $\dfrac{2(4 - t)}{4 + t} > 0$

83. $\dfrac{6}{x - 4} > 2$ **84.** $\dfrac{1}{x + 2} > -3$

In Exercises 85–88, use a graphing utility to solve the inequality. Use the strategies described on page 420 and on page 423.

85. $9 - 0.2(x - 2)^2 > 4$

86. $8x - x^2 < 10$

87. $\dfrac{3x - 4}{x - 4} < -5$

88. $4 - \dfrac{1}{x^2} > 1$

89. *Height of a Projectile*   A projectile is fired straight up from ground level with an initial velocity of 88 feet per second, so that its height at any time t is given by

$$h = -16t^2 + 88t$$

where the height h is measured in feet and the time t is measured in seconds. During what interval of time will the height of the projectile exceed 50 feet?

90. *Company Profits*   The revenue and cost equations for a product are given by

$$R = x(50 - 0.0002x)$$
$$C = 12x + 150,000$$

where R and C are measured in dollars and x represents the number of units sold (see figure). How many units must be sold to obtain a profit of at least $1,650,000?

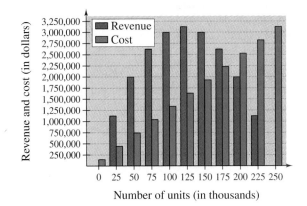

Number of units (in thousands)

91. *Geometry*   You have 64 feet of fencing to enclose a rectangular region, as shown in the figure. Determine the interval for the length such that the area will exceed 240 square feet.

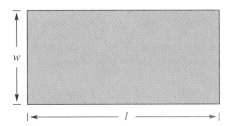

92. *Geometry*   A rectangular playing field with a perimeter of 100 meters is to have an area of at least 500 square meters. Within what bounds must the length of the field lie?

93. *Geometry*   Two circles are tangent to each other (see figure). The distance between their centers is 12 inches.

(a) If x is the radius of one of the circles, express the combined areas of the circles as a function of x. What is the domain of the function?

(b) Use a graphing utility to graph the function found in part (a).

(c) Determine the radii of the circles if the combined areas of the circles must be at least 300 square inches but no more than 400 square inches.

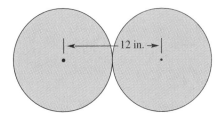

CHAPTER PROJECT: Gravitation and Projectile Paths

In this chapter, you studied position functions for falling objects. On earth, position functions take the form

$$s = -16t^2 + v_0 t + s_0.$$ Position function on earth

At other locations in the universe, position functions take the more general form

$$s = -\tfrac{1}{2}gt^2 + v_0 t + s_0.$$ General position function

On the surface of a planet or moon, the value of g (the acceleration due to gravity) depends on both the mass of the planet or moon and its radius. As discovered by Isaac Newton (1642–1727), the value of g is directly proportional to the mass m and inversely proportional to the square of the radius r. That is,

$$g = \frac{-32m}{r^2}$$ Acceleration due to gravity

where g is measured in feet per second per second, and m and r are the mass and radius *relative to earth's mass and radius.* The table at the left gives the relative masses and radii of earth's moon and the nine planets in our solar system. You are asked to use this data as you investigate the questions below.

Planet/Moon	Mass	Radius
Mercury	0.0532	0.3818
Venus	0.8167	0.9500
Earth	1.0000	1.0000
Moon	0.0123	0.2728
Mars	0.1073	0.5305
Jupiter	317.95	10.949
Saturn	95.066	9.1377
Uranus	14.521	3.6837
Neptune	17.177	3.5654
Pluto	0.002	0.1806

1. Find the acceleration due to gravity on earth's moon and on each of the planets. Organize your results in a table.

2. On which planet is the acceleration due to gravity nearest that on earth? Which planet has the smallest value of g? Which has the largest?

3. You throw a rock straight up with an initial velocity of 50 feet per second and an initial height of 6 feet. Compare the motion of the rock on the moon, earth, and Mars. How long is the rock in motion in each location? What is the maximum height of the rock in each location?

4. The circles below represent the moon and the nine planets from smallest to largest. Trace the circles and label them with their values of g. Does the value of g increase as planet size increases? If not, why?

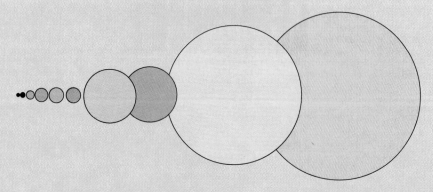

CHAPTER SUMMARY

After studying this chapter, you should have acquired the following skills. These skills are keyed to the Review Exercises that begin on page 430. Answers to odd-numbered Review Exercises are given in the back of the book.

- Solve quadratic equations by factoring. *(Section 6.1)*   **Review Exercises 1–10**

- Solve quadratic equations by extracting square roots. *(Section 6.1)*   **Review Exercises 11–16**

- Solve quadratic equations by completing the square. *(Section 6.2)*   **Review Exercises 17–22**

- Solve quadratic equations using the Quadratic Formula. *(Section 6.3)*   **Review Exercises 23–28**

- Solve quadratic equations by the most convenient method. *(Sections 6.1, 6.2, 6.3)*   **Review Exercises 29–40**

- Solve equations of quadratic form. *(Section 6.1)*   **Review Exercises 41–44**

- Graph functions using a graphing utility. Solve the equations when $y = 0$ and interpret the results. *(Sections 6.1, 6.2, 6.3)*   **Review Exercises 45–50**

- Translate verbal statements into algebraic equations and solve. *(Section 6.4)*   **Review Exercises 51, 52, 57–59, 61–63**

- Solve real-life problems modeled by quadratic equations. *(Sections 6.1, 6.3, 6.4, 6.5)*   **Review Exercises 53, 54, 60, 63, 85**

- Solve problems involving geometry. *(Sections 6.2, 6.3, 6.4)*   **Review Exercises 55, 56**

- Find quadratic equations with the given solutions. *(Section 6.1)*   **Review Exercises 65–68**

- Find quadratic functions that have the given x-intercepts and confirm the results using a graphing utility. *(Sections 6.1, 6.2, 6.3)*   **Review Exercises 69–74**

- Solve inequalities and graph their solutions on the real number line. *(Section 6.5)*   **Review Exercises 75–84**

REVIEW EXERCISES

In Exercises 1–10, solve the equation by factoring.

1. $x^2 + 12x = 0$ **2.** $u^2 - 18u = 0$

3. $3z(z + 10) - 8(z + 10) = 0$

4. $7x(2x - 9) + 4(2x - 9) = 0$

5. $4y^2 - 1 = 0$ **6.** $2z^2 - 72 = 0$

7. $x^2 + \frac{8}{3}x + \frac{16}{9} = 0$ **8.** $4y^2 + 20y + 25 = 0$

9. $2x^2 - 2x = 180$ **10.** $15x^2 - 30x = 45$

In Exercises 11–16, solve the equation by extracting square roots.

11. $x^2 = 10,000$ **12.** $x^2 = 98$

13. $y^2 - 2.25 = 0$ **14.** $y^2 - 8 = 0$

15. $(x - 16)^2 = 400$ **16.** $(x + 3)^2 = 0.04$

In Exercises 17–22, solve the equation by completing the square. Find all real or complex solutions.

17. $x^2 - 6x - 3 = 0$ **18.** $x^2 + 12x + 6 = 0$

19. $x^2 - 3x + 3 = 0$ **20.** $t^2 + \frac{1}{2}t - 1 = 0$

21. $2y^2 + 10y + 3 = 0$ **22.** $3x^2 - 2x + 2 = 0$

In Exercises 23–28, use the Quadratic Formula to solve the equation. Find all real or complex solutions.

23. $y^2 + y - 30 = 0$ **24.** $x^2 - x - 72 = 0$

25. $2y^2 + y - 21 = 0$ **26.** $2x^2 - 3x - 20 = 0$

27. $0.3t^2 - 2t + 5 = 0$ **28.** $-u^2 + 2.5u + 3 = 0$

In Exercises 29–40, solve the equation by the method of your choice. Find all real or complex solutions.

29. $(v - 3)^2 = 250$ **30.** $x^2 - 36x = 0$

31. $-x^2 + 5x + 84 = 0$ **32.** $9x^2 + 6x + 1 = 0$

33. $(x - 9)^2 - 121 = 0$ **34.** $60 - (x - 6)^2 = 0$

35. $z^2 - 6z + 10 = 0$ **36.** $z^2 - 14z + 5 = 0$

37. $2y^2 + 3y + 1 = 0$

38. $0.25y^2 + 0.35y - 0.50 = 0$

39. $\dfrac{1}{x} + \dfrac{1}{x + 1} = \dfrac{1}{2}$

40. $x - 5 = \sqrt{x - 2}$

In Exercises 41–44, solve the equation of quadratic form.

41. $(x^2 - 2x)^2 - 4(x^2 - 2x) - 5 = 0$

42. $\left(\sqrt{x} - 2\right)^2 + 2\left(\sqrt{x} - 2\right) - 3 = 0$

43. $6\left(\dfrac{1}{x}\right)^2 + 7\left(\dfrac{1}{x}\right) - 3 = 0$

44. $\left(\dfrac{x}{x - 2}\right)^2 + 4\left(\dfrac{x}{x - 2}\right) + 3 = 0$

In Exercises 45–48, use a graphing utility to graph the function. Use the graph to approximate any x-intercepts of the graph. Set $y = 0$ and solve the resulting equation. Compare the result with the x-intercepts of the graph.

45. $y = x^2 - 7x$ **46.** $y = 12x^2 + 11x - 15$

47. $y = \dfrac{1}{16}x^4 - x^2 + 3$ **48.** $y = \left(\dfrac{1}{x}\right)^2 + 2\left(\dfrac{1}{x}\right) - 3$

Graphical Reasoning   In Exercises 49 and 50, use a graphing utility to graph the function and observe that the graph has no x-intercepts. Set $y = 0$ and solve the resulting equation. Identify the type of roots of the equation.

49. $y = x^2 - 8x + 17$ **50.** $y = -(x + 3)^2 - 2$

51. Find two consecutive positive integers such that the sum of their squares is 265.

52. Find two consecutive positive integers whose product is 156.

53. *Falling Time* The height h in feet of an object above the ground is given by

$$h = 200 - 16t^2, \quad t \geq 0$$

where t is time in seconds.

(a) How high was the object when it was thrown?

(b) Was the object thrown upward or downward? Explain your reasoning.

(c) Find the time when the object strikes the ground.

54. *Falling Time* The height h in feet of an object above the ground is given by

$$h = -16t^2 + 64t + 192, \quad t \geq 0$$

where t is time in seconds.

(a) How high was the object when it was thrown?

(b) Was the object thrown upward or downward? Explain your reasoning.

(c) Find the time when the object strikes the ground.

55. *Geometry* The perimeter of a rectangle of length l and width w is 48 feet (see figure).

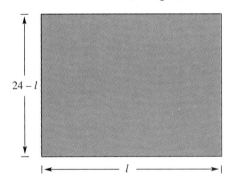

(a) Show that $w = 24 - l$.

(b) Show that the area is $A = lw = l(24 - l)$.

(c) Use the equation in part (b) to complete the table.

l	2	4	6	8	10	12	14	16	18
A									

56. *Geometry* Find the dimensions of a triangle if its height is $1\frac{2}{3}$ times its base and its area is 3000 square inches.

57. *Decreased Price* A Little League baseball team obtains a block of tickets to a ball game for $96. When three more people decide to go to the game, the price per ticket is decreased by $1.60. How many people are going to the game?

58. *Average Speed* A train travels the first 220 miles of a trip at one speed and the last 180 miles at an average speed of 5 miles per hour faster. If the entire trip takes 7 hours, find the higher speed.

59. *Work Rate* Working together, two people can complete a task in 6 hours. Working alone, how long would it take each to do the task if one person takes 3 hours longer than the other?

60. *Path of a Projectile* The path y of a projectile is given by

$$y = -\frac{1}{16}x^2 + 5x$$

where y is the height (in feet) and x is the horizontal distance (in feet) from where the projectile was launched.

(a) Sketch the path of the projectile.

(b) How high is the projectile when it is at its maximum height?

(c) How far from the launch point does the projectile strike the ground?

61. *Reduced Fare* The college wind ensemble charters a bus to attend a concert at a cost of $360. In an attempt to lower the bus fare per person, the ensemble invites nonmembers to go along. When eight nonmembers join the trip, the fare is decreased by $1.50. How many people are going on the excursion?

62. *Think About It* When six nonmembers go along on the excursion described in Exercise 61, the fare is decreased by $16. Describe how it is possible to have fewer nonmembers and a greater decrease in the fare.

63. *Work Rate* Working together, two people can complete a task in 10 hours. Working alone, how long would it take each to do the task if one person takes 2 hours longer than the other?

64. *Solving Graphically, Numerically, and Algebraically*
A launch control team is located 3 miles from the point where a rocket is launched (see figure).

(a) Write the distance d between the rocket and the team as a function of the height h of the rocket. Use a graphing utility to graph the function. Use the graph to approximate the value of h when $d = 4$ miles.

(b) Complete the table. Use the table to approximate the value of h when $d = 4$ miles.

h	1	2	3	4	5	6	7
d							

(c) Find h algebraically when $d = 4$.

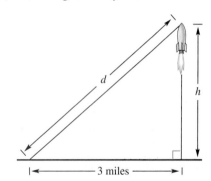

3 miles

In Exercises 65–68, find a quadratic equation with the given solutions.

65. $x = -3, x = 3$ **66.** $x = -\frac{1}{2}, x = 2$

67. $x = -5, x = -1$ **68.** $x = -10, x = 8$

In Exercises 69–72, write a quadratic function that has the given x-intercepts. Use a graphing utility to confirm your result.

69. $(1, 0), (3, 0)$ **70.** $(-2, 0), (2, 0)$

71. $(-4, 0), (-1, 0)$ **72.** $(2, 0), (5, 0)$

73. *Think About It* Find a *different* quadratic equation that has the same solutions as the equation you found in Exercise 65.

74. *Think About It* Find a *different* quadratic equation that has the same solution as the equation you found in Exercise 66.

In Exercises 75–84, solve the inequality and graph its solution on the real number line. (Some of the inequalities have no solution.)

75. $4x - 12 < 0$ **76.** $3(x + 2) > 0$

77. $5x(7 - x) > 0$ **78.** $-2x(x - 10) \le 0$

79. $16 - (x - 2)^2 \le 0$ **80.** $(x - 5)^2 - 36 > 0$

81. $2x^2 + 3x - 20 < 0$ **82.** $3x^2 - 2x - 8 > 0$

83. $\dfrac{x}{2x - 7} \ge 0$ **84.** $\dfrac{2x - 9}{x - 1} \le 0$

85. *College Completion* The percent p of the American population that graduated from college from 1960 to 1990 is approximated by the model

$$p = (2.739 + 0.064t)^2, \quad 0 \le t$$

where $t = 0$ represents 1960 (see figure). According to this model, when will the percent of college graduates exceed 25% of the population? (Source: U.S. Bureau of Census)

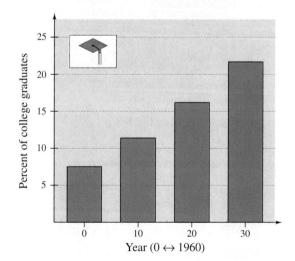

Year (0 ↔ 1960)

CHAPTER TEST

Take this test as you would take a test in class. After you are done, check your work against the answers given in the back of the book.

In Exercises 1 and 2, solve the equation by factoring.

1. $x(x+5) - 10(x+5) = 0$ **2.** $8x^2 - 21x - 9 - 0$

In Exercises 3 and 4, solve the equation by extracting square roots.

3. $(x-2)^2 = 0.09$ **4.** $(x+3)^2 + 81 = 0$

5. Find the real number c such that $x^2 - 3x + c$ is a perfect square trinomial.

6. Solve by completing the square: $2x^2 - 6x + 3 = 0$

7. Find the discriminant and explain how it may be used to determine the type of solutions of the quadratic equation $5x^2 - 12x + 10 = 0$.

In Exercises 8 and 9, solve by the Quadratic Formula.

8. $3x^2 - 8x + 3 = 0$ **9.** $2y(y-2) = 7$

10. Solve, and round the result to two decimal places: $\sqrt{3x+7} - \sqrt{x+12} = 1$

11. Find a quadratic equation having the solutions -4 and 5.

12. Find a quadratic equation that has $i - 1$ as a solution.

In Exercises 13–16, solve the inequality and sketch the solution.

13. $2x(x-3) < 0$ **14.** $16 \le (x-2)^2$

15. $\dfrac{3}{x-2} > 4$ **16.** $\dfrac{3u+2}{u-3} \le 2$

17. Find two consecutive positive integers whose product is 210.

18. The width of a rectangle is 8 feet less than its length. The area of the rectangle is 240 square feet. Find the dimensions of the rectangle.

19. A train traveled the first 125 miles of a trip at one speed and the last 180 miles at an average speed of 5 miles per hour less. If the total time for the trip was $6\frac{1}{2}$ hours, what was the average speed for the first part of the trip?

20. An object is dropped from a height of 75 feet. Its height (in feet) at any time is given by $h = -16t^2 + 75$, where the time t is measured in seconds. Find the time at which the object has fallen to the height of 35 feet.

CUMULATIVE TEST: CHAPTERS 4–6

Take this test as you would take a test in class. After you are done, check your work against the answers given in the back of the book.

In Exercises 1–10, perform the operations and/or simplify.

1. $\dfrac{x^2 + 8x + 16}{18x^2} \cdot \dfrac{2x^4 + 4x^3}{x^2 - 16}$

2. $\dfrac{2}{x} - \dfrac{x}{x^3 + 3x^2} + \dfrac{1}{x + 3}$

3. $\dfrac{\left(\dfrac{x}{y} - \dfrac{y}{x}\right)}{\left(\dfrac{x - y}{xy}\right)}$

4. $\sqrt{-2}\left(\sqrt{-8} + 3\right)$

5. $\dfrac{-4x^{-3}y^4}{6xy^{-2}}$

6. $\left(\dfrac{t^{1/2}}{t^{1/4}}\right)^2$

7. $10\sqrt{20x} + 3\sqrt{125x}$

8. $\left(\sqrt{2x} - 3\right)^2$

9. $\dfrac{6}{\sqrt{10} - 2}$

10. $\dfrac{x^3 + 27}{x + 3}$ (Use synthetic division.)

In Exercises 11 and 12, graph the rational function.

11. $y = \dfrac{4}{x - 2}$

12. $y = \dfrac{4x^2}{x^2 + 1}$

In Exercises 13–16, solve the equation.

13. $x + \dfrac{4}{x} = 4$

14. $\sqrt{x + 10} = x - 2$

15. $(x - 5)^2 + 50 = 0$

16. $3x^2 + 6x + 2 = 0$

17. Use a graphing utility to graph the equation $y = x^2 - 6x - 8$. Use the graph to approximate any x-intercepts of the graph. Set $y = 0$ and solve the resulting equation. Compare the results with the x-intercepts of the graph.

18. Find a quadratic equation having the solutions -2 and 6.

19. Evaluate without the aid of a calculator: $(4 \times 10^3)^2$

20. The volume V of a right circular cylinder is $V = \pi r^2 h$. The two cylinders in the figure have equal volumes. Write r_2 as a function of r_1.

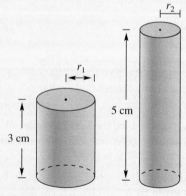

Figure for 20

Lines, Conics, and Variation

7

- Writing Equations of Lines
- Graphs of Linear Inequalities
- Graphs of Quadratic Functions
- Graphs of Second-Degree Equations
- Variation

Mass transportation includes taxis, vans, buses, trains, subways, and airplanes. In 1990, almost four trillion passenger miles were traveled in the United States by these six means of mass transportation. (Passenger miles is the product of the number of vehicle miles and the number of passengers.)

The table at the right shows the number of vehicle miles (in millions) traveled by buses from 1980 through 1991. The data in the table is represented graphically in the scatter plot.

From the scatter plot, you can see that the number of vehicle miles can be approximated by a line. The line that best fits this data is called the *regression line*. It can be used to predict the number of vehicle miles in future years. (Source: Public Transit Association)

t	0	1	2	3	4	5
Miles	1677	1685	1669	1678	1845	1863

t	6	7	8	9	10	11
Miles	2002	2079	2097	2109	2130	2182

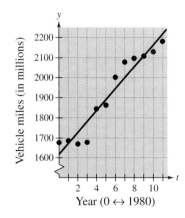

The chapter project related to this information is on page 489.

435

7.1 Writing Equations of Lines

The Point-Slope Equation of a Line ▪
Summary of Equations of Lines ▪ Application

The Point-Slope Equation of a Line

There are two basic types of problems in coordinate geometry.

1. Given an algebraic equation, sketch its graph.

2. Given a description of a graph, write its equation.

The first type of problem can be thought of as moving from algebra to geometry, whereas the second type can be thought of as moving the other way—from geometry to algebra. So far in the text, you have been working primarily with the first type of problem. In this section, you will look at the second problem.

In this section, you will learn that if you know the slope of a line *and* you also know the coordinates of one point on the line, you can find an equation for the line. Before giving a general formula for doing this, let's look at an example.

EXAMPLE 1 *Writing an Equation of a Line*

A line has a slope of $\frac{4}{3}$ and passes through the point $(-2, 1)$. Find an equation of this line.

Solution

Begin by sketching the line, as shown in Figure 7.1. You know that the slope of a line is the same through any two points on the line. Thus, to find an equation of the line, let (x, y) represent *any* point on the line. Using the representative point (x, y) and the given point $(-2, 1)$, it follows that the slope of the line is

$$m = \frac{y - 1}{x - (-2)}.$$

 Difference in y-values
Difference in x-values

Because the slope of the line is $m = \frac{4}{3}$, this equation can be rewritten as follows.

$$\frac{4}{3} = \frac{y - 1}{x + 2} \qquad \text{Slope formula}$$

$$4(x + 2) = 3(y - 1) \qquad \text{Cross-multiply.}$$

$$4x + 8 = 3y - 3 \qquad \text{Distributive Property}$$

$$4x - 3y = -11 \qquad \text{Subtract 8 and } 3y \text{ from both sides.}$$

An equation of the line is $4x - 3y = -11$.

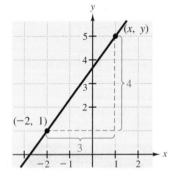

FIGURE 7.1

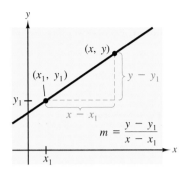

FIGURE 7.2

The procedure in Example 1 can be used to derive a *formula* for the equation of a line, given its slope and a point on the line. In Figure 7.2, let (x_1, y_1) be a given point on the line whose slope is m. If (x, y) is any *other* point on the line, it follows that

$$\frac{y - y_1}{x - x_1} = m.$$

This equation in variables x and y can be rewritten in the form

$$y - y_1 = m(x - x_1)$$

which is called the **point-slope form** of the equation of a line.

Point-Slope Form of the Equation of a Line

The **point-slope form** of the equation of the line that passes through the point (x_1, y_1) and has a slope of m is

$$y - y_1 = m(x - x_1).$$

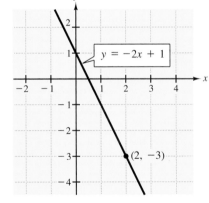

FIGURE 7.3

EXAMPLE 2   The Point-Slope Form of the Equation of a Line

Find an equation of the line that passes through the point $(2, -3)$ and has a slope of -2.

Solution

Use the point-slope form with $(x_1, y_1) = (2, -3)$ and $m = -2$.

$y - y_1 = m(x - x_1)$	Point-slope form
$y - (-3) = -2(x - 2)$	Substitute $y_1 = -3$, $x_1 = 2$, and $m = -2$.
$y + 3 = -2x + 4$	Simplify.
$y = -2x + 1$	Subtract 3 from both sides.

The graph of this line is shown in Figure 7.3.

In Example 2, notice that the final equation is written in slope-intercept form

$$y = mx + b. \qquad \text{Slope-intercept form}$$

You can use this form to check your work. First, observe that the slope is $m = -2$. Then substitute the coordinates of the given point $(2, -3)$ to see that the equation is satisfied.

The point-slope form can be used to find the equation of a line passing through two points (x_1, y_1) and (x_2, y_2). First, use the formula for the slope of the line passing through two points.

$$m = \frac{y_2 - y_1}{x_2 - x_1}$$

Then, once you know the slope, use the point-slope form to obtain the equation

$$y - y_1 = \frac{y_2 - y_1}{x_2 - x_1}(x - x_1). \qquad \text{Two-point form}$$

This is sometimes called the **two-point form** of the equation of a line.

EXAMPLE 3   An Equation of a Line Passing Through Two Points

Find an equation of the line that passes through the points $(4, 2)$ and $(-2, 3)$.

Solution

Let $(x_1, y_1) = (4, 2)$ and $(x_2, y_2) = (-2, 3)$. Then apply the formula for the slope of a line passing through two points as follows.

$$m = \frac{y_2 - y_1}{x_2 - x_1}$$

$$= \frac{3 - 2}{-2 - 4}$$

$$= -\frac{1}{6}$$

Now, using the point-slope form, you can find the equation of the line.

$$y - y_1 = m(x - x_1) \qquad \text{Point-slope form}$$

$$y - 2 = -\frac{1}{6}(x - 4) \qquad \text{Substitute } y_1 = 2,\ x_1 = 4,\ \text{and } m = -\frac{1}{6}.$$

$$y - 2 = -\frac{1}{6}x + \frac{2}{3} \qquad \text{Simplify.}$$

$$y = -\frac{1}{6}x + \frac{8}{3} \qquad \text{Add 2 to both sides.}$$

The graph of this line is shown in Figure 7.4.

In Example 3, it does not matter which of the two points is labeled (x_1, y_1) and which is labeled (x_2, y_2). Try switching these labels to

$$(x_1, y_1) = (-2, 3) \quad \text{and} \quad (x_2, y_2) = (4, 2)$$

and reworking the problem to see that you obtain the same equation.

Programming

The *TI-82* program listed below uses the two-point form to find an equation of a line. (Programs for other calculator models are given in the Appendix.) After the coordinates of two points are entered, the program outputs the slope and y-intercept of the line that passes through the points.

```
PROGRAM: TWOPTFRM
:Disp ''ENTER X1, Y1''
:Input X
:Input Y
:Disp ''Enter X2, Y2''
:Input C
:Input D
:(D-Y)/(C-X)→M
:M*(-X)+Y→B
:Disp ''SLOPE=''
:Disp M
:Disp ''Y-INT=''
:Disp B
```

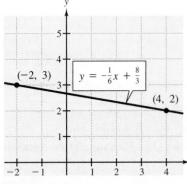

FIGURE 7.4

Summary of Equations of Lines

From the slope-intercept form of the equation of a line, you can see that a horizontal line ($m = 0$) has an equation of the form

$$y = (0)x + b \quad \text{or} \quad y = b. \qquad \text{Horizontal line}$$

This is consistent with the fact that each point on a horizontal line through $(0, b)$ has a y-coordinate of b, as shown in Figure 7.5. Similarly, each point on a vertical line through $(a, 0)$ has an x-coordinate of a, as shown in Figure 7.6. Hence, a vertical line has an equation of the form

$$x = a. \qquad \text{Vertical line}$$

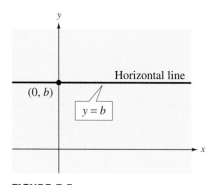

FIGURE 7.5

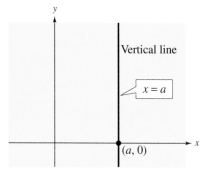

FIGURE 7.6

EXAMPLE 4   *Writing Equations of Horizontal and Vertical Lines*

Write an equation for each line.

a. Vertical line through $(-2, 4)$

b. Line passing through $(-2, 3)$ and $(3, 3)$

c. Line passing through $(-1, 2)$ and $(-1, 3)$

Solution

a. Because the line is vertical and passes through the point $(-2, 4)$, you know that every point on the line has an x-coordinate of -2. Thus, the equation is

$$x = -2.$$

b. The line through $(-2, 3)$ and $(3, 3)$ is horizontal. Thus, its equation is

$$y = 3.$$

c. The line through $(-1, 2)$ and $(-1, 3)$ is vertical. Thus, its equation is

$$x = -1.$$

The equation of a vertical line cannot be written in slope-intercept form because the slope of a vertical line is undefined. However, *every* line has an equation that can be written in the **general form**

$$ax + by + c = 0 \qquad \text{General form}$$

where a and b are not *both* zero.

Summary of Equations of Lines

1. Slope of line through (x_1, y_1) and (x_2, y_2): $m = \dfrac{y_2 - y_1}{x_2 - x_1}$

2. General form of equation of line: $ax + by + c = 0$

3. Equation of vertical line: $x = a$

4. Equation of horizontal line: $y = b$

5. Slope-intercept form of equation of line: $y = mx + b$

6. Point-slope form of equation of line: $y - y_1 = m(x - x_1)$

7. Parallel lines (equal slopes): $m_1 = m_2$

8. Perpendicular lines (negative reciprocal slopes): $m_2 = -\dfrac{1}{m_1}$

EXAMPLE 5   *Parallel and Perpendicular Lines*

Find an equation of the line that passes through the point $(3, -2)$ and is (a) parallel and (b) perpendicular to the line $x - 4y = 6$, as shown in Figure 7.7.

Solution

By writing the given line in slope-intercept form $y = \frac{1}{4}x - \frac{3}{2}$, you can see that it has a slope of $\frac{1}{4}$. Therefore, parallel lines must also have a slope of $\frac{1}{4}$ and perpendicular lines must have a slope of -4.

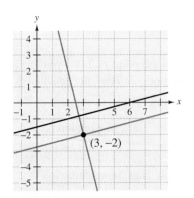

FIGURE 7.7

a. $y - y_1 = m(x - x_1)$ Point-slope form

$y - (-2) = \dfrac{1}{4}(x - 3)$ Substitute $y_1 = -2$, $x_1 = 3$, and $m = \frac{1}{4}$.

$y = \dfrac{1}{4}x - \dfrac{11}{4}$ Equation of parallel line

b. $y - y_1 = m(x - x_1)$ Point-slope form

$y - (-2) = -4(x - 3)$ Substitute $y_1 = -2$, $x_1 = 3$, and $m = -4$.

$y = -4x + 10$ Equation of perpendicular line

Application

EXAMPLE 6   An Application: Total Sales

The total sales of a new computer software company were $500,000 for the second year and $1,000,000 for the fourth year. Using only this information, what would you estimate the total sales to be during the fifth year?

Solution

To solve this problem, use a *linear model*, with y representing the total sales and t representing the year. That is, in Figure 7.8, let $(2, 500)$ and $(4, 1000)$ be two points on the line representing the total sales for the company. The slope of the line passing through these points is

$$m = \frac{1000 - 500}{4 - 2} = 250.$$

Now, using the point-slope form, the equation of the line is

$$y - y_1 = m(t - t_1) \qquad \text{Point-slope form}$$
$$y - 500 = 250(t - 2) \qquad \text{Substitute } y_1 = 500,\ t_1 = 2,\ \text{and } m = 250.$$
$$y = 250t. \qquad \text{Linear model for sales}$$

Finally, estimate the total sales during the fifth year $(t = 5)$ to be

$$y = 250(5) = \$1250 \text{ thousand} = \$1,250,000.$$

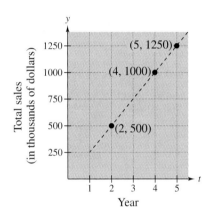

FIGURE 7.8

The estimation method illustrated in Example 6 is called **linear extrapolation.** Note in Figure 7.9 that for linear extrapolation, the estimated point lies to the right of the given points. When the estimated point lies *between* two given points, the procedure is called **linear interpolation.**

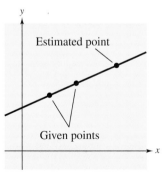

Linear Extrapolation
FIGURE 7.9

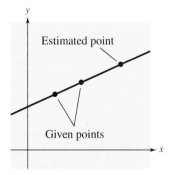

Linear Interpolation

The linear model is the most commonly used type of mathematical model. In the following group activity, you will learn one way to fit a linear model to a set of real data.

Group Activities   Problem Solving

Mathematical Modeling   You can find a linear model that fits a set of real data by first plotting the ordered pairs on a graph known as a **scatter plot.** Using a straightedge, draw a line through the points that seems to have the "best fit." You can then use any of the forms of a linear equation discussed in this section to find the equation of the line.

Suppose your manager asks you to make sense of the following set of data, in which y represents the percent of households with personal computer owners and x represents the year from 1983 to 1994, with $x = 3$ corresponding to 1983. (Source: Electronic Industries Association Consumer Electronic U.S. Sales, 1982–1994)

x	3	4	5	6	7	8	9	10	11	12	13	14
y	7	13	15	16	20	21	22	27	29	34	35	37

You think that a mathematical model will help you understand the trend in the data and may be useful in predicting what could happen in the future. Construct a scatter plot of the data. Do you think that a linear model would represent the data well? If so, find the equation of the best-fitting line. Interpret the meaning of the slope in the context of the data. Use your model to predict the percent of households with personal computer owners in 1999 (assuming that the trend continues).

7.1 Exercises

Discussing the Concepts

1. Can any pair of points on a line be used to determine an equation of the line? Explain.

2. Write the point-slope form, the slope-intercept form, and the general form of an equation of a line.

3. In the equation $y = 2x + 5$, what does the 2 represent? What does the 5 represent?

4. In the equation of a vertical line, the variable y is missing. Explain why.

5. Is it possible to have two perpendicular lines that rise from left to right? Explain.

6. What is the only type of line for which the concept of slope is not defined?

Problem Solving

In Exercises 7–10, match the equation with its graph. Use a graphing utility to verify your result.   [The graphs are labeled (a), (b), (c), and (d).]

(a)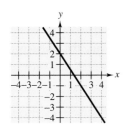

(b)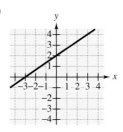

(c)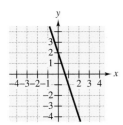

(d)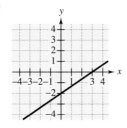

7. $y = \frac{2}{3}x + 2$

8. $y = \frac{2}{3}x - 2$

9. $y = -\frac{3}{2}x + 2$

10. $y = -3x + 2$

In Exercises 11–14, determine the slope of the line and the coordinates of one point through which it passes.

11. $y - 3 = 6(x + 4)$

12. $y = -5x + 12$

13. $3x - 2y = 0$

14. $3x + 4y - 16 = 0$

In Exercises 15–20, find the point-slope form of the equation of the line passing through the given point with the specified slope. Sketch the line.

15. $(0, 0)$, $m = -\frac{1}{2}$

16. $(0, 0)$, $m = \frac{2}{3}$

17. $(0, -4)$, $m = 3$

18. $(0, 6)$, $m = -\frac{3}{4}$

19. $\left(-2, \frac{7}{2}\right)$, $m = -4$

20. $\left(1, -\frac{3}{2}\right)$, $m = \frac{3}{4}$

In Exercises 21–26, find the general form of the equation of the line through the points. Sketch the line.

21. $(0, 0)$, $(2, 3)$

22. $(0, 0)$, $(3, -5)$

23. $(-2, 3)$, $(5, 0)$

24. $(1, -2)$, $(1, 8)$

25. $\left(\frac{3}{2}, 3\right)$, $\left(\frac{9}{2}, -4\right)$

26. $\left(4, \frac{7}{3}\right)$, $\left(-1, \frac{2}{3}\right)$

In Exercises 27–30, write an equation of the line passing through the two points. Use function notation to write y as a function of x. Use a graphing utility to graph the linear function.

27. $(-2, 2)$, $(4, 5)$

28. $(0, 10)$, $(5, 0)$

29. $(-2, 3)$, $(4, 3)$

30. $(-6, -3)$, $(4, 3)$

In Exercises 31–36, write equations of the line through the point that are (a) parallel and (b) perpendicular to the given line.

31. $(2, 1)$, $y = 3x$

32. $(-3, 4)$, $y = -2x$

33. $(-5, 4)$, $x + y = 1$

34. $\left(\frac{5}{8}, \frac{9}{4}\right)$, $5x = 4y$

35. $(-1, 2)$, $y + 5 = 0$

36. $(3, -4)$, $x - 10 = 0$

In Exercises 37–40, use a graphing utility on a square setting to graph both lines on the same screen. Decide whether the lines are parallel, perpendicular, or neither.

37. $y = -0.6x + 1$

$y = \frac{5}{3}x - 2$

38. $y = \frac{1}{2}(4x - 3)$

$y = \frac{1}{4}(4x + 3)$

39. $y = 0.6x + 1$

$y = \frac{1}{5}(3x + 22)$

40. $y = \frac{3}{2}x - 4$

$y = -\frac{2}{3}x + 1$

41. *Cost*   The cost (in dollars) of producing x units of a certain product is given by $C = 20x + 5000$. Use this model to complete the table.

x	0		100		1000
C		6000		15,000	

42. *Temperature Conversion*   The relationship between the Fahrenheit and Celsius temperature scales is given by $C = \frac{5}{9}F - \frac{160}{9}$. Use this equation to complete the table.

F	$-20°$		$20°$		$100°$	
C		$-17.8°$		$0°$		$100°$

Reviewing the Major Concepts

In Exercises 45–48, factor the expression completely.

45. $5x - 20x^2$

46. $64 - (x - 6)^2$

47. $15x^2 - 16x - 15$

48. $8x^3 + 1$

49. *Graphical Reasoning*   Consider the equation

$$2x^3 - 3x^2 - 18x + 27 = (2x - 3)(x + 3)(x - 3).$$

(a) Use a graphing utility to verify the equation by graphing both the left side and right side of the equation. Are the graphs the same?

(b) Verify the equation by multiplying the polynomials on the right side of the equation.

43. *Rental Occupancy*   A real estate office handles an apartment complex with 50 units. When the rent per unit is $450 per month, all 50 units are occupied. However, when the rent is $525 per month, the average number of occupied units drops to 45. Assume the relationship between the monthly rent p and the demand x is linear.

(a) Write the equation of the line giving the demand x in terms of the rent p.

(b) *Linear Extrapolation*   Use the equation from part (a) to predict the number of units occupied when the rent is $570.

(c) *Linear Interpolation*   Use the equation from part (a) to predict the number of units occupied when the rent is $480.

44. *Graphical Estimation*   A sales representative is reimbursed $150 per day for lodging and meals plus $0.34 per mile driven.

(a) Write a linear function giving the daily cost C to the company in terms of x, the number of miles driven.

(b) Use a graphing utility to graphically estimate the daily cost of driving 230 miles. Confirm your estimate algebraically.

(c) Use a graphing utility to graphically estimate the number of miles driven to produce a daily cost of $200. Confirm your estimate algebraically.

50. *Profit*   The cost of producing x units is $C = 12 + 8x$. The revenue from selling x units is

$$R = 16x - \frac{1}{4}x^2, \quad 0 \le x \le 20.$$

The profit is given by $P = R - C$.

(a) Perform the subtraction required to find the polynomial representing profit.

(b) Use a graphing utility to graph the polynomial representing profit.

(c) Determine the profit when $x = 16$ units are produced and sold. Use the graph of part (b) to predict the change in profit if x is some value other than 16.

Additional Problem Solving

In Exercises 51–56, find the slope of the line (if possible) and the coordinates of one point through which it passes.

51. $y + \frac{5}{8} = \frac{3}{4}(x + 2)$

52. $y - \frac{3}{4} = \frac{5}{6}\left(x - \frac{9}{10}\right)$

53. $y = \frac{2}{3}x - 2$

54. $y = -2(6x - 1)$

55. $5x - 2y + 24 = 0$

56. $x + 4 = 0$

In Exercises 57–66, find the point-slope form of the equation of the line passing through the given point with the specified slope.

57. $(0, -6)$, $m = \frac{1}{2}$

58. $(0, 9)$, $m = -\frac{1}{3}$

59. $(-2, 8)$, $m = -2$

60. $(0, 0)$, $m = 3$

61. $(5, 0)$, $m = -\frac{2}{3}$

62. $(-3, -7)$, $m = \frac{5}{4}$

63. $(-8, 5)$, $m = 0$

64. $(2, 1)$, m is undefined.

65. $\left(\frac{3}{4}, \frac{5}{2}\right)$, $m = \frac{4}{3}$

66. $\left(-\frac{3}{2}, \frac{1}{2}\right)$, $m = -3$

In Exercises 67–74, find the general form of the equation of the line passing through the points.

67. $(-5, 2)$, $(5, -2)$

68. $(5, 4)$, $(-3, 5)$

69. $\left(10, \frac{1}{2}\right)$, $\left(\frac{3}{2}, \frac{7}{4}\right)$

70. $(-2, 12)$, $(6, 12)$

71. $(7.5, 2)$, $(7.5, 9)$

72. $(2, -8)$, $\left(6, \frac{8}{3}\right)$

73. $(2, 0.6)$, $(8, -4.2)$

74. $(-5, 0.6)$, $(3, -3.4)$

In Exercises 75–78, write equations of the line through the given point that are (a) parallel and (b) perpendicular to the given line.

75. $(3, 7)$

$4x - y - 3 = 0$

76. $(-5, -10)$

$2x + 5y - 12 = 0$

77. $(6, -4)$

$3x + 10y = 24$

78. $\left(\frac{2}{3}, \frac{4}{3}\right)$

$x - 5 = 0$

In Exercises 79–84, use a graphing utility on a square setting to decide whether the lines are parallel, perpendicular, or neither.

79. $y = 3(x - 2)$

$y = 3x + 2$

80. $y = \frac{3}{2}(2x - 3)$

$y = -\frac{1}{3}(x + 3)$

81. $y = 1 - 1.5x$

$y = \frac{2}{3}x - 4$

82. $y = 2.5x - 4$

$y = -2.5x + 3$

83. $x + 2y - 3 = 0$

$-2x - 4y + 1 = 0$

84. $3x - 4y - 1 = 0$

$4x + 3y + 2 = 0$

Exploration In Exercises 85–88, find an equation of the line with intercepts $(a, 0)$ and $(0, b)$ where the equation is given by

$$\frac{x}{a} + \frac{y}{b} = 1, \quad a \neq 0, \quad b \neq 0.$$

85. x-intercept: $(3, 0)$

y-intercept: $(0, 2)$

86. x-intercept: $(-6, 0)$

y-intercept: $(0, 2)$

87. x-intercept: $\left(-\frac{5}{6}, 0\right)$

y-intercept: $\left(0, -\frac{7}{3}\right)$

88. x-intercept: $\left(-\frac{8}{3}, 0\right)$

y-intercept: $(0, -4)$

89. *Sales Commission* The salary for a sales representative is $1500 per month plus a 3% commission of total monthly sales. Write a linear function giving the salary S in terms of the monthly sales M.

90. *Reimbursed Expenses* A sales representative is reimbursed $125 per day for lodging and meals plus $0.22 per mile driven. Write a linear function giving the daily cost C to the company in terms of x, the number of miles driven.

91. *Discount Price* A store is offering a 30% discount on all items in its inventory.

(a) Write a linear function giving the sale price S for an item in terms of its list price L.

(b) Use the function in part (a) to find the sale price of an item that has a list price of $135.

92. *Straight-Line Depreciation* A small business purchases a photocopier for $7400. After 4 years, its depreciated value will be $1500.

(a) Assuming straight-line depreciation, write a linear function giving the value V of the copier in terms of time t.

(b) Use the function in part (a) to find the value of the copier after 2 years.

93. *College Enrollment*   A small college had an enroll-ment of 1500 students in 1980. During the next 10 years, the enrollment increased by approximately 60 students per year.

(a) Write a linear function giving the enrollment N in terms of the year t. (Let $t = 0$ represent 1980.)

(b) *Linear Extrapolation* Use the function in part (a) to predict the enrollment in the year 2000.

(c) *Linear Interpolation* Use the function in part (a) to estimate the enrollment in 1985.

94. *Soft-Drink Sales*   When soft drinks sold for $0.80 per can at football games, approximately 6000 cans were sold. When the price was raised to $1.00 per can, the demand dropped to 4000. Assume the rela-tionship between the price p and demand x is linear.

(a) Write an equation of the line giving the demand x in terms of the price p.

(b) *Linear Extrapolation* Use the equation in part (a) to predict the number of cans of soft drinks sold if the price is $1.10.

(c) *Linear Interpolation* Use the equation in part (a) to predict the number of cans of soft drinks sold if the price is $0.90.

Graphical Interpretation   In Exercises 95 and 96, use the graph, which shows the cost C (in billions of dollars) of running the Internal Revenue Service from 1980 to 1990.   (Source: Internal Revenue Service)

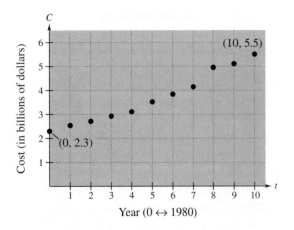

95. Using the costs for 1980 and 1990, write a linear model for the average cost, letting $t = 0$ represent 1980. Estimate the cost of running the IRS in 1995.

96. In 1989, the IRS collected about 9.5 trillion dollars. Use the model to approximate the percent of this amount spent to run the IRS.

97. *Best-Fitting Line*   An instructor gives 20-point quiz-zes and 100-point exams in a mathematics course. The average quiz and test scores for six students are given as ordered pairs (x, y) where x is the average quiz score and y is the average test score. The or-dered pairs are $(18, 87)$, $(10, 55)$, $(19, 96)$, $(16, 79)$, $(13, 76)$, and $(15, 82)$.

(a) Plot the points.

(b) Use a ruler to sketch the "best-fitting" line through the points.

(c) Find an equation for the line sketched in part (b).

(d) Use the equation of part (c) to estimate the aver-age test score for a person with an average quiz score of 17.

98. *Depth Markers*   A swimming pool is 40 feet long, 20 feet wide, 4 feet deep at the shallow end, and 9 feet deep at the deep end. Position the side of the pool on the rectangular coordinate system as shown in the figure and find an equation of the line representing the edge of the inclined bottom of the pool. Use this equation to determine the distances from the deep end at which markers must be placed to indicate each 1-foot change in the depth of the pool.

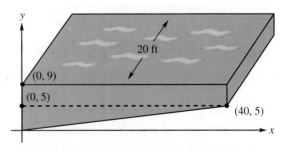

7.2   Graphs of Linear Inequalities

Linear Inequalities in Two Variables ▪
The Graph of a Linear Inequality

Linear Inequalities in Two Variables

A **linear inequality** in variables x and y is an inequality that can be written in one of the following forms.

$$ax + by < c, \quad ax + by > c, \quad ax + by \leq c, \quad \text{and} \quad ax + by \geq c$$

Here are some examples.

$$4x - 3y < 7, \quad x - y > -3, \quad x \leq 2, \quad \text{and} \quad y \geq -4$$

An ordered pair (x_1, y_1) is a **solution** of a linear inequality in x and y if the inequality is true when x_1 and y_1 are substituted for x and y, respectively. For instance, the ordered pair $(3, 2)$ is a solution of the inequality $x - y > 0$ because $3 - 2 > 0$ is a true statement.

EXAMPLE 1   Verifying Solutions of Linear Inequalities

Decide whether each point is a solution of $2x - 3y \geq -2$.

a. $(0, 0)$ **b.** $(2, 2)$ **c.** $(0, 1)$

Solution

a. $2x - 3y \geq -2$ Original inequality

$2(0) - 3(0) \overset{?}{\geq} -2$ Substitute 0 for x and 0 for y.

$0 \geq -2$ Inequality is satisfied. ✓

Because the inequality is satisfied, the point $(0, 0)$ is a solution.

b. $2x - 3y \geq -2$ Original inequality

$2(2) - 3(2) \overset{?}{\geq} -2$ Substitute 2 for x and 2 for y.

$-2 \geq -2$ Inequality is satisfied. ✓

Because the inequality is satisfied, the point $(2, 2)$ is a solution.

c. $2x - 3y \geq -2$ Original inequality

$2(0) - 3(1) \overset{?}{\geq} -2$ Substitute 0 for x and 1 for y.

$-3 \not\geq -2$ Inequality is not satisfied.

Because the inequality is not satisfied, the point $(0, 1)$ is not a solution.

The Graph of a Linear Inequality

The **graph** of an inequality is the collection of all solution points of the inequality. To sketch the graph of a linear inequality such as

$$4x - 3y < 12 \qquad \text{Original inequality}$$

begin by sketching the graph of the *corresponding linear equation*

$$4x - 3y = 12. \qquad \text{Corresponding equation}$$

Use *dashed* lines for the inequalities $<$ and $>$ and *solid* lines for the inequalities $\leq$ and $\geq$. The graph of the equation separates the plane into two **half-planes.** In each half-plane, one of the following must be true.

1. All points in the half-plane are solutions of the inequality.
2. No point in the half-plane is a solution of the inequality.

Thus, you can determine whether the points in an entire half-plane satisfy the inequality by simply testing *one* point in the region.

EXAMPLE 2 *Sketching the Graphs of Linear Inequalities*

Sketch the graph of each linear inequality.

a. $x \geq -3$ **b.** $y < 4$

Solution

a. The graph of the corresponding equation $x = -3$ is a vertical line. The points that satisfy the inequality $x \geq -3$ are those lying on or to the right of this line, as shown in Figure 7.10.

b. The graph of the corresponding equation $y = 4$ is a horizontal line. The points that satisfy the inequality $y < 4$ are those lying below this line, as shown in Figure 7.11.

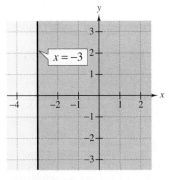

FIGURE 7.10

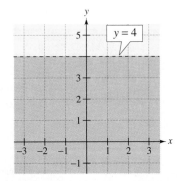

FIGURE 7.11

We summarize the guidelines for sketching the graph of a linear inequality in two variables as follows.

Guidelines for Graphing a Linear Inequality

---

1. Replace the inequality sign by an equal sign, and sketch the graph of the resulting equation. (Use a dashed line for $<$ or $>$, and a solid line for $\leq$ or $\geq$.)

2. Test one point in each of the half-planes formed by the graph in Step 1. If the point satisfies the inequality, shade the entire half-plane to denote that every point in the region satisfies the inequality.

EXAMPLE 3   *Sketching the Graph of a Linear Inequality*

Sketch the graph of the linear inequality

$$x + y > 3. \qquad \text{Original inequality}$$

Solution

The graph of the corresponding equation

$$x + y = 3 \qquad \text{Corresponding equation}$$

is a line, as shown in Figure 7.12. Because the origin $(0, 0)$ does not satisfy the inequality, the graph consists of the half-plane lying above the line. (Try checking a point above the line. Regardless of which point you choose, you will see that it is a solution.)

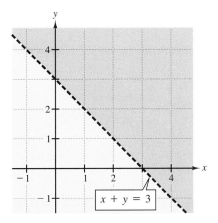

FIGURE 7.12

For a linear inequality in two variables, you can sometimes simplify the graphing procedure by writing the inequality in *slope-intercept form.* For instance, by writing $x + y > 1$ in the form $y > -x + 1$, you can see that the solution points lie *above* the line $y = -x + 1$, as shown in Figure 7.13. Similarly, by writing the inequality $4x - 3y > 12$ in the form

$$y < \frac{4}{3}x - 4$$

you can see that the solutions lie *below* the line $y = \frac{4}{3}x - 4$, as shown in Figure 7.14.

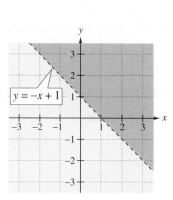

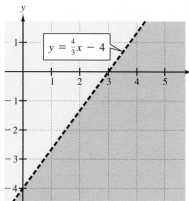

FIGURE 7.13   **FIGURE 7.14**

EXAMPLE 4   *Sketching the Graph of a Linear Inequality*

Use the slope-intercept form of a linear equation as an aid in sketching the graph of the inequality $2x - 3y \leq 15$.

Solution

To begin, rewrite the inequality in slope-intercept form.

$2x - 3y \leq 15$ Original inequality

$-3y \leq -2x + 15$ Subtract $2x$ from both sides.

$y \geq \frac{2}{3}x - 5$ Slope-intercept form

From this form, you can conclude that the solution is the half-plane lying *on or above* the line

$$y = \frac{2}{3}x - 5.$$

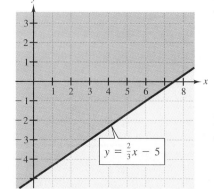

FIGURE 7.15

The graph is shown in Figure 7.15.

Technology

Graphing Utilities and Inequalities

Most graphing utilities can graph inequalities in two variables. For instance, you can sketch the graph of the inequality

$$y \le \tfrac{1}{2}x - 3$$

on a *TI-82* with the following steps.

1. Set the viewing window. For this graph, you can use a standard setting by using the ZOOM feature and the STANDARD option.

2. Graph the inequality by entering the following steps.

    Press DRAW

    Cursor to "Shade("
    Enter the following: Shade(-10,0.5X−3)

    ENTER

    The graph of $y \le \tfrac{1}{2}x - 3$ is shown below.

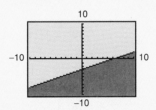

Group Activities   Extending the Concept

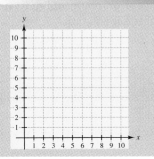

Using Inequalities   Try the following activity. One person picks a point with whole number coordinates on a grid like the one at left without revealing the coordinates. A second person writes the equation of a line passing through the grid region. The first person graphs the line on the grid and indicates whether the secret point lies above, below, or on the line. Continue writing and graphing lines until the second person is able to guess the coordinates of the secret point. Switch roles and try again. What is the fewest number of turns your team required to guess the point?

7.2 **Exercises**

Discussing the Concepts

1. List the four forms of a linear inequality in variables x and y.

2. What is meant by saying (x_1, y_1) is a solution of a linear inequality in x and y?

3. Explain the meaning of the term *half-plane*. Give an example of an inequality whose graph is a half-plane.

4. How does the solution set of $x - y > 1$ differ from the solution set of $x - y \geq 1$?

5. After graphing the boundary, how do you decide which half-plane is the solution set of a linear inequality?

6. Explain the difference between graphing the solution of the inequality $x \leq 3$ on the real number line and graphing it on a rectangular coordinate system.

Problem Solving

In Exercises 7–12, match the inequality with its graph. [The graphs are labeled (a), (b), (c), (d), (e), and (f).]

(a)

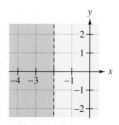

(b)

(c)

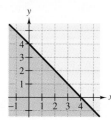

(d)

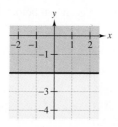

(e)

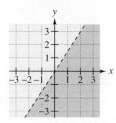

(f)

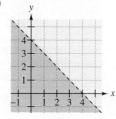

7. $y \geq -2$

8. $x < -2$

9. $3x - 2y < 0$

10. $3x - 2y > 0$

11. $x + y < 4$

12. $x + y \leq 4$

In Exercises 13–16, determine whether the points are solutions of the inequality.

Inequality	*Points*

13. $x - 2y < 4$ (a) $(0, 0)$ (b) $(2, -1)$
          (c) $(3, 4)$ (d) $(5, 1)$

14. $-3x + 5y \geq 6$ (a) $(2, 8)$ (b) $(-10, -3)$
          (c) $(0, 0)$ (d) $(3, 3)$

15. $y > 0.2x - 1$ (a) $(0, 2)$ (b) $(6, 0)$
          (c) $(4, -1)$ (d) $(-2, 7)$

16. $y \geq |x - 3|$ (a) $(0, 0)$ (b) $(1, 2)$
          (c) $(4, 10)$ (d) $(5, -1)$

In Exercises 17–22, sketch the graph of the solution of the linear inequality.

17. $x \geq 2$

18. $y > 2$

19. $y \leq x + 1$

20. $y > 4 - x$

21. $x - 2y \geq 6$

22. $3x + 5y \leq 15$

In Exercises 23–26, use a graphing utility to graph (shade) the solution of the inequality.

23. $y \geq \frac{3}{4}x - 1$

24. $y \leq 9 - 1.5x$

25. $2x + 3y - 12 \leq 0$

26. $x - 3y + 9 \geq 0$

27. Represent all points above the line $y = x$ with an inequality.

28. Represent all points below the line $y = x$ with an inequality.

In Exercises 29–34, write an inequality for the shaded region shown in the figure.

29.

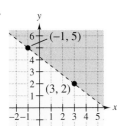

30.

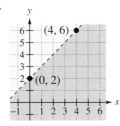

31.

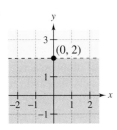

32.

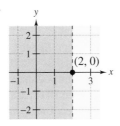

33.

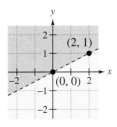

34.

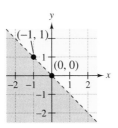

In Exercises 35–38, sketch the region containing the solution points that satisfy both inequalities. Explain your reasoning.

35. $x \geq 2, \ y \leq 4$

36. $y \leq x + 2, \ x \geq 1$

37. $3x + 2y \geq 2, \ y \geq -5$

38. $x + y \leq 5, \ x \geq 2$

In Exercises 39–42, use a graphing utility to graph (shade) the region containing the solution points that satisfy both inequalities.

39. $2x - 3y \leq 6, \ y \leq 4$

40. $6x + 3y \geq 12, \ y \leq 4$

41. $2x + y \leq 2, \ y \geq -4$

42. $x - 2y \geq -6, \ y \leq 6$

43. *Roasting a Turkey* The time t (in minutes) that it takes to roast a turkey weighing p pounds at 350°F is given by the following inequalities.

For a turkey up to 6 pounds: $t \geq 20p$

For a turkey over 6 pounds: $t \geq 15p + 30$

Sketch the graphs of these inequalities. What are the coordinates for a 12-pound turkey that has been roasting for 3 hours and 40 minutes? Is this turkey fully cooked?

44. *Pizza and Soda Pop* You and some friends go out for pizza. Together you have $26. You want to order two large pizzas with cheese at $8 each. Each additional topping costs $0.40, and each small soft drink costs $0.80. Write an inequality that represents the numbers of toppings x and drinks y that your group can afford. Sketch the graph of the inequality. What are the coordinates for an order of six soft drinks and two large pizzas with cheese, each with three additional toppings? Is this a solution of the inequality? (Assume there is no sales tax.)

Reviewing the Major Concepts —————————

In Exercises 45–48, simplify the expression.

45. $(x^3 \cdot x^{-2})^{-3}$

46. $(5x^{-4}y^5)(-3x^2y^{-1})$

47. $\left(\dfrac{2x}{3y}\right)^{-2}$

48. $\left(\dfrac{7u^{-4}}{3v^{-2}}\right)\left(\dfrac{14u}{6v^2}\right)^{-1}$

49. *Geometry* If a television set is advertised as having a 19-inch screen, it means that the diagonal measurement of the screen is 19 inches. If the screen is square, what are its height and width?

50. *Geometry* A tennis court is 36 feet wide and 78 feet long. Find the length of the diagonal of the court.

Additional Problem Solving

In Exercises 51–54, determine whether the points are solutions of the inequality.

Inequality	Points

51. $3x + y \geq 10$ (a) $(1, 3)$ (b) $(-3, 1)$
                     (c) $(3, 1)$ (d) $(2, 15)$

52. $x + y < 3$ (a) $(0, 6)$ (b) $(4, 0)$
                     (c) $(0, -2)$ (d) $(1, 1)$

53. $y \leq 3 - |x|$ (a) $(-1, 4)$ (b) $(2, -2)$
                     (c) $(6, 0)$ (d) $(5, -2)$

54. $y < -3.5x + 7$ (a) $(1, 5)$ (b) $(5, -1)$
                     (c) $(-1, 4)$ (d) $(0, \frac{4}{3})$

In Exercises 55–66, sketch the graph of the solution of the linear inequality.

55. $y < 5$ **56.** $x < -3$

57. $y > \frac{1}{2}x$ **58.** $y \leq 2x$

59. $y - 1 > -\frac{1}{2}(x - 2)$ **60.** $y - 2 < -\frac{2}{3}(x - 1)$

61. $\dfrac{x}{3} + \dfrac{y}{4} \leq 1$ **62.** $\dfrac{x}{2} + \dfrac{y}{6} \geq 1$

63. $3x - 2y \geq 4$ **64.** $x + 3y \leq 5$

65. $0.2x + 0.3y < 2$ **66.** $x - 0.75y > 6$

In Exercises 67–70, use a graphing utility to graph (shade) the solution of the inequality.

67. $y \leq -\frac{2}{3}x + 6$

68. $y \geq \frac{1}{4}x + 3$

69. $x - 2y - 4 \geq 0$

70. $2x + 4y - 3 \leq 0$

In Exercises 71–74, sketch the region showing all points that satisfy both inequalities.

71. $x \leq 5, \ y \geq 2$

72. $y \geq 5 - x, \ x \geq 0$

73. $x + y \geq 4, \ y \leq 4$

74. $3x + y \leq 9, \ x \geq -1$

In Exercises 75–78, use a graphing utility to graph (shade) the region containing the solution points satisfying both inequalities.

75. $2x - 2y \leq 5, \ y \leq 6$

76. $y \geq -2, \ y \leq 8$

77. $2x + 3y \geq 12, \ y \geq 2$

78. $x - y \geq -4, \ y \leq 1$

79. *Geometry* The perimeter of a rectangle of length x and width y cannot exceed 500 feet. Write a linear inequality for this constraint and sketch its graph.

80. *Storage Space* A warehouse for storing chairs and tables has 1000 square feet of floor space. The amounts of space required for each chair and each table are 10 square feet and 15 square feet, respectively. Write a linear inequality for these space constraints if x is the number of chairs and y is the number of tables stored. Sketch a graph of the inequality.

81. *Diet Supplement* A dietitian is asked to design a special diet supplement using two foods. Each ounce of food X contains 30 units of calcium and each ounce of food Y contains 20 units of calcium. The minimum daily requirement in the diet is 300 units of calcium. Write an inequality that represents the different numbers of units of food X and food Y required. Sketch the graph of the inequality. From the graph, find several ordered pairs with positive integer coordinates that are solutions of the inequality.

82. *Weekly Pay* You have two part-time jobs. One is at a grocery store, which pays $9 per hour, and the other is mowing lawns, which pays $6 per hour. Between the two jobs, you want to earn at least $150 a week. Write an inequality that shows the different numbers of hours you can work at each job, and sketch the graph of the inequality. From the graph, find several ordered pairs with positive integer coordinates that are solutions of the inequality.

7.3	**Graphs of Quadratic Functions**

Graphs of Quadratic Functions ▪ Sketching a Parabola ▪
Writing the Equation of a Parabola ▪ Application

Graphs of Quadratic Functions

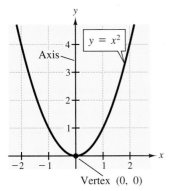

FIGURE 7.16

In this section, you will study graphs of quadratic functions.

$$f(x) = ax^2 + bx + c \qquad \text{Quadratic function}$$

Figure 7.16 shows the graph of a simple quadratic function, $y = x^2$.

Graphs of Quadratic Functions

The graph of $f(x) = ax^2 + bx + c$, $a \neq 0$, is a **parabola**. The completed-square form

$$f(x) = a(x - h)^2 + k \qquad \text{Standard form}$$

is the **standard form** of the function. The **vertex** of the parabola occurs at the point (h, k), and the vertical line passing through the vertex is the **axis** of the parabola.

DISCOVERY

You can use a graphing utility to discover a rule for determining the appearance of a parabola. Graph the equations below.

$$y_1 = x^2 - 3x - 5$$
$$y_2 = 7 - 3x^2$$
$$y_3 = -4 + 6x^2$$
$$y_4 = -x^2 - 6x$$

In your own words, write a rule for determining whether the graph of a parabola opens up or down by just looking at the equation. Does $y = 8 - 2x - 2x^2$ open up or down?

Every parabola is *symmetric* about its axis, which means that if it were folded along its axis, the two parts would match.

If a is positive, the graph of $f(x) = ax^2 + bx + c$ opens up, and if a is negative, the graph opens down, as shown in Figure 7.17.

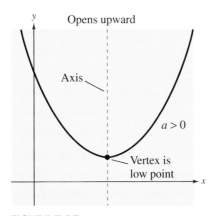

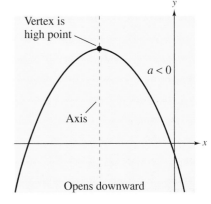

FIGURE 7.17

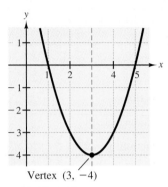

Vertex $(3, -4)$

FIGURE 7.18

EXAMPLE 1   Finding a Vertex by Completing the Square

Find the vertex of the graph of $f(x) = x^2 - 6x + 5$.

Solution

Begin by writing the function in standard form.

$$f(x) = x^2 - 6x + 5 \qquad \text{Original function}$$

$$f(x) = x^2 - 6x + (-3)^2 - (-3)^2 + 5 \qquad \text{Add and subtract } (-3)^2.$$

$$f(x) = (x^2 - 6x + 9) - 9 + 5 \qquad \text{Regroup terms.}$$

$$f(x) = (x - 3)^2 - 4 \qquad \text{Standard form}$$

From the standard form, you can see that the vertex of the parabola occurs at the point $(3, -4)$, as shown in Figure 7.18.

In Example 1, the vertex of the graph was found by *completing the square*. Another approach to finding the vertex is to complete the square once for a general function and then use the resulting formula for the vertex.

$$f(x) = ax^2 + bx + c \quad \Longrightarrow \quad f(x) = a\left(x + \frac{b}{2a}\right)^2 + c - \frac{b^2}{4a}$$

From this form you can see that the vertex occurs when $x = -b/2a$.

EXAMPLE 2   Finding a Vertex with a Formula

Find the vertex of the graph of $f(x) = x^2 + x$.

Solution

From the given function, it follows that $a = 1$ and $b = 1$. Thus, the x-coordinate of the vertex is

$$x = \frac{-b}{2a} = \frac{-1}{2(1)} = -\frac{1}{2}$$

and the y-coordinate is

$$f\left(-\frac{b}{2a}\right) = f\left(-\frac{1}{2}\right) = \left(-\frac{1}{2}\right)^2 + \left(-\frac{1}{2}\right) = \frac{1}{4} - \frac{1}{2} = -\frac{1}{4}.$$

Thus, the vertex of the parabola is $\left(-\frac{1}{2}, -\frac{1}{4}\right)$, and the parabola opens upward, as shown in Figure 7.19.

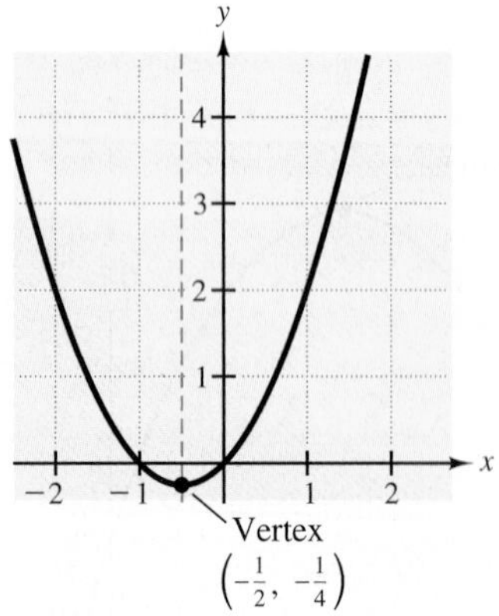

Vertex
$\left(-\frac{1}{2}, -\frac{1}{4}\right)$

FIGURE 7.19

Sketching a Parabola

NOTE The x- and y-intercepts are useful points to plot. Keep this in mind as you study the examples and do the exercises in this section.

Sketching a Parabola

1. Determine the vertex and axis of the parabola by completing the square or by formula.

2. Plot the vertex, axis, and a few additional points on the parabola. (Using the symmetry about the axis can reduce the number of points you need to plot.)

3. Use the fact that the parabola opens upward if $a > 0$ and opens downward if $a < 0$ to complete the sketch.

EXAMPLE 3 *Sketching a Parabola*

Sketch the graph of $x^2 - y + 6x + 8 = 0$.

Solution

Begin by writing the equation in standard form.

$x^2 - y + 6x + 8 = 0$	Original equation
$-y = -x^2 - 6x - 8$	Subtract $x^2 + 6x + 8$ from both sides.
$y = x^2 + 6x + 8$	Multiply both sides by -1.
$y = (x^2 + 6x + 3^2 - 3^2) + 8$	Add and subtract 3^2.
$y = (x^2 + 6x + 9) - 9 + 8$	Regroup terms.
$y = (x + 3)^2 - 1$	Standard form

The vertex occurs at the point $(-3, -1)$ and the axis is given by the line $x = -3$. After plotting this information, calculate a few additional points on the parabola, as shown in the table. Note that the y-intercept is $(0, 8)$ and the x-intercepts are solutions to the equation $x^2 + 6x + 8 = (x + 4)(x + 2) = 0$. The graph of the parabola is shown in Figure 7.20. Note that it opens upward because the leading coefficient (in standard form) is positive.

x-Value	-5	-4	-3	-2	-1
y-Value	3	0	-1	0	3
Solution Point	$(-5, 3)$	$(-4, 0)$	$(-3, -1)$	$(-2, 0)$	$(-1, 3)$

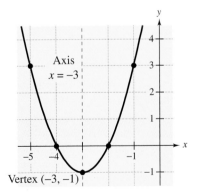

FIGURE 7.20

Writing the Equation of a Parabola

To write the equation of a parabola with a vertical axis, use the fact that its standard equation has the form $y = a(x - h)^2 + k$, where (h, k) is the vertex.

EXAMPLE 4 Writing the Equation of a Parabola

Write the equation of the parabola whose vertex is $(-2, 1)$ and whose y-intercept is $(0, -3)$, as shown in Figure 7.21.

Solution

Because the vertex occurs at $(h, k) = (-2, 1)$, you can write the following.

$$y = a(x - h)^2 + k \qquad \text{Standard form}$$
$$y = a[x - (-2)]^2 + 1 \qquad \text{Substitute } -2 \text{ for } h \text{ and } 1 \text{ for } k.$$
$$y = a(x + 2)^2 + 1 \qquad \text{Simplify.}$$

To find the value of a, use the fact that the y-intercept is $(0, -3)$.

$$y = a(x + 2)^2 + 1 \qquad \text{Standard form}$$
$$-3 = a(0 + 2)^2 + 1 \qquad \text{Substitute 0 for } x \text{ and } -3 \text{ for } y.$$
$$-3 = 4a + 1 \qquad \text{Simplify.}$$
$$-4 = 4a \qquad \text{Subtract 1 from both sides.}$$
$$-1 = a \qquad \text{Divide both sides by 4.}$$

This implies that the standard form of the equation of the parabola is

$$y = -(x + 2)^2 + 1.$$

Vertex $(-2, 1)$

Axis
$x = -2$

$(0, -3)$

FIGURE 7.21

Technology

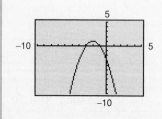

Once you have written an equation for a parabola, you can use a graphing utility to check your equation. For instance, the graph of

$$y = -(x + 2)^2 + 1$$

is shown at the left. By using the TRACE feature of the graphing utility, you can see that the vertex occurs at approximately $(-2, 1)$ and the y-intercept occurs at approximately $(0, -3)$.

Write the equation of the parabola whose vertex is $(3, -2)$ and whose y-intercept is $(0, 7)$. Graph the equation, and use the TRACE feature to check your equation.

Application

EXAMPLE 5   An Application Involving a Minimum Point

A suspension bridge is 100 feet long, as shown in Figure 7.22(a). The bridge is supported by cables attached at the tops of towers at each end of the bridge. Each cable hangs in the shape of a parabola (see Figure 7.22(b)) given by

$$y = 0.01x^2 - x + 35$$

where x and y are both measured in feet. (a) Find the distance between the lowest point of the cable and the roadbed of the bridge. (b) How tall are the towers?

Solution

a. Because $a = 0.01$ and $b = -1$, it follows that the vertex of the parabola occurs when $x = -b/2a = 1/(0.02) = 50$. At this x-value, the value of y is

$$y = 0.01(50)^2 - 50 + 35$$
$$= 10.$$

Thus, the minimum distance between the cable and the roadbed is 10 feet.

b. Because the vertex of the parabola occurs at the midpoint of the bridge, the two towers are located at the points where $x = 0$ and $x = 100$. Substituting an x-value of 0, you can find that the corresponding y-value is

$$y = 0.01(0)^2 - 0 + 35$$
$$= 35.$$

Therefore, the towers are each 35 feet high. (Try substituting $x = 100$ in the equation to see that you obtain the same y-value.)

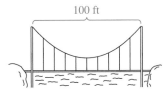

(a)

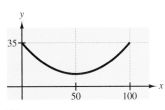

(b)

FIGURE 7.22

Group Activities   Problem Solving

Quadratic Modeling   The data in the table represents the average monthly temperature y in degrees Fahrenheit in Savannah, Georgia for the month x, with $x = 1$ corresponding to November   (Source: National Climate Data Center). Plot the data. Find a quadratic model for the data and use it to find the average temperatures for December and February. The actual average temperature for both December and February is 52°F. How well do you think the model fits the data? Use the model to predict the average temperature for June. How useful do you think the model would be for the whole year?

x	1	3	5
y	59	49	59

7.3 Exercises

Discussing the Concepts

1. In your own words, describe the graph of a quadratic function $f(x) = ax^2 + bx + c$.

2. Explain how to find the vertex of the graph of a quadratic function.

3. Explain how to find the x- and y-intercepts of the graph of a quadratic function.

4. Explain how to determine whether the graph of a quadratic function opens up or down.

5. How is the discriminant related to the graph of a quadratic function?

6. Is it possible for the graph of a quadratic function to have two y-intercepts? Explain.

Problem Solving

In Exercises 7–12, match the equation with its graph. [The graphs are labeled (a), (b), (c), (d), (e), and (f).]

(a)

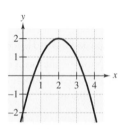

(b)

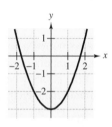

(c)

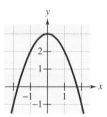

(d)

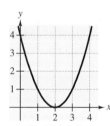

(e)

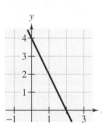

(f)

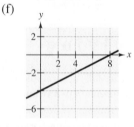

7. $f(x) = 4 - 2x$

8. $f(x) = \frac{1}{2}x - 4$

9. $f(x) = x^2 - 3$

10. $f(x) = -x^2 + 3$

11. $f(x) = (x - 2)^2$

12. $f(x) = 2 - (x - 2)^2$

In Exercises 13–16, state whether the graph opens up or down, and find the vertex.

13. $y = 2(x - 0)^2 + 2$

14. $y = -3(x + 5)^2 - 3$

15. $y = x^2 - 6$

16. $y = -(x + 1)^2$

In Exercises 17–20, find the x- and y-intercepts of the graph.

17. $y = 25 - x^2$

18. $y = x^2 + 4x$

19. $y = x^2 - 3x + 3$

20. $y = x^2 - 3x - 10$

In Exercises 21–24, write the equation in standard form and find the vertex of its graph.

21. $y = x^2 - 4x + 7$

22. $y = x^2 + 6x - 5$

23. $y = -x^2 + 2x - 7$

24. $y = -x^2 - 10x + 10$

In Exercises 25–32, sketch the graph of the equation. Identify the vertex and any x-intercepts. Use a graphing utility to verify your graph.

25. $f(x) = x^2 - 4$

26. $f(x) = -x^2 + 9$

27. $f(x) = -x^2 + 3x$

28. $f(x) = x^2 - 4x$

29. $f(x) = x^2 - 8x + 15$

30. $f(x) = -x^2 + 2x + 8$

31. $f(x) = \frac{1}{5}(3x^2 - 24x + 38)$

32. $f(x) = \frac{1}{5}(2x^2 - 4x + 7)$

In Exercises 33–38, write an equation of the parabola.

33.

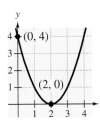

34.

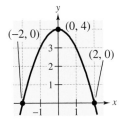

35.

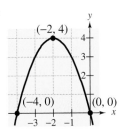

36.

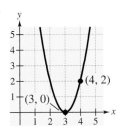

37.

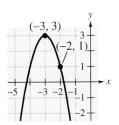

38.

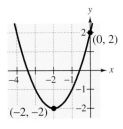

🖩 **39.** *Graphical Interpretation*   The height y (in feet) of a ball thrown by a child is given by

$$y = -\frac{1}{12}x^2 + 2x + 4$$

where x is the horizontal distance (in feet) from where the ball is thrown.

(a) Use a graphing utility to sketch the path of the ball.

(b) How high is the ball when it leaves the child's hand?

(c) How high is the ball when it is at its maximum height?

(d) How far from the child does the ball strike the ground?

40. *Maximum Height of a Diver*   The path of a diver is given by

$$y = -\frac{4}{9}x^2 + \frac{24}{9}x + 10$$

where y is the height in feet and x is the horizontal distance from the end of the diving board in feet (see figure). What is the maximum height of the diver?

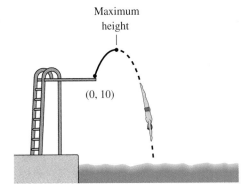

🖩 **41.** *Graphical Interpretation*   A company manufactures radios that cost (the company) $60 each. For buyers who purchase 100 or fewer radios, the purchase price p is $90 per radio. To encourage large orders, the company will reduce the price *per radio* for orders over 100, as follows. If 101 radios are purchased, the price is $89.85 per unit. If 102 radios are purchased, the price is $89.70 per unit. If $(100 + x)$ radios are purchased, the price per unit is

$$p = 90 - x(0.15)$$

where x is the amount over 100 in the order.

(a) Show that the profit for orders over 100 is

$$P = (100 + x)[90 - x(0.15)] - (100 + x)60$$

$$= 3000 + 15x - \frac{3}{20}x^2.$$

(b) Use a graphing utility to graph the profit function.

(c) Find the vertex of the profit curve and determine the order size for maximum profit.

(d) Would you recommend this pricing scheme? Explain your reasoning.

42. *Graphical Interpretation* The advertising revenue for newspapers in the United States for the years 1985 through 1991 is approximated by the model

$$R = -1.03 + 7.11t - 0.38t^2$$

where R is revenue in billions of dollars and t represents the year, with $t = 5$ corresponding to 1985. (Source: McCann-Erickson, Inc.)

(a) Use a graphing utility to graph the revenue function.

(b) Determine the year when advertising revenue was greatest. Approximate the revenue that year.

Think About It In Exercises 43 and 44, write the equation of the parabola.

43.

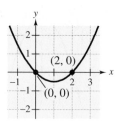

44.

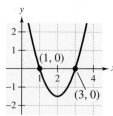

Reviewing the Major Concepts

In Exercises 45–48, solve the inequality and sketch its graph on the real number line.

45. $7 - 3x > 4 - x$ **46.** $2(x + 6) - 20 < 2$

47. $|x - 3| < 2$ **48.** $|x - 5| > 3$

In Exercises 49 and 50, solve the equation.

49. $x^2 - 4x + 3 = 0$ **50.** $2x^2 + 10x + 9 = 0$

Free-Falling Object In Exercises 51 and 52, find the time for an object to reach the ground when it is dropped from a height of s_0 feet. The height h (in feet) is given by

$$h = -16t^2 + s_0$$

where t is the time (in seconds).

51. $s_0 = 80$ **52.** $s_0 = 150$

Additional Problem Solving

In Exercises 53–56, state whether the graph opens up or down, and find the vertex.

53. $y = 4 - (x - 10)^2$ **54.** $y = 2(x - 12)^2 + 3$

55. $y = x^2 - 6x$ **56.** $y = -(x - 3)^2$

In Exercises 57–60, find the x- and y-intercepts of the graph.

57. $y = x^2 - 9x$ **58.** $y = x^2 - 49$

59. $y = 4x^2 - 12x + 9$ **60.** $y = 10 - x - 2x^2$

In Exercises 61–66, write the equation in standard form and find the vertex of its graph.

61. $y = x^2 + 6x + 5$ **62.** $y = x^2 - 4x + 5$

63. $y = -x^2 + 6x - 10$ **64.** $y = 4 - 8x - x^2$

65. $y = 2x^2 + 6x + 2$ **66.** $y = 3x^2 - 3x - 9$

In Exercises 67–82, sketch the graph of the equation. Identify the vertex and any x-intercepts. Use a graphing utility to verify your graph.

67. $f(x) = -x^2 + 4$ **68.** $f(x) = x^2 - 9$

69. $f(x) = x^2 - 3x$ **70.** $f(x) = -x^2 + 4x$

71. $f(x) = (x - 4)^2$ **72.** $f(x) = -(x + 4)^2$

73. $f(x) = 5 - \dfrac{x^2}{3}$ **74.** $f(x) = \dfrac{x^2}{3} - 2$

75. $f(x) = x^2 + 4x + 7$ **76.** $f(x) = x^2 + 4x + 2$

77. $f(x) = 2(x^2 + 6x + 8)$

78. $f(x) = -x^2 + 6x - 7$

79. $f(x) = -(x^2 + 6x + 5)$

80. $f(x) = 3x^2 - 6x + 4$

81. $f(x) = \frac{1}{2}(x^2 - 2x - 3)$

82. $f(x) = -\frac{1}{2}(x^2 - 6x + 7)$

⊞ *Graphical and Algebraic Reasoning* In Exercises 83–86, use a graphing utility to approximate the vertex of the graph. Then check your result algebraically.

83. $y = \frac{1}{6}(2x^2 - 8x + 11)$

84. $y = -\frac{1}{4}(4x^2 - 20x + 13)$

85. $y = -0.7x^2 - 2.7x + 2.3$

86. $y = 0.75x^2 - 7.50x + 23.00$

⊞ In Exercises 87–90, use a graphing utility to graph the two functions on the same screen. Do the graphs intersect? If so, approximate the point or points of intersection.

87. $y_1 = -x^2 + 6, \ y_2 = 2$

88. $y_1 = x^2 - 6x + 8, \ y_2 = 3$

89. $y_1 = \frac{1}{2}x^2 - 3x + \frac{13}{2}, \ y_2 = 3$

90. $y_1 = -2x^2 - 4x, \ y_2 = 1$

In Exercises 91–98, write an equation of the parabola $y = ax^2 + bx + c$ that satisfies the conditions.

91. Vertex: $(2, 1)$; $\quad a = 1$

92. Vertex: $(-3, -3)$; $\quad a = 1$

93. Vertex: $(-3, 4)$; $\quad a = -1$

94. Vertex: $(3, -2)$; $\quad a = -1$

95. Vertex: $(2, -4)$; point on graph: $(0, 0)$

96. Vertex: $(-2, -4)$; point on graph: $(0, 0)$

97. Vertex: $(3, 2)$; point on graph: $(1, 4)$

98. Vertex: $(-1, -1)$; point on graph: $(0, 4)$

⊞ **99.** *Graphical Estimation* The profit (in thousands of dollars) for a company is given by

$$P = 230 + 20s - \frac{1}{2}s^2$$

where s is the amount (in hundreds of dollars) spent on advertising. Use a graphing utility to graph the profit function and approximate the amount of advertising that yields a maximum profit. Verify the maximum algebraically.

100. *Bridge Design*   A bridge is to be constructed over a gorge with the main supporting arch being a parabola (see figure). If the equation of the parabola is

$$y = 4\left(100 - \frac{x^2}{2500}\right)$$

where x and y are measured in feet,

(a) find the length of the road across the gorge.

(b) find the height of the parabolic arch at the center of the span.

(c) find the lengths of the vertical girders at intervals of 100 feet from the center of the bridge.

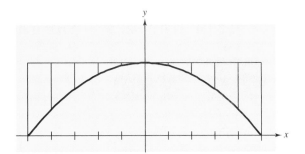

101. *Highway Design*   A highway department engineer must design a parabolic arc to create a turn in a freeway around a city. The vertex of the parabola is placed at the origin, and the parabola must connect with roads represented by the equations

$$y = -0.4x - 100, \quad x < -500$$

$$y = 0.4x - 100, \quad x > 500$$

(see figure). Find an equation of the parabolic arc.

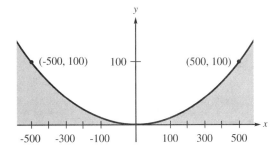

102. *Geometry* The area of a rectangle is given by the equation

$$A = \frac{2}{\pi}(100x - x^2), \quad 0 < x < 100$$

where x is the length of its base in feet. Use a graphing utility to sketch the graph of this equation and use the TRACE feature to approximate the value of x when A is maximum.

103. *Graphical Estimation* The cost of producing x units of a product is given by

$$C = 800 - 10x + \frac{1}{4}x^2, \quad 0 < x < 40.$$

Use a graphing utility to sketch the graph of this equation and use the TRACE feature to approximate the value of x when C is minimum.

Math Matters

Annual Salary Versus Hourly Wage

Here is a quick method that you can use to approximate the annual salary that corresponds to a given hourly wage.

Hourly wage $\times$ 2 $\times$ 1000 $\approx$ annual salary

For example, $5 an hour is approximately $10,000 a year, and $12 hour is approximately $24,000. This approximation assumes that there are only fifty 40-hour weeks in a year. To obtain the exact conversion from hourly wage to annual salary, you should multiply by 40 (hours per week) and then multiply by 52 (weeks per year).

To approximate the hourly wage that corresponds to a given annual salary, use the following rule.

Annual salary $\div$ 1000 $\div$ 2 $\approx$ hourly wage

For example, $12,000 a year is approximately $6 an hour, and $20,000 a year is approximately $10 an hour. To obtain the exact conversion from annual salary to hourly wage, you should divide by 52 (weeks per year) and then divide the result by 40 (hours per week).

Use the quick approximation technique to determine which of the following wages is greater.

(a) $7.50 an hour or $17,500 a year

(b) $12.50 an hour or $22,500 a year

MID-CHAPTER QUIZ

Take this quiz as you would take a quiz in class. After you are done, check your work against the answers given in the back of the book.

In Exercises 1–4, write an equation of the line passing through (x_1, y_1) with slope m.

1. $\left(0, -\frac{3}{2}\right)$, $m = 2$ **2.** $(0, 6)$, $m = -\frac{3}{2}$ **3.** $\left(\frac{5}{2}, 6\right)$, $m = -\frac{3}{4}$ **4.** $(-3.5, -1.8)$, $m = 3$

In Exercises 5–8, write an equation of the line passing through the points.

5. $\left(\frac{1}{3}, 1\right)$, $(4, 5)$ **6.** $(0, 0.8)$, $(3, -2.3)$ **7.** $(3, -1)$, $(10, -1)$ **8.** $\left(4, \frac{5}{3}\right)$, $(4, 8)$

9. Write an equation of the line that passes through the point $(3, 5)$ and is (a) parallel to and (b) perpendicular to the line $2x - 3y = 1$.

10. Decide whether the points are solutions of the inequality $2x - 3y \leq 4$. Explain your reasoning.

    (a) $(5, 2)$ (b) $(-2, 4)$ (c) $(2, -4)$ (d) $(3, 0)$

In Exercises 11 and 12, write an inequality for the shaded region.

11. See figure at right.   **12.** See figure at right.

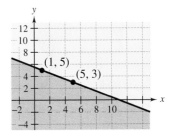

Figure for 11

In Exercises 13 and 14, sketch the region that is determined by both inequalities.

13. $2x + 3y \leq 9$, $y \geq 1$ **14.** $2x - y \leq 4$, $y \leq 4$

In Exercises 15 and 16, write an equation for the indicated parabola.

15. Vertex: $(3, -1)$; passes through the point $(5, 3)$

16. Vertex: $(5, 4)$; passes through the point $(3, 3)$

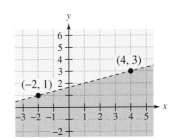

Figure for 12

In Exercises 17 and 18, sketch the graph of the quadratic function. Identify the vertex and the x-intercepts.

17. $y = -\frac{1}{4}(x^2 + 6x + 1)$ **18.** $y = 2x^2 - 4x - 7$

19. You purchase a used car for \$12,400. It is estimated that after 4 years its depreciated value will be \$5000. Assuming straight-line depreciation, write a linear function giving the value V of the car in terms of time t. State the domain of the function.

20. The path of a ball is given by $y = -0.005x^2 + x + 5$. Determine the maximum height of the ball.

7.4 Graphs of Second-Degree Equations

Circles ▪ Ellipses ▪ Hyperbolas ▪ Application

One of the first recognized female mathematicians, Hypatia (370–415 A.D.), wrote a textbook entitled *On the Conics of Apollonius.* Her death marked the end of major mathematical discoveries in Europe for several hundred years.

Circles

In Section 7.3, you learned that the graph of a second-degree equation of the form

$$y = ax^2 + bx + c$$

is a parabola. A parabola is one of four types of **conics** or **conic sections.** The other three types are circles, ellipses, and hyperbolas. All four types have equations that are of second degree. As indicated in Figure 7.23, the name "conic" relates to the fact that each of these figures can be obtained by intersecting a plane with a double-napped cone.

Circle   Parabola   Ellipse   Hyperbola
FIGURE 7.23

Conic sections occur in many practical applications. Reflective surfaces in satellite dishes, flashlights, and telescopes often are of parabolic shape. The orbits of planets are elliptical, and the orbits of comets are usually elliptical or hyperbolic. Ellipses and parabolas are also used in building archways and bridges.

A **circle** in the rectangular coordinate plane consists of all points (x, y) that are a given positive distance r from a fixed point, called the **center** of the circle. The positive distance r is the **radius** of the circle. If the center of the circle is the origin, as shown in Figure 7.24, the relationship between the coordinates of any point (x, y) on the circle and the radius r is given by

$$\text{Radius } = r = \sqrt{(x - 0)^2 + (y - 0)^2} \qquad \text{Distance Formula, center at } (0, 0)$$
$$= \sqrt{x^2 + y^2}$$
$$r = \sqrt{(x - h)^2 + (y - k)^2}. \qquad \text{Distance Formula, center at } (h, k)$$

By squaring both sides of this equation, you obtain the **standard form of the equation of a circle.**

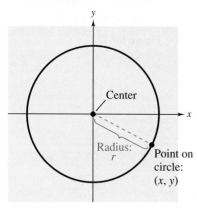

FIGURE 7.24

Standard Form of the Equation of a Circle

The **standard form of the equation of a circle** is

$$x^2 + y^2 = r^2 \qquad \text{Circle with center at } (0, 0)$$
$$(x - h)^2 + (y - k)^2 = r^2. \qquad \text{Circle with center at } (h, k)$$

The positive number r is the **radius** of the circle.

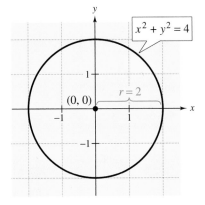

FIGURE 7.25

EXAMPLE 1   *Finding an Equation of a Circle*

Find an equation of the circle whose center is at $(0, 0)$ and whose radius is 2.

Solution

Use the standard form of the equation of a circle with center at the origin.

$$x^2 + y^2 = r^2 \qquad \text{Standard form with center at } (0, 0)$$
$$x^2 + y^2 = 2^2 \qquad \text{Substitute 2 for } r.$$
$$x^2 + y^2 = 4 \qquad \text{Equation of circle}$$

The circle given by this equation is shown in Figure 7.25.

To sketch the circle for a given equation, write the equation in standard form. From the standard form, you can identify the center and radius.

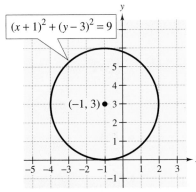

FIGURE 7.26

EXAMPLE 2   *Finding the Center and Radius of a Circle*

Identify the center and radius of the circle given by the equation, and sketch the circle.

$$x^2 + y^2 + 2x - 6y + 1 = 0$$

Solution

$$x^2 + y^2 + 2x - 6y + 1 = 0 \qquad \text{Original equation}$$
$$(x^2 + 2x) + (y^2 - 6y) = -1 \qquad \text{Group terms.}$$
$$(x^2 + 2x + 1) + (y^2 - 6y + 9) = -1 + 1 + 9 \qquad \text{Complete each square.}$$
$$(x + 1)^2 + (y - 3)^2 = 9 \qquad \text{Standard form}$$

From this standard form, you can see that the graph of the equation is a circle whose center is at $(-1, 3)$ and whose radius is 3, as shown in Figure 7.26.

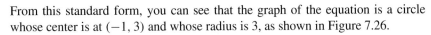

Ellipses

NOTE To trace an ellipse, place two thumbtacks at the foci, as shown in Figure 7.28. If the ends of a fixed length of string are fastened to the thumbtacks and the string is drawn taut with a pencil, the path traced by the pencil will be an ellipse.

An **ellipse** is the set of all points (x, y) such that the sum of the distances between (x, y) and two distinct fixed points is a constant. As shown in Figure 7.27(a), each of the two fixed points is a **focus** of the ellipse. (The plural of focus is *foci.*) In this text, we restrict the study of ellipses to those whose centers are at the origin.

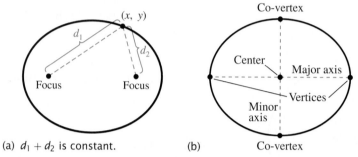

(a) $d_1 + d_2$ is constant. (b)

FIGURE 7.27

FIGURE 7.28

The line through the foci intersects the ellipse at the **vertices,** as shown in Figure 7.27(b). The line segment joining the vertices is the **major axis,** and its midpoint is the **center** of the ellipse. The line segment perpendicular to the major axis at the center is the **minor axis** of the ellipse, and the points at which the minor axis intersects the ellipse are **co-vertices.**

STUDY TIP

If the equation of an ellipse is of the form $\frac{x^2}{a^2} + \frac{y^2}{b^2} = 1$, its major axis is horizontal. Because a is greater than b and its square is the denominator of the x^2 term, you can conclude that the major axis lies along the x-axis. Similarly, if the equation of an ellipse is of the form $\frac{x^2}{b^2} + \frac{y^2}{a^2} = 1$, its major axis is vertical.

Standard Form of the Equation of an Ellipse

The **standard form of the equation of an ellipse** with center at the origin and major and minor axes of lengths $2a$ and $2b$, respectively, is

$$\frac{x^2}{a^2} + \frac{y^2}{b^2} = 1 \quad \text{or} \quad \frac{x^2}{b^2} + \frac{y^2}{a^2} = 1, \qquad 0 < b < a.$$

The vertices lie on the major axis, a units from the center, and the co-vertices lie on the minor axis, b units from the center.

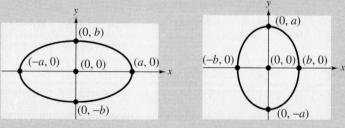

Major axis is horizontal.
Minor axis is vertical.

Major axis is vertical.
Minor axis is horizontal.

EXAMPLE 3   *Finding an Equation of an Ellipse*

Find an equation of the ellipse whose vertices are $(-3, 0)$ and $(3, 0)$ and whose co-vertices are $(0, -2)$ and $(0, 2)$.

Solution

Begin by plotting the vertices and co-vertices, as shown in Figure 7.29. The center of the ellipse is $(0, 0)$, because it is the point that lies halfway between the vertices (and halfway between the co-vertices). Thus, the equation of the ellipse has the form

$$\frac{x^2}{a^2} + \frac{y^2}{b^2} = 1.$$

For this ellipse, the major axis is horizontal. Thus, a is the distance between the center and either vertex, which implies that $a = 3$. Similarly, b is the distance between the center and either co-vertex, which implies that $b = 2$. Thus, the standard equation of the ellipse is

$$\frac{x^2}{3^2} + \frac{y^2}{2^2} = 1.$$

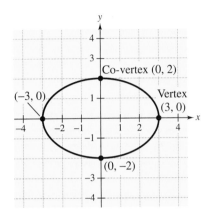

FIGURE 7.29

To sketch an ellipse, it helps to first write its equation in standard form.

EXAMPLE 4   *Sketching an Ellipse*

Sketch the ellipse given by $4x^2 + y^2 = 36$, and identify the vertices and co-vertices.

Solution

Begin by writing the equation in standard form.

$4x^2 + y^2 = 36$	Given equation
$\dfrac{4x^2}{36} + \dfrac{y^2}{36} = \dfrac{36}{36}$	Divide both sides by 36.
$\dfrac{x^2}{9} + \dfrac{y^2}{36} = 1$	Simplify.
$\dfrac{x^2}{3^2} + \dfrac{y^2}{6^2} = 1$	Standard form

Because the denominator of the y^2 term is larger than the denominator of the x^2 term, you can conclude that the major axis is vertical. Moreover, because $a = 6$, the vertices are $(0, -6)$ and $(0, 6)$. Finally, because $b = 3$, the co-vertices are $(-3, 0)$ and $(3, 0)$, as shown in Figure 7.30.

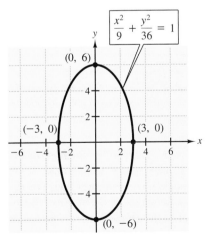

FIGURE 7.30

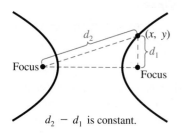

$d_2 - d_1$ is constant.

FIGURE 7.31

Hyperbolas

A **hyperbola** on the rectangular coordinate system consists of all points (x, y) such that the *difference* between (x, y) and two fixed points is a constant, as shown in Figure 7.31. The two fixed points are called the **foci** of the hyperbola. As with ellipses, we will consider only equations of hyperbolas whose foci lie on the x-axis or on the y-axis. The line on which the foci lie is called the **transverse axis** of the hyperbola.

Standard Form of the Equation of a Hyperbola

The **standard form of the equation of a hyperbola** whose center is at the origin is given by

$$\frac{x^2}{a^2} - \frac{y^2}{b^2} = 1 \qquad \text{Transverse axis is horizontal.}$$

$$\frac{y^2}{a^2} - \frac{x^2}{b^2} = 1 \qquad \text{Transverse axis is vertical.}$$

where a and b are positive real numbers. The **vertices** of the hyperbola lie on the transverse axis, a units from the center.

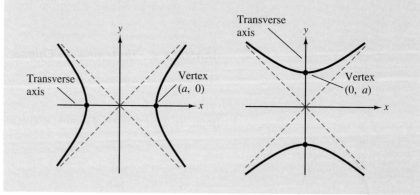

A hyperbola has two disconnected parts, each of which is a **branch** of the hyperbola. The two branches approach a pair of intersecting straight lines called **asymptotes** of the hyperbola. The two asymptotes intersect at the center of the hyperbola.

To sketch a hyperbola, form a **central rectangle** whose center is the origin and whose width and height are $2a$ and $2b$. Note in Figure 7.32 (on page 471) that the asymptotes pass through the corners of the central rectangle and the vertices of the hyperbola lie at the centers of opposite sides of the central rectangle.

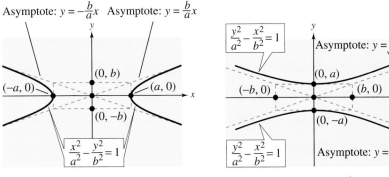

Transverse axis is horizontal.   Transverse axis is vertical.
FIGURE 7.32

EXAMPLE 5   Sketching a Hyperbola

Sketch the hyperbola given by $\dfrac{x^2}{36} - \dfrac{y^2}{16} = 1$.

Solution

From the standard form of the equation

$$\frac{x^2}{6^2} - \frac{y^2}{4^2} = 1$$

you can see that the center of the hyperbola is the origin and the transverse axis is horizontal. Therefore, the vertices lie six units to the left and right of the center at the points $(-6, 0)$ and $(6, 0)$. Because $a = 6$ and $b = 4$, you can sketch the hyperbola by first drawing a central rectangle whose width is $2a = 12$ and whose height is $2b = 8$, as shown in Figure 7.33(a). Next, draw the asymptotes of the hyperbola through the corners of the central rectangle and plot the vertices. Finally, draw the hyperbola, as shown in Figure 7.33(b).

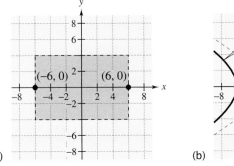

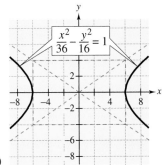

(a)   (b)

FIGURE 7.33

Finding an equation of a hyperbola is a little more difficult than finding equations of the other three types of conics. However, if you know the vertices and the asymptotes, you can find the values of a and b, which enable you to write the equation. Notice in Example 6, that the key to this procedure is knowing that the central rectangle has a width of $2b$ and a height of $2a$.

EXAMPLE 6   Finding the Equation of a Hyperbola

Find an equation of the hyperbola with a vertical transverse axis whose vertices are $(0, 3)$ and $(0, -3)$ and whose asymptotes are given by $y = \frac{3}{5}x$ and $y = -\frac{3}{5}x$.

Solution

To begin, sketch the lines that represent the asymptotes, as shown in Figure 7.34(a). Note that these two lines intersect at the origin, which implies that the center of the hyperbola is $(0, 0)$. Next, plot the two vertices at the points $(0, 3)$ and $(0, -3)$. Because you know where the vertices are located, you can sketch the central rectangle of the hyperbola, as shown in Figure 7.34(a). Note that the corners of the central rectangle occur at the points

$$(-5, 3), \quad (5, 3), \quad (-5, -3), \quad \text{and} \quad (5, -3).$$

Because the width of the central rectangle is $2b = 10$, it follows that $b = 5$. Similarly, because the height of the central rectangle is $2a = 6$, it follows that $a = 3$. Now that you know the values of a and b, you can conclude that the standard form of the equation of the hyperbola is

$$\frac{y^2}{3^2} - \frac{x^2}{5^2} = 1.$$

The graph is shown in Figure 7.34(b).

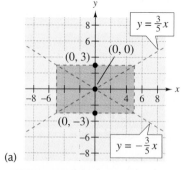
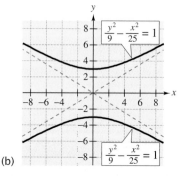

(a)

(b)

FIGURE 7.34

Application

EXAMPLE 7   An Application Involving an Ellipse

You are responsible for designing a semielliptical archway, as shown in Figure 7.35. The height of the archway is 10 feet and its width is 30 feet. Find an equation of the ellipse and use the equation to sketch an accurate diagram of the archway.

Solution

To make the equation simple, place the origin at the center of the ellipse. This means that the standard form of the equation is

$$\frac{x^2}{a^2} + \frac{y^2}{b^2} = 1.$$

Because the major axis is horizontal, it follows that $a = 15$ and $b = 10$, which implies that the equation is

$$\frac{x^2}{15^2} + \frac{y^2}{10^2} = 1.$$

In order to make an accurate sketch of the semiellipse, it is helpful to solve this equation for y as follows.

$$\frac{x^2}{15^2} + \frac{y^2}{10^2} = 1$$

$$\frac{x^2}{225} + \frac{y^2}{100} = 1$$

$$\frac{y^2}{100} = 1 - \frac{x^2}{225}$$

$$y^2 = 100\left(1 - \frac{x^2}{225}\right)$$

$$y = 10\sqrt{1 - \frac{x^2}{225}}$$

Next, calculate several y-values for the archway, as shown in the table. Then use the values in the table to sketch the archway, as shown in Figure 7.36.

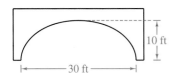

FIGURE 7.35

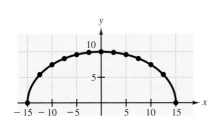

FIGURE 7.36

x-Value	±15	±12.5	±10	±7.5	±5	±2.5	0
y-Value	0	5.53	7.45	8.66	9.43	9.86	10

Technology

Graphing Conic Sections

Most graphing utilities are designed to sketch the graphs of equations in which y is isolated on the left side of the equation. Thus, if you want to sketch the graph of the circle given by $x^2 + y^2 = 36$, you should solve the equation for y and obtain the following *two* equations.

$$y = \sqrt{36 - x^2} \qquad \text{and} \qquad y = -\sqrt{36 - x^2}$$

The first of these two equations represents the upper half of the circle and the second represents the lower half of the circle. Try graphing both the upper and lower halves of the circle on the same screen.

With a *standard* setting, the graph should appear as shown in the figure on the left. The reason the graph does not look like a circle is that with the standard setting the tick marks on the x-axis are farther apart than the tick marks on the y-axis. To correct this, choose a *square* setting, as shown in the figure on the right.

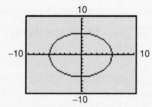 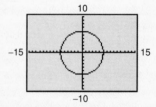

Group Activities   Extending the Concept

Identifying Conic Sections   Cut cone-shaped pieces of styrofoam to demonstrate how to obtain each type of conic section: circle, parabola, ellipse, and hyperbola. Discuss how you could write directions for someone else to form each conic section. Compile a list of real-life situations and/or everyday objects in which conic sections may be seen.

7.4 **Exercises**

Discussing the Concepts

1. Name the four types of conics.

2. Define a circle and give the standard form of the equation of a circle centered at the origin.

3. Define an ellipse and give the standard form of the equation of an ellipse centered at the origin.

4. Define a hyperbola and give the standard form of the equation of a hyperbola centered at the origin.

5. From its equation, how can you determine the lengths of the axes of an ellipse?

6. Explain how the central rectangle of a hyperbola can be used to sketch its asymptotes.

Problem Solving

In Exercises 7–12, match the equation with its graph. [The graphs are labeled (a), (b), (c), (d), (e), and (f).]

(a)
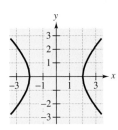

(b)
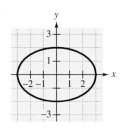

(c)
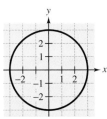

(d)
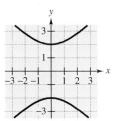

(e)
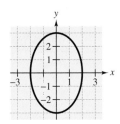

(f)
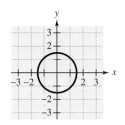

7. $x^2 + y^2 = 9$

8. $4x^2 + 4y^2 = 9$

9. $\dfrac{x^2}{4} + \dfrac{y^2}{9} = 1$

10. $\dfrac{x^2}{9} + \dfrac{y^2}{4} = 1$

11. $x^2 - y^2 = 4$

12. $x^2 - y^2 = -4$

In Exercises 13–16, find an equation of the circle with center at $(0, 0)$ that satisfies the given criteria.

13. Radius: 5

14. Radius: $\frac{5}{2}$

15. Passes through the point $(5, 2)$

16. Passes through the point $(-1, -4)$

In Exercises 17–20, identify the center and radius of the circle and sketch its graph.

17. $x^2 + y^2 = 16$

18. $x^2 + y^2 = 25$

19. $25x^2 + 25y^2 - 144 = 0$

20. $\dfrac{x^2}{4} + \dfrac{y^2}{4} - 1 = 0$

In Exercises 21–24, write the standard form of the equation of the ellipse, centered at the origin.

	Vertices	*Co-vertices*
21.	$(-4, 0), (4, 0)$	$(0, -3), (0, 3)$
22.	$(-4, 0), (4, 0)$	$(0, -1), (0, 1)$
23.	$(0, -4), (0, 4)$	$(-3, 0), (3, 0)$
24.	$(0, -5), (0, 5)$	$(-1, 0), (1, 0)$

25. Write the standard form of the equation of the ellipse, centered at the origin, with a horizontal major axis of length 20 and a vertical minor axis of length 12.

26. Write the standard form of the equation of the ellipse, centered at the origin, with a horizontal minor axis of length 30 and a vertical major axis of length 50.

In Exercises 27–30, sketch the ellipse. Identify its vertices and co-vertices.

27. $\dfrac{x^2}{16} + \dfrac{y^2}{4} = 1$ 28. $\dfrac{x^2}{9} + \dfrac{y^2}{25} = 1$

29. $4x^2 + y^2 - 4 = 0$ 30. $4x^2 + 9y^2 - 36 = 0$

In Exercises 31–34, sketch the hyperbola. Identify its vertices and asymptotes.

31. $\dfrac{x^2}{9} - \dfrac{y^2}{25} = 1$ 32. $\dfrac{x^2}{4} - \dfrac{y^2}{9} = 1$

33. $\dfrac{y^2}{9} - \dfrac{x^2}{25} = 1$ 34. $\dfrac{y^2}{4} - \dfrac{x^2}{9} = 1$

In Exercises 35–38, find an equation of the hyperbola centered at the origin.

	Vertices	*Asymptotes*	
35.	$(-4, 0), (4, 0)$	$y = 2x$	$y = -2x$
36.	$(-2, 0), (2, 0)$	$y = \frac{1}{3}x$	$y = -\frac{1}{3}x$
37.	$(0, -4), (0, 4)$	$y = \frac{1}{2}x$	$y = -\frac{1}{2}x$
38.	$(0, -2), (0, 2)$	$y = 3x$	$y = -3x$

In Exercises 39 and 40, sketch the circle and identify its center and radius.

39. $(x - 2)^2 + (y - 3)^2 = 4$

40. $(x + 4)^2 + (y - 3)^2 = 25$

In Exercises 41 and 42, write the equation in completed-square form. Then sketch its graph.

41. $x^2 + y^2 - 4x - 2y + 1 = 0$

42. $x^2 + y^2 + 6x - 4y - 3 = 0$

In Exercises 43–46, use a graphing utility to graph the equation.

43. $x^2 + y^2 = 30$ 44. $4x^2 + 4y^2 = 45$

45. $6x^2 - y^2 = 40$ 46. $x^2 + 4y^2 = 18$

47. *Satellite Orbit*   Find an equation of the circular orbit of a satellite 500 miles above the surface of the earth. Place the origin of the rectangular coordinate system at the center of the earth and assume the radius of the earth to be 4000 miles.

48. *Architecture*   The top portion of a stained-glass window is in the form of a pointed Gothic arch (see figure). Each side of the arch is an arc of a circle that has a radius of 12 feet and a center at the base of the opposite arch. Find an equation of one of the circles and use it to determine the height of the point of the arch above the horizontal base of the window.

|←——— 12 ft ———→|

49. *Height of an Arch*   A *semielliptical* arch for a tunnel under a river has a width of 100 feet and a height of 40 feet (see figure). Determine the height of the arch 5 feet from the edge of the tunnel.

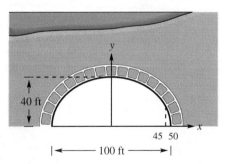

50. *Hyperbolic Mirror*   A hyperbolic mirror (used in some telescopes) has the property that a light ray directed at the first focus will be reflected to the second focus. The foci of a hyperbolic mirror are $(\pm 12, 0)$. Find the vertex of the mirror if its mount has coordinates $(12, 12)$.

Reviewing the Major Concepts

In Exercises 51–54, sketch the graph of the equation.

51. $y = 2x - 3$

52. $y = -\frac{3}{4}x + 2$

53. $y = x^2 - 4x + 4$

54. $y = 9 - x^2$

55. *Geometry*   Approximate the radius of a circle whose area is 3 square feet.

56. *Geometry*   The perimeter of a rectangle is 72 inches and the length of the diagonal is $12\sqrt{5}$ inches. Find the dimensions of the rectangle.

Additional Problem Solving

In Exercises 57–60, find an equation of the circle with center at $(0, 0)$ that satisfies the given criteria.

57. Radius: $\frac{2}{3}$ **58.** Radius: 7

59. Passes through the point $(0, 8)$

60. Passes through the point $(-2, 0)$

In Exercises 61–64, identify the center and radius of the circle and sketch its graph.

61. $x^2 + y^2 = 36$ **62.** $x^2 + y^2 = 10$

63. $4x^2 + 4y^2 = 1$ **64.** $9x^2 + 9y^2 = 64$

In Exercises 65–70, write the standard form of the equation of the ellipse, centered at the origin.

	Vertices	*Co-vertices*
65.	$(-2, 0), (2, 0)$	$(0, -1), (0, 1)$
66.	$(-10, 0), (10, 0)$	$(0, -4), (0, 4)$
67.	$(0, -2), (0, 2)$	$(-1, 0), (1, 0)$
68.	$(0, -8), (0, 8)$	$(-4, 0), (4, 0)$

69. The major axis is vertical with length 10, and the minor axis has length 6.

70. The major axis is horizontal with length 24, and the minor axis has length 10.

In Exercises 71–74, sketch the ellipse. Identify its vertices and co-vertices.

71. $\dfrac{x^2}{4} + \dfrac{y^2}{16} = 1$ **72.** $\dfrac{x^2}{25} + \dfrac{y^2}{9} = 1$

73. $\dfrac{x^2}{25/9} + \dfrac{y^2}{16/9} = 1$ **74.** $\dfrac{x^2}{1} + \dfrac{y^2}{1/4} = 1$

In Exercises 75–80, sketch the hyperbola. Identify its vertices and asymptotes.

75. $x^2 - y^2 = 9$ **76.** $x^2 - y^2 = 1$

77. $y^2 - x^2 = 9$ **78.** $y^2 - x^2 = 1$

79. $4y^2 - x^2 + 16 = 0$ **80.** $4y^2 - 9x^2 - 36 = 0$

In Exercises 81–84, write an equation of the hyperbola centered at the origin.

	Vertices	*Asymptotes*	
81.	$(-9, 0), (9, 0)$	$y = \frac{3}{2}x$	$y = -\frac{3}{2}x$
82.	$(-1, 0), (1, 0)$	$y = \frac{1}{2}x$	$y = -\frac{1}{2}x$
83.	$(0, -1), (0, 1)$	$y = 2x$	$y = -2x$
84.	$(0, -5), (0, 5)$	$y = x$	$y = -x$

In Exercises 85–94, identify the graph of the equation as a line, circle, parabola, ellipse, or hyperbola.

85. $y = 2x^2 - 8x + 2$ **86.** $y = 10 - \frac{3}{2}x$

87. $4x^2 + 9y^2 = 36$ **88.** $4x^2 + 4y^2 = 36$

89. $4x^2 - 9y^2 = 36$ **90.** $x^2 - 4y + 2x = 0$

91. $x^2 + y^2 - 1 = 0$ **92.** $2x^2 + 2y^2 = 9$

93. $3x + 2 = 0$ **94.** $y^2 = x^2 + 2$

In Exercises 95–98, use a graphing utility to graph the equation.

95. $x^2 - 2y^2 = 4$ **96.** $9x^2 + 9y^2 = 64$

97. $3x^2 + y^2 - 12 = 0$ **98.** $5x^2 - 2y^2 + 10 = 0$

99. A rectangle centered at the origin with sides parallel to the coordinate axes is placed in a circle of radius 25 inches centered at the origin (see figure). The length of the rectangle is $2x$ inches.

(a) Show that the width and area of the rectangle are given by $2\sqrt{625 - x^2}$ and $4x\sqrt{625 - x^2}$, respectively.

(b) Use a graphing utility to graph the area function. Approximate the value of x for which the area is maximum.

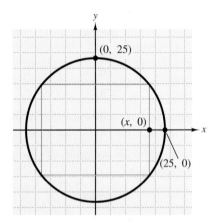

100. *Height of an Arch*   A *semicircular* arch for a tunnel under a river has a diameter of 100 feet (see figure). Determine the height of the arch 5 feet from the edge of the tunnel.

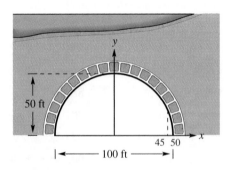

101. *Area*   The area A of the ellipse

$$\frac{x^2}{a^2} + \frac{y^2}{b^2} = 1$$

is given by $A = \pi ab$. Find the equation of an ellipse with area 301.59 square units and $a + b = 20$.

102. Sketch a graph of the ellipse that consists of all points (x, y) such that the sum of the distances between (x, y) and two fixed points is 15 units and for which the foci are located at the centers of the two sets of concentric circles in the figure.

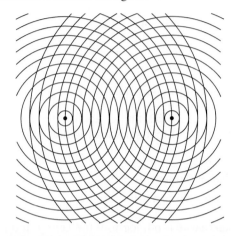

7.5	**Variation**
	Direct Variation ▪ Inverse Variation ▪ Joint Variation

Direct Variation

In the mathematical model for direct variation, y is a *linear* function of x. Specifically, $y = kx$.

STUDY TIP

To use this mathematical model in applications involving direct variation, you are usually given specific values of x and y, which then enable you to find the value for the constant k.

Direct Variation

The following statements are equivalent. The number k is the **constant of proportionality.**

1. y **varies directly** as x.

2. y is **directly proportional** to x.

3. $y = kx$ for some constant k.

EXAMPLE 1   Direct Variation

Assume that the total revenue R (in dollars) obtained from selling x units of a product is directly proportional to the number of units sold. When 10,000 units are sold, the total revenue is $142,500.

a. Find a model that relates the total revenue R to the number of units sold x.

b. Find the total revenue obtained from selling 12,000 units.

Solution

a. Because the total revenue is directly proportional to the number of units sold, the linear model is $R = kx$. To find the value of the constant k, substitute 142,500 for R and 10,000 for x

$$142,500 = k(10,000) \qquad \text{Substitute for } R \text{ and } x.$$

which implies that $k = 142,500/10,000 = 14.25$. Thus, the equation relating the total revenue to the total number of units sold is

$$R = 14.25x. \qquad \text{Direct variation model}$$

b. When $x = 12,000$, the total revenue is

$$R = 14.25(12,000) = \$171,000.$$

EXAMPLE 2   Direct Variation

Hooke's Law for springs states that the distance a spring is stretched (or compressed) is proportional to the force on the spring. A force of 20 pounds stretches a particular spring 5 inches.

a. Find a mathematical model that relates the distance the spring is stretched to the force applied to the spring.

b. How far will a force of 30 pounds stretch the spring?

Solution

a. For this problem, let d represent the distance (in inches) that the spring is stretched and let F represent the force (in pounds) that is applied to the spring. Because the distance d is proportional to the force F, the model is

$$d = kF.$$

To find the value of the constant k, use the fact that $d = 5$ when $F = 20$. Substituting these values into the given model produces

$$5 = k(20) \qquad\qquad \text{Substitute 5 for } d \text{ and 20 for } F.$$

implying that $k = \frac{5}{20} = \frac{1}{4}$. Thus, the equation relating distance and force is

$$d = \frac{1}{4}F. \qquad\qquad \text{Direct variation model}$$

b. When $F = 30$, the distance is

$$d = \frac{1}{4}(30) = 7.5 \text{ inches.} \qquad\qquad \text{See Figure 7.37.}$$

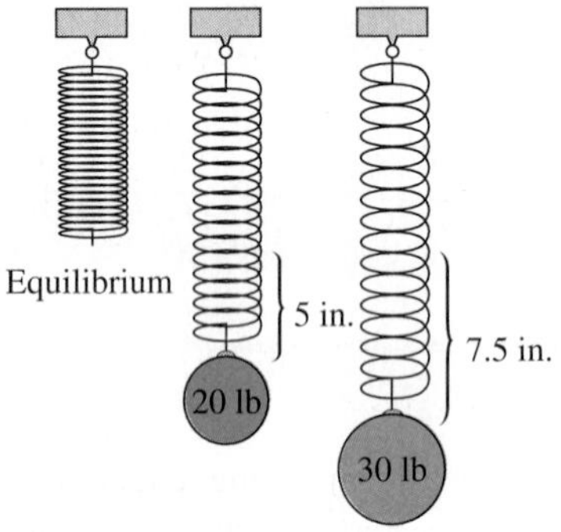

Equilibrium

5 in.

20 lb

7.5 in.

30 lb

FIGURE 7.37

In Example 2, you can get a clearer understanding of Hooke's Law by using the model $d = \frac{1}{4}F$ to create a table. From the table, you can see what it means for the distance to be "proportional to the force."

Force, F	10 lb	20 lb	30 lb	40 lb	50 lb	60 lb
Distance, d	2.5 in.	5.0 in.	7.5 in.	10.0 in.	12.5 in.	15.0 in.

In Examples 1 and 2, the direct variations were such that an *increase* in one variable corresponded to an *increase* in the other variable. There are, however, other applications of direct variation in which an increase in one variable corresponds to a *decrease* in the other variable. For instance, in the model $y = -2x$, an increase in x will yield a decrease in y.

A second type of direct variation relates one variable to a *power* of another.

Direct Variation as *n*th Power

The following statements are equivalent.

1. *y* **varies directly as the *n*th power** of *x*.

2. *y* is **directly proportional to the *n*th power** of *x*.

3. $y = kx^n$ for some constant *k*.

EXAMPLE 3   *Direct Variation as a Power*

The distance a ball rolls down an inclined plane is directly proportional to the square of the time it rolls. Assume that, during the first second, a ball rolls down a particular plane a distance of 6 feet.

a.   Find a mathematical model that relates the distance traveled to the time.

b.   How far will the ball roll during the first 2 seconds?

Solution

a.   Letting *d* be the distance (in feet) that the ball rolls and letting *t* be the time (in seconds), we obtain the model

$$d = kt^2.$$

Because $d = 6$ when $t = 1$, it follows that $k = 6$. Therefore, the equation relating distance to time is

$$d = 6t^2.$$ Direct variation as 2nd power model

b.   When $t = 2$, the distance traveled is

$$d = 6(2)^2 = 6(4) = 24 \text{ feet.}$$ See Figure 7.38.

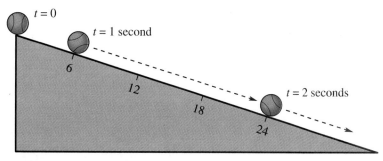

FIGURE 7.38

Inverse Variation

A third type of variation is called **inverse variation.** For this type of variation, we say that one of the variables is inversely proportional to the other variable.

NOTE If x and y are related by an equation of the form $y = k/x^n$, we say that y varies inversely as the nth power of x (or y is inversely proportional to the nth power of x).

Inverse Variation

The following statements are equivalent.

1. y **varies inversely** as x.

2. y is **inversely proportional** to x.

3. $y = \dfrac{k}{x}$ for some constant k.

EXAMPLE 4   *Inverse Variation*

The marketing department of a large company has found that the demand for one of its products varies inversely as the price of the product. (When the price is low, more people are willing to buy the product than when the price is high.) When the price of the product is $7.50, the monthly demand is 50,000 units. Approximate the monthly demand if the price is reduced to $6.00.

Solution

Let x represent the number of units that are sold each month (the demand), and let p represent the price per unit (in dollars). Because the demand is inversely proportional to the price, the model is

$$x = \frac{k}{p}.$$

By substituting $x = 50{,}000$ when $p = 7.50$, you obtain $k = (7.5)(50{,}000) = 375{,}000$. Thus, the model is

$$x = \frac{375{,}000}{p}. \qquad \text{Inverse variation model}$$

To find the demand that corresponds to a price of $6.00, substitute $p = 6$ into the equation and obtain a demand of

$$x = \frac{375{,}000}{6} = 62{,}500 \text{ units.}$$

Thus, if the price were lowered from $7.50 per unit to $6.00 per unit, the monthly demand could be expected to increase from 50,000 units to 62,500 units.

Some applications of variation involve problems with *both* direct and inverse variation in the same model.

EXAMPLE 5   *Direct and Inverse Variation*

A company determines that the demand for one of its products is directly proportional to the amount spent on advertising and inversely proportional to the price of the product. When $40,000 is spent on advertising and the price per unit is $20, the monthly demand is 10,000 units.

a. If the amount of advertising were increased to $50,000, how much could the price be increased to maintain a monthly demand of 10,000 units?

b. If you were in charge of the advertising department, would you recommend this increased expense in advertising?

Solution

a. Let x represent the number of units that are sold each month (the demand), let a represent the amount spent on advertising (in dollars), and let p represent the price per unit (in dollars). Because the demand is directly proportional to the advertising and inversely proportional to the price, the model is

$$x = \frac{ka}{p}.$$

By substituting $x = 10,000$ when $a = 40,000$ and $p = 20$, you obtain $k = (10,000)(20)/(40,000) = 5$. Thus, the model is

$$x = \frac{5a}{p}. \qquad \text{Direct and inverse variation model}$$

To find the price that corresponds to a demand of 10,000 and an advertising expense of $50,000, substitute $x = 10,000$ and $a = 50,000$ into the model and solve for p.

$$10,000 = \frac{5(50,000)}{p} \qquad \Longrightarrow \qquad p = \frac{5(50,000)}{10,000} = \$25$$

b. The total revenue from selling 10,000 units at $20 is $200,000, and the revenue from selling 10,000 units at $25 is $250,000. Thus, increasing the advertising expense from $40,000 to $50,000 would increase the revenue by $50,000. This implies that you should recommend the increased expense in advertising.

Amount of Advertising	Price	Revenue
$40,000	$20.00	$10,000 \times 20 = \$200,000$
$50,000	$25.00	$10,000 \times 25 = \$250,000$

Joint Variation

Joint Variation
The following statements are equivalent.
1. z **varies jointly** as x and y.
2. z is **jointly proportional** to x and y.
3. $z = kxy$ for some constant k.

EXAMPLE 6   Joint Variation

The *simple interest* for a certain savings account is jointly proportional to the time and the principal. After one quarter (3 months), the interest for a principal of $6000 is $120. How much interest would a principal of $7500 earn in 5 months?

Solution

To begin, let I represent the interest earned (in dollars), let P represent the principal (in dollars), and let t represent the time (in years). Because the interest is jointly proportional to the time and the principal, the model is

$$I = ktP.$$

Because $I = 120$ when $P = 6000$ and $t = \frac{1}{4}$, it follows that $k = 120/\left(6000 \cdot \frac{1}{4}\right) = 0.08$. Thus, the model that relates interest to time and principal is

$$I = 0.08tP. \qquad \text{Joint variation model}$$

To find the interest earned on a principal of $7500 over a 5-month period of time, substitute $P = 7500$ and $t = \frac{5}{12}$ into the model and obtain an interest of $I = 0.08(\frac{5}{12})(7500) = \$250.00$.

Group Activities   Exploring with Technology

Investigating Variation Models   Use a graphing utility to sketch the graphs of the direct variation, direct variation as nth power, and inverse variation models for different values of k (and n as appropriate). Summarize the basic characteristics of the graph for each type of model. Does understanding the basic graphs of these models shed light on the models themselves?

7.5 Exercises

Discussing the Concepts

1. Suppose the constant of proportionality is positive and y varies directly as x. If one of the variables increases, how will the other change? Explain.

2. Suppose the constant of proportionality is positive and y varies inversely as x. If one of the variables increases, how will the other change? Explain.

3. If y varies directly as the square of x and x is doubled, how will y change? Use the properties of exponents to explain your answer.

4. If y varies inversely as the square of x and x is doubled, how will y change? Use the properties of exponents to explain your answer.

Problem Solving

In Exercises 5–10, write a model for the statement.

5. I varies directly as V.

6. V varies directly as the cube of x.

7. p varies inversely as d.

8. S is inversely proportional to the square of v.

9. *Boyle's Law*   If the temperature of a gas is constant, its absolute pressure P is inversely proportional to its volume V.

10. *Newton's Law of Universal Gravitation*   The gravitational attraction F between two particles of masses m_1 and m_2 is proportional to the product of the masses and inversely proportional to the square of the distance r between the particles.

In Exercises 11–14, write a sentence using variation terminology to describe the formula.

11. *Area of a Triangle:* $A = \frac{1}{2}bh$

12. *Area of a Circle:* $A = \pi r^2$

13. *Volume of a Right Circular Cylinder:* $V = \pi r^2 h$

14. *Average Speed:* $r = d/t$

In Exercises 15–22, find the constant of proportionality and give the equation relating the variables.

15. s varies directly as t, and $s = 20$ when $t = 4$.

16. h is directly proportional to r, and $h = 28$ when $r = 12$.

17. F is directly proportional to the square of x, and $F = 500$ when $x = 40$.

18. v varies directly as the square root of s, and $v = 24$ when $s = 16$.

19. n varies inversely as m, and $n = 32$ when $m = 1.5$.

20. q is inversely proportional to p, and $q = \frac{3}{2}$ when $p = 50$.

21. d varies directly as the square of x and inversely with r, and $d = 3000$ when $x = 10$ and $r = 4$.

22. z is directly proportional to x and inversely proportional to the square root of y, and $z = 720$ when $x = 48$ and $y = 81$.

23. *Hooke's Law*   A baby weighing $10\frac{1}{2}$ pounds compresses the spring of a baby scale 7 millimeters (see figure). What weight will compress the spring 12 millimeters?

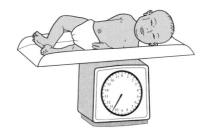

24. *Hooke's Law*   A force of 50 pounds compresses the spring of a scale 1.5 inches. Find the distance the spring will be compressed when a 20-pound object is placed on the scale.

25. *Stopping Distance* The stopping distance d of an automobile is directly proportional to the square of its speed s. On a certain road surface, a car requires 75 feet to stop when its speed is 30 miles per hour. Estimate the stopping distance if the brakes are applied when the car is traveling at 50 miles per hour under similar road conditions.

26. *Distance* Neglecting air resistance, the distance d that an object falls varies directly as the square of the time t it has been falling. If an object falls 64 feet in 2 seconds, determine the distance it will fall in 6 seconds.

27. *Power Generation* The power P generated by a wind turbine varies directly as the cube of the wind speed w. The turbine generates 750 watts of power in a 25-mile-per-hour wind. Find the power it generates in a 40-mile-per-hour wind.

28. *Velocity of a Stream* The diameter d of the largest particle that can be moved by a stream is directly proportional to the square of the velocity v of the stream. A stream with a velocity of $\frac{1}{4}$ mile per hour can move coarse sand particles about 0.02 inch in diameter. What must the velocity be to carry particles with a diameter of 0.12 inch?

29. *Demand Function* A company has found that the daily demand x for its product is inversely proportional to the price p. When the price is \$5, the demand is 800 units. Approximate the demand if the price is increased to \$6.

30. *Weight of an Astronaut* The gravitational force F with which an object is attracted to the earth is inversely proportional to its distance r from the center of the earth. If an astronaut weighs 190 pounds on the surface of the earth ($r \approx 4000$ miles), what will the astronaut weigh 1000 miles above the earth's surface?

31. *Amount of Illumination* The illumination I from a light source varies inversely as the square of the distance d from the light source. If you raise a lamp from 18 inches to 36 inches over your desk (see figure), the illumination will change by what factor?

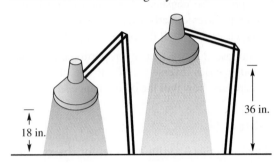

18 in.   36 in.

32. *Frictional Force* The frictional force F between the tires and the road required to keep a car on a curved section of a highway is directly proportional to the square of the speed s of the car. If the speed of the car is doubled, the required frictional force will change by what factor?

33. *Load of a Beam* The load that can be safely supported by a horizontal beam varies jointly as the width of the beam and the square of its depth and inversely as the length of the beam. A beam with width 3 inches, depth 8 inches, and length 10 feet can safely support 2000 pounds. Determine the safe load of a beam made from the same material if its depth is increased to 10 inches.

34. *Best Buy* The prices of 9-inch, 12-inch, and 15-inch diameter pizzas at a certain pizza shop are \$6.78, \$9.78, and \$12.18, respectively. You might expect that the price of a certain size of pizza is directly proportional to its area. Is this the case for this pizza shop? If not, which size of pizza is the best buy?

Reviewing the Major Concepts

In Exercises 35–38, solve for y in terms of x.

35. $3x + 4y - 5 = 0$

36. $-2x - 3y + 6 = 0$

37. $-2x^2 + 3y + 2 = 0$

38. $x^2 - 2y = 9$

39. Find two positive consecutive odd integers whose product is 255.

40. Find two positive consecutive even integers, the sum of whose squares is 452.

Additional Problem Solving

In Exercises 41–48, write a model for the statement.

41. V is directly proportional to t.

42. C varies directly as r.

43. u is directly proportional to the square of v.

44. s varies directly as the cube of t.

45. P is inversely proportional to the square root of $1+r$.

46. A varies inversely as the fourth power of t.

47. A varies jointly as l and w.

48. V varies jointly as h and the square of r.

In Exercises 49–52, write a sentence using variation terminology to describe the formula.

49. *Volume of a Sphere:* $V = \frac{4}{3}\pi r^3$

50. *Surface Area of a Sphere:* $A = 4\pi r^2$

51. *Area of an Ellipse:* $A = \pi ab$

52. *Height of a Cylinder:* $h = V/(\pi r^2)$

In Exercises 53–58, find the constant of proportionality and give the equation relating the variables.

53. H is directly proportional to u, and $H = 100$ when $u = 40$.

54. M varies directly as the cube of n, and $M = 0.012$ when $n = 0.2$.

55. g varies inversely as the square root of z, and $g = \frac{4}{5}$ when $z = 25$.

56. u varies inversely as the square of v, and $u = 40$ when $v = \frac{1}{2}$.

57. F varies jointly as x and y, and $F = 500$ when $x = 15$ and $y = 8$.

58. V varies jointly as h and the square of b, and $V = 288$ when $h = 6$ and $b = 12$.

59. *Revenue* The total revenue R is directly proportional to the number of units sold x. When 500 units are sold, the revenue is \$3875.

(a) Find the revenue when 635 units are sold.

(b) Interpret the constant of proportionality.

60. *Hooke's Law* A force of 50 pounds stretches a spring 3 inches.

(a) How far will a 20-pound force stretch the spring?

(b) What force will stretch the spring 1.5 inches?

61. *Free-Falling Object* The velocity v of a free-falling object is proportional to the time that it has fallen. The constant of proportionality is the acceleration due to gravity. Find the acceleration due to gravity if the velocity of a falling object is 96 feet per second after the object has fallen 3 seconds.

62. *Free-Falling Object* The distance d that a free-falling object has fallen is proportional to the square of the time that it has fallen. An object falls 144 feet in 3 seconds. Find the constant of proportionality.

63. *Travel Time* The travel time between two cities is inversely proportional to the average speed. If a train travels between two cities in 3 hours at an average speed of 65 miles per hour, how long would it take at an average speed of 80 miles per hour? What does the constant of proportionality measure in this problem?

64. *Predator-Prey* The number N of prey t months after a natural predator is introduced into the test area is inversely proportional to $t + 1$. If $N = 500$ when $t = 0$, find N when $t = 4$.

65. *Weight* A person's weight on the moon varies directly with his or her weight on earth. Neil Armstrong, the first man on the moon, weighed 360 pounds on earth, including his equipment. On the moon he weighed only 60 pounds, with equipment. If the first woman in space, Valentina V. Tereshkova, had landed on the moon and weighed 54 pounds, with equipment, how much would she have weighed on earth?

66. *Snowshoes* When a person walks, the pressure P on each sole varies inversely with the area A of the sole. Denise is trudging through deep snow, wearing boots that have a sole area of 29 square inches each. The sole pressure is 4 pounds per square inch. If Denise were wearing snowshoes, each with an area 11 times that of her boot soles, what would be the pressure on each snowshoe? The constant of variation is Denise's weight. What does she weigh?

67. *Reading a Graph*   The graph shows the percent p of oil that remained in Chedabucto Bay, Nova Scotia after an oil spill. The cleaning of the spill was left primarily to natural actions such as wave motion, evaporation, photochemical decomposition, and bacterial decomposition. After about a year, the percent that remained varied inversely with time. Find a model that relates p and t, where t is the number of years since the spill. Then use the model to find the amount of oil that remained $6\frac{1}{2}$ years after the spill.

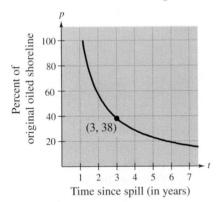

Time since spill (in years)

68. *Reading a Graph*   The graph shows the temperature of the water in the north central Pacific Ocean. At depths greater than 900 meters, the water temperature varies inversely with the water depth. Find a model that relates the temperature T and the depth d. What is the temperature at a depth of 4385 meters?

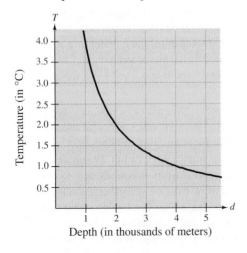

Depth (in thousands of meters)

In Exercises 69–72, complete the table and plot the resulting points.

x	2	4	6	8	10
$y = kx^2$					

69. $k = 1$ **70.** $k = 2$

71. $k = \frac{1}{2}$ **72.** $k = \frac{1}{4}$

In Exercises 73–76, complete the table and plot the resulting points.

x	2	4	6	8	10
$y = \dfrac{k}{x^2}$					

73. $k = 2$ **74.** $k = 5$

75. $k = 10$ **76.** $k = 20$

In Exercises 77–80, determine whether the variation model is of the form $y = kx$ or $y = k/x$, and find k.

77.

x	10	20	30	40	50
y	$\frac{2}{5}$	$\frac{1}{5}$	$\frac{2}{15}$	$\frac{1}{10}$	$\frac{2}{25}$

78.

x	10	20	30	40	50
y	2	4	6	8	10

79.

x	10	20	30	40	50
y	-3	-6	-9	-12	-15

80.

x	10	20	30	40	50
y	60	30	20	15	12

CHAPTER PROJECT: Mass Transportation

Mass transportation includes taxis, vans, buses, trains, subways, and airplanes. In 1991, approximately $800 billion was spent on mass transportation in the U.S.

The table below shows the amounts (in billions of dollars) spent on mass transportation for taxis and buses from 1978 through 1991. In this table, t represents the year, with $t = 0$ representing 1970. (Source: Eno Foundation for Transportation)

Year, t	8	9	10	11	12	13	14
Amount Spent on Taxis	4.4	4.5	5.2	5.2	5.6	5.3	5.5
Amount Spent on Buses	6.7	7.5	9.3	10.3	10.6	11.6	13.2

Year, t	15	16	17	18	19	20	21
Amount Spent on Taxis	5.6	6.0	6.4	6.9	7.1	7.5	7.9
Amount Spent on Buses	13.5	14.5	15.1	15.7	15.9	16.7	17.5

Use the information in the table to investigate the following questions.

1. *Linear Modeling*   Plot the points in the table to form scatter plots for the amounts spent on taxis and buses. Draw the line that you think best approximates each scatter plot. Then find the equation of each line.

2. *Linear Modeling*   Use a graphing utility or computer program to find the equation of the line that best fits each of the data sets. To do this with a *TI-82*, use the following steps.

   • Activate the STAT mode and clear previously entered data.

   • Enter the data for taxis or buses.

   • Use the linear regression program to find the best-fitting line.

   How do the equations for the regression lines compare with the ones you found in Question 1?

3. *Graphical Interpretation*   What are the slopes of the linear models found in Question 1 or 2? Interpret these slopes in the context of the data. Which slope is greater? What real-life implications does this have?

4. *Prediction*   Use the linear models found in Question 1 or 2 to predict the amounts of money that was spent on taxis and buses in 1994.

5. *Correlation*   Which of the linear models that you found fits its set of data better? How can you tell?

6. *Research Project*   Use your school's library or some other reference source to find the revenues for two other forms of mass transportation. Repeat Questions 1 and 2 for the data. Write a paragraph that interprets the data.

CHAPTER SUMMARY

After studying this chapter, you should have acquired the following skills. These skills are keyed to the Review Exercises that begin on page 491. Answers to odd-numbered Review Exercises are given in the back of the book.

- Write the equation of a line given its slope and a point on the line. *(Section 7.1)*

  Review Exercises 1–8

- Write the equation of a line passing through two points. *(Section 7.1)*

  Review Exercises 9–14

- Write the equation of a line through a given point that is parallel or perpendicular to a given line. *(Section 7.1)*

  Review Exercises 15–18

- Sketch pairs of lines using a graphing utility and determine if the lines are parallel, perpendicular, or neither. *(Section 7.1)*

  Review Exercises 19–22

- Translate real-life situations into algebraic equations or inequalities. *(Sections 7.1, 7.2, 7.5)*

  Review Exercises 23, 25, 26, 39, 71–74

- Solve real-life problems modeled by equations. *(Sections 7.1, 7.3)*

  Review Exercises 24, 57–60

- Sketch the graphs of inequalities. *(Section 7.2)*

  Review Exercises 27–32

- Sketch, with and without a graphing utility, the graphs of solution points that satisfy pairs of inequalities. *(Section 7.2)*

  Review Exercises 33–38

- Solve problems involving geometry. *(Section 7.2)*

  Review Exercise 40

- Identify and sketch conics represented by equations. *(Sections 7.3, 7.4)*

  Review Exercises 41–50

- Determine the equations of conics given points, dimensions, and/or other characteristics of the conics. *(Sections 7.3, 7.4)*

  Review Exercises 51–56

- Match equations with their graphs. *(Sections 7.1, 7.2, 7.3, 7.4)*

  Review Exercises 61–66

- Find the constant of proportionality and write an equation that relates the variables. *(Section 7.5)*

  Review Exercises 67–70

REVIEW EXERCISES

In Exercises 1–8, write an equation of the line passing through the point with the specified slope.

1. $(1, -4)$, $m = 2$

2. $(-5, -5)$, $m = 3$

3. $(-1, 4)$, $m = -4$

4. $(5, -2)$, $m = -2$

5. $\left(\frac{5}{2}, 4\right)$, $m = -\frac{2}{3}$

6. $\left(-2, -\frac{4}{3}\right)$, $m = \frac{3}{2}$

7. $(7, 8)$, m is undefined.

8. $(-6, 5)$, $m = 0$

In Exercises 9–14, write an equation of the line passing through the points.

9. $(-6, 0)$, $(0, -3)$

10. $(-2, -3)$, $(4, 6)$

11. $(0, 10)$, $(6, 10)$

12. $(-10, 2)$, $(4, -7)$

13. $\left(\frac{4}{3}, \frac{1}{6}\right)$, $\left(4, \frac{7}{6}\right)$

14. $\left(\frac{5}{2}, 0\right)$, $\left(\frac{5}{2}, 5\right)$

In Exercises 15–18, find equations of the line passing through the point that are (a) parallel and (b) perpendicular to the given line.

15. $\left(\frac{3}{5}, -\frac{4}{5}\right)$, $3x + y = 2$

16. $(-1, 5)$, $2x + 4y = 1$

17. $(12, 1)$, $5x = 3$

18. $\left(\frac{3}{8}, 3\right)$, $4x - 3y = 12$

In Exercises 19–22, use a graphing utility on a square setting to sketch the lines. Decide whether the lines are parallel, perpendicular, or neither.

19. $y = 2x - 3$
$y = \frac{1}{2}(4 - x)$

20. $y = \frac{2}{3}x + 3$
$y = \frac{2}{9}(3x - 4)$

21. $2x - 3y - 5 = 0$
$x + 2y - 6 = 0$

22. $4x + 3y - 6 = 0$
$3x - 4y - 8 = 0$

23. *Cost and Profit*   A company produces a product for which the variable cost is $8.55 per unit and the fixed costs are $25,000. The product is sold for $12.60, and the company can sell all it produces.

(a) Write the cost C as a linear function of x, the number of units produced.

(b) Write the profit P as a linear function of x, the number of units sold.

24. *Velocity*   The velocity of a ball thrown upward from ground level is given by $v = -32t + 80$, where t is time in seconds and v is velocity in feet per second.

(a) Find the velocity when $t = 2$.

(b) Find the time when the ball reaches its maximum height. (*Hint:* Find the time when $v = 0$.)

(c) Find the velocity when $t = 3$.

Rocker Arm Construction   In Exercises 25 and 26, consider the rocker arm shown in the figure.

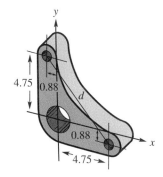

25. Find an equation of the line through the centers of the two small bolt holes in the rocker arm.

26. Find the distance d between the centers of the two small bolt holes in the rocker arm.

In Exercises 27–32, sketch the graph of the inequality.

27. $x - 2 \geq 0$

28. $y + 3 < 0$

29. $2x + y < 1$

30. $3x - 4y > 2$

31. $x \leq 4y - 2$

32. $(y - 3) \geq \frac{2}{3}(x - 5)$

In Exercises 33–36, sketch a graph of the solution points that satisfy both inequalities.

33. $x \geq 5$, $y \leq 2$

34. $y \leq \frac{1}{2}x + 1$, $x \geq 2$

35. $x - 2y \geq 2$, $y \geq -1$

36. $x + y \geq 2$, $x \leq 2$

In Exercises 37 and 38, use a graphing utility to graph the region containing the solution points that satisfy both inequalities.

37. $x + y \geq 0$, $y \leq 5$　　　**38.** $4x - 3y \geq 2$, $y \geq 1$

39. *Weekly Pay* You have two part-time jobs. One is at a grocery store, which pays $8 per hour, and the other is mowing lawns, which pays $10 per hour. Between the two jobs, you want to earn at least $200 a week. Write an inequality that shows the different numbers of hours you can work at each job, and sketch the graph of the inequality. From the graph, find several ordered pairs with positive integer coordinates that are solutions of the inequality.

40. *Geometry* The perimeter of a rectangle of length x and width y cannot exceed 800 feet. Write a linear inequality for this constraint and sketch its graph.

In Exercises 41–50, identify and sketch the conic.

41. $x^2 - 2y = 0$　　　　　**42.** $x^2 + y^2 = 64$

43. $x^2 - y^2 = 64$　　　　**44.** $x^2 + 4y^2 = 64$

45. $y = x(x - 6)$　　　　　**46.** $y = 9 - (x - 3)^2$

47. $\dfrac{x^2}{25} + \dfrac{y^2}{4} = 1$　　　**48.** $\dfrac{x^2}{25} - \dfrac{y^2}{4} = -1$

49. $4x^2 + 4y^2 - 9 = 0$　　**50.** $x^2 + 9y^2 - 9 = 0$

In Exercises 51–56, find the general form of the equation for the conic meeting the given criteria.

51. *Parabola:* vertex: $(5, 0)$;
passes through the point $(1, 1)$

52. *Parabola:* vertex: $(-2, 5)$;
passes through the point $(0, 1)$

53. *Ellipse:* vertices: $(0, -5)$, $(0, 5)$;
co-vertices: $(-2, 0)$, $(2, 0)$

54. *Ellipse:* vertices: $(-10, 0)$, $(10, 0)$;
co-vertices: $(0, -6)$, $(0, 6)$

55. *Circle:* center: $(0, 0)$;
radius: 20

56. *Hyperbola:* vertices: $(0, -4)$, $(0, 4)$;
asymptotes: $y = 2x$, $y = -2x$

57. *Graphical Estimation* The profit (in thousands of dollars) for a certain product is given by

$$P = 320 + 10s - \frac{1}{2}s^2$$

where s is the amount (in hundreds of dollars) spent on advertising. Use a graphing utility to graph the profit function and approximate the amount of advertising that yields a maximum profit. Verify the maximum algebraically.

58. *Graphical Interpretation* The height y (in feet) of a ball thrown by a child is given by

$$y = -\frac{1}{10}x^2 + 3x + 6$$

where x is the horizontal distance (in feet) from where the ball is thrown.

(a) Use a graphing utility to graph the path of the ball.

(b) How high is the ball when it leaves the child's hand?

(c) What is the maximum height of the ball?

(d) How far from the child does the ball hit the ground?

59. *Graphical Estimation* The enrollment E (in millions) in public elementary schools in the United States for the years 1970 through 1990 is approximated by the model

$$E = 27.611 - 0.586t + 0.026t^2$$

where t represents the calendar year, with $t = 0$ corresponding to 1970. (Source: U.S. National Center for Education Statistics)

(a) Use a graphing utility to graph the enrollment function.

(b) Use the graph of part (a) to approximate the year when the enrollment was minimum.

(c) Predict the enrollment in 1999 if this trend continues.

60. *Graphical Estimation* The number N of opera- ble nuclear power units generating electricity in the United States for the years 1985 through 1992 is ap- proximated by the model

$$N = 47.601 + 12.613t - 0.625t^2$$

where t represents the calendar year, with $t = 5$ corresponding to 1985. (Source: U.S. Energy In- formation Administration)

(a) Use a graphing utility to graph the function.

(b) Use the graph of part (a) to approximate the year when the number of operable generating units was maximum.

(c) What was the number of operable generating units in 1995?

In Exercises 61–66, match the equation with its graph. [The graphs are labeled (a), (b), (c), (d), (e), and (f).]

(a)
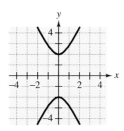

(b)
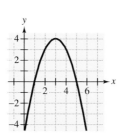

(c)
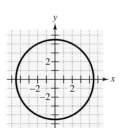

(d)
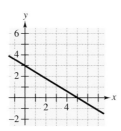

(e)
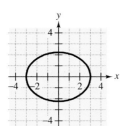

(f)
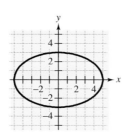

61. $4x^2 + 4y^2 = 81$

62. $3x + 5y = 15$

63. $\dfrac{y^2}{4} - x^2 = 1$

64. $\dfrac{x^2}{25} + \dfrac{y^2}{9} = 1$

65. $y = -x^2 + 6x - 5$

66. $\dfrac{x^2}{9} + \dfrac{y^2}{25} = 1$

In Exercises 67–70, find the constant of proportion- ality and write an equation that relates the variables.

67. y varies directly as the cube root of x, and $y = 12$ when $x = 9$.

68. r varies inversely as s, and $r = 45$ when $s = \frac{3}{5}$.

69. T varies jointly as r and the square of s, and $T = 5000$ when $r = 0.09$ and $s = 1000$.

70. D is directly proportional to the cube of x and in- versely proportional to y, and $D = 810$ when $x = 3$ and $y = 25$.

71. *Hooke's Law* A force of 100 pounds stretches a spring 4 inches. Find the force required to stretch the same spring 6 inches.

72. *Stopping Distance* The stopping distance d of an automobile is directly proportional to the square of its speed s. How will the stopping distance be changed by doubling the speed of the car?

73. *Demand Function* A company has found that the daily demand x for its product varies inversely as the square root of the price p. When the price is $25, the demand is approximately 1000 units. Approximate the demand if the price is increased to $28.

74. *Weight of an Astronaut* The gravitational force F with which an object is attracted to the earth is in- versely proportional to its distance r from the center of the earth. If an astronaut weighs 200 pounds on the surface of the earth ($r \approx 4000$ miles), what will the astronaut weigh 500 miles above the earth's surface?

CHAPTER TEST

Take this test as you would take a test in class. After you are done, check your work against the answers given in the back of the book.

In Exercises 1–4, write an equation of the line.

1. The line has a slope of -2 and passes through $(2, -4)$.

2. The line passes through $(25, -15)$ and $(75, 10)$.

3. The line is horizontal and passes through $(5, -1)$.

4. The line is vertical and passes through $(-2, 4)$.

5. Find the slope of a line perpendicular to the line given by $5x + 3y - 9 = 0$.

6. After 4 years, a \$26,000 car will have depreciated to a value of \$10,000. Write a linear equation that gives the value V in terms of t, the number of years. When will the car be worth \$16,000? Explain your reasoning.

7. Sketch the graph of the inequality $x + 2y \leq 4$.

8. Sketch the graph of the inequality $3x - y \geq 6$.

9. Sketch the graph of $y = -2(x - 2)^2 + 8$. Label its vertex and intercepts.

10. Write an equation of the parabola shown at the right.

11. The revenue R for a chartered bus trip is given by $R = -\frac{1}{20}(n^2 - 240n)$, where $80 \leq n \leq 160$ and n is the number of passengers. How many passengers will produce a maximum revenue? Explain your reasoning.

12. Write an equation of the circle shown at the right.

13. Write the standard form of the equation of the ellipse centered at the origin with vertices $(0, -10)$ and $(0, 10)$ and co-vertices $(-3, 0)$ and $(3, 0)$.

14. Write the standard form of the equation of the hyperbola centered at the origin with vertices $(-3, 0)$ and $(3, 0)$ and asymptotes $y = \frac{1}{2}x$ and $y = -\frac{1}{2}x$.

In Exercises 15–18, sketch the graph of the equation.

15. $x^2 + y^2 = 9$

16. $\dfrac{x^2}{9} + \dfrac{y^2}{16} = 1$

17. $\dfrac{x^2}{9} - \dfrac{y^2}{16} = 1$

18. $\dfrac{x}{3} - \dfrac{y}{4} = 1$

19. Write a mathematical model for the statement, "S varies directly as the square of x and inversely as y."

20. Find the constant of proportionality if v varies directly as the square root of u and $v = \frac{3}{2}$ when $u = 36$.

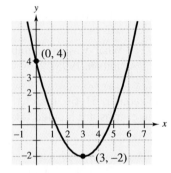

Figure for 10

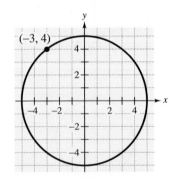

Figure for 12

Systems of Equations

8

- Introduction to Systems of Equations
- Systems of Linear Equations in Two Variables
- Systems of Linear Equations in Three Variables
- Matrices and Systems of Linear Equations
- Linear Systems and Determinants

Diagnostic, Inc. and Invacare Corporation are two American healthcare companies. The annual revenue R (in millions of dollars) for Diagnostic, Inc. from 1988 through 1993 can be modeled by

$$R = -564.34 + 79.67t$$

where $t = 0$ represents 1980.

The annual revenue R (in millions of dollars) for Invacare Corporation during that same period can be modeled by

$$R = -179.49 + 40.41t$$

where $t = 0$ represents 1980.

From the graph at the right, notice that the revenue for Diagnostic, Inc. was increasing at a faster rate than the revenue for Invacare Corporation. In 1990, both companies earned about $230 million. The table at the right confirms this estimate.

t	8	9	10	11	12	13
Diagnostic	71.42	150.89	230.36	309.83	389.30	468.77
Invacare	150.79	191.20	231.61	272.02	312.43	352.84

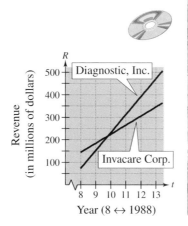

The chapter project related to this information is on page 556.

495

8.1   Introduction to Systems of Equations

Systems of Equations ▪ Solving Systems of Equations by Substitution ▪ Solving Systems of Equations by Graphing ▪ Applications

Systems of Equations

Many problems in business and science involve **systems of equations** that consist of two or more equations, each involving two or more variables.

$$ax + by = c \qquad \text{Equation 1}$$
$$dx + ey = f \qquad \text{Equation 2}$$

A **solution** of such a system is an ordered pair (x, y) of real numbers that satisfies *each* equation in the system. When you find the set of all solutions of the system of equations, you are **solving the system of equations.**

EXAMPLE 1   Checking Solutions of a System of Equations

Which of the ordered pairs is a solution of the system: (a) (3, 3) or (b) (4, 2)?

$$x + y = 6 \qquad \text{Equation 1}$$
$$2x - 5y = -2 \qquad \text{Equation 2}$$

Solution

a. To determine whether the ordered pair (3, 3) is a solution of the system of equations, you should substitute $x = 3$ and $y = 3$ into *each* of the equations. Substituting into Equation 1 produces

$$3 + 3 = 6. \checkmark \qquad \text{Substitute 3 for } x \text{ and 3 for } y.$$

Similarly, substituting into Equation 2 produces

$$2(3) - 5(3) \neq -2. \ ✗ \qquad \text{Substitute 3 for } x \text{ and 3 for } y.$$

Because the ordered pair (3, 3) fails to check in *both* equations, you can conclude that it *is not* a solution of the original system of equations.

b. By substituting the coordinates of the ordered pair (4, 2) into the original equations, you can determine that it is a solution of the first equation

$$4 + 2 = 6 \checkmark \qquad \text{Substitute 4 for } x \text{ and 2 for } y.$$

and is also a solution of the second equation

$$2(4) - 5(2) = -2. \checkmark \qquad \text{Substitute 4 for } x \text{ and 2 for } y.$$

Thus, (4, 2) *is* a solution of the original system of equations.

Solving Systems of Equations by Substitution

One way to solve a system of two equations in two variables is to convert the system to *one* equation in *one* variable by an appropriate substitution.

EXAMPLE 2 The Method of Substitution: One-Solution Case

Solve the system of equations.

$-x + y = 3$	Equation 1
$3x + y = -1$	Equation 2

Solution

Begin by solving for y in Equation 1.

$y = x + 3$	Solve for y in Equation 1.

Next, substitute this expression for y in Equation 2.

$3x + y = -1$	Equation 2
$3x + (x + 3) = -1$	Substitute $x + 3$ for y.
$4x = -4$	Simplify.
$x = -1$	Divide both sides by 4.

At this point, you know that the x-coordinate of the solution is -1. To find the y-coordinate, *back-substitute* the x-value into the revised Equation 1.

$y = x + 3$	Revised Equation 1
$y = -1 + 3$	Substitute -1 for x.
$y = 2$	Simplify.

The solution is $(-1, 2)$. Check this in the original system of equations.

NOTE The term **back-substitute** implies that you work backwards. After finding a value for one of the variables, substitute that value back into one of the equations in the original (or revised) system to find the value of the other variable.

When you use substitution, it does not matter which variable you solve for first. Whether you solve for y first or x first, you will obtain the same solution. When making your choice, you should choose the variable that is easier to work with. For instance, in the system

$3x - 2y = 1$	Equation 1
$x + 4y = 3$	Equation 2

it is easier to solve for x first (in the second equation). But in the system

$2x + y = 5$	Equation 1
$3x - 2y = 11$	Equation 2

it is easier to solve for y first (in the first equation).

The steps for using substitution are summarized as follows.

The Method of Substitution

---

1. Solve one of the equations for one variable in terms of the other.

2. Substitute the expression found in Step 1 into the other equation to obtain an equation in one variable.

3. Solve the equation obtained in Step 2.

4. Back-substitute the solution from Step 3 into the expression obtained in Step 1 to find the value of the other variable.

5. Check the solution in the original system.

Both of the equations in Example 2 are linear. That is, the variables x and y appear in the first power only. The method of substitution can also be used to solve a system of equations in which one or both of the equations is nonlinear.

EXAMPLE 3   The Method of Substitution: Two-Solution Case

Solve the system of equations.

$$x^2 + y^2 = 25 \qquad \text{Equation 1}$$
$$-x + y = -1 \qquad \text{Equation 2}$$

Solution

$$y = x - 1 \qquad \text{Solve for } y \text{ in Equation 2.}$$
$$x^2 + y^2 = 25 \qquad \text{Equation 1}$$
$$x^2 + (x - 1)^2 = 25 \qquad \text{Substitute } x - 1 \text{ for } y.$$
$$2x^2 - 2x - 24 = 0 \qquad \text{Simplify.}$$
$$2(x - 4)(x + 3) = 0 \qquad \text{Factor.}$$
$$x = 4, \ -3 \qquad \text{Solve for } x.$$

Finally, back-substitute these values of x into the revised second equation to solve for y. For $x = 4$, you have

$$y = 4 - 1 = 3$$

which implies that $(4, 3)$ is a solution. For $x = -3$, you have

$$y = -3 - 1 = -4$$

which implies that $(-3, -4)$ is a solution. Check these in the original system.

Solving Systems of Equations by Graphing

A system of two equations in two variables can have exactly one solution, more than one solution, or no solution. In practice, you can gain insight about the location and number of solutions of a system of equations by sketching the graph of each equation in the same coordinate plane. The solutions of the system correspond to the **points of intersection** of the graphs. For instance, Figure 8.1 shows the graphs of three pairs (systems) of equations.

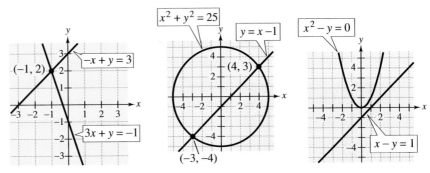

FIGURE 8.1

EXAMPLE 4   *The Graphical Method of Solving a System*

Use the graphical method to solve the system of equations.

$$2x + 3y = 7 \qquad \text{Equation 1}$$
$$2x - 5y = -1 \qquad \text{Equation 2}$$

Solution

Because both equations in the system are linear, you know that they have graphs that are straight lines. To sketch these lines, write each equation in slope-intercept form, as follows.

$$y = -\frac{2}{3}x + \frac{7}{3} \qquad \text{Equation 1}$$

$$y = \frac{2}{5}x + \frac{1}{5} \qquad \text{Equation 2}$$

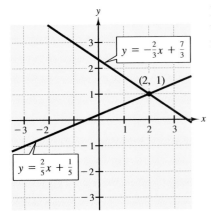

FIGURE 8.2

The lines corresponding to these two equations are shown in Figure 8.2. From this figure, it appears that the two lines intersect in a single point, and that the coordinates of the point are approximately (2, 1). You can check these as follows.

$$2(2) + 3(1) = 7 \qquad \text{Solution checks in Equation 1.} \checkmark$$
$$2(2) - 5(1) = -1 \qquad \text{Solution checks in Equation 2.} \checkmark$$

Applications

To model a real-life situation with a system of equations, you can use the same basic problem-solving strategy that has been used throughout the text.

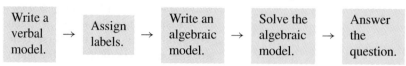

| Write a verbal model. | → | Assign labels. | → | Write an algebraic model. | → | Solve the algebraic model. | → | Answer the question. |

After answering the question, remember to check the answer in the original statement of the problem.

EXAMPLE 5   An Application

In 1990, there were about 90 thousand construction companies that specialized in single-family construction. These companies had an average of 4 employees each.

A roofing contractor bought 30 bundles of shingles and four rolls of roofing paper for $528. A second purchase (at the same prices) cost $140 for eight bundles of shingles and one roll of roofing paper. Find the price per bundle of shingles and the price per roll of roofing paper.

Solution

Verbal Model:

$$\boxed{\text{Cost of 30 bundles}} + \boxed{\text{Cost of 4 rolls}} = \boxed{528}$$

$$\boxed{\text{Cost of 8 bundles}} + \boxed{\text{Cost of 1 roll}} = \boxed{140}$$

Labels: Price of bundle of shingles $= x$ (dollars)
            Price of roll of roofing paper $= y$ (dollars)

System: $30x + 4y = 528$ Equation 1
            $8x + \ y = 140$ Equation 2

Solving the second equation for y produces $y = 140 - 8x$, and substituting this expression into the first equation produces the following.

$$30x + 4(140 - 8x) = 528$$
$$30x + 560 - 32x = 528$$
$$-2x = -32$$
$$x = 16$$

Back-substituting $x = 16$ into the revised second equation produces

$$y = 140 - 8(16)$$
$$= 12.$$

Thus, you can conclude that the price of shingles is $16 per bundle and the price of roofing paper is $12 per roll. Check this in the original statement of the problem.

The total cost C of producing x units of a product usually has two components—the initial cost and the cost per unit. When enough units have been sold so that the total revenue R equals the total cost, the sales are said to have reached the **break-even point.** You can find this break-even point by setting C equal to R and solving for x. In other words, the break-even point corresponds to the point of intersection of the cost and revenue graphs.

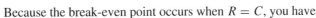

EXAMPLE 6   An Application: Break-Even Analysis

A small business invests \$14,000 in equipment to produce a product. Each unit of the product costs \$0.80 to produce and is sold for \$1.50. How many items must be sold before the business breaks even?

Solution

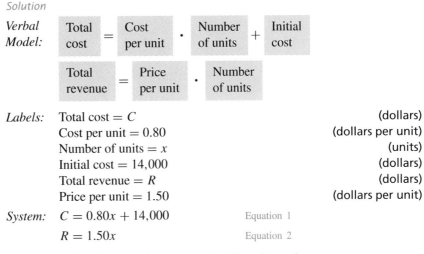

Verbal Model:

$$\boxed{\text{Total cost}} = \boxed{\text{Cost per unit}} \cdot \boxed{\text{Number of units}} + \boxed{\text{Initial cost}}$$

$$\boxed{\text{Total revenue}} = \boxed{\text{Price per unit}} \cdot \boxed{\text{Number of units}}$$

Labels:
Total cost = C (dollars)
Cost per unit = 0.80   (dollars per unit)
Number of units = x (units)
Initial cost = 14,000   (dollars)
Total revenue = R (dollars)
Price per unit = 1.50   (dollars per unit)

System:
$$C = 0.80x + 14{,}000 \qquad \text{Equation 1}$$
$$R = 1.50x \qquad \text{Equation 2}$$

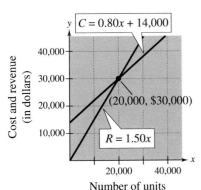

FIGURE 8.3

Because the break-even point occurs when $R = C$, you have

$$1.5x = 0.8x + 14{,}000$$
$$0.7x = 14{,}000$$
$$x = 20{,}000.$$

Thus, it follows that the business must sell 20,000 units before it breaks even. Note in Figure 8.3 that sales less than the break-even point correspond to a loss for the business, whereas sales greater than the break-even point correspond to a profit for the business. The following table helps confirm this conclusion.

Units, x	0	5000	10,000	15,000	20,000	25,000
Profit, P	−\$14,000	−\$10,500	−\$7000	−\$3500	\$0	\$3500

EXAMPLE 7 *An Interest Rate Problem*

A total of $12,000 is invested in two funds paying 6% and 8% simple interest. If the interest for 1 year is $880, how much of the $12,000 was invested in each fund?

Solution

Verbal Model:

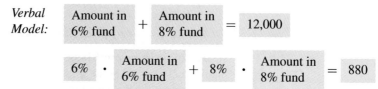

Labels: Amount in 6% fund $= x$ (dollars)
            Amount in 8% fund $= y$ (dollars)

System: $x + \quad y = 12,000$ Equation 1
            $0.06x + 0.08y = \quad 880$ Equation 2

To solve this system with a graphing utility, solve each equation for y to get

$$y = 12,000 - x$$ Equation 1
$$y = 11,000 - 0.75x.$$ Equation 2

Graph both equations on the same screen, as shown in Figure 8.4. You can estimate that the two graphs intersect when

$$x = 4000 \quad \text{and} \quad y = 8000.$$

Check this in the original statement of the problem.

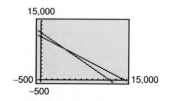

FIGURE 8.4

Group Activities You Be the Instructor

Problem Posing   You want to create several systems of equations with relatively simple solutions that students can use for practice. Discuss how to create a system of equations that has a given solution. Illustrate your method by creating a system of linear equations that has one of the following solutions: $(1, 4)$, $(-2, 5)$, $(-3, 1)$, or $(4, 2)$. Trade your system for that of another group member, and verify that each other's systems have the desired solutions.

8.1 Exercises

Discussing the Concepts

1. What is meant by a solution of a system of equations in two variables?

2. List and explain the basic steps in solving a system of equations by substitution.

3. When solving a system of equations by substitution, how do you recognize that the system has no solution?

4. What does it mean to *back-substitute* when solving a system of equations?

5. Give a geometric description of the solution of a system of equations in two variables.

6. Describe any advantages of substitution over the graphical method of solving a system of equations.

Problem Solving

In Exercises 7–10, determine whether each ordered pair is a solution of the system of equations.

7. $x + 2y = 9$ (a) $(1, 4)$
   $-2x + 3y = 10$ (b) $(3, -1)$

8. $5x - 4y = 34$ (a) $(0, 3)$
   $x - 2y = 8$ (b) $(6, -1)$

9. $-2x + 7y = 46$ (a) $(-3, 2)$
   $3x + y = 0$ (b) $(-2, 6)$

10. $-5x - 2y = 23$ (a) $(-3, -4)$
    $x + 4y = -19$ (b) $(3, 7)$

In Exercises 11–20, solve the system by substitution.

11. $x - 2y = 0$ 12. $x - 3y = -2$
    $3x + 2y = 8$ $5x + 3y = 17$

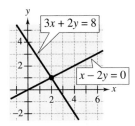

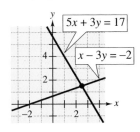

13. $x + y = 2$ 14. $2x + y = 10$
    $x^2 - y = 0$ $x^2 + y^2 = 25$

15. $x + y = 3$ 16. $-x + y = 5$
    $2x - y = 0$ $x - 4y = 0$

17. $4x - 14y = -15$ 18. $5x - 24y = -12$
    $18x - 12y = 9$ $17x - 24y = 36$

19. $x^2 + y^2 = 25$ 20. $x^2 - y^2 = 16$
    $2x - y = -5$ $3x - y = 12$

In Exercises 21–26, use the graphs of the equations to determine whether the system has any solutions. Find any solutions that exist.

21. $x + y = 4$ 22. $-x + y = 5$
    $x + y = -1$ $x + 2y = 4$

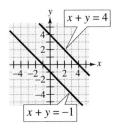

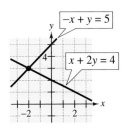

23. $5x - 3y = 4$ 24. $2x - y = 4$
    $2x + 3y = 3$ $-4x + 2y = -12$

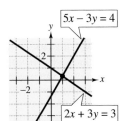

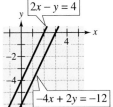

25. $x - 2y = 4$
    $x^2 - y = 0$

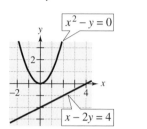

26. $y = \sqrt{x - 2}$
    $x - 2y = 1$

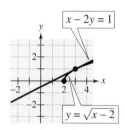

In Exercises 27–32, use a graphing utility to graph the equations and approximate any solutions of the system of equations.

27. $-2x + y = 1$
    $x - 3y = 2$

28. $5x - 6y = -30$
    $5x + 4y = 20$

29. $x^2 - y^2 = 12$
    $x - 2y = 0$

30. $x^2 + y^2 = 20$
    $x + 3y = 10$

31. $y = x^3$
    $y = x^3 - 3x^2 + 3x$

32. $y = \frac{1}{5}(24 - x)$
    $y = \sqrt{64 - x^2}$

In Exercises 33 and 34, use a graphing utility to graph each equation in the system. The graphs appear parallel. Yet, from the slope-intercept forms of the lines, you find that the slopes are not equal and thus the graphs intersect. Find the point of intersection of the two lines.

33. $x - 100y = -200$
    $3x - 275y = 198$

34. $35x - 33y = 0$
    $12x - 11y = 92$

35. *Break-Even Analysis* A small business invests $8000 in equipment to produce a product. Each unit of the product costs $1.20 to produce and is sold for $2.00. How many items must be sold before the business breaks even?

36. *Break-Even Analysis* A business invests $50,000 in equipment to produce a product. Each unit of the product costs $19.25 to produce and is sold for $35.95. How many items must be sold before the business breaks even?

37. *Hyperbolic Mirror* In a hyperbolic mirror, light rays directed to one focus are reflected to the other focus. The mirror illustrated below has the equation

$$\frac{x^2}{36} - \frac{y^2}{64} = 1.$$

At which point on the mirror will light from the point $(0, 10)$ reflect to the focus?

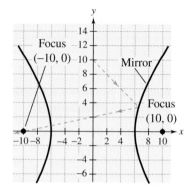

38. *Miniature Golf* You are playing miniature golf. Your golf ball is at the point $(-15, 25)$ (see figure). The wall at the end of the enclosed area is part of a hyperbola whose equation is

$$\frac{x^2}{19} - \frac{y^2}{81} = 1.$$

Using the reflective property of hyperbolas given in Exercise 37, at which point on the wall must your ball hit for it to go into the hole? (The ball bounces off a wall only once.)

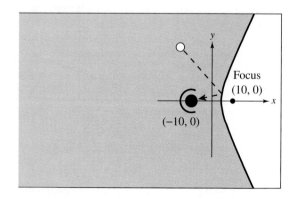

Reviewing the Major Concepts

In Exercises 39–42, find the general form of the equation of the line through the two points.

39. $(-1, -2)$, $(3, 6)$

40. $(1, 5)$, $(6, 0)$

41. $\left(\frac{3}{2}, 8\right)$, $\left(\frac{11}{2}, \frac{5}{2}\right)$

42. $(0, 2)$, $(7.3, 15.4)$

43. *Work Rate* Machine A can complete a job in 4 hours and Machine B can complete it in 6 hours. How long will it take both machines working together to complete the job?

44. *Geometry* Draw a rectangle whose diagonal has a length of 5 units. Do all such rectangles have the same dimensions? Explain.

Additional Problem Solving

In Exercises 45–66, solve the system by substitution.

45.
$$x \qquad\; = 4$$
$$x - 2y = -2$$

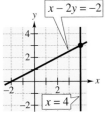

46.
$$y = 2$$
$$x - 6y = -6$$

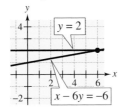

47. $7x + 8y = 24$
$$x - 8y = 8$$

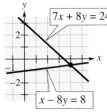

48.
$$x - y = 0$$
$$5x - 2y = 6$$

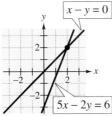

49. $y = \sqrt{8 - x}$
$$y = -\frac{1}{5}(x - 14)$$

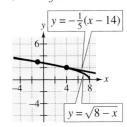

50. $9x^2 + 4y^2 = 36$
$$3x - 2y = -6$$

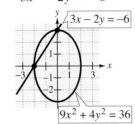

51. $x + y = 2$
$$x - 4y = 12$$

52. $x - 2y = -1$
$$x - 5y = 2$$

53. $x + 6y = 19$
$$x - 7y = -7$$

54. $x - 5y = -6$
$$4x - 3y = 10$$

55. $2x + 5y = 29$
$$5x + 2y = 13$$

56. $-13x + 16y = 10$
$$5x + 16y = -26$$

57. $y = 2x^2$
$$y = -2x + 12$$

58. $y = 5x^2$
$$y = -15x - 10$$

59. $x^2 + y = 9$
$$x - y = -3$$

60. $x - y^2 = 0$
$$x - y = 2$$

61. $3x + 2y = 90$
$$xy = 300$$

62. $x + 2y = 40$
$$xy = 150$$

63. $x^2 + y^2 = 100$
$$x = 12$$

64. $x^2 + y^2 = 169$
$$x + y = 7$$

65. $y = \sqrt{4 - x}$
$$x + 3y = 6$$

66. $x^2 + 2y = 6$
$$x - y = -4$$

In Exercises 67–78, use a graphing utility to graph the equations and approximate any solutions of the system.

67. $2x - 5y = 20$
$$4x - 5y = 40$$

68. $5x + 3y = 24$
$$x - 2y = 10$$

69. $x - y = -3$
$$2x - y = 6$$

70. $x = 4$
$$y = 3$$

71. $-5x + 3y = 15$
$x + y = 1$

72. $9x + 4y = 8$
$7x - 4y = 8$

73. $y = x^2$
$y = 4x - x^2$

74. $y = 8 - x^2$
$y = 6 - x$

75. $\sqrt{x} - y = 0$
$\phantom{\sqrt{}}x - 5y = -6$

76. $x^2 + y = 4$
$x + y = 6$

77. $16x^2 + 9y^2 = 144$
$4x + 3y = 12$

78. $x^2 - y^2 = 1$
$\frac{1}{2}x^2 + y^2 = 1$

In Exercises 79–82, find two positive integers that satisfy the given requirements.

79. The sum of two numbers is 80 and their difference is 18.

80. The sum of a larger number and twice a smaller number is 61 and the difference between the numbers is 7.

81. The sum of two numbers is 52 and the larger number is 8 less than twice the smaller number.

82. The sum of two numbers is 160 and the larger number is three times the smaller number.

Geometry In Exercises 83–86, find the dimensions of the rectangle meeting the specified conditions.

	Perimeter	*Condition*
83.	50 feet	The length is 5 feet greater than the width.
84.	320 inches	The width is 20 inches less than the length.
85.	68 yards	The width is $\frac{7}{10}$ of the length.
86.	90 meters	The length is $1\frac{1}{2}$ times the width.

87. *Simple Interest*   A combined total of $20,000 is invested in two bonds that pay 8% and 9.5% simple interest. The annual interest is $1675. How much is invested in each bond?

88. *Simple Interest*   A combined total of $12,000 is invested in two bonds that pay 8.5% and 10% simple interest. The annual interest is $1140. How much is invested in each bond?

89. *Busing Boundary*   To be eligible to ride the school bus to East High School, a student must live at least 1 mile from the school (see figure). Describe the portion of Clarke Street from which the residents are *not* eligible to ride the school bus. Use a coordinate system in which the school is at (0, 0) and each unit represents 1 mile.

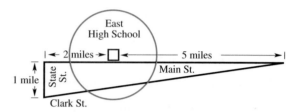

90. *Dimensions of a Corral*   You have 250 feet of fencing to enclose two corrals of equal size (see figure). The combined area of the corrals is 2400 square feet. Find the dimensions of each corral.

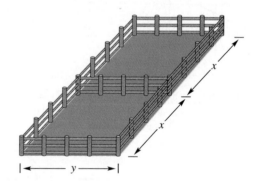

91. *Comparing Populations*   From 1982 to 1988, the northeastern part of the United States grew at a slower rate than the western part. Two models that represent the populations of the two regions are

$$P = 49{,}094.5 + 106.9t + 9.7t^2 \qquad \text{Northeast}$$
$$P = 43{,}331.9 + 907.8t \qquad \text{West}$$

where P is the population in thousands and t is the calendar year with $t = 3$ corresponding to 1983. Use a graphing utility to determine when the population of the West overtook the population of the Northeast. (Source: U.S. Bureau of Census)

92. *Geometry*   A theorem from geometry states that if a triangle is inscribed in a circle so that one side of the triangle is a diameter of the circle, the triangle is a right triangle (see figure). Show that this theorem is true for the circle

$$x^2 + y^2 = 100$$

and the triangle formed by the lines

$$y = 0, \quad y = \frac{1}{2}x + 5, \quad \text{and} \quad y = -2x + 20.$$

(Find the vertices of the triangle and verify that it is a right triangle.)

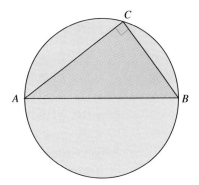

Math Matters   Circle Problem of Apollonius

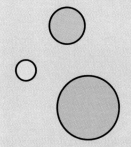

Apollonius and Euclid are the two most famous mathematicians of the classical Greek period. Euclid lived about 300 B.C. and Apollonius lived about 100 years later. In one of the writings of Apollonius, he poses the following problem.

Suppose you are given three circles that have no points in common and no circle is inside either of the other two circles, as shown in the figure at the left. Is it always possible to find a fourth circle that is tangent to each of the given circles? The answer is yes. In fact, it can be shown that the problem always has eight distinct solutions. Four are shown here. Can you find the other four? (The answer is given in the back of the book.)

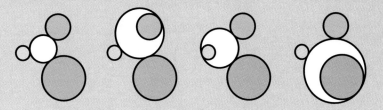

8.2 Systems of Linear Equations in Two Variables

The Method of Elimination ▪ Graphical Interpretation of Solutions ▪ Applications

The Method of Elimination

In Section 8.1, you studied two ways of solving a system of equations—substitution and graphing. In this section, you will study a third way—the **method of elimination.**

The key step in the method of elimination is to obtain, for one of the variables, coefficients that differ only in sign so that when the two equations are *added* this variable is eliminated. Notice how this is accomplished in Example 1—when the two equations are added, the *y*-terms are eliminated.

EXAMPLE 1   The Method of Elimination

Solve the system of linear equations.

$$3x + 2y = 4 \qquad \text{Equation 1}$$
$$5x - 2y = 8 \qquad \text{Equation 2}$$

Solution

Begin by noting that the coefficients of *y* differ only in sign. By adding the two equations, you can eliminate *y*.

$$
\begin{array}{llr}
3x + 2y = & 4 & \text{Equation 1} \\
5x - 2y = & 8 & \text{Equation 2} \\
\hline
8x = & 12 & \text{Add equations}
\end{array}
$$

Thus, $x = \frac{3}{2}$. By back-substituting this value into the first equation, you can solve for *y* as follows.

$$3x + 2y = 4 \qquad \text{Equation 1}$$
$$3\left(\frac{3}{2}\right) + 2y = 4 \qquad \text{Substitute } \tfrac{3}{2} \text{ for } x.$$
$$2y = -\frac{1}{2} \qquad \text{Simplify.}$$
$$y = -\frac{1}{4} \qquad \text{Divide both sides by 2.}$$

The solution is $\left(\frac{3}{2}, -\frac{1}{4}\right)$. Check this solution in the original system of linear equations.

Try using substitution to solve the system given in Example 1. Which method do you think is easier: substitution or elimination? Many people find that the method of elimination is more efficient.

The Method of Elimination

1. Obtain coefficients for x (or y) that differ only in sign by multiplying all terms of one or both equations by suitably chosen constants.

2. Add the equations to eliminate one variable and solve the resulting equation.

3. Back-substitute the value obtained in Step 2 into either of the original equations and solve for the other variable.

4. Check your solution in both of the original equations.

To obtain coefficients (for one of the variables) that differ only in sign, you often need to multiply one or both of the equations by a suitable constant.

EXAMPLE 2   The Method of Elimination

Solve the system of linear equations.

$$4x - 5y = 13 \qquad \text{Equation 1}$$
$$3x - y = 7 \qquad \text{Equation 2}$$

Solution

To obtain coefficients of y that differ only in sign, multiply Equation 2 by -5.

$$
\begin{array}{rcl}
4x - 5y &=& 13 \\
3x - y &=& 7
\end{array}
\implies
\begin{array}{rcll}
4x - 5y &=& 13 & \text{Equation 1} \\
-15x + 5y &=& -35 & \text{Multiply Equation 2 by } -5. \\
\hline
-11x &=& -22 & \text{Add equations.}
\end{array}
$$

Thus, $x = 2$. Back-substitute this value into Equation 2 and solve for y.

$$3x - y = 7 \qquad \text{Equation 2}$$
$$3(2) - y = 7 \qquad \text{Substitute 2 for } x.$$
$$y = -1 \qquad \text{Solve for } y.$$

The solution is $(2, -1)$. Check this in the original system of equations.

$$4(2) - 5(-1) = 13 \qquad \text{Solution checks in Equation 1.} \checkmark$$
$$3(2) - (-1) = 7 \qquad \text{Solution checks in Equation 2.} \checkmark$$

Graphical Interpretation of Solutions

Rewrite each of the following systems of equations in slope-intercept form and graph the equations using a graphing utility. What is the relationship between the slopes of the two lines and the number of points of intersection?

a. $2x + 4y = 8$
$4x - 3y = -6$

b. $-x + 5y = 15$
$2x - 10y = 7$

c. $x - y = 9$
$2x - 2y = 18$

As you observed in Section 8.1, it is possible for a *general* system of equations to have exactly one solution, two or more solutions, or no solution. For a system of *linear* equations, the second of these three possibilities is different.

Specifically, if a system of linear equations has two different solutions, it must have an *infinite* number of solutions. To see why this is true, consider the following graphical interpretations of a system of two linear equations in two variables. (Remember that the graph of a linear equation in two variables is a straight line.)

Graphical Interpretation of Solutions

For a system of two linear equations in two variables, the number of solutions is given by one of the following.

Number of Solutions	*Graphical Interpretation*
1. Exactly one solution	The two lines intersect at one point.
2. Infinitely many solutions	The two lines are identical.
3. No solution	The two lines are parallel.

These three possibilities are shown in Figure 8.5.

Consistent   Inconsistent

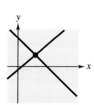 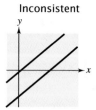

Two lines that intersect: single point of intersection

Two lines that coincide: infinitely many points of intersection

Two parallel lines: no point of intersection

FIGURE 8.5

A system of linear equations is **consistent** if it has at least one solution, and it is **inconsistent** if it has no solution. Here are some examples.

Consistent System	*Consistent System*	*Inconsistent System*
$x + y = 2$	$x + y = 2$	$x + y = 2$
$x + 2y = 4$	$2x + 2y = 4$	$x + y = 4$

Try using a graphing utility to determine how many solutions each of these systems has.

Example 3 shows how the method of elimination can be used to determine that a system of linear equations has no solution.

EXAMPLE 3   The Method of Elimination: No-Solution Case

Solve the system of linear equations.

$$3x + 9y = 8 \qquad \text{Equation 1}$$
$$2x + 6y = 7 \qquad \text{Equation 2}$$

Solution

To obtain coefficients of x that differ only in sign, multiply the first equation by 2 and multiply the second equation by -3.

$$3x + 9y = 8 \qquad\Longrightarrow\qquad 6x + 18y = 16 \qquad \text{Multiply Equation 1 by 2.}$$
$$2x + 6y = 7 \qquad\Longrightarrow\qquad \underline{-6x - 18y = -21} \qquad \text{Multiply Equation 2 by } -3.$$
$$0 = -5 \qquad \text{False statement}$$

Because there are no values of x and y for which $0 = -5$, you can conclude that the system is inconsistent and has no solution. The lines corresponding to the two equations given in this system are shown in Figure 8.6. Note that the two lines are parallel, and therefore have no point of intersection.

Example 4 shows how the method of elimination works with a system that has infinitely many solutions.

EXAMPLE 4   The Method of Elimination: Many-Solutions Case

Solve the system of linear equations.

$$-2x + 6y = 3 \qquad \text{Equation 1}$$
$$4x - 12y = -6 \qquad \text{Equation 2}$$

Solution

To obtain coefficients of x that differ only in sign, multiply the first equation by 2.

$$-2x + 6y = 3 \qquad\Longrightarrow\qquad -4x + 12y = 6 \qquad \text{Multiply Equation 1 by 2.}$$
$$4x - 12y = -6 \qquad\Longrightarrow\qquad \underline{4x - 12y = -6} \qquad \text{Equation 2}$$
$$0 = 0 \qquad \text{Add equations.}$$

Because the two equations turn out to be equivalent, the system has infinitely many solutions. The solution set consists of all points (x, y) lying on the line $-2x + 6y = 3$, as shown in Figure 8.7.

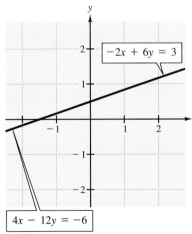

FIGURE 8.6

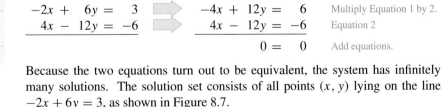

FIGURE 8.7

Programming

The general solution of the linear system

$$ax + by = c$$
$$dx + ey = f$$

is $x = (ce - bf)/(ae - db)$ and $y = (af - cd)/(ae - db)$. If $ae - db = 0$, the system does not have a unique solution. A *TI-82* program for solving such a system is given below. (Programs for other calculator models are given in the Appendix.) Try using this program to solve the systems in Examples 5 and 6.

```
PROGRAM: SOLVE
:Prompt A
:Prompt B
:Prompt C
:Prompt D
:Prompt E
:Prompt F
:If AE - DB = 0
:Then
:Disp ''NO SOLUTION''
:Else
:(CE - BF)/(AE - DB) → X
:(AF - CD)/(AE - DB) → Y
:Disp X
:Disp Y
:End
```

Applications

To determine which real-life problems can be solved using a system of linear equations, consider the following. (1) Does the problem involve more than one unknown quantity? (2) Are there two (or more) equations or conditions to be satisfied? If one or both of these conditions occur, the appropriate mathematical model for the problem may be a system of linear equations.

EXAMPLE 5 A Mixture Problem

A company with two stores buys six large delivery vans and five small delivery vans. The first store receives four of the large vans and two of the small vans for a total cost of $160,000. The second store receives two of the large vans and three of the small vans for a total cost of $128,000. What is the cost of each type of van?

Solution

The two unknowns in this problem are the costs of the two types of vans.

Verbal Model:

$$4 \left(\begin{array}{c} \text{Cost of} \\ \text{large van} \end{array} \right) + 2 \left(\begin{array}{c} \text{Cost of} \\ \text{small van} \end{array} \right) = \$160,000$$

$$2 \left(\begin{array}{c} \text{Cost of} \\ \text{large van} \end{array} \right) + 3 \left(\begin{array}{c} \text{Cost of} \\ \text{small van} \end{array} \right) = \$128,000$$

Labels: Cost of large van $= x$ (dollars)
Cost of small van $= y$ (dollars)

System: $4x + 2y = 160,000$ Equation 1
$2x + 3y = 128,000$ Equation 2

To solve this system of linear equations, use the method of elimination. To obtain coefficients of x that differ only in sign, multiply the second equation by -2.

$$\begin{array}{l} 4x + 2y = 160,000 \\ 2x + 3y = 128,000 \end{array} \implies \begin{array}{r} 4x + 2y = 160,000 \\ -4x - 6y = -256,000 \\ \hline -4y = -96,000 \end{array}$$

The cost of each small van is $y = \$24,000$. Back-substitute this value into Equation 1 to find the cost of each large van.

$$4x + 2y = 160,000 \qquad \text{Equation 1}$$
$$4x + 2(24,000) = 160,000 \qquad \text{Substitute 24,000 for } y.$$
$$4x = 112,000 \qquad \text{Simplify.}$$
$$x = 28,000 \qquad \text{Divide both sides by 4.}$$

The cost of each large van is $x = \$28,000$. Check this solution in the original statement of the problem.

EXAMPLE 6 *An Application Involving Two Speeds*

You take a motorboat trip on a river (18 miles upstream and 18 miles back downstream). You run the motor at the same speed going up and down the river, but because of the current of the river, the trip takes longer going upstream than it does downstream. You do not know the speed of the river's current, but you know that the trip upstream takes $1\frac{1}{2}$ hours and the trip downstream takes only 1 hour. From this information, can you determine the speed of the current?

Solution

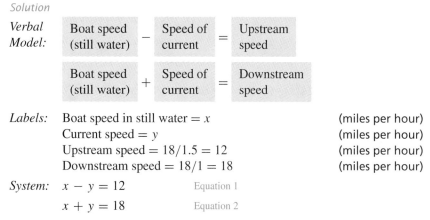

Verbal Model:

| Boat speed (still water) | − | Speed of current | = | Upstream speed |

| Boat speed (still water) | + | Speed of current | = | Downstream speed |

Labels: Boat speed in still water $= x$ (miles per hour)
Current speed $= y$ (miles per hour)
Upstream speed $= 18/1.5 = 12$ (miles per hour)
Downstream speed $= 18/1 = 18$ (miles per hour)

System: $x - y = 12$ Equation 1
$x + y = 18$ Equation 2

To solve this system of linear equations, use the method of elimination to obtain $x = 15$ miles per hour and $y = 3$ miles per hour. Check this in the original statement of the problem.

Group Activities — Problem Solving

Fitting a Line to Data   The median existing home price in the United States in 1988 was $90,600, and in 1994 it was $107,600. Suppose you wish to fit a linear equation to the points representing this data: $(8, 90,600)$ and $(14, 107,600)$. You remember that one form of an equation for a line is $y = mx + b$, but you can't remember how to use the methods for finding the equation of a line that you used in previous chapters. Write a system of equations that could be solved to find the values of m and b. Solve the system and write the linear equation that fits the data. Interpret the slope in the context of the data. By your model, what was the median existing home price in 1991? The actual median price in 1991 was $99,700. How does the value obtained from your model compare?   (Source: National Association of Realtors)

8.2 Exercises

Discussing the Concepts

1. Give a geometric description of the solution of a system of linear equations in two variables.

2. Explain what is meant by an *inconsistent* system of linear equations.

3. Is it possible for a consistent system of linear equations to have exactly two solutions? Explain.

4. In your own words, explain how to solve a system of linear equations by elimination.

5. How can you recognize that a system of linear equations has no solution? Give an example.

6. When might substitution be better than elimination for solving a system of linear equations?

Problem Solving

In Exercises 7–10, use a graphing utility to graph the equations in the system. Use the graphs to determine whether the system is consistent or inconsistent. If it is consistent, determine the number of solutions.

7. $\frac{1}{3}x - \frac{1}{2}y = 1$
   $-2x + 3y = 6$

8. $x + y = 5$
   $x - y = 5$

9. $-2x + 3y = 6$
   $x - y = -1$

10. $2x - 4y = 9$
    $x - 2y = 4.5$

In Exercises 11 and 12, determine whether the system is consistent or inconsistent.

11. $x + 2y = 6$
    $x + 2y = 3$

12. $x - 2y = 3$
    $2x - 4y = 6$

In Exercises 13–22, solve the system of linear equations by elimination. Identify and label each line with its equation and label the point of intersection (if any).

13. $-x + 2y = 1$
    $x - y = 2$

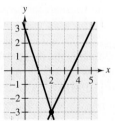

14. $3x + y = 3$
    $2x - y = 7$

15. $x + y = 0$
    $3x - 2y = 10$

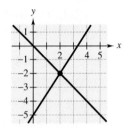

16. $-x + 2y = 2$
    $3x + y = 15$

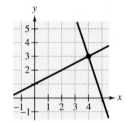

17. $x - y = 1$
    $-3x + 3y = 8$

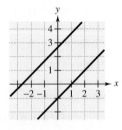

18. $3x + 4y = 2$
    $0.6x + 0.8y = 1.6$

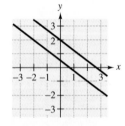

19. $x - 3y = 5$
    $-2x + 6y = -10$

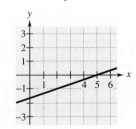

20. $x - 4y = 5$
    $5x + 4y = 7$

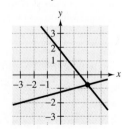

21. $2x - 8y = -11$
$5x + 3y = 7$

22. $3x + 4y = 0$
$9x - 5y = 17$

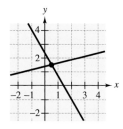
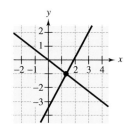

In Exercises 23–28, solve the system of linear equations by the method of elimination.

23. $3x - 2y = 5$
$x + 2y = 7$

24. $7r - s = -25$
$2r + 5s = 14$

25. $12x - 5y = 2$
$-24x + 10y = 6$

26. $-2x + 3y = 9$
$6x - 9y = -27$

27. $2x + 3y = 0$
$3x + 5y = -1000$

28. $0.4u + 1v = 800$
$0.7u + 2v = 1850$

In Exercises 29–32, solve the system by any convenient method.

29. $\frac{3}{2}x + 2y = 12$
$\frac{1}{4}x + y = 4$

30. $4x + y = -2$
$-6x + y = 18$

31. $y = 5x - 3$
$y = -2x + 11$

32. $3x + 2y = 5$
$y = 2x + 13$

In Exercises 33 and 34, find a system of linear equations that has the given solution. (There are many correct answers.)

33. $\left(3, -\frac{3}{2}\right)$

34. $(-8, 12)$

35. *Break-Even Analysis* You are planning to open a restaurant. You need an initial investment of $85,000. Each week your costs will be about $7400. If your weekly revenue is $8100, how many weeks will it take to break even?

36. *Investing in Two Funds* A total of $4500 is invested in two funds paying 4% and 5% annual interest. The combined annual interest is $210. How much of the $4500 is invested in each fund?

37. *Average Speed* A truck travels for 4 hours at an average speed of 42 miles per hour. How much longer must the truck travel at an average speed of 55 miles per hour so that the average speed for the entire trip will be 50 miles per hour?

38. *Air Speed* An airplane flying into a headwind travels the 3000-mile flying distance between two cities in 6 hours and 15 minutes. On the return flight, the distance is traveled in 5 hours. Find the speed of the plane in still air and the speed of the wind, assuming that both remain constant throughout the round trip.

39. *Rope Length* You must cut a rope that is 160 inches long into two pieces so that one piece is four times as long as the other. Find the length of each piece.

40. *Driving Distance* Two people share the driving on a trip of 300 miles. One person drives three times as far as the other. Find the distance that each person drives.

41. *Vietnam Veterans Memorial* "The Wall" in Washington, D.C., designed by Maya Ling Lin when she was a student at Yale University, has two vertical, triangular sections of black granite with a common side (see figure). The top of each section is level with the ground. The bottoms of the two sections can be modeled by the equations $y = \frac{2}{25}x - 10$ and $y = -\frac{5}{61}x - 10$ when the x-axis is superimposed on the top of the wall. Each unit on the coordinate system represents 1 foot. How deep is the memorial at the point where the two sections meet? How long is each section?

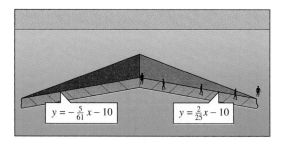

42. *Mathematical Modeling* The total employment (in thousands) in aircraft industries in the United States from 1990 through 1992 is given in the table. (Source: U.S. Bureau of Labor Statistics)

Year	1990	1991	1992
Employment	898	838	758

(a) Plot the data given in the table, where $x = 0$ corresponds to 1990.

(b) The line $y = mx + b$ that best fits the data is given by the solution of the following system.

$$3b + 3m = 2494$$
$$3b + 5m = 2354$$

Solve the system and find the equation of the required line. Graph the line on the coordinate system of part (a).

(c) Explain the meaning of the slope of the line in the context of this problem.

Reviewing the Major Concepts

In Exercises 43–46, find an equation of the line through the two points.

43. $(-2, 7), (5, 5)$

44. $(3.5, 4), (10, 6)$

45. $(6, 3), (10, 3)$

46. $(4, -2), (4, 5)$

In Exercises 47 and 48, find the number of units x that produces a maximum revenue R. Confirm your result graphically with a graphing utility.

47. $R = 900x - 0.1x^2$

48. $R = 400x - x^2$

Additional Problem Solving

In Exercises 49–66, solve the system of linear equations by the method of elimination.

49. $2x + y = 9$
$3x - y = 16$

50. $-x + 2y = 9$
$x + 3y = 16$

51. $4x + y = -3$
$-4x + 3y = 23$

52. $-3x + 5y = -23$
$2x - 5y = 22$

53. $x - 3y = 2$
$3x - 7y = 4$

54. $2s - t = 9$
$3s + 4t = -14$

55. $2u + 3v = 8$
$3u + 4v = 13$

56. $4x - 3y = 25$
$-3x + 8y = 10$

57. $\frac{2}{3}r - s = 0$
$10r + 4s = 19$

58. $x - y = -\frac{1}{2}$
$4x - 48y = -35$

59. $0.7u - v = -0.4$
$0.3u - 0.8v = 0.2$

60. $0.15x - 0.35y = -0.5$
$-0.12x + 0.25y = 0.1$

61. $5x + 7y = 25$
$x + 1.4y = 5$

62. $12b - 13m = 2$
$-6b + 6.5m = -2$

63. $\frac{3}{2}x - y = 4$
$-x + \frac{2}{3}y = -1$

64. $12x - 3y = 6$
$4x - y = 2$

65. $2x = 25$
$4x - 10y = 0.52$

66. $6x - 6y = 25$
$3y = 11$

In Exercises 67–70, solve the system by any convenient method.

67. $2x - y = 20$
$-x + y = -5$

68. $3x - 2y = -20$
$5x + 6y = 32$

69. $3y = 2x + 21$
$x = 50 - 4y$

70. $x + 2y = 4$
$\frac{1}{2}x + \frac{1}{3}y = 1$

In Exercises 71 and 72, determine the value of k such that the system of linear equations is inconsistent.

71. $5x - 10y = 40$
$-2x + ky = 30$

72. $12x - 18y = 5$
$-18x + ky = 10$

In Exercises 73–78, decide whether the system is consistent or inconsistent.

73. $\quad 4x - 5y = 3$
$\quad\quad -8x + 10y = -6$

74. $\quad 4x - 5y = 3$
$\quad\quad -8x + 10y = 14$

75. $-2x + 5y = 3$
$\quad\quad 5x + 2y = 8$

76. $\quad x + 10y = 12$
$\quad\quad -2x + 5y = 2$

77. $-10x + 15y = 25$
$\quad\quad 2x - 3y = -24$

78. $\quad 4x - 5y = 28$
$\quad\quad -2x + 2.5y = -14$

In Exercises 79–82, use a graphing utility to sketch the graphs of the two equations. Use the graphs to approximate the solution of the system.

79. $5x + 4y = 35$
$\quad\quad -x + 3y = 12$

80. $\quad 5x - 4y = 0$
$\quad\quad -3x + 8y = 14$

81. $4x - y = 3$
$\quad\quad 6x + 2y = 1$

82. $\quad x - 6y = 2$
$\quad\quad 2x + 3y = 9$

Geometry In Exercises 83 and 84, find the dimensions of the rectangle meeting the specified conditions.

Perimeter	Condition
83. 220 meters	The length is 120% of the width.
84. 280 feet	The width is 75% of the length.

In Exercises 85–88, determine how many coins of each type will yield the given value.

Coins	Value
85. 15 dimes and quarters	$3.00
86. 32 dimes and quarters	$6.95
87. 25 nickels and dimes	$1.95
88. 50 nickels and quarters	$7.50

89. *Gasoline Mixture* Twelve gallons of regular unleaded gasoline plus 8 gallons of premium unleaded gasoline cost $23.08. The price of premium unleaded is 11 cents more per gallon than the price of regular unleaded. Find the price per gallon for each grade of gasoline.

90. *Ticket Sales* Eight hundred tickets were sold for a theater production and the receipts for the performance were $8600. The tickets for adults and students sold for $12.50 and $7.50, respectively. How many of each kind of ticket were sold?

91. *Nut Mixture* Ten pounds of mixed nuts sell for $6.95 per pound. The mixture is obtained from two kinds of nuts, with one variety priced at $5.65 per pound and the other priced at $8.95 per pound. How many pounds of each variety of nuts were used in the mixture?

92. *Hay Mixture* How many tons of hay at $125 per ton must be mixed with hay at $75 per ton to have 100 tons of hay with a value of $90 per ton?

93. *Alcohol Mixture* How many liters of a 40% alcohol solution must be mixed with a 65% solution to obtain 20 liters of a 50% solution?

94. *Acid Mixture* Fifty gallons of a 70% acid solution is obtained by mixing an 80% solution with a 50% solution. How many gallons of each solution must be used to obtain the desired mixture?

95. *Best-Fitting Line* The slope and y-intercept of the line $y = mx + b$ that "best fits" the three noncollinear points $(0, 0)$, $(1, 1)$, and $(2, 3)$ are given by the following system of linear equations.

$$3b + 3m = 4$$
$$3b + 6m = 7$$

(a) Solve the system and find the equation of the "best-fitting" line.

(b) Plot the three points and sketch the graph of the "best-fitting" line.

96. *Best-Fitting Line* The slope and y-intercept of the line $y = mx + b$ that "best fits" the three points $(0, 3)$, $(1, 2)$, and $(2, 1)$ are given by the following system of linear equations.

$$3b + 3m = 6$$
$$3b + 5m = 4$$

(a) Solve the system and find the equation of the "best-fitting" line.

(b) Plot the three points and sketch the graph of the "best-fitting" line.

8.3 Systems of Linear Equations in Three Variables

Row-Echelon Form ▪ The Method of Elimination ▪ Applications

Row-Echelon Form

The method of elimination can be applied to a system of linear equations in more than two variables. In fact, this method easily adapts to computer use for solving systems of linear equations with dozens of variables.

When the method of elimination is used to solve a system of linear equations, the goal is to rewrite the system in a form to which back-substitution can be applied. For instance, consider the following two systems of linear equations.

$$
\begin{aligned}
x - 2y + 2z &= 9 \\
-x + 3y &= -4 \\
2x - 5y + z &= 10
\end{aligned}
\qquad\qquad
\begin{aligned}
x - 2y + 2z &= 9 \\
y + 2z &= 5 \\
z &= 3
\end{aligned}
$$

The system on the right is said to be in **row-echelon form,** which means that it has a "stair-step" pattern with leading coefficients of 1. Which of these two systems do you think is easier to solve? After comparing the two systems, it should be clear that it is easier to solve the system on the right.

EXAMPLE 1 Using Back-Substitution

Solve the system of linear equations.

$$
\begin{aligned}
x - 2y + 2z &= 9 &&\text{Equation 1}\\
y + 2z &= 5 &&\text{Equation 2}\\
z &= 3 &&\text{Equation 3}
\end{aligned}
$$

Solution

From Equation 3, you know the value of z. To solve for y, substitute $z = 3$ into Equation 2 to obtain

$$
\begin{aligned}
y + 2(3) &= 5 &&\text{Substitute 3 for } z.\\
y &= -1. &&\text{Solve for } y.
\end{aligned}
$$

Finally, substitute $y = -1$ and $z = 3$ into Equation 1 to obtain

$$
\begin{aligned}
x - 2(-1) + 2(3) &= 9 &&\text{Substitute } -1 \text{ for } y \text{ and 3 for } z.\\
x &= 1. &&\text{Solve for } x.
\end{aligned}
$$

The solution is $x = 1$, $y = -1$, and $z = 3$, which can also be written as the **ordered triple** $(1, -1, 3)$. Check this in the original system of equations.

The Method of Elimination

Two systems of equations are **equivalent** if they have the same solution set. To solve a system that is not in row-echelon form, first convert it to an *equivalent* system that is in row-echelon form. To see how this is done, let's take another look at the method of elimination, as applied to a system of two linear equations.

EXAMPLE 2   *The Method of Elimination*

NOTE   As shown in Example 2, rewriting a system of linear equations in row-echelon form usually involves a *chain* of equivalent systems, each of which is obtained by using one of the three basic row operations. This process is called **Gaussian elimination,** after the German mathematician Carl Friedrich Gauss (1777–1855).

Solve the system of linear equations.

$$3x - 2y = -1 \qquad \text{Equation 1}$$
$$x - y = 0 \qquad \text{Equation 2}$$

Solution

$$x - y = 0$$
$$3x - 2y = -1$$

You can interchange two equations in the system.

$$-3x + 3y = 0$$

Multiply the first equation by -3.

$$-3x + 3y = 0$$
$$3x - 2y = -1$$

You can add the multiple of the first equation to the second equation to obtain a new second equation.

$$\underline{}$$
$$y = -1$$

$$x - y = 0$$
$$y = -1$$

New system in row-echelon form

Now, using back-substitution, you can determine that the solution is $y = -1$ and $x = -1$, which can be written as the ordered pair $(-1, -1)$. Check this in the original system of equations.

Operations That Produce Equivalent Systems

Each of the following **row operations** on a system of linear equations produces an *equivalent* system of linear equations.

1. Interchange two equations.

2. Multiply one of the equations by a nonzero constant.

3. Add a multiple of one of the equations to another equation to replace the latter equation.

EXAMPLE 3 Using Elimination to Solve a System

Solve the system of linear equations.

$$x - 2y + 2z = 9 \qquad \text{Equation 1}$$
$$-x + 3y = -4 \qquad \text{Equation 2}$$
$$2x - 5y + z = 10 \qquad \text{Equation 3}$$

Solution

There are many ways to begin, but we suggest working from the upper left corner, saving the x in the upper left position and eliminating the other x's from the first column.

$$x - 2y + 2z = 9$$
$$y + 2z = 5$$
$$2x - 5y + z = 10$$

Adding the first equation to the second equation produces a new second equation.

$$x - 2y + 2z = 9$$
$$y + 2z = 5$$
$$-y - 3z = -8$$

Adding −2 times the first equation to the third equation produces a new third equation.

Now that all but the first x have been eliminated from the first column, go to work on the second column. (You need to eliminate y from the third equation.)

$$x - 2y + 2z = 9$$
$$y + 2z = 5$$
$$-z = -3$$

Adding the second equation to the third equation produces a new third equation.

Finally, you need a coefficient of 1 for z in the third equation.

$$x - 2y + 2z = 9$$
$$y + 2z = 5$$
$$z = 3$$

Multiplying the third equation by −1 produces a new third equation.

This is the same system that was solved in Example 1, and, as in that example, you can conclude that the solution is

$$x = 1, \quad y = -1, \quad \text{and} \quad z = 3.$$

In Example 3, you can check the solution by substituting $x = 1$, $y = -1$, and $z = 3$ into each equation, as follows.

Equation 1: $(1) - 2(-1) + 2(3) = 9$ ✓
Equation 2: $-(1) + 3(-1) = -4$ ✓
Equation 3: $2(1) - 5(-1) + (3) = 10$ ✓

EXAMPLE 4   *Using Elimination to Solve a System*

Solve the following system of linear equations.

$$4x + y - 3z = 11 \qquad \text{Equation 1}$$
$$2x - 3y + 2z = 9 \qquad \text{Equation 2}$$
$$x + y + z = -3 \qquad \text{Equation 3}$$

Solution

$$x + y + z = -3$$
$$2x - 3y + 2z = 9$$
$$4x + y - 3z = 11$$

> Interchange the first and third equations.

$$x + y + z = -3$$
$$- 5y = 15$$
$$4x + y - 3z = 11$$

> Adding -2 times the first equation to the second equation produces a new second equation.

$$x + y + z = -3$$
$$- 5y = 15$$
$$- 3y - 7z = 23$$

> Adding -4 times the first equation to the third equation produces a new third equation.

$$x + y + z = -3$$
$$y = -3$$
$$- 3y - 7z = 23$$

> Multiplying the second equation by $-\frac{1}{5}$ produces a new second equation.

$$x + y + z = -3$$
$$y = -3$$
$$- 7z = 14$$

> Adding 3 times the second equation to the third equation produces a new third equation.

$$x + y + z = -3$$
$$y = -3$$
$$z = -2$$

> Multiplying the third equation by $-\frac{1}{7}$ produces a new third equation.

Now that the system of equations is in row-echelon form, you can see that $z = -2$ and $y = -3$. Moreover, by back-substituting these values into Equation 1, you can determine that $x = 2$. Therefore, the solution is

$$x = 2, \qquad y = -3, \qquad \text{and} \qquad z = -2.$$

You can check this solution as follows.

Equation 1: $4(2) + (-3) - 3(-2) = 11$ ✓

Equation 2: $2(2) - 3(-3) + 2(-2) = 9$ ✓

Equation 3: $(2) + (-3) + (-2) = -3$ ✓

The next example involves an inconsistent system—one that has no solution. The key to recognizing an inconsistent system is that at some stage in the elimination process, you obtain an absurdity such as $0 = -2$.

EXAMPLE 5 An Inconsistent System

Solve the system of linear equations.

$$
\begin{aligned}
x - 3y + z &= 1 \qquad &\text{Equation 1} \\
2x - y - 2z &= 2 \qquad &\text{Equation 2} \\
x + 2y - 3z &= -1 \qquad &\text{Equation 3}
\end{aligned}
$$

Solution

$$
\begin{aligned}
x - 3y + z &= 1 \\
5y - 4z &= 0 \\
x + 2y - 3z &= -1
\end{aligned}
$$

> Adding −2 times the first equation to the second equation produces a new second equation.

$$
\begin{aligned}
x - 3y + z &= 1 \\
5y - 4z &= 0 \\
5y - 4z &= -2
\end{aligned}
$$

> Adding −1 times the first equation to the third equation produces a new third equation.

$$
\begin{aligned}
x - 3y + z &= 1 \\
5y - 4z &= 0 \\
0 &= -2
\end{aligned}
$$

> Adding −1 times the second equation to the third equation produces a new third equation.

Because the third "equation" is impossible, you can conclude that this system is inconsistent and therefore has no solution. Moreover, because this system is equivalent to the original system, you can conclude that the original system also has no solution.

As with a system of linear equations in two variables, the solution(s) of a system of linear equations in more than two variables must fall into one of three categories.

The Number of Solutions of a Linear System

For a system of linear equations, exactly one of the following is true.

1. There is exactly one solution.

2. There are infinitely many solutions.

3. There is no solution.

EXAMPLE 6   *A System with Infinitely Many Solutions*

Solve the system of linear equations.

$$
\begin{aligned}
x + y - 3z &= -1 & \text{Equation 1}\\
y - z &= 0 & \text{Equation 2}\\
-x + 2y &= 1 & \text{Equation 3}
\end{aligned}
$$

Solution

Begin by rewriting the system in row-echelon form.

$$
\begin{aligned}
x + y - 3z &= -1\\
y - z &= 0\\
3y - 3z &= 0
\end{aligned}
$$

> Adding the first equation to the third equation produces a new third equation.

$$
\begin{aligned}
x + y - 3z &= -1\\
y - z &= 0\\
0 &= 0
\end{aligned}
$$

> Adding -3 times the second equation to the third equation produces a new third equation.

This means that Equation 3 depends on Equations 1 and 2 in the sense that it gives us no additional information about the variables. Thus, the original system is equivalent to the system

$$
\begin{aligned}
x + y - 3z &= -1\\
y - z &= 0.
\end{aligned}
$$

In this last equation, solve for y in terms of z to obtain $y = z$. Back-substituting for y in the previous equation produces $x = 2z - 1$. Finally, letting $z = a$, the solutions to the given system are all of the form

$$
x = 2a - 1, \quad y = a, \quad \text{and} \quad z = a
$$

where a is a real number. Thus, every ordered triple of the form

$$
(2a - 1, a, a), \qquad a \text{ is a real number}
$$

is a solution of the system.

In Example 6, there are other ways to write the same infinite set of solutions. For instance, the solutions could have been written as

$$
\left(b, \tfrac{1}{2}(b + 1), \tfrac{1}{2}(b + 1)\right), \qquad b \text{ is a real number}.
$$

Try convincing yourself of this by substituting $a = 0$, $a = 1$, $a = 2$, and $a = 3$ into the solution listed in Example 6. Then substitute $b = -1$, $b = 1$, $b = 3$, and $b = 5$ into the solution listed above. In both cases, you should obtain the same ordered triples. Thus, when comparing descriptions of an infinite solution set, keep in mind that there is more than one way to describe the set.

Applications

EXAMPLE 7   An Application: Moving Object

The height at time t of an object that is moving in a (vertical) line with constant acceleration a is given by the **position equation**

$$s = \tfrac{1}{2}at^2 + v_0t + s_0.$$

The height s is measured in feet, the acceleration a is measured in feet per second squared, the time t is measured in seconds, v_0 is the initial velocity (at time $t = 0$), and s_0 is the initial height. Find the values of a, v_0, and s_0, if $s = 164$ feet at 1 second, $s = 180$ feet at 2 seconds, and $s = 164$ feet at 3 seconds.

Solution

By substituting the three values of t and s into the position equation, you obtain three linear equations in a, v_0, and s_0.

When $t = 1$, $s = 164$: $\tfrac{1}{2}a(1)^2 + v_0(1) + s_0 = 164$

When $t = 2$, $s = 180$: $\tfrac{1}{2}a(2)^2 + v_0(2) + s_0 = 180$

When $t = 3$, $s = 164$: $\tfrac{1}{2}a(3)^2 + v_0(3) + s_0 = 164$

By multiplying the first and third equations by 2, this system can be rewritten

$$
\begin{aligned}
a + 2v_0 + 2s_0 &= 328 &\quad& \text{Equation 1}\\
2a + 2v_0 + s_0 &= 180 &\quad& \text{Equation 2}\\
9a + 6v_0 + 2s_0 &= 328 &\quad& \text{Equation 3}
\end{aligned}
$$

and you can apply elimination to obtain

$$
\begin{aligned}
a + 2v_0 + 2s_0 &= 328\\
-2v_0 - 3s_0 &= -476\\
2s_0 &= 232.
\end{aligned}
$$

From the third equation, $s_0 = 116$, so that back-substitution into the second equation yields

$$
\begin{aligned}
-2v_0 - 3(116) &= -476\\
-2v_0 &= -128\\
v_0 &= 64.
\end{aligned}
$$

Finally, back-substituting $s_0 = 116$ and $v_0 = 64$ into the first equation yields

$$
\begin{aligned}
a + 2(64) + 2(116) &= 328\\
a &= -32.
\end{aligned}
$$

Thus, the position equation for this object is $s = -16t^2 + 64t + 116$.

EXAMPLE 8   Data Analysis: Curve-Fitting

Find a quadratic equation $y = ax^2 + bx + c$ whose graph passes through the points $(-1, 3)$, $(1, 1)$, and $(2, 6)$.

Solution

Because the graph of $y = ax^2 + bx + c$ passes through the points $(-1, 3)$, $(1, 1)$, and $(2, 6)$, you can write the following.

When $x = -1$, $y = 3$: $a(-1)^2 + b(-1) + c = 3$

When $x = 1$, $y = 1$: $a(1)^2 + b(1) + c = 1$

When $x = 2$, $y = 6$: $a(2)^2 + b(2) + c = 6$

This produces the following system of linear equations.

$$
\begin{aligned}
a - b + c &= 3 && \text{Equation 1}\\
a + b + c &= 1 && \text{Equation 2}\\
4a + 2b + c &= 6 && \text{Equation 3}
\end{aligned}
$$

The solution of this system is $a = 2$, $b = -1$, and $c = 0$. Thus, the equation of the parabola is $y = 2x^2 - x$, as shown in Figure 8.8.

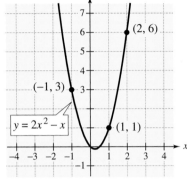

FIGURE 8.8

Group Activities   Problem Solving

Fitting a Quadratic Model   The data in the table represents the United States government's annual net receipts y (in billions of dollars) from individual income taxes for the year x from 1990 through 1992, where $x = 0$ corresponds to 1990.   (Source: U.S. Department of the Treasury)

x	0	1	2
y	467	468	476

Use a system of three linear equations to find a quadratic model that fits the data. According to your model, what were the annual net receipts from individual income taxes in 1993? The actual annual net receipts for 1993 were $510 billion. How does the value obtained from your quadratic model compare? Suppose you had been involved in planning the 1993 federal budget and had used this model to estimate how much federal income could be expected from 1993 individual income taxes. When you review the actual 1993 tax receipts and see that the model wasn't completely accurate, how do you evaluate the model's prediction performance? Are you satisfied with it? Why or why not?

8.3 Exercises

Discussing the Concepts

1. Give an example of a system of linear equations that is in row-echelon form.

2. Show how to use back-substitution to solve the system you found for Exercise 1.

3. Describe the row operations that are performed on a system of linear equations to produce an equivalent system of equations.

4. Are the following two systems of equations equivalent? Give reasons for your answer.

$$\begin{aligned} x + 3y - z &= 6 \\ 2x - y + 2z &= 1 \\ 3x + 2y - z &= 2 \end{aligned} \qquad \begin{aligned} x + 3y - z &= 6 \\ -7y + 4z &= 1 \\ -7y - 4z &= -16 \end{aligned}$$

Problem Solving

In Exercises 5–8, use back-substitution to solve the system of linear equations.

5.
$$\begin{aligned} x - 2y + 4z &= 4 \\ 3y - z &= 2 \\ z &= -5 \end{aligned}$$

6.
$$\begin{aligned} 5x + 4y - z &= 0 \\ 10y - 3z &= 11 \\ z &= 3 \end{aligned}$$

7.
$$\begin{aligned} x - 2y + 4z &= 4 \\ y &= 3 \\ y + z &= 2 \end{aligned}$$

8.
$$\begin{aligned} x &= 10 \\ 3x + 2y &= 2 \\ x + y + 2z &= 0 \end{aligned}$$

In Exercises 9–18, use Gaussian elimination to solve the system of linear equations.

9.
$$\begin{aligned} x + z &= 4 \\ y &= 2 \\ 4x + z &= 7 \end{aligned}$$

10.
$$\begin{aligned} x - y + 2z &= -4 \\ 3x + y - 4z &= -6 \\ 2x + 3y - 4z &= 4 \end{aligned}$$

11.
$$\begin{aligned} x &= 3 \\ -x + 3y &= 3 \\ y + 2z &= 4 \end{aligned}$$

12.
$$\begin{aligned} x + y + z &= -3 \\ 4x + y - 3z &= 11 \\ 2x - 3y + 2z &= 9 \end{aligned}$$

13.
$$\begin{aligned} x + 2y + 6z &= 5 \\ -x + y - 2z &= 3 \\ x - 4y - 2z &= 1 \end{aligned}$$

14.
$$\begin{aligned} 2x + y + 3z &= 1 \\ 2x + 6y + 8z &= 3 \\ 6x + 8y + 18z &= 5 \end{aligned}$$

15.
$$\begin{aligned} 0.2x + 1.3y + 0.6z &= 0.1 \\ 0.1x + 0.3z &= 0.7 \\ 2x + 10y + 8z &= 8 \end{aligned}$$

16.
$$\begin{aligned} 0.3x - 0.1y + 0.2z &= 0.35 \\ 2x + y - 2z &= -1 \\ 2x + 4y + 3z &= 10.5 \end{aligned}$$

17.
$$\begin{aligned} x + 4y - 2z &= 2 \\ -3x + y + z &= -2 \\ 5x + 7y - 5z &= 6 \end{aligned}$$

18.
$$\begin{aligned} 2x + z &= 1 \\ 5y - 3z &= 2 \\ 6x + 20y - 9z &= 11 \end{aligned}$$

Curve-Fitting   In Exercises 19 and 20, find the equation of the parabola

$$y = ax^2 + bx + c$$

that passes through the given points.

19. $(0, 5), (1, 6), (2, 5)$

20. $(0, -4), (1, 1), (2, 10)$

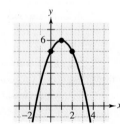

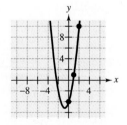

Curve-Fitting In Exercises 21 and 22, find the equation of the circle

$$x^2 + y^2 + Dx + Ey + F = 0$$

that passes through the given points.

21. $(0, 0), (2, -2), (4, 0)$ **22.** $(0, 0), (0, 6), (-3, 3)$

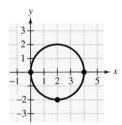   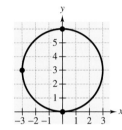

Vertical Motion In Exercises 23–26, find the position equation, $s = \frac{1}{2}at^2 + v_0t + s_0$, for an object that has the indicated heights at the specified times.

23. $s = 128$ feet at $t = 1$ second
$s = 80$ feet at $t = 2$ seconds
$s = 0$ feet at $t = 3$ seconds

24. $s = 48$ feet at $t = 1$ second
$s = 64$ feet at $t = 2$ seconds
$s = 48$ feet at $t = 3$ seconds

25. $s = 32$ feet at $t = 1$ second
$s = 32$ feet at $t = 2$ seconds
$s = 0$ feet at $t = 3$ seconds

26. $s = 10$ feet at $t = 0$ seconds
$s = 54$ feet at $t = 1$ second
$s = 46$ feet at $t = 3$ seconds

27. *Crop Spraying* A mixture of 12 gallons of chemical A, 16 gallons of chemical B, and 26 gallons of chemical C is required to kill a certain destructive crop insect. Commercial spray X contains 1, 2, and 2 parts, respectively, of these chemicals. Commercial spray Y contains only chemical C. Commercial spray Z contains only chemicals A and B in equal amounts. How much of each type of commercial spray is needed to get the desired mixture?

28. *Chemistry* A chemist needs 10 liters of a 25% acid solution. The solution is to be mixed from three solutions whose concentrations are 10%, 20%, and 50%. How many liters of each solution should the chemist use to satisfy the following?

(a) Use as little as possible of the 50% solution.

(b) Use as much as possible of the 50% solution.

(c) Use 2 liters of the 50% solution.

29. *School Orchestra* The table shows the percents of each section of the North High School orchestra that were chosen to participate in the city orchestra, the county orchestra, and the state orchestra. Thirty members of the city orchestra, 17 members of the county orchestra, and 10 members of the state orchestra are from North. How many members are in each section of North High's orchestra?

Orchestra	String	Wind	Percussion
City Orchestra	40%	30%	50%
County Orchestra	20%	25%	25%
State Orchestra	10%	15%	25%

30. *Diagonals of a Polygon* The total numbers of sides and diagonals of regular polygons with three, four, and five sides are three, six, and ten, as shown in the figure. Find a quadratic function $y = ax^2 + bx + c$ that fits this data. Then check to see if it gives the correct answer for a polygon with six sides.

3

6

10

15

Reviewing the Major Concepts

In Exercises 31–34, find the domain of the function.

31. $f(x) = x^3 - 2x$

32. $g(x) = \sqrt[3]{x}$

33. $h(x) = \sqrt{16 - x^2}$

34. $A(x) = \dfrac{3}{(36 - x^2)}$

35. *Predator-Prey* The number N of prey animals t months after a predator is introduced into a test area is inversely proportional to $t + 1$. When $t = 0$, $N = 300$. Find N when $t = 5$.

36. *Geometry* Find the area A of the rectangle in the figure as a function of x.

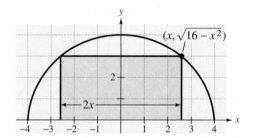

Additional Problem Solving

In Exercises 37 and 38, perform the row operation and write the equivalent system.

37. Add Equation 1 to Equation 2.

$$\begin{aligned} x - 2y + 3z &= 5 \quad &\text{Equation 1}\\ -x + y + 5z &= 4 \quad &\text{Equation 2}\\ 2x \phantom{{}+y} - 3z &= 0 \quad &\text{Equation 3} \end{aligned}$$

38. Add -2 times Equation 1 to Equation 3.

$$\begin{aligned} x - 2y + 3z &= 5 \quad &\text{Equation 1}\\ -x + y + 5z &= 4 \quad &\text{Equation 2}\\ 2x \phantom{{}+y} - 3z &= 0 \quad &\text{Equation 3} \end{aligned}$$

In Exercises 39 and 40, decide whether each ordered triple is a solution of the system.

39.
$$\begin{aligned} x + 3y + 2z &= 1\\ 5x - y + 3z &= 16\\ -3x + 7y + z &= -14 \end{aligned}$$

(a) $(0, 3, -2)$ (b) $(12, 5, -13)$

(c) $(1, -2, 3)$ (d) $(-2, 5, -3)$

40.
$$\begin{aligned} 3x - y + 4z &= -10\\ -x + y + 2z &= 6\\ 2x - y + z &= -8 \end{aligned}$$

(a) $(-2, 4, 0)$ (b) $(0, -3, 10)$

(c) $(1, -1, 5)$ (d) $(7, 19, -3)$

In Exercises 41–56, use Gaussian elimination to solve the system of linear equations.

41.
$$\begin{aligned} x + y + z &= 6\\ 2x - y + z &= 3\\ 3x \phantom{{}+y} - z &= 0 \end{aligned}$$

42.
$$\begin{aligned} x + y + z &= 2\\ -x + 3y + 2z &= 8\\ 4x + y \phantom{{}+ 2z} &= 4 \end{aligned}$$

43.
$$\begin{aligned} x + y + 8z &= 3\\ 2x + y + 11z &= 4\\ x \phantom{{}+y} + 3z &= 0 \end{aligned}$$

44.
$$\begin{aligned} x + 6y + 2z &= 9\\ 3x - 2y + 3z &= -1\\ 5x - 5y + 2z &= 7 \end{aligned}$$

45.
$$\begin{aligned} 2x \phantom{{}+3y} + 2z &= 2\\ 5x + 3y \phantom{{}+4z} &= 4\\ 3y - 4z &= 4 \end{aligned}$$

46.
$$\begin{aligned} 6y + 4z &= -12\\ 3x + 3y \phantom{{}+4z} &= 9\\ 2x \phantom{{}+3y} - 3z &= 10 \end{aligned}$$

47.
$$\begin{aligned} 3x - y - 2z &= 5\\ 2x + y + 3z &= 6\\ 6x - y - 4z &= 9 \end{aligned}$$

48.
$$\begin{aligned} 2x - 4y + z &= 0\\ 3x \phantom{{}+4y} + 2z &= -1\\ -6x + 3y + 2z &= -10 \end{aligned}$$

49.
$$\begin{aligned} 5x + 2y \phantom{{}+z} &= -8\\ z &= 5\\ 3x - y + z &= 9 \end{aligned}$$

50.
$$\begin{aligned} y + z &= 5\\ 2x \phantom{{}+y} + 4z &= 4\\ 2x - 3y \phantom{{}+z} &= -14 \end{aligned}$$

51.
$$\begin{aligned} x + 2y - 2z &= 4\\ 2x + 5y - 7z &= 5\\ 3x + 7y - 9z &= 10 \end{aligned}$$

52.
$$\begin{aligned} 2x + 6y - 4z &= 8\\ 3x + 10y - 7z &= 12\\ -2x - 6y + 5z &= -3 \end{aligned}$$

53. $2x + y - z = 4$
$\quad\quad y + 3z = 2$
$\quad 3x + 2y = 4$

54. $x - 2y - z = 3$
$\quad 2x + y - 3z = 1$
$\quad x + 8y - 3z = -7$

55. $3x + y + z = 2$
$\quad 4x + 2z = 1$
$\quad 5x - y + 3z = 0$

56. $2x + 3z = 4$
$\quad 5x + y + z = 2$
$\quad 11x + 3y - 3z = 0$

Curve-Fitting   In Exercises 57–60, find the equation of the parabola $y = ax^2 + bx + c$ that passes through the given points.

57. $(1, 0), (2, -1), (3, 0)$ **58.** $(1, 2), (2, 1), (3, -4)$

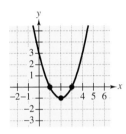
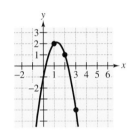

59. $(-1, -3), (1, 1), (2, 0)$
60. $(-1, -1), (1, 1), (2, -4)$

Curve-Fitting   In Exercises 61–64, find the equation of the circle $x^2 + y^2 + Dx + Ey + F = 0$ that passes through the given points.

61. $(0, 0), (0, 2), (3, 0)$ **62.** $(3, -1), (-2, 4), (6, 8)$

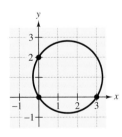
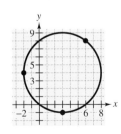

63. $(-3, 5), (4, 6), (5, 5)$
64. $(5, 13), (17, 5), (10, 12)$

65. *Graphical Estimation*   The table gives the amounts y, in millions of short tons, of newsprint produced in the years 1990 through 1992 in the United States. (Source: American Paper Institute)

Year	1990	1991	1992
y	6.6	6.8	7.1

(a) Find the equation of the parabola $y = at^2 + bt + c$ that passes through the three points, letting $t = 0$ correspond to 1990.

(b) Use a graphing utility to graph the model found in part (a).

(c) Use the model of part (a) to predict newsprint production in 1999 if the trend continues.

66. *Rewriting a Fraction*   The fraction $1/(x^3 - x)$ can be written as a sum of three fractions as follows.

$$\frac{1}{x^3 - x} = \frac{A}{x} + \frac{B}{x+1} + \frac{C}{x-1}$$

The numbers A, B, and C are the solutions of the following system.

$$A + B + C = 0$$
$$-B + C = 0$$
$$-A = 1$$

Solve the system and verify that the sum of the three resulting fractions is the original fraction.

In Exercises 67 and 68, find a system of linear equations in three variables with integer coefficients that has the given point as a solution. (*Note:* There are many correct answers.)

67. $(4, -3, 2)$ **68.** $(5, 7, -10)$

MID-CHAPTER QUIZ

Take this quiz as you would take a quiz in class. After you are done, check your work against the answers given in the back of the book.

1. Which is the solution of the system $5x - 12y = 2$ and $2x + 1.5y = 26$: $(1, -2)$ or $(10, 4)$? Explain your reasoning.

In Exercises 2–4, graph the equations in the system. Use the graphs to determine the number of solutions of the system.

2. $-6x + 9y = 9$
$\quad 2x - 3y = 6$

3. $x - 2y = -4$
$\quad 3x - 2y = 4$

4. $y = x - 1$
$\quad y = 1 + 2x - x^2$

In Exercises 5–8, solve the system of equations graphically.

5. $x \quad = 4$
$\quad 2x - y = 6$

6. $y = \frac{1}{3}(1 - 2x)$
$\quad y = \frac{1}{3}(5x - 13)$

7. $2x + 7y = 16$
$\quad 3x + 2y = 24$

8. $7x - 17y = -169$
$\quad x^2 + y^2 = 169$

In Exercises 9–12, use substitution to solve the system. Use a graphing utility to check the solution.

9. $2x - 3y = 4$
$\quad y = 2$

10. $y = 5 - x^2$
$\quad y = 2(x + 1)$

11. $5x - y = 32$
$\quad 6x - 9y = 18$

12. $0.2x + 0.7y = 8$
$\quad -x + 2y = 15$

In Exercises 13–16, use Gaussian elimination to solve the linear system.

13. $x + 10y = 18$
$\quad 5x + 2y = 42$

14. $3x + 11y = 38$
$\quad 7x - 5y = -34$

15. $a + b + c = 1$
$\quad 4a + 2b + c = 2$
$\quad 9a + 3b + c = 4$

16. $x + 4z = 17$
$\quad -3x + 2y - z = -20$
$\quad x - 5y + 3z = 19$

In Exercises 17 and 18, find a system of linear equations that has the unique solution. (There are many correct answers.)

17. $(10, -12)$

18. $(1, 3, -7)$

19. Twenty gallons of a 30% brine solution is obtained by mixing a 20% solution with a 50% solution. Let x represent the number of gallons of the 20% solution and let y represent the number of gallons of the 50% solution. Write a system of equations that models this problem and solve the system.

20. Find the equation of the parabola $y = ax^2 + bx + c$ that passes through the points $(1, 2)$, $(-1, -4)$, and $(2, 8)$.

8.4   **Matrices and Systems of Linear Equations**

Matrices ▪ Elementary Row Operations ▪
Solving a System of Linear Equations

Matrices

In this section, you will study a streamlined technique for solving systems of linear equations. This technique involves the use of a rectangular array of real numbers called a **matrix.** (The plural of matrix is *matrices.*) Here is an example of a matrix.

$$\begin{bmatrix} 3 & -2 & 4 & 1 \\ 0 & 1 & -1 & 2 \\ 2 & 0 & -3 & 0 \end{bmatrix}$$

This matrix has three rows and four columns, which means that its **order** is 3×4, which is read as "3 by 4." Each number in the matrix is an **entry** of the matrix.

EXAMPLE 1   *Examples of Matrices*

The following matrices have the indicated orders.

a. Order: 2×3 **b.** Order: 2×2 **c.** Order: 3×2

$$\begin{bmatrix} 1 & -2 & 4 \\ 0 & 1 & -2 \end{bmatrix} \qquad \begin{bmatrix} 0 & 0 \\ 0 & 0 \end{bmatrix} \qquad \begin{bmatrix} 1 & -3 \\ -2 & 0 \\ 4 & -2 \end{bmatrix}$$

A matrix with the same number of rows as columns is called a **square matrix.** For instance, the 2×2 matrix in part (b) is square.

A matrix derived from a system of linear equations (each written in standard form) is the **augmented matrix** of the system. Moreover, the matrix derived from the coefficients of the system (but that does not include the constant terms) is the **coefficient matrix** of the system. Here is an example.

System	Coefficient Matrix	Augmented Matrix

$$\begin{array}{r} x - 4y + 3z = 5 \\ -x + 3y - z = -3 \\ 2x \quad\;\; - 4z = 6 \end{array} \qquad \begin{bmatrix} 1 & -4 & 3 \\ -1 & 3 & -1 \\ 2 & 0 & -4 \end{bmatrix} \qquad \begin{bmatrix} 1 & -4 & 3 & \vdots & 5 \\ -1 & 3 & -1 & \vdots & -3 \\ 2 & 0 & -4 & \vdots & 6 \end{bmatrix}$$

Note the use of 0 for the missing y-variable in the third equation, and also note the fourth column of constant terms in the augmented matrix.

When forming either the coefficient matrix or the augmented matrix of a system, you should begin by vertically aligning the variables in the equations.

Given System	*Align Variables*	*Form Augmented Matrix*
$x + 3y = 9$	$x + 3y \quad = 9$	
$-y + 4z = -2$	$-y + 4z = -2$	$\begin{bmatrix} 1 & 3 & 0 & \vdots & 9 \\ 0 & -1 & 4 & \vdots & -2 \\ 1 & 0 & -5 & \vdots & 0 \end{bmatrix}$
$x - 5z = 0$	$x \quad\quad - 5z = 0$	

EXAMPLE 2 Forming Coefficient and Augmented Matrices

Form the coefficient matrix and the augmented matrix for each system of linear equations.

a. $\quad -x + 5y = 2$ **b.** $\quad 3x + 2y - z = 1$ **c.** $\quad x = 3y - 1$

$\quad\quad 7x - 2y = -6$ $x + 2z = -3$ $2y - 5 = 9x$

$\quad\quad\quad\quad\quad\quad\quad\quad\quad\quad\quad\quad -2x - y = 4$

Solution

	System	*Coefficient Matrix*	*Augmented Matrix*
a.	$-x + 5y = 2$ $7x - 2y = -6$	$\begin{bmatrix} -1 & 5 \\ 7 & -2 \end{bmatrix}$	$\begin{bmatrix} -1 & 5 & \vdots & 2 \\ 7 & -2 & \vdots & -6 \end{bmatrix}$
b.	$3x + 2y - z = 1$ $x + 2z = -3$ $-2x - y = 4$	$\begin{bmatrix} 3 & 2 & -1 \\ 1 & 0 & 2 \\ -2 & -1 & 0 \end{bmatrix}$	$\begin{bmatrix} 3 & 2 & -1 & \vdots & 1 \\ 1 & 0 & 2 & \vdots & -3 \\ -2 & -1 & 0 & \vdots & 4 \end{bmatrix}$
c.	$x - 3y = -1$ $-9x + 2y = 5$	$\begin{bmatrix} 1 & -3 \\ -9 & 2 \end{bmatrix}$	$\begin{bmatrix} 1 & -3 & \vdots & -1 \\ -9 & 2 & \vdots & 5 \end{bmatrix}$

EXAMPLE 3 Forming Linear Systems from Their Matrices

Write systems of linear equations that are represented by the following matrices.

a. $\begin{bmatrix} 3 & -5 & \vdots & 4 \\ -1 & 2 & \vdots & 0 \end{bmatrix}$ **b.** $\begin{bmatrix} 1 & 3 & \vdots & 2 \\ 0 & 1 & \vdots & -3 \end{bmatrix}$ **c.** $\begin{bmatrix} 2 & 0 & -8 & \vdots & 1 \\ -1 & 1 & 1 & \vdots & 2 \\ 5 & -1 & 7 & \vdots & 3 \end{bmatrix}$

Solution

a. $\quad 3x - 5y = 4$ **b.** $\quad x + 3y = 2$ **c.** $\quad 2x \quad\quad - 8z = 1$

$\quad\quad -x + 2y = 0$ $y = -3$ $-x + y + z = 2$

$\quad\quad\quad\quad\quad\quad\quad\quad\quad\quad\quad\quad\quad\quad 5x - y + 7z = 3$

Elementary Row Operations

In Section 8.3, you studied three operations that can be used on a system of linear equations to produce an equivalent system: (1) interchange two rows, (2) multiply a row by a nonzero constant, and (3) add a multiple of a row to another row. In matrix terminology, these three operations correspond to **elementary row operations**.

Elementary Row Operations

Any of the following **elementary row operations** performed on an augmented matrix will produce a matrix that is row-equivalent to the original matrix. Two matrices are **row-equivalent** if one can be obtained from the other by a sequence of elementary row operations.

1. Interchange two rows.

2. Multiply a row by a nonzero constant.

3. Add a multiple of a row to another row.

EXAMPLE 4   Elementary Row Operations

a. Interchange the first and second rows.

Original Matrix

$$\begin{bmatrix} 0 & 1 & 3 & 4 \\ -1 & 2 & 0 & 3 \\ 2 & -3 & 4 & 1 \end{bmatrix}$$

New Row-Equivalent Matrix

$$\begin{matrix} R_2 \\ R_1 \end{matrix} \begin{bmatrix} -1 & 2 & 0 & 3 \\ 0 & 1 & 3 & 4 \\ 2 & -3 & 4 & 1 \end{bmatrix}$$

b. Multiply the first row by $\frac{1}{2}$.

Original Matrix

$$\begin{bmatrix} 2 & -4 & 6 & -2 \\ 1 & 3 & -3 & 0 \\ 5 & -2 & 1 & 2 \end{bmatrix}$$

New Row-Equivalent Matrix

$$\tfrac{1}{2} R_1 \rightarrow \begin{bmatrix} 1 & -2 & 3 & -1 \\ 1 & 3 & -3 & 0 \\ 5 & -2 & 1 & 2 \end{bmatrix}$$

c. Add -2 times the first row to the third row.

Original Matrix

$$\begin{bmatrix} 1 & 2 & -4 & 3 \\ 0 & 3 & -2 & -1 \\ 2 & 1 & 5 & -2 \end{bmatrix}$$

New Row-Equivalent Matrix

$$-2R_1 + R_3 \rightarrow \begin{bmatrix} 1 & 2 & -4 & 3 \\ 0 & 3 & -2 & -1 \\ 0 & -3 & 13 & -8 \end{bmatrix}$$

In Section 8.3, Gaussian elimination was used with back-substitution to solve a system of linear equations. Example 5 demonstrates the matrix version of Gaussian elimination. The two methods are essentially the same. The basic difference is that with matrices you do not need to keep writing the variables.

EXAMPLE 5 Solving a System of Linear Equations

Linear System

$$x - 2y + 2z = 9$$
$$-x + 3y \quad\quad = -4$$
$$2x - 5y + z = 10$$

Associated Augmented Matrix

$$\left[\begin{array}{ccc:c} 1 & -2 & 2 & 9 \\ -1 & 3 & 0 & -4 \\ 2 & -5 & 1 & 10 \end{array}\right]$$

Add the first equation to the second equation.

Add the first row to the second row $(R_1 + R_2)$.

$$x - 2y + 2z = 9$$
$$y + 2z = 5$$
$$2x - 5y + z = 10$$

$$R_1 + R_2 \rightarrow \left[\begin{array}{ccc:c} 1 & -2 & 2 & 9 \\ 0 & 1 & 2 & 5 \\ 2 & -5 & 1 & 10 \end{array}\right]$$

Add -2 times the first equation to the third equation.

Add -2 times the first row to the third row $(-2R_1 + R_3)$.

$$x - 2y + 2z = 9$$
$$y + 2z = 5$$
$$-y - 3z = -8$$

$$-2R_1 + R_3 \rightarrow \left[\begin{array}{ccc:c} 1 & -2 & 2 & 9 \\ 0 & 1 & 2 & 5 \\ 0 & -1 & -3 & -8 \end{array}\right]$$

Add the second equation to the third equation.

Add the second row to the third row $(R_2 + R_3)$.

$$x - 2y + 2z = 9$$
$$y + 2z = 5$$
$$-z = -3$$

$$R_2 + R_3 \rightarrow \left[\begin{array}{ccc:c} 1 & -2 & 2 & 9 \\ 0 & 1 & 2 & 5 \\ 0 & 0 & -1 & -3 \end{array}\right]$$

Multiply the third equation by -1.

Multiply the third row by -1.

$$x - 2y + 2z = 9$$
$$y + 2z = 5$$
$$z = 3$$

$$-R_3 \rightarrow \left[\begin{array}{ccc:c} 1 & -2 & 2 & 9 \\ 0 & 1 & 2 & 5 \\ 0 & 0 & 1 & 3 \end{array}\right]$$

At this point, you can use back-substitution to find that the solution is $x = 1$, $y = -1$, and $z = 3$.

The last matrix in Example 5 is in **row-echelon form.** The term *echelon* refers to the stair-step pattern formed by the nonzero elements of the matrix.

Solving a System of Linear Equations

> ### Gaussian Elimination with Back-Substitution
>
> To use matrices and Gaussian elimination to solve a system of linear equations, use the following steps.
>
> 1. Write the augmented matrix of the system of linear equations.
>
> 2. Use elementary row operations to rewrite the augmented matrix in row-echelon form.
>
> 3. Write the system of linear equations corresponding to the matrix in row-echelon form, and use back-substitution to find the solution.

When you perform Gaussian elimination with back-substitution, we suggest that you operate from *left to right by columns,* using elementary row operations to obtain zeros in all entries directly below the leading 1's.

EXAMPLE 6   *Gaussian Elimination with Back-Substitution*

Solve the system of linear equations.

$$2x - 3y = -2$$
$$x + 2y = 13$$

Solution

$$\begin{bmatrix} 2 & -3 & \vdots & -2 \\ 1 & 2 & \vdots & 13 \end{bmatrix}$$ Augmented matrix for system of linear equations.

$$\begin{matrix} R_2 \\ R_1 \end{matrix} \begin{bmatrix} 1 & 2 & \vdots & 13 \\ 2 & -3 & \vdots & -2 \end{bmatrix}$$ First column has leading 1 in upper left corner.

$$-2R_1 + R_2 \rightarrow \begin{bmatrix} 1 & 2 & \vdots & 13 \\ 0 & -7 & \vdots & -28 \end{bmatrix}$$ First column has a zero under its leading 1.

$$-\tfrac{1}{7}R_2 \rightarrow \begin{bmatrix} 1 & 2 & \vdots & 13 \\ 0 & 1 & \vdots & 4 \end{bmatrix}$$ Second column has leading 1 in second row.

The system of linear equations that corresponds to the (row-echelon) matrix is

$$x + 2y = 13$$
$$y = 4.$$

Using back-substitution, you can find that the solution of the system is $x = 5$ and $y = 4$. Check this solution in the original system of linear equations.

EXAMPLE 7 *Gaussian Elimination with Back-Substitution*

Solve the system of linear equations.

$$
\begin{aligned}
3x + 3y &= 9 \\
2x - 3z &= 10 \\
6y + 4z &= -12
\end{aligned}
$$

Solution

$$
\begin{bmatrix}
3 & 3 & 0 & \vdots & 9 \\
2 & 0 & -3 & \vdots & 10 \\
0 & 6 & 4 & \vdots & -12
\end{bmatrix}
$$
Augmented matrix for system of linear equations.

$$
\tfrac{1}{3}R_1 \rightarrow
\begin{bmatrix}
1 & 1 & 0 & \vdots & 3 \\
2 & 0 & -3 & \vdots & 10 \\
0 & 6 & 4 & \vdots & -12
\end{bmatrix}
$$
First column has leading 1 in upper left corner.

$$
-2R_1 + R_2 \rightarrow
\begin{bmatrix}
1 & 1 & 0 & \vdots & 3 \\
0 & -2 & -3 & \vdots & 4 \\
0 & 6 & 4 & \vdots & -12
\end{bmatrix}
$$
First column has zeros under its leading 1.

$$
-\tfrac{1}{2}R_2 \rightarrow
\begin{bmatrix}
1 & 1 & 0 & \vdots & 3 \\
0 & 1 & \tfrac{3}{2} & \vdots & -2 \\
0 & 6 & 4 & \vdots & -12
\end{bmatrix}
$$
Second column has leading 1 in second row.

$$
\begin{bmatrix}
1 & 1 & 0 & \vdots & 3 \\
0 & 1 & \tfrac{3}{2} & \vdots & -2 \\
0 & 0 & -5 & \vdots & 0
\end{bmatrix}
$$
$-6R_2 + R_3 \rightarrow$

Second column has zero under its leading 1.

$$
\begin{bmatrix}
1 & 1 & 0 & \vdots & 3 \\
0 & 1 & \tfrac{3}{2} & \vdots & -2 \\
0 & 0 & 1 & \vdots & 0
\end{bmatrix}
$$
$-\tfrac{1}{5}R_3 \rightarrow$

Third column has leading 1 in third row.

The system of linear equations that corresponds to this (row-echelon) matrix is

$$
\begin{aligned}
x + y &= 3 \\
y + \tfrac{3}{2}z &= -2 \\
z &= 0.
\end{aligned}
$$

Using back-substitution, you can find that the solution is

$$
x = 5, \quad y = -2, \quad \text{and} \quad z = 0.
$$

Check this in the original system, as follows.

$$
\begin{aligned}
3(5) + 3(-2) &= 9 && \text{Substitute in Equation 1.} \;\checkmark \\
2(5) - 3(0) &= 10 && \text{Substitute in Equation 2.} \;\checkmark \\
6(-2) + 4(0) &= -12 && \text{Substitute in Equation 3.} \;\checkmark
\end{aligned}
$$

EXAMPLE 8   A System with No Solution

Solve the system of linear equations.

$$6x - 10y = -4$$
$$9x - 15y = 5$$

Solution

$$\begin{bmatrix} 6 & -10 & \vdots & -4 \\ 9 & -15 & \vdots & 5 \end{bmatrix}$$
Augmented matrix for system of linear equations.

$$\tfrac{1}{6}R_1 \rightarrow \begin{bmatrix} 1 & -\tfrac{5}{3} & \vdots & -\tfrac{2}{3} \\ 9 & -15 & \vdots & 5 \end{bmatrix}$$
First column has leading 1 in upper left corner.

$$-9R_1 + R_2 \rightarrow \begin{bmatrix} 1 & -\tfrac{5}{3} & \vdots & -\tfrac{2}{3} \\ 0 & 0 & \vdots & 11 \end{bmatrix}$$
First column has a zero under its leading 1.

The "equation" that corresponds to the second row of this matrix is $0 = 11$. Because this does not make sense, the system of equations has no solution.

EXAMPLE 9   A System with Infinitely Many Solutions

Solve the system of linear equations.

$$12x - 6y = -3$$
$$-8x + 4y = 2$$

Solution

$$\begin{bmatrix} 12 & -6 & \vdots & -3 \\ -8 & 4 & \vdots & 2 \end{bmatrix}$$
Augmented matrix for system of linear equations.

$$\tfrac{1}{12}R_1 \rightarrow \begin{bmatrix} 1 & -\tfrac{1}{2} & \vdots & -\tfrac{1}{4} \\ -8 & 4 & \vdots & 2 \end{bmatrix}$$
First column has leading 1 in upper left corner.

$$8R_1 + R_2 \rightarrow \begin{bmatrix} 1 & -\tfrac{1}{2} & \vdots & -\tfrac{1}{4} \\ 0 & 0 & \vdots & 0 \end{bmatrix}$$
First column has a zero under its leading 1.

Because the second row of the matrix is all zeros, you can conclude that the system of equations has an infinite number of solutions, represented by all points (x, y) on the line $x - \tfrac{1}{2}y = -\tfrac{1}{4}$. Because this line can be written as

$$x = -\frac{1}{4} + \frac{1}{2}y$$

you can write the solution set as

$$\left(-\frac{1}{4} + \frac{1}{2}a, \ a\right), \qquad \text{where } a \text{ is any real number.}$$

Technology

Graphing Utilities and Matrices

Most graphing utilities are capable of working with matrices. For example, to write the matrix shown at the left in row-echelon form, use the following steps.

1. Use the MATRIX feature to enter the matrix.

2. To use a *TI-82* to write the matrix in row-echelon form, use the elementary row operations that appear in the MATRIX MATH menu, as follows.

row+([A],1,2) ENTER	Add row 1 to row 2.
STO ▷ [A] ENTER	Store the result as matrix A.
*row+(−2,[A],1,3) ENTER	Multiply row 1 by −2 and add to row 3.
STO ▷ [A] ENTER	Store the result as matrix A.
row+([A],2,3) ENTER	Add row 2 to row 3.
STO ▷ [A] ENTER	Store the result as matrix A.
*row(−1,[A],3) ENTER	Multiply row 3 by −1.
STO ▷ [A] ENTER	Store the result as matrix A.

3. The result that is displayed on the screen should be as follows.

$$\begin{bmatrix} [& 1 & -2 & 2 & 9 &] \\ [& 0 & 1 & 2 & 5 &] \\ [& 0 & 0 & 1 & 3 &] \end{bmatrix}$$

At this point, you can use back-substitution to solve the corresponding system of equations.

Group Activities   Exploring with Technology

Analyzing Solutions to Systems of Equations   Use a graphing utility to graph each system of equations given in Example 6, Example 8, and Example 9. Verify the solution given in each example and explain how you may reach the same conclusion by using the graph. Summarize how you may conclude that a system has a unique solution, no solution, or infinitely many solutions when you use Gaussian elimination.

8.4 Exercises

Discussing the Concepts

1. Describe the three elementary row operations that can be performed on an augmented matrix.

2. What is the relationship between the three elementary row operations on an augmented matrix and the row operations on a system of linear equations?

3. What is meant by saying that two augmented matrices are *row-equivalent*?

4. Give an example of a matrix in *row-echelon form.*

5. Describe the row-echelon form of an augmented matrix that corresponds to a system of linear equations that is inconsistent.

6. Describe the row-echelon form of an augmented matrix that corresponds to a system of linear equations that has an infinite number of solutions.

Problem Solving

In Exercises 7–10, state the order of the matrix.

7. $\begin{bmatrix} 3 & -2 \\ -4 & 0 \\ 2 & -7 \end{bmatrix}$

8. $\begin{bmatrix} 4 & 0 & -5 \\ -1 & 8 & 9 \\ 0 & -3 & 4 \end{bmatrix}$

9. $\begin{bmatrix} 5 & -8 \\ 7 & 15 \end{bmatrix}$

10. $\begin{bmatrix} -2 & 5 \\ 0 & -1 \end{bmatrix}$

In Exercises 11–14, write the augmented matrix for the system of linear equations.

11. $4x - 5y = -2$
$-x + 8y = 10$

12. $8x + 3y = 25$
$3x - 9y = 12$

13. $x + 10y - 3z = 2$
$5x - 3y + 4z = 0$
$2x + 4y = 6$

14. $9x - 3y + z = 13$
$12x - 8z = 5$

In Exercises 15–18, write the system of linear equations represented by the augmented matrix. (Use variables x, y, z, and w.)

15. $\begin{bmatrix} 4 & 3 & \vdots & 8 \\ 1 & -2 & \vdots & 3 \end{bmatrix}$

16. $\begin{bmatrix} 9 & -4 & \vdots & 0 \\ 6 & 1 & \vdots & -4 \end{bmatrix}$

17. $\begin{bmatrix} 1 & 0 & 2 & \vdots & -10 \\ 0 & 3 & -1 & \vdots & 5 \\ 4 & 2 & 0 & \vdots & 3 \end{bmatrix}$

18. $\begin{bmatrix} 5 & 8 & 2 & 0 & \vdots & -1 \\ -2 & 15 & 5 & 1 & \vdots & 9 \\ 1 & 6 & -7 & 0 & \vdots & -3 \end{bmatrix}$

In Exercises 19 and 20, fill in the blank(s) by using elementary row operations to form a row-equivalent matrix.

19. $\begin{bmatrix} 1 & 4 & 3 \\ 2 & 10 & 5 \end{bmatrix}$

$\begin{bmatrix} 1 & 4 & 3 \\ 0 & \blacksquare & -1 \end{bmatrix}$

20. $\begin{bmatrix} 2 & 4 & 8 & 6 \\ 1 & -1 & -3 & 2 \\ 2 & 6 & 4 & 9 \end{bmatrix}$

$\begin{bmatrix} 1 & \blacksquare & \blacksquare & \blacksquare \\ 1 & -1 & -3 & 2 \\ 2 & 6 & 4 & 9 \end{bmatrix}$

$\begin{bmatrix} 1 & 2 & 4 & 3 \\ 0 & \blacksquare & -7 & -1 \\ 0 & 2 & \blacksquare & \blacksquare \end{bmatrix}$

In Exercises 21–24, convert the matrix to row-echelon form. (*Note:* There is more than one correct answer.)

21. $\begin{bmatrix} 1 & 2 & 3 \\ 2 & -1 & -4 \end{bmatrix}$

22. $\begin{bmatrix} 10 & 30 & 60 & 10 \\ -4 & -9 & 3 & 5 \end{bmatrix}$

23. $\begin{bmatrix} 1 & 1 & -1 & 3 \\ 2 & 1 & 2 & 5 \\ 3 & 2 & 1 & 8 \end{bmatrix}$

24. $\begin{bmatrix} 1 & -3 & -2 & -8 \\ 1 & 3 & -2 & 17 \\ 1 & 2 & -2 & -5 \end{bmatrix}$

In Exercises 25 and 26, write the system of linear equations represented by the augmented matrix. Then use back-substitution to find the solution. (Use variables x, y, and z.)

25. $\begin{bmatrix} 1 & 5 & \vdots & 3 \\ 0 & 1 & \vdots & -2 \end{bmatrix}$

26. $\begin{bmatrix} 1 & 5 & -3 & \vdots & 0 \\ 0 & 1 & 0 & \vdots & 6 \\ 0 & 0 & 1 & \vdots & -5 \end{bmatrix}$

In Exercises 27–34, use matrices to solve the system of linear equations.

27. $6x - 4y = 2$
$5x + 2y = 7$

28. $2x - y = -0.1$
$3x + 2y = 1.6$

29. $x - 2y - z = 6$
$y + 4z = 5$
$4x + 2y + 3z = 8$

30. $x - 3z = -2$
$3x + y - 2z = 5$
$2x + 2y + z = 4$

31. $x + y - 5z = 3$
$x - 2z = 1$
$2x - y - z = 0$

32. $2x + 3z = 3$
$4x - 3y + 7z = 5$
$8x - 9y + 15z = 9$

33. $2x + 4z = 1$
$x + y + 3z = 0$
$x + 3y + 5z = 0$

34. $3x + y - 2z = 2$
$6x + 2y - 4z = 1$
$-3x - y + 2z = 1$

35. *Simple Interest* A corporation borrowed $1,500,000 to expand its product line. Some of the money was borrowed at 8%, some at 9%, and the remainder at 12%. The total annual interest payment to the lenders was $133,000. If the amount borrowed at 8% was four times the amount borrowed at 12%, how much was borrowed at each rate?

36. *Nut Mixture* A grocer wishes to mix three kinds of nuts priced at $3.50, $4.50, and $6.00 per pound to obtain 50 pounds of a mixture priced at $4.95 per pound. How many pounds of each variety should the grocer use if half the mixture is to be composed of the two cheapest varieties?

Curve-Fitting In Exercises 37 and 38, find the equation of the parabola $y = ax^2 + bx + c$ that passes through the given points.

37.

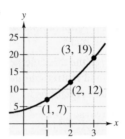

38.

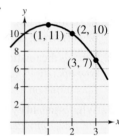

39. *Mathematical Modeling* A videotape of the path of a ball thrown by a baseball player was analyzed on a television set with a grid on the screen (see figure). The tape was paused three times and the position of the coordinates of the ball were measured each time. The coordinates were approximately $(0, 6)$, $(25, 18.5)$, and $(50, 26)$. (The x-coordinate measures the horizontal distance from the player and the y-coordinate is the height of the ball above the ground. Both are measured in feet.)

(a) Find the equation of the parabola $y = ax^2 + bx + c$ that passes through the three points.

(b) Use a graphing utility to graph the parabola. Approximate the maximum height of the ball and the point at which the ball struck the ground.

(c) Find algebraically the maximum height of the ball and the point at which it struck the ground.

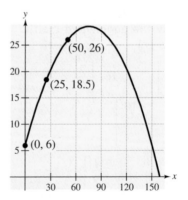

40. *Reading a Graph*   The bar graph at the right gives the gross private savings y (in billions of dollars) in the years 1990 through 1992 in the United States. (Source: U. S. Bureau of Economic Analysis)

(a) Find the equation of the parabola $y = at^2 + bt + c$ that passes through the three points, letting $t = 0$ correspond to 1990.

(b) Use a graphing utility to graph the model found in part (a).

(c) Use the model of part (a) to predict gross private savings in 1999 if the trend continues.

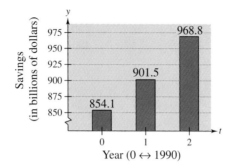

Reviewing the Major Concepts

In Exercises 41–44, evaluate the function.

41. $f(x) = \frac{1}{3}x^2$

(a) $f(6)$

(b) $f\left(\frac{3}{4}\right)$

42. $f(x) = 3 - 2x$

(a) $f(5)$

(b) $f(x + 3) - f(3)$

43. $g(x) = \dfrac{x}{x + 10}$

(a) $g(5)$

(b) $g(c - 6)$

44. $h(x) = \sqrt{x - 4}$

(a) $h(16)$

(b) $h(t + 3)$

45. *Cost*   The inventor of a new game believes that the variable cost for producing the game is \$5.75 per unit and the fixed costs are \$12,000. Let x represent the number of games produced. Express the total cost C as a function of x.

46. *Geometry*   The length of a rectangle is $1\frac{1}{2}$ times its width. Express the length L of the diagonal of the rectangle as a function of its width w.

Additional Problem Solving

In Exercises 47–50, fill in the blanks by using elementary row operations to form a row-equivalent matrix.

47. $\begin{bmatrix} 9 & -18 & 6 \\ 2 & 8 & 15 \end{bmatrix}$

$\begin{bmatrix} 1 & & \\ 2 & 8 & 15 \end{bmatrix}$

48. $\begin{bmatrix} 3 & 6 & 8 \\ 4 & -3 & 6 \end{bmatrix}$

$\begin{bmatrix} 3 & 6 & 8 \\ 1 & -9 & \end{bmatrix}$

49. $\begin{bmatrix} 1 & 1 & 4 & -1 \\ 3 & 8 & 10 & 3 \\ -2 & 1 & 12 & 6 \end{bmatrix}$

$\begin{bmatrix} 1 & 1 & 4 & -1 \\ 0 & 5 & & \\ 0 & 3 & & \end{bmatrix}$

$\begin{bmatrix} 1 & 1 & 4 & -1 \\ 0 & 1 & -\frac{2}{5} & \frac{6}{5} \\ 0 & 3 & & \end{bmatrix}$

50. $\begin{bmatrix} 2 & 3 & -5 & 6 \\ 5 & -7 & 12 & 9 \\ -4 & 6 & 9 & 5 \end{bmatrix}$

$\begin{bmatrix} 2 & 3 & -5 & 6 \\ 5 & -7 & 12 & 9 \\ 0 & 12 & & \end{bmatrix}$

In Exercises 51–56, convert the matrix to row-echelon form. (*Note:* There is more than one correct answer.)

51. $\begin{bmatrix} 4 & 6 & 1 \\ -2 & 2 & 5 \end{bmatrix}$

52. $\begin{bmatrix} 3 & 2 & 6 \\ 2 & 3 & -3 \end{bmatrix}$

53. $\begin{bmatrix} 1 & 1 & 0 & 5 \\ -2 & -1 & 2 & -10 \\ 3 & 6 & 7 & 14 \end{bmatrix}$

54. $\begin{bmatrix} 1 & 2 & -1 & 3 \\ 3 & 7 & -5 & 14 \\ -2 & -1 & -3 & 8 \end{bmatrix}$

55. $\begin{bmatrix} 1 & -1 & -1 & 1 \\ 4 & -4 & 1 & 8 \\ -6 & 8 & 18 & 0 \end{bmatrix}$

56. $\begin{bmatrix} 1 & -3 & 0 & -7 \\ -3 & 10 & 1 & 23 \\ 4 & -10 & 2 & -24 \end{bmatrix}$

In Exercises 57–60, write the system of linear equations represented by the augmented matrix. Then use back-substitution to find the solution. (Use variables x, y, and z.)

57. $\begin{bmatrix} 1 & -2 & \vdots & 4 \\ 0 & 1 & \vdots & -3 \end{bmatrix}$

58. $\begin{bmatrix} 1 & 5 & \vdots & 0 \\ 0 & 1 & \vdots & -1 \end{bmatrix}$

59. $\begin{bmatrix} 1 & -1 & 2 & \vdots & 4 \\ 0 & 1 & -1 & \vdots & 2 \\ 0 & 0 & 1 & \vdots & -2 \end{bmatrix}$

60. $\begin{bmatrix} 1 & 2 & -2 & \vdots & -1 \\ 0 & 1 & 1 & \vdots & 9 \\ 0 & 0 & 1 & \vdots & -3 \end{bmatrix}$

In Exercises 61–74, use matrices to solve the system of linear equations.

61. $x + 2y = 7$
$3x + y = 8$

62. $2x + 6y = 16$
$2x + 3y = 7$

63. $-x + 2y = 1.5$
$2x - 4y = 3$

64. $x - 3y = 5$
$-2x + 6y = -10$

65. $x - 3y + 2z = 8$
$2y - z = -4$
$x + z = 3$

66. $2y + z = 3$
$-4y - 2z = 0$
$x + y + z = 2$

67. $2x - y + 3z = 24$
$2y - z = 14$
$7x - 5y = 6$

68. $2x + 4y = 10$
$2x + 2y + 3z = 3$
$-3x + y + 2z = -3$

69. $x + 3y = 2$
$2x + 6y = 4$
$2x + 5y + 4z = 3$

70. $2x + 2y + z = 8$
$2x + 3y + z = 7$
$6x + 8y + 3z = 22$

71. $2x + 4y + 5z = 5$
$x + 3y + 3z = 2$
$2x + 4y + 4z = 2$

72. $-2x - 2y - 15z = 0$
$x + 2y + 2z = 18$
$3x + 3y + 22z = 2$

73. $3x + 3y + z = 4$
$2x + 6y + z = 5$
$-x - 3y + 2z = -5$

74. $2x + y - 2z = 4$
$3x - 2y + 4z = 6$
$-4x + y + 6z = 12$

Investment Portfolio   In Exercises 75 and 76, consider an investor with a portfolio totaling $500,000 that is to be allocated among the following types of investments: certificates of deposit, municipal bonds, blue-chip stocks, and growth or speculative stocks. How much should be allocated to each type of investment?

75. The certificates of deposit pay 10% annually, and the municipal bonds pay 8% annually. Over a 5-year period, the investor expects the blue-chip stocks to return 12% annually, and expects the growth stocks to return 13% annually. The investor wants a combined annual return of 10% and also wants to have only one-fourth of the portfolio invested in stocks.

76. The certificates of deposit pay 9% annually, and the municipal bonds pay 5% annually. Over a 5-year period, the investor expects the blue-chip stocks to return 12% annually, and expects the growth stocks to return 14% annually. The investor wants a combined annual return of 10% and also wants to have only one-fourth of the portfolio invested in stocks.

77. *Curve-Fitting*   Find the equation of the parabola
$$y = ax^2 + bx + c$$
that passes through the points $(1, 1)$, $(-3, 17)$, and $(2, -\frac{1}{2})$.

78. *Curve-Fitting*   Find the equation of the circle
$$x^2 + y^2 + Dx + Ey + F = 0$$
that passes through the points $(1, 1)$, $(3, 3)$, and $(4, 2)$.

79. The sum of three positive numbers is 33. The second number is 3 greater than the first, and the third is four times the first. Find the three numbers.

80. *Investments*   An inheritance of $16,000 was divided among three investments yielding a total of $990 in simple interest per year. The interest rates for the three investments were 5%, 6%, and 7%. Find the amount placed in each investment if the 5% and 6% investments were $3000 and $2000 less than the 7% investment, respectively.

81. *Rewriting a Fraction*   The fraction

$$\frac{2x^2 - 9x}{(x-2)^3}$$

can be written as a sum of three fractions as follows.

$$\frac{2x^2 - 9x}{(x-2)^3} = \frac{A}{x-2} + \frac{B}{(x-2)^2} + \frac{C}{(x-2)^3}$$

The numbers A, B, and C are the solutions of the system

$$\begin{aligned}
4A - 2B + C &= 0 \\
-4A + B &= -9 \\
A &= 2.
\end{aligned}$$

Solve the system and verify that the sum of the three resulting fractions is the original fraction.

82. *Rewriting a Fraction*   The fraction

$$\frac{x+1}{x(x^2+1)}$$

can be written as a sum of two fractions as follows.

$$\frac{x+1}{x(x^2+1)} = \frac{A}{x} + \frac{Bx+C}{x^2+1}.$$

The numbers A, B, and C are the solutions of the system

$$\begin{aligned}
2A + B + C &= 2 \\
2A + B - C &= 0 \\
5A + 4B + 2C &= 3.
\end{aligned}$$

Solve the system and verify that the sum of the two resulting fractions is the original fraction.

CAREER INTERVIEW

Richard S. Schroeder
Electrical Engineer
Lorain Products
Lorain, OH 44502

Lorain Products designs and manufactures power systems for the telecommunications industry. These power supplies convert the power delivered by an electric utility into a form that a telecommunication system can use. I work in Lorain Products' Custom Power unit, and it is my responsibility to design power supplies that meet customers' specifications, can be manufactured easily, and meet the quoted cost. A typical power supply design takes about one year to go from an idea to production, and a typical supply will have a production run of five years.

Because power supplies are vital to business communications throughout the world, it is my job as a designer to ensure that the power systems that Lorain Products supplies are reliable. To ensure reliability, I often use a circuit analysis technique that results in a system of linear equations. In a recent design I needed to find the currents in a certain circuit to check that none of the circuit's components would operate beyond their limits. My analysis of this circuit resulted in the following system of two equations in two unknowns.

$$\begin{aligned}
5.99I_1 + I_2 &= 2.5 \\
-I_1 + 101I_2 &= 14.3
\end{aligned}$$

This system can easily be solved to show that $I_1 = 0.393$ amps and $I_2 = 0.146$ amps. Knowing the currents I_1 and I_2 and the current ratings for each part of the circuit, I can tell if the circuit will last a long time. Systems of equations are quite useful in my line of work.

8.5 Linear Systems and Determinants

The Determinant of a Matrix ▪ Cramer's Rule ▪
Applications of Determinants

The Determinant of a Matrix

Associated with each square matrix is a real number called its **determinant.** The use of determinants arose from special number patterns that occur during the solution of systems of linear equations. For instance, the system

$$a_1 x + b_1 y = c_1$$
$$a_2 x + b_2 y = c_2$$

has a solution given by

$$x = \frac{c_1 b_2 - c_2 b_1}{a_1 b_2 - a_2 b_1} \quad \text{and} \quad y = \frac{a_1 c_2 - a_2 c_1}{a_1 b_2 - a_2 b_1}$$

provided that $a_1 b_2 - a_2 b_1 \neq 0$. Note that the denominator of each fraction is the same. We call this denominator the **determinant** of the coefficient matrix of the system.

Coefficient Matrix   *Determinant*

$$A = \begin{bmatrix} a_1 & b_1 \\ a_2 & b_2 \end{bmatrix} \qquad \det(A) = a_1 b_2 - a_2 b_1$$

The determinant of the matrix A can also be denoted by vertical bars on both sides of the matrix, as indicated in the following definition.

Arthur Cayley (1821–1895) is credited with creating the theory of matrices. Determinants had been studied as rectangular arrays of numbers since the middle of the 18th century. Thus, the use and basic properties of matrices were well established when Cayley first published articles introducing matrices as distinct entities.

Definition of the Determinant of a 2 x 2 Matrix

$$\det(A) = \begin{vmatrix} a_1 & b_1 \\ a_2 & b_2 \end{vmatrix} = a_1 b_2 - a_2 b_1$$

A convenient method for remembering the formula for the determinant of a 2×2 matrix is shown in the following diagram.

$$\det(A) = \begin{vmatrix} a_1 & b_1 \\ a_2 & b_2 \end{vmatrix} = a_1 b_2 - a_2 b_1$$

Note that the determinant is given by the difference of the products of the two diagonals of the matrix.

EXAMPLE 1   The Determinant of a 2 x 2 Matrix

Find the determinant of each matrix.

a. $A = \begin{bmatrix} 2 & -3 \\ 1 & 4 \end{bmatrix}$ **b.** $B = \begin{bmatrix} -1 & 2 \\ 2 & -4 \end{bmatrix}$ **c.** $C = \begin{bmatrix} 1 & 3 \\ 2 & 5 \end{bmatrix}$

Solution

NOTE   Notice in Example 1 that the determinant of a matrix can be positive, zero, or negative.

a. $\det(A) = \begin{vmatrix} 2 & -3 \\ 1 & 4 \end{vmatrix} = 2(4) - 1(-3) = 8 + 3 = 11$

b. $\det(B) = \begin{vmatrix} -1 & 2 \\ 2 & -4 \end{vmatrix} = (-1)(-4) - 2(2) = 4 - 4 = 0$

c. $\det(C) = \begin{vmatrix} 1 & 3 \\ 2 & 5 \end{vmatrix} = 1(5) - 2(3) = 5 - 6 = -1$

One way to evaluate the determinant of a 3×3 matrix, called **expanding by minors,** allows you to write the determinant of a 3×3 matrix in terms of three 2×2 determinants. The **minor** of an entry in a 3×3 matrix is the determinant of the 2×2 matrix that remains after deletion of the row and column in which the entry occurs. Here are two examples.

Technology

A graphing utility can be used to evaluate the determinant of a square matrix. To find the determinant of

$\begin{bmatrix} 2 & -3 \\ 1 & 4 \end{bmatrix}$

first use the MATRIX feature to enter the matrix, and then evaluate the determinant, as follows.

det [A] ENTER

Check the result against Example 1(a). Evaluate the determinant of the 3×3 matrix at the right using a graphing utility. Then finish the evaluation of the determinant by expanding by minors to check the result.

Given Determinant	*Entry*	*Minor of Entry*	*Value of Minor*
$\begin{vmatrix} 1 & -1 & 3 \\ 0 & 2 & 5 \\ -2 & 4 & -7 \end{vmatrix}$	1	$\begin{vmatrix} 2 & 5 \\ 4 & -7 \end{vmatrix}$	$2(-7) - 4(5) = -34$
$\begin{vmatrix} 1 & -1 & 3 \\ 0 & 2 & 5 \\ -2 & 4 & -7 \end{vmatrix}$	-1	$\begin{vmatrix} 0 & 5 \\ -2 & -7 \end{vmatrix}$	$0(-7) - (-2)(5) = 10$

Expanding by Minors

$$\det(A) = \begin{vmatrix} a_1 & b_1 & c_1 \\ a_2 & b_2 & c_2 \\ a_3 & b_3 & c_3 \end{vmatrix}$$

$$= a_1(\text{minor of } a_1) - b_1(\text{minor of } b_1) + c_1(\text{minor of } c_1)$$

$$= a_1 \begin{vmatrix} b_2 & c_2 \\ b_3 & c_3 \end{vmatrix} - b_1 \begin{vmatrix} a_2 & c_2 \\ a_3 & c_3 \end{vmatrix} + c_1 \begin{vmatrix} a_2 & b_2 \\ a_3 & b_3 \end{vmatrix}$$

This pattern is called **expanding by minors** along the first row. A similar pattern can be used to expand by minors along any row or column.

$$\begin{bmatrix} + & - & + \\ - & + & - \\ + & - & + \end{bmatrix}$$

FIGURE 8.9 Sign Pattern for
3×3 Matrix

The *signs* of the terms used in expanding by minors follows the alternating pattern shown in Figure 8.9. For instance, the signs used to expand by minors along the second row are $-, +, -$, as follows.

$$\det(A) = \begin{vmatrix} a_1 & b_1 & c_1 \\ a_2 & b_2 & c_2 \\ a_3 & b_3 & c_3 \end{vmatrix}$$

$$= -a_2(\text{minor of } a_2) + b_2(\text{minor of } b_2) - c_2(\text{minor of } c_2)$$

EXAMPLE 2   Finding the Determinant of a 3 x 3 Matrix

Find the determinant of $A = \begin{bmatrix} -1 & 1 & 2 \\ 0 & 2 & 3 \\ 3 & 4 & 2 \end{bmatrix}$.

Solution

By expanding by minors along the *first column*, you obtain the following.

$$\det(A) = \begin{vmatrix} -1 & 1 & 2 \\ 0 & 2 & 3 \\ 3 & 4 & 2 \end{vmatrix}$$

$$= (-1)\begin{vmatrix} 2 & 3 \\ 4 & 2 \end{vmatrix} - (0)\begin{vmatrix} 1 & 2 \\ 4 & 2 \end{vmatrix} + (3)\begin{vmatrix} 1 & 2 \\ 2 & 3 \end{vmatrix}$$

$$= (-1)(4 - 12) - (0)(2 - 8) + (3)(3 - 4)$$

$$= 8 - 0 - 3$$

$$= 5$$

STUDY TIP

Note in the expansion in Example 2 that a zero entry will always yield a zero term when expanding by minors. Thus, when you are evaluating the determinant of a matrix, you should choose to expand along the row or column that has the most zero entries.

EXAMPLE 3   Finding the Determinant of a 3 x 3 Matrix

Find the determinant of $A = \begin{bmatrix} 1 & 2 & 1 \\ 3 & 0 & 2 \\ 4 & 0 & -1 \end{bmatrix}$.

Solution

By expanding by minors along the *second column*, you obtain the following.

$$\det(A) = \begin{vmatrix} 1 & 2 & 1 \\ 3 & 0 & 2 \\ 4 & 0 & -1 \end{vmatrix}$$

$$= -(2)\begin{vmatrix} 3 & 2 \\ 4 & -1 \end{vmatrix} + (0)\begin{vmatrix} 1 & 1 \\ 4 & -1 \end{vmatrix} - (0)\begin{vmatrix} 1 & 1 \\ 3 & 2 \end{vmatrix}$$

$$= -(2)(-3 - 8) + 0 - 0$$

$$= 22$$

Cramer's Rule

So far in this chapter, you have studied three methods for solving a system of linear equations: substitution, elimination (with equations), and elimination (with matrices). We now look at one more method, called **Cramer's Rule,** which is named after Gabriel Cramer (1704–1752). This rule uses determinants to write the solution of a system of linear equations.

Cramer's Rule

---

1. For the system of linear equations

$$a_1x + b_1y = c_1$$
$$a_2x + b_2y = c_2$$

the solution is given by

$$x = \frac{D_x}{D} = \frac{\begin{vmatrix} c_1 & b_1 \\ c_2 & b_2 \end{vmatrix}}{\begin{vmatrix} a_1 & b_1 \\ a_2 & b_2 \end{vmatrix}}, \qquad y = \frac{D_y}{D} = \frac{\begin{vmatrix} a_1 & c_1 \\ a_2 & c_2 \end{vmatrix}}{\begin{vmatrix} a_1 & b_1 \\ a_2 & b_2 \end{vmatrix}}$$

provided that $D \neq 0$.

2. For the system of linear equations

$$a_1x + b_1y + c_1z = d_1$$
$$a_2x + b_2y + c_2z = d_2$$
$$a_3x + b_3y + c_3z = d_3$$

the solution is given by

$$x = \frac{D_x}{D} = \frac{\begin{vmatrix} d_1 & b_1 & c_1 \\ d_2 & b_2 & c_2 \\ d_3 & b_3 & c_3 \end{vmatrix}}{\begin{vmatrix} a_1 & b_1 & c_1 \\ a_2 & b_2 & c_2 \\ a_3 & b_3 & c_3 \end{vmatrix}}, \qquad y = \frac{D_y}{D} = \frac{\begin{vmatrix} a_1 & d_1 & c_1 \\ a_2 & d_2 & c_2 \\ a_3 & d_3 & c_3 \end{vmatrix}}{\begin{vmatrix} a_1 & b_1 & c_1 \\ a_2 & b_2 & c_2 \\ a_3 & b_3 & c_3 \end{vmatrix}},$$

$$z = \frac{D_z}{D} = \frac{\begin{vmatrix} a_1 & b_1 & d_1 \\ a_2 & b_2 & d_2 \\ a_3 & b_3 & d_3 \end{vmatrix}}{\begin{vmatrix} a_1 & b_1 & c_1 \\ a_2 & b_2 & c_2 \\ a_3 & b_3 & c_3 \end{vmatrix}}$$

provided that $D \neq 0$.

EXAMPLE 4   Using Cramer's Rule for a 2 x 2 System

Use Cramer's Rule to solve the system of linear equations.

$$4x - 2y = 10$$
$$3x - 5y = 11$$

Solution

Begin by finding the determinant of the coefficient matrix, $D = -14$.

$$x = \frac{D_x}{D} = \frac{\begin{vmatrix} 10 & -2 \\ 11 & -5 \end{vmatrix}}{-14} = \frac{(-50) - (-22)}{-14} = \frac{-28}{-14} = 2$$

$$y = \frac{D_y}{D} = \frac{\begin{vmatrix} 4 & 10 \\ 3 & 11 \end{vmatrix}}{-14} = \frac{44 - 30}{-14} = \frac{14}{-14} = -1$$

The solution is $(2, -1)$. Check this in the original system of equations.

EXAMPLE 5   Using Cramer's Rule for a 3 x 3 System

Use Cramer's Rule to solve the system of linear equations.

$$-x + 2y - 3z = 1$$
$$2x + \qquad z = 0$$
$$3x - 4y + 4z = 2$$

Solution

The determinant of the coefficient matrix is $D = 10$.

NOTE When using Cramer's Rule, remember that the method *does not* apply if the determinant of the coefficient matrix is zero. For instance, the system

$$6x - 10y = -4$$
$$9x - 15y = 5$$

has no solution (see Example 8 in Section 8.4), and the determinant of the coefficient matrix of this system is zero.

$$x = \frac{D_x}{D} = \frac{\begin{vmatrix} 1 & 2 & -3 \\ 0 & 0 & 1 \\ 2 & -4 & 4 \end{vmatrix}}{10} = \frac{8}{10} = \frac{4}{5}$$

$$y = \frac{D_y}{D} = \frac{\begin{vmatrix} -1 & 1 & -3 \\ 2 & 0 & 1 \\ 3 & 2 & 4 \end{vmatrix}}{10} = \frac{-15}{10} = -\frac{3}{2}$$

$$z = \frac{D_z}{D} = \frac{\begin{vmatrix} -1 & 2 & 1 \\ 2 & 0 & 0 \\ 3 & -4 & 2 \end{vmatrix}}{10} = \frac{-16}{10} = -\frac{8}{5}$$

The solution is $\left(\frac{4}{5}, -\frac{3}{2}, -\frac{8}{5}\right)$. Check this in the original system of equations.

Applications of Determinants

In addition to Cramer's Rule, determinants have many other practical applications. For instance, you can use a determinant to find the area of a triangle whose vertices are given by three points on a rectangular coordinate system.

Area of a Triangle

The area of a triangle with vertices (x_1, y_1), (x_2, y_2), and (x_3, y_3) is

$$\text{Area} = \pm \frac{1}{2} \begin{vmatrix} x_1 & y_1 & 1 \\ x_2 & y_2 & 1 \\ x_3 & y_3 & 1 \end{vmatrix}$$

where the symbol $(\pm)$ indicates that the appropriate sign should be chosen to yield a positive area.

EXAMPLE 6   *Finding the Area of a Triangle*

Find the area of the triangle whose vertices are $(2, 0)$, $(1, 3)$, and $(3, 2)$, as shown in Figure 8.10.

Solution

Choose $(x_1, y_1) = (2, 0)$, $(x_2, y_2) = (1, 3)$, and $(x_3, y_3) = (3, 2)$. To find the area of the triangle, evaluate the determinant

$$\begin{vmatrix} x_1 & y_1 & 1 \\ x_2 & y_2 & 1 \\ x_3 & y_3 & 1 \end{vmatrix} = \begin{vmatrix} 2 & 0 & 1 \\ 1 & 3 & 1 \\ 3 & 2 & 1 \end{vmatrix}$$

$$= 2 \begin{vmatrix} 3 & 1 \\ 2 & 1 \end{vmatrix} - 0 \begin{vmatrix} 1 & 1 \\ 3 & 1 \end{vmatrix} + 1 \begin{vmatrix} 1 & 3 \\ 3 & 2 \end{vmatrix}$$

$$= 2(1) - 0 + 1(-7)$$

$$= -5.$$

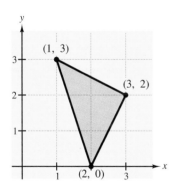

FIGURE 8.10

Using this value, you can conclude that the area of the triangle is

$$\text{Area} = -\frac{1}{2} \begin{vmatrix} 2 & 0 & 1 \\ 1 & 3 & 1 \\ 3 & 2 & 1 \end{vmatrix} = -\frac{1}{2}(-5) = \frac{5}{2}.$$

NOTE   To see the benefit of the "determinant formula," try finding the area of the triangle in Example 6 using the standard formula: $\text{Area} = \frac{1}{2}(\text{base})(\text{height})$.

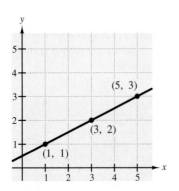

FIGURE 8.11

Suppose the three points in Example 6 had been on the same line. What would have happened had we applied the area formula to three such points? The answer is that the determinant would have been zero. Consider for instance, the three collinear points $(1, 1)$, $(3, 2)$, and $(5, 3)$, as shown in Figure 8.11. The area of the "triangle" that has these three points as vertices is

$$\frac{1}{2}\begin{vmatrix} 1 & 1 & 1 \\ 3 & 2 & 1 \\ 5 & 3 & 1 \end{vmatrix} = \frac{1}{2}\left(1\begin{vmatrix} 2 & 1 \\ 3 & 1 \end{vmatrix} - 1\begin{vmatrix} 3 & 1 \\ 5 & 1 \end{vmatrix} + 1\begin{vmatrix} 3 & 2 \\ 5 & 3 \end{vmatrix}\right)$$

$$= \frac{1}{2}[-1 - (-2) + (-1)]$$

$$= 0.$$

This result is generalized as follows.

Test for Collinear Points

Three points (x_1, y_1), (x_2, y_2), and (x_3, y_3) are collinear (lie on the same line) if and only if

$$\begin{vmatrix} x_1 & y_1 & 1 \\ x_2 & y_2 & 1 \\ x_3 & y_3 & 1 \end{vmatrix} = 0.$$

EXAMPLE 7 Testing for Collinear Points

Determine whether the points $(-2, -2)$, $(1, 1)$, and $(7, 5)$ lie on the same line. (See Figure 8.12.)

Solution

Letting $(x_1, y_1) = (-2, -2)$, $(x_2, y_2) = (1, 1)$, and $(x_3, y_3) = (7, 5)$, you have

$$\begin{vmatrix} x_1 & y_1 & 1 \\ x_2 & y_2 & 1 \\ x_3 & y_3 & 1 \end{vmatrix} = \begin{vmatrix} -2 & -2 & 1 \\ 1 & 1 & 1 \\ 7 & 5 & 1 \end{vmatrix}$$

$$= -2\begin{vmatrix} 1 & 1 \\ 5 & 1 \end{vmatrix} - (-2)\begin{vmatrix} 1 & 1 \\ 7 & 1 \end{vmatrix} + 1\begin{vmatrix} 1 & 1 \\ 7 & 5 \end{vmatrix}$$

$$= -2(-4) - (-2)(-6) + 1(-2)$$

$$= -6.$$

Because the value of this determinant *is not* zero, you can conclude that the three points *do not* lie on the same line.

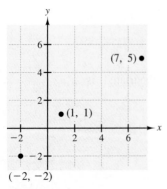

FIGURE 8.12

Two-Point Form of the Equation of a Line

An equation of the line passing through the distinct points (x_1, y_1) and (x_2, y_2) is given by

$$\begin{vmatrix} x & y & 1 \\ x_1 & y_1 & 1 \\ x_2 & y_2 & 1 \end{vmatrix} = 0.$$

EXAMPLE 8   *Finding an Equation of a Line*

Find an equation of the line passing through $(-2, 1)$ and $(3, -2)$.

Solution

$$\begin{vmatrix} x & y & 1 \\ -2 & 1 & 1 \\ 3 & -2 & 1 \end{vmatrix} = 0$$

$$x \begin{vmatrix} 1 & 1 \\ -2 & 1 \end{vmatrix} - y \begin{vmatrix} -2 & 1 \\ 3 & 1 \end{vmatrix} + 1 \begin{vmatrix} -2 & 1 \\ 3 & -2 \end{vmatrix} = 0$$

$$3x + 5y + 1 = 0$$

Therefore, an equation of the line is $3x + 5y + 1 = 0$.

Group Activities   Extending the Concept

Determinant of a 3 x 3 Matrix   There is an alternative method for evaluating the determinant of a 3×3 matrix A. (This method works *only* for 3×3 matrices.) To apply this method, copy the first and second columns of A to form fourth and fifth columns. The determinant of A is then obtained by adding the products of three diagonals and subtracting the products of three diagonals.

$$|A| = \begin{vmatrix} 0 & 2 & 1 \\ 3 & 1 & 2 \\ 4 & -4 & 1 \end{vmatrix} \begin{matrix} 0 & 2 \\ 3 & -1 \\ 4 & -4 \end{matrix} = 0 + 16 - 12 - (-4) - 0 - 6 = 2$$

Try using this technique to find the determinants of the matrices in Examples 2 and 3. Do you think this method is easier than expanding by minors?

8.5 Exercises

Discussing the Concepts

1. Explain the difference between a square matrix and its determinant.

2. Is it possible to find the determinant of a 2×3 matrix? Explain your answer.

3. What is meant by the *minor* of an entry of a square matrix?

4. What conditions must be met in order to use Cramer's Rule to solve a system of linear equations?

Problem Solving

In Exercises 5–8, find the determinant of the matrix.

5. $\begin{bmatrix} 2 & 1 \\ 3 & 4 \end{bmatrix}$

6. $\begin{bmatrix} -3 & 1 \\ 5 & 2 \end{bmatrix}$

7. $\begin{bmatrix} -7 & 6 \\ \frac{1}{2} & 3 \end{bmatrix}$

8. $\begin{bmatrix} -1.2 & 4.5 \\ 0.4 & -0.9 \end{bmatrix}$

In Exercises 9 and 10, evaluate the determinant of the matrix six different ways by expanding by minors along each row and column.

9. $\begin{bmatrix} 2 & 3 & -1 \\ 6 & 0 & 0 \\ 4 & 1 & 1 \end{bmatrix}$

10. $\begin{bmatrix} 10 & 2 & -4 \\ 8 & 0 & -2 \\ 4 & 0 & 2 \end{bmatrix}$

In Exercises 11–14, evaluate the determinant of the matrix. Expand by minors on the row or column that appears to make the computation easiest.

11. $\begin{bmatrix} 2 & 4 & 6 \\ 0 & 3 & 1 \\ 0 & 0 & -5 \end{bmatrix}$

12. $\begin{bmatrix} 3 & 2 & 2 \\ 2 & 2 & 2 \\ -4 & 4 & 3 \end{bmatrix}$

13. $\begin{bmatrix} 6 & 8 & -7 \\ 0 & 0 & 0 \\ 4 & -6 & 22 \end{bmatrix}$

14. $\begin{bmatrix} 0.1 & 0.2 & 0.3 \\ -0.3 & 0.2 & 0.2 \\ 5 & 4 & 4 \end{bmatrix}$

In Exercises 15–18, use a graphing utility to evaluate the determinant of the matrix.

15. $\begin{bmatrix} 5 & -3 & 2 \\ 7 & 5 & -7 \\ 0 & 6 & -1 \end{bmatrix}$

16. $\begin{bmatrix} -\frac{1}{2} & -1 & 6 \\ 8 & -\frac{1}{4} & -4 \\ 1 & 2 & 1 \end{bmatrix}$

17. $\begin{bmatrix} 0.2 & 0.8 \\ -8 & -5 \end{bmatrix}$

18. $\begin{bmatrix} 250 & -125 \\ 60 & -50 \end{bmatrix}$

In Exercises 19–26, use Cramer's Rule to solve the system of equations. (If it is not possible, state the reason.)

19. $3x + 4y = -2$
    $5x + 3y = 4$

20. $18x + 12y = 13$
    $30x + 24y = 23$

21. $-0.4x + 0.8y = 1.6$
    $2x - 4y = 5$

22. $-0.4x + 0.8y = 1.6$
    $0.2x + 0.3y = 2.2$

23. $4x - y + z = -5$
    $2x + 2y + 3z = 10$
    $5x - 2y + 6z = 1$

24. $4x - 2y + 3z = -2$
    $2x + 2y + 5z = 16$
    $8x - 5y - 2z = 4$

25. $3x + 4y + 4z = 11$
    $4x - 4y + 6z = 11$
    $6x - 6y = 3$

26. $2x + 3y + 5z = 4$
    $3x + 5y + 9z = 7$
    $5x + 9y + 17z = 13$

In Exercises 27–30, use a graphing utility and Cramer's Rule to solve the system of equations.

27. $-3x + 10y = 22$
    $9x - 3y = 0$

28. $3x + 7y = 3$
    $7x + 25y = 11$

29. $x + y - z = 2$
    $6x + 4y + 3z = 4$
    $3x + 6z = -3$

30. $3x + 2y - 5z = -10$
$ 6x - z = 8$
$ -y + 3z = -2$

Geometry   In Exercises 31 and 32, use a determinant to find the area of the triangle with the given vertices.

31. $(0, 3), (4, 0), (8, 5)$

32. $(-4, 2), (1, 5), (4, -4)$

33. *Geometry*   A large region of forest has been infested with gypsy moths. The region is roughly triangular, as shown in the figure. Approximate the number of square miles in this region.

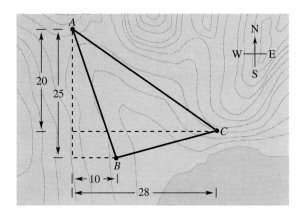

34. *Geometry*   You have purchased a triangular tract of land, as shown in the figure. How many square feet are there in the tract of land?

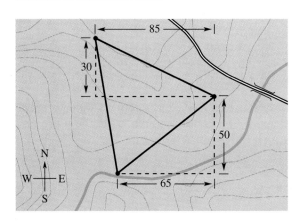

Collinear Points   In Exercises 35 and 36, decide whether the points are collinear.

35. $(-1, -5), (1, -1), (4, 5)$

36. $(-1, 8), (1, 2), (2, 0)$

Equation of a Line   In Exercises 37 and 38, find an equation of the line through the points.

37. $\left(-2, \frac{3}{2}\right), (3, -3)$ **38.** $(2, 3.6), (8, 10)$

Curve-Fitting   In Exercises 39 and 40, use Cramer's Rule to find the equation of the parabola that passes through the points.

39. $(-2, 6), (2, -2), (4, 0)$

40. $(-1, 0), (1, 4), (4, -5)$

41. *Mathematical Modeling*   The table gives the merchandise exports y_1 and the merchandise imports y_2 (in billions of dollars) for the years 1990 through 1992 in the United States.   (Source: U.S. Bureau of Census)

Year	1990	1991	1992
y_1	393.6	421.7	448.2
y_2	495.3	487.1	532.5

(a) Find a quadratic model for exports. Let $t = 0$ represent 1990.

(b) Find a quadratic model for imports. Let $t = 0$ represent 1990.

(c) Use a graphing utility to graph the models found in parts (a) and (b).

(d) Find a model for the merchandise trade balance (merchandise exports − merchandise imports).

42. (a) Use Cramer's Rule to solve the system of linear equations.

$$kx + (1 - k)y = 1$$
$$(1 - k)x + ky = 3$$

(b) For what value(s) of k will the system be inconsistent?

Reviewing the Major Concepts

In Exercises 43–46, sketch a graph of the equation.

43. $y = 3 - \frac{1}{2}x$

44. $4x^2 + 4y^2 = 25$

45. $\dfrac{y^2}{25} + x^2 = 1$

46. $\dfrac{x^2}{9} - \dfrac{y^2}{25} = 1$

47. *Defective Units*   A quality control engineer for a certain buyer found two defective units in a sample of 75. At that rate, what is the expected number of defective units in a shipment of 10,000 units?

48. *Geometry*   Find the perimeter and area of a right triangle whose legs are 4 inches and 5 inches.

Additional Problem Solving

In Exercises 49–56, find the determinant.

49. $\begin{bmatrix} 5 & 2 \\ -6 & 3 \end{bmatrix}$

50. $\begin{bmatrix} 2 & -2 \\ 4 & 3 \end{bmatrix}$

51. $\begin{bmatrix} 5 & -4 \\ -10 & 8 \end{bmatrix}$

52. $\begin{bmatrix} 4 & -3 \\ 0 & 0 \end{bmatrix}$

53. $\begin{bmatrix} 2 & 6 \\ 0 & 3 \end{bmatrix}$

54. $\begin{bmatrix} -2 & 3 \\ 6 & -9 \end{bmatrix}$

55. $\begin{bmatrix} 0.3 & 0.5 \\ 0.5 & 0.3 \end{bmatrix}$

56. $\begin{bmatrix} \frac{2}{3} & \frac{5}{6} \\ 14 & -2 \end{bmatrix}$

In Exercises 57–60, evaluate the determinant of the matrix six different ways by expanding by minors along each row and column.

57. $\begin{bmatrix} 2 & -5 & 0 \\ 4 & 7 & 0 \\ -7 & 25 & 3 \end{bmatrix}$

58. $\begin{bmatrix} 8 & 7 & 6 \\ -4 & 0 & 0 \\ 5 & 1 & 4 \end{bmatrix}$

59. $\begin{bmatrix} 1 & 1 & 2 \\ 3 & 1 & 0 \\ -2 & 0 & 3 \end{bmatrix}$

60. $\begin{bmatrix} 2 & 1 & 3 \\ 1 & 4 & 4 \\ 1 & 0 & 2 \end{bmatrix}$

In Exercises 61–72, evaluate the determinant of the matrix. Expand by minors on the row or column that appears to make the computation easiest.

61. $\begin{bmatrix} -2 & 2 & 3 \\ 1 & -1 & 0 \\ 0 & 1 & 4 \end{bmatrix}$

62. $\begin{bmatrix} 2 & 3 & 1 \\ 0 & 5 & -2 \\ 0 & 0 & -2 \end{bmatrix}$

63. $\begin{bmatrix} 1 & 4 & -2 \\ 3 & 6 & -6 \\ -2 & 1 & 4 \end{bmatrix}$

64. $\begin{bmatrix} 2 & -1 & 0 \\ 4 & 2 & 1 \\ 4 & 2 & 1 \end{bmatrix}$

65. $\begin{bmatrix} -3 & 2 & 1 \\ 4 & 5 & 6 \\ 2 & -3 & 1 \end{bmatrix}$

66. $\begin{bmatrix} -3 & 4 & 2 \\ 6 & 3 & 1 \\ 4 & -7 & -8 \end{bmatrix}$

67. $\begin{bmatrix} 1 & 4 & -2 \\ 3 & 2 & 0 \\ -1 & 4 & 3 \end{bmatrix}$

68. $\begin{bmatrix} -0.4 & 0.4 & 0.3 \\ 0.2 & 0.2 & 0.2 \\ 0.3 & 0.2 & 0.2 \end{bmatrix}$

69. $\begin{bmatrix} 0.4 & 0.3 & 0.3 \\ -0.2 & 0.6 & 0.6 \\ 3 & 1 & 1 \end{bmatrix}$

70. $\begin{bmatrix} \frac{1}{2} & \frac{3}{2} & \frac{1}{2} \\ 4 & 8 & 10 \\ -2 & -6 & 12 \end{bmatrix}$

71. $\begin{bmatrix} x & y & 1 \\ 3 & 1 & 1 \\ -2 & 0 & 1 \end{bmatrix}$

72. $\begin{bmatrix} x & y & 1 \\ -2 & -2 & 1 \\ 1 & 5 & 1 \end{bmatrix}$

In Exercises 73–76, use a graphing utility to evaluate the determinant of the matrix.

73. $\begin{bmatrix} 35 & 15 & 70 \\ -8 & 20 & 3 \\ -5 & 6 & 20 \end{bmatrix}$

74. $\begin{bmatrix} 3 & -1 & 2 \\ 1 & -1 & 2 \\ -2 & 3 & 10 \end{bmatrix}$

75. $\begin{bmatrix} 0.3 & -0.2 & 0.5 \\ 0.6 & 0.4 & -0.3 \\ 1.2 & 0 & 0.7 \end{bmatrix}$

76. $\begin{bmatrix} \frac{3}{2} & -\frac{3}{4} & 1 \\ 10 & 8 & 7 \\ 12 & -4 & 12 \end{bmatrix}$

In Exercises 77–86, use Cramer's Rule to solve the system. (If it is not possible, state the reason.)

77. $\begin{aligned} x + 2y &= 5 \\ -x + y &= 1 \end{aligned}$

78. $\begin{aligned} 2x - y &= -10 \\ 3x + 2y &= -1 \end{aligned}$

79. $20x + 8y = 11$
$12x - 24y = 21$

80. $13x - 6y = 17$
$26x - 12y = 8$

81. $3u + 6v = 5$
$6u + 14v = 11$

82. $3x_1 + 2x_2 = 1$
$2x_1 + 10x_2 = 6$

83. $3a + 3b + 4c = 1$
$3a + 5b + 9c = 2$
$5a + 9b + 17c = 4$

84. $14x_1 - 21x_2 - 7x_3 = 10$
$-4x_1 + 2x_2 - 2x_3 = 4$
$56x_1 - 21x_2 + 7x_3 = 5$

85. $5x - 3y + 2z = 2$
$2x + 2y - 3z = 3$
$x - 7y + 8z = -4$

86. $3x + 2y + 5z = 4$
$4x - 3y - 4z = 1$
$-8x + 2y + 3z = 0$

In Exercises 87–90, use a graphing utility and Cramer's Rule to solve the system of equations.

87. $4x - y = -2$
$-2x + y = 3$

88. $4x + 8y = 6$
$8x + 26y = 19$

89. $3x + y + z = 6$
$x - 4y + 2z = -1$
$x - 3y + z = 0$

90. $3x - 2y + 3z = 8$
$x + 3y + 6z = -3$
$x + 2y + 9z = -5$

Geometry   In Exercises 91–94, use a determinant to find the area of the triangle with the given vertices.

91. $(0, 0)$, $(3, 1)$, $(1, 5)$

92. $(-2, -3)$, $(2, -3)$, $(0, 4)$

93. $(-2, 1)$, $(3, -1)$, $(1, 6)$

94. $\left(0, \frac{1}{2}\right)$, $\left(\frac{5}{2}, 0\right)$, $(4, 3)$

Geometry   In Exercises 95 and 96, find the area of the shaded region of the figure.

95.

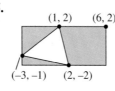

96.

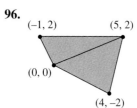

Collinear Points   In Exercises 97–100, decide whether the points are collinear.

97. $(-1, 11)$, $(0, 8)$, $(2, 2)$

98. $(-1, -1)$, $(1, 9)$, $(2, 13)$

99. $\left(-2, \frac{1}{3}\right)$, $(2, 1)$, $\left(3, \frac{1}{5}\right)$

100. $\left(0, \frac{1}{2}\right)$, $\left(1, \frac{7}{6}\right)$, $\left(9, \frac{13}{2}\right)$

Equation of a Line   In Exercises 101–104, find an equation of the line through the points.

101. $(0, 0)$, $(5, 3)$

102. $(-4, 3)$, $(2, 1)$

103. $(10, 7)$, $(-2, -7)$

104. $\left(-\frac{1}{2}, 3\right)$, $\left(\frac{5}{2}, 1\right)$

Curve-Fitting   In Exercises 105–108, use Cramer's Rule to find the equation of the parabola that passes through the points.

105. $(0, 1)$, $(1, -3)$, $(-2, 21)$

106. $(2, 3)$, $\left(-1, \frac{9}{2}\right)$, $(-2, 9)$

107. $(1, -1)$, $(-1, -5)$, $\left(\frac{1}{2}, \frac{1}{4}\right)$

108. $(-2, 6)$, $(1, 9)$, $(3, 1)$

In Exercises 109 and 110, solve the equation.

109. $\begin{vmatrix} 5 - x & 4 \\ 1 & 2 - x \end{vmatrix} = 0$

110. $\begin{vmatrix} 4 - x & -2 \\ 1 & 1 - x \end{vmatrix} = 0$

CHAPTER PROJECT: Healthcare

The table below gives models for the annual revenues of four companies in the medical supply industry from 1988 through 1993. The revenue R is listed in millions of dollars and t represents the year, with $t = 8$ corresponding to 1988. (Source: SciMed Life Systems, Nellcor, St. Jude Medical, and Diagnostic Products)

Company	Model Revenue Equation	Years
SciMed Life Systems, Inc.	$R = -371.43 + 49.31t$	$8 \le t \le 13$
Nellcor, Inc.	$R = -103.67 + 24.64t$	$8 \le t \le 13$
St. Jude Medical, Inc.	$R = -110.69 + 28.63t$	$8 \le t \le 13$
Diagnostic Products Corp.	$R = -52.13 + 12.64t$	$8 \le t \le 13$

Each of these models was created using a linear regression program on a graphing utility. Use the information in the table to answer the following questions.

1. **Graphical Interpretation**   Which of the equations in the table are represented in the graph at the left? What is the point of intersection of the lines? What does the point of intersection represent?

2. **Graphical Interpretation**   Use a graphing utility to graph the equations for Nellcor, Inc. and St. Jude Medical, Inc. for the given years. Do the equations appear parallel? What does this tell you about the two companies from 1988 through 1993?

3. **Equal Revenues**   During which year did Diagnostic Products Corp. and St. Jude Medical, Inc. have about the same revenue? Does the answer make sense? Explain your reasoning.

4. **A System with Three Equations**   Sketch the graph of the following system of equations. Is there a single point of intersection?

$$2x + 2y =\ \ 6$$
$$x + 2y =\ \ 5$$
$$-2x +\ \ y = -5$$

5. **Numerical and Graphical Reasoning**   Create a table showing the revenues for each year from 1988 to 1993 for SciMed Life Systems, Inc., Nellcor, Inc., and St. Jude Medical, Inc. Graph the three equations in the same coordinate plane. Write a paragraph relating the information in the table to the positions and intersections of the lines of the graph.

6. **Research Project**   Use your school's library or some other reference source to find the 1994 revenues for the four companies listed above. How closely did the models predict the 1994 revenues?

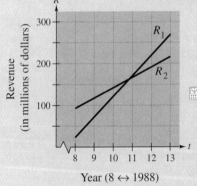

Revenue (in millions of dollars)

R_1

R_2

Year (8 ↔ 1988)

CHAPTER SUMMARY

After studying this chapter, you should have acquired the following skills. These skills are keyed to the Review Exercises that begin on page 558. Answers to odd-numbered Review Exercises are given in the back of the book.

- Solve systems of equations using the method of substitution. *(Sections 8.1, 8.2)* — **Review Exercises 1–10**

- Solve systems of equations graphically. *(Section 8.1)* — **Review Exercises 11–16**

- Solve systems of equations using a graphing utility. *(Sections 8.1, 8.2)* — **Review Exercises 17–20**

- Solve systems of linear equations using the method of elimination. *(Sections 8.2, 8.3)* — **Review Exercises 21–24**

- Solve systems of linear equations using matrices and elementary row operations. *(Section 8.4)* — **Review Exercises 25–28**

- Evaluate the determinant of a 2 × 2 matrix. *(Section 8.5)* — **Review Exercises 29, 30**

- Evaluate the determinant of a 3 × 3 matrix by expanding by minors. *(Section 8.5)* — **Review Exercises 31, 32**

- Evaluate the determinant of a 3 × 3 matrix using any appropriate method. *(Section 8.5)* — **Review Exercises 33, 34**

- Solve systems of linear equations using Cramer's Rule. *(Section 8.5)* — **Review Exercises 35–40**

- Evaluate determinants to find equations of lines and areas of triangles. *(Section 8.5)* — **Review Exercises 41–48**

- Create systems of equations having given solutions. *(Sections 8.1, 8.2)* — **Review Exercises 49, 50**

- Model real-life situations with equations and solve. *(Sections 8.1, 8.2, 8.3, 8.4)* — **Review Exercises 51–56, 61**

- Find equations of parabolas and circles passing through points. *(Sections 8.3, 8.4, 8.5)* — **Review Exercises 57–60**

REVIEW EXERCISES

In Exercises 1–10, use substitution to solve the system.

1. $2x + 3y = 1$
$x + 4y = -2$

2. $3x - 7y = 10$
$-2x + y = -14$

3. $-5x + 2y = 4$
$10x - 4y = 7$

4. $5x + 2y = 3$
$2x + 3y = 10$

5. $3x - 7y = 5$
$5x - 9y = -5$

6. $24x - 4y = 20$
$6x - y = 5$

7. $y = 5x^2$
$y = -15x - 10$

8. $y^2 = 16x$
$4x - y = -24$

9. $x^2 + y^2 = 1$
$x + y = -1$

10. $x^2 + y^2 = 100$
$x + y = 0$

In Exercises 11–16, solve the system graphically.

11. $2x - y = 0$
$-x + y = 4$

12. $x = y + 3$
$x = y + 1$

13. $\dfrac{x^2}{16} + \dfrac{y^2}{4} = 1$
$y = x + 2$

14. $\dfrac{x^2}{100} + \dfrac{y^2}{25} = 1$
$y = -x - 5$

15. $\dfrac{x^2}{25} + \dfrac{y^2}{9} = 1$
$\dfrac{x^2}{25} - \dfrac{y^2}{9} = 1$

16. $x^2 + y^2 = 16$
$-x^2 + \dfrac{y^2}{16} = 1$

In Exercises 17–20, use a graphing utility to solve the system.

17. $5x - 3y = 3$
$2x + 2y = 14$

18. $8x + 5y = 1$
$3x - 4y = 18$

19. $x^2 + y^2 = 25$
$y^2 - x^2 = 7$

20. $x^2 + y = 9$
$x^2 - y^2 = 7$

In Exercises 21–24, use elimination to solve the system.

21. $x + y = 0$
$2x + y = 0$

22. $4x + y = 1$
$x - y = 4$

23. $-x + y + 2z = 1$
$2x + 3y + z = -2$
$5x + 4y + 2z = 4$

24. $2x + 3y + z = 10$
$2x - 3y - 3z = 22$
$4x - 2y + 3z = -2$

In Exercises 25–28, use matrices and elementary row operations to solve the system.

25. $5x + 4y = 2$
$-x + y = -22$

26. $2x - 5y = 2$
$3x - 7y = 1$

27. $x + 2y + 6z = 4$
$-3x + 2y - z = -4$
$4x + 2z = 16$

28. $2x_1 + 3x_2 + 3x_3 = 3$
$6x_1 + 6x_2 + 12x_3 = 13$
$12x_1 + 9x_2 - x_3 = 2$

In Exercises 29 and 30, find the determinant of the matrix.

29. $\begin{bmatrix} 7 & 10 \\ 10 & 15 \end{bmatrix}$

30. $\begin{bmatrix} -3.4 & 1.2 \\ -5 & 2.5 \end{bmatrix}$

In Exercises 31 and 32, evaluate the determinant of the matrix six different ways by expanding by minors along each row and column.

31. $\begin{bmatrix} 8 & 6 & 3 \\ 6 & 3 & 0 \\ 3 & 0 & 2 \end{bmatrix}$

32. $\begin{bmatrix} 7 & -1 & 10 \\ -3 & 0 & -2 \\ 12 & 1 & 1 \end{bmatrix}$

In Exercises 33 and 34, find the determinant of the matrix.

33. $\begin{bmatrix} 8 & 3 & 2 \\ 1 & -2 & 4 \\ 6 & 0 & 5 \end{bmatrix}$

34. $\begin{bmatrix} 4 & 0 & 10 \\ 0 & 10 & 0 \\ 10 & 0 & 34 \end{bmatrix}$

In Exercises 35–40, solve the system of linear equations by using Cramer's Rule. (If it is not possible, state the reason.)

35. $7x + 12y = 63$
     $2x + \ 3y = 15$

36. $12x + 42y = -17$
     $30x - 18y = \ \ \ 19$

37. $\ \ 3x - 2y = \ 16$
     $12x - 8y = -5$

38. $\ \ \ 4x + 24y = \ 20$
     $-3x + 12y = -5$

39. $-x + \ y + 2z = \ \ \ 1$
     $2x + 3y + \ z = -2$
     $5x + 4y + 2z = \ \ \ 4$

40. $2x_1 + \ x_2 + 2x_3 = 4$
     $2x_1 + 2x_2 \ \ \ \ \ \ \ \ = 5$
     $2x_1 - \ x_2 + 6x_3 = 2$

In Exercises 41–44, use a determinant to find the equation of the line through the points.

41. $(-4, 0), (4, 4)$

42. $(2, 5), (6, -1)$

43. $\left(-\frac{5}{2}, 3\right), \left(\frac{7}{2}, 1\right)$

44. $(-0.8, 0.2), (0.7, 3.2)$

In Exercises 45–48, use a determinant to find the area of the triangle with the given vertices.

45. $(1, 0), (5, 0), (5, 8)$

46. $(-4, 0), (4, 0), (0, 6)$

47. $(1, 2), (4, -5), (3, 2)$

48. $\left(\frac{3}{2}, 1\right), \left(4, -\frac{1}{2}\right), (4, 2)$

In Exercises 49 and 50, create a system of equations having the given solution.

49. $\left(\frac{2}{3}, -4\right)$

50. $(-10, \ 12)$

51. *Break-Even Analysis* A small business invests $25,000 in equipment to produce a product. Each unit of the product costs $3.75 to produce and is sold for $5.25. How many items must be sold before the business breaks even?

52. *Geometry* The perimeter of a rectangle is 480 meters and its length is 150% of its width. Find the dimensions of the rectangle.

53. *Acid Mixture* One hundred gallons of a 60% acid solution is obtained by mixing a 75% solution with a 50% solution. How many gallons of each solution must be used to obtain the desired mixture?

54. *Rope Length* You must cut a rope that is 128 inches long into two pieces such that one piece is three times as long as the other. Find the length of each piece.

55. *Flying Speeds* Two planes leave Pittsburgh and Philadelphia at the same time, each going to the other city. Because of the wind, one plane flies 25 miles per hour faster than the other. Find the ground speed of each plane if the cities are 275 miles apart and the planes pass one another after 40 minutes.

56. *Investments* An inheritance of $20,000 was divided among three investments yielding $1780 in interest per year. The interest rates for the three investments were 7%, 9%, and 11%. Find the amount placed in each investment if the second and third were $3000 and $1000 less than the first, respectively.

Curve-Fitting In Exercises 57 and 58, find an equation of the parabola passing through the points.

57. $(0, -6), (1, -3), (2, 4)$

58. $(-5, 0), (1, -6), (2, 14)$

Curve-Fitting In Exercises 59 and 60, find an equation of the circle $x^2 + y^2 + Dx + Ey + F = 0$ passing through the points.

59. $(2, 2), (5, -1), (-1, -1)$

60. $(4, 2), (1, 3), (-2, -6)$

61. *Mathematical Modeling* A child throws a softball over a garage. The location of the eaves and the peak of the roof are given by $(0, 10)$, $(15, 15)$, and $(30, 10)$.

(a) Find the equation of the parabola for the path of the ball if the ball follows a path 1 foot over the eaves and the peak of the roof.

(b) Use a graphing utility to graph the path.

(c) How far from the edge of the garage is the child standing if the ball is at a height of 5 feet when it leaves the child's hand?

CHAPTER TEST

Take this test as you would take a test in class. After you are done, check your work against the answers given in the back of the book.

1. Which ordered pair is the solution of the system at the right: $(3, -4)$ or $\left(1, \frac{1}{2}\right)$?

$$2x - 2y = 1$$
$$-x + 2y = 0$$

System for 1

In Exercises 2–13, use the indicated method to solve the system.

2. *Substitution:* $\begin{aligned} 5x - y &= 6 \\ 4x - 3y &= -4 \end{aligned}$

3. *Substitution:* $\begin{aligned} x + y &= 8 \\ xy &= 12 \end{aligned}$

4. *Graphical:* $\begin{aligned} x - 2y &= -1 \\ 2x + 3y &= 12 \end{aligned}$

5. *Elimination:* $\begin{aligned} 3x - 4y &= -14 \\ -3x + y &= 8 \end{aligned}$

6. *Elimination:* $\begin{aligned} 8x + 3y &= 3 \\ 4x - 6y &= -1 \end{aligned}$

7. *Elimination:* $\begin{aligned} x + 2y - 4z &= 0 \\ 3x + y - 2z &= 5 \\ 3x - y + 2z &= 7 \end{aligned}$

8. *Matrices:* $\begin{aligned} x - 3z &= -10 \\ -2y + 2z &= 0 \\ x - 2y &= -7 \end{aligned}$

9. *Matrices:* $\begin{aligned} x - 3y + z &= -3 \\ 3x + 2y - 5z &= 18 \\ y + z &= -1 \end{aligned}$

10. *Cramer's Rule:* $\begin{aligned} 2x - 7y &= 7 \\ 3x + 7y &= 13 \end{aligned}$

11. *Graphical:* $\begin{aligned} x - 2y &= -3 \\ 2x + 3y &= 22 \end{aligned}$

12. *Any Method:* $\begin{aligned} 3x - 2y + z &= 12 \\ x - 3y &= 2 \\ -3x - 9z &= -6 \end{aligned}$

13. *Any Method:* $\begin{aligned} 4x + y + 2z &= -4 \\ 3y + z &= 8 \\ -3x + y - 3z &= 5 \end{aligned}$

14. Describe the number of possible solutions of a system of linear equations.

15. Evaluate the determinant of A, as shown at the right.

16. Find the value of a such that the system at the right is inconsistent.

17. Find a system of linear equations with integer coefficients that has the solution $(5, -3)$. (The problem has many correct answers.)

18. Two people share the driving on a 200-mile trip. One person drives four times as far as the other. Write a system of linear equations that models the problem. Find the distance each person drives.

19. Find the equation of the parabola $y = ax^2 + bx + c$ that passes through the points $(0, 4)$, $(1, 3)$, and $(2, 6)$.

20. Find the area of the triangle with vertices $(0, 0)$, $(5, 4)$, and $(6, 0)$.

$$A = \begin{bmatrix} 3 & -2 & 0 \\ -1 & 5 & 3 \\ 2 & 7 & 1 \end{bmatrix}$$

Matrix for 15

$$5x - 8y = 3$$
$$3x + ay = 0$$

System for 16

Exponential and Logarithmic Functions

9

- Exponential Functions
- Composite and Inverse Functions
- Logarithmic Functions
- Properties of Logarithms
- Solving Exponential and Logarithmic Equations
- Applications

Radiocarbon, also known as carbon 14, is a radioactive isotope of carbon. Plants and animals absorb radiocarbon into their systems while they are alive. Once the organism dies, it no longer absorbs the isotope. From that time on, the radiocarbon decays exponentially.

The *half-life* of radiocarbon, which is the amount of time it takes for half of the radiocarbon to decay, is 5700 years. By measuring the amounts of radiocarbon remaining in prehistoric plants and animals that lived up to 50,000 years ago, scientists can determine their ages.

The table and graph at the right show the time it takes for 16 grams of radiocarbon to decay to 1 gram. Notice that after 11,400 years, one-fourth of the radiocarbon is still present.

Grams	16	8	4	2	1
Years	0	5700	11,400	17,100	22,800

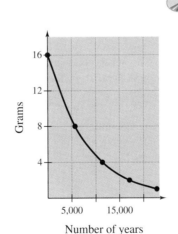

Number of years

The chapter project related to this information is on page 626.

9.1 Exponential Functions

Exponential Functions ▪ Graphs of Exponential Functions ▪
The Natural Exponential Function ▪ Compound Interest

Exponential Functions

In this section, you will study a new type of function called an **exponential function.** Whereas polynomial and rational functions have terms with variable bases and constant exponents, exponential functions have terms with *constant bases* and *variable exponents.* Here are some examples.

Polynomial or Rational Function

Constant Exponent

$$x^2, \quad x^{-3}$$

Variable Base

Exponential Function

Variable Exponent

$$2^x, \quad 3^{-x}$$

Constant Base

NOTE   The base $a = 1$ is excluded because $f(x) = 1^x = 1$ is a constant function, *not* an exponential function.

Definition of Exponential Function

The **exponential function** f with base a is denoted by

$$f(x) = a^x$$

where $a > 0$, $a \neq 1$, and x is any real number.

In Chapter 5, you learned to evaluate a^x for integer and rational values of x. For example, you know that

$$a^3 = a \cdot a \cdot a \qquad \text{and} \qquad a^{2/3} = \left(\sqrt[3]{a}\right)^2.$$

However, to evaluate a^x for any real number x, you need to interpret forms with *irrational* exponents. For the purpose of this text, it is sufficient to think of a number such as

$$a^{\sqrt{2}}$$

where $\sqrt{2} \approx 1.414214$, as the number that has the successively closer approximations

$$a^{1.4},\ a^{1.41},\ a^{1.414},\ a^{1.4142},\ a^{1.41421},\ a^{1.414214},\ \dots\ .$$

The properties of exponents that were discussed in Section 5.1 can be extended to cover exponential functions, as described on page 563.

Properties of Exponential Functions

1. $a^x \cdot a^y = a^{x+y}$ **2.** $(a^x)^y = a^{xy}$

3. $\dfrac{a^x}{a^y} = a^{x-y}$ **4.** $a^{-x} = \dfrac{1}{a^x} = \left(\dfrac{1}{a}\right)^x$

To evaluate exponential functions with a calculator, you can use the exponential key $\boxed{y^x}$ (where y is the base and x is the exponent) or $\boxed{\wedge}$. For example, to evaluate $3^{-1.3}$, you can use the following keystrokes.

Keystrokes	Display	
3 $\boxed{y^x}$ 1.3 $\boxed{+/-}$ $\boxed{=}$	0.239741	Scientific
3 $\boxed{\wedge}$ $\boxed{(}$ $\boxed{(-)}$ 1.3 $\boxed{)}$ $\boxed{\text{ENTER}}$	0.239741	Graphing

---

EXAMPLE 1   *Evaluating Exponential Functions*

Evaluate each function at the indicated values of x. Use a calculator only if it is necessary or more efficient.

Function	*Values*
a. $f(x) = 2^x$	$x = 3,\ x = -4,\ x = \pi$
b. $g(x) = 12^x$	$x = 3,\ x = -0.1,\ x = \frac{5}{7}$
c. $h(x) = (1.085)^x$	$x = 0,\ x = -3$

Solution

Evaluation	*Comment*
a. $f(3) = 2^3 = 8$	Calculator is not necessary.
$f(-4) = 2^{-4} = \dfrac{1}{2^4} = \dfrac{1}{16}$	Calculator is not necessary.
$f(\pi) = 2^{\pi} \approx 8.825$	Calculator is necessary.
b. $g(3) = 12^3 = 1728$	Calculator is more efficient.
$g(-0.1) = 12^{-0.1} \approx 0.7800$	Calculator is necessary.
$g\left(\dfrac{5}{7}\right) = 12^{5/7} \approx 5.900$	Calculator is necessary.
c. $h(0) = (1.085)^0 = 1$	Calculator is not necessary.
$h(-3) = (1.085)^{-3} \approx 0.7829$	Calculator is more efficient.

Graphs of Exponential Functions

The basic nature of the graph of an exponential function can be determined by the point-plotting method or by using a graphing utility.

EXAMPLE 2 The Graphs of Exponential Functions

In the same coordinate plane, sketch the graphs of the following functions. Determine the domains and ranges.

a. $f(x) = 2^x$ **b.** $g(x) = 4^x$

Solution

The table lists some values of each function, and Figure 9.1 shows their graphs. From the graphs, you can see that the domain of each function is the set of all real numbers and that the range of each function is the set of all positive real numbers.

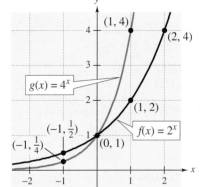

FIGURE 9.1

x	-2	-1	0	1	2	3
2^x	$\frac{1}{4}$	$\frac{1}{2}$	1	2	4	8
4^x	$\frac{1}{16}$	$\frac{1}{4}$	1	4	16	64

You know from your study of functions in Chapter 2 that the graph of $h(x) = f(-x) = 2^{-x}$ is a reflection of the graph of $f(x) = 2^x$ in the y-axis. This is reinforced in the following example.

EXAMPLE 3 The Graphs of Exponential Functions

In the same coordinate plane, sketch the graph of each function.

a. $f(x) = 2^{-x}$ **b.** $g(x) = 4^{-x}$

Solution

The table lists some values of each function, and Figure 9.2 shows their graphs.

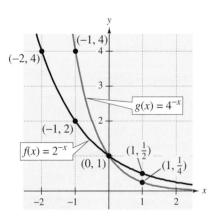

FIGURE 9.2

x	-3	-2	-1	0	1	2
2^{-x}	8	4	2	1	$\frac{1}{2}$	$\frac{1}{4}$
4^{-x}	64	16	4	1	$\frac{1}{4}$	$\frac{1}{16}$

Examples 2 and 3 suggest that for $a > 1$, the graph of $y = a^x$ increases and the graph of $y = a^{-x}$ decreases. The graphs shown in Figure 9.3 are typical of the graphs of exponential functions. Note that each has a y-intercept at $(0, 1)$.

Graph of $y = a^x$

- Domain: $(-\infty, \infty)$
- Range: $(0, \infty)$
- Intercept: $(0, 1)$
- Increasing

Graph of $y = a^{-x}$

- Domain: $(-\infty, \infty)$
- Range: $(0, \infty)$
- Intercept: $(0, 1)$
- Decreasing

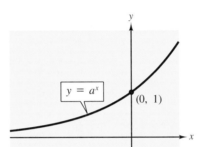

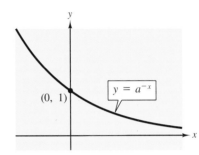

FIGURE 9.3   Characteristics of the Exponential Functions a^x and $a^{-x} (a > 1)$

EXAMPLE 4 An Application: Radioactive Decay

Let y represent the mass of a particular radioactive element whose half-life is 25 years. The initial mass is 10 grams. After t years, the mass (in grams) is given by

$$y = 10 \left(\tfrac{1}{2} \right)^{t/25}, \quad t \geq 0.$$

How much of the initial mass remains after 120 years?

Solution

When $t = 120$, the mass is given by

$$y = 10 \left(\tfrac{1}{2} \right)^{120/25} \qquad \text{Substitute 120 for } t.$$

$$= 10 \left(\tfrac{1}{2} \right)^{4.8} \qquad \text{Simplify.}$$

$$\approx 0.359 \text{ gram.} \qquad \text{Use a calculator.}$$

Thus, after 120 years, the mass has decayed from an initial amount of 10 grams to only 0.359 gram. Note in Figure 9.4 that the graph of the function shows the 25-year half-life. That is, after 25 years the mass is 5 grams (half of the original), after another 25 years the mass is 2.5 grams, and so on.

FIGURE 9.4

The Natural Exponential Function

So far, we have used integers or rational numbers as bases of exponential functions. In many applications of exponential functions, the convenient choice for a base is the following irrational number, denoted by the letter "*e*."

$$e \approx 2.71828\dots\dots \qquad \text{Natural base}$$

This number is called the **natural base.** The function

$$f(x) = e^x \qquad \text{Natural exponential function}$$

is called the **natural exponential function.** Be sure you understand that for this function, *e* is the constant number 2.71828 . . . , and *x* is a variable. To evaluate the natural exponential function, you need a calculator, preferably one having a natural exponential key $\boxed{e^x}$. Here are some examples of how to use such a calculator to evaluate the natural exponential function.

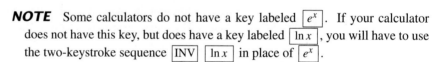

Value	Keystrokes	Display	
e^2	2 $\boxed{e^x}$	7.3890561	Scientific
e^2	$\boxed{e^x}$ 2 $\boxed{\text{ENTER}}$	7.3890561	Graphing
e^{-3}	3 $\boxed{+/-}$ $\boxed{e^x}$	0.049787	Scientific
e^{-3}	$\boxed{e^x}$ $\boxed{(}$ $\boxed{(-)}$ 3 $\boxed{)}$ $\boxed{\text{ENTER}}$	0.049787	Graphing
$e^{0.32}$	.32 $\boxed{e^x}$	1.3771278	Scientific
$e^{0.32}$	$\boxed{e^x}$ .32 $\boxed{\text{ENTER}}$	1.3771278	Graphing

NOTE   Some calculators do not have a key labeled $\boxed{e^x}$. If your calculator does not have this key, but does have a key labeled $\boxed{\ln x}$, you will have to use the two-keystroke sequence $\boxed{\text{INV}}$ $\boxed{\ln x}$ in place of $\boxed{e^x}$.

After evaluating the natural exponential function at several values, as shown in the table, you can sketch its graph, as shown in Figure 9.5.

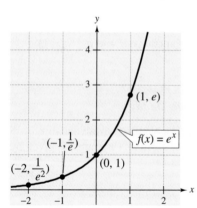

FIGURE 9.5

x	-1.5	-1.0	-0.5	0.0	0.5	1.0	1.5
$f(x) = e^x$	0.223	0.368	0.607	1.000	1.649	2.718	4.482

From the graph, notice the following properties of the natural exponential function.

- The domain is the set of all real numbers.
- The range is the set of positive real numbers.
- The *y*-intercept is (0, 1).

Compound Interest

One of the most familiar uses of exponential functions involves **compound interest.** Suppose a principal P is invested at an annual interest rate r (in decimal form), compounded once a year. If the interest is added to the principal at the end of the year, the balance is

$$A = P + Pr = P(1 + r).$$

This pattern of multiplying the previous principal by $(1 + r)$ is then repeated each successive year, as shown below.

Time in Years	Balance at Given Time
0	$A = P$
1	$A = P(1 + r)$
2	$A = P(1 + r)(1 + r) = P(1 + r)^2$
3	$A = P(1 + r)^2(1 + r) = P(1 + r)^3$
$\vdots$	$\vdots$
t	$A = P(1 + r)^t$

To account for more frequent compounding of interest (such as quarterly or monthly compounding), let n be the number of compoundings per year and let t be the number of years. Then the rate per compounding is r/n and the account balance after t years is

$$A = P\left(1 + \frac{r}{n}\right)^{nt}.$$

EXAMPLE 5   *Finding the Balance for Compound Interest*

A sum of $10,000 is invested at an annual interest rate of 7.5%, compounded monthly. Find the balance in the account after 10 years.

Solution

Using the formula for compound interest, with $P = 10,000$, $r = 0.075$, $n = 12$ (for monthly compounding), and $t = 10$, you obtain the following balance.

$$A = 10,000\left(1 + \frac{0.075}{12}\right)^{12(10)} \approx \$21,120.65$$

A second method that banks use to compute interest is called **continuous compounding.** The formula for the balance for this type of compounding is

$$A = Pe^{rt}.$$

The formulas for both types of compounding are summarized on page 568.

Formulas for Compound Interest

After t years, the balance A in an account with principal P and annual interest rate r (in decimal form) is given by the following formulas.

1. For n compoundings per year: $A = P\left(1 + \dfrac{r}{n}\right)^{nt}$

2. For continuous compounding: $A = Pe^{rt}$

EXAMPLE 6   Comparing Two Types of Compounding

A total of $15,000 is invested at an annual interest rate of 8%. Find the balance after 6 years if the interest is compounded

a. quarterly   and   **b.** continuously.

Solution

a. Letting $P = 15{,}000$, $r = 0.08$, $n = 4$, and $t = 6$, the balance after 6 years at quarterly compounding is

$$A = 15{,}000\left(1 + \frac{0.08}{4}\right)^{4(6)}$$

$$= \$24{,}126.56.$$

NOTE   Example 6 illustrates the following general rule. For a given principal, interest rate, and time, the more often the interest is compounded per year, the greater the balance will be. Moreover, the balance obtained by continuous compounding is larger than the balance obtained by compounding n times per year.

b. Letting $P = 15{,}000$, $r = 0.08$, and $t = 6$, the balance after 6 years at continuous compounding is

$$A = 15{,}000e^{0.08(6)}$$

$$= \$24{,}241.12.$$

Note that the balance is greater with continuous compounding than with quarterly compounding.

Group Activities   Exploring with Technology

Finding a Pattern   Use a graphing utility to investigate the function $f(x) = k^x$ for different values of k. Discuss the effect that k has on the shape of the graph.

9.1 Exercises

Discussing the Concepts

1. Describe some differences between exponential functions and polynomial or rational functions.

2. Explain why 1^x is not an exponential function.

3. Compare the graphs of $f(x) = 3^x$ and $g(x) = \left(\frac{1}{3}\right)^x$.

4. Describe some applications of the exponential functions $f(x) = a^x$ and $g(x) = a^{-x}$.

5. *True or False?* $e = 271,801/99,990$. Explain.

6. Without using a calculator, how do you know that $2^{\sqrt{2}}$ is greater than 2, but less than 4?

Problem Solving

In Exercises 7–10, simplify the expression.

7. $2^x \cdot 2^{x-1}$

8. $\dfrac{3^{2x+3}}{3^{x+1}}$

9. $(2e^x)^3$

10. $\sqrt{4e^{6x}}$

In Exercises 11–14, evaluate the function as indicated. Use a calculator only if it is necessary or more efficient. (Round to three decimal places.)

11. $f(x) = 3^x$
 (a) $x = -2$
 (b) $x = 0$
 (c) $x = 1$

12. $F(x) = 3^{-x}$
 (a) $x = -2$
 (b) $x = 0$
 (c) $x = 1$

13. $g(x) = 10e^{-0.5x}$
 (a) $x = -4$
 (b) $x = 4$
 (c) $x = 8$

14. $f(z) = \dfrac{100}{1 + e^{-0.05z}}$
 (a) $z = 0$
 (b) $z = 10$
 (c) $z = 20$

In Exercises 15–20, match the function with its graph. [The graphs are labeled (a), (b), (c), (d), (e), and (f).]

(a)

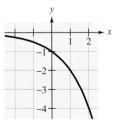

(b)

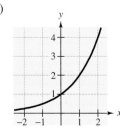

(c)

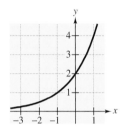

(d)

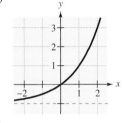

(e)

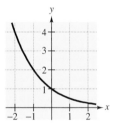

(f)

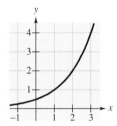

15. $f(x) = 2^x$

16. $f(x) = -2^x$

17. $f(x) = 2^{-x}$

18. $f(x) = 2^x - 1$

19. $f(x) = 2^{x-1}$

20. $f(x) = 2^{x+1}$

In Exercises 21–28, sketch the graph of the function.

21. $f(x) = 3^x$

22. $f(x) = 3^{-x} = \left(\frac{1}{3}\right)^x$

23. $g(x) = 3^x - 2$

24. $g(x) = 3^x + 1$

25. $f(t) = 2^{-t^2}$

26. $f(t) = 2^{t^2}$

27. $f(x) = -2^{0.5x}$

28. $h(t) = -2^{-0.5t}$

In Exercises 29–34, use a graphing utility to graph the function.

29. $y = 5^{x/3}$

30. $y = 5^{(x-2)/3}$

31. $f(x) = e^{0.2x}$

32. $f(x) = e^{-0.2x}$

33. $P(t) = 100e^{-0.1t}$

34. $A(t) = 1000e^{0.08t}$

35. *Depreciation*   After t years, the value of a car that cost \$16,000 is given by

$$V(t) = 16,000 \left(\tfrac{3}{4}\right)^t.$$

Sketch a graph of the function and determine the value of the car 2 years after it was purchased.

36. *Depreciation*   Suppose straight-line depreciation is used to determine the value of the car in Exercise 35. If the car depreciates \$3000 per year, the model for its value after t years is given by

$$V(t) = 16,000 - 3000t.$$

(a) Sketch the graph of this line on the same coordinate axes used for the graph in Exercise 35.

(b) If you were selling the car after owning it 2 years, which depreciation model would you prefer?

(c) If you sell the car after 4 years, which model would be to your advantage?

Creating a Table   In Exercises 37 and 38, complete the table for P dollars invested at rate r for t years and compounded n times per year.

n	1	4	12	365	Continuous
A					

	Principal	Rate	Time
37.	$P = \$100$	$r = 8\%$	$t = 20$ years
38.	$P = \$400$	$r = 8\%$	$t = 50$ years

Creating a Table   In Exercises 39 and 40, complete the table to find the principal P that yields a balance of A dollars when invested at rate r for t years, compounded n times per year.

n	1	4	12	365	Continuous
P					

	Balance	Rate	Time
39.	$A = \$5000$	$r = 7\%$	$t = 10$ years
40.	$A = \$100,000$	$r = 9\%$	$t = 20$ years

41. *Graphical Interpretation*   An investment of \$500 in two different accounts with respective interest rates of 6% and 8% is compounded continuously. The balances in the accounts after t years are modeled by

$$A_1 = 500e^{0.06t} \quad \text{and} \quad A_2 = 500e^{0.08t}.$$

(a) Use a graphing utility to graph each of the models.

(b) Use a graphing utility to graph the function $A_2 - A_1$ in the same window as the graphs of part (a).

(c) Use the graphs to discuss the rates of increase of the balances in the two accounts.

42. *Graphical Estimation*   From 1970 through 1991, the number of recreational boats B (in millions) in the United States can be modeled by

$$B = 6.10(1.04)^t + 2.62$$

where $t = 0$ represents 1970. (Source: National Marine Manufacturers Association)

(a) Use a graphing utility to graph the model.

(b) Approximate the number of recreational boats owned in the United States in 1990.

Reviewing the Major Concepts

In Exercises 43–46, simplify the expression.

43. $(4x + 3y) - 3(5x + y)$

44. $(-15u + 4v) + 5(3u - 9v)$

45. $2x^2 + (2x - 3)^2 + 12x$

46. $y^2 - (y + 2)^2 + 4y$

47. *Geometry*   A circle has a circumference of 1 inch.

(a) Find the radius of the circle.

(b) Find the area of the circle.

48. *Geometry*   A square has a perimeter of 1 inch. What is its area?

Additional Problem Solving

In Exercises 49–52, evaluate the expression. (Round to three decimal places.)

49. $4^{\sqrt{3}}$

50. $6^{-\pi}$

51. $e^{1/3}$

52. $e^{-1/3}$

In Exercises 53–60, evaluate the function as indicated. Use a calculator only if it is necessary or more efficient. (Round to three decimal places.)

53. $g(x) = 5^x$
  (a) $x = -1$
  (b) $x = 1$
  (c) $x = 3$

54. $G(x) = 5^{-x}$
  (a) $x = -1$
  (b) $x = 1$
  (c) $x = \sqrt{3}$

55. $f(t) = 500 \left(\frac{1}{2}\right)^t$
  (a) $t = 0$
  (b) $t = 1$
  (c) $t = \pi$

56. $g(s) = 1200 \left(\frac{2}{3}\right)^s$
  (a) $s = 0$
  (b) $s = 2$
  (c) $s = 4$

57. $f(x) = 1000(1.05)^{2x}$
  (a) $x = 0$
  (b) $x = 5$
  (c) $x = 10$

58. $P(t) = \dfrac{10,000}{(1.01)^{12t}}$
  (a) $t = 2$
  (b) $t = 10$
  (c) $t = 20$

59. $f(x) = e^x$
  (a) $x = -1$
  (b) $x = 0$
  (c) $x = \frac{1}{2}$

60. $A(t) = 200e^{0.1t}$
  (a) $t = 10$
  (b) $t = 20$
  (c) $t = 40$

In Exercises 61–66, match the function with its graph. [The graphs are labeled (a), (b), (c), (d), (e), and (f).]

(a)

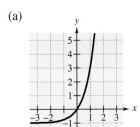

(b)

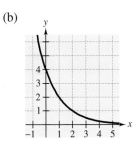

(c)

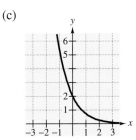

(d)

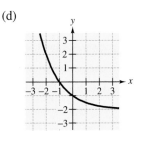

(e)

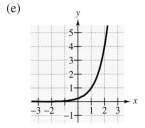

(f)

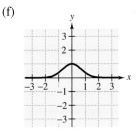

61. $f(x) = \left(\frac{1}{2}\right)^x - 2$

62. $f(x) = \left(\frac{1}{2}\right)^{x-2}$

63. $f(x) = 4^x - 1$

64. $f(x) = 4^{x-1}$

65. $f(x) = e^{-x^2}$

66. $f(x) = 2e^{-x}$

In Exercises 67–78, sketch the graph of the function.

67. $h(x) = \frac{1}{2}(3^x)$

68. $h(x) = \frac{1}{2}(3^{-x})$

69. $h(x) = 2^{0.5x}$

70. $g(t) = 2^{-0.5t}$

71. $f(x) = 4^{x-5}$

72. $f(x) = 4^{x+1}$

73. $g(x) = 4^x - 5$

74. $g(x) = 4^x + 1$

75. $f(x) = -\left(\frac{1}{3}\right)^x$

76. $f(x) = \left(\frac{3}{4}\right)^x + 1$

77. $g(t) = 200 \left(\frac{1}{2}\right)^t$

78. $h(y) = 27 \left(\frac{2}{3}\right)^y$

In Exercises 79–88, use a graphing utility to graph the function.

79. $y = 3^{-x/2}$

80. $y = 3^{-x/2} + 2$

81. $y = 5^{-x/3}$

82. $y = 5^{-x/3} + 2$

83. $y = 500(1.06)^t$

84. $y = 100(1.06)^{-t}$

85. $y = 3e^{0.2x}$

86. $y = 50e^{-0.05x}$

87. $y = 6e^{-x^2/3}$

88. $g(t) = \dfrac{10}{1 + e^{-0.5t}}$

89. *Population Growth* The population of the United States (in recent years) can be approximated by the exponential function

$$P(t) = 203(1.0118)^{t-1970}$$

where t is the year and P is the population in millions. Use the model to approximate the population in the years (a) 1995 and (b) 2000.

90. *Property Value* Suppose the value of a piece of property doubles every 15 years. If you buy the property for $64,000, its value t years after the date of purchase should be

$$V(t) = 64,000(2)^{t/15}.$$

Use the model to approximate the value of the property (a) 5 years and (b) 20 years after it is purchased.

91. *Inflation Rate* Suppose the annual rate of inflation averages 5% over the next 10 years. With this rate of inflation, the approximate cost C of goods or services during any year in that decade will be given by

$$C(t) = P(1.05)^t, \quad 0 \le t \le 10$$

where t is time in years and P is the present cost. If the price of an oil change for your car is presently $19.95, estimate the price 10 years from now.

92. *Price and Demand* The daily demand x and price p for a certain product are related by

$$p = 25 - 0.4e^{0.02x}.$$

Find the prices for demands of (a) $x = 100$ units and (b) $x = 125$ units.

Creating a Table In Exercises 93–96, complete the table for P dollars invested at rate r for t years, compounded n times per year.

n	1	4	12	365	Continuous
A					

	Principal	Rate	Time
93.	$P = \$2000$	$r = 9\%$	$t = 10$ years
94.	$P = \$1500$	$r = 7\%$	$t = 2$ years

	Principal	Rate	Time
95.	$P = \$5000$	$r = 10\%$	$t = 40$ years
96.	$P = \$10,000$	$r = 9.5\%$	$t = 30$ years

Creating a Table In Exercises 97 and 98, complete the table to determine the principal P that will yield a balance of A dollars when invested at rate r for t years, compounded n times per year.

n	1	4	12	365	Continuous
P					

	Balance	Rate	Time
97.	$A = \$1,000,000$	$r = 10.5\%$	$t = 40$ years
98.	$A = \$2500$	$r = 7.5\%$	$t = 2$ years

99. On the same set of coordinate axes, sketch the graph of the following functions. Which functions are exponential?

(a) $f(x) = 2x$ (b) $f(x) = 2x^2$
(c) $f(x) = 2^x$ (d) $f(x) = 2^{-x}$

100. *Savings Plan* Suppose you decide to start saving pennies according to the following pattern. You save 1 penny the first day, 2 pennies the second day, 4 the third day, 8 the fourth day, and so on. Each day you save twice the number of pennies as the previous day. Which function in Exercise 99 models this problem, and how many pennies do you save on the thirtieth day? (In the next chapter you will learn how to find the total number saved.)

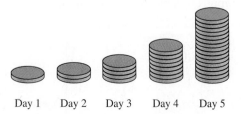

Day 1   Day 2   Day 3   Day 4   Day 5

101. *Graphical Estimation*   A parachutist jumps from a plane and opens the parachute at a height of 2000 feet (see figure). The height of the parachutist is

$$h = 1950 + 50e^{-1.6t} - 20t$$

where h is the height in feet and t is the time in seconds. (The time $t = 0$ corresponds to the time when the parachute is opened.)

(a) Use a graphing utility to graph the function.

(b) Find the height of the parachutist when $t = 0$, 25, 50, and 75.

(c) Approximate the time when the parachutist will reach the ground.

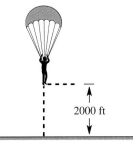

2000 ft

102. *Reading a Graph*   The popularity of 8-track cartridges increased until about 1980. From 1980 through 1985, the number n (in millions) of 8-track cartridges sold each year can be modeled by

$$n = 88\left(\frac{10}{21}\right)^t$$

where $t = 0$ corresponds to 1980 (see figure). How many were sold in 1985? (Source: Recording Industry Association of America)

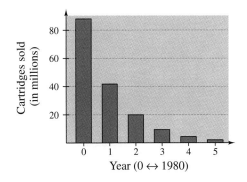

Year (0 ↔ 1980)

103. *Creating a Table*   The median price of a home in the United States for the years 1987 through 1992 is given in the following table. (Source: Chicago Title Insurance Company)

Year	1987	1988	1989
Price	$99,260	$121,920	$129,800

Year	1990	1991	1992
Price	$131,200	$134,300	$141,000

A model for this data is given by

$$y = 131{,}368e^{0.0102t^3}$$

where t is time in years, with $t = 0$ representing 1990.

(a) Use the model to complete the table and compare the results with the actual data.

Year	1987	1988	1989	1990	1991	1992
Price						

(b) Use a graphing utility to graph the model.

(c) If the model were used to predict home prices in the years ahead, would the predictions be increasing at a higher rate or a lower rate? Do you think the model would be reliable for predicting the future prices of homes? Explain.

104. *Identifying Graphs*   Identify the graphs of $y_1 = e^{0.2x}$, $y_2 = e^{0.5x}$, and $y_3 = e^x$ in the accompanying figure. Describe the effect on the graph of $y = e^{kx}$ when $k > 0$ is changed.

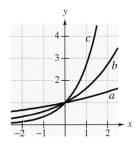

9.2	**Composite and Inverse Functions**
	Composition of Functions ▪ Inverse and One-to-One Functions ▪ Finding Inverse Functions ▪ Graphs of Inverse Functions

Composition of Functions

Two functions can be combined to form another function called the **composition** of the two functions. For instance, if $f(x) = 2x^2$ and $g(x) = x - 1$, the composition of f with g is given by

$$f(g(x)) = f(x - 1) = 2(x - 1)^2.$$

This composition is denoted by $f \circ g$.

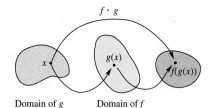

$f \circ g$

Domain of g Domain of f

FIGURE 9.6

Definition of Composition of Two Functions

The **composition** of the functions f and g is given by

$$(f \circ g)(x) = f(g(x)).$$

The domain of the **composite function** $(f \circ g)$ is the set of all x in the domain of g such that $g(x)$ is in the domain of f. (See Figure 9.6.)

EXAMPLE 1   *Forming the Composition of Two Functions*

Find the composition of f with g. Evaluate the composite function when $x = 1$.

$$f(x) = 2x + 4 \quad \text{and} \quad g(x) = 3x - 1$$

Solution

The composition of f with g is given by

$$\begin{aligned}
(f \circ g)(x) &= f(g(x)) \\
&= f(3x - 1) \\
&= 2(3x - 1) + 4 \\
&= 6x - 2 + 4 \\
&= 6x + 2.
\end{aligned}$$

When $x = 1$, the value of this function is

$$(f \circ g)(1) = 6(1) + 2 = 8.$$

The composition of f with g is generally *not* the same as the composition of g with f. This is illustrated in Example 2.

EXAMPLE 2   Comparing the Compositions of Functions

Given $f(x) = 2x - 3$ and $g(x) = x^2 + 1$, find each of the following.

a. $(f \circ g)(x)$ **b.** $(g \circ f)(x)$

Solution

a. The composition of f with g is as follows.

$$
\begin{aligned}
(f \circ g)(x) &= f(g(x)) && \text{Definition of } f \circ g \\
&= f(x^2 + 1) && \text{Definition of } g(x) \\
&= 2(x^2 + 1) - 3 && \text{Definition of } f(x) \\
&= 2x^2 + 2 - 3 && \text{Simplify.} \\
&= 2x^2 - 1 && \text{Simplify.}
\end{aligned}
$$

b. The composition of g with f is as follows.

$$
\begin{aligned}
(g \circ f)(x) &= g(f(x)) && \text{Definition of } g \circ f \\
&= g(2x - 3) && \text{Definition of } f(x) \\
&= (2x - 3)^2 + 1 && \text{Definition of } g(x) \\
&= 4x^2 - 12x + 9 + 1 && \text{Simplify.} \\
&= 4x^2 - 12x + 10 && \text{Simplify.}
\end{aligned}
$$

Note that $(f \circ g)(x) \neq (g \circ f)(x)$.

EXAMPLE 3   Finding the Domain of a Composite Function

Find the domain of the composition of $f(x) = x^2$ with $g(x) = \sqrt{x}$.

Solution

The composition of f with g is given by

$$
\begin{aligned}
(f \circ g)(x) &= f(g(x)) \\
&= f\left(\sqrt{x}\right) \\
&= \left(\sqrt{x}\right)^2 \\
&= x, \qquad x \geq 0.
\end{aligned}
$$

The domain of g consists of all nonnegative real numbers, $[0, \infty)$. This implies that the domain of the composition of f with g is this same set. That is, the domain of $f \circ g$ is $[0, \infty)$.

Inverse and One-to-One Functions

In Section 2.5, you learned that a function can be represented by a set of ordered pairs. For instance, the function $f(x) = x + 2$ from the set $A = \{1, 2, 3, 4\}$ to the set $B = \{3, 4, 5, 6\}$ can be written as follows.

$$f(x) = x + 2: \quad \{(1, 3), (2, 4), (3, 5), (4, 6)\}$$

By interchanging the first and second coordinates of each of these ordered pairs, you can form another function that is called the **inverse function** of f. The inverse function is denoted by f^{-1}. It is a function from the set B to the set A, and can be written as follows.

$$f^{-1}(x) = x - 2: \quad \{(3, 1), (4, 2), (5, 3), (6, 4)\}$$

Interchanging the ordered pairs for a function f will only produce another function when f is **one-to-one.** A function f is **one-to-one** if each value of the dependent variable corresponds to exactly one value of the independent variable.

Horizontal Line Test for Inverse Functions

A function f has an inverse function if and only if f is one-to-one. Graphically, a function f has an inverse function if and only if no *horizontal* line intersects the graph of f at more than one point.

EXAMPLE 4   *Applying the Horizontal Line Test*

a. The function $f(x) = x^3 - 1$ has an inverse because no horizontal line intersects its graph at more than one point, as shown in Figure 9.7.

b. The function $f(x) = x^2 - 1$ does not have an inverse because it is possible to find a horizontal line that intersects the graph of f at more than one point, as shown in Figure 9.8.

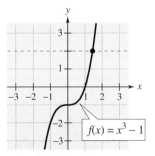

FIGURE 9.7

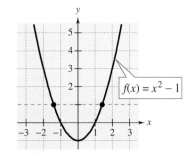

FIGURE 9.8

Definition of the Inverse of a Function

Let f and g be two functions such that

$$f(g(x)) = x \qquad \text{for every } x \text{ in the domain of } g$$

and

$$g(f(x)) = x \qquad \text{for every } x \text{ in the domain of } f.$$

The function g is the **inverse** of the function f, and is denoted by f^{-1} (read "f-inverse"). Thus, $f(f^{-1}(x)) = x$ and $f^{-1}(f(x)) = x$. The domain of f is equal to the range of f^{-1}, and vice versa.

If the function g is the inverse of the function f, it must also be true that the function f is the inverse of the function g. For this reason, you can refer to the functions f and g as being *inverses of each other*.

EXAMPLE 5   Verifying Inverse Functions

Show that each function is an inverse of the other.

$$f(x) = x^3 + 1 \qquad \text{and} \qquad g(x) = \sqrt[3]{x - 1}$$

Solution

Begin by noting that the domain and range of both functions is the entire set of real numbers. To show that f and g are inverses of each other, you need to show that $f(g(x)) = x$ and $g(f(x)) = x$, as follows.

$$
\begin{aligned}
f(g(x)) &= f\left(\sqrt[3]{x - 1}\right) \\
&= \left(\sqrt[3]{x - 1}\right)^3 + 1 \\
&= (x - 1) + 1 \\
&= x \\
g(f(x)) &= g(x^3 + 1) \\
&= \sqrt[3]{(x^3 + 1) - 1} \\
&= \sqrt[3]{x^3} \\
&= x
\end{aligned}
$$

Note that the two functions f and g "undo" each other in the following verbal sense. The function f first cubes the input x and then adds 1, whereas the function g first subtracts 1, and then takes the cube root of the result.

You can use a graphing utility to determine if two functions are inverses of each other. Graph each function and its inverse in the same viewing rectangle.

Example 5

$$f(x) = x^3 + 1$$

$$f^{-1}(x) = \sqrt[3]{x - 1}$$

Example 6

$$f(x) = 2x + 3$$

$$f^{-1}(x) = \frac{x - 3}{2}$$

Use the TRACE feature to find several solutions of the functions. Can you discover a rule for determining whether two functions are inverses of each other by studying their graphs? Algebraically find the inverse of $f(x) = \frac{1}{4}(1 - x)$. Graph the two functions using a graphing utility to check your answer.

Finding Inverse Functions

You can find the inverse of a simple function by inspection. For instance, the inverse of $f(x) = 10x$ is $f^{-1}(x) = x/10$. For more complicated functions, however, it is best to use the following steps for finding the inverse of a function. The key step in these guidelines is switching the roles of x and y. This step corresponds to the fact that inverse functions have ordered pairs with the coordinates reversed.

Finding the Inverse of a Function

1. In the equation for $f(x)$, replace $f(x)$ by y.

2. Interchange the roles of x and y.

3. If the new equation does not represent y as a function of x, the function f does not have an inverse function. If the new equation does represent y as a function of x, solve the new equation for y.

4. Replace y by $f^{-1}(x)$.

EXAMPLE 6 Finding the Inverse of a Function

Find the inverse of $f(x) = 2x + 3$.

Solution

$$f(x) = 2x + 3 \qquad \text{Original function}$$

$$y = 2x + 3 \qquad \text{Replace } f(x) \text{ by } y.$$

$$x = 2y + 3 \qquad \text{Interchange } x \text{ and } y.$$

$$y = \frac{x - 3}{2} \qquad \text{Solve for } y.$$

$$f^{-1}(x) = \frac{x - 3}{2} \qquad \text{Replace } y \text{ by } f^{-1}(x).$$

Thus, the inverse of $f(x) = 2x + 3$ is

$$f^{-1}(x) = \frac{x - 3}{2}.$$

NOTE Note in Step 3 of the guidelines for finding the inverse of a function that it is possible that a function has no inverse. For instance, the function $f(x) = x^2$ has no inverse.

Graphs of Inverse Functions

The graphs of a function f and its inverse f^{-1} are related to each other in the following way. If the point (a, b) lies on the graph of f, the point (b, a) must lie on the graph of f^{-1}, and vice versa. This means that the graph of f^{-1} is a reflection of the graph of f in the line $y = x$, as shown in Figure 9.9.

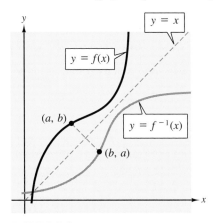

FIGURE 9.9

EXAMPLE 7 The Graphs of f and f^{-1}

Sketch the graphs of the inverse functions

$$f(x) = 2x - 3 \qquad \text{and} \qquad f^{-1}(x) = \frac{1}{2}(x + 3)$$

on the same rectangular coordinate system and show that the graphs are reflections of each other in the line $y = x$.

Solution

The graphs of f and f^{-1} are shown in Figure 9.10. Visually, it appears that the graphs are reflections of each other. You can further verify this by testing a few points on each graph. Note in the following list that if the point (a, b) is on the graph of f, the point (b, a) is on the graph of f^{-1}.

$f(x) = 2x - 3$	$f^{-1}(x) = \frac{1}{2}(x + 3)$
$(-1, -5)$	$(-5, -1)$
$(0, -3)$	$(-3, 0)$
$(1, -1)$	$(-1, 1)$
$(2, 1)$	$(1, 2)$
$(3, 3)$	$(3, 3)$

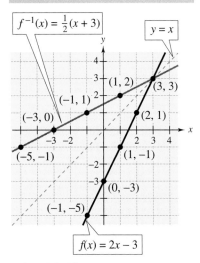

FIGURE 9.10

At the bottom of page 578, we mentioned that the function $f(x) = x^2$ has no inverse. What we mean is that *assuming the domain of* f *is the entire real line*, the function $f(x) = x^2$ has no inverse. If, however, you restrict the domain of f to the nonnegative real numbers, f does have an inverse.

EXAMPLE 8 The Graphs of f and f⁻¹

Graphically show that each function is an inverse of the other.

$$f(x) = x^2, \quad x \geq 0, \quad \text{and} \quad f^{-1}(x) = \sqrt{x}$$

Solution

The graphs of f and f^{-1} are shown in Figure 9.11. Visually, it appears that the graphs are reflections of each other in the line $y = x$. You can further verify this by testing a few points on each graph. Note in the following list that if the point (a, b) is on the graph of f, the point (b, a) is on the graph of f^{-1}.

$f(x) = x^2, \quad x \geq 0$	$f^{-1}(x) = \sqrt{x}$
$(0, 0)$	$(0, 0)$
$(1, 1)$	$(1, 1)$
$(2, 4)$	$(4, 2)$
$(3, 9)$	$(9, 3)$

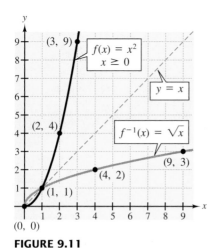

FIGURE 9.11

Group Activities   You Be the Instructor

Error Analysis   Suppose you are an algebra instructor and one of your students hands in the following solutions. Find and correct the errors and discuss how you can help your student avoid such errors in the future.

a. If $f(x) = 2x - 1$ and $g(x) = x^3 + 1$, find $(f \circ g)(2)$.

$$(f \circ g)(2) = (2 \cdot 2 - 1)(2^3 + 1)$$
$$= (4 - 1)(8 + 1)$$
$$= (3)(9)$$
$$= 27$$

b. If $f(x) = 3x^2 + x$ and $g(x) = x - 2$, find $(f \circ g)(1)$.

$$(f \circ g)(1) = f(1) - 2$$
$$= [3(1)^2 + 1] - 2$$
$$= (3 + 1) - 2$$
$$= 2$$

9.2 Exercises

Discussing the Concepts

1. Give an example showing that the composite functions $(f \circ g)(x)$ and $(g \circ f)(x)$ are not necessarily the same.

2. Describe how to find the inverse function of a function given by a set of ordered pairs. Give an example.

3. Describe how to find the inverse function of a function given by an equation in x and y. Give an example.

4. Give an example of a function that does not have an inverse function.

5. Explain the Horizontal Line Test. What is the relationship between this test and a function being one-to-one?

6. Describe the relationship between the graph of a function and its inverse.

Problem Solving

In Exercises 7–10, find the indicated composites.

7. $f(x) = x - 3$, $g(x) = x^2$

(a) $(f \circ g)(4)$ (b) $(g \circ f)(7)$

(c) $(f \circ g)(x)$ (d) $(g \circ f)(x)$

8. $f(x) = |x|$, $g(x) = 2x + 5$

(a) $(f \circ g)(-2)$ (b) $(g \circ f)(-4)$

(c) $(f \circ g)(x)$ (d) $(g \circ f)(x)$

9. $f(x) = \sqrt{x}$, $g(x) = x + 5$

(a) $(f \circ g)(4)$ (b) $(g \circ f)(9)$

(c) $(f \circ g)(x)$ (d) $(g \circ f)(x)$

10. $f(x) = \dfrac{4}{x^2 - 4}$, $g(x) = \dfrac{1}{x}$

(a) $(f \circ g)(-2)$ (b) $(g \circ f)(1)$

(c) $(f \circ g)(x)$ (d) $(g \circ f)(x)$

In Exercises 11–14, use the functions f and g to find the indicated values.

$f = \{(-2, 3), \ (-1, 1), \ (0, 0), \ (1, -1), \ (2, -3)\}$,

$g = \{(-3, 1), \ (-1, -2), \ (0, 2), \ (2, 2), \ (3, 1)\}$

11. (a) $f(1)$

(b) $g(-1)$

(c) $(g \circ f)(1)$

12. (a) $g(0)$

(b) $f(2)$

(c) $(f \circ g)(0)$

13. (a) $(f \circ g)(-3)$

(b) $(g \circ f)(-2)$

14. (a) $(f \circ g)(2)$

(b) $(g \circ f)(2)$

In Exercises 15 and 16, find the domain of the compositions (a) $f \circ g$ and (b) $g \circ f$.

15. $f(x) = \sqrt{x}$

$g(x) = x - 2$

16. $f(x) = \dfrac{x}{x - 4}$

$g(x) = \sqrt{x}$

17. *Sales Bonus* You are a sales representative for a clothing manufacturer. You are paid an annual salary plus a bonus of 2% of your sales over $200,000. Consider the two functions

$f(x) = x - 200{,}000$ and $g(x) = 0.02x$.

If x is greater than $200,000, which of the following represents your bonus? Explain.

(a) $f(g(x))$ (b) $g(f(x))$

18. *Geometry* You are standing on a bridge over a calm pond and drop a pebble, causing ripples of concentric circles in the water. The radius (in feet) of the outer ripple is given by

$r(t) = 0.6t$

where t is time in seconds after the pebble hits the water. The area of the circle is given by the function

$A(r) = \pi r^2$.

Find an equation for the composite function $A(r(t))$. Interpret this composite function in the context of the problem.

In Exercises 19–22, find the inverse of the function f informally.

19. $f(x) = 5x$

20. $f(x) = x - 5$

21. $f(x) = x^7$

22. $f(x) = x^{1/5}$

In Exercises 23–26, verify algebraically that the functions f and g are inverses of each other.

23. $f(x) = x + 15$
    $g(x) = x - 15$

24. $f(x) = 2x - 1$
    $g(x) = \frac{1}{2}(x + 1)$

25. $f(x) = \sqrt[3]{x + 1}$
    $g(x) = x^3 - 1$

26. $f(x) = \dfrac{1}{x - 3}$
    $g(x) = 3 + \dfrac{1}{x}$

In Exercises 27–30, match the graph with the graph of its inverse. [The graphs of the inverse functions are labeled (a), (b), (c), and (d).]

(a)
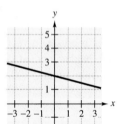

(b)
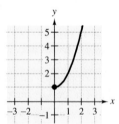

(c)
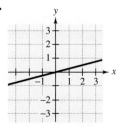

(d)
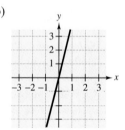

27.
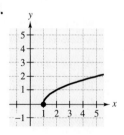

28.
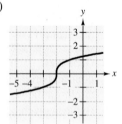

29.
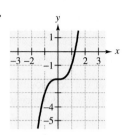

30.
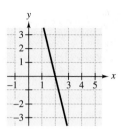

In Exercises 31–34, use a graphing utility to verify that the functions are inverses of each other.

31. $f(x) = \frac{1}{3}x$
    $g(x) = 3x$

32. $f(x) = \frac{1}{5}x - 1$
    $g(x) = 5x + 5$

33. $f(x) = \sqrt[3]{x + 2}$
    $g(x) = x^3 - 2$

34. $f(x) = \sqrt{x + 1}$
    $g(x) = x^2 - 1, \ x \geq 0$

In Exercises 35–38, find the inverse of the function.

35. $g(x) = 3 - 4x$

36. $h(x) = \sqrt{x + 5}$

37. $f(t) = t^3 - 1$

38. $f(s) = \dfrac{2}{3 - s}$

In Exercises 39–42, decide whether the function has an inverse.

39. $f(x) = x^2 - 2$

40. $f(x) = \frac{1}{5}x$

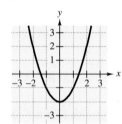

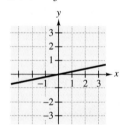

41. $g(x) = \sqrt{25 - x^2}$

42. $g(x) = |x - 4|$

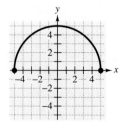

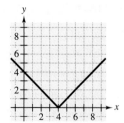

In Exercises 43–46, place a restriction on the domain of the function so that the graph of the restricted function satisfies the Horizontal Line Test. Then find the inverse of the restricted function. (*Note:* There is more than one correct answer.)

43. $f(x) = x^4$

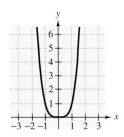

44. $f(x) = 9 - x^2$

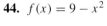
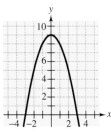

45. $f(x) = (x - 2)^2$

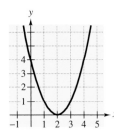

46. $f(x) = |x - 2|$

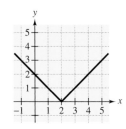

Graphical Reasoning In Exercises 47 and 48, use the graph of f to sketch the graph of f^{-1}. Explain the relationship between the graph of f and the graph of f^{-1}.

47.

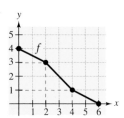

48.

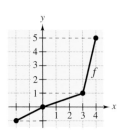

True or False? In Exercises 49 and 50, is the statement true or false? If true, explain your reasoning. If false, give an example to show why.

49. If the inverse of f exists, the y-intercept of f is an x-intercept of f^{-1}.

50. There exists no function f such that $f = f^{-1}$.

Reviewing the Major Concepts

In Exercises 51–54, solve the equation by the specified method.

51. $3x^2 + 9x - 12 = 0$ *Factoring*

52. $x^2 + 3x - 2 = 0$ *Completing the square*

53. $4x^2 - 7 = -6x$ *Quadratic Formula*

54. $x^2 - 10 = 0$ *Extracting square roots*

In Exercises 55 and 56, use $y = -x^2 + 4x$.

55. (a) Does the graph open up or down? Explain.

  (b) Find the x-intercepts algebraically.

  (c) Find the coordinates of the vertex of the parabola.

56. Use a graphing utility to graph the equation and verify the results of Exercise 55 graphically.

Additional Problem Solving

In Exercises 57–62, find the indicated composites.

57. $f(x) = x + 1$, $g(x) = 2x$

  (a) $(f \circ g)(3)$ (b) $(g \circ f)(3)$

  (c) $(f \circ g)(x)$ (d) $(g \circ f)(x)$

58. $f(x) = x^2 - 3x$, $g(x) = 5x + 3$

  (a) $(f \circ g)(-1)$ (b) $(g \circ f)(3)$

  (c) $(f \circ g)(x)$ (d) $(g \circ f)(x)$

59. $f(x) = |x - 3|, \quad g(x) = 3x$

(a) $(f \circ g)(1)$ (b) $(g \circ f)(2)$

(c) $(f \circ g)(x)$ (d) $(g \circ f)(x)$

60. $f(x) = x + 5, \quad g(x) = x^3$

(a) $(f \circ g)(2)$ (b) $(g \circ f)(-3)$

(c) $(f \circ g)(x)$ (d) $(g \circ f)(x)$

61. $f(x) = \dfrac{1}{x - 3}, \quad g(x) = \sqrt{x}$

(a) $(f \circ g)(49)$ (b) $(g \circ f)(12)$

(c) $(f \circ g)(x)$ (d) $(g \circ f)(x)$

62. $f(x) = \sqrt{x + 6}, \quad g(x) = 2x - 3$

(a) $(f \circ g)(3)$ (b) $(g \circ f)(-2)$

(c) $(f \circ g)(x)$ (d) $(g \circ f)(x)$

In Exercises 63–66, use the functions f and g to find the indicated values.

$f = \{(0, 1), \ (1, 2), \ (2, 5), \ (3, 10), \ (4, 17)\},$

$g = \{(5, 4), \ (10, 1), \ (2, 3), \ (17, 0), \ (1, 2)\}$

63. (a) $f(3)$ **64.** (a) $g(2)$

    (b) $(g \circ f)(3)$ (b) $(f \circ g)(10)$

65. (a) $(g \circ f)(4)$ **66.** (a) $(f \circ g)(1)$

    (b) $(f \circ g)(2)$ (b) $(g \circ f)(0)$

In Exercises 67–70, find the domain of the compositions (a) $f \circ g$ and (b) $g \circ f$.

67. $f(x) = x^2 + 1$ **68.** $f(x) = 2 - 3x$

   $g(x) = 2x$ $g(x) = 5x + 3$

69. $f(x) = \dfrac{9}{x + 9}$ **70.** $f(x) = \sqrt{x - 5}$

   $g(x) = x^2$ $g(x) = x^2$

71. *Production Cost* The daily cost of producing x units of a product is $C(x) = 8.5x + 300$. The number of units produced in t hours during a day is given by $x(t) = 12t, \ 0 \le t \le 8$. Find, simplify, and interpret $(C \circ x)(t)$.

72. *Rebate and Discount* The suggested retail price of a new car is p dollars. The dealership is advertising a factory rebate of $2000 and a 5% discount.

(a) Write a function R in terms of p, giving the cost of the car after receiving the rebate from the factory.

(b) Write a function S in terms of p, giving the cost of the car after receiving the dealership discount.

(c) Form the composite functions $(R \circ S)(p)$ and $(S \circ R)(p)$ and interpret each.

(d) Find $(R \circ S)(26,000)$ and $(S \circ R)(26,000)$. Which yields the smaller cost for the car? Explain.

In Exercises 73–80, find the inverse of the function f informally.

73. $f(x) = 6x$ **74.** $f(x) = \frac{1}{3}x$

75. $f(x) = x + 10$ **76.** $f(x) = 10 - x$

77. $f(x) = \sqrt[3]{x}$ **78.** $f(x) = x^5$

79. $f(x) = 2x - 1$ **80.** $f(x) = \frac{1}{2}x + 3$

In Exercises 81–88, verify algebraically that the functions f and g are inverses of each other.

81. $f(x) = 10x$ **82.** $f(x) = \frac{2}{3}x$

   $g(x) = \frac{1}{10}x$ $g(x) = \frac{3}{2}x$

83. $f(x) = 1 - 2x$ **84.** $f(x) = 3 - x$

   $g(x) = \frac{1}{2}(1 - x)$ $g(x) = 3 - x$

85. $f(x) = 2 - 3x$ **86.** $f(x) = -\frac{1}{4}x + 3$

   $g(x) = \frac{1}{3}(2 - x)$ $g(x) = -4(x - 3)$

87. $f(x) = 1/x$ **88.** $f(x) = x^7$

   $g(x) = 1/x$ $g(x) = \sqrt[7]{x}$

In Exercises 89–92, use a graphing utility to verify that the functions are inverses of each other.

89. $f(x) = 3x + 4$ **90.** $f(x) = |x - 2|, \ x \ge 2$

   $g(x) = \frac{1}{3}(x - 4)$ $g(x) = x + 2, \ x \ge 0$

91. $f(x) = \frac{1}{8}x^3$

$g(x) = 2\sqrt[3]{x}$

92. $f(x) = \sqrt{4 - x}$

$g(x) = 4 - x^2, \quad x \geq 0$

In Exercises 93–102, find the inverse of the function.

93. $f(x) = 8x$

94. $f(x) = \dfrac{x}{10}$

95. $g(x) = x + 25$

96. $f(x) = 7 - x$

97. $g(t) = -\frac{1}{4}t + 2$

98. $g(t) = 6t + 1$

99. $h(x) = \sqrt{x}$

100. $h(t) = t^5$

101. $g(s) = \dfrac{5}{s}$

102. $f(x) = \dfrac{4}{x - 2}$

In Exercises 103 and 104, find the inverse of the function f. Use a graphing utility to sketch the graph of f and f^{-1}.

103. $f(x) = x^3 + 1$

104. $f(x) = \sqrt{x^2 - 4}, \quad x \geq 2$

In Exercises 105–114, use a graphing utility to graph the function and determine whether the function is one-to-one.

105. $f(x) = \frac{1}{4}x^3$

106. $f(x) = x^2 - 2$

107. $f(t) = \sqrt[3]{5 - t}$

108. $h(t) = 4 - \sqrt[3]{t}$

109. $g(x) = x^4$

110. $f(x) = (x + 2)^5$

111. $h(t) = \dfrac{5}{t}$

112. $g(t) = \dfrac{5}{t^2}$

113. $f(s) = \dfrac{4}{s^2 + 1}$

114. $f(x) = \dfrac{1}{x - 2}$

In Exercises 115 and 116, use the graph of f to sketch the graph of f^{-1}.

115.

116.

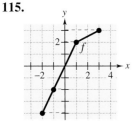

117. Consider the function $f(x) = 3 - 2x$.

(a) Find $f^{-1}(x)$.

(b) Find $(f^{-1})^{-1}(x)$.

118. Consider the functions $f(x) = 4x$ and $g(x) = x + 6$.

(a) Find $(f \circ g)(x)$.

(b) Find $(f \circ g)^{-1}(x)$.

(c) Find $f^{-1}(x)$ and $g^{-1}(x)$.

(d) Find $(g^{-1} \circ f^{-1})(x)$ and compare the result with that of part (b).

119. *Hourly Wage* Your wage is $9.00 per hour plus $0.65 for each unit produced per hour. Thus, your hourly wage y in terms of the number of units produced is given by $y = 9 + 0.65x$.

(a) Determine the inverse of the function.

(b) What does each variable represent in the inverse function?

(c) Determine the number of units produced when your hourly wage averages $14.20.

120. *Cost* Suppose you need 100 pounds of two commodities costing $0.50 and $0.75 per pound.

(a) Verify that your total cost y is given by $y = 0.50x + 0.75(100 - x)$, where x is the number of pounds of the less expensive commodity.

(b) Find the inverse of the cost function. What does each variable represent in the inverse function?

(c) Use the context of the problem to determine the domain of the inverse function.

(d) Determine the number of pounds of the less expensive commodity purchased if the total cost is $60.

True or False? In Exercises 121 and 122, decide whether the statement is true or false. If true, explain your reasoning. If false, give an example to show why.

121. If the inverse of f exists, the domains of f and f^{-1} are the same.

122. If the inverse of f exists and its graph passes through the point $(2, 2)$, the graph of f^{-1} also passes through the point $(2, 2)$.

9.3 | Logarithmic Functions

Logarithmic Functions ■ Graphs of Logarithmic Functions ■ The Natural Logarithmic Function ■ Change of Base

Logarithmic Functions

In Section 9.2, you were introduced to the concept of the inverse of a function. Moreover, you saw that if a function has the property that no horizontal line intersects the graph of the function more than once, the function must have an inverse. By looking back at the graphs of the exponential functions introduced in Section 9.1, you will see that every function of the form

$$f(x) = a^x \qquad \text{Exponential functions have inverses.}$$

passes the horizontal line test, and therefore must have an inverse. This inverse function is called the **logarithmic function with base *a*.**

Definition of Logarithmic Function

Let *a* and *x* be positive real numbers such that $a \neq 1$. The **logarithm of *x* with base *a*** is denoted by $\log_a x$ and is defined as follows.

$$y = \log_a x \qquad \text{if and only if} \qquad x = a^y$$

The function $f(x) = \log_a x$ is the **logarithmic function with base *a*.**

From the definition of a logarithmic function, you can see that the equations $y = \log_a x$ and $x = a^y$ are equivalent.

Logarithmic Equation	*Exponential Equation*
$y = \log_a x$	$x = a^y$

The first equation is in *logarithmic* form and the second is in *exponential* form. From these equivalent equations it should also be clear that *a logarithm is an exponent.* For instance, because the exponent in the expression

$$2^3 = 8 \qquad \text{The exponent of } 2^3 \text{ is 3.}$$

is 3, the value of the logarithm $\log_2 8$ is 3. That is,

$$\log_2 8 = 3. \qquad \text{A logarithm is an exponent.}$$

Therefore, to evaluate the logarithmic expression $\log_a x$, you need to ask the question, "To what power must *a* be raised to obtain *x*?"

EXAMPLE 1   *Evaluating Logarithms*

Evaluate each logarithm.

a. $\log_2 16$ **b.** $\log_3 9$ **c.** $\log_4 2$

Solution

In each case you should answer the question, "To what power must the base be raised to obtain the given number?"

a. The power to which 2 must be raised to obtain 16 is 4. That is,

$$2^4 = 16 \qquad \Longrightarrow \qquad \log_2 16 = 4.$$

b. The power to which 3 must be raised to obtain 9 is 2. That is,

$$3^2 = 9 \qquad \Longrightarrow \qquad \log_3 9 = 2.$$

c. The power to which 4 must be raised to obtain 2 is $\frac{1}{2}$. That is,

$$4^{1/2} = 2 \qquad \Longrightarrow \qquad \log_4 2 = \frac{1}{2}.$$

Each of the logarithms in Example 2 involves a special important case. Study this example carefully.

EXAMPLE 2   *Evaluating Logarithms*

Evaluate each logarithm.

a. $\log_5 1$ **b.** $\log_{10} \dfrac{1}{10}$ **c.** $\log_3(-1)$ **d.** $\log_4 0$

Solution

a. The power to which 5 must be raised to obtain 1 is 0. That is,

$$5^0 = 1 \qquad \Longrightarrow \qquad \log_5 1 = 0.$$

b. The power to which 10 must be raised to obtain $\frac{1}{10}$ is -1. That is,

$$10^{-1} = \frac{1}{10} \qquad \Longrightarrow \qquad \log_{10} \frac{1}{10} = -1.$$

c. There is no power to which 3 can be raised to obtain -1. The reason for this is that for any value of x, 3^x is a positive number. Thus, $\log_3(-1)$ is undefined.

d. There is no power to which 4 can be raised to obtain 0. Thus, $\log_4 0$ is undefined.

The following properties of logarithms follow directly from the definition of the logarithmic function with base a.

Properties of Logarithms

Let a and x be positive real numbers such that $a \neq 1$. Then the following properties are true.

1. $\log_a 1 = 0$ because $a^0 = 1$.

2. $\log_a a = 1$ because $a^1 = a$.

3. $\log_a a^x = x$ because $a^x = a^x$.

The logarithmic function with base 10 is called the **common logarithmic function.** On most calculators, this function can be evaluated with the common logarithmic key $\boxed{\log}$, as illustrated in the next example.

EXAMPLE 3   *Evaluating Common Logarithms*

Evaluate each logarithm. Use a calculator only if necessary.

a. $\log_{10} 100$ **b.** $\log_{10} 0.01$ **c.** $\log_{10} 5$

Solution

a. Because $10^2 = 100$, it follows that

$$\log_{10} 100 = 2.$$

b. Because $10^{-2} = \frac{1}{100} = 0.01$, it follows that

$$\log_{10} 0.01 = -2.$$

c. There is no simple power to which 10 can be raised to obtain 5, so you should use a calculator to evaluate $\log_{10} 5$.

Keystrokes	*Display*	
5 $\boxed{\log}$	0.69897	Scientific
$\boxed{\log}$ 5 $\boxed{\text{ENTER}}$	0.69897	Graphing

Thus, rounded to three decimal places, $\log_{10} 5 \approx 0.699$.

NOTE   Be sure you see that a logarithm can be zero or negative, *but* you cannot take the logarithm of zero or a negative number.

Graphs of Logarithmic Functions

To sketch the graph of $y = \log_a x$, we can use the fact that the graphs of inverse functions are reflections of each other in the line $y = x$.

EXAMPLE 4   *Graphs of Exponential and Logarithmic Functions*

On the same rectangular coordinate system, sketch the graphs of the following.

a. $f(x) = 2^x$ **b.** $g(x) = \log_2 x$

Solution

a. Begin by making a table of values for $f(x) = 2^x$.

x	-2	-1	0	1	2	3
$f(x) = 2^x$	$\frac{1}{4}$	$\frac{1}{2}$	1	2	4	8

By plotting these points and connecting them with a smooth curve, you obtain the graph shown in Figure 9.12.

b. Because $g(x) = \log_2 x$ is the inverse function of $f(x) = 2^x$, the graph of g is obtained by reflecting the graph of f in the line $y = x$, as shown in Figure 9.12.

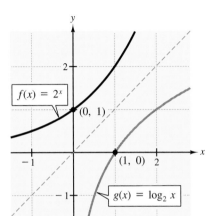

FIGURE 9.12   Inverse Functions

Notice from the graph of $g(x) = \log_2 x$, shown in Figure 9.12, that the domain of the function is the set of positive numbers and the range is the set of all real numbers. The basic characteristics of the graph of a logarithmic function are summarized in Figure 9.13. In this figure, note that the graph has one x-intercept, at $(1, 0)$. Also note that the y-axis is a vertical asymptote of the graph.

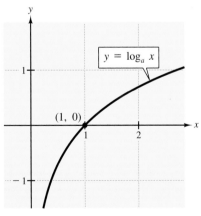

Graph of $y = \log_a x$, $a > 1$
- Domain: $(0, \infty)$
- Range: $(-\infty, \infty)$
- Intercept: $(1, 0)$
- Increasing

FIGURE 9.13

The Natural Logarithmic Function

As with exponential functions, the most widely used base for logarithmic functions is the number e. The logarithmic function with base e is the **natural logarithmic function** and is denoted by the special symbol $\ln x$, which is read as "el en of x."

The Natural Logarithmic Function

The function defined by

$$f(x) = \log_e x = \ln x$$

where $x > 0$, is called the **natural logarithmic function.**

NOTE On most calculators, the natural logarithm key is denoted by $\boxed{\ln}$. For instance, on a scientific calculator, you can evaluate $\ln 2$ as

$$2 \boxed{\ln}$$

and on a graphing calculator, you can evaluate it as

$$\boxed{\ln} \quad 2 \quad \boxed{\text{ENTER}}.$$

In either case, you should obtain a display of 0.6931472.

The three properties of logarithms listed earlier in this section are also valid for natural logarithms.

Properties of Natural Logarithms

Let x be a positive real number. Then the following properties are true.

1. $\ln 1 = 0$ because $e^0 = 1$.

2. $\ln e = 1$ because $e^1 = e$.

3. $\ln e^x = x$ because $e^x = e^x$.

EXAMPLE 5   *Evaluating the Natural Logarithmic Function*

Evaluate each of the following. Then incorporate the results into a graph of the natural logarithmic function.

a. $\ln e^2$ **b.** $\ln \dfrac{1}{e}$

Solution

Using the property that $\ln e^x = x$, you obtain the following.

a. $\ln e^2 = 2$

b. $\ln \dfrac{1}{e} = \ln e^{-1} = -1$

Using the points, $(1/e, -1)$, $(1, 0)$, $(e, 1)$, and $(e^2, 2)$, you can sketch the graph of the natural logarithmic function, as shown in Figure 9.14.

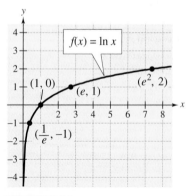

FIGURE 9.14

Change of Base

Although 10 and e are the most frequently used bases, you occasionally need to evaluate logarithms with other bases. In such cases the following **change-of-base formula** is useful.

Change-of-Base Formula

Let a, b, and x be positive real numbers such that $a \neq 1$ and $b \neq 1$. Then $\log_a x$ is given as follows.

$$\log_a x = \frac{\log_b x}{\log_b a} \quad \text{or} \quad \log_a x = \frac{\ln x}{\ln a}$$

The usefulness of this change-of-base formula is that you can use a calculator that has only the common logarithm key $\boxed{\log}$ and the natural logarithm key $\boxed{\ln}$ to evaluate logarithms to any base.

EXAMPLE 6 *Changing the Base to Evaluate Logarithms*

a. Use *common* logarithms to evaluate $\log_3 5$.

b. Use *natural* logarithms to evaluate $\log_6 2$.

Solution

Using the change-of-base formula, you can convert to common and natural logarithms by writing

$$\log_3 5 = \frac{\log_{10} 5}{\log_{10} 3} \quad \text{and} \quad \log_6 2 = \frac{\ln 2}{\ln 6}.$$

Now, use the following keystrokes.

a.

Keystrokes	Display	
5 $\boxed{\log}$ $\boxed{\div}$ 3 $\boxed{\log}$ $\boxed{=}$	1.4649735	Scientific
$\boxed{\log}$ 5 $\boxed{\div}$ $\boxed{\log}$ 3 $\boxed{\text{ENTER}}$	1.4649735	Graphing

Thus, $\log_3 5 \approx 1.465$.

b.

Keystrokes	Display	
2 $\boxed{\ln}$ $\boxed{\div}$ 6 $\boxed{\ln}$ $\boxed{=}$	0.3868528	Scientific
$\boxed{\ln}$ 2 $\boxed{\div}$ $\boxed{\ln}$ 6 $\boxed{\text{ENTER}}$	0.3868528	Graphing

Thus, $\log_6 2 \approx 0.387$.

At this point, you have been introduced to all the basic types of functions that are covered in this course: polynomial functions, radical functions, rational functions, exponential functions, and logarithmic functions. The only other common types of functions are *trigonometric functions*, which you will study if you go on to take a course in trigonometry or precalculus.

Group Activities   Reviewing the Major Concepts

Comparing Models   Suppose you work for a research and development firm that deals with a wide variety of disciplines. Your supervisor has asked your group to give a presentation to your department on four basic kinds of mathematical models. Identify each of the models shown below. Develop a presentation describing the types of data sets that each model would best represent. Include distinctions in domain, range, and intercepts and a discussion of the types of applications to which each model is suited.

a.   b.

c.   d.

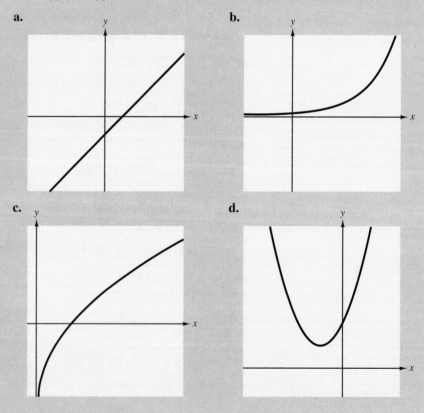

9.3 Exercises

Discussing the Concepts

1. Write "logarithm of x with base 5" symbolically.

2. Explain the relationship between $f(x) = 2^x$ and $g(x) = \log_2 x$.

3. Explain why $\log_a a = 1$.

4. Explain why $\log_a a^x = x$.

5. What are common and natural logarithms?

6. Describe how to use a calculator to find the logarithm of a number if the base is not 10 or e.

Problem Solving

In Exercises 7–10, write the logarithmic equation in exponential form.

7. $\log_5 25 = 2$

8. $\log_3 \frac{1}{243} = -5$

9. $\log_{36} 6 = \frac{1}{2}$

10. $\log_{32} 4 = \frac{2}{5}$

In Exercises 11–14, write the exponential equation in logarithmic form.

11. $3^{-2} = \frac{1}{9}$

12. $5^4 = 625$

13. $8^{2/3} = 4$

14. $81^{3/4} = 27$

In Exercises 15–22, evaluate the expression without a calculator. (If it is not possible, state the reason.)

15. $\log_2 8$

16. $\log_3 27$

17. $\log_2 \frac{1}{4}$

18. $\log_3 \frac{1}{9}$

19. $\log_2 -3$

20. $\log_3 1$

21. $\log_9 3$

22. $\log_{25} 125$

In Exercises 23–26, use the properties of natural logarithms to evaluate the expression.

23. $\ln e^2$

24. $\ln e^{-4}$

25. $\ln e$

26. $\ln(e^2 \cdot e^4)$

In Exercises 27–30, use a calculator to evaluate the logarithm. Round to four decimal places.

27. $\log_{10} 31$

28. $\log_{10} \dfrac{\sqrt{3}}{2}$

29. $\ln 0.75$

30. $\ln(\sqrt{3} + 1)$

In Exercises 31–36, match the function with its graph. [The graphs are labeled (a), (b), (c), (d), (e), and (f).]

(a)

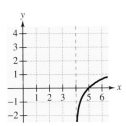

(b)
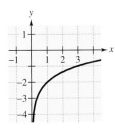

(c)
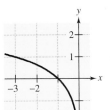

(d)
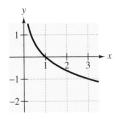

(e)
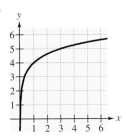

(f)
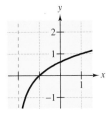

31. $f(x) = 4 + \log_3 x$

32. $f(x) = -2 + \log_3 x$

33. $f(x) = -\log_3 x$

34. $f(x) = \log_3(-x)$

35. $f(x) = \log_3(x - 4)$

36. $f(x) = \log_3(x + 2)$

In Exercises 37 and 38, describe the relationship between the graphs of f and g.

37. $f(x) = \log_3 x$
   $g(x) = 3^x$

38. $f(x) = \log_4 x$
   $g(x) = 4^x$

In Exercises 39–42, sketch the graph of the function.

39. $f(x) = \log_5 x$

40. $g(x) = \log_8 x$

41. $f(x) = 3 + \log_2 x$

42. $f(x) = -2 + \log_3 x$

In Exercises 43–46, use a graphing utility to sketch the graph of the function.

43. $y = 5 \log_{10} x$

44. $y = 5 \log_{10}(x - 3)$

45. $f(x) = 3 \ln x$

46. $h(x) = 2 + \ln x$

In Exercises 47–50, use a calculator to evaluate the logarithm by means of the change-of-base formula.

47. $\log_8 132$

48. $\log_5 510$

49. $\log_2 0.72$

50. $\log_3(1 + e^2)$

51. *Creating a Table*   The time t in years for an investment to double in value when compounded continuously at annual interest rate r is given by

$$t = \frac{\ln 2}{r}.$$

Complete the table, which shows the "doubling time" for several annual interest rates.

r	0.07	0.08	0.09	0.10	0.11	0.12
t						

52. *Intensity of Sound*   The relationship between the number of decibels B and the intensity of a sound I in watts per meter squared is given by

$$B = 10 \log_{10}\left(\frac{I}{10^{-16}}\right).$$

Determine the number of decibels of a sound with an intensity of 10^{-4} watts per centimeter squared.

Reviewing the Major Concepts

In Exercises 53 and 54, use a graphing utility to graph each equation and use the graph to approximate any points of intersection. Find the solution of the system of equations algebraically.

53. $x + 3y = 11$
   $x^2 - 3y = -5$

54. $x + 2y = -7$
   $-x - y^2 = -17$

Geometry   In Exercises 55 and 56, find the area of the shaded region.

55.

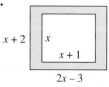

56.

In Exercises 57 and 58, use the following information. From 1986 through 1990, the number of compact discs C (in millions) and the number of record albums R (in millions) sold in the United States can be approximated by the models

$$C = 4.3t^2 - 9.8t - 42.7 \text{ and } R = -28.8t + 301$$

where $t = 6$ represents 1986. (Source: Recording Industry Association of America)

57. How many compact discs and how many records were sold in 1986?

58. During which year was the number of compact discs sold approximately equal to the number of record albums sold?

Additional Problem Solving

In Exercises 59–64, write the logarithmic equation in exponential form.

59. $\log_6 36 = 2$

60. $\log_{10} 10,000 = 4$

61. $\log_4 \frac{1}{16} = -2$

62. $\log_8 \frac{1}{8} = -1$

63. $\log_8 4 = \frac{2}{3}$

64. $\log_{16} 8 = \frac{3}{4}$

In Exercises 65–70, write the exponential equation in logarithmic form.

65. $7^2 = 49$

66. $6^4 = 1296$

67. $25^{-1/2} = \frac{1}{5}$

68. $6^{-3} = \frac{1}{216}$

69. $4^0 = 1$

70. $10^{0.12} \approx 1.318$

In Exercises 71–82, evaluate the expression without a calculator. (If it is not possible, state the reason.)

71. $\log_4 1$

72. $\log_5 (-6)$

73. $\log_{10} 10$

74. $\log_8 8$

75. $\log_{10} 1000$

76. $\log_{10} \frac{1}{100}$

77. $\log_4 (-4)$

78. $\log_2 0$

79. $\log_{16} 8$

80. $\log_{144} 12$

81. $\log_7 7^4$

82. $\log_5 5^3$

In Exercises 83–86, use the properties of natural logarithms to evaluate the expression.

83. $\ln 1$

84. $\ln 8^0$

85. $\ln \frac{e^3}{e^2}$

86. $\ln \frac{1}{e^3}$

In Exercises 87–92, use a calculator to evaluate the logarithm. Round to four decimal places.

87. $\log_{10} (\sqrt{2} + 4)$

88. $\log_{10} 5310$

89. $\log_{10} 0.85$

90. $\log_{10} 0.345$

91. $\ln 25$

92. $\ln 6.57$

In Exercises 93 and 94, describe the relationship between the graphs of f and g.

93. $f(x) = \log_6 x$

$g(x) = 6^x$

94. $f(x) = \log_{(1/2)} x$

$g(x) = \left(\frac{1}{2}\right)^x$

In Exercises 95–100, sketch the graph of the function.

95. $g(t) = -\log_2 t$

96. $h(s) = -2\log_3 s$

97. $g(x) = \log_2 (x - 3)$

98. $h(x) = \log_3 (x + 1)$

99. $f(x) = \log_{10}(10x)$

100. $g(x) = \log_4(4x)$

In Exercises 101–108, use a graphing utility to sketch the graph of the function.

101. $y = -3 + 5\log_{10} x$

102. $y = 5\log_{10}(3x)$

103. $h(t) = -2\ln t$

104. $f(x) = 4\ln x$

105. $g(x) = \ln(x + 6)$

106. $h(x) = -\ln(x - 2)$

107. $f(x) = \ln(-x)$

108. $g(x) = -\frac{3}{2}\ln x$

In Exercises 109–116, use a calculator to evaluate the logarithm by means of the change-of-base formula. Use (a) the common logarithm key and (b) the natural logarithm key.

109. $\log_3 7$

110. $\log_7 4$

111. $\log_{15} 1250$

112. $\log_{20} 125$

113. $\log_4 \sqrt{42}$

114. $\log_{12} 0.6$

115. $\log_{(1/2)} 4$

116. $\log_{(1/3)}(0.015)$

117. *American Elk*   The antler spread a (in inches) and shoulder height h (in inches) of an adult male American elk are related by the model

$$h = 116\log_{10}(a + 40) - 176.$$

Approximate the shoulder height of a male American elk with an antler spread of 55 inches.

118. *Tornadoes* Most tornadoes last less than 1 hour and travel less than 20 miles. The speed of the wind S (in miles per hour) near the center of the tornado is related to the distance the tornado travels d (in miles) by the model

$$S = 93 \log_{10} d + 65.$$

On March 18, 1925, a large tornado struck portions of Missouri, Illinois, and Indiana, covering a distance of 220 miles. Approximate the speed of the wind near the center of this tornado.

In Exercises 119–124, answer the question for the function $f(x) = \log_{10} x$. (Do not use a calculator.)

119. What is the domain of f?

120. Find the inverse function of f.

121. Describe the values of $f(x)$ for $1000 \le x \le 10{,}000$.

122. Describe the values of x, given that $f(x)$ is negative.

123. By what amount will x increase, given that $f(x)$ is increased by one unit?

124. Find the ratio of a to b, given that $f(a) = 3f(b)$.

Math Matters

Animal Species

There are approximately 6 million different species of living organisms on earth—animals, plants, single-cell organisms, fungi, and so on. Of those that are animals, by far most are insects. What percent of the different species on earth are insects? (The answer is given in the back of the book.)

Type of Animal	Examples	Number of Species
Insect	Ant, Butterfly, Honey Bee	950,000
Arachnid	Spider, Scorpion, Mite	110,000
Mollusk	Snail, Clam, Squid	45,000
Fish	Tuna, Salmon, Trout	30,000
Crustacean	Shrimp, Lobster, Crab	26,000
Bird	Robin, Ostrich, Eagle	8,650
Worm	Earthworm, Flatworm, Roundworm	6,700
Echinoderm	Starfish, Sea Urchin, Sea Cucumber	6,000
Reptile	Snake, Lizard, Alligator	5,180
Mammal	Human, Mouse, Whale	4,230
Amphibian	Frog, Toad, Salamander	3,000

MID-CHAPTER QUIZ

Take this quiz as you would take a quiz in class. After you are done, check your work against the answers given in the back of the book.

1. Given $f(x) = \left(\frac{4}{3}\right)^x$, find (a) $f(2)$, (b) $f(0)$, (c) $f(-1)$, and (d) $f(1.5)$.

2. Find the domain and range of $g(x) = 2^{-0.5x}$.

In Exercises 3–6, sketch the graph of the function.

3. $y = \frac{1}{2}(4^x)$ 4. $y = 5(2^{-x})$ 5. $f(t) = 12e^{-0.4t}$ 6. $g(x) = 100(1.08)^x$

7. You deposit $750 at $7\frac{1}{2}\%$ interest, compounded n times per year or continuously. Find the balance A after 20 years.

n	1	4	12	365	Continuous compounding
A					

8. A gallon of milk costs $2.23 now. If the price increases by 4% each year, what will the price be after 5 years?

9. Given $f(x) = 2x - 3$ and $g(x) = x^3$, find the indicated composition.
   (a) $(f \circ g)(-2)$ (b) $(g \circ f)(4)$ (c) $(f \circ g)(x)$ (d) $(g \circ f)(x)$

10. Verify algebraically and graphically that $f(x) = 3 - 5x$ and $g(x) = \frac{1}{5}(3 - x)$ are inverses of each other.

In Exercises 11 and 12, find the inverse of the function.

11. $h(x) = 10x + 3$ 12. $g(t) = \frac{1}{2}t^3 + 2$

13. Write the logarithmic equation $\log_4\left(\frac{1}{16}\right) = -2$ in exponential form.

14. Write the exponential equation $3^4 = 81$ in logarithmic form.

15. Evaluate $\log_5 125$ without the aid of a calculator.

16. Write a paragraph comparing the graphs of $f(x) = \log_5 x$ and $g(x) = 5^x$.

In Exercises 17 and 18, use a graphing utility to sketch the graph of the function.

17. $f(t) = \frac{1}{2}\ln t$ 18. $h(x) = 3 - \ln x$

19. Use the graph of f at the right to determine h and k if $f(x) = \log_5(x - h) + k$.

20. Use a calculator and the change-of-base formula to evaluate $\log_6 450$.

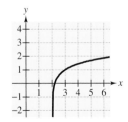

Figure for 19

9.4   Properties of Logarithms

Properties of Logarithms ▪ Rewriting Logarithmic Expressions ▪ Application

Properties of Logarithms

Before electronic hand-held calculators became available in the 1970s, mathematicians, engineers, and scientists relied on a tool called the slide rule. Created by Edmund Gunter (1581–1626), the slide rule uses logarithms to quickly multiply and divide numbers.

You know from the previous section that the logarithmic function with base a is the *inverse* of the exponential function with base a. Thus, it makes sense that each property of exponents should have a corresponding property of logarithms. For instance, the exponential property

$$a^0 = 1 \qquad \text{Exponential property}$$

has the corresponding logarithmic property

$$\log_a 1 = 0. \qquad \text{Corresponding logarithmic property}$$

In this section you will study the logarithmic properties that correspond to the following three exponential properties:

Base a	Natural Base
1. $a^n a^m = a^{n+m}$	$e^n e^m = e^{n+m}$
2. $\dfrac{a^n}{a^m} = a^{n-m}$	$\dfrac{e^n}{e^m} = e^{n-m}$
3. $(a^n)^m = a^{nm}$	$(e^n)^m = e^{nm}$

Properties of Logarithms

Let a be a positive real number such that $a \neq 1$, and let n be a real number. If u and v are real numbers, variables, or algebraic expressions such that $u > 0$ and $v > 0$, the following properties are true.

Logarithm with Base a	Natural Logarithm
1. $\log_a(uv) = \log_a u + \log_a v$	$\ln(uv) = \ln u + \ln v$
2. $\log_a \dfrac{u}{v} = \log_a u - \log_a v$	$\ln \dfrac{u}{v} = \ln u - \ln v$
3. $\log_a u^n = n \log_a u$	$\ln u^n = n \ln u$

NOTE   There is no general property of logarithms that can be used to simplify $\log_a(u + v)$. Specifically,

$$\log_a(u + v) \ \textit{does not equal} \ \log_a u + \log_a v.$$

EXAMPLE 1   Using Properties of Logarithms

Use the fact that $\ln 2 \approx 0.693$, $\ln 3 \approx 1.099$, and $\ln 5 \approx 1.609$ to approximate each of the following.

a. $\ln \dfrac{2}{3}$ **b.** $\ln 10$ **c.** $\ln 30$

Solution

a. $\ln \dfrac{2}{3} = \ln 2 - \ln 3$ Property 2

           $\approx 0.693 - 1.099$ Substitute for $\ln 2$ and $\ln 3$.

           $= -0.406$ Simplify.

b. $\ln 10 = \ln(2 \cdot 5)$ Factor.

           $= \ln 2 + \ln 5$ Property 1

           $\approx 0.693 + 1.609$ Substitute for $\ln 2$ and $\ln 5$.

           $= 2.302$ Simplify.

c. $\ln 30 = \ln(2 \cdot 3 \cdot 5)$ Factor.

           $= \ln 2 + \ln 3 + \ln 5$ Property 1

           $\approx 0.693 + 1.099 + 1.609$ Substitute for $\ln 2$, $\ln 3$, and $\ln 5$.

           $= 3.401$ Simplify.

NOTE   When using the properties of logarithms, it helps to state the properties *verbally*. For instance, the verbal form of the property $\ln(uv) = \ln u + \ln v$ is: *The log of a product is the sum of the logs of the factors.* Similarly, the verbal form of the property $\ln(u/v) = \ln u - \ln v$ is: *The log of a quotient is the difference of the logs of the numerator and denominator.*

EXAMPLE 2   Using Properties of Logarithms

Use the properties of logarithms to verify that $-\ln 2 = \ln \frac{1}{2}$.

Solution

Using Property 3, you can write the following.

       $-\ln 2 = (-1) \ln 2$ Rewrite coefficient as -1.

           $= \ln 2^{-1}$ Property 3

           $= \ln \dfrac{1}{2}$ Rewrite 2^{-1} as $\frac{1}{2}$.

Rewriting Logarithmic Expressions

In Examples 1 and 2, the properties of logarithms were used to rewrite logarithmic expressions involving the log of a *constant*. A more common use of the properties is to rewrite the log of a *variable expression*.

EXAMPLE 3 Rewriting the Logarithm of a Product

Use the properties of logarithms to rewrite $\log_{10} 7x^3$.

Solution

$$\log_{10} 7x^3 = \log_{10} 7 + \log_{10} x^3 \qquad \text{Property 1}$$
$$= \log_{10} 7 + 3\log_{10} x \qquad \text{Property 3}$$

When you rewrite a logarithmic expression as in Example 3, you are **expanding** the expression. The reverse procedure is demonstrated in Example 4, and is called **condensing** a logarithmic expression.

EXAMPLE 4 Condensing a Logarithmic Expression

Use the properties of logarithms to condense $\ln x - \ln 3$.

Solution

Using Property 2, you can write

$$\ln x - \ln 3 = \ln \frac{x}{3}. \qquad \text{Property 2}$$

NOTE In Example 4, try confirming the result by substituting values of x.

EXAMPLE 5 Expanding a Logarithmic Expression

Expand the logarithmic expression.

$$\log_2 3xy^2, \qquad x > 0, \quad y > 0$$

Solution

$$\log_2 3xy^2 = \log_2 3 + \log_2 x + \log_2 y^2 \qquad \text{Property 1}$$
$$= \log_2 3 + \log_2 x + 2\log_2 y \qquad \text{Property 3}$$

Sometimes expanding or condensing logarithmic expressions involves several steps. In the next example, be sure that you can justify each step in the solution. Also, notice how different the expanded expression is from the original.

EXAMPLE 6   Expanding a Logarithmic Expression

Expand the logarithmic expression.

$$\ln \sqrt{x^2 - 1}, \qquad x > 1$$

Solution

$$
\begin{aligned}
\ln \sqrt{x^2 - 1} &= \ln(x^2 - 1)^{1/2} && \text{Rewrite using fractional exponent.}\\
&= \tfrac{1}{2} \ln(x^2 - 1) && \text{Property 3}\\
&= \tfrac{1}{2} \ln(x - 1)(x + 1) && \text{Factor.}\\
&= \tfrac{1}{2}[\ln(x - 1) + \ln(x + 1)] && \text{Property 1}\\
&= \tfrac{1}{2} \ln(x - 1) + \tfrac{1}{2} \ln(x + 1) && \text{Distributive Property}
\end{aligned}
$$

EXAMPLE 7   Condensing a Logarithmic Expression

Use the properties of logarithms to condense the expression.

$$\ln 2 - 2 \ln x$$

Solution

$$
\begin{aligned}
\ln 2 - 2 \ln x &= \ln 2 - \ln x^2, && x > 0 && \text{Property 3}\\
&= \ln \frac{2}{x^2}, && x > 0 && \text{Property 2}
\end{aligned}
$$

When you expand or condense a logarithmic expression, it is possible to change the domain of the expression. For instance, the domain of the function

$$f(x) = 2 \ln x \qquad\qquad \text{Domain is the set of positive real numbers.}$$

is the set of positive real numbers, whereas the domain of

$$g(x) = \ln x^2 \qquad\qquad \text{Domain is the set of nonzero real numbers.}$$

is the set of nonzero real numbers. Thus, when you expand or condense a logarithmic expression, you should check to see whether the rewriting has changed the domain of the expression. In such cases, you should restrict the domain appropriately. For instance, you can write

$$f(x) = 2 \ln x = \ln x^2, \qquad x > 0.$$

Application

EXAMPLE 8   An Application: Human Memory Model

Students participating in a psychological experiment attended several lectures on a subject. Every month for a year after that, the students were tested to see how much of the material they remembered. The average scores for the group were given by the **human memory model**

$$f(t) = 80 - \ln(t + 1)^9, \qquad 0 \le t \le 12$$

where t is the time in months. Find the average scores for the group after 2 months and 8 months.

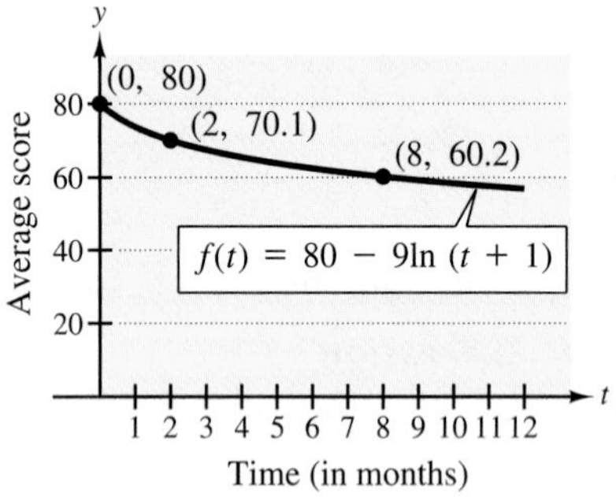

Solution

To make the calculations easier, rewrite the model as

$$f(t) = 80 - 9\ln(t + 1), \qquad 0 \le t \le 12.$$

After 2 months, the average score will be

$$f(2) = 80 - 9\ln 3 \approx 70.1 \qquad \text{Average score after 2 months}$$

and after 8 months, the average score will be

$$f(8) = 80 - 9\ln 9 \approx 60.2. \qquad \text{Average score after 8 months}$$

The graph of the function is shown in Figure 9.15.

FIGURE 9.15   Human Memory Model

The graph shows average score versus time (in months), with points (0, 80), (2, 70.1), (8, 60.2), and the labeled curve $f(t) = 80 - 9\ln(t + 1)$.

Group Activities

Problem Solving

x	5	7	9	11	13
y	522	1682	3370	5375	12,789

Mathematical Modeling   The data in the table represents the annual number y of new AIDS cases in females reported in the United States for the year x from 1985 through 1993, with $x = 5$ corresponding to 1998. (Source: National Center for Health Statistics)

a. Plot the data. Would a linear model fit the points well?

b. Add a third row to the table giving the values of $\ln y$.

c. Plot the coordinate pairs $(x, \ln y)$. Would a linear model fit these points well? If so, draw the best-fitting line and find its equation.

d. Describe the shapes of the two scatter plots. Using your knowledge of logarithms, explain why the second scatter plot is so different from the first.

9.4 Exercises

Discussing the Concepts

1. State the properties of logarithms verbally.

2. Is it true that $\log_4(x^2 + 9) = \log_4 x^2 + \log_4 9$? Explain.

3. If $f(x) = \log_a x$, does $f(ax) = 1 + f(x)$? Explain.

4. If $f(x) = \log_a x$, does $f(a^n) = n$?

Problem Solving

In Exercises 5–8, evaluate the expression without using a calculator. (If it is not possible, state the reason.)

5. $\log_5 5^2$

6. $\log_6 2 + \log_6 3$

7. $\log_2 \frac{1}{8}$

8. $\log_3(-3)$

In Exercises 9–14, use $\log_{10} 3 \approx 0.477$ and $\log_{10} 12 \approx 1.079$ to approximate the value of the expression. Use a calculator to verify your results.

9. $\log_{10} 9$

10. $\log_{10} \frac{1}{4}$

11. $\log_{10} 36$

12. $\log_{10} 144$

13. $\log_{10} \sqrt{36}$

14. $\log_{10} 5^0$

In Exercises 15–22, use the properties of logarithms to expand the expression as a sum, difference, or multiple of logarithms.

15. $\log_2 3x$

16. $\log_3 x^3$

17. $\log_5 x^{-2}$

18. $\log_2 \sqrt{s}$

19. $\ln \dfrac{5}{x-2}$

20. $\ln \sqrt[3]{\dfrac{x^2}{x+1}}$

21. $\ln \sqrt{x(x+2)}$

22. $\ln \left(\dfrac{x+1}{x-1}\right)^2$

In Exercises 23–30, use the properties of logarithms to condense the expression.

23. $\log_{12} x - \log_{12} 3$

24. $\log_6 12 + \log_6 y$

25. $-2\log_5 2x$

26. $10 \log_4 z$

27. $5\ln 2 - \ln x + 3\ln y$

28. $4\ln 2 + 2\ln x - \frac{1}{2}\ln y$

29. $\frac{1}{2}(\ln 8 + \ln 2x)$

30. $5\left[\ln x - \frac{1}{2}\ln(x+4)\right]$

In Exercises 31–36, simplify the expression.

31. $\log_4 \dfrac{4}{x}$

32. $\log_3(3^2 \cdot 4)$

33. $\log_5 \sqrt{50}$

34. $\log_2 \sqrt{22}$

35. $\ln 3e^2$

36. $\ln \dfrac{6}{e^5}$

In Exercises 37 and 38, use a graphing utility to verify that the expressions are equivalent.

37. $\ln\left[10/(x^2+1)\right]^2$ and $2\left[\ln 10 - \ln(x^2+1)\right]$

38. $\ln \sqrt{x(x+1)}$ and $\frac{1}{2}[\ln x + \ln(x+1)]$

39. *Think About It*   Explain how you could show that $(\ln x)/(\ln y) \neq \ln(x/y)$.

40. *Think About It*   Approximate the natural logarithms of as many integers as possible between 1 and 20 using $\ln 2 \approx 0.6931$, $\ln 3 \approx 1.0986$, and $\ln 5 \approx 1.6094$. (Do not use a calculator.)

Biology   In Exercises 41 and 42, use the following information. The energy E (in kilocalories per gram molecule) required to transport a substance from the outside to the inside of a living cell is

$$E = 1.4(\log_{10} C_2 - \log_{10} C_1)$$

where C_1 is the concentration of the substance outside the cell and C_2 is the concentration inside.

41. Condense the expression.

42. The concentration of a particular substance inside a cell is twice the concentration outside the cell. How much energy is required to transport the substance from outside to inside the cell?

Reviewing the Major Concepts

In Exercises 43–48, use the properties of exponents to simplify the expression.

43. $(x^2 \cdot x^3)^4$

44. $4^{-2} \cdot x^2$

45. $\dfrac{15y^{-3}}{10y^2}$

46. $\left(\dfrac{3x^2}{2y}\right)^{-2}$

47. $\dfrac{3x^2 y^3}{18x^{-1} y^2}$

48. $(x^2 + 1)^0$

49. *Ticket Prices* A service organization paid \$288 for a block of tickets to a game. The block contained three more tickets than the organization needed for its members. By inviting three more people to share in the cost, the organization lowered the price per ticket by \$8. How many people will attend?

50. *Geometry* Sketch a diagram of a right triangle whose perimeter is 12 inches.

Additional Problem Solving

In Exercises 51–58, evaluate the expression without using a calculator. (If it is not possible, state the reason.)

51. $\log_4 2 + \log_4 8$

52. $\log_5 50 - \log_5 2$

53. $\log_3 9$

54. $\log_6 \sqrt{6}$

55. $\ln e^5 - \ln e^2$

56. $\ln e^4$

57. $\ln 1$

58. $\log_{10} 1$

In Exercises 59–70, approximate the logarithm given that $\log_4 2 = 0.5000$ and $\log_4 3 \approx 0.7925$. Do not use a calculator.

59. $\log_4 4$

60. $\log_4 8$

61. $\log_4 6$

62. $\log_4 24$

63. $\log_4 \frac{3}{2}$

64. $\log_4 \frac{9}{2}$

65. $\log_4 \sqrt{2}$

66. $\log_4 \sqrt[3]{9}$

67. $\log_4 3 \cdot 2^4$

68. $\log_4 \sqrt{3 \cdot 2^5}$

69. $\log_4 3^0$

70. $\log_4 4^3$

In Exercises 71–84, use the properties of logarithms to expand the expression as a sum, difference, or multiple of logarithms.

71. $\log_3 11x$

72. $\ln 5x$

73. $\ln y^3$

74. $\log_7 x^2$

75. $\log_2 \dfrac{z}{17}$

76. $\log_{10} \dfrac{7}{y}$

77. $\log_3 \sqrt[3]{x+1}$

78. $\log_4 \dfrac{1}{\sqrt{t}}$

79. $\ln 3x^2 y$

80. $\ln y(y-1)^2$

81. $\log_2 \dfrac{x^2}{x-3}$

82. $\log_5 \sqrt{\dfrac{x}{y}}$

83. $\ln \sqrt[3]{x(x+5)}$

84. $\ln\left[3x(x-5)\right]^2$

In Exercises 85–98, use the properties of logarithms to condense the expression.

85. $\log_2 3 + \log_2 x$

86. $\log_5 2x + \log_5 3y$

87. $\log_{10} 4 - \log_{10} x$

88. $\ln 10x - \ln z$

89. $4 \ln b$

90. $-5 \ln (x+3)$

91. $\frac{1}{3} \ln (2x+1)$

92. $-\frac{1}{2} \log_3 5y$

93. $\log_3 2 + \frac{1}{2} \log_3 y$

94. $\ln 6 - 3 \ln z$

95. $2 \ln x + 3 \ln y - \ln z$

96. $4 \ln 3 - 2 \ln x - \ln y$

97. $4(\ln x + \ln y)$

98. $2[\ln x - \ln (x+1)]$

True or False? In Exercises 99–104, determine whether the equation is true or false. Explain.

99. $\ln e^{2-x} = 2 - x$

100. $\log_2 8x = 3 + \log_2 x$

101. $\log_8 4 + \log_8 16 = 2$

102. $\log_3 (u + v) = \log_3 u + \log_3 v$

103. $\log_3 (u + v) = \log_3 u \cdot \log_3 v$

104. $\dfrac{\log_6 10}{\log_6 3} = \log_6 10 - \log_6 3$

In Exercises 105 and 106, use a graphing utility to graph the two equations in the same viewing rectangle. Use the graphs to verify that the expressions are equivalent.

105. $y_1 = \ln\left[x^2(x+2)\right], \quad x > 0$

$y_2 = 2\ln x + \ln(x+2)$

106. $y_1 = \ln\left(\dfrac{\sqrt{x}}{x-3}\right)$

$y_2 = \dfrac{1}{2}\ln x - \ln(x-3)$

107. *Intensity of Sound* The relationship between the number of decibels B and the intensity of a sound I in watts per meter squared is given by

$$B = 10\log_{10}\left(\dfrac{I}{10^{-16}}\right).$$

Use the properties of logarithms to write the formula in simpler form, and determine the number of decibels of a sound with an intensity of 10^{-10} watts per centimeter squared.

108. *Human Memory Model* Students participating in a psychological experiment attended several lectures on a subject. Every month for a year after that, the students were tested to see how much of the material they remembered. The average scores for the group were given by the human memory model

$$f(t) = 80 - \log_{10}(t+1)^{12}, \quad 0 \le t \le 12$$

where t is the time in months.

(a) Find the average scores for the group after 2 months and 8 months.

(b) Use a graphing utility to graph the function.

True or False? In Exercises 109–114, determine if the statement is true or false given that $f(x) = \ln x$. If false, state why or give an example to show that it is false.

109. $f(0) = 0$

110. $f(2x) = \ln 2 + \ln x, \quad x > 0$

111. $f(x-3) = \ln x - \ln 3, \quad x > 3$

112. $\sqrt{f(x)} = \frac{1}{2}\ln x$

113. If $f(u) = 2f(v)$, then $v = u^2$.

114. If $f(x) > 0$, then $x > 1$.

CAREER INTERVIEW

Mary Kay Brown

Research Assistant

Dartmouth Medical School

Hanover, NH 03755

A question that is often asked in the lab is "How much drug does it take to kill 50% of a cell population?" To find out, we set up an assay using different drug concentrations to treat the same number of cells. After a period of several days, the percent of cells that survive at each drug concentration is calculated and graphed. However, there is such a wide range of concentrations (from 0.005 to 100 units in this example) that a linear-scale graph is difficult to read. So we graph this data using a *logarithmic scale*, which gives the same results as graphing $\log_{10}$ of each concentration. The logarithmic scale spreads the points out in a nice curve, and it is easy to see that a drug concentration between one and two units killed 50% of the cell population.

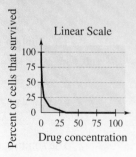

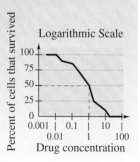

9.5	**Solving Exponential and Logarithmic Equations**
	Exponential and Logarithmic Equations ▪ Solving Exponential Equations ▪ Solving Logarithmic Equations ▪ Application

Exponential and Logarithmic Equations

So far in this chapter, we have focused on the definitions, graphs, and properties of exponential and logarithmic functions. In this section, you will study procedures for *solving equations* that involve exponential or logarithmic expressions. As a simple example, consider the exponential equation $2^x = 16$. By rewriting this equation in the form $2^x = 2^4$, you can see that the solution is $x = 4$. To solve this equation, you can use one of the following properties.

Properties of Exponential and Logarithmic Equations

Let a be a positive real number such that $a \neq 1$, and let x and y be real numbers. Then the following properties are true.

1. $a^x = a^y$ if and only if $x = y$.

2. $\log_a x = \log_a y$ if and only if $x = y$ $(x > 0, \ y > 0)$.

EXAMPLE 1   *Solving Exponential and Logarithmic Equations*

Solve each equation.

a. $4^{x+2} = 4^5$ **b.** $\ln(2x - 3) = \ln 11$

Solution

a. $\quad 4^{x+2} = 4^5$ Original equation

$\quad\quad x + 2 = 5$ Property 1

$\quad\quad\quad\quad x = 3$ Subtract 2 from both sides.

The solution is 3. Check this in the original equation.

b. $\ln(2x - 3) = \ln 11$ Original equation

$\quad\quad\quad 2x - 3 = 11$ Property 2

$\quad\quad\quad\quad\quad 2x = 14$ Add 3 to both sides.

$\quad\quad\quad\quad\quad\quad x = 7$ Divide both sides by 2.

The solution is 7. Check this in the original equation.

Solving Exponential Equations

In Example 1(a), you were able to solve the given equation because both sides of the equation were written in exponential form (with the same base). However, if only one side of the equation is written in exponential form, it is more difficult to solve the equation. For example, how would you solve the following equation?

$$2^x = 7$$

To solve this equation, you must find the power to which 2 can be raised to obtain 7. To do this, you can use one of the following inverse properties of exponents and logarithms.

<table>
<tr><td colspan="2">**Inverse Properties of Exponents and Logarithms**</td></tr>
<tr><td>*Base a*</td><td>*Natural Base e*</td></tr>
<tr><td>**1.** $\log_a(a^x) = x$</td><td>$\ln(e^x) = x$</td></tr>
<tr><td>**2.** $a^{(\log_a x)} = x$</td><td>$e^{(\ln x)} = x$</td></tr>
</table>

EXAMPLE 2   Solving an Exponential Equation

Solve $2^x = 7$.

Solution

To isolate the x, take the $\log_2$ of both sides of the equation, as follows.

$$2^x = 7 \qquad\qquad \text{Original equation}$$
$$x = \log_2 7 \qquad\qquad \text{Inverse property}$$

The solution is $x = \log_2 7 \approx 2.807$. Check this in the original equation.

EXAMPLE 3   Solving an Exponential Equation

Solve $2e^x = 10$.

Solution

$$2e^x = 10 \qquad\qquad \text{Original equation}$$
$$e^x = 5 \qquad\qquad \text{Divide both sides by 2.}$$
$$x = \ln 5 \qquad\qquad \text{Inverse property}$$

The solution is $x = \ln 5 \approx 1.609$. Check this in the original equation.

Technology

You can use logarithmic and exponential functions to derive other functions. Use the definitions of the logarithmic and natural logarithmic functions to show that $e^{(\ln x)} = x$.

Technology

Graphical Check of Solutions

Remember that you can use a graphing utility to solve equations graphically or check solutions that are obtained algebraically. For instance, to check the solutions in Examples 2 and 3, graph both sides of the equations, as shown below.

Graph $y = 2^x$ and $y = 7$. Then approximate the intersection of the two graphs to be $x \approx 2.807$.

Graph $y = 2e^x$ and $y = 10$. Then approximate the intersection of the two graphs to be $x \approx 1.609$.

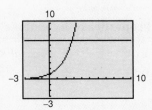

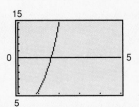

EXAMPLE 4   Solving an Exponential Equation

Solve $5 + e^{x+1} = 20$.

Solution

$$
\begin{aligned}
5 + e^{x+1} &= 20 && \text{Original equation} \\
e^{x+1} &= 15 && \text{Subtract 5 from both sides.} \\
x + 1 &= \ln 15 && \text{Inverse property} \\
x &= -1 + \ln 15 && \text{Subtract 1 from both sides.}
\end{aligned}
$$

The solution is $x = -1 + \ln 15 \approx 1.708$. Check this in the original equation, as follows.

Check

$$
\begin{aligned}
5 + e^{x+1} &= 20 && \text{Original equation} \\
5 + e^{1.708+1} &\overset{?}{=} 20 && \text{Substitute 1.708 for } x. \\
5 + e^{2.708} &\overset{?}{=} 20 && \text{Simplify.} \\
5 + 14.999 &\approx 20 && \text{Solution checks.} \quad
\end{aligned}
$$

Solving Logarithmic Equations

You know how to solve an exponential equation by *taking the logarithms of both sides*. To solve a logarithmic equation, you need to **exponentiate** both sides. For instance, to solve a logarithmic equation such as

$$\ln x = 2$$

you can exponentiate both sides of the equation as follows.

$\ln x = 2$	Original equation
$x = e^2$	Inverse property

This procedure is demonstrated in the next three examples. We suggest the following guidelines for solving exponential and logarithmic equations.

Solving Exponential and Logarithmic Equations

1. To solve an exponential equation, first isolate the exponential expression, then **take the logarithms of both sides of the equation** and solve for the variable.

2. To solve a logarithmic equation, first isolate the logarithmic expression, then **exponentiate both sides of the equation** and solve for the variable.

EXAMPLE 5   *Solving a Logarithmic Equation*

Solve $2 \ln x = 5$.

Solution

$2 \ln x = 5$	Original equation
$\ln x = \dfrac{5}{2}$	Divide both sides by 2.
$x = e^{5/2}$	Inverse property

The solution is $x = e^{5/2} \approx 12.182$. Check this in the original equation, as follows.

Check

$2 \ln x = 5$	Original equation
$2 \ln(12.182) \overset{?}{=} 5$	Substitute 12.182 for x.
$2(2.49996) \overset{?}{=} 5$	Use a calculator.
$4.99992 \approx 5$	Solution checks.

EXAMPLE 6   Solving a Logarithmic Equation

Solve $3\log_{10} x = 6$.

Solution

$$3\log_{10} x = 6 \qquad\qquad \text{Original equation}$$
$$\log_{10} x = 2 \qquad\qquad \text{Divide both sides by 3.}$$
$$x = 100 \qquad\qquad \text{Inverse property}$$

The solution is $x = 100$. Check this in the original equation.

EXAMPLE 7   Solving a Logarithmic Equation

Solve $20\ln 0.2x = 30$.

Solution

$$20\ln 0.2x = 30 \qquad\qquad \text{Original equation}$$
$$\ln 0.2x = 1.5 \qquad\qquad \text{Divide both sides by 20.}$$
$$0.2x = e^{1.5} \qquad\qquad \text{Inverse property}$$
$$x = 5e^{1.5} \qquad\qquad \text{Divide both sides by 0.2.}$$

The solution is $x = 5e^{1.5} \approx 22.408$. Check this in the original equation.

The next example uses logarithmic properties as part of the solution.

EXAMPLE 8   Solving a Logarithmic Equation

Solve $\log_{10} 2x - \log_{10}(x - 3) = 1$.

Solution

$$\log_{10} 2x - \log_{10}(x - 3) = 1 \qquad\qquad \text{Original equation}$$
$$\log_{10} \frac{2x}{x - 3} = 1 \qquad\qquad \text{Condense the left side.}$$
$$\frac{2x}{x - 3} = 10 \qquad\qquad \text{Inverse property}$$
$$2x = 10x - 30 \qquad\qquad \text{Multiply both sides by } x - 3.$$
$$-8x = -30 \qquad\qquad \text{Subtract } 10x \text{ from both sides.}$$
$$x = \frac{15}{4} \qquad\qquad \text{Divide both sides by } -8.$$

The solution is $x = \frac{15}{4}$. Check this in the original equation.

Application

EXAMPLE 9   An Application: Compound Interest

A deposit of $5000 is placed in a savings account for 2 years. The interest for the account is compounded continuously. At the end of 2 years, the balance in the account is $5867.56. What is the annual interest rate for this account?

Solution

Using the formula for continuously compounded interest, $A = Pe^{rt}$, you have the following solution.

Formula:	$A = Pe^{rt}$	
Labels:	Principal $= P = 5000$	(dollars)
	Amount $= A = 5867.56$	(dollars)
	Time $= t = 2$	(years)
	Annual interest rate $= r$	(percent in decimal form)

$$\textit{Equation:} \quad 5867.56 = 5000e^{2r}$$

$$\frac{5867.56}{5000} = e^{2r}$$

$$1.1735 \approx e^{2r}$$

$$\ln 1.1735 \approx \ln(e^{2r})$$

$$0.16 \approx 2r$$

$$0.08 \approx r$$

The rate is 8%. Check this in the original statement of the problem.

Group Activities   R e v i e w i n g   t h e   M a j o r   C o n c e p t s

Solving Equations   Solve each equation.

a. $x^2 - 3x - 4 = 0$

b. $e^{2x} - 3e^x - 4 = 0$

c. $(\ln x)^2 - 3\ln x - 4 = 0$

Explain your strategy. What are the similarities among the three equations? One of the equations has only one solution. Explain why.

9.5 Exercises

Discussing the Concepts

1. State the three basic properties of logarithms.

2. Which equation requires logarithms for its solution:

$2^{x-1} = 32$ or $2^{x-1} = 30$?

3. Explain how to solve $10^{2x-1} = 5316$.

4. In your own words, state the guidelines for solving exponential and logarithmic equations.

Problem Solving

In Exercises 5–10, determine whether the x-values are solutions of the equation.

5. $3^{2x-5} = 27$
   (a) $x = 1$
   (b) $x = 4$

6. $4^{x+3} = 16$
   (a) $x = -1$
   (b) $x = 0$

7. $e^{x+5} = 45$
   (a) $x = -5 + \ln 45$
   (b) $x = -5 + e^{45}$

8. $2^{3x-1} = 324$
   (a) $x \approx 3.1133$
   (b) $x \approx 2.4327$

9. $\log_9(6x) = \frac{3}{2}$
   (a) $x = 27$
   (b) $x = \frac{9}{2}$

10. $\ln(x + 3) = 2.5$
    (a) $x = -3 + e^{2.5}$
    (b) $x \approx 9.1825$

In Exercises 11–18, solve the equation.

11. $3^{x-1} = 3^7$

12. $4^{2x} = 64$

13. $2^{x+2} = \frac{1}{16}$

14. $3^{2-x} = 9$

15. $\log_3(2 - x) = 2$

16. $\log_4(x-4) = \log_4 12$

17. $\ln(2x - 3) = \ln 15$

18. $\ln 3x = \ln 24$

In Exercises 19–22, simplify the expression.

19. $\ln e^{2x-1}$

20. $\log_3 3^{x^2}$

21. $10^{\log_{10} 2x}$

22. $e^{\ln(x+1)}$

In Exercises 23–32, solve the exponential equation. (Round the solution to two decimal places.)

23. $4^x = 8$

24. $2^x = 1.5$

25. $\frac{1}{2}e^{3x} = 20$

26. $6e^{-x} = 3$

27. $5(2)^{3x} - 4 = 13$

28. $-16 + 0.2(10)^x = 35$

29. $8 - 12e^{-x} = 7$

30. $4 - 2e^x = -23$

31. $\dfrac{8000}{(1.03)^t} = 6000$

32. $\dfrac{500}{1 + e^{-0.1t}} = 400$

In Exercises 33–42, solve the logarithmic equation. (Round the solution to two decimal places.)

33. $\log_2 x = 4.5$

34. $\log_4(25x) = 7$

35. $4 \log_3 x = 28$

36. $5 \log_{10}(x + 2) = 18$

37. $16 \ln x = 30$

38. $\ln\left(\frac{1}{2}t\right) = \frac{1}{4}$

39. $1 - 2\ln x = -4$

40. $-5 + 2\ln 3x = 5$

41. $\log_2(x - 1) + \log_2(x + 3) = 3$

42. $\log_{10}(25x) - \log_{10}(x - 1) = 2$

In Exercises 43–46, use a graphing utility to approximate the x-intercept of the graph.

43. $y = 10^{x/2} - 5$

44. $y = 2e^x - 21$

45. $y = 6\ln(0.4x) - 13$

46. $y = 5\log_{10}(x + 1) - 3$

47. *Doubling Time* At 9% interest, compounded continuously, how long does it take an investment to double?

48. *Human Memory Model* The average score A for students who took a test t months after the completion of a course is given by the memory model

$$A = 80 - \log_{10}(t + 1)^{12}.$$

How long after completing the course will the average score fall to $A = 72$?

49. *Friction*   To restrain an untrained horse, a person partially wraps the rope around a cylindrical post in the corral (see figure).   If the horse is pulling on the rope with a force of 200 pounds, the force F in pounds required by the person is

$$F = 200e^{-0.2\pi\theta/180}$$

where θ is the angle of wrap in degrees.   Find the smallest value of θ if F cannot exceed 80 pounds.

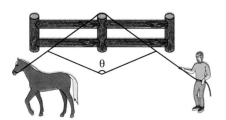

50. *Military Personnel*   The number N (in thousands) of United States military personnel on active duty for the years 1988 through 1992 is modeled by the equation

$$N = 2123.53 - 44.15e^t, \quad -2 \le t \le 2$$

where t is time in years, with $t = 0$ corresponding to 1990.   (Source: U.S. Department of Defense)

(a) Use a graphing utility to graph the equation over the specified domain.

(b) Use the graph of part (a) to estimate the value of t when $N = 2000$.

(c) Answer the question in part (b) algebraically. Compare your result with the result obtained graphically.

Reviewing the Major Concepts

In Exercises 51–54, simplify the expression.

51. $\sqrt{24x^2y^3}$

52. $\sqrt[3]{9} \cdot \sqrt[3]{15}$

53. $(12a^{-4}b^6)^{1/2}$

54. $(16^{1/3})^{3/4}$

55. *Balance*   \$4000 is deposited at 6%, compounded monthly. Find the account balance after 5 years.

56. *Demand for a Product*   The demand per day x and price per unit p for a certain product are related by

$$p = 30 - \sqrt{0.5(x - 1)}.$$

Find the demand if the price is \$26.76.

Additional Problem Solving

In Exercises 57–72, solve the equation. (Do not use a calculator.)

57. $2^x = 2^5$

58. $5^x = 5^3$

59. $3^{x+4} = 3^{12}$

60. $10^{1-x} = 10^4$

61. $4^{x-1} = 16$

62. $3^{2x} = 81$

63. $5^x = \frac{1}{125}$

64. $4^{x+1} = \frac{1}{64}$

65. $\log_5 2x = \log_5 36$

66. $\log_2(x + 3) = \log_2 7$

67. $\ln 5x = \ln 22$

68. $\ln(2x - 3) = \ln 17$

69. $\log_3 x = 4$

70. $\log_5 x = 3$

71. $\log_{10} 2x = 6$

72. $\log_2(3x - 1) = 5$

In Exercises 73–92, solve the exponential equation. (Round the solution to two decimal places.)

73. $2^x = 45$

74. $5^x = 212$

75. $10^{2y} = 52$

76. $12^{x-1} = 1500$

77. $\frac{1}{5}4^{x+2} = 300$

78. $3(2^{t+4}) = 350$

79. $4 + e^{2x} = 150$

80. $500 - e^{x/2} = 35$

81. $23 - 5e^{x+1} = 3$

82. $2e^x + 5 = 115$

83. $300e^{x/2} = 9000$

84. $1000^{0.12x} = 25,000$

85. $6000e^{-2t} = 1200$

86. $10,000e^{-0.1t} = 4000$

87. $32(1.5)^x = 640$

88. $250(1.04)^x = 1000$

89. $\dfrac{1600}{(1.1)^x} = 200$

90. $\dfrac{5000}{(1.05)^x} = 250$

91. $4(1 + e^{x/3}) = 84$

92. $50(3 - e^{2x}) = 125$

In Exercises 93–108, solve the logarithmic equation. (Round the solution to two decimal places.)

93. $\log_{10} x = 0$

94. $\ln x = 1$

95. $\log_{10} x = 3$

96. $\log_{10} x = -2$

97. $\log_{10} 4x = \frac{3}{2}$

98. $\log_{10}(x + 3) = \frac{5}{3}$

99. $\ln x = 2.1$

100. $\ln 2x = 3$

101. $\frac{2}{3}\ln(x + 1) = -1$

102. $8\ln(3x - 2) = 1.5$

103. $2\log_{10}(x + 5) = 15$

104. $-1 + 3\log_{10}\dfrac{x}{2} = 8$

105. $\ln x^2 = 6$

106. $\ln\sqrt{x} = 6.5$

107. $\log_{10} x + \log_{10}(x - 3) = 1$

108. $\log_{10} x + \log_{10}(x + 1) = 0$

In Exercises 109–112, use a graphing utility to approximate the point of intersection of the graphs.

109. $y_1 = 2$
$y_2 = e^x$

110. $y_1 = 2$
$y_2 = \ln x$

111. $y_1 = 3$
$y_2 = 2\ln(x + 3)$

112. $y_1 = 200$
$y_2 = 1000e^{-x/2}$

113. *Intensity of Sound* The relationship between the number of decibels B and the intensity of a sound I in watts per meter squared is given by

$$B = 10\log_{10}\left(\dfrac{I}{10^{-16}}\right).$$

Determine the intensity of a sound I if it registers 75 decibels on an intensity meter.

114. *Doubling Time* Solve the exponential equation

$$10{,}000 = 5000e^{10r}$$

for r to determine the interest rate required for an investment of $5000 to double in value when compounded continuously for 10 years.

115. *Muon Decay* A muon is an elementary particle that is similar to an electron, but much heavier. Muons are unstable—they quickly decay to form electrons and other particles. In an experiment conducted in 1943, the number of muon decays m (of an original 5000 muons) was related to the time T (in microseconds) by the model

$$T = 15.7 - 2.48\ln m.$$

How many decays were recorded when $T = 2.5$?

116. *Oceanography* Oceanographers use the density d (in grams per cubic centimeter) of seawater to obtain information about the circulation of water masses and the rates at which waters of different densities mix. For water with a salinity of 30%, the water temperature T (in degrees Celsius) is related to the density by

$$T = 7.9\ln(1.0245 - d) + 61.84.$$

Find the densities of the subantarctic water and the antarctic bottom water shown in the figure.

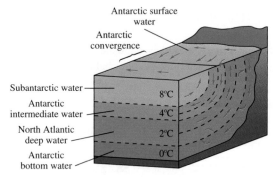

This cross section shows complex currents at various depths in the South Atlantic Ocean off Antarctica.

9.6   **Applications**

Compound Interest ▪ Growth and Decay ▪ Intensity Models

Compound Interest

In Section 9.1, you are introduced to the following two formulas for compound interest. In these formulas, A is the balance, P is the principal, r is the annual interest rate (in decimal form), and t is the time in years.

n Compoundings per Year

$$A = P \left(1 + \frac{r}{n} \right)^{nt}$$

Continuous Compounding

$$A = Pe^{rt}$$

EXAMPLE 1   *Finding the Annual Interest Rate*

An investment of $50,000 is made in an account that compounds interest quarterly. After 4 years, the balance in the account is $71,381.07. What is the annual interest rate for this account?

Solution

Formula: $A = P \left(1 + \frac{r}{n} \right)^{nt}$

Labels: Principal = P = 50,000   (dollars)
Amount = A = 71,381.07   (dollars)
Time = t = 4   (years)
Number of compoundings per year = n = 4
Annual interest rate = r (percent in decimal form)

Equation: $71{,}381.07 = 50{,}000 \left(1 + \frac{r}{4} \right)^{(4)(4)}$

$$\frac{71{,}381.07}{50{,}000} = \left(1 + \frac{r}{4} \right)^{16}$$

$$1.42762 \approx \left(1 + \frac{r}{4} \right)^{16}$$

$(1.42762)^{1/16} \approx 1 + \frac{r}{4}$ Raise both sides to $\frac{1}{16}$ power.

$$1.0225 \approx 1 + \frac{r}{4}$$

$$0.09 \approx r$$

NOTE   Notice in Example 1 that you can "get rid of" the 16th power on the right side of the equation by raising both sides of the equation to the $\frac{1}{16}$ power.

The annual interest rate is approximately 9%. Check this in the original problem.

EXAMPLE 2 Doubling Time for Continuous Compounding

An investment is made in a trust fund at an annual interest rate of 8.75%, compounded continuously. How long will it take for the investment to double?

Solution

$$A = Pe^{rt}$$
Formula for continuous compounding

$$2P = Pe^{0.0875t}$$
Substitute known values.

$$2 = e^{0.0875t}$$
Divide both sides by P.

$$\ln 2 = 0.0875t$$
Inverse property

$$\frac{\ln 2}{0.0875} = t$$
Divide both sides by 0.0875.

$$7.92 \approx t$$

NOTE In Example 2, note that you do not need to know the principal to find the doubling time. In other words, the doubling time is the same for *any* principal.

It will take approximately 7.92 years for the investment to double. Check this in the original problem.

EXAMPLE 3 Finding the Type of Compounding

You deposit $1000 in an account. At the end of 1 year your balance is $1077.63. If the bank tells you that the annual interest rate for the account is 7.5%, how was the interest compounded?

Solution

If the interest had been compounded continuously at 7.5%, the balance would have been

$$A = 1000e^{(0.075)(1)} = \$1077.88.$$

Because the actual balance is slightly less than this, you should use the formula for interest that is compounded n times per year.

$$A = 1000 \left(1 + \frac{0.075}{n}\right)^n = 1077.63$$

At this point, it is not clear what you should do to solve the equation for n. However, by completing a table, you can see that $n = 12$. Thus, the interest was compounded monthly.

n	1	4	12	365
$\left(1 + \dfrac{0.075}{n}\right)^n$	1.075	1.07714	1.07763	1.07788

In Example 3, notice that an investment of $1000 compounded monthly produced a balance of $1077.63 at the end of 1 year. Because $77.63 of this amount is interest, the **effective yield** for the investment is

$$\text{Effective yield} = \frac{\text{year's interest}}{\text{amount invested}} = \frac{77.63}{1000} = 0.07763 = 7.763\%.$$

In other words, the effective yield for an investment collecting compound interest is the *simple interest rate* that would yield the same balance at the end of 1 year.

EXAMPLE 4 *Finding the Effective Yield*

An investment is made in an account that pays 6.75% interest, compounded continuously. What is the effective yield for this investment?

Solution

Notice that you do not have to know the principal or the time that the money will be left in the account. Instead, you can choose an arbitrary principal, such as $1000. Then, because effective yield is based on the balance at the end of 1 year, you can use the following formula.

$$A = Pe^{rt} = 1000e^{0.0675(1)} = 1069.83$$

Now, because the account would earn $69.83 in interest after 1 year for a principal of $1000, you can conclude that the effective yield is

$$\text{Effective yield} = \frac{69.83}{1000} = 0.06983 = 6.983\%.$$

Growth and Decay

The balance in an account earning *continuously* compounded interest is one example of a quantity that increases over time according to the **exponential growth model** $y = Ce^{kt}$.

Exponential Growth and Decay

The mathematical model for exponential growth or decay is given by

$$y = Ce^{kt}.$$

For this model, t is the time, C is the original amount of the quantity, and y is the amount after the time t. The number k is a constant that is determined by the rate of growth. If $k > 0$, the model represents **exponential growth,** and if $k < 0$, it represents **exponential decay.**

One common application of exponential growth is in modeling the growth of a population, as shown in Example 5.

EXAMPLE 5   Population Growth

A country's population was 2 million in 1980 and 3 million in 1990. What would you predict the population of the country to be in the year 2000?

Solution

If you assumed a *linear growth model*, you would simply predict the population in the year 2000 to be 4 million. If, however, you assumed an *exponential growth model*, the model would have the form

$$y = Ce^{kt}.$$

In this model, let $t = 0$ represent the year 1980. The given information about the population can be described by the following table.

t (years)	0	10	20
Ce^{kt} (million)	$Ce^{k(0)} = 2$	$Ce^{k(10)} = 3$	$Ce^{k(20)} = ?$

To find the population when $t = 20$, you must first find the values of C and k. From the table, you can use the fact that $Ce^{k(0)} = Ce^0 = 2$ to conclude that $C = 2$. Then, using this value of C, you solve for k as follows.

$$Ce^{k(10)} = 3 \qquad \text{From table}$$

$$2e^{10k} = 3 \qquad \text{Substitute value of } C.$$

$$e^{10k} = \frac{3}{2} \qquad \text{Divide both sides by 2.}$$

$$10k = \ln \frac{3}{2} \qquad \text{Inverse property}$$

$$k = \frac{1}{10} \ln \frac{3}{2} \qquad \text{Divide both sides by 10.}$$

$$k \approx 0.0405 \qquad \text{Simplify.}$$

Finally, you can use this value of k to conclude that the population in the year 2000 is given by

$$2e^{0.0405(20)} \approx 2(2.25) = 4.5 \text{ million.}$$

Figure 9.16 graphically compares the exponential growth model with a linear growth model.

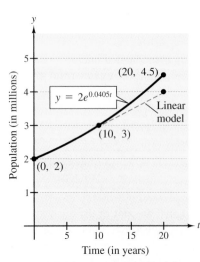

FIGURE 9.16   Population Models

EXAMPLE 6 *Radioactive Decay*

Radioactive iodine is a by-product of some types of nuclear reactors. Its **half-life** is 60 days. That is, after 60 days, a given amount of radioactive iodine will have decayed to half the original amount. Suppose a nuclear accident occurs and releases 20 grams of radioactive iodine. How long will it take for the radioactive iodine to decay to a level of 1 gram?

Solution

To solve this problem, use the model for exponential decay.

$$y = Ce^{kt}$$

Next, use the information given in the problem to set up the following table.

t (days)	0	60	?
Ce^{kt} (grams)	$Ce^{k(0)} = 20$	$Ce^{k(60)} = 10$	$Ce^{k(t)} = 1$

Because $Ce^{k(0)} = Ce^0 = 20$, you can conclude that $C = 20$. Then, using this value of C, you can solve for k, as follows.

$Ce^{k(60)} = 10$	From table
$20e^{60k} = 10$	Substitute value of C.
$e^{60k} = \dfrac{1}{2}$	Divide both sides by 20.
$60k = \ln \dfrac{1}{2}$	Inverse property
$k = \dfrac{1}{60} \ln \dfrac{1}{2} \approx -0.01155$	Divide both sides by 60 and simplify.

Finally, you can use this value of k to find the time when the amount is 1 gram, as follows.

$Ce^{kt} = 1$	From table
$20e^{-0.01155t} = 1$	Substitute values of C and k.
$e^{-0.01155t} = \dfrac{1}{20}$	Divide both sides by 20.
$-0.01155t = \ln \dfrac{1}{20}$	Inverse property
$t = \dfrac{1}{-0.01155} \ln \dfrac{1}{20} \approx 259.4 \text{ days}$	Divide both sides by -0.01155 and simplify.

Thus, 20 grams of radioactive iodine will have decayed to 1 gram after about 259.4 days. This solution is shown graphically in Figure 9.17.

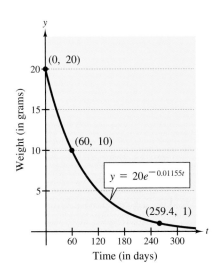

FIGURE 9.17 Radioactive Decay

Earthquakes take place along faults in the earth's crust. The 1989 earthquake in California took place along the San Andreas Fault.

Intensity Models

On the **Richter scale,** the magnitude R of an earthquake can be measured by the **intensity model**

$$R = \log_{10} I$$

where I is the intensity of the shock wave.

EXAMPLE 7 Earthquake Intensity

In 1906, San Francisco experienced an earthquake that measured 8.6 on the Richter scale. In 1989, another earthquake, which measured 7.7 on the Richter scale, struck the same area. Compare the intensities of these two earthquakes.

Solution

The intensity of the 1906 earthquake is given as follows.

$$8.6 = \log_{10} I \qquad \text{Given}$$
$$10^{8.6} = I \qquad \text{Inverse property}$$

The intensity of the 1989 earthquake can be found in a similar way.

$$7.7 = \log_{10} I \qquad \text{Given}$$
$$10^{7.7} = I \qquad \text{Inverse property}$$

The ratio of these two intensities is

$$\frac{I \text{ for } 1906}{I \text{ for } 1989} = \frac{10^{8.6}}{10^{7.7}} = 10^{8.6-7.7} = 10^{0.9} \approx 7.94.$$

Thus, the 1906 earthquake had an intensity that was about eight times greater than the 1989 earthquake.

Group Activities   You Be the Instructor

Problem Posing   Write a problem that could be answered by investigating the exponential growth model $y = 10e^{0.08t}$ or the exponential decay model $y = 5e^{-0.25t}$. Exchange your problem for that of another group member, and solve one another's problems.

9.6 Exercises

Discussing the Concepts

1. If the equation $y = Ce^{kt}$ models exponential growth, what must be true about k?

2. If the equation $y = Ce^{kt}$ models exponential decay, what must be true about k?

3. The formulas for periodic and continuous compounding have the four variables A, P, r, and t in common. Explain what each variable measures.

4. What is meant by the effective yield of an investment? Explain how it is computed.

5. In your own words, explain what is meant by the half-life of a radioactive isotope.

6. If the reading on the Richter scale is increased by 1, the intensity of the earthquake is increased by what factor?

Problem Solving

Annual Interest Rate   In Exercises 7–10, find the annual interest rate.

	Principal	Balance	Time	Compounding
7.	$1000	$36,581.00	40 years	Daily
8.	$200	$314.85	5 years	Yearly
9.	$750	$8267.38	30 years	Continuous
10.	$2000	$4234.00	10 years	Continuous

Doubling Time   In Exercises 11–14, find the time for an investment to double. Explain how to use a graphing utility to graphically check your result.

	Principal	Rate	Compounding
11.	$6000	8%	Quarterly
12.	$500	$5\frac{1}{4}\%$	Monthly
13.	$2000	10.5%	Daily
14.	$100	6%	Continuous

Type of Compounding   In Exercises 15–18, determine the type of compounding.

	Principal	Balance	Time	Rate
15.	$750	$1587.75	10 years	7.5%
16.	$10,000	$73,890.56	20 years	10%
17.	$100	$141.48	5 years	7%
18.	$4000	$4788.76	2 years	9%

In Exercises 19–22, find the effective yield.

	Rate	Compounding
19.	8%	Continuous
20.	9.5%	Daily
21.	7%	Monthly
22.	8%	Yearly

In Exercises 23–26, find the principal that must be deposited in an account to obtain the given balance.

	Balance	Rate	Time	Compounding
23.	$10,000	9%	20	Continuous
24.	$5000	8%	5	Continuous
25.	$750	6%	3	Daily
26.	$3000	7%	10	Monthly

Exponential Growth and Decay   In Exercises 27–30, find the constant k such that the graph of $y = Ce^{kt}$ passes through the given points.

27.

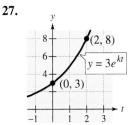

$y = 3e^{kt}$

28.

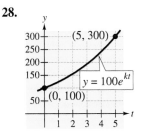

$y = 100e^{kt}$

29.

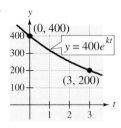

30.

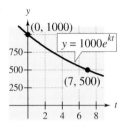

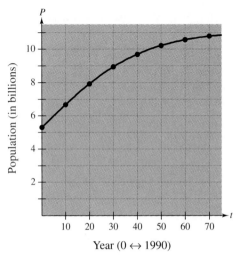

Figure for 36

Population of a City In Exercises 31–34, the population of a city for the year 1992 and the predicted population for the year 2000 are given. Find the constants C and k to obtain the exponential growth model $y = Ce^{kt}$ for the population growth. (Let $t = 0$ correspond to the year 1992.) Use the model to predict the population of the city in the year 2005. (Source: U.S. Bureau of the Census)

City	1992	2000
31. Los Angeles	10.1 million	10.7 million
32. Houston	2.4 million	2.7 million
33. Dhaka, Bangladesh	4.6 million	6.5 million
34. Lagos, Nigeria	8.5 million	12.5 million

35. *Rate of Growth*

  (a) Compare the values of k in Exercises 31 and 33. Which is larger? Explain.

  (b) What variable in the continuous compounding interest formula is equivalent to k in the model for population growth? Use your answer to give an interpretation of k.

36. *Graphical Estimation* The figure shows the population P (in billions) of the world as projected by the Population Reference Bureau. The Bureau's projection can be modeled by

$$P = \frac{11.14}{1 + 1.101e^{-0.051t}}$$

where $t = 0$ represents 1990.

  (a) Estimate the world population in 2020.

  (b) Use a graphing utility to estimate the year when the population will be twice the 1990 population.

37. *Radioactive Decay* Radioactive radium (Ra^{226}) has a half-life of 1620 years. If you start with 5 grams of the isotope, how much remains after 1000 years?

38. *Radioactive Decay* The isotope Pu^{230} has a half-life of 24,360 years. If you start with 10 grams of the isotope, how much remains after 10,000 years?

39. *Earthquake Intensity* On March 27, 1964, Alaska had an earthquake that measured 8.4 on the Richter scale. On February 9, 1971, an earthquake in the San Fernando Valley measured 6.6 on the Richter scale. Compare the intensities of these two earthquakes.

40. *Earthquake Intensity* On March 10, 1933, Long Beach, California had an earthquake that measured 6.2 on the Richter scale. On December 3, 1988, an earthquake in Pasadena, California measured 5.0 on the Richter scale. Compare the intensities of these two earthquakes.

Acidity Model In Exercises 41 and 42, use the acidity model $pH = -\log_{10} [H^+]$, where acidity (pH) is a measure of the hydrogen ion concentration $[H^+]$ (measured in moles of hydrogen per liter) of solution.

41. Find the pH of a solution that has a hydrogen ion concentration of 9.2×10^{-8}.

42. Compute the hydrogen ion concentration if the pH of a solution is 4.7.

Reviewing the Major Concepts

In Exercises 43–46, simplify the compound fraction.

43. $\dfrac{\left(\dfrac{9}{x}\right)}{\left(\dfrac{6}{x}+2\right)}$

44. $\dfrac{\left(1+\dfrac{2}{x}\right)}{\left(x-\dfrac{4}{x}\right)}$

45. $\dfrac{\left(\dfrac{4}{x^2-9}+\dfrac{2}{x-2}\right)}{\left(\dfrac{1}{x+3}+\dfrac{1}{x-3}\right)}$

46. $\dfrac{\left(\dfrac{1}{x+1}+\dfrac{1}{2}\right)}{\left(\dfrac{3}{2x^2+4x+2}\right)}$

47. *Equal Parts*   Find two real numbers that divide the real number line between $x/2$ and $4x/3$ into three equal parts.

48. *Capacitance*   When two capacitors, with capacitance C_1 and C_2, respectively, are connected in series, the equivalent capacitance is given by

$$\frac{1}{\left(\dfrac{1}{C_1}+\dfrac{1}{C_2}\right)}.$$

Simplify this compound fraction.

Additional Problem Solving

Annual Interest Rate   In Exercises 49–54, find the annual interest rate.

	Principal	Balance	Time	Compounding
49.	$500	$1004.83	10 years	Monthly
50.	$3000	$21,628.70	20 years	Quarterly
51.	$5000	$22,405.68	25 years	Daily
52.	$10,000	$110,202.78	30 years	Daily
53.	$1500	$24,666.97	40 years	Continuous
54.	$7500	$15,877.50	15 years	Continuous

Doubling Time   In Exercises 55–60, find the time for an investment to double.

	Principal	Rate	Compounding
55.	$300	5%	Yearly
56.	$10,000	9.5%	Yearly
57.	$6000	7%	Quarterly
58.	$500	9%	Daily
59.	$1500	7.5%	Continuous
60.	$12,000	4%	Continuous

In Exercises 61–66, find the effective yield.

	Rate	Compounding
61.	6%	Quarterly
62.	9%	Quarterly
63.	8%	Monthly
64.	$5\frac{1}{4}\%$	Daily
65.	7.5%	Continuous
66.	4%	Continuous

67. *Doubling Time*   Is it necessary to know the principal P to find the doubling time in Exercises 11–14 and Exercises 55–60? Explain.

68. *Effective Yield*

(a) Is it necessary to know the principal P to find the effective yield in Exercises 61–66? Explain.

(b) When the interest is compounded more frequently, what inference can you make about the difference between the effective yield and the annual interest rate?

In Exercises 69–74, find the principal that must be deposited in an account to obtain the given balance.

	Balance	Rate	Time	Compounding
69.	$25,000	7%	30 years	Monthly
70.	$8000	6%	2 years	Monthly
71.	$1000	5%	1 year	Daily
72.	$100,000	9%	40 years	Daily
73.	$500,000	8%	25 years	Continuous
74.	$1,000,000	10%	50 years	Continuous

Balance After Monthly Deposits In Exercises 75–78, you make monthly deposits of P dollars in a savings account at an annual interest rate r, compounded continuously. Find the balance A after t years given that

$$A = \frac{P(e^{rt} - 1)}{e^{r/12} - 1}.$$

Principal	Interest Rate	Time
75. $P = 30$	$r = 8\%$	$t = 10$ years
76. $P = 100$	$r = 9\%$	$t = 30$ years
77. $P = 50$	$r = 10\%$	$t = 40$ years
78. $P = 20$	$r = 7\%$	$t = 20$ years

Reading a Graph In Exercises 79 and 80, you make monthly deposits of $30 in a savings account at an annual interest rate of 8%, compounded continuously (see figure).

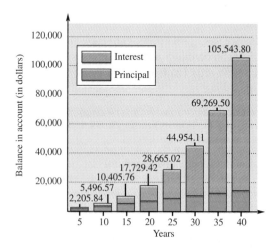

79. Find the total amount that you deposited in the account in 20 years and the total interest earned.

80. Find the total amount that you deposited in the account in 40 years and the total interest earned.

81. *Radioactive Decay* The isotope Pu^{230} has a half-life of 24,360 years. If you start with 2 grams of this isotope, how much remains after 30,000 years?

82. *Radioactive Decay* The isotope C^{14} has a half-life of 5730 years. If you start with 5 grams of this isotope, how much remains after 1000 years?

Population of a City In Exercises 83–86, the population of a city for the year 1992 and the predicted population for the year 2000 are given. Find the constants C and k to obtain the exponential growth model $y = Ce^{kt}$ for the population growth. (Let $t = 0$ correspond to the year 1992.) Use the model to predict the population of the city in the year 2005. (Source: U.S. Bureau of the Census)

City	1992	2000
83. Chicago	6.5 million	6.6 million
84. San Francisco	4.0 million	4.2 million
85. Mexico City, Mexico	21.6 million	27.9 million
86. Sao Paulo, Brazil	19.3 million	25.4 million

87. *Graphical Interpretation* The population p of a certain species t years after it is introduced into a new habitat is given by

$$p(t) = \frac{5000}{1 + 4e^{-t/6}}.$$

(a) Use a graphing utility to graph the population function.

(b) Determine the population size that was introduced into the habitat.

(c) Determine the population size after 9 years.

(d) After how many years will the population be 2000?

88. *Carbon 14 Dating* C^{14} dating assumes that the carbon dioxide on earth today has the same radioactive content as it did centuries ago. If this is true, the amount of C^{14} absorbed by a tree that grew several centuries ago should be the same as the amount of C^{14} absorbed by a tree growing today. A piece of ancient charcoal contains only 15% as much of the radioactive carbon as a piece of modern charcoal. How long ago did the tree burn to make the ancient charcoal if the half-life of C^{14} is 5730 years? (Round your answer to the nearest 100 years.)

89. *Depreciation* A car that cost $22,000 new has a depreciated value of $16,500 after 1 year. Find the value of the car when it is 3 years old by using the exponential model $y = Ce^{kt}$.

90. *Graphical Estimation* After x years, the value y of a truck that cost $32,000$ is given by $y = 32,000(0.8)^x$.

(a) Use a graphing utility to graph the equation.

(b) Graphically approximate the value after 1 year.

(c) Graphically approximate the time when the truck's value will be $16,000.

91. *Graphical Estimation* Annual sales y of a product x years after it is introduced are approximated by

$$y = \frac{2000}{1 + 4e^{-x/2}}.$$

(a) Use a graphing utility to graph the equation.

(b) Graphically approximate sales when $x = 4$.

(c) Graphically approximate the time when annual sales will be 1100 units.

(d) Graphically estimate the maximum level annual sales will approach.

92. *Advertising Effect* The sales S (in thousands of units) of a product after spending x hundred dollars in advertising are given by

$$S = 10(1 - e^{kx}).$$

(a) Find S as a function of x if 2500 units are sold when $500 is spent on advertising.

(b) How many units will be sold if advertising expenditures are raised to $700?

Earthquake Intensity In Exercises 93 and 94, compare the earthquake intensities.

93. On August 16, 1906, Chile had an earthquake that measured 8.6 on the Richter scale. On December 7, 1988, an earthquake in Armenia, USSR measured 6.8 on the Richter scale.

94. On September 19, 1985, Mexico City, Mexico had an earthquake that measured 8.1 on the Richter scale. On August 20, 1988, an earthquake in Nepal measured 6.5 on the Richter scale.

Acidity Model In Exercises 95 and 96, use the acidity model $pH = -\log_{10}[H^+]$, where acidity (pH) is a measure of the hydrogen ion concentration $[H^+]$ (measured in moles of hydrogen per liter) of solution.

95. A certain fruit has a pH of 2.5 and an antacid tablet has a pH of 9.5. The hydrogen ion concentration of the fruit is how many times the concentration of the tablet?

96. If the pH of a solution is decreased by one unit, the hydrogen ion concentration is increased by what factor?

97. *Comparing Models* The figure gives the earnings per share of common stock for the years 1981 through 1993 for Automatic Data Processing, Inc. A list of models ($t = 1$ represents 1981) for the data is also given. For each of the models, (a) use a graphing utility to obtain its graph, and (b) find the sum of the squares of the differences between the actual data and the approximations given by the model. Use this sum to determine which model "best fits" the data. (Source: Automatic Data Processing, Inc.)

Linear: $E = 0.146t + 0.007$

Quadratic: $E = 0.009t^2 + 0.018t + 0.325$

Exponential: $E = 0.301(1.165)^t$

Exponential: $E = 0.301e^{0.153t}$

Logarithmic: $E = 0.201 + 0.228t - 0.444\ln t$

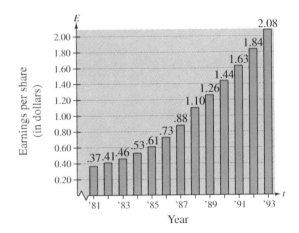

CHAPTER PROJECT: Radioactivity

Radioactive elements occur naturally. Some of these elements have proven to be useful. For instance, carbon 14 dating enables people to determine the ages of prehistoric animals and plants. Modern medicine uses iodine 131 to diagnose thyroid conditions, iron 59 to study red blood cells, technetium 99 to assess heart damage, and sodium 24 to study the circulatory system.

In large quantities, however, radioactive elements can cause serious illness. For instance, radon 222 is a radioactive gas that is produced by the decay of uranium 238. Overexposure to radon 222 can lead to lung cancer.

The nucleus of a radioactive element decays into other elements until the nucleus of the new element becomes stable. The uranium 238 decay series includes 14 radioactive elements. When the last radioactive element in this series, polonium 210, decays, it forms lead 206, a stable nonradioactive element.

Investigate the following questions.

1. Consider the exponential decay model $y = Ce^{kt}$, where y is the amount of radioactive substance left after time t, C is the beginning amount of the substance, and k is the rate constant of the radioactive element. The half-life of iodine 131 is 8.1 days. Determine its rate constant k.

2. Radon 222 is a radioactive gas that can be found in the basements of homes. It is one of the radioactive elements produced by the decay of uranium 238. Use the model for exponential decay $y = Ce^{kt}$ and the information in the table to determine the half-life of radon 222.

t (days)	0	2	?
Ce^{kt} (grams)	$Ce^{kt} = 10$	$Ce^{kt} = 6.9565$	$Ce^{kt} = 5$

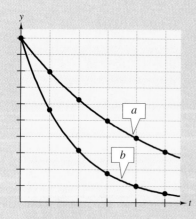

3. Technetium 99 has a half-life of 6.0 hours and sodium 24 has a half-life of 14.8 hours. Decay models for both are graphed on the same figure at the left. Determine the scales used for the x- and y-axes and write a paragraph describing how to determine which graph is the model for technetium 99 and which graph is the model for sodium 24.

4. The radioactive element iron 59 has a half-life of 45.1 days. Determine an algebraic decay model for 10 grams of the element. Use the model to complete the table below and draw a graph of the model.

t (days)	0	45.1	90.2	?	?
y (grams)	10	?	?	2.150	0.462

CHAPTER SUMMARY

After studying this chapter, you should have acquired the following skills. These skills are keyed to the Review Exercises that begin on page 628. Answers to odd-numbered Review Exercises are given in the back of the book.

- Evaluate exponential and logarithmic functions for given values of the variable. *(Sections 9.1, 9.3)* — **Review Exercises 1–10**

- Match exponential and logarithmic functions with their graphs. *(Sections 9.1, 9.3)* — **Review Exercises 11–16**

- Sketch the graphs of exponential and logarithmic functions. *(Sections 9.1, 9.3)* — **Review Exercises 17–26**

- Graph exponential and logarithmic functions using a graphing utility. *(Sections 9.1, 9.3)* — **Review Exercises 27–34**

- Find composite functions. *(Sections 9.2, 9.5)* — **Review Exercises 35–38**

- Find composite functions and determine their domains. *(Section 9.2)* — **Review Exercises 39, 40**

- Determine if functions have inverses using a graphing utility. *(Section 9.2)* — **Review Exercises 41–44**

- Find the inverses of functions. *(Section 9.2)* — **Review Exercises 45–50**

- Write exponential equations in logarithmic form and logarithmic equations in exponential form. *(Section 9.3)* — **Review Exercises 51–54**

- Evaluate logarithmic expressions. *(Sections 9.3, 9.4)* — **Review Exercises 55–62**

- Rewrite logarithmic expressions in expanded form using the properties of logarithms. *(Section 9.4)* — **Review Exercises 63–68**

- Rewrite logarithmic expressions in condensed form using the properties of logarithms. *(Section 9.4)* — **Review Exercises 69–74**

- Graphically verify that two logarithmic expressions are equivalent using a graphing utility. *(Section 9.4)* — **Review Exercises 75, 76**

- Decide whether exponential and logarithmic equations are true or false. *(Section 9.4)* — **Review Exercises 77–82**

- Approximate values of logarithmic expressions given the values of specific logarithms. *(Section 9.4)* — **Review Exercises 83–88**

- Evaluate logarithmic expressions using the change-of-base formula. *(Section 9.3)* — **Review Exercises 89–92**

- Solve exponential and logarithmic equations. *(Section 9.5)* — **Review Exercises 93–110**

- Solve real-life problems involving exponential and logarithmic functions. *(Sections 9.1, 9.3, 9.4, 9.5, 9.6)* — **Review Exercises 111–122**

REVIEW EXERCISES

In Exercises 1–10, evaluate the function as indicated.

1. $f(x) = 2^x$

   (a) $x = -3$ (b) $x = 1$ (c) $x = 2$

2. $g(x) = 2^{-x}$

   (a) $x = -2$ (b) $x = 0$ (c) $x = 2$

3. $g(t) = e^{-t/3}$

   (a) $t = -3$ (b) $t = \pi$ (c) $t = 6$

4. $h(s) = 1 - e^{0.2s}$

   (a) $s = 0$ (b) $s = 2$ (c) $s = \sqrt{10}$

5. $f(x) = \log_3 x$

   (a) $x = 1$ (b) $x = 27$ (c) $x = 0.5$

6. $g(x) = \log_{10} x$

   (a) $x = 0.01$ (b) $x = 0.1$ (c) $x = 30$

7. $f(x) = \ln x$

   (a) $x = e$ (b) $x = \frac{1}{3}$ (c) $x = 10$

8. $h(x) = \ln x$

   (a) $x = e^2$ (b) $x = \frac{5}{4}$ (c) $x = 1200$

9. $g(x) = \ln e^{3x}$

   (a) $x = -2$ (b) $x = 0$ (c) $x = 7.5$

10. $f(x) = \log_2 \sqrt{x}$

   (a) $x = 4$ (b) $x = 64$ (c) $x = 5.2$

In Exercises 11–16, match the function with its graph. [The graphs are labeled (a), (b), (c), (d), (e), and (f).]

11. $f(x) = 2^x$ **12.** $f(x) = 2^{-x}$

13. $f(x) = -2^x$ **14.** $f(x) = 2^x + 1$

15. $f(x) = \log_2 x$ **16.** $f(x) = \log_2(x - 1)$

(a)

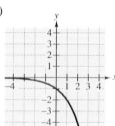

(b)

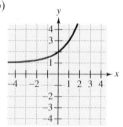

(c)

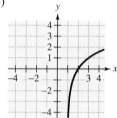

(d)

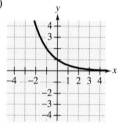

(e)

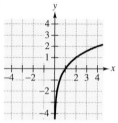

(f)

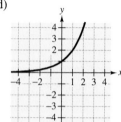

In Exercises 17–26, sketch the graph of the function.

17. $y = 3^{x/2}$ **18.** $y = 3^{x/2} - 2$

19. $f(x) = 3^{-x/2}$ **20.** $f(x) = -3^{-x/2}$

21. $f(x) = 3^{-x^2}$ **22.** $g(t) = 3^{|t|}$

23. $f(x) = -2 + \log_3 x$ **24.** $f(x) = 2 + \log_3 x$

25. $y = \log_2(x - 4)$ **26.** $y = 2\log_4(x + 1)$

In Exercises 27–34, use a graphing utility to graph the function.

27. $y = 5e^{-x/4}$ **28.** $y = 6 - e^{x/2}$

29. $f(x) = e^{x+2}$ **30.** $h(t) = \dfrac{8}{1 + e^{-t/5}}$

31. $g(x) = \ln 2x$ **32.** $f(x) = 3 + \ln x$

33. $g(t) = 2 - \ln(t - 1)$ **34.** $g(x) = \ln(x - 5)$

In Exercises 35–38, find $(f \circ g)(x)$ and $(g \circ f)(x)$.

35. $f(x) = x + 2$, $g(x) = x^2$

36. $f(x) = \sqrt[3]{x}$, $g(x) = x + 2$

37. $f(x) = \sqrt{x + 1}$, $g(x) = x^2 - 1$

38. $f(x) = e^x$, $g(x) = \ln x$

In Exercises 39 and 40, find the domains of the compositions (a) $(f \circ g)$ and (b) $(g \circ f)$.

39. $f(x) = \sqrt{x - 4}, \quad g(x) = 2x$

40. $f(x) = \dfrac{2}{x - 4}, \quad g(x) = x^2$

In Exercises 41–44, use a graphing utility to decide if the function has an inverse. Explain your reasoning.

41. $f(x) = x^2 - 25$ **42.** $f(x) = \frac{1}{4}x^3$

43. $h(x) = 4\sqrt[3]{x}$ **44.** $g(x) = \sqrt{9 - x^2}$

In Exercises 45–50, find the inverse of the function. (If it is not possible, state the reason.)

45. $f(x) = \frac{1}{4}x$ **46.** $f(x) = 2x - 3$

47. $h(x) = \sqrt{x}$ **48.** $g(x) = x^2 + 2, \quad x \ge 0$

49. $f(t) = |t + 3|$ **50.** $h(t) = t$

In Exercises 51 and 52, write the exponential equation in logarithmic form.

51. $4^3 = 64$ **52.** $25^{3/2} = 125$

In Exercises 53 and 54, write the logarithmic equation in exponential form.

53. $\ln e = 1$ **54.** $\log_3 \frac{1}{9} = -2$

In Exercises 55–62, evaluate the expression.

55. $\log_{10} 1000$ **56.** $\log_9 3$

57. $\log_3 \frac{1}{9}$ **58.** $\log_4 \frac{1}{16}$

59. $\ln e^7$ **60.** $\log_a \dfrac{1}{a}$

61. $\ln 1$ **62.** $\ln e^{-3}$

In Exercises 63–68, use the properties of logarithms to expand the expression.

63. $\log_4 6x^4$ **64.** $\log_{10} 2x^{-3}$

65. $\log_5 \sqrt{x + 2}$ **66.** $\ln \sqrt[3]{\frac{1}{5}x}$

67. $\ln \dfrac{x + 2}{x - 2}$ **68.** $\ln x(x - 3)^2$

In Exercises 69–74, use properties of logarithms to condense the expression.

69. $\log_4 x - \log_4 10$ **70.** $5 \log_2 y$

71. $4(1 + \ln x + \ln x)$ **72.** $\log_8 16x + \log_8 2x^2$

73. $-2(\ln 2x - \ln 3)$ **74.** $-\frac{2}{3} \ln 3y$

In Exercises 75 and 76, use a graphing utility to graphically verify that the expressions are equivalent.

75. $y_1 = \ln \left(\dfrac{x + 1}{x - 1}\right)^2, \quad x > 1$

$y_2 = 2[\ln(x + 1) - \ln(x - 1)]$

76. $y_1 = \ln \sqrt{x^2 - 4}, \quad x > 2$

$y_2 = \frac{1}{2}[\ln(x + 2) + \ln(x - 2)]$

True or False? In Exercises 77–82, decide whether the equation is true or false. Explain your reasoning.

77. $\log_2 4x = 2 \log_2 x$ **78.** $\dfrac{\ln 5x}{\ln 10x} = \ln \dfrac{1}{2}$

79. $\log_{10} 10^{2x} = 2x$ **80.** $e^{\ln t} = t$

81. $\log_4 \dfrac{16}{x} = 2 - \log_4 x$

82. $e^{2x} - 1 = (e^x + 1)(e^x - 1)$

In Exercises 83–88, approximate the logarithm given that $\log_5 2 \approx 0.43068$ and $\log_5 3 \approx 0.6826$.

83. $\log_5 18$ **84.** $\log_5 \sqrt{6}$

85. $\log_5 \frac{1}{2}$ **86.** $\log_5 \frac{2}{3}$

87. $\log_5 (12)^{2/3}$ **88.** $\log_5 (5^2 \cdot 6)$

In Exercises 89–92, evaluate the logarithm using the change-of-base formula. Round each logarithm to three decimal places.

89. $\log_4 9$ **90.** $\log_{1/2} 5$

91. $\log_{12} 200$ **92.** $\log_3 0.28$

In Exercises 93–98, solve the equation.

93. $2^x = 64$ **94.** $3^{x-2} = 81$

95. $4^{x-3} = \frac{1}{16}$

96. $\log_2 2x = \log_2 100$

97. $\log_3 x = 5$

98. $\log_5(x - 10) = 2$

In Exercises 99–110, solve the equation. (Round your answer to two decimal places.)

99. $3^x = 500$

100. $8^x = 1000$

101. $2e^{x/2} = 45$

102. $100e^{-0.6x} = 20$

103. $\dfrac{500}{(1.05)^x} = 100$

104. $25(1 - e^t) = 12$

105. $\log_{10} 2x = 1.5$

106. $\frac{1}{3}\log_2 x + 5 = 7$

107. $\ln x = 7.25$

108. $\ln x = -0.5$

109. $\log_2 2x = -0.65$

110. $\log_5(x + 1) = 4.8$

Creating a Table In Exercises 111–114, complete the table to determine the balance A for P dollars invested at rate r for t years, and compounded n times per year.

n	1	4	12	365	Continuous
A					

	Principal	Interest Rate	Time
111.	$P = \$500$	$r = 7\%$	$t = 30$ years
112.	$P = \$100$	$r = 5\frac{1}{4}\%$	$t = 60$ years
113.	$P = \$10,000$	$r = 10\%$	$t = 20$ years
114.	$P = \$2500$	$r = 8\%$	$t = 1$ year

Creating a Table In Exercises 115 and 116, complete the table to determine the principal P that will yield a balance of A dollars when invested at rate r for t years and compounded n times per year.

n	1	4	12	365	Continuous
P					

	Balance	Interest Rate	Time
115.	$A = \$50,000$	$r = 8\%$	$t = 40$ years
116.	$A = \$1000$	$r = 6\%$	$t = 1$ year

117. *Inflation Rate* If the annual rate of inflation averages 5% over the next 10 years, the approximate cost C of goods or services during any year in that decade will be given by

$$C(t) = P(1.05)^t, \quad 0 \le t \le 10$$

where t is the time in years and P is the present cost. If the price of an oil change is presently $19.95, when will it cost $25.00?

118. *Doubling Time* Find the time for an investment of $1000 to double in value when invested at 8% compounded monthly.

119. *Product Demand* The demand x and price p for a product are related by

$$p = 25 - 0.4e^{0.02x}.$$

Approximate the demand when the price is $16.97.

120. *Sound Intensity* The relationship between the number of decibels B and the intensity of a sound I in watts per meter squared is given by

$$B = 10\log_{10}\left(\frac{I}{10^{-16}}\right).$$

Determine the intensity of a sound in watts per meter squared if the decibel level is 125.

121. *Graphical Interpretation* The population p of a certain species t years after it is introduced into a new habitat is given by

$$p(t) = \frac{600}{1 + 2e^{-0.2t}}.$$

Use a graphing utility to graph the function. Use the graph to determine the limiting size of the population in this habitat.

122. *Deer Herd* The state Parks and Wildlife Department releases 100 deer into a wilderness area. The population P of the herd can be modeled by

$$P = \frac{500}{1 + 4e^{-0.36t}}$$

where t is measured in years.

(a) Find the population after 5 years.

(b) After how many years will the population be 250?

CHAPTER TEST

Take this test as you would take a test in class. After you are done, check your work against the answers given in the back of the book.

1. Evaluate $f(t) = 54 \left(\frac{2}{3}\right)^t$ when $t = -1, 0, \frac{1}{2}$, and 2.

2. Sketch a graph of the function $f(x) = 2^{x/3}$.

3. Write the logarithmic equation $\log_5 125 = 3$ in exponential form.

4. Write the exponential equation $4^{-2} = \frac{1}{16}$ in logarithmic form.

5. Evaluate $\log_8 2$ without the aid of a calculator.

6. Describe the relationship between the graphs of $f(x) = \log_5 x$ and $g(x) = 5^x$.

7. Use the properties of logarithms to expand $\log_4 \left(5x^2/\sqrt{y}\right)$.

8. Use the properties of logarithms to condense $\ln x - 4 \ln y$.

9. Simplify $\log_5 5^3 \cdot 6$.

In Exercises 10–13, solve the equation.

10. $\log_4 x = 64$

11. $10^{3y} = 832$

12. $400e^{0.08t} = 1200$

13. $3 \ln(2x - 3) = 10$

14. Determine the balance after 20 years if $2000 is invested at 7% compounded (a) quarterly and (b) continuously.

15. Determine the principal that will yield $100,000 when invested at 9% compounded quarterly for 25 years.

16. A principal of $500 yields a balance of $1006.88 in 10 years when the interest is compounded continuously. What is the annual interest rate?

17. A car that cost $18,000 new has a depreciated value of $14,000 after 1 year. Find the value of the car when it is 3 years old by using the exponential model $y = Ce^{kt}$.

In Exercises 18–20, the population of a certain species t years after it is introduced into a new habitat is given by

$$p(t) = \frac{2400}{1 + 3e^{-t/4}}.$$

18. Determine the population size that was introduced into the habitat.

19. Determine the population after 4 years.

20. After how many years will the population be 1200?

CUMULATIVE TEST: CHAPTERS 7–9

Take this test as you would take a test in class. After you are done, check your work against the answers given in the back of the book.

1. Find an equation of the line through $(-4, 0)$ and $(4, 6)$.

In Exercises 2–4, sketch the graph of the equation.

2. $x^2 + y^2 = 8$ 3. $\dfrac{x^2}{1} + \dfrac{y^2}{4} = 1$ 4. $\dfrac{x^2}{1} - \dfrac{y^2}{4} = 1$

5. Graph the inequality $5x + 2y > 10$.

6. Find an equation of the parabola shown at the right.

7. A semicircular arch is positioned over the roadway onto the grounds of an estate. The roadway is 10 feet wide and the arch is sitting on pillars that are 8 feet tall. Find the maximum height of a truck that can be driven onto the estate if its width is 8 feet.

8. The stopping distance d of a car is directly proportional to the square of its speed s. On a certain type of pavement, a car requires 50 feet to stop when its speed is 25 miles per hour. Estimate the stopping distance when the speed of the car is 40 miles per hour. Explain your reasoning.

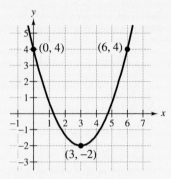

Figure for 6

In Exercises 9–12, solve the system of equations by the specified method.

9. *Graphical:* $x - y = 1$
                   $2x + y = 5$

10. *Substitution:* $4x + 2y = 8$
                      $x - 5y = 13$

11. *Elimination:* $4x - 3y = 8$
                     $-2x + y = -6$

12. *Cramer's Rule:* $2x - y = 4$
                       $3x + y = -5$

13. The sum of three positive numbers is 44. The second number is 4 greater than the first, and the third number is three times the first. Find the numbers.

14. Graph $y = 4e^{-x^2/4}$.

15. Evaluate $\log_4 \frac{1}{16}$ without using a calculator.

16. Describe the relationship between the graphs of $f(x) = e^x$ and $g(x) = \ln x$.

17. Use the properties of logarithms to condense $3(\log_2 x + \log_2 y) - \log_2 z$.

18. Solve each equation.

    (a) $\log_x \left(\frac{1}{9}\right) = -2$ (b) $4 \ln x = 10$

    (c) $500(1.08)^t = 2000$ (d) $3(1 + e^{2x}) = 20$

19. Determine the effective yield of an 8% interest rate compounded continuously.

20. Determine the length of time for an investment of $1000 to quadruple in value if the investment earns 9% compounded continuously.

Additional Topics in Algebra

- Sequences
- Arithmetic Sequences
- Geometric Sequences
- The Binomial Theorem
- Counting Principles
- Probability

In 1991, 73.2% of the households in the United States held interest-earning accounts such as savings accounts, money market deposit accounts, certificates of deposit, and interest-earning checking accounts at financial institutions. Other interest-earning assets include money market funds, government securities, corporate and municipal bonds, stocks and mutual fund shares, U.S. savings bonds, and IRA and Keogh accounts.

The balance A of a $1000 savings account with an interest rate of 3%, compounded quarterly, increases in value, as long as there are no withdrawals, according to the formula

$$A = 1000 \left(1 + \frac{0.03}{4} \right)^{4t}.$$

The increase in the balance of the account is shown in the table and graph at the right.

Year	1	5	10	15
Balance	$1030.34	$1161.18	$1348.35	$1565.70

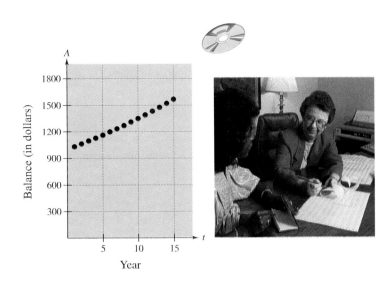

The chapter project related to this information is on page 689.

10.1	**Sequences**
	Sequences ▪ Factorial Notation ▪ Sigma Notation

Sequences

Suppose you were given the following choices of a contract offer for the next 5 years of employment.

Contract A $20,000 the first year and a $2200 raise each year
Contract B $20,000 the first year and a 10% raise each year

Year	Contract A	Contract B
1	$20,000	$20,000
2	$22,200	$22,000
3	$24,400	$24,200
4	$26,600	$26,620
5	$28,800	$29,282
Total	$122,000	$122,102

Which contract offers the largest salary over the 5-year period? The salaries for each contract are shown at the left. The salaries for contract A represent the first five terms of an **arithmetic sequence,** and the salaries for contract B represent the first five terms of a **geometric sequence.** Notice that after 5 years the geometric sequence represents a better contract offer than the arithmetic sequence.

A mathematical **sequence** is simply an ordered list of numbers. Each number in the list is a **term** of the sequence. A sequence can have a finite number of terms or an infinite number of terms. For instance, the sequence of positive odd integers that are less than 15 is a *finite* sequence

$$1, \ 3, \ 5, \ 7, \ 9, \ 11, \ 13 \qquad \text{Finite sequence}$$

whereas the sequence of positive odd integers is an *infinite* sequence.

$$1, \ 3, \ 5, \ 7, \ 9, \ 11, \ 13, \ . \ . \ . \qquad \text{Infinite sequence}$$

Note that the three dots indicate that the sequence continues and has an infinite number of terms.

Definition of an Infinite Sequence

An **infinite sequence** is an ordered list of real numbers.

$$a_1, \ a_2, \ a_3, \ a_4, \ a_5, \ . \ . \ . \ , \ a_n, \ . \ . \ .$$

Each number in the sequence is a **term** of the sequence, and the sequence consists of an infinite number of terms.

NOTE   Sometimes it is convenient to begin subscripting an infinite sequence with 0 instead of 1. In such cases, the terms of the sequence are denoted by

$$a_0, \ a_1, \ a_2, \ a_3, \ a_4, \ a_5, \ . \ . \ . \ , \ a_n, \ . \ . \ . \ .$$

Technology

Most graphing utilities have a "sequence graphing mode" that allows you to plot the terms of a sequence as points on a rectangular coordinate system. For instance, the graph of the first six terms of the sequence given by

$$a_n = n^2 - 1$$

is shown below.

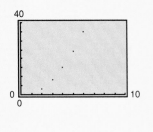

EXAMPLE 1   *Finding the Terms of a Sequence*

Write the first six terms of the sequence whose nth term is

$$a_n = n^2 - 1. \qquad\qquad \text{Begin sequence with } n = 1.$$

Solution

$$a_1 = (1)^2 - 1 = 0 \qquad a_2 = (2)^2 - 1 = 3 \qquad a_3 = (3)^2 - 1 = 8$$
$$a_4 = (4)^2 - 1 = 15 \qquad a_5 = (5)^2 - 1 = 24 \qquad a_6 = (6)^2 - 1 = 35$$

To represent the entire sequence, you can write the following.

$$0, \ 3, \ 8, \ 15, \ 24, \ 35, \ \ldots, \ n^2 - 1, \ \ldots$$

EXAMPLE 2   *Finding the Terms of a Sequence*

Write the first six terms of the sequence whose nth term is

$$a_n = 3(2^n). \qquad\qquad \text{Begin sequence with } n = 0.$$

Solution

$$a_0 = 3(2^0) = 3 \cdot 1 = 3 \qquad a_1 = 3(2^1) = 3 \cdot 2 = 6$$
$$a_2 = 3(2^2) = 3 \cdot 4 = 12 \qquad a_3 = 3(2^3) = 3 \cdot 8 = 24$$
$$a_4 = 3(2^4) = 3 \cdot 16 = 48 \qquad a_5 = 3(2^5) = 3 \cdot 32 = 96$$

The entire sequence can be written as follows.

$$3, \ 6, \ 12, \ 24, \ 48, \ 96, \ \ldots, \ 3(2^n), \ \ldots$$

EXAMPLE 3   *A Sequence Whose Terms Alternate in Sign*

Write the first six terms of the sequence whose nth term is

$$a_n = \frac{(-1)^n}{2n - 1}. \qquad\qquad \text{Begin sequence with } n = 1.$$

Solution

$$a_1 = \frac{(-1)^1}{2(1) - 1} = -\frac{1}{1} \qquad a_2 = \frac{(-1)^2}{2(2) - 1} = \frac{1}{3} \qquad a_3 = \frac{(-1)^3}{2(3) - 1} = -\frac{1}{5}$$
$$a_4 = \frac{(-1)^4}{2(4) - 1} = \frac{1}{7} \qquad a_5 = \frac{(-1)^5}{2(5) - 1} = -\frac{1}{9} \qquad a_6 = \frac{(-1)^6}{2(6) - 1} = \frac{1}{11}$$

The entire sequence can be written as follows.

$$-1, \ \frac{1}{3}, \ -\frac{1}{5}, \ \frac{1}{7}, \ -\frac{1}{9}, \ \frac{1}{11}, \ \ldots, \ \frac{(-1)^n}{2n - 1}, \ \ldots$$

Factorial Notation

Some very important sequences in mathematics involve terms that are defined with special types of products called **factorials.**

Definition of Factorial

If n is a positive integer, **n factorial** is defined as

$$n! = 1 \cdot 2 \cdot 3 \cdot 4 \cdots \cdots (n-1) \cdot n.$$

As a special case, zero factorial is defined as $0! = 1$.

The first several factorial values are as follows.

$0! = 1$ $1! = 1$

$2! = 1 \cdot 2 = 2$ $3! = 1 \cdot 2 \cdot 3 = 6$

$4! = 1 \cdot 2 \cdot 3 \cdot 4 = 24$ $5! = 1 \cdot 2 \cdot 3 \cdot 4 \cdot 5 = 120$

Many calculators have a factorial key, denoted by $\boxed{n!}$. If your calculator has such a key, try using it to evaluate $n!$ for several values of n. You will see that the value of n does not have to be very large before the value of $n!$ is huge. For instance,

$$10! = 3{,}628{,}800.$$

EXAMPLE 4   *A Sequence Involving Factorials*

Write the first six terms of the sequence whose nth term is

$$a_n = \frac{1}{n!}.$$

Begin sequence with $n = 0$.

Solution

$$a_0 = \frac{1}{0!} = \frac{1}{1} = 1 \qquad\qquad a_1 = \frac{1}{1!} = \frac{1}{1} = 1$$

$$a_2 = \frac{1}{2!} = \frac{1}{2} \qquad\qquad a_3 = \frac{1}{3!} = \frac{1}{1 \cdot 2 \cdot 3} = \frac{1}{6}$$

$$a_4 = \frac{1}{4!} = \frac{1}{1 \cdot 2 \cdot 3 \cdot 4} = \frac{1}{24} \qquad\qquad a_5 = \frac{1}{5!} = \frac{1}{1 \cdot 2 \cdot 3 \cdot 4 \cdot 5} = \frac{1}{120}$$

The entire sequence can be written as follows.

$$1,\ 1,\ \frac{1}{2},\ \frac{1}{6},\ \frac{1}{24},\ \frac{1}{120},\ \cdots,\ \frac{1}{n!},\ \cdots$$

Factorials follow the same conventions for order of operation as do exponents. For instance, $2n!$ means $2(n!)$, not $(2n)!$. Notice how these conventions are used in the next example.

EXAMPLE 5   A Sequence Involving Factorials

Write the first six terms of the sequence whose nth term is

$$a_n = \frac{2n!}{(2n)!}.$$

Begin sequence with $n = 0$.

Solution

$$a_0 = \frac{2(0!)}{(2 \cdot 0)!} = \frac{2(0!)}{0!} = \frac{2}{1} = 2$$

$$a_1 = \frac{2(1!)}{(2 \cdot 1)!} = \frac{2(1!)}{2!} = \frac{2}{2} = 1$$

$$a_2 = \frac{2(2!)}{(2 \cdot 2)!} = \frac{2(2!)}{4!} = \frac{4}{24} = \frac{1}{6}$$

$$a_3 = \frac{2(3!)}{(2 \cdot 3)!} = \frac{2(3!)}{6!} = \frac{12}{720} = \frac{1}{60}$$

$$a_4 = \frac{2(4!)}{(2 \cdot 4)!} = \frac{2(4!)}{8!} = \frac{48}{40,320} = \frac{1}{840}$$

$$a_5 = \frac{2(5!)}{(2 \cdot 5)!} = \frac{2(5!)}{10!} = \frac{240}{3,628,800} = \frac{1}{15,120}$$

The entire sequence can be written as follows.

$$2, \ 1, \ \frac{1}{6}, \ \frac{1}{60}, \ \frac{1}{840}, \ \frac{1}{15,120}, \ \cdots \ , \ \frac{2n!}{(2n)!}, \ \cdots$$

In Example 5, the numerators and denominators were multiplied before reducing the fractions. When you are finding the terms of a sequence, reducing is often easier if you leave the numerator and denominator in factored form. For instance, notice the cancellation in the following fraction.

$$a_5 = \frac{2(5!)}{10!}$$

$$= \frac{2 \cdot 1 \cdot 2 \cdot 3 \cdot 4 \cdot 5}{1 \cdot 2 \cdot 3 \cdot 4 \cdot 5 \cdot 6_3 \cdot 7 \cdot 8 \cdot 9 \cdot 10}$$

$$= \frac{1}{3 \cdot 7 \cdot 8 \cdot 9 \cdot 10}$$

$$= \frac{1}{15,120}$$

Sigma Notation

Many applications involve finding the sum of the first n terms of a sequence. A convenient shorthand notation for such a sum is **sigma notation.** This name comes from the use of the uppercase Greek letter sigma, written as Σ.

Definition of Sigma Notation

The sum of the first n terms of the sequence whose nth term is a_n is

$$\sum_{i=1}^{n} a_i = a_1 + a_2 + a_3 + a_4 + \cdots + a_n$$

where i is the **index of summation,** n is the **upper limit of summation,** and 1 is the **lower limit of summation.**

NOTE In Example 6, the index of summation is i and the summation begins with $i = 1$. Any letter can be used as the index of summation, and the summation can begin with any integer. For instance, in Example 7, the index of summation is k and the summation begins with $k = 0$.

EXAMPLE 6   Sigma Notation for Sums

Find the sum $\displaystyle\sum_{i=1}^{6} 2i$.

Solution

$$\sum_{i=1}^{6} 2i = 2(1) + 2(2) + 2(3) + 2(4) + 2(5) + 2(6)$$

$$= 2 + 4 + 6 + 8 + 10 + 12$$

$$= 42$$

EXAMPLE 7   Sigma Notation for Sums

Find the sum $\displaystyle\sum_{k=0}^{8} \frac{1}{k!}$.

Solution

$$\sum_{k=0}^{8} \frac{1}{k!} = \frac{1}{0!} + \frac{1}{1!} + \frac{1}{2!} + \frac{1}{3!} + \frac{1}{4!} + \frac{1}{5!} + \frac{1}{6!} + \frac{1}{7!} + \frac{1}{8!}$$

$$= 1 + 1 + \frac{1}{2} + \frac{1}{6} + \frac{1}{24} + \frac{1}{120} + \frac{1}{720} + \frac{1}{5040} + \frac{1}{40,320}$$

$$\approx 2.71828$$

Note that this sum is approximately $e = 2.71828 \ldots$.

EXAMPLE 8   Sigma Notation for Sums

a. $\displaystyle\sum_{i=1}^{4} 5 = 5 + 5 + 5 + 5 = 20$

b. $\displaystyle\sum_{i=0}^{6} \frac{1}{2^i} = \frac{1}{2^0} + \frac{1}{2^1} + \frac{1}{2^2} + \frac{1}{2^3} + \frac{1}{2^4} + \frac{1}{2^5} + \frac{1}{2^6}$

$$= \frac{1}{1} + \frac{1}{2} + \frac{1}{4} + \frac{1}{8} + \frac{1}{16} + \frac{1}{32} + \frac{1}{64}$$

$$= \frac{127}{64}$$

EXAMPLE 9   Writing a Sum in Sigma Notation

Write the sum in sigma notation.

$$\frac{2}{2} + \frac{2}{3} + \frac{2}{4} + \frac{2}{5} + \frac{2}{6}$$

Solution

To write this sum in sigma notation, you must find a pattern for the terms. After examining the terms, you can see that they have numerators of 2 and denominators that range over the integers from 2 to 6. Thus, one possible sigma notation is

$$\sum_{i=1}^{5} \frac{2}{i+1} = \frac{2}{2} + \frac{2}{3} + \frac{2}{4} + \frac{2}{5} + \frac{2}{6}.$$

Group Activities

Communicating Mathematically

Finding a Pattern   You learned in this section that a sequence is an ordered list of numbers. Study the following sequence and see if you can guess what its next term should be.

Z, O, T, T, F, F, S, S, E, N, T, E, T, . . .

(*Hint:* you might try to figure out what numbers the letters represent.) Construct another sequence with letters. Can the other members of your group guess the next term?

10.1 Exercises

Discussing the Concepts

1. Give an example of an infinite sequence.

2. State the definition of n factorial.

3. The nth term of a sequence is $a_n = (-1)^n n$. Which terms of the sequence are negative?

In Exercises 4–6, decide whether the statement is true. Explain your reasoning.

4. $\displaystyle\sum_{i=1}^{4} (i^2 + 2i) = \sum_{i=1}^{4} i^2 + \sum_{i=1}^{4} 2i$

5. $\displaystyle\sum_{k=1}^{4} 3k = 3 \sum_{k=1}^{4} k$

6. $\displaystyle\sum_{j=1}^{4} 2^j = \sum_{j=3}^{6} 2^{j-2}$

Problem Solving

In Exercises 7–14, write the first five terms of the sequence. (Begin with $n = 1$.)

7. $a_n = 2n$

8. $a_n = (-1)^{n+1} 3n$

9. $a_n = \left(-\dfrac{1}{2}\right)^n$

10. $a_n = \left(\dfrac{2}{3}\right)^{n-1}$

11. $a_n = \dfrac{2n}{3n + 2}$

12. $a_n = \dfrac{5n}{4n + 3}$

13. $a_n = \dfrac{2^n}{n!}$

14. $a_n = \dfrac{n!}{(n-1)!}$

In Exercises 15–18, match the sequence with the graph of its first 10 terms. [The graphs are labeled (a), (b), (c), and (d).]

15. $a_n = \dfrac{6}{n + 1}$

16. $a_n = \dfrac{6n}{n + 1}$

17. $a_n = (0.6)^{n-1}$

18. $a_n = \dfrac{3^n}{n!}$

(a)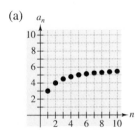

(b)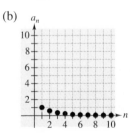

(c)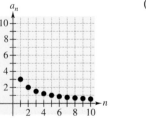

(d)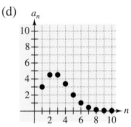

In Exercises 19 and 20, use a graphing utility to graph the first 10 terms of the sequence.

19. $a_n = (-0.8)^{n-1}$

20. $a_n = \dfrac{2n^2}{n^2 + 1}$

In Exercises 21 and 22, find the indicated term of the sequence.

21. $a_n = (-1)^n (5n - 3)$

$a_{15} = $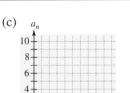

22. $a_n = \dfrac{n^2}{n!}$

$a_{12} = $

In Exercises 23–26, simplify the expression.

23. $\dfrac{25!}{27!}$

24. $\dfrac{20!}{15! \cdot 5!}$

25. $\dfrac{(n + 1)!}{(n - 1)!}$

26. $\dfrac{(3n)!}{(3n + 2)!}$

In Exercises 27–34, find the sum.

27. $\displaystyle\sum_{k=1}^{6} 3k$

28. $\displaystyle\sum_{k=1}^{4} 5k$

29. $\displaystyle\sum_{i=0}^{6} (2i+5)$

30. $\displaystyle\sum_{j=3}^{7} (6j-10)$

31. $\displaystyle\sum_{m=2}^{6} \frac{2m}{2(m-1)}$

32. $\displaystyle\sum_{k=1}^{5} \frac{10k}{k+2}$

33. $\displaystyle\sum_{i=0}^{4} (i!+4)$

34. $\displaystyle\sum_{k=1}^{6} \left(\frac{1}{2k}-\frac{1}{2k-1}\right)$

In Exercises 35–40, write the sum using sigma notation. (Begin with $k=0$ or $k=1$.)

35. $2+4+6+8+10$

36. $24+30+36+42$

37. $\dfrac{4}{1+3}+\dfrac{4}{2+3}+\dfrac{4}{3+3}+\cdots+\dfrac{4}{20+3}$

38. $\dfrac{1}{2^3}-\dfrac{1}{4^3}+\dfrac{1}{6^3}-\dfrac{1}{8^3}+\cdots+\dfrac{1}{14^3}$

39. $\frac{2}{4}+\frac{4}{5}+\frac{6}{6}+\frac{8}{7}+\cdots+\frac{40}{23}$

40. $\left(2+\frac{1}{1}\right)+\left(2+\frac{1}{2}\right)+\left(2+\frac{1}{3}\right)+\cdots+\left(2+\frac{1}{25}\right)$

41. *Compound Interest*   A deposit of $500 is made in an account that earns 7% interest compounded yearly. The balance in the account after N years is given by

$$A_N = 500(1+0.07)^N, \quad N=1,\ 2,\ 3,\ \ldots.$$

(a) Compute the first eight terms of this sequence.

(b) Find the balance in this account after 40 years by computing A_{40}.

(c) Use a graphing utility to graph the first 40 terms of the sequence.

(d) The terms of the sequence are increasing. Is the rate of growth of the terms increasing? Explain.

42. *Depreciation*   At the end of each year, the value of a car with an initial cost of $16,000 is three-fourths what it was at the beginning of the year. Thus, after n years its value is given by

$$a_n = 16{,}000\left(\frac{3}{4}\right)^n, \quad n=1,\ 2,\ 3,\ \ldots.$$

(a) Find the value of the car 3 years after it was purchased by computing a_3.

(b) Find the value of the car 6 years after it was purchased by computing a_6. Is this value half of what it was after 3 years?

Reviewing the Major Concepts

In Exercises 43–46, simplify the expression.

43. $(x+10)^2, \ x \neq -10$

44. $\dfrac{18(x-3)^5}{(x-3)^2}$

45. $(a^2)^{-4}, \ a \neq 0$

46. $(8x^3)^{1/3}$

47. Find an equation of the line through $(2,3)$ and $(5,6)$.

48. *Volleyball Court*   A volleyball court is 60 feet long and 30 feet wide. To be assured that the court is rectangular, you check the diagonals of the court. How long should each be?

Additional Problem Solving

In Exercises 49–60, write the first five terms of the sequence. (Begin with $n=1$.)

49. $a_n = (-1)^n 2n$

50. $a_n = 3n$

51. $a_n = \left(\frac{1}{2}\right)^n$

52. $a_n = \left(\frac{1}{3}\right)^n$

53. $a_n = \dfrac{1}{n+1}$

54. $a_n = \dfrac{3}{2n+1}$

55. $a_n = \dfrac{(-1)^n}{n^2}$

56. $a_n = \dfrac{1}{\sqrt{n}}$

57. $a_n = 5 - \dfrac{1}{2^n}$

58. $a_n = 7 + \dfrac{1}{3^n}$

59. $a_n = 2 + (-2)^n$

60. $a_n = \dfrac{1+(-1)^n}{n^2}$

In Exercises 61–64, use a graphing utility to graph the first 10 terms of the sequence.

61. $a_n = \frac{1}{2}n$

62. $a_n = 10\left(\frac{3}{4}\right)^{n-1}$

63. $a_n = 3 - \frac{4}{n}$

64. $a_n = \frac{n+2}{n}$

In Exercises 65–72, simplify the expression.

65. $\frac{5!}{4!}$

66. $\frac{18!}{17!}$

67. $\frac{10!}{12!}$

68. $\frac{5!}{8!}$

69. $\frac{n!}{(n+1)!}$

70. $\frac{(n+2)!}{n!}$

71. $\frac{(2n)!}{(2n-1)!}$

72. $\frac{(2n+2)!}{(2n)!}$

In Exercises 73–88, find the sum.

73. $\displaystyle\sum_{i=0}^{4} (2i+3)$

74. $\displaystyle\sum_{i=2}^{7} (4i-1)$

75. $\displaystyle\sum_{j=1}^{5} \frac{(-1)^{j+1}}{j}$

76. $\displaystyle\sum_{j=0}^{3} \frac{1}{j^2+1}$

77. $\displaystyle\sum_{k=1}^{6} (-8)$

78. $\displaystyle\sum_{n=3}^{12} 10$

79. $\displaystyle\sum_{i=1}^{8} \left(\frac{1}{i} - \frac{1}{i+1}\right)$

80. $\displaystyle\sum_{k=1}^{5} \left(\frac{2}{k} - \frac{2}{k+2}\right)$

81. $\displaystyle\sum_{n=0}^{5} \left(-\frac{1}{3}\right)^n$

82. $\displaystyle\sum_{n=0}^{6} \left(\frac{3}{2}\right)^n$

83. $\displaystyle\sum_{n=1}^{6} n(n+1)$

84. $\displaystyle\sum_{n=0}^{5} 2n^2$

85. $\displaystyle\sum_{j=2}^{6} (j!-j)$

86. $\displaystyle\sum_{j=0}^{4} \frac{6}{j!}$

87. $\displaystyle\sum_{k=1}^{6} \ln k$

88. $\displaystyle\sum_{k=2}^{4} \frac{k}{\ln k}$

In Exercises 89–100, write the sum using sigma notation. (Begin with $k = 0$ or $k = 1$.)

89. $1 + 2 + 3 + 4 + 5$

90. $8 + 9 + 10 + 11 + 12 + 13$

91. $\frac{1}{2(1)} + \frac{1}{2(2)} + \frac{1}{2(3)} + \frac{1}{2(4)} + \cdots + \frac{1}{2(10)}$

92. $\frac{3}{1+1} + \frac{3}{1+2} + \frac{3}{1+3} + \frac{3}{1+4} + \cdots + \frac{3}{1+50}$

93. $\frac{1}{1^2} + \frac{1}{2^2} + \frac{1}{3^2} + \frac{1}{4^2} + \cdots + \frac{1}{20^2}$

94. $\frac{1}{2^0} + \frac{1}{2^1} + \frac{1}{2^2} + \frac{1}{2^3} + \cdots + \frac{1}{2^{12}}$

95. $\frac{1}{3^0} - \frac{1}{3^1} + \frac{1}{3^2} - \frac{1}{3^3} + \cdots - \frac{1}{3^9}$

96. $\left(-\frac{2}{3}\right)^0 + \left(-\frac{2}{3}\right)^1 + \left(-\frac{2}{3}\right)^2 + \cdots + \left(-\frac{2}{3}\right)^{20}$

97. $\frac{1}{2} + \frac{2}{3} + \frac{3}{4} + \frac{4}{5} + \frac{5}{6} + \cdots + \frac{11}{12}$

98. $\frac{2}{4} + \frac{4}{7} + \frac{6}{10} + \frac{8}{13} + \frac{10}{16} + \cdots + \frac{20}{31}$

99. $1 + 1 + 2 + 6 + 24 + 120 + 720$

100. $1 + 1 + \frac{1}{2} + \frac{1}{6} + \frac{1}{24} + \frac{1}{120} + \frac{1}{720}$

Arithmetic Mean In Exercises 101–104, find the arithmetic mean of the set. The *arithmetic mean* $\bar{x}$ of a set of n measurements $x_1, x_2, x_3, \ldots, x_n$ is

$$\bar{x} = \frac{1}{n} \sum_{i=1}^{n} x_i.$$

101. 1, 2, 3, 4, 5

102. 66, 70, 74, 78

103. 5, 8, 11, 14, 17, 20

104. 17, 22, 27, 32, 37

105. *Soccer Ball* The number of degrees a_n in each angle of a regular n-sided polygon is

$$a_n = \frac{180(n-2)}{n}, \quad n \geq 3.$$

The surface of a soccer ball is made of regular hexagons and pentagons. If a soccer ball is taken apart and flattened, as shown in the figure at the top of the next page, the sides of the hexagons do not meet each other. Use the terms a_5 and a_6 to explain why there are gaps between adjacent hexagons.

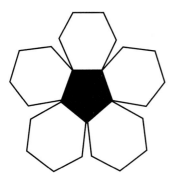

Figure for 105

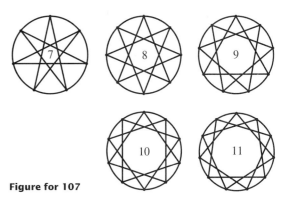

Figure for 107

106. *Stars* The number of degrees d_n in each tip of the n-pointed stars in the figure is given by

$$d_n = \frac{180(n-4)}{n}, \quad n \geq 5.$$

Write the first six terms of this sequence.

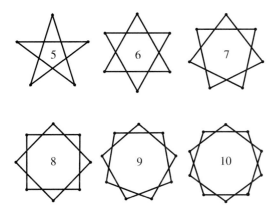

107. *Stars* The stars in Exercise 106 are formed by placing n equally spaced points on a circle and connecting each point with the second point from it on the circle. The stars in the figure at the top of the next column are formed in a similar way except each point is connected with the third point from it. For these stars, the number of degrees in a tip is

$$d_n = \frac{180(n-6)}{n}, \quad n \geq 7.$$

Write the first five terms of this sequence.

10.2	**Arithmetic Sequences**
	Arithmetic Sequences · The Sum of an Arithmetic Sequence · Application

Arithmetic Sequences

A sequence whose consecutive terms have a common difference is called an **arithmetic sequence.**

Definition of an Arithmetic Sequence

A sequence is called **arithmetic** if the differences between consecutive terms are the same. Thus, the sequence

$$a_1, \ a_2, \ a_3, \ a_4, \ \ldots, \ a_n, \ \ldots$$

is arithmetic if there is a number d such that

$$a_2 - a_1 = d, \quad a_3 - a_2 = d, \quad a_4 - a_3 = d$$

and so on. The number d is the **common difference** of the sequence.

EXAMPLE 1   *Examples of Arithmetic Sequences*

a. The sequence whose nth term is $3n + 2$ is arithmetic. For this sequence, the common difference between consecutive terms is 3.

$$5, \ 8, \ 11, \ 14, \ \ldots, \ 3n + 2, \ \ldots$$
$$8 - 5 = 3$$

b. The sequence whose nth term is $7 - 5n$ is arithmetic. For this sequence, the common difference between consecutive terms is -5.

$$2, \ -3, \ -8, \ -13, \ \ldots, \ 7 - 5n, \ \ldots$$
$$-3 - 2 = -5$$

c. The sequence whose nth term is $\frac{1}{4}(n + 3)$ is arithmetic. For this sequence, the common difference between consecutive terms is $\frac{1}{4}$.

$$1, \ \frac{5}{4}, \ \frac{3}{2}, \ \frac{7}{4}, \ \ldots, \ \frac{n + 3}{4}, \ \ldots$$
$$\tfrac{5}{4} - 1 = \tfrac{1}{4}$$

The nth Term of an Arithmetic Sequence

The nth term of an arithmetic sequence has the form

$$a_n = a_1 + (n - 1)d$$

where d is the common difference between the terms of the sequence, and a_1 is the first term.

EXAMPLE 2   Finding the nth Term of an Arithmetic Sequence

Find a formula for the nth term of the arithmetic sequence whose common difference is 2 and whose first term is 5.

Solution

You know that the formula for the nth term is of the form $a_n = a_1 + (n - 1)d$. Moreover, because the common difference is $d = 2$, and the first term is $a_1 = 5$, the formula must have the form

$$a_n = 5 + 2(n - 1).$$

Thus, the formula for the nth term is

$$a_n = 2n + 3.$$

The sequence therefore has the following form.

$$5, \ 7, \ 9, \ 11, \ 13, \ \ldots, \ 2n + 3, \ \ldots$$

If you know the nth term and the common difference of an arithmetic sequence, you can find the $(n+1)$th term by using the following **recursion formula.**

$$a_{n+1} = a_n + d$$

EXAMPLE 3   Using a Recursion Formula

The 12th term of an arithmetic sequence is 52 and the common difference is 3. What is the 13th term of the sequence?

Solution

$$a_{13} = a_{12} + 3 = 52 + 3 = 55$$

The Sum of an Arithmetic Sequence

The sum of the first n terms of an arithmetic sequence is called the **nth partial sum** of the sequence. For instance, the 5th partial sum of the arithmetic sequence whose nth term is $3n + 4$ is

$$\sum_{i=1}^{5} (3i + 4) = 7 + 10 + 13 + 16 + 19 = 65.$$

A formula for the nth partial sum of an arithmetic sequence is given below.

STUDY TIP

You can use the formula for the nth partial sum of an arithmetic sequence to find the sum of consecutive numbers. For instance, the sum of the integers from 1 to 100 is

$$\sum_{i=1}^{100} i = \frac{100}{2}(1 + 100)$$
$$= 50(101)$$
$$= 5050.$$

The nth Partial Sum of an Arithmetic Sequence

The nth partial sum of the arithmetic sequence whose nth term is a_n is

$$\sum_{i=1}^{n} a_i = a_1 + a_2 + a_3 + a_4 + \cdots + a_n$$
$$= n\left(\frac{a_1 + a_n}{2}\right).$$

In other words, to find the sum of the first n terms of an arithmetic sequence, find the average of the first and nth terms, and multiply by n.

NOTE This summation form of an arithmetic sequence is also called a **series.**

EXAMPLE 4 Finding the nth Partial Sum

Find the sum of the first 20 terms of the arithmetic sequence whose nth term is $4n + 1$.

Solution

The first term of this sequence is $a_1 = 4(1) + 1 = 5$ and the 20th term is $a_{20} = 4(20) + 1 = 81$. Therefore, the sum of the first 20 terms is given by

$$\sum_{i=1}^{n} a_i = \frac{n}{2}(a_1 + a_n)$$
$$\sum_{i=1}^{20} (4i + 1) = \frac{20}{2}(a_1 + a_{20})$$
$$= 10(5 + 81)$$
$$= 10(86)$$
$$= 860.$$

Application

EXAMPLE 5   An Application: Total Sales

Your business sells $100,000 worth of products during its first year. You have a goal of increasing annual sales by $25,000 each year for 9 years. If you meet this goal, how much will you sell during your first 10 years of business?

Solution

The annual sales during the first 10 years form the following arithmetic sequence.

$100,000, $125,000, $150,000, $175,000, $200,000,
$225,000, $250,000, $275,000, $300,000, $325,000

Using the formula for the nth partial sum of an arithmetic sequence, you find the total sales during the first 10 years as follows.

$$\text{Total sales} = \frac{10}{2}(100,000 + 325,000) = 5(425,000) = \$2,125,000$$

From the bar graph shown in Figure 10.1, notice that the annual sales for this company follows a *linear growth* pattern. In other words, saying that a quantity increases arithmetically is the same as saying that it increases linearly.

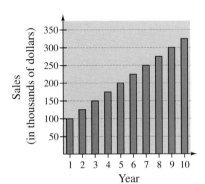

FIGURE 10.1

Group Activities   Extending the Concept

6	1	8
7	5	3
2	9	4

Using Arithmetic Sequences   A magic square is a square table of positive integers in which each row, column, and diagonal adds up to the same number. One example is shown at left. In addition, the values in the middle row, in the middle column, and along both diagonals form arithmetic sequences. See if you can complete the following magic squares.

a.

	11	14
	10	
		15

b.

8		
	9	
	13	

c.

		20
	13	
6		

10.2 Exercises

Discussing the Concepts

1. In your own words, explain what makes a sequence arithmetic.

2. The second and third terms of an arithmetic sequence are 12 and 15, respectively. What is the first term?

3. Explain how the first two terms of an arithmetic sequence can be used to find the nth term.

4. Explain what is meant by the recursion formula.

5. Explain what is meant by the nth partial sum of a sequence.

6. Explain how to find the sum of the integers from 100 to 200.

Problem Solving

In Exercises 7–10, find the common difference of the arithmetic sequence.

7. 10, 22, 34, 46, 58, . . .

8. 4, $\frac{9}{2}$, 5, $\frac{11}{2}$, 6, . . .

9. $\frac{7}{2}$, $\frac{9}{4}$, 1, $-\frac{1}{4}$, $-\frac{3}{2}$, . . .

10. $\frac{5}{2}$, $\frac{11}{6}$, $\frac{7}{6}$, $\frac{1}{2}$, $-\frac{1}{6}$, . . .

In Exercises 11–18, determine whether the sequence is arithmetic. If it is, find the common difference.

11. 10, 8, 6, 4, 2, . . .

12. 1, 2, 4, 8, 16, . . .

13. 3, $\frac{5}{2}$, 2, $\frac{3}{2}$, 1, . . .

14. $\frac{1}{3}$, $\frac{2}{3}$, $\frac{4}{3}$, $\frac{8}{3}$, $\frac{16}{3}$, . . .

15. -12, -8, -4, 0, 4, . . .

16. $\frac{9}{4}$, 2, $\frac{7}{4}$, $\frac{3}{2}$, $\frac{5}{4}$, . . .

17. $\ln 4$, $\ln 8$, $\ln 12$, $\ln 16$, . . .

18. e, e^2, e^3, e^4, . . .

In Exercises 19–22, write the first five terms of the arithmetic sequence. (Begin with $n = 1$.)

19. $a_n = -5n + 45$

20. $a_n = 3n + 1$

21. $a_n = \frac{3}{5}n + 1$

22. $a_n = \frac{3}{4}(n + 1) - 2$

In Exercises 23 and 24, write the first five terms of the arithmetic sequence defined recursively.

23. $a_1 = 25$
$a_{k+1} = a_k + 3$

24. $a_1 = 12$
$a_{k+1} = a_k - 6$

In Exercises 25–28, match the sequence with its graph. [The graphs are labeled (a), (b), (c), and (d).]

(a)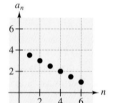

(b)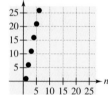

(c)

(d)

25. $a_n = -\frac{1}{3}n + 2$

26. $a_n = 5n - 4$

27. $a_n = 2n - 3$

28. $a_n = -\frac{1}{2}n + 4$

In Exercises 29–34, find a formula for the nth term of the arithmetic sequence.

29. $a_1 = 3, d = \frac{3}{2}$

30. $a_1 = 12, d = -3$

31. $a_3 = 20, d = -4$

32. $a_6 = 5, d = \frac{3}{2}$

33. $a_1 = 5, a_5 = 13$

34. $a_2 = 93, a_6 = 65$

In Exercises 35 and 36, find the sum.

35. $\displaystyle\sum_{k=1}^{10} 5k$

36. $\displaystyle\sum_{n=1}^{30} \left(\frac{1}{2}n + 2\right)$

In Exercises 37 and 38, use a graphing utility to find the sum.

37. $\displaystyle\sum_{j=1}^{25}(750-30j)$

38. $\displaystyle\sum_{i=1}^{60}\left(300-\tfrac{8}{3}i\right)$

In Exercises 39–42, find the nth partial sum of the arithmetic sequence.

39. 2, 8, 14, 20, . . . , $n=25$

40. 500, 480, 460, 440, . . . , $n=20$

41. 0.5, 0.9, 1.3, 1.7, . . . , $n=10$

42. $a_1=15$, $a_{100}=312$, $n=100$

43. Find the sum of the multiples of 6 from 12 to 240.

44. *Clock Chimes* A clock chimes once at 1:00, twice at 2:00, three times at 3:00, and so on. How many times does the clock chime in a 12-hour period?

45. *Free-Falling Object* A free-falling object will fall 16 feet during the first second, 48 more feet during the second, 80 more feet during the third, and so on. What is the total distance the object will fall in 8 seconds if this pattern continues?

46. *Pile of Logs* Logs are stacked in a pile as shown in the figure. The top row has 15 logs and the bottom row has 21 logs. How many logs are in the stack?

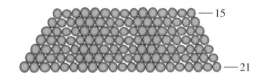

Reviewing the Major Concepts

In Exercises 47–50, solve the system of equations.

47. $y=x^2$
$-3x+2y=2$

48. $x-y^3=0$
$x-2y^2=0$

49. $-x+y=1$
$x+2y-2z=3$
$3x-y+2z=3$

50. $2x+y-2z=1$
$x-z=1$
$3x+3y+z=12$

51. *Ticket Sales* Twelve hundred tickets are sold for a total of $21,120. Adult tickets cost $20 and children's tickets cost $12.50. How many of each type of ticket were sold?

52. *Best-Fitting Line* The slope and y-intercept of the line $y=mx+b$ that "best fits" the four points $(-1,5)$, $(0,3)$, $(2,3)$, and $(4,0)$ are given by the solution of the system of linear equations.
$4b+5m=11$
$5b+21m=1$

(a) Solve the system and find the equation of the best-fitting line.

(b) Plot the three points and sketch the graph of the best-fitting line.

Additional Problem Solving

In Exercises 53–58, find the common difference of the arithmetic sequence.

53. 2, 5, 8, 11, . . .

54. −8, 0, 8, 16, . . .

55. 100, 94, 88, 82, . . .

56. 3200, 2800, 2400, 2000, . . .

57. 1, $\tfrac{5}{3}$, $\tfrac{7}{3}$, 3, . . .

58. $\tfrac{1}{2}$, $\tfrac{5}{4}$, 2, $\tfrac{11}{4}$, . . .

In Exercises 59–70, determine whether the sequence is arithmetic. If it is, find the common difference.

59. 2, 4, 6, 8, . . .

60. 2, 6, 10, 14, . . .

61. 2, $\tfrac{7}{2}$, 5, $\tfrac{13}{2}$, . . .

62. 5, 13, 21, 29, 37, . . .

63. 32, 16, 8, 4, . . .

64. 32, 16, 0, −16, . . .

65. $\frac{1}{3}, \frac{1}{2}, \frac{2}{3}, \frac{5}{6}, 1, \ldots$

66. $\frac{1}{3}, \frac{2}{3}, \frac{4}{3}, \frac{8}{3}, \frac{16}{3}, \ldots$

67. $3.2, 4, 4.8, 5.6, \ldots$

68. $8, 4, 2, 1, 0.5, 0.25, \ldots$

69. $1, \sqrt{2}, \sqrt{3}, 2, \sqrt{5}, \ldots$

70. $1, 4, 9, 16, 25, \ldots$

In Exercises 71–78, write the first five terms of the arithmetic sequence. (Begin with $n = 1$.)

71. $a_n = 3n + 4$ **72.** $a_n = 5n - 4$

73. $a_n = -2n + 8$ **74.** $a_n = -10n + 100$

75. $a_n = \frac{5}{2}n - 1$ **76.** $a_n = \frac{2}{3}n + 2$

77. $a_n = -\frac{1}{4}(n - 1) + 4$ **78.** $a_n = 4(n + 2) + 24$

In Exercises 79–82, use a graphing utility to graph the first 10 terms of the sequence.

79. $a_n = -2n + 21$ **80.** $a_n = \frac{3}{2}n + 1$

81. $a_n = \frac{3}{5}n + \frac{3}{2}$ **82.** $a_n = -25n + 500$

In Exercises 83–90, find a formula for the nth term of the arithmetic sequence.

83. $a_1 = 3, d = \frac{1}{2}$ **84.** $a_1 = -1, d = 1.2$

85. $a_1 = 64, d = -8$ **86.** $a_1 = 1000, d = -25$

87. $a_3 = 16, a_4 = 20$ **88.** $a_5 = 30, a_4 = 25$

89. $a_1 = 50, a_3 = 30$ **90.** $a_{10} = 32, a_{12} = 48$

In Exercises 91–94, write the first five terms of the arithmetic sequence defined recursively.

91. $a_1 = 9$

$a_{k+1} = a_k - 3$

92. $a_1 = 8$

$a_{k+1} = a_k + 7$

93. *First term:* -10

(k+1)st term: $a_k + 6$

94. *First term:* -20

(k+1)st term: $a_k + 7$

In Exercises 95–100, match the sequence with its graph. [The graphs are labeled (a), (b), (c), (d), (e), and (f).]

(a)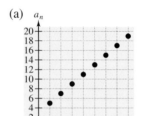

(b)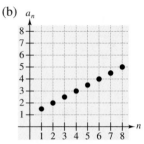

(c)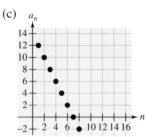

(d)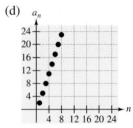

(e)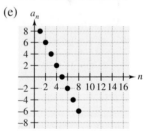

(f)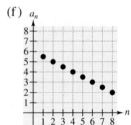

95. $a_n = \frac{1}{2}n + 1$ **96.** $a_n = -\frac{1}{2}n + 6$

97. $a_n = -2n + 10$ **98.** $a_n = 2n + 3$

99. $a_1 = 12$ **100.** $a_1 = 2$

$\;a_{k+1} = a_k - 2$ $\;a_{k+1} = a_k + 3$

In Exercises 101–106, find the sum.

101. $\displaystyle\sum_{n=1}^{20} n$ **102.** $\displaystyle\sum_{n=1}^{30} 4n$

103. $\displaystyle\sum_{n=1}^{50} (2n + 3)$ **104.** $\displaystyle\sum_{n=1}^{100} (4n - 1)$

105. $\displaystyle\sum_{n=1}^{500} \frac{n}{2}$ **106.** $\displaystyle\sum_{n=1}^{600} \frac{2n}{3}$

In Exercises 107 and 108, use a graphing utility to find the sum.

107. $\displaystyle\sum_{n=1}^{40}(1000-25n)$ **108.** $\displaystyle\sum_{n=1}^{20}(500-10n)$

In Exercises 109–116, find the nth partial sum of the arithmetic sequence.

109. 5, 12, 19, 26, 33, . . . , $n=12$

110. 2, 12, 22, 32, 42, . . . , $n=20$

111. 200, 175, 150, 125, 100, . . . , $n=8$

112. 800, 785, 770, 755, 740, . . . , $n=25$

113. $-50, -38, -26, -14, -2, \ldots,$ $n=50$

114. $-16, -8, 0, 8, 16, \ldots,$ $n=30$

115. 1, 4.5, 8, 11.5, 15, . . . , $n=12$

116. 2.2, 2.8, 3.4, 4.0, 4.6, . . . , $n=12$

117. Find the sum of the first 75 positive integers.

118. Find the sum of the integers from 35 to 100.

119. Find the sum of the first 50 even positive integers.

120. Find the sum of the first 100 positive odd integers.

121. *Salary Increases*   In your new job you are told that your starting salary will be $36,000 with an increase of $2000 at the end of each 5 years. How much will you be paid through the end of your first 6 years of employment with the company?

122. *Daily Wages*   Suppose that you receive 25 cents on the first day of the month, 50 cents the second day, 75 cents the third day, and so on. Determine the total amount that you will receive during a month with 30 days.

123. *Ticket Prices*   There are 20 rows of seats on the main floor of a concert hall—20 seats in the first row, 21 seats in the second row, 22 seats in the third row, and so on. How much should you charge per ticket in order to obtain $15,000 for the sale of all of the seats on the main floor?

124. *Baling Hay*   In the first two trips baling hay around a large field (see figure), a farmer obtains 93 bales and 89 bales, and the farmer estimates that the same pattern will continue. Estimate the total number of bales made if there are another 6 trips around the field.

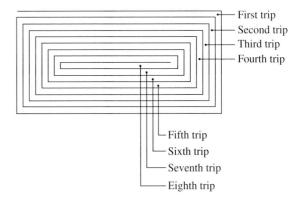

125. *Pattern Recognition*

(a) Compute the sums of positive odd integers.

$1+3 = $

$1+3+5 = $

$1+3+5+7 = $

$1+3+5+7+9 = $

$1+3+5+7+9+11 = $

(b) Use the sums of part (a) to make a conjecture about the sums of positive odd integers. Check your conjecture for the sum

$1+3+5+7+9+11+13 = $

(c) Verify your conjecture algebraically.

126. *Think About It*   Each term of an arithmetic sequence is multiplied by a constant C. Is the resulting sequence arithmetic? If so, how does the common difference compare with the common difference of the original sequence?

10.3	**Geometric Sequences**
	Geometric Sequences ▪ The Sum of a Geometric Sequence ▪ Applications

Geometric Sequences

In Section 10.2, you studied sequences whose consecutive terms have a common *difference*. In this section, you will study sequences whose consecutive terms have a common *ratio*.

Definition of a Geometric Sequence

A sequence is called **geometric** if the ratios of consecutive terms are the same. Thus, the sequence a_1, a_2, a_3, a_4, . . . , a_n, . . . is geometric if there is a number r, $r \neq 0$, such that

$$\frac{a_2}{a_1} = r, \quad \frac{a_3}{a_2} = r, \quad \frac{a_4}{a_3} = r$$

and so on. The number r is the **common ratio** of the sequence.

EXAMPLE 1   Examples of Geometric Sequences

a. The sequence whose nth term is 2^n is geometric. For this sequence, the common ratio between consecutive terms is 2.

$$2, \; 4, \; 8, \; 16, \; . . . , \; 2^n, \; . . .$$

$\frac{4}{2} = 2$

b. The sequence whose nth term is $4(3^n)$ is geometric. For this sequence, the common ratio between consecutive terms is 3.

$$12, \; 36, \; 108, \; 324, \; . . . , \; 4(3^n), \; . . .$$

$\frac{36}{12} = 3$

c. The sequence whose nth term is $\left(-\frac{1}{3}\right)^n$ is geometric. For this sequence, the common ratio between consecutive terms is $-\frac{1}{3}$.

$$-\frac{1}{3}, \frac{1}{9}, \; -\frac{1}{27}, \frac{1}{81}, \; . . . , \; \left(-\frac{1}{3}\right)^n, \; . . .$$

$\frac{1/9}{-1/3} = -\frac{1}{3}$

	The nth Term of a Geometric Sequence

NOTE If you know the nth term of a geometric sequence, the $(n + 1)$th term can be found by multiplying by r. That is,

$$a_{n+1} = ra_n.$$

The nth term of a geometric sequence has the form

$$a_n = a_1 r^{n-1}$$

where r is the common ratio of consecutive terms of the sequence. Thus, every geometric sequence can be written in the following form.

$$a_1, \ a_1 r, \ a_1 r^2, \ a_1 r^3, \ a_1 r^4, \ \ldots, \ a_1 r^{n-1}, \ \ldots$$

EXAMPLE 2   Finding the nth Term of a Geometric Sequence

Find a formula for the nth term of the geometric sequence whose common ratio is 3 and whose first term is 1. What is the eighth term of this sequence?

Solution

The formula for the nth term is of the form $a_1 r^{n-1}$. Moreover, because the common ratio is $r = 3$, and the first term is $a_1 = 1$, the formula must have the form

$$a_n = a_1 r^{n-1} = (1)(3)^{n-1} = 3^{n-1}.$$

The sequence therefore has the following form.

$$1, \ 3, \ 9, \ 27, \ 81, \ \ldots, \ 3^{n-1}, \ \ldots$$

The eighth term of the sequence is $a_8 = 3^{8-1} = 3^7 = 2187$.

EXAMPLE 3   Finding the nth Term of a Geometric Sequence

Find a formula for the nth term of the geometric sequence whose first two terms are 4 and 2.

Solution

Because the common ratio is

$$\frac{a_2}{a_1} = \frac{2}{4} = \frac{1}{2}$$

the formula for the nth term must be

$$a_n = a_1 r^{n-1} = 4\left(\frac{1}{2}\right)^{n-1}.$$

The sequence therefore has the following form.

$$4, \ 2, \ 1, \ \frac{1}{2}, \ \frac{1}{4}, \ \ldots, \ 4\left(\frac{1}{2}\right)^{n-1}, \ \ldots$$

The Sum of a Geometric Sequence

The nth Partial Sum of a Geometric Sequence

The nth partial sum of the geometric sequence whose nth term is $a_n = a_1 r^{n-1}$ is given by

$$\sum_{i=1}^{n} a_1 r^{i-1} = a_1 + a_1 r + a_1 r^2 + a_1 r^3 + \cdots + a_1 r^{n-1} = a_1\left(\frac{r^n - 1}{r - 1}\right).$$

Some of the early work in representing functions by series was done by the French mathematician Joseph Fourier (1768–1830). Fourier's work is important in the history of calculus, partly because it forced 18th-century mathematicians to question the then-prevailing narrow concept of a function. Both Cauchy and Dirichlet were motivated by Fourier's work in series, and in 1837 Dirichlet published the general definition of a function that is used today.

EXAMPLE 4   Finding the nth Partial Sum

Find the sum.

$$1 + 2 + 4 + 8 + 16 + 32 + 64 + 128$$

Solution

This is a geometric sequence whose common ratio is $r = 2$. Because the first term of the sequence is $a_1 = 1$, it follows that the sum is

$$\sum_{i=1}^{8} 2^{i-1} = (1)\left(\frac{2^8 - 1}{2 - 1}\right) = \frac{256 - 1}{2 - 1} = 255.$$

EXAMPLE 5   Finding the nth Partial Sum

Find the sum of the first five terms of the geometric sequence whose nth term is $a_n = \left(\frac{2}{3}\right)^n$.

Solution

$$\sum_{i=1}^{5} \left(\frac{2}{3}\right)^i = \frac{2}{3}\left[\frac{(2/3)^5 - 1}{(2/3) - 1}\right] \qquad \text{Substitute } \tfrac{2}{3} \text{ for } a_1 \text{ and } \tfrac{2}{3} \text{ for } r.$$
$$= \frac{2}{3}\left[\frac{(32/243) - 1}{-1/3}\right]$$
$$= \frac{2}{3}\left(-\frac{211}{243}\right)(-3)$$
$$= \frac{422}{243}$$
$$\approx 1.737$$

Applications

EXAMPLE 6   An Application: A Lifetime Salary

You have accepted a job that pays a salary of $28,000 the first year. During the next 39 years, suppose you receive a 6% raise each year. What will your total salary be over the 40-year period?

Solution

Using a geometric sequence, your salary during the first year will be

$$a_1 = 28,000.$$

Then, with a 6% raise, your salary during the next 2 years will be as follows.

$$a_2 = 28,000 + 28,000(0.06) = 28,000(1.06)^1$$

$$a_3 = 28,000(1.06) + 28,000(1.06)(0.06) = 28,000(1.06)^2$$

From this pattern, you can see that the common ratio of the geometric sequence is $r = 1.06$. Using the formula for the nth partial sum of a geometric sequence, you will find that the total salary over the 40-year period is given by

$$\text{Total salary} = a_1 \left(\frac{r^n - 1}{r - 1} \right)$$

$$= 28,000 \left[\frac{(1.06)^{40} - 1}{1.06 - 1} \right]$$

$$= 28,000 \left[\frac{(1.06)^{40} - 1}{0.06} \right]$$

$$\approx \$4,333,335.$$

The bar graph in Figure 10.2 illustrates your salary during the 40-year period.

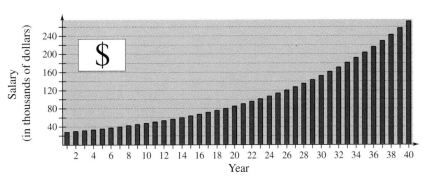

FIGURE 10.2

EXAMPLE 7 An Application: Increasing Annuity

You deposit $100 in an account each month for 2 years. The account pays an annual interest rate of 9%, compounded monthly. What is your balance at the end of 2 years? (This type of savings plan is called an **increasing annuity.**)

Solution

The first deposit would earn interest for the full 24 months, the second deposit would earn interest for 23 months, the third deposit would earn interest for 22 months, and so on. Using the formula for compound interest, you can see that the total of the 24 deposits would be

$$
\begin{aligned}
\text{Total} &= a_1 + a_2 + \cdots + a_{24} \\
&= 100(1.0075)^1 + 100(1.0075)^2 + \cdots + 100(1.0075)^{24} \\
&= 100(1.0075)\left(\frac{1.0075^{24} - 1}{1.0075 - 1}\right) \qquad a_1\left(\frac{r^n - 1}{r - 1}\right) \\
&= \$2638.49.
\end{aligned}
$$

Group Activities   Extending the Concept

Annual Revenue   The two bar graphs below show the annual revenues for two companies. One company's revenue grew at an arithmetic rate, whereas the other grew at a geometric rate. Which company had the greatest revenue during the 10-year period? Which company would you rather own? Explain.

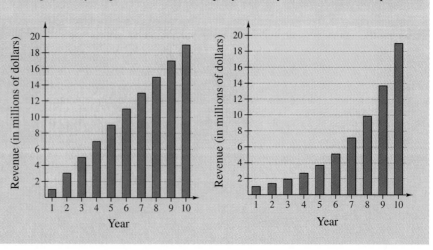

10.3 Exercises

Discussing the Concepts

1. In your own words, explain what makes a sequence geometric.

2. What is the general formula for the nth term of a geometric sequence?

3. The second and third terms of a geometric sequence are 6 and 3, respectively. What is the first term?

4. Give an example of a geometric sequence whose terms alternate in sign.

5. Explain why the terms of a geometric sequence decrease when $0 < r < 1$.

6. In your own words, describe an increasing annuity.

Problem Solving

In Exercises 7–10, find the common ratio of the geometric sequence.

7. $1, -3, 9, -27, \ldots$

8. $5, -\frac{5}{2}, \frac{5}{4}, -\frac{5}{8}, \ldots$

9. $1, \pi, \pi^2, \pi^3, \ldots$

10. $50(1.06), 50(1.06)^2, 50(1.06)^3, 50(1.06)^4, \ldots$

In Exercises 11–14, determine whether the sequence is geometric. If it is, find the common ratio.

11. $5, 10, 20, 40, \ldots$

12. $54, -18, 6, -2, \ldots$

13. $1, 8, 27, 64, 125, \ldots$

14. $12, 7, 2, -3, -8, \ldots$

In Exercises 15–18, write the first five terms of the geometric sequence.

15. $a_1 = 4, r = -\frac{1}{2}$

16. $a_1 = 4, r = \frac{3}{2}$

17. $a_1 = 20, r = 1.07$

18. $a_1 = 100, r = \dfrac{1}{1.05}$

In Exercises 19–22, find the nth term of the geometric sequence.

19. $a_1 = 120, r = -\frac{1}{3}, a_{10} = $ ▩

20. $a_1 = 5, r = \sqrt{3}, a_9 = $ ▩

21. $a_1 = 200, r = 1.2, a_{12} = $ ▩

22. $a_1 = 1200, a_2 = 1000, a_6 = $ ▩

In Exercises 23–26, find the formula for the nth term of the geometric sequence. (Begin with $n = 1$.)

23. $a_1 = 25, r = 4$

24. $a_1 = 12, r = -\frac{4}{3}$

25. $1, \frac{3}{2}, \frac{9}{4}, \frac{27}{8}, \ldots$

26. $4, -6, 9, -\frac{27}{2}, \ldots$

In Exercises 27–30, match the sequence with its graph. [The graphs are labeled (a), (b), (c), and (d).]

(a)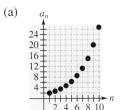

(b)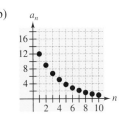

(c)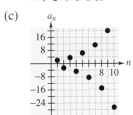

(d)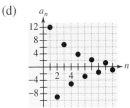

27. $a_n = 12\left(\frac{3}{4}\right)^{n-1}$

28. $a_n = 12\left(-\frac{3}{4}\right)^{n-1}$

29. $a_n = 2\left(\frac{4}{3}\right)^{n-1}$

30. $a_n = 2\left(-\frac{4}{3}\right)^{n-1}$

▦ In Exercises 31 and 32, use a graphing utility to graph the first 10 terms of the sequence.

31. $a_n = 20(-0.6)^{n-1}$

32. $a_n = 4(1.4)^{n-1}$

In Exercises 33–36, find the sum.

33. $\displaystyle\sum_{i=1}^{12} 4(-2)^{i-1}$

34. $\displaystyle\sum_{i=1}^{20} 16\left(\frac{1}{2}\right)^{i-1}$

35. $\displaystyle\sum_{i=1}^{8} 6(0.1)^{i-1}$

36. $\displaystyle\sum_{i=1}^{24} 1000(1.06)^{i-1}$

In Exercises 37 and 38, use a graphing utility to find the sum.

37. $\displaystyle\sum_{i=1}^{30} 100(0.75)^{i-1}$

38. $\displaystyle\sum_{i=1}^{24} 5000(1.08)^{-(i-1)}$

In Exercises 39–42, find the *n*th partial sum of the geometric sequence.

39. 4, 12, 36, 108, . . . , $n = 8$

40. $\frac{1}{36}$, $-\frac{1}{12}$, $\frac{1}{4}$, $-\frac{3}{4}$, . . . , $n = 20$

41. 60, -15, $\frac{15}{4}$, $-\frac{15}{16}$, . . . , $n = 12$

42. 50, $50(1.04)$, $50(1.04)^2$, $50(1.04)^3$, . . . , $n = 18$

43. *Depreciation* Your company buys a machine for $250,000. During the next 5 years, the machine depreciates at the rate of 25% per year. (That is, at the end of each year, the depreciated value is 75% of what it was at the beginning of the year.)

(a) Find a formula for the *n*th term of the geometric sequence that gives the value of the machine *n* full years after it was purchased.

(b) Find the depreciated value of the machine at the end of 5 full years.

(c) During which year did the machine depreciate the most?

44. *Population Increase* A city of 500,000 people is growing at a rate of 1% per year. (That is, at the end of each year, the population is 1.01 times the population at the beginning of the year.)

(a) Find a formula for the *n*th term of the geometric sequence that gives the population *t* years from now.

(b) Estimate the population 20 years from now.

Increasing Annuity In Exercises 45 and 46, find the balance in an increasing annuity when *P* dollars is invested each month for *t* years, compounded monthly at rate *r*.

45. $P = \$75$, $t = 30$ years, $r = 6\%$

46. $P = \$100$, $t = 25$ years, $r = 8\%$

47. *Geometry* A square has 12-inch sides. A new square is formed by connecting the midpoints of the sides of the square, and two of the triangles are shaded (see figure). This process is repeated five more times. What is the total area of the shaded region?

48. *Bungee Jumping* A bungee jumper stretches a cord 100 feet. Successive bounces stretch the cord 75% of each previous length (see figure). Find the total distance traveled by the bungee jumper during 10 bounces.

$$100 + 2(100)(0.75) + \cdots + 2(100)(0.75)^{10}$$

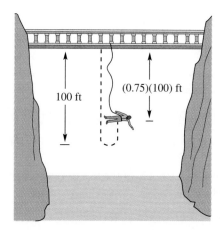

Reviewing the Major Concepts

In Exercises 49–52, evaluate the determinant.

49. $\begin{bmatrix} 10 & 25 \\ 6 & -5 \end{bmatrix}$ **50.** $\begin{bmatrix} 3 & 7 \\ -2 & 6 \end{bmatrix}$

51. $\begin{bmatrix} 3 & -2 & 1 \\ 0 & 5 & 3 \\ 6 & 1 & 1 \end{bmatrix}$ **52.** $\begin{bmatrix} 4 & 3 & 5 \\ 3 & 2 & -2 \\ 5 & -2 & 0 \end{bmatrix}$

53. *Geometry*   Use a determinant to find the area of the triangle with vertices $(-5, 8)$, $(10, 0)$, and $(3, -4)$.

54. Solve the system of linear equations.

$$3x - 2y + z = 1$$
$$x + 5y - 6z = 4$$
$$4x - 3y + 2z = 2$$

Additional Problem Solving

In Exercises 55–62, find the common ratio of the geometric sequence.

55. 2, 6, 18, 54, . . .   **56.** 5, −10, 20, −40, . . .

57. 54, 18, 6, 2, . . .   **58.** 12, −6, 3, $-\frac{3}{2}$, . . .

59. 1, $-\frac{3}{2}$, $\frac{9}{4}$, $-\frac{27}{8}$, . . .

60. 9, 6, 4, $\frac{8}{3}$, . . .

61. e, e^2, e^3, e^4, . . .

62. 1.1, $(1.1)^2$, $(1.1)^3$, $(1.1)^4$, . . .

In Exercises 63–70, determine whether the sequence is geometric. If it is, find the common ratio.

63. 10, 15, 20, 25, . . .

64. 10, 20, 40, 80, . . .

65. 64, 32, 16, 8, . . .

66. 64, 32, 0, −32, . . .

67. 1, $-\frac{2}{3}$, $\frac{4}{9}$, $-\frac{8}{27}$, . . .

68. $\frac{1}{3}$, $-\frac{2}{3}$, $\frac{4}{3}$, $-\frac{8}{3}$, . . .

69. $10(1 + 0.02)$, $10(1 + 0.02)^2$, $10(1 + 0.02)^3$, . . .

70. 1, 0.2, 0.04, 0.008, . . .

In Exercises 71–78, write the first five terms of the geometric sequence.

71. $a_1 = 4, r = 2$ **72.** $a_1 = 3, r = 4$

73. $a_1 = 6, r = \frac{1}{3}$ **74.** $a_1 = 4, r = \frac{1}{2}$

75. $a_1 = 1, r = -\frac{1}{2}$ **76.** $a_1 = 32, r = -\frac{3}{4}$

77. $a_1 = 1000, r = 1.01$ **78.** $a_1 = 4000, r = \dfrac{1}{1.01}$

In Exercises 79–86, find the nth term of the geometric sequence.

79. $a_1 = 6, r = \frac{1}{2}, a_{10} =$

80. $a_1 = 8, r = \frac{3}{4}, a_8 =$

81. $a_1 = 3, r = \sqrt{2}, a_{10} =$

82. $a_1 = 500, r = 1.06, a_{40} =$

83. $a_1 = 4, a_2 = 3, a_5 =$

84. $a_1 = 1, a_2 = 9, a_7 =$

85. $a_1 = 1, a_3 = \frac{9}{4}, a_6 =$

86. $a_2 = 12, a_3 = 16, a_4 =$

In Exercises 87–94, find the formula for the nth term of the geometric sequence. (Begin with $n = 1$.)

87. $a_1 = 2, r = 3$ **88.** $a_1 = 5, r = 4$

89. $a_1 = 1, r = 2$ **90.** $a_1 = 1, r = -5$

91. $a_1 = 4, r = -\frac{1}{2}$ **92.** $a_1 = 9, r = \frac{2}{3}$

93. $a_1 = 8, a_2 = 2$ **94.** $a_1 = 18, a_2 = 8$

In Exercises 95–100, find the sum.

95. $\displaystyle\sum_{i=1}^{10} 2^{i-1}$ **96.** $\displaystyle\sum_{i=1}^{6} 3^{i-1}$

97. $\displaystyle\sum_{i=1}^{12} 3\left(\frac{3}{2}\right)^{i-1}$ **98.** $\displaystyle\sum_{i=1}^{20} 12\left(\frac{2}{3}\right)^{i-1}$

99. $\displaystyle\sum_{i=1}^{15} 3\left(-\frac{1}{3}\right)^{i-1}$ **100.** $\displaystyle\sum_{i=1}^{8} 8\left(-\frac{1}{4}\right)^{i-1}$

In Exercises 101 and 102, use a graphing utility to find the sum.

101. $\displaystyle\sum_{i=1}^{20} 100(1.1)^i$ **102.** $\displaystyle\sum_{i=1}^{40} 50(1.07)^i$

In Exercises 103–110, find the nth partial sum of the geometric sequence.

103. $1, -3, 9, -27, 81, \ldots, \quad n = 10$

104. $3, -6, 12, -24, 48, \ldots, \quad n = 12$

105. $8, 4, 2, 1, \frac{1}{2}, \ldots, \quad n = 15$

106. $9, 6, 4, \frac{8}{3}, \frac{16}{9}, \ldots, \quad n = 10$

107. $1, \sqrt{2}, 2, 2\sqrt{2}, 4, \ldots, \quad n = 12$

108. $40, -10, \frac{5}{2}, -\frac{5}{8}, \frac{5}{32}, \ldots, \quad n = 10$

109. $3, 3(1.06), 3(1.06)^2, 3(1.06)^3, \ldots, n = 20$

110. $10, 10(1.08), 10(1.08)^2, 10(1.08)^3, \ldots, n = 40$

111. *Power Supply*   The electrical power for an implanted medical device decreases by 0.1% each day.

(a) Find a formula for the nth term of the geometric sequence that gives the power n days after the device is implanted.

(b) What percent of the initial power is still available 1 year after the device is implanted?

(c) The power supply needs to be changed when half the power is depleted. Use a graphing utility to graph the first 750 terms and estimate when the power source should be changed.

112. *Cooling*   The temperature of an item is 70° when it is placed in a freezer. Its temperature n hours after being placed in the freezer is 20% less than 1 hour earlier.

(a) Find a formula for the nth term of the geometric sequence that gives the temperature of the item n hours after being placed in the freezer.

(b) Find the temperature of the item 6 hours after being placed in the freezer.

(c) Use a graphing utility to estimate the time when the item freezes. Explain your reasoning.

113. *Salary Increases*   You accept a job that pays a salary of $30,000 the first year. During the next 39 years, you receive a 5% raise each year. What is your *total* salary over the 40-year period?

114. *Salary Increases*   You accept a job that pays a salary of $30,000 the first year. During the next 39 years, you receive a 5.5% raise each year. What is your *total* salary over the 40-year period?

Increasing Annuity   In Exercises 115–118, find the balance in an increasing annuity when P dollars is invested each month for t years, compounded monthly at rate r.

115. $P = \$100, t = 10$ years, $r = 9\%$

116. $P = \$50, t = 5$ years, $r = 7\%$

117. $P = \$30, t = 40$ years, $r = 8\%$

118. $P = \$200, t = 30$ years, $r = 10\%$

119. *Salary*   You start work at a company that pays $0.01 for the first day, $0.02 for the second day, $0.04 for the third day, and so on. If the daily wage keeps doubling, what would your total income be for working (a) 29 days and (b) 30 days?

120. *Salary*   You start work at a company that pays $0.01 for the first day, $0.03 for the second day, $0.09 for the third day, and so on. If the daily wage keeps tripling, what would your total income be for working (a) 25 days and (b) 26 days?

121. *Geometry*   A square has 12-inch sides. The square is divided into nine smaller squares and the center square is shaded (see figure). Each of the eight unshaded squares is then divided into nine smaller squares and each center square is shaded. This process is repeated four more times. What is the total area of the shaded region?

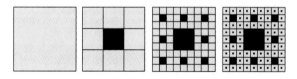

122. *Number of Ancestors*   The number of direct ancestors a person has had is as follows.

$$2 + 2^2 + 2^3 + 2^4 + \cdots + 2^n + \cdots.$$

This formula is valid provided the person has no common ancestors. (A common ancestor is one to whom you are related in more than one way.) During the past 2000 years, suppose your ancestry can be traced through 66 generations. During that time, your total number of ancestors would be

$$2 + 2^2 + 2^3 + 2^4 + \cdots + 2^{66}.$$

Considering the total, do you think that you have had no common ancestors in the past 2000 years?

Math Matters   Duo–Dice*

Ordinary six-sided dice

Twelve-sided Duo-Dice

Ordinary playing dice are made in the shape of cubes. On the sides of each cube, the dice are imprinted with one, two, three, four, five, or six dots, as shown in the figure. On normal playing dice, these dots are usually recessed to prevent the paint from wearing off. These recessions, however, cause the dice to be unbalanced. The heaviest side is the side with only one recession, and the lightest side is the side with six recessions. In practice, this unbalancing is enough to create an increased probability that the lightest side (the side with six recessions) will turn up.

To prevent the unbalancing caused by recessed dots, you can make the dice in the shape of dodecahedrons, as shown in the figure. On two opposing sides, one dot is recessed. On two other opposing sides, two dots are recessed, and so on. Dice made with this pattern are perfectly balanced (because each pair of opposing sides has the same number of recessed dots). When two of these dice are used in any game that requires two six-sided dice, the probability of rolling a given total is the same as if you used two perfectly balanced, six-sided dice.

*Duo-Dice are patented by Roland E. Larson (one of the authors of this text). (Patent Number: 4,465,279, August 14, 1984) If you would like a free pair of these dice, write to Roland E. Larson, Department of Mathematics, Pennsylvania State University, Erie, Pennsylvania 16563.

MID-CHAPTER QUIZ

Take this quiz as you would take a quiz in class. After you are done, check your work against the answers given in the back of the book.

In Exercises 1 and 2, write the first five terms of the sequence.

1. $a_n = 32 \left(\dfrac{1}{4}\right)^{n-1}$ (Begin with $n = 1$.) **2.** $a_n = \dfrac{(-3)^n n}{n + 4}$ (Begin with $n = 1$.)

In Exercises 3–6, find the sum.

3. $\displaystyle\sum_{k=1}^{4} 10k$ **4.** $\displaystyle\sum_{i=1}^{10} 4$ **5.** $\displaystyle\sum_{j=1}^{5} \dfrac{60}{j + 1}$ **6.** $\displaystyle\sum_{n=1}^{8} 8 \left(-\dfrac{1}{2}\right)$

In Exercises 7 and 8, write the sum using sigma notation.

7. $\dfrac{2}{3(1)} + \dfrac{2}{3(2)} + \dfrac{2}{3(3)} + \cdots + \dfrac{2}{3(20)}$ **8.** $\dfrac{1}{1^3} - \dfrac{1}{2^3} + \dfrac{1}{3^3} - \cdots + \dfrac{1}{25^3}$

In Exercises 9 and 10, find the common difference of the arithmetic sequence.

9. $1, \frac{3}{2}, 2, \frac{5}{2}, 3, \ldots$ **10.** $100, 94, 88, 82, 76, \ldots$

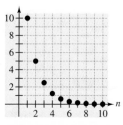

In Exercises 11 and 12, find a formula for a_n.

11. Arithmetic, $a_1 = 20$, $a_4 = 11$ **12.** Geometric, $a_1 = 32$, $r = -\frac{1}{4}$

In Exercises 13–16, find the sum.

13. $\displaystyle\sum_{n=1}^{50} (3n + 5)$ **14.** $\displaystyle\sum_{n=1}^{300} \dfrac{n}{5}$

15. $\displaystyle\sum_{i=1}^{8} 9 \left(\dfrac{2}{3}\right)^{i-1}$ **16.** $\displaystyle\sum_{j=1}^{20} 500(1.06)^{j-1}$

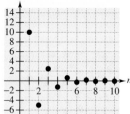

17. Find the 12th term of $625, -250, 100, -40, \ldots$.

18. Match $a_n = 10 \left(\frac{1}{2}\right)^{n-1}$ and $b_n = 10 \left(-\frac{1}{2}\right)^{n-1}$ with the graphs at the right.

Figure for 18

19. The temperature of a coolant decreases by 25.75°F the first hour. For each subsequent hour, the temperature decreases by 2.25°F less than it decreased the previous hour. How much does the temperature decrease during the 10th hour?

20. The sequence given by $a_n = 2^{n-1}$ is geometric. Describe the sequence given by $b_n = \ln a_n$.

10.4	**The Binomial Theorem**
	Binomial Coefficients ▪ Pascal's Triangle ▪ Binomial Expansions

Binomial Coefficients

Recall that a **binomial** is a polynomial that has two terms. In this section, you will study a formula that provides a quick method of raising a binomial to a power. To begin, let's look at the expansion of $(x + y)^n$ for several values of n.

$$(x + y)^0 = 1$$
$$(x + y)^1 = x + y$$
$$(x + y)^2 = x^2 + 2xy + y^2$$
$$(x + y)^3 = x^3 + 3x^2y + 3xy^2 + y^3$$
$$(x + y)^4 = x^4 + 4x^3y + 6x^2y^2 + 4xy^3 + y^4$$
$$(x + y)^5 = x^5 + 5x^4y + 10x^3y^2 + 10x^2y^3 + 5xy^4 + y^5$$

There are several observations you can make about these expansions.

1. In each expansion, there are $n + 1$ terms.

2. In each expansion, x and y have symmetrical roles. The powers of x decrease by 1 in successive terms, whereas the powers of y increase by 1.

3. The sum of the powers of each term is n. For instance, in the expansion of $(x + y)^5$, the sum of the powers of each term is 5.

$$\overbrace{4 + 1 = 5} \quad \overbrace{3 + 2 = 5}$$
$$(x + y)^5 = x^5 + 5x^4y^1 + 10x^3y^2 + 10x^2y^3 + 5xy^4 + y^5$$

4. The coefficients increase and then decrease in a symmetric pattern.

The coefficients of a binomial expansion are called **binomial coefficients.** To find them, you can use the following theorem.

The Binomial Theorem

In the expansion of $(x + y)^n$

$$(x+y)^n = x^n + nx^{n-1}y + \cdots + {}_nC_r x^{n-r}y^r + \cdots + nxy^{n-1} + y^n$$

the coefficient of $x^{n-r}y^r$ is given by

$$_nC_r = \frac{n!}{(n-r)!r!}.$$

EXAMPLE 1   Finding Binomial Coefficients

Find each binomial coefficient.

a. $_8C_2$ **b.** $_{10}C_3$ **c.** $_7C_0$ **d.** $_8C_8$

Solution

a. $_8C_2 = \dfrac{8!}{6! \cdot 2!} = \dfrac{(8 \cdot 7) \cdot \cancel{6!}}{\cancel{6!} \cdot 2!} = \dfrac{8 \cdot 7}{2 \cdot 1} = 28$

b. $_{10}C_3 = \dfrac{10!}{7! \cdot 3!} = \dfrac{(10 \cdot 9 \cdot 8) \cdot \cancel{7!}}{\cancel{7!} \cdot 3!} = \dfrac{10 \cdot 9 \cdot 8}{3 \cdot 2 \cdot 1} = 120$

c. $_7C_0 = \dfrac{7!}{7! \cdot 0!} = 1$

d. $_8C_8 = \dfrac{8!}{0! \cdot 8!} = 1$

NOTE   When $r \neq 0$ and $r \neq n$, as in parts (a) and (b) above, there is a simple pattern for evaluating binomial coefficients.

$$\overbrace{}^{\text{2 factors}} \qquad\qquad \overbrace{}^{\text{3 factors}}$$

$$_8C_2 = \underbrace{\dfrac{8 \cdot 7}{2 \cdot 1}}_{\text{2 factorial}} \quad \text{and} \quad _{10}C_3 = \underbrace{\dfrac{10 \cdot 9 \cdot 8}{3 \cdot 2 \cdot 1}}_{\text{3 factorial}}$$

EXAMPLE 2   Finding Binomial Coefficients

Find each binomial coefficient.

a. $_7C_3$ **b.** $_7C_4$ **c.** $_{12}C_1$ **d.** $_{12}C_{11}$

Solution

a. $_7C_3 = \dfrac{7 \cdot 6 \cdot 5}{3 \cdot 2 \cdot 1} = 35$

b. $_7C_4 = \dfrac{7 \cdot 6 \cdot 5 \cdot 4}{4 \cdot 3 \cdot 2 \cdot 1} = 35$

c. $_{12}C_1 = \dfrac{12!}{11!1!} = \dfrac{(12) \cdot \cancel{11!}}{\cancel{11!} \cdot 1!} = \dfrac{12}{1} = 12$

d. $_{12}C_{11} = \dfrac{12!}{1! \cdot 11!} = \dfrac{(12) \cdot \cancel{11!}}{1! \cdot \cancel{11!}} = \dfrac{12}{1} = 12$

NOTE   In Example 2, it is not a coincidence that the results in parts (a) and (b) are the same, *and* the results in parts (c) and (d) are the same. In general it is true that

$$_nC_r = {}_nC_{n-r}.$$

This shows the symmetric property of binomial coefficients.

Pascal's Triangle

There is a convenient way to remember a pattern for binomial coefficients. By arranging the coefficients in a triangular pattern, you obtain the following array, which is called **Pascal's Triangle.** This triangle is named after the famous French mathematician Blaise Pascal (1623–1662).

$$
\begin{array}{ccccccccccccccc}
 & & & & & & & 1 & & & & & & & \\
 & & & & & & 1 & & 1 & & & & & & \\
 & & & & & 1 & & 2 & & 1 & & & & & \\
 & & & & 1 & & 3 & & 3 & & 1 & & & & \\
 & & & 1 & & 4 & & 6 & & 4 & & 1 & & & \\
 & & 1 & & 5 & & 10 & & 10 & & 5 & & 1 & & \\
 & 1 & & 6 & & 15 & & 20 & & 15 & & 6 & & 1 & \\
1 & & 7 & & 21 & & 35 & & 35 & & 21 & & 7 & & 1
\end{array}
$$

NOTE   The top row in Pascal's Triangle is called the *zero row* because it corresponds to the binomial expansion

$$(x + y)^0 = 1.$$

Similarly, the next row is called the *first row* because it corresponds to the binomial expansion

$$(x + y)^1 = 1(x) + 1(y).$$

In general, the *nth row* in Pascal's Triangle gives the coefficients of $(x + y)^n$.

The first and last numbers in each row of Pascal's Triangle are 1. Every other number in each row is formed by adding the two numbers immediately above the number. Pascal noticed that numbers in this triangle are precisely the same numbers that are the coefficients of binomial expansions, as follows.

$$(x + y)^0 = 1$$
$$(x + y)^1 = 1x + 1y$$
$$(x + y)^2 = 1x^2 + 2xy + 1y^2$$
$$(x + y)^3 = 1x^3 + 3x^2y + 3xy^2 + 1y^3$$
$$(x + y)^4 = 1x^4 + 4x^3y + 6x^2y^2 + 4xy^3 + 1y^4$$
$$(x + y)^5 = 1x^5 + 5x^4y + 10x^3y^2 + 10x^2y^3 + 5xy^4 + 1y^5$$
$$(x + y)^6 = 1x^6 + 6x^5y + 15x^4y^2 + 20x^3y^3 + 15x^2y^4 + 6xy^5 + 1y^6$$
$$(x + y)^7 = 1x^7 + 7x^6y + 21x^5y^2 + 35x^4y^3 + 35x^3y^4 + 21x^2y^5 + 7xy^6 + 1y^7$$

EXAMPLE 3   Using Pascal's Triangle

You can use the seventh row of Pascal's Triangle to find the eighth row.

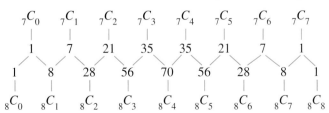

Binomial Expansions

As mentioned at the beginning of this section, when you write out the coefficients for a binomial that is raised to a power, you are **expanding a binomial.** The formulas for binomial coefficients give you an easy way to expand binomials, as demonstrated in the next three examples.

EXAMPLE 4   Expanding a Binomial

Write the expansion for the expression.

$$(x + 1)^3$$

Solution

The binomial coefficients from the third row of Pascal's Triangle are

$$1, 3, 3, 1.$$

Therefore, the expansion is as follows.

$$(x + 1)^3 = (1)x^3 + (3)x^2(1) + (3)x(1^2) + (1)(1^3)$$
$$= x^3 + 3x^2 + 3x + 1$$

To expand binomials representing *differences*, rather than sums, you alternate signs. Here are two examples.

$$(x - 1)^3 = x^3 - 3x^2 + 3x - 1$$
$$(x - 1)^4 = x^4 - 4x^3 + 6x^2 - 4x + 1$$

EXAMPLE 5   Expanding a Binomial

Write the expansion for the expression.

$$(x - 3)^4$$

Solution

The binomial coefficients from the fourth row of Pascal's Triangle are

$$1, 4, 6, 4, 1.$$

Therefore, the expansion is as follows.

$$(x - 3)^4 = (1)x^4 - (4)x^3(3) + (6)x^2(3^2) - (4)x(3^3) + (1)(3^4)$$
$$= x^4 - 12x^3 + 54x^2 - 108x + 81$$

EXAMPLE 6   *Expanding a Binomial*

Write the expansion for $(x - 2y)^4$.

Solution

Use the fourth row of Pascal's Triangle, as follows.

$$(x - 2y)^4 = (1)x^4 - (4)x^3(2y) + (6)x^2(2y)^2 - (4)x(2y)^3 + (1)(2y)^4$$
$$= x^4 - 8x^3y + 24x^2y^2 - 32xy^3 + 16y^4$$

EXAMPLE 7   *Finding a Term in the Binomial Expansion*

Find the sixth term of $(a + 2b)^8$.

Solution

From the Binomial Theorem, you can see that the $(r + 1)$th term is $_nC_rx^{n-r}y^r$. So in this case, $6 = r + 1$ means that $r = 5$. Because $n = 8$, $x = a$, and $y = 2b$, the sixth term in the binomial expansion is

$$_8C_5a^{8-5}(2b)^5 = 56 \cdot a^3 \cdot (2b)^5$$
$$= 56(2^5)a^3b^5$$
$$= 1792a^3b^5$$

Group Activities   Extending the Concept

Finding a Pattern   By adding the terms in each of the rows of Pascal's Triangle, you obtain the following.

Row 0: $1 = 1$
Row 1: $1 + 1 = 2$
Row 2: $1 + 2 + 1 = 4$
Row 3: $1 + 3 + 3 + 1 = 8$
Row 4: $1 + 4 + 6 + 4 + 1 = 16$

Find a pattern for this sequence. Then use the pattern to find the sum of the terms in the 10th row of Pascal's Triangle. Finally, check your answer by actually adding the terms of the 10th row.

10.4 **Exercises**

Discussing the Concepts

1. How many terms are in the expansion of $(x + y)^n$?

2. Describe the pattern for the exponents with base x in the expansion of $(x + y)^n$.

3. How do the expansions of $(x + y)^n$ and $(x - y)^n$ differ?

4. What is the relationship between $_nC_r$ and $_nC_{n-r}$?

5. Which of the following is equal to $_{11}C_5$? Explain.

(a) $\dfrac{11 \cdot 10 \cdot 9 \cdot 8 \cdot 7}{5 \cdot 4 \cdot 3 \cdot 2 \cdot 1}$ (b) $\dfrac{11 \cdot 10 \cdot 9 \cdot 8 \cdot 7}{6 \cdot 5 \cdot 4 \cdot 3 \cdot 2 \cdot 1}$

6. In your own words, explain how to form the rows of Pascal's Triangle.

Problem Solving

In Exercises 7–10, evaluate $_nC_r$.

7. $_6C_4$ **8.** $_7C_3$

9. $_{20}C_{20}$ **10.** $_{200}C_1$

In Exercises 11–14, use a graphing utility to evaluate $_nC_r$.

11. $_{30}C_6$ **12.** $_{25}C_{10}$

13. $_{52}C_5$ **14.** $_{100}C_8$

In Exercises 15–18, evaluate the binomial coefficient $_nC_r$. Also, evaluate its symmetric coefficient $_nC_{n-r}$.

15. $_{15}C_3$ **16.** $_9C_4$

17. $_{25}C_5$ **18.** $_{30}C_3$

19. Find the ninth row of Pascal's Triangle.

20. Use Pascal's Triangle to evaluate $_nC_r$.

(a) $_6C_2$ (b) $_9C_3$

In Exercises 21–24, use the Binomial Theorem to expand the expression.

21. $(x + 3)^6$ **22.** $(x - 5)^4$

23. $(u - 2v)^3$ **24.** $(2x + y)^5$

In Exercises 25–28, use Pascal's Triangle to expand the expression.

25. $(x + y)^8$ **26.** $(r - s)^7$

27. $(x - 2)^6$ **28.** $(2x + 3)^5$

In Exercises 29–32, find the coefficient of the given term of the expression.

	Expression	*Term*
29.	$(x + 1)^{10}$	x^7
30.	$(x + 3)^{12}$	x^9
31.	$(x - y)^{15}$	x^4y^{11}
32.	$(x - 3y)^{14}$	x^3y^{11}

Probability In Exercises 33–36, use the Binomial Theorem to expand the expression. In the study of probability, it is sometimes necessary to use the expansion $(p + q)^n$, where $p + q = 1$.

33. $\left(\dfrac{1}{2} + \dfrac{1}{2}\right)^5$ **34.** $\left(\dfrac{2}{3} + \dfrac{1}{3}\right)^4$

35. $\left(\dfrac{1}{4} + \dfrac{3}{4}\right)^4$ **36.** $(0.4 + 0.6)^6$

In Exercises 37–40, use the Binomial Theorem to approximate the quantity accurate to two decimal places. For example,

$(1.02)^{10} \approx 1 + 10(0.02) + 45(0.02)^2 \approx 1.22.$

37. $(1.02)^8$ **38.** $(2.005)^{10}$

39. $(2.99)^{12}$ **40.** $(1.98)^9$

Reviewing the Major Concepts

In Exercises 41–44, rewrite the logarithmic equation in exponential form.

41. $\log_4 64 = 3$

42. $\log_3 \frac{1}{81} = -4$

43. $\ln 1 = 0$

44. $\ln 5 \approx 1.6094$

45. *Graphical Estimation*   After t years, the value of a car is given by $V(t) = 22{,}000(0.8)^t$. Graphically determine when the value of the car will be $15,000.

46. *Interest*   Find the balance when $10,000 is invested at $7\frac{1}{2}\%$ compounded monthly for 15 years.

Additional Problem Solving

In Exercises 47–54, evaluate $_nC_r$.

47. $_{10}C_5$

48. $_{12}C_9$

49. $_{18}C_{18}$

50. $_{15}C_0$

51. $_{50}C_{48}$

52. $_{75}C_1$

53. $_{25}C_4$

54. $_{18}C_5$

In Exercises 55–60, use a graphing utility to evaluate $_nC_r$.

55. $_{12}C_7$

56. $_{40}C_5$

57. $_{200}C_{10}$

58. $_{500}C_6$

59. $_{25}C_{12}$

60. $_{1000}C_2$

In Exercises 61–66, evaluate the binomial coefficient $_nC_r$. Also, evaluate its symmetric coefficient $_nC_{n-r}$.

61. $_5C_2$

62. $_8C_6$

63. $_{12}C_5$

64. $_{14}C_8$

65. $_{10}C_0$

66. $_{25}C_{25}$

In Exercises 67–70, use Pascal's Triangle to evaluate the binomial coefficient.

67. $_7C_3$

68. $_9C_5$

69. $_8C_4$

70. $_{10}C_6$

In Exercises 71–78, use the Binomial Theorem to expand the expression.

71. $(x + 1)^5$

72. $(x + 2)^5$

73. $(x - 4)^6$

74. $(x - 8)^4$

75. $(x + y)^4$

76. $(u + v)^6$

77. $(x - y)^5$

78. $(4t - 1)^4$

In Exercises 79–84, use Pascal's Triangle to expand the expression.

79. $(a + 2)^3$

80. $(x + 3)^5$

81. $(2x - 1)^4$

82. $(4 - 3y)^3$

83. $(2y + z)^6$

84. $(2t - s)^5$

In Exercises 85–90, find the coefficient of the given term of the expression.

Expression	Term
85. $(x - 1)^{10}$	x^7
86. $(x - 2)^8$	x^4
87. $(2x + y)^{12}$	$x^3 y^9$
88. $(x + y)^{10}$	$x^7 y^3$
89. $(x^2 - 3)^4$	x^4
90. $(3 - y^3)^5$	y^9

91. *Patterns in Pascal's Triangle*   Use each encircled group of numbers to form a 2×2 matrix. Find the determinant of each matrix. Describe the pattern.

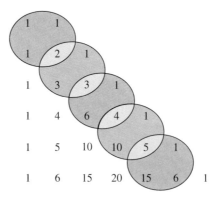

10.5	**Counting Principles**
	Simple Counting Problems ▪ Counting Principles ▪ Permutations ▪ Combinations

Simple Counting Problems

The last two sections of this chapter contain a brief introduction to some of the basic counting principles and their application to probability. In the next section, you will see that much of probability has to do with counting the number of ways an event can occur. Examples 1, 2, and 3 describe some simple cases.

EXAMPLE 1 A Random Number Generator

A random number generator (on a computer) selects an integer from 1 to 30. Find the number of ways each event can occur.

a. An even integer is selected.

b. A number that is less than 12 is selected.

c. A prime number is selected.

d. A perfect square is selected.

Solution

a. Because half of the numbers from 1 to 30 are even, this event can occur in 15 different ways.

b. The positive integers that are less than 12 are as follows.

$$\{1,\ 2,\ 3,\ 4,\ 5,\ 6,\ 7,\ 8,\ 9,\ 10,\ 11\}$$

Because this set has 11 members, you can conclude that there are 11 different ways this event can happen.

c. The prime numbers between 1 and 30 are as follows.

$$\{2,\ 3,\ 5,\ 7,\ 11,\ 13,\ 17,\ 19,\ 23,\ 29\}$$

Because this set has 10 members, you can conclude that there are 10 different ways this event can happen.

d. The perfect square numbers between 1 and 30 are as follows.

$$\{1,\ 4,\ 9,\ 16,\ 25\}$$

Because this set has five members, you can conclude that there are five different ways this event can happen.

EXAMPLE 2   *Selecting Pairs of Numbers at Random*

Eight pieces of paper are numbered from 1 to 8 and placed in a box. One piece of paper is drawn from the box, its number is written down, and the piece of paper is replaced in the box. Then, a second piece of paper is drawn from the box, and its number is written down. Finally, the two numbers are added together. How many different ways can a total of 12 be obtained?

Solution

To solve this problem, count the different ways that a total of 12 can be obtained using two numbers between 1 and 8.

$$\boxed{\text{First number}} \;+\; \boxed{\text{Second number}} \;=\; \boxed{12}$$

After considering the various possibilities, you can see that this equation can be solved in the following five ways.

$$
\begin{array}{lccccc}
\textit{First Number:} & 4 & 5 & 6 & 7 & 8 \\
\textit{Second Number:} & 8 & 7 & 6 & 5 & 4
\end{array}
$$

Thus, a total of 12 can be obtained in five different ways.

Solving counting problems can be tricky. Often, seemingly minor changes in the statement of a problem can affect the answer. For instance, compare the counting problem in the next example with that given in Example 2.

EXAMPLE 3   *Selecting Pairs of Numbers at Random*

Eight pieces of paper are numbered from 1 to 8 and placed in a box. Two pieces of paper are drawn from the box, and the numbers on them are written down and totaled. How many different ways can a total of 12 be obtained?

NOTE   The difference between the counting problems in Examples 2 and 3 can be distinguished by saying that the random selection in Example 2 occurs *with replacement,* whereas the random selection in Example 3 occurs *without replacement,* which eliminates the possibility of choosing two 6's.

Solution

To solve this problem, count the different ways that a total of 12 can be obtained *using two different numbers* between 1 and 8.

$$\boxed{\text{First number}} \;+\; \boxed{\text{Second number}} \;=\; \boxed{12}$$

After considering the various possibilities, you can see that this equation can be solved in the following four ways.

$$
\begin{array}{lcccc}
\textit{First Number:} & 4 & 5 & 7 & 8 \\
\textit{Second Number:} & 8 & 7 & 5 & 4
\end{array}
$$

Thus, a total of 12 can be obtained in four different ways.

Counting Principles

The first three examples in this section are considered simple counting problems in which you can *list* each possible way that an event can occur. When it is possible, this is always the best way to solve a counting problem. However, some events can occur in so many different ways that it is not feasible to write out the entire list. In such cases, you must rely on formulas and counting principles. The most important of these is called the **Fundamental Counting Principle.**

NOTE The Fundamental Counting Principle can be extended to three or more events. For instance, the number of ways that three events E_1, E_2, and E_3 can occur is $m_1 \cdot m_2 \cdot m_3$.

Fundamental Counting Principle

Let E_1 and E_2 be two events. The first event E_1 can occur in m_1 different ways. After E_1 has occurred, E_2 can occur in m_2 different ways. The number of ways that the two events can occur is

$$m_1 \cdot m_2.$$

EXAMPLE 4   *Applying the Fundamental Counting Principle*

The English alphabet contains 26 letters. Thus, the number of possible "two-letter words" is $26 \cdot 26 = 676$.

EXAMPLE 5   *Applying the Fundamental Counting Principle*

Telephone numbers in the United States have ten digits. The first three are the *area code* and the next seven are the *local telephone number*. How many different telephone numbers are possible within each area code? (A telephone number cannot have 0 or 1 as its first or second digit.)

Solution

There are only eight choices for the first and second digits because neither can be 0 or 1. For each of the other digits, there are 10 choices.

In 1991, there were 131 million active telephone numbers in use in the United States.

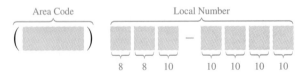

Thus, by the Fundamental Counting Principle, the number of local telephone numbers that are possible within each area code is

$$8 \cdot 8 \cdot 10 \cdot 10 \cdot 10 \cdot 10 \cdot 10 = 6,400,000.$$

Permutations

One important application of the Fundamental Counting Principle is in determining the number of ways that n elements can be arranged (in order). An ordering of n elements is called a **permutation** of the elements.

Definition of Permutation

A **permutation** of n different elements is an ordering of the elements such that one element is first, one is second, one is third, and so on.

EXAMPLE 6 Listing Permutations

The six possible permutations of the letters A, B, and C are as follows.

A, B, C	B, A, C	C, A, B
A, C, B	B, C, A	C, B, A

EXAMPLE 7 Finding the Number of Permutations of n Elements

How many permutations are possible for the letters A, B, C, D, E, and F?

Solution

1st position:	Any of the *six* letters.
2nd position:	Any of the remaining *five* letters.
3rd position:	Any of the remaining *four* letters.
4th position:	Any of the remaining *three* letters.
5th position:	Any of the remaining *two* letters.
6th position:	The *one* remaining letter.

Thus, the number of choices for the six positions are as follows.

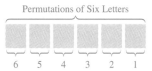

Permutations of Six Letters

By the Fundamental Counting Principle, the total number of permutations of the six letters is

$$6 \cdot 5 \cdot 4 \cdot 3 \cdot 2 \cdot 1 = 6! = 720.$$

The result obtained in Example 7 is generalized below.

Number of Permutations of n Elements

The number of permutations of n elements is given by

$$n \cdot (n-1) \cdot \cdots \cdot 4 \cdot 3 \cdot 2 \cdot 1 = n!.$$

In other words, there are $n!$ different ways that n elements can be ordered.

EXAMPLE 8 Finding the Number of Permutations

How many ways can you form a four-digit number using each of the digits 1, 3, 5, and 7 exactly once?

Solution

One way to solve this problem is simply to list the number of ways.

1357, 1375, 1537, 1573, 1735, 1753
3157, 3175, 3517, 3571, 3715, 3751
5137, 5173, 5317, 5371, 5713, 5731
7135, 7153, 7315, 7351, 7513, 7531

Another way to solve the problem is to use the formula for the number of permutations of four elements. By that formula, there are $4! = 24$ permutations.

EXAMPLE 9 Finding the Number of Permutations

You are a supervisor for eleven different employees. One of your responsibilities is to perform an annual evaluation for each employee, and then rank the eleven different performances. How many different rankings are possible?

Solution

Because there are eleven different employees, you have eleven choices for first ranking. After choosing the first ranking, you can choose any of the remaining ten for second ranking, and so on.

Rankings of Eleven Employees

11 10 9 8 7 6 5 4 3 2 1

Thus, the number of different rankings is $11! = 39,916,800$.

Combinations

When one counts the number of possible permutations of a set of elements, order is important. The final topic in this section is a method of selecting subsets of a larger set in which order is *not important.* Such subsets are called **combinations of *n* elements taken *r* at a time.** For instance, the combinations

{A, B, C}   and   {B, A, C}

are equivalent because both sets contain the same three elements, and the order in which the elements are listed is *not important.* Hence, you would count only one of the two sets. A common example of how a combination occurs is a card game in which the player is free to reorder the cards after they have been dealt.

EXAMPLE 10   *Combination of n Elements Taken r at a Time*

In how many different ways can three letters be chosen from the letters A, B, C, D, and E? (The order of the three letters is not important.)

Solution

The following subsets represent the different combinations of three letters that can be chosen from five letters.

{A, B, C}	{A, B, D}
{A, B, E}	{A, C, D}
{A, C, E}	{A, D, E}
{B, C, D}	{B, C, E}
{B, D, E}	{C, D, E}

From this list, you can conclude that there are 10 different ways that three letters can be chosen from five letters. Because order is not important, the set {B, C, A} is not chosen. It is represented by the set {A, B, C}.

The formula for the number of combinations of *n* elements taken *r* at a time is as follows.

Number of Combinations of *n* Elements Taken *r* at a Time

The number of combinations of *n* elements taken *r* at a time is

$$_nC_r = \frac{n!}{(n-r)!r!}.$$

When solving problems involving counting principles, you need to be able to distinguish among the various counting principles in order to determine which is necessary to solve the problem correctly. To do this, ask yourself the following questions.

1. Is the order of the elements important? *Permutation*

2. Are the chosen elements a subset of a larger set in which order is not important? *Combination*

3. Does the problem involve two or more separate events? *Fundamental Counting Principle*

Note that the formula for $_nC_r$ is the same one given for binomial coefficients. To see how this formula is used, consider the counting problem given in Example 10. In that problem, you need to find the number of combinations of five elements taken three at a time. Thus, $n = 5$, $r = 3$, and the number of combinations is

$$_5C_3 = \frac{5!}{2!3!}$$
$$= \frac{5 \cdot 4 \cdot 3}{3 \cdot 2 \cdot 1}$$
$$= 10$$

which is the same as the answer obtained in Example 10.

EXAMPLE 11 Combinations of n Elements Taken r at a Time

A standard poker hand consists of five cards dealt from a deck of 52. How many different poker hands are possible? (After the cards are dealt, the player may reorder them, and therefore order is not important.)

Solution

Use the formula for the number of combinations of 52 elements taken five at a time, as follows.

$$_{52}C_5 = \frac{52!}{47!5!}$$
$$= \frac{52 \cdot 51 \cdot 50 \cdot 49 \cdot 48}{5 \cdot 4 \cdot 3 \cdot 2 \cdot 1}$$
$$= 2,598,960$$

Thus, there are almost 2.6 million different hands.

Group Activities   Problem Solving

Applying Counting Methods   The Boston Market restaurant chain offers individual rotisserie chicken meals with two side items. Customers can choose either dark or white meat and can select side items from a list of 15. How many different meals are available if two different side items are to be ordered? Of the 15 side items, nine are hot and six are cold. How many different meals are available if a customer wishes to order one hot and one cold side item? (Source: Boston Market, Inc.)

10.5 Exercises

Discussing the Concepts

1. State the Fundamental Counting Principle.

2. When you use the Fundamental Counting Principle, what are you counting?

3. Give examples of a permutation and a combination.

4. Without calculating the numbers, determine which of the following is greater. Explain.

   (a) The combination of 10 elements taken 6 at a time

   (b) The permutation of 10 elements taken 6 at a time

Problem Solving

Random Selection   In Exercises 5–10, find the number of ways the specified event can occur when one or more marbles are selected from a bowl containing 10 marbles numbered 0 through 9.

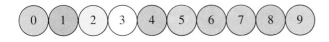

5. One number is drawn and it is even.

6. One number is drawn and it is prime.

7. Two marbles are drawn one after the other. The first is replaced before the second is drawn. The sum of the numbers is 10.

8. Two marbles are drawn one after the other. The first is replaced before the second is drawn. The sum of the numbers is 7.

9. Two marbles are drawn without replacement. The sum of the numbers is 10.

10. Two marbles are drawn without replacement. The sum of the numbers is 7.

11. *Staffing Choices*   A small grocery store needs to open another checkout line. Three people who can run the cash register are available and two people are available to bag groceries. How many different ways can the additional checkout line be staffed?

12. *Computer System*   You are in the process of purchasing a new computer system. You must choose one of three monitors, one of two computers, and one of two keyboards. How many different configurations of the system are available to you?

13. *Taking a Trip*   Five people are taking a long trip in a car. Two sit in the front seat and three in the back seat. Three of the people agree to share the driving. In how many different arrangements can the five people sit?

14. *Aircraft Boarding*   Eight people are boarding an aircraft. Three have tickets for first class and board before those in the economy class. In how many different ways can the eight people board the aircraft?

15. *Permutations*   List all the permutations of the letters X, Y, and Z.

16. *Permutations*   List all the permutations of the letters A, B, C, and D.

17. *Seating Arrangement*   In how many ways can five children be seated in a single row of chairs?

18. *Combination Lock*   A combination lock will open when the right choice of three numbers (from 1 to 40, inclusive) is selected. How many different lock combinations are possible?

19. *Work Assignments*   Eight workers are assigned to eight different tasks. In how many ways can this be done assuming there are no restrictions in making the assignments?

20. *Work Assignments*   Eight workers are available for five different tasks. In how many ways can five workers be selected from the eight and assigned to the tasks if there are no restrictions in making the assignments?

21. *Number of Subsets*   List all the subsets with two elements that can be formed from the set of letters {A, B, C, D, E, F}.

22. *Number of Subsets*   List all the subsets with three elements that can be formed from the set of letters {A, B, C, D, E, F}.

23. *Relationships*   As the size of a group increases, the number of relationships increases dramatically (see figure). Determine the number of two-person relationships in a group that has the following numbers.

(a) 3   (b) 4

(c) 6   (d) 8

(e) 10   (f) 12

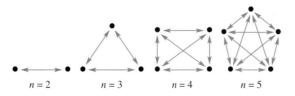

$n = 2$ $n = 3$ $n = 4$ $n = 5$

24. *Committee Selection*   Three students are selected from a class of 20 to form a fundraising committee. In how many ways can the committee be formed?

25. *Number of Triangles*   Eight points are located in the coordinate plane such that no three are collinear. How many different triangles can be formed having their vertices as three of the eight points?

26. *Test Questions*   A student is required to answer any nine questions from 12 questions on an exam. In how many ways can the student select the questions?

27. *Defective Units*   A shipment of 10 microwave ovens contains two defective units. In how many ways can a vending company purchase three of these units and receive (a) all good units, (b) two good units, and (c) one good unit?

28. *Job Applicants*   An employer interviews six people for four openings in the company. Four of the six people are women. If all six are qualified, in how many ways can the employer fill the four positions if (a) the selection is random and (b) exactly two are women?

Reviewing the Major Concepts

In Exercises 29–32, solve the equation. (Round your answer to two decimal places.)

29. $\sqrt{x-5} = 6$ **30.** $\dfrac{4}{t} + \dfrac{3}{2t} = 1$

31. $\log_2(x-5) = 6$ **32.** $e^{x/2} = 8$

33. Write the equation of the line that passes through $(3, 5)$ and $(6, 7)$.

34. *Radioactive Decay*   Carbon 14 has a half-life of 5730 years. If you start with 10 grams of this isotope, how much remains after 3000 years?

Additional Problem Solving

Random Selection   In Exercises 35–44, determine the number of ways the specified event can occur when one or more marbles are selected from a bowl containing 20 marbles numbered 1 through 20.

35. One number is drawn, and it is odd.

36. One number is drawn, and it is even.

37. One number is drawn, and it is prime.

38. One number is drawn, and it is greater than 12.

39. One number is drawn, and it is divisible by 3.

40. One number is drawn, and it is divisible by 6.

41. Two marbles are drawn one after the other. The first is replaced before the second is drawn. The sum of the numbers is 8.

42. Two marbles are drawn one after the other. The first is replaced before the second is drawn. The sum of the numbers is 15.

43. Two marbles are drawn without replacement. The sum of the numbers is 8.

44. Three marbles are drawn, one after another. The first and second are replaced before drawing the second and third. The sum of the numbers is 15.

45. *License Plates*   How many distinct automobile license plates can be formed by using a four-digit number followed by two letters?

46. *Task Assignment*   Four people are assigned to four different tasks. In how many ways can the assignments be made if one of the four is not qualified for the first task?

47. *Permutations*   List all the permutations of two letters selected from the letters A, B, C, and D.

48. *Permutations*   List all the permutations of two letters selected from the letters A, B, and C.

49. *Posing for a Photograph*   In how many ways can four children line up in one row to have their picture taken?

50. *Seating Arrangement*   In how many ways can six people be seated in a six-passenger car?

51. *Choosing Officers*   From a pool of 10 candidates, the offices of president, vice-president, secretary, and treasurer will be filled. In how many ways can the offices be filled if each of the 10 can hold any one of the offices?

52. *Time Management Study*   There are eight steps in accomplishing a certain task and these steps can be performed in any order. Management wants to test each possible order to determine which is least time consuming.

  (a) How many different orders will have to be tested?

  (b) How many different orders will have to be tested if one step in accomplishing the task must be done first? (The other seven steps can be performed in any order.)

53. *Number of Subsets*   List all the subsets with three elements that can be formed from the set of letters {A, B, C, D, E}.

54. *Number of Subsets*   List all the subsets with two elements that can be formed from the set of letters {A, B, C, D, E, F}.

55. *Identification Numbers*   In a statistical study, each participant is given an identification label consisting of a letter of the alphabet followed by a single digit. How many distinct labels are possible?

56. *Test Questions*   A student may answer any seven questions from a total of 10 questions on an exam. In how many different ways can the student select the questions?

57. *Committee Selection*   In how many ways can a committee of five be formed from a group of 30 people?

58. *Menu Selection*   A group of four people go out to dinner at a restaurant. There are nine entrees on the menu and the four people decide that no two will order the same thing. How many ways can the four order from the nine entrees?

59. *Basketball Lineup*   A high school basketball team has 15 players. In how many different ways can the coach choose the starting lineup of five? (Assume each player can play each position.)

60. *Softball League*   Six churches form a softball league. If each team must play every other team twice during the season, what is the total number of league games played?

61. *Group Selection*   Four people are to be selected from four couples. In how many ways can this be done if

  (a) there are no restrictions?

  (b) one person from each couple must be selected?

62. *Geometry*   Three points that are not on a line determine three lines. How many lines are determined by seven points, no three of which are on a line?

63. *Diagonals of a Polygon*   Find the number of diagonals of each polygon. (A line segment connecting any two nonadjacent vertices of a polygon is called a *diagonal* of the polygon.)

  (a) Pentagon   (b) Hexagon   (c) Octagon

10.6   Probability

The Probability of an Event ▪
Using Counting Methods to Find Probabilities

The Probability of an Event

The **probability of an event** is a number from 0 to 1 that indicates the likelihood that the event will occur. An event that is certain to occur has a probability of 1. An event that cannot occur has a probability of 0. An event that is equally likely to occur or not occur has a probability of $\frac{1}{2}$ or 0.5.

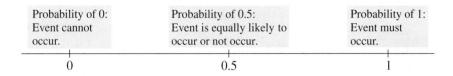

| Probability of 0:
Event cannot
occur. | Probability of 0.5:
Event is equally likely to
occur or not occur. | Probability of 1:
Event must
occur. |

0   0.5   1

The Probability of an Event

Consider a **sample space** S that is composed of a finite number of outcomes, each of which is equally likely to occur. A subset E of the sample space is an **event.** The probability that an outcome E will occur is

$$P = \frac{\text{number of outcomes in event}}{\text{number of outcomes in sample space}}.$$

EXAMPLE 1   *Finding the Probability of an Event*

a. You are dialing a friend's phone number but cannot remember the last digit. If you choose a digit at random, the probability that it is correct is

$$P = \frac{\text{number of correct digits}}{\text{number of possible digits}} = \frac{1}{10}.$$

b. On a multiple-choice test, you know that the answer to a question is not (a) or (d), but you are not sure about (b), (c), and (e). If you guess, the probability that you are wrong is

$$P = \frac{\text{number of wrong answers}}{\text{number of possible answers}} = \frac{2}{3}.$$

EXAMPLE 2   Conducting a Poll

In 1990, the Center for Disease Control took a survey of 11,631 high school students. The students were asked whether they considered themselves to be a good weight, underweight, or overweight. The results are shown in Figure 10.3.

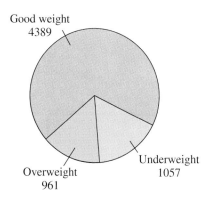

FIGURE 10.3

a. If you choose a female at random from those surveyed, the probability that she said she was underweight is

$$P = \frac{\text{number of females who answered "underweight"}}{\text{number of females in survey}}$$

$$= \frac{366}{3082 + 366 + 1776}$$

$$= \frac{366}{5224}$$

$$\approx 0.07.$$

b. If you choose a person who answered "underweight" from those surveyed, the probability that the person is female is

$$P = \frac{\text{number of females who answered "underweight"}}{\text{number in survey who answered "underweight"}}$$

$$= \frac{366}{366 + 1057}$$

$$= \frac{366}{1423}$$

$$\approx 0.26.$$

Polls such as the one described in Example 2 are often used to make inferences about a population that is larger than the sample. For instance, from Example 2, you might infer that 7% of *all* high school girls consider themselves to be underweight. When you make such an inference, it is important that those surveyed are representative of the entire population.

EXAMPLE 3   Using Area to Find Probability

You have just stepped into the tub to take a shower when one of your contact lenses falls out. (You have not yet turned on the water.) Assuming the lens is equally likely to land anywhere on the bottom of the tub, what is the probability that it lands in the drain? Use the dimensions in Figure 10.4 to answer the question.

Solution

Because the area of the tub bottom is $(26)(50) = 1300$ square inches and the area of the drain is

$$\pi(1^2) = \pi \qquad \text{Area of drain}$$

square inches, the probability that the lens lands in the drain is about

$$P = \frac{\pi}{1300} \approx 0.0024.$$

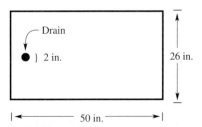

FIGURE 10.4

EXAMPLE 4   The Probability of Inheriting Certain Genes

Common parakeets have genes that can produce any one of four feather colors:

Green: BBCC, BBCc, BbCC, BbCc
Blue: BBcc, Bbcc
Yellow: bbCC, bbCc
White: bbcc

Use the *Punnett square* in Figure 10.5 to find the probability that an offspring to two green parents (both with BbCc feather genes) will be yellow. Note that each parent passes along a B or b gene and a C or c gene.

Solution

The probability that an offspring will be yellow is

$$P = \frac{\text{number of yellow possibilities}}{\text{number of possibilities}}$$

$$= \frac{3}{16}.$$

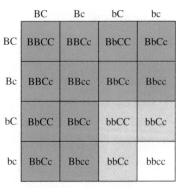

FIGURE 10.5

Using Counting Methods to Find Probabilities

Standard 52-Card Deck

A♠	A♥	A♦	A♣
K♠	K♥	K♦	K♣
Q♠	Q♥	Q♦	Q♣
J♠	J♥	J♦	J♣
10♠	10♥	10♦	10♣
9♠	9♥	9♦	9♣
8♠	8♥	8♦	8♣
7♠	7♥	7♦	7♣
6♠	6♥	6♦	6♣
5♠	5♥	5♦	5♣
4♠	4♥	4♦	4♣
3♠	3♥	3♦	3♣
2♠	2♥	2♦	2♣

FIGURE 10.6

EXAMPLE 5   The Probability of a Royal Flush

Five cards are dealt at random from a standard deck of 52 playing cards (see Figure 10.6). What is the probability that the cards are 10-J-Q-K-A of the same suit?

Solution

On page 676, you saw that the number of possible five-card hands from a deck of 52 cards is $_{52}C_5 = 2,598,960$. Because only four of these five-card hands are 10-J-Q-K-A of the same suit, the probability that the hand contains these cards is

$$P = \frac{4}{2,598,960} = \frac{1}{649,740}.$$

EXAMPLE 6   Conducting a Survey

In 1990, a survey was conducted of 500 adults who had worn Halloween costumes. Each person was asked how he or she acquired a Halloween costume: created it, rented it, bought it, or borrowed it. The results are shown in Figure 10.7. What is the probability that the first four people who were polled all created their costumes?

Solution

To answer this question, you need to use the formula for the number of combinations *twice*. First, find the number of ways to choose four people from 360 who created their own costumes.

$$_{360}C_4 = \frac{360 \cdot 359 \cdot 358 \cdot 357}{4 \cdot 3 \cdot 2 \cdot 1} = 688,235,310$$

Next, find the number of ways to choose four people from the 500 who were surveyed.

$$_{500}C_4 = \frac{500 \cdot 499 \cdot 498 \cdot 497}{4 \cdot 3 \cdot 2 \cdot 1} = 2,573,031,125$$

The probability that all of the first four people surveyed created their own costumes is the ratio of these two numbers.

$$P = \frac{\text{number of ways to choose 4 from 360}}{\text{number of ways to choose 4 from 500}}$$

$$= \frac{688,235,310}{2,573,031,125}$$

$$\approx 0.267$$

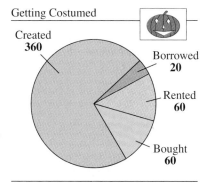

Getting Costumed

Created 360
Borrowed 20
Rented 60
Bought 60

FIGURE 10.7

EXAMPLE 7 Forming a Committee

To obtain input from 200 company employees, the management of a company selected a committee of five. Of the 200 employees, 56 were from minority groups. None of the 56, however, was selected to be on the committee. Does this indicate that the management's selection was biased?

Solution

Part of the solution is similar to that of Example 6. If the five committee members were selected at random, the probability that all five would be nonminority is

$$P = \frac{\text{number of ways to choose 5 from 144 nonminority employees}}{\text{number of ways to choose 5 from 200 employees}}$$

$$= \frac{_{144}C_5}{_{200}C_5}$$

$$= \frac{481,008,528}{2,535,650,040}$$

$$\approx 0.19.$$

Thus, if the committee were chosen at random (that is, without bias), the likelihood that it would have no minority members is about 0.19. Although this does not *prove* that there was bias, it does suggest it.

Group Activities Extending the Concept

Probability of Guessing Correctly   You are taking a chemistry test and are asked to arrange the first 10 elements in the order in which they appear on the periodic table of elements. Suppose that you have no idea of the correct order and simply guess. Does the following computation represent the probability that you guess correctly?

Solution

You have 10 choices for the first element, nine choices for the second, eight choices for the third, and so on. The number of different orders is $10! = 3,628,800$, which means that your probability of guessing correctly is

$$P = \frac{1}{3,628,800}.$$

10.6 Exercises

Discussing the Concepts

1. The probability of an event must be a real number in what interval? Is the interval open or closed?

2. The probability of an event is $\frac{3}{4}$. What is the probability that the event *does not* occur? Explain.

3. What is the sum of the probabilities of all elements in a sample space? Explain.

4. The weather forecast indicates that the probability of rain is 40%. Explain what this means.

Problem Solving

In Exercises 5–8, determine the sample space for the experiment.

5. One letter from the alphabet is chosen.

6. A six-sided die is tossed twice and the sum is recorded.

7. Two county supervisors are selected from five supervisors, A, B, C, D, and E.

8. A salesperson makes a presentation in three homes. In each home there may be a sale (denote by Y) or there may be no sale (denote by N).

In Exercises 9 and 10, you are given the probability that an event will occur. Find the probability that the event will not occur.

9. $p = 0.35$ **10.** $p = 0.8$

Coin Tossing   In Exercises 11–14, a coin is tossed three times. Find the probability of the specified event. Use the following sample space.

HHH, HHT, HTH, THH
HTT, THT, TTH, TTT

11. The event of getting two heads

12. The event of getting a tail on the second toss

13. The event of getting at least one head

14. The event of getting no more than two heads

Reading a Graph   In Exercises 15 and 16, use the circle graph, which shows the number of people in the United States in 1990 with each blood type. (Source: American Association of Blood Banks)

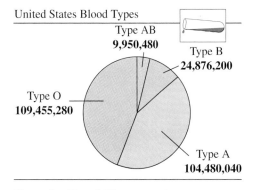

United States Blood Types

Type AB
9,950,480

Type B
24,876,200

Type O
109,455,280

Type A
104,480,040

Figure for 15 and 16

15. A person is selected at random from the United States population. What is the probability that the person does not have blood type B?

16. What is the probability that a person selected at random from the United States population does have blood type B? How is this probability related to the probability found in Exercise 15?

17. *Random Selection*   Twenty marbles numbered 1 through 20 are placed in a bag, and one is selected. Find the probability of the specified event.

(a) The number is 12.

(b) The number is prime.

(c) The number is odd.

(d) The number is less than 6.

18. *Class Election*   Three people are running for class president. It is estimated that the probability that candidate A will win is 0.5 and the probability that candidate B will win is 0.3. What is the probability that the third candidate will win?

19. *Meteorites*   The largest meteorite that ever landed in the United States was found in the Willamette Valley of Oregon in 1902. Earth contains 57,510,000 square miles of land and 139,440,000 square miles of water. What is the probability that a meteorite that hits the earth will fall onto land? What is the probability that a meteorite that hits the earth will fall into water?

20. *Estimating Pi*   A coin of diameter d is dropped onto a paper that contains a grid of squares d units on a side (see figure).

    (a) Find the probability that the coin covers a vertex of one of the squares in the grid.

    (b) Repeat the experiment 100 times and use the results to approximate π.

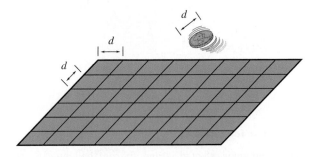

In Exercises 21–26, the sample spaces are large, and therefore you should use the counting principles discussed in Section 10.5.

21. *Game Show*   On a game show, you are given five digits to arrange in the proper order to form the price of a car. If you arrange them correctly, you win the car. Find the probability of winning if you know the correct position of only one digit and must guess the positions of the other digits.

22. *Shelving Books*   A parent instructs a young child to place a five-volume set of books on a bookshelf. Find the probability that the books are in correct order if the child places them on the shelf at random.

23. *Defective Units*   A shipment of 10 food processors to a certain store contained two defective units. If you purchase two of these food processors as birthday gifts for friends, determine the probability that you get both defective units.

24. *Book Selection*   Four books are selected at random from a shelf containing six novels and four autobiographies. Find the probability that the four autobiographies are selected.

25. *Card Selection*   Five cards are selected from a standard deck of 52 cards. Find the probability that the four aces are selected.

26. *Card Selection*   Five cards are selected from a standard deck of 52 cards. Find the probability of getting all hearts.

Reviewing the Major Concepts

In Exercises 27–30, describe the relationship between the graphs of f and g.

27. $g(x) = f(x) - 4$ **28.** $g(x) = f(x - 4)$

29. $g(x) = -f(x)$ **30.** $g(x) = f(-x)$

In Exercises 31 and 32, use a graphing utility to solve the system of equations.

31. $\begin{aligned} 5x - 2y &= -25 \\ -3x + 7y &= 44 \end{aligned}$ **32.** $\begin{aligned} 6x + 2y &= 20 \\ 3x - y &= 14 \end{aligned}$

Additional Problem Solving

In Exercises 33–36, determine the sample space for the experiment.

33. A taste tester must taste and rank three brands of yogurt, A, B, and C, according to preference.

34. A coin and a die are tossed.

35. A basketball tournament between two teams consists of three games. For each game, your team may win (denote by W) or lose (denote by L).

36. Two students are randomly selected from four students, A, B, C, and D.

In Exercises 37 and 38, you are given the probability that an event will not occur. Find the probability that the event will occur.

37. $p = 0.82$ **38.** $p = 0.13$

Playing Cards   In Exercises 39–42, a card is drawn from a standard deck of playing cards. Find the probability of drawing the indicated card.

39. A red card   **40.** A queen
41. A face card   **42.** A black face card

Tossing a Die   In Exercises 43–46, a six-sided die is tossed. Find the probability of the specified event.

43. The number is a 5.   **44.** The number is a 7.
45. The number is no more than 5.
46. The number is at least 1.

Reading a Graph   In Exercises 47–50, use the circle graphs, which show for a certain college the numbers of incoming freshmen in each average high school grade category for the years 1970 and 1992.

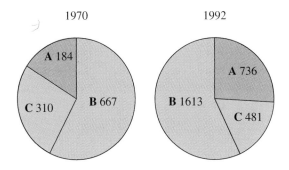

47. A person is selected at random from the 1970 freshman class. What is the probability that the person's high school average was an A?
48. A person is selected at random from the 1992 freshman class. What is the probability that the person's high school average was an A?

49. What is the probability that a person selected from the 1970 freshman class did not have a high school average grade of C?
50. What is the probability that a person selected from the 1992 freshman class did not have a high school average grade of B?

Geometry   A child uses a spring-loaded device to shoot a marble into the square box shown in the figure. The base of the square is horizontal and the marble has an equal likelihood of coming to rest at any point on the base. In Exercises 51–54, find the probability that the marble comes to rest in the specified region.

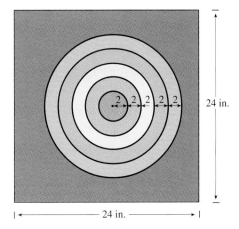

51. The red center
52. The blue ring
53. The purple border
54. Not in the yellow ring
55. *Multiple-Choice Test*   A student takes a multiple-choice test in which there are five choices for each question. Find the probability that the first question is answered correctly given the following conditions.
  (a) The student has no idea of the answer and guesses at random.
  (b) The student can eliminate two of the choices and guesses from the remaining choices.
  (c) The student knows the answer.

56. *Multiple-Choice Test* A student takes a multiple-choice test in which there are four choices for each question. Find the probability that the first question is answered correctly given the following conditions.

(a) The student has no idea of the answer and guesses at random.

(b) The student can eliminate two of the choices and guesses from the remaining choices.

(c) The student knows the answer.

57. *Girl or Boy?* The genes that determine the sex of a human baby are denoted by XX (female) and XY (male). Complete the Punnett square below. Then use the result to explain why it is equally likely that a newborn baby will be a boy or a girl.

Female

	X	X
X	?	?
Y	?	?

Male

58. *Blood Types* There are four basic human blood types: A (AA or Ao), B (BB or Bo), AB (AB), and O (oo). Complete the Punnett square below. What is the blood type of each parent? What is the probability that their offspring will have blood type A? B? AB? O?

	A	o
B	?	?
o	?	?

59. *Continuing Education* In a high school graduating class of 325 students, 255 are going to continue their education. What is the probability that a student selected at random from the class will not be furthering his or her education?

60. *Study Questions* An instructor gives his class a list of four study questions for the next exam. Two of the four study questions will be on the exam. Find the probability that a student who knows the material relating to three of the four questions will be able to answer both questions selected for the exam.

In Exercises 61–66, the sample spaces are large, and therefore you should use the counting principles discussed in Section 10.5.

61. *Lottery* You buy a lottery ticket inscribed with a five-digit number. On the designated day, five digits are randomly selected from the digits 0 through 9, inclusive. What is the probability that you have a winning ticket?

62. *Game Show* On a game show, you are given four digits to arrange in proper order to form the price of a grandfather clock. What is the probability of winning given the following conditions?

(a) You guess the position of each digit.

(b) You know the first digit, but must guess the remaining three.

63. *Preparing for a Test* An instructor gives her class a list of 10 study problems, from which she will select eight to be answered on an exam. If you know how to solve eight of the problems, find the probability you will be able to answer all eight test questions.

64. *Committee Selection* A committee of three students is to be selected from a group of three girls and five boys. Find the probability that the committee is composed entirely of girls.

65. *Defective Units* A shipment of 12 compact disc players contains two defective units. A husband and wife buy three of these players to give to their children as gifts.

(a) What is the probability that all three are good players?

(b) What is the probability that they buy at least one defective player?

66. *Card Selection* Five cards are selected from a standard deck of 52 cards. Find the probability that two aces and three queens are selected.

CHAPTER PROJECT: Savings Account Interest

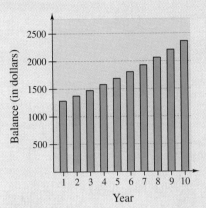

Figure for 1

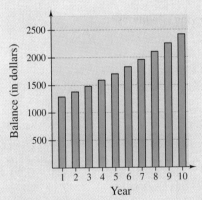

Figure for 2

Saving money requires careful planning and determination. Six factors that should be considered when devising a savings plan are: present assets, present debts, present income, present expenses, future income, and future goals. The type of savings account depends on the amount of deposit and the length of time the money will be in the account.

A compound interest formula to determine a savings account balance is

$$A = P \left(1 + \frac{r}{n} \right)^{nt}$$

where A is the balance in the account after t years, P is the principal, and r is the annual interest rate compounded n times per year. Use this information to investigate the following questions.

1. Use the formula for compound interest to find the balance for each of the first 10 years for a savings account with a principal of $1200 and 7% interest, compounded annually.

2. Use the formula for compound interest to find the balance for each of the first 10 years for a savings account with a principal of $1200 and 7% interest, compounded monthly.

3. In Questions 1 and 2, each savings account had a principal of $1200. Write a paragraph that answers the following questions.

   • What was the difference in the balances of the two accounts at the end of 10 years?

   • How do you account for this difference?

   • Which would you prefer: an account that compounds yearly, an account that compounds monthly, or an account that compounds daily? What other factors should you consider when choosing an account?

4. Use the information from Questions 1 and 2 to create a double bar graph of the annual account balances for the two savings accounts. For each account, in which year was the earned interest the least? For each account, in which year was the earned interest the greatest? How can you tell this from the graph?

5. Use the formula $R = I/P$, where R is the *annual interest rate,* I is the earned interest, and P is the principal, to find the annual interest rate that takes into consideration the effect of compounding for the third year for each account in Questions 1 and 2. Which account has a higher rate?

6. Use the formula for compound interest to find the balances in a savings account for each of the first five years with a principal of $10,000 and 8.5% interest compounded yearly, quarterly, and monthly. Create a triple bar graph to display the information.

CHAPTER SUMMARY

After studying this chapter, you should have acquired the following skills. These skills are keyed to the Review Exercises that begin on page 691. Answers to odd-numbered Review Exercises are given in the back of the book.

- Write sums in sigma notation. *(Section 10.1)*   **Review Exercises 1–4**

- Simplify expressions involving factorials. *(Section 10.1)*   **Review Exercises 5–8**

- Write the first several terms of an arithmetic sequence and of a geometric sequence. *(Sections 10.2, 10.3)*   **Review Exercises 9–12, 15–18**

- Find the common difference of an arithmetic sequence and the common ratio of a geometric sequence. *(Sections 10.2, 10.3)*   **Review Exercises 13, 14, 19, 20**

- Find formulas for the nth terms of an arithmetic sequence and a geometric sequence. *(Sections 10.2, 10.3)*   **Review Exercises 21–30**

- Match arithmetic and geometric sequences with their graphs. *(Sections 10.2, 10.3)*   **Review Exercises 31–36**

- Graph the first several terms of sequences using a graphing utility. *(Sections 10.1, 10.2, 10.3)*   **Review Exercises 37–40**

- Evaluate sums expressed in sigma notation. *(Sections 10.1, 10.2, 10.3)*   **Review Exercises 41–56**

- Find and evaluate the sums of arithmetic and geometric sequences from verbal statements. *(Sections 10.2, 10.3)*   **Review Exercises 57–62**

- Evaluate binomial coefficients using the Binomial Theorem or Pascal's Triangle. *(Section 10.4)*   **Review Exercises 63–66**

- Evaluate binomial coefficients using a graphing utility. *(Section 10.4)*   **Review Exercises 67, 68**

- Expand binomial expressions using the Binomial Theorem. *(Section 10.4)*   **Review Exercises 69–74**

- Find the coefficients of specified terms of binomial expressions. *(Section 10.4)*   **Review Exercises 75, 76**

- Calculate the numbers of ways events can occur using the Fundamental Counting Principle, permutations, or combinations. *(Section 10.5)*   **Review Exercises 77–80**

- Find the probabilities of the occurrences of specified events. *(Section 10.6)*   **Review Exercises 81–86**

REVIEW EXERCISES

In Exercises 1–4, use sigma notation to write the sum.

1. $[5(1) - 3] + [5(2) - 3] + [5(3) - 3] + [5(4) - 3]$

2. $[9 - 2(1)] + [9 - 2(2)] + [9 - 2(3)] + [9 - 2(4)]$

3. $\dfrac{1}{3(1)} + \dfrac{1}{3(2)} + \dfrac{1}{3(3)} + \dfrac{1}{3(4)} + \dfrac{1}{3(5)} + \dfrac{1}{3(6)}$

4. $\left(-\tfrac{1}{3}\right)^0 + \left(-\tfrac{1}{3}\right)^1 + \left(-\tfrac{1}{3}\right)^2 + \left(-\tfrac{1}{3}\right)^3 + \left(-\tfrac{1}{3}\right)^4$

In Exercises 5–8, simplify the expression.

5. $\dfrac{20!}{18!}$

6. $\dfrac{50!}{53!}$

7. $\dfrac{n!}{(n-3)!}$

8. $\dfrac{(n-1)!}{(n+1)!}$

In Exercises 9–12, write the first five terms of the arithmetic sequence. (Begin with $n = 1$.)

9. $a_n = 132 - 5n$

10. $a_n = 2n + 3$

11. $a_n = \tfrac{3}{4}n + \tfrac{1}{2}$

12. $a_n = -\tfrac{3}{5}n + 1$

In Exercises 13 and 14, find the common difference of the arithmetic sequence.

13. $30, 27.5, 25, 22.5, 20, \ldots$

14. $9, 12, 15, 18, 21, \ldots$

In Exercises 15–18, write the first five terms of the geometric sequence.

15. $a_1 = 10, \ r = 3$

16. $a_1 = 2, \ r = -5$

17. $a_1 = 100, \ r = -\tfrac{1}{2}$

18. $a_1 = 12, \ r = \tfrac{1}{6}$

In Exercises 19 and 20, find the common ratio of the geometric sequence.

19. $8, 12, 18, 27, \tfrac{81}{2}, \ldots$

20. $81, -54, 36, -24, 16, \ldots$

In Exercises 21–30, find a formula for the nth term of the specified sequence.

21. Arithmetic sequence: $a_1 = 10, \ d = 4$

22. Arithmetic sequence: $a_1 = 32, \ d = -2$

23. Arithmetic sequence: $a_1 = 1000, \ a_2 = 950$

24. Arithmetic sequence: $a_1 = 12, \ a_2 = 20$

25. Geometric sequence: $a_1 = 1, \ r = -\tfrac{2}{3}$

26. Geometric sequence: $a_1 = 100, \ r = 1.07$

27. Geometric sequence: $a_1 = 24, \ a_2 = 48$

28. Geometric sequence: $a_1 = 16, \ a_2 = -4$

29. Geometric sequence: $a_1 = 12, \ a_4 = -\tfrac{3}{2}$

30. Geometric sequence: $a_2 = 1, \ a_3 = \tfrac{1}{3}$

In Exercises 31–36, match the sequence with its graph. [The graphs are labeled (a), (b), (c), (d), (e), and (f).]

(a)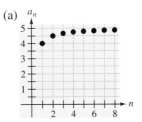

(b)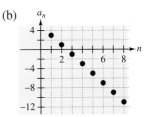

(c)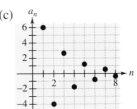

(d)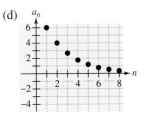

(e)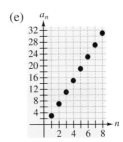

(f)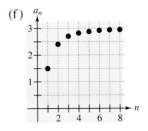

31. $a_n = 5 - \dfrac{1}{n}$

32. $a_n = \dfrac{3n^2}{n^2 + 1}$

33. $a_n = 5 - 2n$

34. $a_n = 4n - 1$

35. $a_n = 6\left(\dfrac{2}{3}\right)^{n-1}$

36. $a_n = 6\left(-\dfrac{2}{3}\right)^{n-1}$

In Exercises 37–40, use a graphing utility to graph the first 10 terms of the sequence.

37. $a_n = \dfrac{3n}{n+1}$

38. $a_n = \dfrac{3}{n+1}$

39. $a_n = 5\left(\dfrac{3}{4}\right)^{n-1}$

40. $a_n = 5\left(-\dfrac{3}{4}\right)^{n-1}$

In Exercises 41–56, evaluate the sum.

41. $\displaystyle\sum_{k=1}^{4} 7$

42. $\displaystyle\sum_{k=1}^{4} \dfrac{(-1)^k}{k}$

43. $\displaystyle\sum_{n=1}^{4} \left(\dfrac{1}{n} - \dfrac{1}{n+1}\right)$

44. $\displaystyle\sum_{n=1}^{4} \left(\dfrac{1}{n} - \dfrac{1}{n+2}\right)$

45. $\displaystyle\sum_{k=1}^{12} (7k - 5)$

46. $\displaystyle\sum_{k=1}^{10} (100 - 10k)$

47. $\displaystyle\sum_{j=1}^{100} \dfrac{j}{4}$

48. $\displaystyle\sum_{j=1}^{50} \dfrac{3j}{2}$

49. $\displaystyle\sum_{n=1}^{12} 2^n$

50. $\displaystyle\sum_{n=1}^{12} (-2)^n$

51. $\displaystyle\sum_{k=1}^{8} 5\left(-\dfrac{3}{4}\right)^k$

52. $\displaystyle\sum_{k=1}^{10} 4\left(\dfrac{3}{2}\right)^k$

53. $\displaystyle\sum_{i=1}^{8} (1.25)^{i-1}$

54. $\displaystyle\sum_{i=1}^{8} (-1.25)^{i-1}$

55. $\displaystyle\sum_{n=1}^{120} 500(1.01)^n$

56. $\displaystyle\sum_{n=1}^{40} 1000(1.1)^n$

57. Find the sum of the first 50 positive integers that are multiples of 4.

58. Find the sum of the integers from 225 to 300.

59. *Auditorium Seating* Each row in a small auditorium has three more seats than the preceding row. Find the seating capacity of the auditorium if the front row seats 22 people and there are 12 rows of seats.

60. *Depreciation* A company pays $120,000 for a machine. During the next 5 years, the machine depreciates at a rate of 30% per year. (That is, at the end of each year, the depreciated value is 70% of what it was at the beginning of the year.)

    (a) Find a formula for the nth term of the geometric sequence that gives the value of the machine n full years after it was purchased.

    (b) Find the depreciated value of the machine at the end of 5 full years.

61. *Population Increase* A city of 85,000 people is growing at the rate of 1.2% per year. (That is, at the end of each year, the population is 1.012 times the population at the beginning of the year.)

    (a) Find a formula for the nth term of the geometric sequence that gives the population n years from now.

    (b) Estimate the population 50 years from now.

62. *Salary Increase* You accept a job that pays a salary of $32,000 the first year. During the next 39 years, you receive a 5.5% raise each year. What is your total salary over the 40-year period?

In Exercises 63–66, evaluate $_nC_r$.

63. $_8C_3$

64. $_{12}C_2$

65. $_{12}C_0$

66. $_{100}C_1$

In Exercises 67 and 68, use a graphing utility to evaluate $_nC_r$.

67. $_{40}C_4$

68. $_{15}C_9$

In Exercises 69–74, use the Binomial Theorem to expand the expression. Simplify your answer.

69. $(x + 1)^{10}$

70. $(u - v)^9$

71. $(y - 2)^6$

72. $(x + 3)^5$

73. $\left(\dfrac{1}{2} - x\right)^8$

74. $(3x - 2y)^4$

In Exercises 75 and 76, find the coefficient of the given term of the expression.

Expression	Term
75. $(x - 3)^{10}$	x^5
76. $(2x - 3y)^5$	$x^2 y^3$

77. *Morse Code*   In Morse code, all characters are transmitted using a sequence of *dots* and *dashes*. How many different characters can be formed by using a sequence of three dots and dashes? (These can be repeated. For example, dash-dot-dot represents the letter *d*.)

78. *Forming Line Segments*   How many straight line segments can be formed by five points of which no three are collinear?

79. *Committee Selection*   Determine the number of ways a committee of five people can be formed from a group of 15 people.

80. *Program Listing*   There are seven participants in a piano recital. In how many orders can their names be listed in the program?

81. *Rolling a Die*   Find the probability of obtaining a number greater than 4 when a single six-sided die is rolled.

82. *Coin Tossing*   Find the probability of obtaining at least one head when a coin is tossed four times.

83. *Book Selection*   A child who does not know how to read carries a four-volume set of books to a bookshelf. Find the probability that the child will put the books on the shelf in the correct order.

84. *Rolling a Die*   Are the chances of rolling a 3 with one six-sided die the same as the chances of rolling a total of 6 with two six-sided dice? If not, which has the greater probability of occurring?

85. *Hospital Inspection*   As part of a monthly inspection at a hospital, the inspection team randomly selects reports from eight of the 84 nurses who are on duty. What is the probability that none of the reports selected will be from the 10 most experienced nurses on duty?

86. *Target Shooting*   An archer shoots an arrow at the target shown in the figure. Suppose that the arrow is equally likely to hit any point on the target. What is the probability that the arrow hits the bull's-eye? What is the probability that the arrow hits the blue ring?

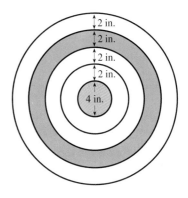

CHAPTER TEST

Take this test as you would take a test in class. After you are done, check your work against the answers given in the back of the book.

1. Write the first five terms of the sequence $a_n = \left(-\frac{2}{3}\right)^{n-1}$. (Begin with $n = 1$.)

2. Evaluate: $\displaystyle\sum_{j=0}^{4} (3j + 1)$ 3. Evaluate: $\displaystyle\sum_{n=1}^{5} (3 - 4n)$

4. Use sigma notation to write $\dfrac{2}{3(1) + 1} + \dfrac{2}{3(2) + 1} + \cdots + \dfrac{2}{3(12) + 1}$.

5. Write the first five terms of the arithmetic sequence whose first term is $a_1 = 12$ and whose common difference is $d = 4$.

6. Find a formula for the nth term of the arithmetic sequence whose first term is $a_1 = 5000$ and whose common difference is $d = -100$.

7. Find the sum of the first 50 positive integers that are multiples of 3.

8. Find the common ratio of the geometric sequence: $2, \ -3, \ \frac{9}{2}, \ -\frac{27}{4}, \ \ldots$.

9. Find a formula for the nth term of the geometric sequence whose first term is $a_1 = 4$ and whose common ratio is $r = \frac{1}{2}$.

10. Evaluate: $\displaystyle\sum_{n=1}^{8} 2(2^n)$ 11. Evaluate: $\displaystyle\sum_{n=1}^{10} 3\left(\tfrac{1}{2}\right)^n$

12. Fifty dollars is deposited each month in an increasing annuity that pays 8%, compounded monthly. What is the balance after 25 years?

13. Evaluate: $\ _{20}C_3$

14. Explain how to use Pascal's Triangle to expand $(x - 2)^5$.

15. Find the coefficient of the term $x^3 y^5$ in the expansion of $(x + y)^8$.

16. How many license plates can consist of one letter followed by three digits?

17. Four students are randomly selected from a class of 25 to answer questions from a reading assignment. In how many ways can the four be selected?

18. The weather report indicates that the probability of snow tomorrow is 0.75. What is the probability that it will not snow?

19. A card is drawn from a standard deck of playing cards. Find the probability that it is a red face card.

20. Suppose two spark plugs require replacement in a four-cylinder engine. If the mechanic randomly removes two plugs, find the probability that they are the two defective plugs.

Appendices

SECTION A.1 — Introduction to Logic
Statements • *Truth Tables*

Statements

In everyday speech and in mathematics we make inferences that adhere to common **laws of logic**. These laws (or methods of reasoning) allow us to build an algebra of statements by using logical operations to form compound statements from simpler ones. One of the primary goals of logic is to determine the truth value (true or false) of a compound statement knowing the truth value of its simpler component statements. For instance, we will learn that the compound statement "The temperature is below freezing and it is snowing" is true only if both component statements are true.

Definition of a Statement	
	1. A **statement** is a sentence to which only one truth value (either true or false) can be meaningfully assigned.
	2. An **open statement** is a sentence that contains one or more variables and becomes a statement when each variable is replaced by a specific item from a designated set.

NOTE: In the above definition, the word *statement* can be replaced by the word *proposition*.

EXAMPLE 1 ■ Statements, Nonstatements, and Open Statements

Statement	*Truth Value*
A square is a rectangle.	T
-3 is less than -5.	F

Nonstatement	*Truth Value*
Do your homework.	No truth value can be meaningfully assigned.
Did you call the police?	No truth value can be meaningfully assigned.

Open Statement	*Truth Value*
x is an irrational number.	We need a value of x.
She is a computer science major.	We need a specific person. ■

Symbolically, we represent statements by lowercase letters p, q, r, and so on. Statements can be changed or combined to form **compound statements** by means of the three logical operations **and**, **or**, and **not**, which we represent by $\wedge$ (and), $\vee$ (or), and $\sim$ (not). In logic we use the word *or* in the *inclusive* sense (meaning "and/or" in everyday language). That is, the statement "p or q" is true if p is true, q is true, or both p and q are true. The following list summarizes the terms and symbols used with these three operations of logic.

Operations of Logic	Operation	Verbal Statement	Symbolic Form	Name of Operation
	$\sim$	not p	$\sim p$	**Negation**
	$\wedge$	p and q	$p \wedge q$	**Conjunction**
	$\vee$	p or q	$p \vee q$	**Disjunction**

Compound statements can be formed using more than one logical operation, as demonstrated in Example 2.

EXAMPLE 2 ■ Forming Negations and Compound Statements

The statements p and q are as follows.

> $p:$ The temperature is below freezing.
> $q:$ It is snowing.

Write the verbal form for each of the following.

(a) $p \wedge q$ (b) $\sim p$

(c) $\sim(p \vee q)$ (d) $\sim p \wedge \sim q$

Solution

(a) The temperature is below freezing and it is snowing.

(b) The temperature is not below freezing.

(c) It is not true that the temperature is below freezing or it is snowing.

(d) The temperature is not below freezing and it is not snowing. ■

EXAMPLE 3 ■ Forming Compound Statements

The statements p and q are as follows.

> $p:$ The temperature is below freezing.
> $q:$ It is snowing.

(a) Write the symbolic form for: *The temperature is not below freezing or it is not snowing.*

(b) Write the symbolic form for: *It is not true that the temperature is below freezing and it is snowing.*

Solution

(a) The symbolic form is: $\sim p \vee \sim q$

(b) The symbolic form is: $\sim(p \wedge q)$ ■

Truth Tables

To determine the truth value of a compound statement, we use charts called **truth tables**. The following tables represent the three basic logical operations.

TABLE 1 Negation

p	q	$\sim p$	$\sim q$
T	T	F	F
T	F	F	T
F	T	T	F
F	F	T	T

TABLE 2 Conjunction

p	q	$p \wedge q$
T	T	T
T	F	F
F	T	F
F	F	F

TABLE 3 Disjunction

p	q	$p \vee q$
T	T	T
T	F	T
F	T	T
F	F	F

For the sake of uniformity, all truth tables with two component statements will have T and F values for p and q assigned in the order shown in the first column of these three tables. Truth tables for several operations can be combined into one chart by using the same two first columns. For each operation a new column is added. Such an arrangement is especially useful with compound statements that involve more than one logical operation and for showing that two statements are logically equivalent.

Logical Equivalence

Two compound statements are **logically equivalent** if they have identical truth tables. Symbolically, we denote the equivalence of the statements p and q by writing $p \equiv q$.

EXAMPLE 4 ■ Logical Equivalence

Use a truth table to show the logical equivalence of the statements $\sim p \wedge \sim q$ and $\sim(p \vee q)$.

Solution

TABLE 4

p	q	$\sim p$	$\sim q$	$\sim p \wedge \sim q$	$p \vee q$	$\sim(p \vee q)$
T	T	F	F	F	T	F
T	F	F	T	F	T	F
F	T	T	F	F	T	F
F	F	T	T	T	F	T

$\uparrow$ —— Identical —— $\uparrow$

Since the fifth and seventh columns in Table 4 are identical, the two given statements are logically equivalent. ■

The equivalence established in Example 4 is one of two well-known rules in logic called **DeMorgan's Laws**. Verification of the second of DeMorgan's Laws is left as an exercise.

| **DeMorgan's Laws** | 1. $\sim(p \lor q) \equiv \sim p \land \sim q$ |
| | 2. $\sim(p \land q) \equiv \sim p \lor \sim q$ |

Compound statements that are true, no matter what the truth values of component statements, are called **tautologies**. One simple example is the statement "p or not p," as shown in Table 5.

TABLE 5 $p \lor \sim p$ **is a tautology**

p	$\sim p$	$p \lor \sim p$
T	F	T
F	T	T

A.1   EXERCISES

In Exercises 1–12, classify the sentence as a statement, a nonstatement, or an open statement.

1. All dogs are brown.

2. Can I help you?

3. That figure is a circle.

4. Substitute 4 for x.

5. x is larger than 4.

6. 8 is larger than 4.

7. $x + y = 10$

8. $12 + 3 = 14$

9. Hockey is fun to watch.

10. One mile is greater than one kilometer.

11. It is more than one mile to the school.

12. Come to the party.

In Exercises 13–20, determine whether the open statement is true for the given values of x.

Open Statement	Values of x	
13. $x^2 - 5x + 6 = 0$	(a) $x = 2$	(b) $x = -2$
14. $x^2 - x - 6 = 0$	(a) $x = 2$	(b) $x = -2$
15. $x^2 \le 4$	(a) $x = -2$	(b) $x = 0$
16. $|x - 3| = 4$	(a) $x = -1$	(b) $x = 7$
17. $4 - |x| = 2$	(a) $x = 0$	(b) $x = 1$
18. $\sqrt{x^2} = x$	(a) $x = 3$	(b) $x = -3$
19. $\dfrac{x}{x} = 1$	(a) $x = -4$	(b) $x = 0$
20. $\sqrt[3]{x} = -2$	(a) $x = 8$	(b) $x = -8$

In Exercises 21–24, write the verbal form for each of the following.

(a) $\sim p$ (b) $\sim q$ (c) $p \wedge q$ (d) $p \vee q$

21. p: The sun is shining.
  q: It is hot.

22. p: The car has a radio.
  q: The car is red.

23. p: Lions are mammals.
  q: Lions are carnivorous.

24. p: Twelve is less than fifteen.
  q: Seven is a prime number.

In Exercises 25–28, write the verbal form for each of the following.

(a) $\sim p \wedge q$ (b) $\sim p \vee q$ (c) $p \wedge \sim q$ (d) $p \vee \sim q$

25. p: The sun is shining.
  q: It is hot.

26. p: The car has a radio.
  q: The car is red.

27. p: Lions are mammals.
  q: Lions are carnivorous.

28. p: Twelve is less than fifteen.
  q: Seven is a prime number.

In Exercises 29–32, write the symbolic form of the given compound statement. In each case let p represent the statement "It is four o'clock," and let q represent the statement "It is time to go home."

29. It is four o'clock and it is not time to go home.

30. It is not four o'clock or it is not time to go home.

31. It is not four o'clock or it is time to go home.

32. It is four o'clock and it is time to go home.

In Exercises 33–36, write the symbolic form of the given compound statement. In each case let p represent the statement "The dog has fleas," and let q represent the statement "The dog is scratching."

33. The dog does not have fleas or the dog is not scratching.

34. The dog has fleas and the dog is scratching.

35. The dog does not have fleas and the dog is scratching.

36. The dog has fleas or the dog is not scratching.

In Exercises 37–42, write the negation of the given statement.

37. The bus is not blue.

38. Frank is not six feet tall.

39. x is equal to 4.

40. x is not equal to 4.

41. The Earth is not flat.

42. The Earth is flat.

In Exercises 43–48, construct a truth table for the given compound statement.

43. $\sim p \wedge q$

44. $\sim p \vee q$

45. $\sim p \vee \sim q$

46. $\sim p \wedge \sim q$

47. $p \vee \sim q$

48. $p \wedge \sim q$

In Exercises 49–54, use a truth table to determine whether the given statements are logically equivalent.

49. $\sim p \wedge q, \quad p \vee \sim q$

50. $\sim(p \wedge \sim q), \quad \sim p \vee q$

51. $\sim(p \vee \sim q), \quad \sim p \wedge q$

52. $\sim(p \vee q), \quad \sim p \vee \sim q$

53. $p \wedge \sim q, \quad \sim(\sim p \vee q)$

54. $p \wedge \sim q, \quad \sim(\sim p \wedge q)$

In Exercises 55–58, determine whether the statements are logically equivalent.

55. (a) The house is red and it is not made of wood.
(b) The house is red or it is not made of wood.

56. (a) It is not true that the tree is not green.
(b) The tree is green.

57. (a) The statement that the house is white or blue is not true.
(b) The house is not white and it is not blue.

58. (a) I am not twenty-five years old and I am not applying for this job.
(b) The statement that I am twenty-five years old and applying for this job is not true.

In Exercises 59–62, use a truth table to determine whether the given statement is a tautology.

59. $\sim p \wedge p$

60. $\sim p \vee p$

61. $\sim(\sim p) \vee \sim p$

62. $\sim(\sim p) \wedge \sim p$

63. Use a truth table to verify the second of DeMorgan's Laws:

$$\sim(p \wedge q) \equiv \sim p \vee \sim q$$

Implications, Quantifiers, and Venn Diagrams
Implications • Logical Quantifiers • Venn Diagrams

Implications

A statement of the form "If p, then q," is called an **implication** (or a conditional statement) and is denoted by

$$p \rightarrow q.$$

We call p the **hypothesis** and q the **conclusion**. There are many different ways to express the implication $p \rightarrow q$, as shown in the following list.

Different Ways of Stating Implications

The implication $p \rightarrow q$ has the following equivalent verbal forms.

1. If p, then q.   4. q follows from p.

2. p implies q.   5. q is necessary for p.

3. p, only if q.   6. p is sufficient for q.

Normally, we think of the implication $p \rightarrow q$ as having a cause-and-effect relationship between the hypothesis p and the conclusion q. However, you should be careful not to confuse the truth value of the component statements with the truth value of the implication. The following truth table should help you keep this distinction in mind.

TABLE 6   Implication

p	q	$p \rightarrow q$
T	T	T
T	F	F
F	T	T
F	F	T

Note in Table 6 that the implication $p \rightarrow q$ is false only when p is true and q is false. This is like a promise. Suppose you promise a friend that "If the sun shines, I will take you fishing." The only way you can break your promise is for the sun to shine (p is true) and you do not take your friend fishing (q is false). If the sun doesn't shine (p is false), you have no obligation to go fishing, and hence, the promise cannot be broken.

EXAMPLE 1 ■   Finding Truth Values of Implications

Give the truth value of each implication.

(a) If 3 is odd, then 9 is odd.

(b) If 3 is odd, then 9 is even.

(c) If 3 is even, then 9 is odd.

(d) If 3 is even, then 9 is even.

Solution

	Hypothesis	*Conclusion*	*Implication*
(a)	T	T	T
(b)	T	F	F
(c)	F	T	T
(d)	F	F	T

■

The next example shows how to write an implication as a disjunction.

EXAMPLE 2 ■   Identifying Equivalent Statements

Use a truth table to show the logical equivalence of the following statements.

(a) If I get a raise, I will take my family on a vacation.

(b) I will not get a raise *or* I will take my family on a vacation.

Solution

We let p represent the statement "I will get a raise," and let q represent the statement "I will take my family on a vacation." Then, we can represent the statement in part (a) as $p \rightarrow q$ and the statement in part (b) as $\sim p \vee q$. The logical equivalence of these two statements is shown in the following truth table.

TABLE 7 $p \rightarrow q \equiv \sim p \vee q$

p	q	$\sim p$	$\sim p \vee q$	$p \rightarrow q$
T	T	F	T	T
T	F	F	F	F
F	T	T	T	T
F	F	T	T	T

└— Identical —┘

Because the fourth and fifth columns of the truth table are identical, we can conclude that the two statements $p \rightarrow q$ and $\sim p \vee q$ are equivalent.   ■

From Table 7 and the fact that $\sim(\sim p) \equiv p$, we can write the **negation of an implication**. That is, since $p \rightarrow q$ is equivalent to $\sim p \vee q$, it follows that the negation of $p \rightarrow q$ must be $\sim(\sim p \vee q)$, which by DeMorgan's Laws can be written as follows.

$$\sim(p \rightarrow q) \equiv p \wedge \sim q$$

For the implication $p \rightarrow q$, there are three important associated implications.

1. The **converse** of $p \rightarrow q$: $q \rightarrow p$

2. The **inverse** of $p \rightarrow q$: $\sim p \rightarrow \sim q$

3. The **contrapositive** of $p \rightarrow q$: $\sim q \rightarrow \sim p$

From Table 8 you can see that these four statements yield two pairs of logically equivalent implications.

TABLE 8

p	q	$\sim p$	$\sim q$	$p \rightarrow q$	$\sim q \rightarrow \sim p$	$q \rightarrow p$	$\sim p \rightarrow \sim q$
T	T	F	F	T	T	T	T
T	F	F	T	F	F	T	T
F	T	T	F	T	T	F	F
F	F	T	T	T	T	T	T

Identical — — — Identical

NOTE: The connective "$\rightarrow$" is used to determine the truth values in the last three columns of Table 8.

EXAMPLE 3   ■   Writing the Converse, Inverse, and Contrapositive

Write the converse, inverse, and contrapositive for the implication "If I get a B on my test, then I will pass the course."

Solution

(a) *Converse:* If I pass the course, then I got a B on my test.

(b) *Inverse:* If I do not get a B on my test, then I will not pass the course.

(c) *Contrapositive:* If I do not pass the course, then I did not get a B on my test. ■

NOTE: In Example 3 be sure you see that neither the converse nor the inverse are logically equivalent to the original implication. To see this, consider that the original implication simply states that if you get a B on your test, then you will pass the course. The converse is not true because knowing that you passed the course does not imply that you got a B on the test. After all, you might have gotten an A on the test!

A **biconditional statement**, denoted by $p \leftrightarrow q$, is the conjunction of the implications $p \rightarrow q$ and $q \rightarrow p$. We often write a biconditional statement as "p if and only if q," or in shorter form as "p iff q." A biconditional statement is true when both components are true and when both components are false, as shown in the following truth table.

TABLE 9 Biconditional Statement: p if and only if q

p	q	$p \rightarrow q$	$q \rightarrow p$	$p \leftrightarrow q$	$(p \rightarrow q) \wedge (q \rightarrow p)$
T	T	T	T	T	T
T	F	F	T	F	F
F	T	T	F	F	F
F	F	T	T	T	T

The following list summarizes some of the laws of logic that we have discussed up to this point.

Laws of Logic

1. For every statement p, either p is true or p is false.   Law of Excluded Middle
2. $\sim(\sim p) \equiv p$ Law of Double Negation
3. $\sim(p \vee q) \equiv \sim p \wedge \sim q$ DeMorgan's Law
4. $\sim(p \wedge q) \equiv \sim p \vee \sim q$ DeMorgan's Law
5. $p \rightarrow q \equiv \sim p \vee q$ Law of Implication
6. $p \rightarrow q \equiv \sim q \rightarrow \sim p$ Law of Contraposition

Logical Quantifiers

Logical quantifiers are words such as *some*, *all*, *every*, *each*, *one*, and *none*. Here are some examples of statements with quantifiers.

Some isosceles triangles are right triangles.

Every painting on display is for sale.

Not all corporations have male chief executive officers.

All squares are parallelograms.

Being able to recognize the negation of a statement involving a quantifier is one of the most important skills in logic. For instance, consider the statement "All dogs are brown." In order for this statement to be false, we do not have to show that *all* dogs are not brown, we must simply find at least one dog that is not brown. Thus, the negation of the statement is "Some dogs are brown."

Next we list some of the more common negations involving quantifiers.

Negating Statements with Quantifiers

Statement	Negation
1. All *p* are *q*.	Some *p* are not *q*.
2. Some *p* are *q*.	No *p* is *q*.
3. Some *p* are not *q*.	All *p* are *q*.
4. No *p* is *q*.	Some *p* are *q*.

When using logical quantifiers, the word *all* can be replaced by the words *each* or *every*. For instance, the following are equivalent.

All *p* are *q*.   Each *p* is *q*.   Every *p* is *q*.

Similarly, the word *some* can be replaced by the words *at least one*. For instance, the following are equivalent.

Some *p* are *q*.   At least one *p* is *q*.

EXAMPLE 4 ■ Negating Quantifying Statements

Write the negation of each of the following.
(a) All students study.
(b) Not all prime numbers are odd.
(c) At least one mammal can fly.
(d) Some bananas are not yellow.

Solution
(a) Some students do not study.
(b) All prime numbers are odd.
(c) No mammals can fly.
(d) All bananas are yellow.

Venn Diagrams

Venn diagrams are figures that are used to show relationships between two or more sets of objects. They can help us interpret quantifying statements. Study the following Venn diagrams in which the circle marked *A* represents people over six feet tall and the circle marked *B* represents the basketball players.

1. All basketball players are over six feet tall.

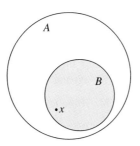

2. Some basketball players are over six feet tall.

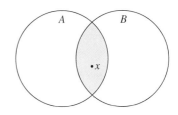

3. Some basketball players are not over six feet tall.

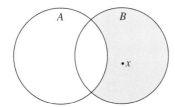

4. No basketball player is over six feet tall.

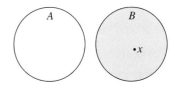

A.2   EXERCISES

In Exercises 1–4, write the verbal form for each of the following.

(a) $p \rightarrow q$ (b) $q \rightarrow p$ (c) $\sim q \rightarrow \sim p$ (d) $p \rightarrow \sim q$

1. *p:* The engine is running.
   q: The engine is wasting gasoline.

2. *p:* The student is at school.
   q: It is nine o'clock.

3. *p:* The integer is even.
   q: It is divisible by two.

4. *p:* The person is generous.
   q: The person is rich.

In Exercises 5–10, write the symbolic form of the compound statement. Let p represent the statement "The economy is expanding," and let q represent the statement "Interest rates are low."

5. If interest rates are low, then the economy is expanding.

6. If interest rates are not low, then the economy is not expanding.

7. An expanding economy implies low interest rates.

8. Low interest rates are sufficient for an expanding economy.

9. Low interest rates are necessary for an expanding economy.

10. The economy will expand only if interest rates are low.

In Exercises 11–20, give the truth value of the implication.

11. If 4 is even, then 12 is even.

12. If 4 is even, then 2 is odd.

13. If 4 is odd, then 3 is odd.

14. If 4 is odd, then 2 is odd.

15. If $2n$ is even, then $2n + 2$ is odd.

16. If $2n + 1$ is even, then $2n + 2$ is odd.

17. $3 + 11 > 16$ only if $2 + 3 = 5$.

18. $\frac{1}{6} < \frac{2}{3}$ is necessary for $\frac{1}{2} > 0$.

19. $x = -2$ follows from $2x + 3 = x + 1$.

20. If $2x = 224$, then $x = 10$.

In Exercises 21–26, write the converse, inverse, and contrapositive of the statement.

21. If the sky is clear, then you can see the eclipse.

22. If the person is nearsighted, then he is ineligible for the job.

23. If taxes are raised, then the deficit will increase.

24. If wages are raised, then the company's profits will decrease.

25. It is necessary to have a birth certificate to apply for the visa.

26. The number is divisible by three only if the sum of its digits is divisible by three.

In Exercises 27–40, write the negation of the statement.

27. Paul is a junior or senior.

28. Jack is a senior and he plays varsity basketball.

29. If the temperature increases, then the metal rod will expand.

30. If the test fails, then the project will be halted.

31. We will go to the ocean only if the weather forecast is good.

32. Completing the pass on this play is necessary if we are going to win the game.

33. Some students are in extracurricular activities.

34. Some odd integers are not prime numbers.

35. All contact sports are dangerous.

36. All members must pay their dues prior to June 1.

37. No child is allowed at the concert.

38. No contestant is over the age of twelve.

39. At least one of the $20 bills is counterfeit.

40. At least one unit is defective.

In Exercises 41–48, construct a truth table for the compound statement.

41. $\sim(p \to \sim q)$

42. $\sim q \to (p \to q)$

43. $\sim(q \to p) \wedge q$

44. $p \to (\sim p \vee q)$

45. $[(p \vee q) \wedge (\sim p)] \to q$

46. $[(p \to q) \wedge (\sim q)] \to p$

47. $(p \leftrightarrow \sim q) \to \sim p$

48. $(p \vee \sim q) \leftrightarrow (q \to \sim p)$

In Exercises 49–56, use a truth table to show the logical equivalence of the two statements.

49. $q \to p$ $\qquad\qquad$ $\sim p \to \sim q$

50. $\sim p \to q$ $\qquad\qquad$ $p \vee q$

51. $\sim(p \to q)$ $\qquad\qquad$ $p \wedge \sim q$

52. $(p \vee q) \to q$ $\qquad\qquad$ $p \to q$

53. $(p \to q) \vee \sim q$ $\qquad$ $p \vee \sim p$

54. $q \to (\sim p \vee q)$ $\qquad$ $q \vee \sim q$

55. $p \to (\sim p \wedge q)$ $\qquad$ $\sim p$

56. $\sim(p \wedge q) \to \sim q$ $\qquad$ $p \vee \sim q$

57. Select the statement that is logically equivalent to the statement "If a number is divisible by six, then it is divisible by two."
(a) If a number is divisible by two, then it is divisible by six.
(b) If a number is not divisible by six, then it is not divisible by two.
(c) If a number is not divisible by two, then it is not divisible by six.
(d) Some numbers are divisible by six and not divisible by two.

58. Select the statement that is logically equivalent to the statement "It is not true that Pam is a conservative and a Democrat."
(a) Pam is a conservative and a Democrat.
(b) Pam is not a conservative and not a Democrat.
(c) Pam is not a conservative or she is not a Democrat.
(d) If Pam is not a conservative, then she is a Democrat.

59. Select the statement that is *not* logically equivalent to the statement "Every citizen over the age of 18 has the right to vote."
(a) Some citizens over the age of 18 have the right to vote.
(b) Each citizen over the age of 18 has the right to vote.
(c) All citizens over the age of 18 have the right to vote.
(d) No citizen over the age of 18 can be restricted from voting.

60. Select the statement that is *not* logically equivalent to the statement "It is necessary to pay the registration fee to take the course."
(a) If you take the course, then you must pay the registration fee.
(b) If you do not pay the registration fee, then you cannot take the course.
(c) If you pay the registration fee, then you may take the course.
(d) You may take the course only if you pay the registration fee.

In Exercises 61–70, sketch a Venn diagram and shade the region that illustrates the given statement. Let A be a circle that represents people who are happy, and let B be a circle that represents college students.

61. All college students are happy.

62. All happy people are college students.

63. No college students are happy.

64. No happy people are college students.

65. Some college students are not happy.

66. Some happy people are not college students.

67. At least one college student is happy.

68. At least one happy person is not a college student.

69. Each college student is sad.

70. Each sad person is not a college student.

In Exercises 71–74, state whether the statement follows from the given Venn diagram. Assume that each area shown in the Venn diagram is non-empty. (*Note:* Use only the information given in the diagram. Do not be concerned with whether the statement is actually true or false.)

71. (a) All toads are green.
(b) Some toads are green.

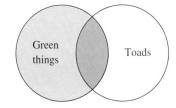

72. (a) All men are company presidents.
(b) Some company presidents are women.

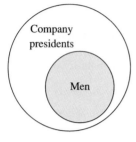

73. (a) All blue cars are old.
(b) Some blue cars are not old.

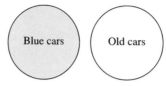

74. (a) No football players are over six feet tall.
(b) Every football player is over six feet tall.

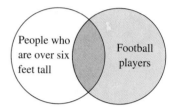

SECTION A.3	**Logical Arguments**
	Arguments • *Venn Diagrams and Arguments* • *Proofs*

Arguments

An **argument** is a collection of statements, listed in order. The last statement is called the **conclusion** and the other statements are called the **premises**. An argument is **valid** if the conjunction of all the premises implies the conclusion. The most common type of argument takes the following form.

Premise #1: $p \rightarrow q$
Premise #2: p
Conclusion: q

This form of argument is called the **Law of Detachment** or *Modus Ponens*. It is illustrated in the following example.

EXAMPLE 1 ■ A Valid Argument

Show that the following argument is valid.

Premise #1:   If Sean is a freshman, then he is taking algebra.
Premise #2:   Sean is a freshman.
Conclusion: Therefore, Sean is taking algebra.

Solution

We let p represent the statement "Sean is a freshman," and q represent the statement "Sean is taking algebra." Then the argument fits the Law of Detachment, which can be written as follows.

$$[(p \rightarrow q) \wedge p] \rightarrow q$$

The validity of this argument is shown in the following truth table.

TABLE 10   Law of Detachment

p	q	$p \rightarrow q$	$(p \rightarrow q) \wedge p$	$[(p \rightarrow q) \wedge p] \rightarrow q$
T	T	T	T	T
T	F	F	F	T
F	T	T	F	T
F	F	T	F	T

Keep in mind that the validity of an argument has nothing to do with the truthfulness of the premises or conclusion. For instance, the following argument is valid—the fact that it is fanciful does not alter its validity.

Premise #1:   If I snap my fingers, elephants will stay out of my house.
Premise #2:   I am snapping my fingers.
Conclusion: Therefore, elephants will stay out of my house.

We have discussed the most common form of logical argument. This and three other commonly used forms of valid arguments are summarized in the following list.

Four Types of Valid Arguments

Name	Pattern	
1. **Law of Detachment** or *Modus Ponens*	Premise #1:	$p \rightarrow q$
	Premise #2:	p
	Conclusion:	q
2. **Law of Contraposition** or *Modus Tollens*	Premise #1:	$p \rightarrow q$
	Premise #2:	$\sim q$
	Conclusion:	$\sim p$
3. **Law of Transitivity** or *Syllogism*	Premise #1:	$p \rightarrow q$
	Premise #2:	$q \rightarrow r$
	Conclusion:	$p \rightarrow r$
4. **Law of Disjunctive Syllogism**	Premise #1:	$p \vee q$
	Premise #2:	$\sim p$
	Conclusion:	q

EXAMPLE 2   ■   An Invalid Argument

Determine whether the following argument is valid.

Premise #1:	If John is elected, the income tax will be increased.
Premise #2:	The income tax was increased.
Conclusion:	Therefore, John was elected.

Solution

This argument has the following form.

Pattern   *Implication*

Premise #1:	$p \rightarrow q$	$[(p \rightarrow q) \wedge q] \rightarrow p$
Premise #2:	q	
Conclusion:	p	

This is not one of the four valid forms of arguments that we listed. We can construct a truth table to verify that the argument is invalid, as follows.

TABLE 11   **An Invalid Argument**

p	q	$p \rightarrow q$	$(p \rightarrow q) \wedge q$	$[(p \rightarrow q) \wedge q] \rightarrow p$
T	T	T	T	T
T	F	F	F	T
F	T	T	T	F
F	F	T	F	T

An invalid argument, like the one in Example 2, is called a **fallacy**. Other common fallacies are given in the following example.

EXAMPLE 3   ■   Common Fallacies

Each of the following arguments is invalid.

(a) *Arguing from the Converse:* If the football team wins the championship, then students will skip classes. The students skipped classes. Therefore, the football team won the championship.

(b) *Arguing from the Inverse:* If the football team wins the championship, then students will skip classes. The football team did not win the championship. Therefore, the students did not skip classes.

(c) *Arguing from False Authority:* Wheaties are best for you because Joe Montana eats them.

(d) *Arguing from an Example:* Beta Brand products are not reliable because my Beta Brand snowblower does not start in cold weather.

(e) *Arguing from Ambiguity:* If automobile carburetors are modified, the automobile will pollute. Brand X automobiles have modified carburetors. Therefore, Brand X automobiles pollute.

(f) *Arguing by False Association:* Joe was running through the alley when the fire alarm went off. Therefore, Joe started the fire.   ■

EXAMPLE 4   ■   A Valid Argument

Determine whether the following argument is valid.

Premise #1:   You like strawberry pie or you like chocolate pie.
Premise #2:   You do not like strawberry pie.
Conclusion: Therefore, you like chocolate pie.

Solution

This argument has the following form.

Premise #1: $p \lor q$
Premise #2: $\sim p$
Conclusion: q

This argument is a disjunctive syllogism, which is one of the four common types of valid arguments.   ■

In a valid argument, the conclusion drawn from the premise is called a **valid conclusion**.

EXAMPLE 5   ■   Making Valid Conclusions

Given the following two premises, which of the conclusions are valid?

Premise #1:   If you like boating, then you like swimming.
Premise #2:   If you like swimming, then you are a scholar.

(a) Conclusion: If you like boating, then you are a scholar.

(b) Conclusion: If you do not like boating, then you are not a scholar.

(c) Conclusion: If you are not a scholar, then you do not like boating.

Solution

(a) This conclusion is valid. It follows from the Law of Transitivity (or syllogism).

(b) This conclusion is invalid. The fallacy stems from arguing from the inverse.

(c) This conclusion is valid. It follows from the Law of Contraposition.   ■

FIGURE A.1

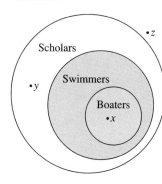

Venn Diagrams and Arguments

Venn diagrams can be used to informally test the validity of an argument. For instance, a Venn diagram for the premises in Example 5 is shown in Figure A.1. In this figure the validity of Conclusion (a) is seen by choosing a boater x in all three sets. Conclusion (b) is seen to be invalid by choosing a person y who is a scholar but does not like boating. Finally, person z indicates the validity of Conclusion (c).

Venn diagrams work well for testing arguments that involve quantifiers, as shown in the next two examples.

EXAMPLE 6 ■ Using a Venn Diagram to Show that an Argument Is Not Valid

Use a Venn diagram to test the validity of the following argument.

Premise #1:   Some plants are green.
Premise #2:   All lettuce is green.
Conclusion: Therefore, lettuce is a plant.

Solution

FIGURE A.2

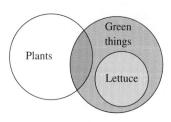

From the Venn diagram shown in Figure A.2, we can see that this is not a valid argument. Remember that even though the conclusion is true (lettuce is a plant), this does not imply that the argument is true. ■

NOTE: When you are using Venn diagrams, you must remember to draw the most general case. For example, in Figure A.2 the circle representing plants is not drawn entirely within the circle representing green things because we are told that only *some* plants are green.

EXAMPLE 7 ■ Using a Venn Diagram to Show that an Argument Is Valid

Use a Venn diagram to test the validity of the following argument.

Premise #1:   All good tennis players are physically fit.
Premise #2:   Some golfers are good tennis players.
Conclusion: Therefore, some golfers are physically fit.

FIGURE A.3

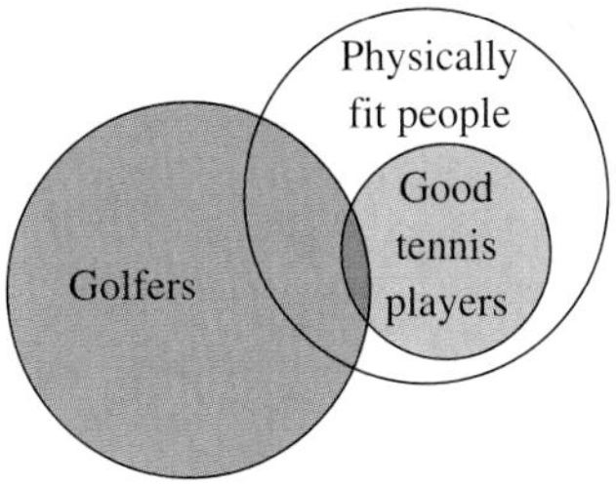

Solution

Since the set of golfers intersects the set of good tennis players, we see from Figure A.3 that the set of golfers must also intersect the set of physically fit people. Therefore, the argument is valid. ∎

Proofs

What does the word *proof* mean to you? In mathematics we use the word *proof* to simply mean a valid argument. Many proofs involve more than two premises and a conclusion. For instance, the proof in Example 8 involves three premises and a conclusion.

EXAMPLE 8   ■   **A Proof by Contraposition**

Use the following three premises to prove that "It is not snowing today."

Premise #1:   If it is snowing today, Greg will go skiing.
Premise #2:   If Greg is skiing today, then he is not studying.
Premise #3:   Greg is studying today.

Solution

We let p represent the statement "It is snowing today," let q represent "Greg is skiing," and let r represent "Greg is studying today." Thus, the given premises have the following form.

Premise #1: $p \to q$
Premise #2: $q \to \sim r$
Premise #3: r

By noting that $r \equiv \sim(\sim r)$, reordering the premises, and writing the contrapositives of the first and second premises, we can obtain the following valid argument.

Premise #3: r
Contrapositive of Premise #2: $r \to \sim q$
Contrapositive of Premise #1: $\sim q \to \sim p$
Conclusion: $\sim p$

Thus, we can conclude $\sim p$. That is, "It is not snowing today." ∎

A.3   EXERCISES

In Exercises 1–4, use a truth table to show that the given argument is valid.

1. Premise #1: $p \rightarrow \sim q$
Premise #2: q
Conclusion: $\sim p$

2. Premise #1: $p \leftrightarrow q$
Premise #2: p
Conclusion: q

3. Premise #1: $p \vee q$
Premise #2: $\sim p$
Conclusion: q

4. Premise #1: $p \wedge q$
Premise #2: $\sim p$
Conclusion: q

In Exercises 5–8, use a truth table to show that the given argument is invalid.

5. Premise #1: $\sim p \rightarrow q$
Premise #2: p
Conclusion: $\sim q$

6. Premise #1: $p \rightarrow q$
Premise #2: $\sim p$
Conclusion: $\sim q$

7. Premise #1: $p \vee q$
Premise #2: q
Conclusion: p

8. Premise #1: $\sim(p \wedge q)$
Premise #2: q
Conclusion: p

In Exercises 9–22, determine whether the argument is valid or invalid.

9. Premise #1:   If taxes are increased, then businesses will leave the state.
Premise #2:   Taxes are increased.
Conclusion: Therefore, businesses will leave the state.

10. Premise #1:   If a student does the homework, then a good grade is certain.
Premise #2:   Liza does the homework.
Conclusion: Therefore, Liza will receive a good grade for the course.

11. Premise #1:   If taxes are increased, then businesses will leave the state.
Premise #2:   Businesses are leaving the state.
Conclusion: Therefore, taxes were increased.

12. Premise #1:   If a student does the homework, then a good grade is certain.
Premise #2:   Liza received a good grade for the course.
Conclusion: Therefore, Liza did her homework.

13. Premise #1:   If the doors are kept locked, then the car will not be stolen.
Premise #2:   The car was stolen.
Conclusion: Therefore, the car doors were unlocked.

14. Premise #1:   If Jan passes the exam, she is eligible for the position.
Premise #2:   Jan is not eligible for the position.
Conclusion: Therefore, Jan did not pass the exam.

15. Premise #1:   All cars manufactured by the Ford Motor Company are reliable.
    Premise #2:   Lincolns are manufactured by Ford.
    Conclusion: Therefore, Lincolns are reliable cars.

16. Premise #1:   Some cars manufactured by the Ford Motor Company are reliable.
    Premise #2:   Lincolns are manufactured by Ford.
    Conclusion: Therefore, Lincolns are reliable.

17. Premise #1:   All federal income tax forms are subject to the Paperwork Reduction Act of 1980.
    Premise #2:   The 1040 Schedule A form is subject to the Paperwork Reduction Act of 1980.
    Conclusion: Therefore, the 1040 Schedule A form is a federal income tax form.

18. Premise #1:   All integers divisible by six are divisible by three.
    Premise #2:   Eighteen is divisible by six.
    Conclusion: Therefore, eighteen is divisible by three.

19. Premise #1:   Eric is at the store or the handball court.
    Premise #2:   He is not at the store.
    Conclusion: Therefore, he must be at the handball court.

20. Premise #1:   The book must be returned within two weeks or you pay a fine.
    Premise #2:   The book was not returned within two weeks.
    Conclusion: Therefore, you must pay a fine.

21. Premise #1:   It is not true that it is a diamond and it sparkles in the sunlight.
    Premise #2:   It does sparkle in the sunlight.
    Conclusion: Therefore, it is a diamond.

22. Premise #1:   Either I work tonight or I pass the mathematics test.
    Premise #2:   I'm going to work tonight.
    Conclusion: Therefore, I will fail the mathematics test.

In Exercises 23–30, determine which conclusion is valid from the given premises.

23. Premise #1:   If seven is a prime number, then seven does not divide evenly into twenty-one.
    Premise #2:   Seven divides evenly into twenty-one.
    (a) Conclusion: Therefore, seven is a prime number.
    (b) Conclusion: Therefore, seven is not a prime number.
    (c) Conclusion: Therefore, twenty-one divided by seven is three.

24. Premise #1:   If the fuel is shut off, then the fire will be extinguished.
    Premise #2:   The fire continues to burn.
    (a) Conclusion: Therefore, the fuel was not shut off.
    (b) Conclusion: Therefore, the fuel was shut off.
    (c) Conclusion: Therefore, the fire becomes hotter.

25. Premise #1:   It is necessary that interest rates be lowered for the economy to improve.
Premise #2:   Interest rates were not lowered.
(a) Conclusion: Therefore, the economy will improve.
(b) Conclusion: Therefore, interest rates are irrelevant to the performance of the economy.
(c) Conclusion: Therefore, the economy will not improve.

26. Premise #1:   It will snow only if the temperature is below 32° at some level of the atmosphere.
Premise #2:   It is snowing.
(a) Conclusion: Therefore, the temperature is below 32° at ground level.
(b) Conclusion: Therefore, the temperature is above 32° at some level of the atmosphere.
(c) Conclusion: Therefore, the temperature is below 32° at some level of the atmosphere.

27. Premise #1:   Smokestack emissions must be reduced or acid rain will continue as an environmental problem.
Premise #2:   Smokestack emissions have not decreased.
(a) Conclusion: Therefore, the ozone layer will continue to be depleted.
(b) Conclusion: Therefore, acid rain will continue as an environmental problem.
(c) Conclusion: Therefore, stricter automobile emission standards must be enacted.

28. Premise #1:   The library must upgrade its computer system or service will not improve.
Premise #2:   Service at the library has improved.
(a) Conclusion: Therefore, the computer system was upgraded.
(b) Conclusion: Therefore, more personnel were hired for the library.
(c) Conclusion: Therefore, the computer system was not upgraded.

29. Premise #1:   If Rodney studies, then he will make good grades.
Premise #2:   If he makes good grades, then he will get a good job.
(a) Conclusion: Therefore, Rodney will get a good job.
(b) Conclusion: Therefore, if Rodney doesn't study, then he won't get a good job.
(c) Conclusion: Therefore, if Rodney doesn't get a good job, then he didn't study.

30. Premise #1:   It is necessary to have a ticket and an ID card to get into the arena.
Premise #2:   Janice entered the arena.
(a) Conclusion: Therefore, Janice does not have a ticket.
(b) Conclusion: Therefore, Janice has a ticket and an ID card.
(c) Conclusion: Therefore, Janice has an ID card.

In Exercises 31–34, use a Venn diagram to test the validity of the argument.

31. Premise #1:   All numbers divisible by ten are divisible by five.
Premise #2:   Fifty is divisible by ten.
Conclusion: Therefore, fifty is divisible by five.

32. Premise #1:   All human beings require adequate rest.
Premise #2:   All infants are human beings.
Conclusion: Therefore, all infants require adequate rest.

33. Premise #1:   No person under the age of eighteen is eligible to vote.
    Premise #2:   Some college students are eligible to vote.
    Conclusion: Therefore, some college students are under the age of eighteen.

34. Premise #1:   Every amateur radio operator has a radio license.
    Premise #2:   Jackie has a radio license.
    Conclusion: Therefore, Jackie is an amateur radio operator.

In Exercises 35–38, use the premises to prove the given conclusion.

35. Premise #1:   If Sue drives to work, then she will stop at the grocery store.
    Premise #2:   If she stops at the grocery store, then she'll buy milk.
    Premise #3:   Sue drove to work today.
    Conclusion: Therefore, Sue will get milk.

36. Premise #1:   If Bill is patient, then he will succeed.
    Premise #2:   Bill will get bonus pay if he succeeds.
    Premise #3:   Bill did not get bonus pay.
    Conclusion: Therefore, Bill is not patient.

37. Premise #1:   If this is a good product, then we should buy it.
    Premise #2:   Either it was made by XYZ Corporation, or we will not buy it.
    Premise #3:   It is not made by XYZ Corporation.
    Conclusion: Therefore, it is not a good product.

38. Premise #1:   If the book is returned within two weeks, then there is no fine.
    Premise #2:   You pay a fine or you may not check out another book.
    Premise #3:   You are allowed to check out another book.
    Conclusion: Therefore, the book was not returned within two weeks.

Further Concepts in Statistics

Stem-and-Leaf Plots • *Histograms and Frequency Distributions* • *Line Graphs* •
Choosing an Appropriate Graph • *Scatter Plots* • *Fitting a Line to Data* •

Stem-and-Leaf Plots

Statistics is the branch of mathematics that studies techniques for collecting, organizing, and interpreting data. In this section, you will study several ways to organize and interpret data.

One type of plot that can be used to organize sets of numbers by hand is a **stem-and-leaf plot.** A set of test scores and the corresponding stem-and-leaf plot are shown below.

Test Scores		*Stems*	*Leaves*
93, 70, 76, 58, 86, 93, 82, 78, 83, 86,		5	8
64, 78, 76, 66, 83, 83, 96, 74, 69, 76,		6	4 4 6 9
64, 74, 79, 76, 88, 76, 81, 82, 74, 70		7	0 0 4 4 4 6 6 6 6 6 8 8 9
		8	1 2 2 3 3 3 6 6 8
		9	3 3 6

Note that the *leaves* represent the units digits of the numbers and the *stems* represent the tens digits. Stem-and-leaf plots can also be used to compare two sets of data, as shown in the following example.

EXAMPLE 1   ■   Comparing Two Sets of Data

Use a stem-and-leaf plot to compare the test scores given above with the following test scores. Which set of test scores is better?

90, 81, 70, 62, 64, 73, 81, 92, 73, 81, 92, 93, 83, 75, 76,
83, 94, 96, 86, 77, 77, 86, 96, 86, 77, 86, 87, 87, 79, 88

Solution

Begin by ordering the second set of scores.

62, 64, 70, 73, 73, 75, 76, 77, 77, 77, 79, 81, 81, 81, 83,
83, 86, 86, 86, 86, 87, 87, 88, 90, 92, 92, 93, 94, 96, 96

Now that the data has been ordered, you can construct a *double* stem-and-leaf plot by letting the leaves to the right of the stems represent the units digits for the first group of test scores and letting the leaves to the left of the stems represent the units digits for the second group of test scores.

Leaves (2nd Group)	Stems	Leaves (1st Group)
	5	8
4 2	6	4 4 6 9
9 7 7 7 6 5 3 3 0	7	0 0 4 4 4 6 6 6 6 6 8 8 9
8 7 7 6 6 6 6 3 3 1 1 1	8	1 2 2 3 3 3 6 6 8
6 6 4 3 2 2 0	9	3 3 6

By comparing the two sets of leaves, you can see that the second group of test scores is better than the first group. ■

EXAMPLE 2   ■   Using a Stem-and-Leaf Plot

Table B.1 shows the percent of the population of each state and the District of Columbia that was at least 65 years old in 1989. Use a stem-and-leaf plot to organize the data. (*Source:* U.S. Bureau of Census)

TABLE B.1

AK	4.1	AL	12.7	AR	14.8	AZ	13.1	CA	10.6
CO	9.8	CT	13.6	DC	12.5	DE	11.8	FL	18.0
GA	10.1	HI	10.7	IA	15.1	ID	11.9	IL	12.3
IN	12.4	KS	13.7	KY	12.7	LA	11.1	MA	13.8
MD	10.8	ME	13.4	MI	11.9	MN	12.6	MO	13.9
MS	12.4	MT	13.2	NC	12.1	ND	13.9	NE	13.9
NH	11.4	NJ	13.2	NM	10.5	NV	10.9	NY	13.0
OH	12.8	OK	13.3	OR	13.9	PA	15.1	RI	14.8
SC	11.1	SD	14.4	TN	12.6	TX	10.1	UT	8.6
VA	10.8	VT	11.9	WA	11.9	WI	13.4	WV	14.6
WY	9.8								

Solution

Begin by ordering the numbers, as shown below.

4.1, 8.6, 9.8, 9.8, 10.1, 10.1, 10.5, 10.6, 10.7, 10.8, 10.8,
10.9, 11.1, 11.1, 11.4, 11.8, 11.9, 11.9, 11.9, 11.9, 12.1, 12.3,
12.4, 12.4, 12.5, 12.6, 12.6, 12.7, 12.7, 12.8, 13.0, 13.1, 13.2,
13.2, 13.3, 13.4, 13.4, 13.6, 13.7, 13.8, 13.9, 13.9, 13.9, 13.9,
14.4, 14.6, 14.8, 14.8, 15.1, 15.1, 18.0

Next construct the stem-and-leaf plot using the leaves to represent the digits to the right of the decimal points.

Stems	Leaves	
4.	1	Alaska has the lowest percent.
5.		
6.		
7.		
8.	6	
9.	8 8	
10.	1 1 5 6 7 8 8 9	
11.	1 1 4 8 9 9 9 9	
12.	1 3 4 4 5 6 6 7 7 8	
13.	0 1 2 2 3 4 4 6 7 8 9 9 9 9	
14.	4 6 8 8	
15.	1 1	
16.		
17.		
18.	0	Florida has the highest percent.

∎

Histograms and Frequency Distributions

With data such as that given in Example 2, it is useful to group the numbers into intervals and plot the frequency of the data in each interval. For instance, the **frequency distribution** and **histogram** shown in Figure B.1 represent the data given in Example 2.

Frequency Distribution

Interval	Tally
[4, 6)	I
[6, 8)	
[8, 10)	III
[10, 12)	JHT JHT JHT I
[12, 14)	JHT JHT JHT JHT IIII
[14, 16)	JHT I
[16, 18)	
[18, 20)	I

FIGURE B.1

Histogram

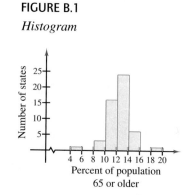

A histogram has a portion of a real number line as its horizontal axis. A **bar graph** is similar to a histogram, except that the rectangles (bars) can be either horizontal or vertical and the labels of the bars are not necessarily numbers.

Another difference between a bar graph and a histogram is that the bars in a bar graph are usually separated by spaces, whereas the bars in a histogram are not separated by spaces.

EXAMPLE 3   ■   Constructing a Bar Graph
The data below shows the average monthly precipitation (in inches) in Houston, Texas. Construct a bar graph for this data. What can you conclude? (*Source:* PC USA)

January	3.2	February	3.3	March	2.7
April	4.2	May	4.7	June	4.1
July	3.3	August	3.7	September	4.9
October	3.7	November	3.4	December	3.7

Solution
To create a bar graph, begin by drawing a vertical axis to represent the precipitation and a horizontal axis to represent the months. The bar graph is shown in Figure B.2. From the graph, you can see that Houston receives a fairly consistent amount of rain throughout the year—the driest month tends to be March and the wettest month tends to be September.

FIGURE B.2

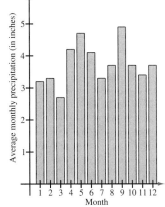

NOTE   Bar graphs are used in exercises and examples throughout this book. For example, see Exercises 77 and 78 on page 92 and Exercise 74 on page 417. The material presented here allows for a more in-depth discussion.

Line Graphs

A **line graph** is similar to a standard coordinate graph. Line graphs are usually used to show trends over periods of time.

EXAMPLE 4   ■   Constructing a Line Graph

The following data shows the number of immigrants (in thousands) to the United States per decade. Construct a line graph of the data. What can you conclude?

Decade	Number	Decade	Number
1851–1860	2598	1861–1870	2315
1871–1880	2812	1881–1890	5247
1891–1900	3688	1901–1910	8795
1911–1920	5736	1921–1930	4107
1931–1940	528	1941–1950	1035
1951–1960	2515	1961–1970	3322
1971–1980	4493	1981–1990	6447

NOTE Line graphs are used in exercises throughout this book. For example, see Exercises 99 and 101 on page 167 and Exercises 55 and 56 on page 178. The material presented here allows for a more in-depth discussion.

Solution

Begin by drawing a vertical axis to represent the number of immigrants in thousands. Then label the horizontal axis with decades and plot the points shown in the table. Finally, connect the points with line segments, as shown in Figure B.3. From the line graph, you can see that the number of immigrants hit a low point during the depression of the 1930's. Since then the numbers have steadily increased.

FIGURE B.3

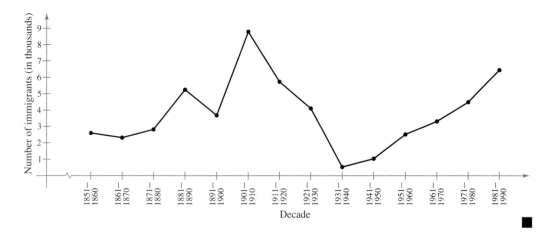

Choosing an Appropriate Graph

Line graphs and bar graphs are commonly used for displaying data. When you are using a graph to organize and present data, you must first decide which type of graph to use.

EXAMPLE 5   ■   Organizing Data with a Graph

Listed below are the daily average numbers of miles walked by people while working at their jobs. Organize the data graphically. (*Source:* American Podiatry Association)

Occupation	Miles Walked per Day
Mail Carrier	4.4
Medical Doctor	3.5
Nurse	3.9
Police Officer	6.8
Television Reporter	4.2

STUDY TIP

Here are some guidelines to use when you must decide which type of graph to use.

1. Use a bar graph when the data falls into distinct categories and you want to compare totals.
2. Use a line graph when you want to show the relationship between consecutive amounts or data over time.

Solution

You can use a bar graph because the data falls into distinct categories, and it would be useful to compare totals. The bar graph shown in Figure B.4 is horizontal. This makes it easier to label each bar. Also notice that the occupations are listed in order of the number of miles walked.

FIGURE B.4

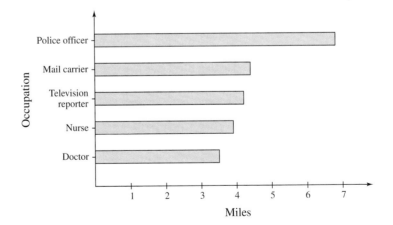

t	P
0	109
1	110
2	112
3	113
4	115
5	117

t	P
6	120
7	122
8	123
9	126
10	126

NOTE Scatter plots are discussed on page 442 and used in chapter projects throughout this book. The idea of correlation is included in several chapter projects. For example, see the Chapter 2 project on page 191 and the Chapter 7 project on page 489. The material presented here allows for a more in-depth discussion.

Scatter Plots

Many real-life situations involve finding relationships between two variables, such as the year and the number of people in the labor force. In a typical situation, data is collected and written as a set of ordered pairs. The graph of such a set is called a **scatter plot.**

FIGURE B.5

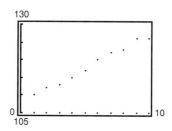

From the scatter plot in Figure B.5, it appears that the points describe a relationship that is nearly linear. (The relationship is not *exactly* linear because the labor force did not increase by precisely the same amount each year.) A mathematical equation that approximates the relationship between t and P is called a *mathematical model.* When developing a mathematical model, you strive for two (often conflicting) goals—accuracy and simplicity.

Consider a collection of ordered pairs of the form (x, y). If y tends to increase as x increases, the collection is said to have a **positive correlation.** If y tends to decrease as x increases, the collection is said to have a **negative correlation.** Figure B.6 shows three examples: one with a positive correlation, one with a negative correlation, and one with no (discernible) correlation.

FIGURE B.6

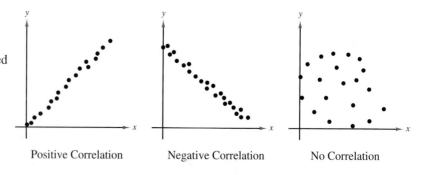

Positive Correlation Negative Correlation No Correlation

EXAMPLE 6 ◼ Interpreting Scatter Plots

FIGURE B.7

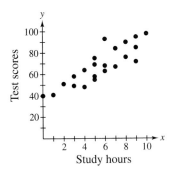

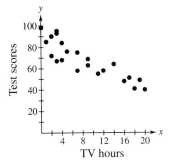

On a Friday, 22 students in a class were asked to keep track of the numbers of hours they spent studying for a test on Monday and the numbers of hours they spent watching television. The numbers are shown below. Construct a scatter plot for each set of data. Then determine whether the points are positively correlated, negatively correlated, or have no discernible correlation. What can you conclude? (The first coordinate is the number of hours and the second coordinate is the score obtained on Monday's test.)

Study Hours: (0, 40), (1, 41), (2, 51), (3, 58), (3, 49), (4, 48), (4, 64), (5, 55), (5, 69), (5, 58), (5, 75), (6, 68), (6, 63), (6, 93), (7, 84), (7, 67), (8, 90), (8, 76), (9, 95), (9, 72), (9, 85), (10, 98)

TV Hours: (0, 98), (1, 85), (2, 72), (2, 90), (3, 67), (3, 93), (3, 95), (4, 68), (4, 84), (5, 76), (7, 75), (7, 58), (9, 63), (9, 69), (11, 55), (12, 58), (14, 64), (16, 48), (17, 51), (18, 41), (19, 49), (20, 40)

Solution

Scatter plots for the two sets of data are shown in Figure B.7. The scatter plot relating study hours and test scores has a positive correlation. This means that the more a student studied, the higher his or her score tended to be. The scatter plot relating television hours and test scores has a negative correlation. This means that the more time a student spent watching television, the lower his or her score tended to be. ◼

Fitting a Line to Data

Finding a linear model that represents the relationship described by a scatter plot is called **fitting a line to data.** You can do this graphically by simply sketching the line that appears to fit the points, finding two points on the line, and then finding the equation of the line that passes through the two points.

EXAMPLE 7 ◼ Fitting a Line to Data

Find a linear model that relates the year with the number of people P (in millions) who were part of the United States labor force from 1980 through 1990. In Table B.2, t represents the year, with $t = 0$ corresponding to 1980. (*Source:* U.S. Bureau of Labor Statistics)

TABLE B.2

t	0	1	2	3	4	5	6	7	8	9	10
P	109	110	112	113	115	117	120	122	123	126	126

Solution

After plotting the data from Table B.2, draw the line that you think best represents the data, as shown in Figure B.8. Two points that lie on this line are (0, 109) and (9, 126). Using the point-slope form, you can find the equation of the line to be

$$P = \frac{17}{9}t + 109.$$ Linear model

FIGURE B.8

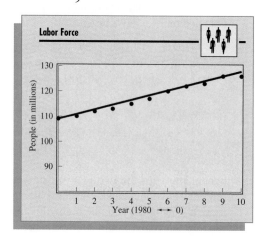

NOTE Fitting a line to data is incorporated throughout this book. For example, see the group activities on pages 442 and 513, and Exercises 95–97 on page 446. The material presented here allows for a more in-depth discussion.

Once you have found a model, you can measure how well the model fits the data by comparing the actual values with the values given by the model, as shown in Table B.3.

TABLE B.3

t	0	1	2	3	4	5	6	7	8	9	10
Actual → P	109	110	112	113	115	117	120	122	123	126	126
Model → P	109	110.9	112.8	114.7	116.6	118.4	120.3	122.2	124.1	126	127.9

The sum of the squares of the differences between the actual values and the model's values is the **sum of the squared differences.** The model that has the least sum is called the **least squares regression line** for the data. For the model in Example 7, the sum of the squared differences is 13.81. The least squares regression line for the data is

$$P = 1.864t + 108.2.$$ Best-fitting linear model

Its sum of squared differences is 4.7.

Many calculators have "built-in" least squares regression programs. If your calculator has such a program, enter the data in Table B.2 and use it to find the least squares regression line.

NOTE Built-in regression features of calculators are mentioned in the chapter projects throughout this book. For example, see the Chapter 2 project on page 191.

B. EXERCISES

1. Construct a stem-and-leaf plot for the following exam for a class of 30 students. The scores are for a 100-point exam.

77, 100, 77, 70, 83, 89, 87, 85, 81, 84, 81, 78, 89, 78, 88, 85, 90, 92, 75, 81, 85, 100, 98, 81, 78, 75, 85, 89, 82, 75

2. *Education Expenses*   The table shows the per capita expenditures for public elementary and secondary education in the 50 states and the District of Columbia in 1991. Use a stem-and-leaf plot to organize the data. (*Source:* National Education Association)

AK	1626	AL	694	AR	668	AZ	892	CA	918
CO	841	CT	1151	DC	1010	DE	891	FL	862
GA	859	HI	784	IA	846	ID	725	IL	788
IN	925	KS	906	KY	725	LA	758	MA	866
MD	944	ME	1062	MI	926	MN	990	MO	742
MS	671	MT	983	NC	813	ND	719	NE	757
NH	881	NJ	1223	NM	915	NV	1004	NY	1186
OH	861	OK	776	OR	925	PA	889	RI	892
SC	835	SD	716	TN	618	TX	905	UT	828
VA	941	VT	992	WA	1095	WI	928	WV	883
WY	1178								

In Exercises 3 and 4, use the following set of data, which lists students' scores on a 100-point exam.

93, 84, 100, 92, 66, 89, 78, 52, 71, 85, 83, 95, 98, 99, 93, 81, 80, 79, 67, 59, 90, 55, 77, 62, 90, 78, 66, 63, 93, 87, 74, 96, 72, 100, 70 ,73

3. Use a stem-and-leaf plot to organize the data.

4. Draw a histogram to represent the data.

5. Complete the following frequency distribution table and draw a histogram to represent the data.

44, 33, 17, 23, 16, 18, 44, 47, 18, 20, 25, 27, 18, 29, 29, 28, 27, 18, 36, 22, 32, 38, 33, 41, 49, 48, 45, 38, 49, 15

Interval	Tally
[15, 22)	
[22, 29)	
[29, 36)	
[36, 43)	
[43, 50)	

6. *Snowfall*   The data below shows the seasonal snowfall (in inches) at Erie, Pennsylvania, for the years 1960 through 1989 (the amounts are listed in order by year). How would you organize this data? Explain your reasoning. (*Source:* National Oceanic and Atmospheric Administration)

69.6, 42.5, 75.9, 115.9, 92.9, 84.8, 68.6, 107.9, 79.7, 85.6, 120.0, 92.3, 53.7, 68.6, 66.7, 66.0, 111.5, 142.8, 76.5, 55.2, 89.4, 71.3, 41.2, 110.0, 106.3, 124.9, 68.2, 103.5, 76.5, 114.9

7. *Travel to the United States*   The data below gives the places of origin and the numbers of travelers (in millions) to the United States in 1991. Construct a horizontal bar graph for this data. (*Source:* U.S. Travel and Tourism Administration)

Canada   18.9   Mexico 7.0   Europe 7.4
Latin America   2.0   Other   6.8

8. *Fruit Crops*   The data below shows the cash receipts (in millions of dollars) from fruit crops for farmers in 1990. Construct a bar graph for this data. (*Source:* U.S. Department of Agriculture)

Apples	1159	Peaches	365
Grapefruit	317	Pears	266
Grapes	1668	Plums and Prunes	293
Lemons	278	Strawberries	560
Oranges	1707		

Handling Garbage   In Exercises 9–14, use the line graph given below. (*Source:* Franklin Associates)

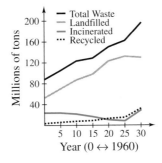

9. Estimate the total waste in 1980 and 1990.

10. Estimate the amount of incinerated garbage in 1970.

11. Which quantities increased every year?

12. During which time period did the amount of incinerated garbage decrease?

13. *Reasoning*   What is the relationship between the four quantities in the line graph?

14. *Think About It*   Why do you think landfill garbage is decreasing?

15. *College Attendance*   The following table shows the enrollment in a liberal arts college. Construct a line graph for the data.

Year	1985	1986	1987	1988
Enrollment	1675	1704	1710	1768

Year	1989	1990	1991	1992
Enrollment	1833	1918	1967	1972

16. *Oil Imports*   The table shows the crude oil imports into the United States in millions of barrels for the years 1982 through 1990. Construct a line graph for the data and state what information it reveals. (*Source:* U.S. Energy Information Administration)

Year	1982	1983	1984	1985	1986
Oil Imports	1273	1215	1254	1168	1525

Year	1987	1988	1989	1990
Oil Imports	1706	1869	2133	2145

17. *Stock Market*   The list below shows stock prices for selected companies in March of 1994. Draw a graph that best represents the data. Explain why you chose that type of graph.

Company	Stock Price
Sears, Roebuck	$48
Wal-Mart Stores	$28
JC Penney	$55
K Mart Corp.	$19
The Gap, Inc.	$45

18. *Net Profit*   The table shows the net profits (in millions of dollars) of Blockbuster Entertainment for the years 1988 through 1993. Draw a graph that best represents the net profit and explain why you chose that type of graph.

Year	1988	1989	1990	1991	1992	1993
Net Profit	15.5	44.2	68.7	93.7	142.0	243.6

19. *Videocassette Recorders*   The average numbers (out of 100) of people who owned videocassette recorders in selected years from 1980 to 1992 are given in the table. Organize the data graphically. Explain your reasoning. (*Source:* The Roper Organization)

Year	1980	1983	1985	1987	1988	1990	1992
Number	3	10	19	50	64	65	68

20. *Owning Cats*   The average numbers (out of 100) of cat owners who state various reasons for owning a cat are listed below. Organize the data graphically. Explain your reasoning. (*Source:* Gallup Poll)

Reason for Owning a Cat	Number
Have a pet to play with	93
Companionship	84
Help children learn responsibility	78
Have a pet to communicate with	62
Security	51

Interpreting a Scatter Plot   In Exercises 21–24, use the scatter plot shown. The scatter plot compares the number of hits *x* made by 30 softball players during the first half of the season with the number of runs batted in *y*.

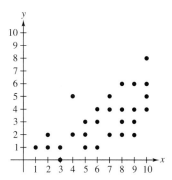

21. Do *x* and *y* have a positive correlation, a negative correlation, or no correlation?

22. Why does the scatter plot show only 28 points?

23. From the scatter plot, does it appear that players with more hits tend to have more runs batted in?

24. Can a player have more runs batted in than hits? Explain.

In Exercises 25–28, decide whether a scatter plot relating the two quantities would tend to have a positive, negative, or no correlation. Explain.

25. The age and value of a car

26. A student's study time and test scores

27. The height and age of a pine tree

28. A student's height and test scores

In Exercises 29–32, use the data in the table, which shows the relationship between the altitude *A* (in thousands of feet) and the air pressure *P* (in pounds per square foot).

A	0	5	10	15	20	25
P	14.7	12.3	10.2	8.4	6.8	5.4

A	30	35	40	45	50
P	4.5	3.5	2.8	2.1	1.8

29. Sketch a scatter plot of the data.

30. How are *A* and *P* related?

31. Estimate the air pressure at 42,500 feet.

32. Estimate the altitude at which the air pressure is 5.0 pounds per square foot.

Crop Yield   In Exercises 33–36, use the data in the table, where *x* is the number of units of fertilizer applied to sample plots and *y* is the yield (in bushels) of a crop.

x	0	1	2	3	4	5	6	7	8
y	58	60	59	61	63	66	65	67	70

33. Sketch a scatter plot of the data.

34. Determine whether the points are positively correlated, are negatively correlated, or have no discernible correlation.

35. Sketch a linear model that you think best represents the data. Find an equation of the line you sketched. Use the line to predict the yield if 10 units of fertilizer are used.

36. Can the model found in Exercise 35 be used to predict yields for arbitrarily large values of *x*? Explain.

Speed of Sound   In Exercises 37–40, use the data in the table, where *h* is altitude in thousands of feet and *v* is the speed of sound in feet per second.

h	0	5	10	15	20	25	30	35
v	1116	1097	1077	1057	1036	1015	995	973

37. Sketch a scatter plot of the data.

38. Determine whether the points are positively correlated, are negatively correlated, or have no discernible correlation.

39. Sketch a linear model that you think best represents the data. Find an equation of the line you sketched. Use the line to predict the speed of sound at an altitude of 27,000 feet.

40. The speed of sound at an altitude of 70,000 feet is approximately 971 feet per second. What does this suggest about the validity of using the model in Exercise 39 to extrapolate beyond the data given in the table?

In Exercises 41–44, use a graphing utility to find the least squares regression line for the data. Sketch a scatter plot and the regression line.

41. $(0, 23)$, $(1, 20)$, $(2, 19)$, $(3, 17)$, $(4, 15)$, $(5, 11)$, $(6, 10)$

42. $(4, 52.8)$, $(5, 54.7)$, $(6, 55.7)$, $(7, 57.8)$, $(8, 60.2)$, $(9, 63.1)$, $(10, 66.5)$

43. $(-10, 5.1)$, $(-5, 9.8)$, $(0, 17.5)$, $(2, 25.4)$, $(4, 32.8)$, $(6, 38.7)$, $(8, 44.2)$, $(10, 50.5)$

44. $(-10, 213.5)$, $(-5, 174.9)$, $(0, 141.7)$, $(5, 119.7)$, $(8, 102.4)$, $(10, 87.6)$

45. *Advertising* The management of a department store ran an experiment to determine if a relationship existed between sales S (in thousands of dollars) and the amount spent on advertising x (in thousands of dollars). The following data was collected.

x	1	2	3	4	5	6	7	8
S	405	423	455	466	492	510	525	559

Use a graphing utility to find the least squares regression line. Use the equation to estimate sales if $4500 is spent on advertising. Make a scatter plot of the data and sketch the graph of the regression line.

Programs

Programs for the *TI–82* are given in several sections of the text. This appendix contains additional programming hints for the *TI–82* and translations of the text programs for the following graphics calculators from Texas Instruments, Casio, and Sharp:

- *TI–80*
- *TI–81*
- *TI–82*
- *TI–85*
- *Sharp EL–9200*
- *Sharp EL–9300*
- *Casio fx–6300G*
- *Casio fx–7700G*
- *Casio fx–7700GE*
- *Casio fx–9700GE*
- *Casio CFX–9800G.*

Similar programs can be written for other brands and models of graphics calculators.

Enter a program in your calculator, then refer to the text discussion and apply the program as appropriate. Section references are provided to help you locate the text discussion of the programs and their uses.

To illustrate the power and versatility of programmable calculators, a variety of types of programs is presented, including a simulation program (Graph Reflection Program) and a tutorial program (Reflections and Shifts Program).

Evaluating an Algebraic Expression (Section P.4)

The program, shown in the marginal Programming note on page 36, can be used to evaluate an algebraic expression in one variable at several values of the variable.

Note: On the *TI–82*, the "Lbl" and "Goto" commands may be entered through the "CTL" menu accessed by pressing the PRGM key. The "Disp" and "Input" commands may be entered through the "I/O" menu accessed by pressing the PRGM key. The symbol "Y_1" may be entered through the "Function" menu accessed by pressing the Y-VARS key. Keystroke sequences required for similar commands on other calculators will vary. Consult the user's manual for your calculator.

TI–80

```
PROGRAM:EVALUATE
:Lbl A
:Input "ENTER X ",X
:Disp Y₁
:Goto A
```

To use this program, enter an expression in Y_1. Expressions may also be evaluated directly on the *TI–80*'s home screen.

TI–81

```
Prgm1: EVALUATE
:Lbl 1
:Disp "ENTER X"
:Input X
:Disp Y1
:Goto 1
```

To use this program, enter an expression in Y_1. Expressions may also be evaluated on the *TI–81*'s home screen.

TI–82

```
PROGRAM:EVALUATE
:Lbl A
:Input "ENTER X ",X
:Disp Y₁
:Goto A
```

To use this program, enter an expression in Y_1. Expressions may also be evaluated directly on the *TI–82*'s home screen.

TI–85

```
PROGRAM:
:Lbl A
:Input "Enter x ",x
:Disp y1
:Goto A
```

To use this program, enter an expression in Y_1.

Sharp EL 9200
Sharp EL 9300

```
Evaluate
------------REAL
Goto top
Label eqtn
Y=f(X)
Return
Label top
Input X
Gosub equation
Print Y
Goto top
End
```

To use this program, replace $f(X)$ with your expression in X.

Casio fx–6300G

```
EVALUATE:
Lbl 1:
"X="?→X:
"F(X)="◢Prog 0◢
Goto 1
```

To use this program, write the expression as Prog 0.

Casio fx–7700G

EVALUATE
Lbl 1
"X="?→X
"F(X)=":f$_1$ ◢
Goto 1

To use this program, enter an expression in f_1.

Casio fx–7700GE
Casio fx–9700GE
Casio CFX–9800G

EVALUATE↵
Lbl 1↵
"X="?→X↵
"F (X) =":Y1 ◢
Goto 1

To use this program, enter an expression in Y_1.

Simple Interest Program (Section 1.3)

The program, shown in the marginal Programming note on page 85, can be used to find the amount of simple interest earned by a given principal at a given annual interest rate for a certain amount of time.

Note: The program requires a line of code that restricts the displayed result to two decimal places. For instance, the *TI–82* program requires "Fix 2" as the first line. To do this, press the MODE key, cursor to "2" on the line beginning with "Float," and press ENTER. Similarly, the last line of the program may be entered by pressing the MODE key, cursoring to "Float," and pressing ENTER. On the *TI–82,* the → symbol represents pressing the STO▷ key. For additional keystroke instructions, see previous programs in this appendix. Keystroke sequences required for similar commands on other calculators will vary. Consult the user's manual for your calculator.

TI–80
TI–81
TI–82

```
SIMPINT
:Fix 2
:Disp "PRINCIPAL"
:Input P
:Disp "INTEREST RATE"
:Disp "IN DECIMAL FORM"
:Input R
:Disp "NO. OF YEARS"
:Input T
:PRT→I
:Disp "THE INTEREST IS"
:Disp I
:Float
```

TI–85

```
SimpInt
:Disp "Principal"
:Input P
:Disp "Interest rate"
:Disp "in decimal form"
:Input R
:Disp "No. of years"
:Input T
:P*R*T→I
:Disp "The interest is"
:Disp I
:Float
```

Sharp EL 9200
Sharp EL 9300

```
SimpInt
Input principal
Print "Interest rate
Print "in decimal form
Input rate
Print "No. of years
Input time
interest=principal*rate*time
Print interest
```

Casio fx–7700G
Casio fx–7700GE
Casio fx–9700GE
Casio CFX–9800G

```
SIMPINT↵
Fix 2↵
"PRINCIPAL"?→P↵
"INTEREST RATE"↵
"IN DECIMAL FORM"?→R↵
"NO. OF YEARS"?→T↵
PRT→I↵
"THE INTEREST IS":I↵
Norm
```

Casio fx–6300G

```
SIMPINT:Fix 2:"PRINCIPAL"?→P:
"INTEREST"◢"RATE AS"◢"DECIMAL"?→R:
"NO. YEARS"?→T:PRT→I:
"INTEREST IS"◢I:Norm
```

Reflections and Shifts Program (Section 2.6)

The program, referenced in the marginal Programming note on page 186, will sketch a graph of the function $y = R(x + H)^2 + V$, where $R = \pm 1$, H is an integer between -6 and 6, and V is an integer between -3 and 3. This program gives you practice working with reflections, horizontal shifts, and vertical shifts.

Note: On the *TI–82,* the "int" and "rand" commands may be entered through the "NUM" and "PRB" menus, respectively, accessed by pressing the MATH key. The "=" and "<" symbols may be entered through the "TEST" menu accessed by pressing the TEST key. Other commands such as "If," "Then," "Else," and "End" may be entered through the "CTL" menu accessed by pressing the PRGM key. The commands "Xmin," "Xmax," "Xscl," "Ymin," "Ymax," and "Yscl" may be entered through the "Window" menu accessed by pressing the VARS key. The commands "DispGraph" and "Pause" may be entered through the "I/O" and "CTL" menus, respectively, accessed by pressing the PRGM key. For additional keystroke instructions, see previous programs in this appendix. Keystroke sequences for similar commands on other calculators will vary. Consult the user's manual for your calculator.

TI–80

PROGRAM:PARABOLA
:-6 + int (12rand)→H
:-3 + int (6rand)→V
:rand→R
:If R < .5
:Then
:-1→R
:Else
:1→R
:End
:"R(X + H)^2 + V"→Y$_1$
:-9→Xmin
:9→Xmax
:1→Xscl
:-6→Ymin
:6→Ymax
:1→Yscl
:DispGraph
:Pause
:Disp "R = ",R
:Disp "H = ",H
:Disp "V = ",V

Press ENTER after the graph
to display the coordinates.

TI–81

Prgm2: PARABOLA
:Rand→H
:-6+Int (12H) → H
:Rand→V
:-3+Int (6V) → V
:Rand→R
:If R<.5
:-1 → R
:If R>.49
:1 → R
:"R(X+H)^2+V"→Y$_1$
:-9 → Xmin
:9 → Xmax
:1 → Xscl
:-6 → Ymin
:6 → Ymax
:1 → Yscl
:DispGraph
:Pause
:Disp "Y=R(X+H)^2+V"
:Disp "R="
:Disp R
:Disp "H="
:Disp H
:Disp "V="
:Disp V
:End

Press ENTER after the graph
to display the coordinates.

TI–82

PROGRAM:PARABOLA
:-6 + int (12rand)→H
:-3 + int (6rand)→V
:rand→R
:If R < .5
:Then
:-1→R
:Else
:1→R
:End
:"R(X + H)^2 + V"→Y$_1$
:-9→Xmin
:9→Xmax
:1→Xscl
:-6→Ymin
:6→Ymax
:1→Yscl
:DispGraph
:Pause
:Disp "R = ",R
:Disp "H = ",H
:Disp "V = ",V

Press ENTER after the graph
to display the coordinates.

TI–85

```
 PROGRAM:
:rand→H
: -6+int(12H)→H
:rand→V
: -3+int(6V)→V
:rand→R
:If R<.5
: -1→R
:If R>.49
: 1→R
:y1=R(x+H)²+V
: -9→xMin
:9→xMax
:1→xScl
: -6→yMin
:6→yMax
:1→yScl
:DispG
:Pause
:Disp "Y=R(X+H)²+V"
:Disp "R=",R
:Disp "H=",H
:Disp "V=",V
```

Press ENTER after the graph to display the coefficients.

Sharp EL 9200
Sharp EL 9300

```
Parabola
-----------REAL
H=int (random*12)-6
V=int (random*6)-3
S=(random*2)-1
R=S/abs S
Range -9,9,1,-6,6,1
Graph R(X+H)²+V
Wait
Print "Y=R(X+H)²+V"
Print R
Print H
Print V
End
```

Press ENTER after the graph to display the coefficients.

Casio fx–6300G

R(X+H)²+V:
-6+Int (12Ran#)→H:
-3+Int (6Ran#)→V:
-1→R:Ran#<0.5⇒1→R:
Range -9,9,1,-6,6,1:
Graph Y=R(X+H)²+V◢
"Y=R(X+H)²+V"◢
"R="◢R◢
"H="◢H◢
"V="◢V

Casio 7700

R(X+H)²+V
-6+Int (12Ran#)→H
-3+Int (6Ran#)→V
-1→R:Ran#<0.5⇒1→R
Range -9,9,1,-6,6,1
Graph Y=R(X+H)²+V◢
"Y=R(X+H)²+V"
"R=":R◢
"H=":H◢
"V=":V

Casio fx–7700GE
Casio fx–9700GE
Casio CFX–9800G

```
R(X+H)²+V↵
 -6+Int (12Ran#)→H↵
 -3+Int (6Ran#)→V↵
Ran#→R↵
R<.5⇒-1→R↵
R≥.5⇒1→R↵
Range -9,9,1,-6,6,1↵
Graph Y=R(X+H)²+V◢
"Y=R(X+H)²+V"↵
"R=":R◢
"H=":H◢
"V=":V
```

Press ENTER after the graph to display the coordinates.

Quadratic Formula Program (Section 6.3)

The program, shown in the marginal Programming note on page 402, will display the solutions to quadratic equations or the words "No Real Solution." To use the program, write the quadratic equation in standard form and enter the values of a, b, and c.

Note: On the *TI–82,* the "Prompt" command may be entered through the "I/O" menu acccessed by pressing the PRGM key. For additional keystroke instructions, see previous programs in this appendix. Keystroke sequences for similar commands on other calculators will vary. Consult the user's manual for your calculator.

TI–80

```
PROGRAM:QUADRAT
:Disp "AX²+BX+C=0"
:Input "ENTER A ", A
:Input "ENTER B ", B
:Input "ENTER C ", C
:B²−4AC→D
:If D≥0
:Then
:(−B+√D)/(2A)→M
:Disp M
:(−B−√D)/(2A)→N
:Disp N
:Else
:Disp "NO REAL SOLUTION"
:End
```

TI–81

```
Prgm4: QUADRAT
:Disp "ENTER A"
:Input A
:Disp "ENTER B"
:Input B
:Disp "ENTER C"
:Input C
:B²−4AC→D
:If D<0
:Goto 1
:((−B+√D)/(2A))→S
:Disp S
:((−B−√D)/(2A))→S
:Disp S
:End
:Lbl 1
:Disp "NO REAL"
:Disp "SOLUTION"
:End
```

TI–82

```
PROGRAM:QUADRAT
:Disp "AX²+BX+C=0"
:Prompt A
:Prompt B
:Prompt C
:B²−4AC→D
:If D≥0
:Then
:(−B+√D)/(2A)→M
:Disp M
:(−B−√D)/(2A)→N
:Disp N
:Else
:Disp "NO REAL SOLUTION"
:End
```

TI–85

```
PROGRAM:
:Input "ENTER A ",A
:Input "ENTER B ",B
:Input "ENTER C ",C
:B²-4*A*C→D
:Disp (-B+√D)/(2A)
:Disp (-B-√D)/(2A)
```

This program gives both real and complex answers. Solutions to quadratic equations are also available directly by using the *TI–85* POLY function.

Sharp EL 9200
Sharp EL 9300

```
Quadratic
----------COMPLEX
Input A
Input B
Input C
D=B²-4AC
x1=(-B+√ D)/(2A)
x2=(-B-√ D)/(2A)
Print x1
Print x2
X=x1
Y=x2
End
```

This program is written in the program's complex mode, so both real and complex answers are given. The answers are also stored under variables X and Y so they can be used in the calculator mode.

Casio fx–6300G

```
QUADRATICS:
"AX²+BX+C=0"◢
"A="?→A:
"B="?→B:
"C="?→C:
B²-4AC→D:
D<0⇒Goto 1:
"X="◢(-B+√D)÷2A ◢
"OR X="◢(-B-√D)÷2A:
Goto 2:
Lbl 1:
"NO REAL SOLUTION"
Lbl 2
```

Casio fx–7700G

```
QUADRATICS
"AX²+BX+C=0"
"A="?→A
"B="?→B
"C="?→C
B²-4AC→D
D<0⇒Goto 1
"X=":(-B+√D)÷2A ◢
"OR X=":(-B-√D)÷2A
Goto 2
Lbl 1
"NO REAL SOLUTION"
Lbl 2
```

Casio fx–7700GE
Casio fx–9700GE
Casio CFX–9800G

```
QUADRATIC↵
"AX²+BX+C=0"↵
"A="?→A↵
"B="?→B↵
"C="?→C↵
B²-4AC→D↵
(-B+√D)÷2A ◢
(-B-√D)÷2A
```

Both real and complex answers are given. Solutions to quadratic equations are also available directly from the Casio calculator's EQUATION MENU.

Two-Point Form of a Line (Section 7.1)

The program, shown in the marginal Programming note on page 438, will display the slope and y-intercept for the line that passes through two points, (x_1, y_1) and (x_2, y_2), entered by the user.

Note: For help with *TI–82* keystrokes, see previous programs in this appendix.

TI–80
TI–81
TI–82

```
TWOPTFM
:Disp "ENTER X1,Y1"
:Input X
:Input Y
:Disp "ENTER X2,Y2"
:Input C
:Input D
:(D-Y)/(C-X)→M
:M*(-X)+Y→B
:Disp "SLOPE ="
:Disp M
:Disp "Y-INT ="
:Disp B
```

TI–85

```
TwoPtFrm
:Disp "Enter X1,Y1"
:Input X
:Input Y
:Disp "Enter X2,Y2"
:Input C
:Input D
:(D-Y)/(C-X)→M
:M*(-X)+Y→B
:Disp "Slope ="
:Disp M
:Disp "Y-int ="
:Disp B
```

Sharp EL 9200
Sharp EL 9300

```
TwoPtForm
Print "Enter X1,Y1
Input x
c=x
Input y
d=y
Print "Enter X2,Y2
Input x
Input y
m=(d-y)÷(c-x)
b=m*(-x)+y
Print "Slope
Print m
Print "Y-int
Print b
```

Casio fx–6300G

```
TWOPTFORM:"ENTER X1,Y1"?→X:?→Y:
"ENTER X2,Y2"?→C:?→D:
(D-Y)÷(C-X)→M:M×(-X)+Y→B:
"SLOPE ="◢M◢"Y-INT ="◢B
```

Casio fx–7700G
Casio fx–7700GE
Casio fx–9700GE
Casio CFX–9800G

```
TWOPTFORM↵
"ENTER X1,Y1"?→X:?→Y↵
"ENTER X2,Y2"?→C:?→D↵
(D-Y)÷(C-X)→M↵
M×(-X)+Y→B↵
"SLOPE =":M◢
"Y-INT =":B
```

Systems of Linear Equations (Section 8.2)

The program, shown in the marginal Programming note on page 512, will display the solution of a system of two linear equations in two variables of the form

$$ax + by = c$$

$$dx + ey = f$$

if a unique solution exists.

Note: For help with *TI–82* keystrokes, see previous programs in this appendix.

TI–80

PROGRAM:SOLVE
:Disp "AX + BY = C"
:Input "ENTER A ", A
:Input "ENTER B ", B
:Input "ENTER C ", C
:Disp "DX + EY = F"
:Input "ENTER D ", D
:Input "ENTER E ", E
:Input "ENTER F ", F
:If AE−DB = 0
:Then
:Disp "NO SOLUTION"
:Else
:(CE−BF)/(AE−DB)→X
:(AF−CD)/(AE−DB)→Y
:Disp X
:Disp Y
:End

TI–81

Prgm5: SOLVE
:Disp "ENTER A,B,C,D,E,F"
:Input A
:Input B
:Input C
:Input D
:Input E
:Input F
:If AE−DB=0
:Goto 1
:(CE−BF)/(AE−DB) → X
:(AF−CD)/(AE−DB) → Y
:Disp X
:Disp Y
:Lbl 1
:End

TI–82

PROGRAM:SOLVE
:Disp "AX + BY = C"
:Prompt A
:Prompt B
:Prompt C
:Disp "DX + EY = F"
:Prompt D
:Prompt E
:Prompt F
:If AE−DB = 0
:Then
:Disp "NO SOLUTION"
:Else
:(CE−BF)/(AE−DB)→X
:(AF−CD)/(AE−DB)→Y
:Disp X
:Disp Y
:End

TI–85

```
PROGRAM:
:Disp "ax+by=c"
:Input "Enter a ",A
:Input "Enter b ",B
:Input "Enter c ",C
:Disp "dx+ey=f"
:Input "Enter d ",D
:Input "Enter e ",E
:Input "Enter f ",F
:If A*E-D*B=0
:Goto A
:(C*E-B*F)/(A*E-D*B)→X
:(A*F-C*D)/(A*E-D*B)→Y
:Disp X
:Disp Y
:Lbl A
```

Sharp EL 9200
Sharp EL 9300

```
Solve
----------REAL
Input A
Input B
Input C
Input D
Input E
Input F
If A*E-D*B=0 Goto 1
X=(C*E-B*F)/(A*E-D*B)
Y=(A*F-C*D)/(A*E-D*B)
Print X
Print Y
End
Label 1
Print "No solution
End
```

Equations must be entered in the form: $Ax + By = C$; $Dx + Ey = F$. Uppercase letters are used so that the values can be accessed in the calculation mode of the calculator.

Casio fx–7700GE
Casio fx–9700GE
Casio CFX–9800G

```
SOLVE↵
"ENTER A,B,C,D,E,F"↵
"A":?→A↵
"B":?→B↵
"C":?→C↵
"D":?→D↵
"E":?→E↵
"F":?→F↵
AE-DB=0⇨Goto 1↵
"X=":(CE-BF)÷(AE-DB)◢
"Y=":(AF-CD)÷(AE-DB)↵
Goto 2↵
Lbl 1↵
"NO UNIQUE SOLUTION"↵
Lbl 2
```

Solutions to systems of linear equations are also available directly from the Casio calculator's EQUATION MENU.

Casio fx–7700G

```
SOLVE
"ENTER A,B,C,D,E,F"
"A="?→A
"B="?→B
"C="?→C
"D="?→D
"E="?→E
"F="?→F
AE-DB=0⇒Goto 1
"X=":(CE-BF)÷(AE-DB)◢
"Y=":(AF-CD)÷(AE-DB)
Goto 2
Lbl 1
"NO UNIQUE SOLUTION"
Lbl 2
```

Casio fx–6300G

```
SOLVE:
"A="?→A:
"B="?→B:
"C="?→C:
"D="?→D:
"E="?→E:
"F="?→F:
AE-DB=0⇒Goto 1:
"X="◢(CE-BF)÷(AE-DB)◢
"Y="◢(AF-CD)÷(AE-DB):
Goto 2:
Lbl 1:
"NO UNIQUE SOLUTION":
Lbl 2
```

Graph Reflection Program (Section 9.2)

The program, shown in the marginal Programming note on page 579, will graph a function f and its reflection in the line $y = x$.

Note: On the *TI–82,* the "While" command may be entered through the "CTL" menu accessed by pressing the PRGM key. The "Pt-On(" command may be entered through the "POINTS" menu accessed by pressing the DRAW key. For additional keystroke instructions, see previous programs in this appendix. Keystroke sequences required for similar commands on other calculators will vary. Consult the user's manual for your calculator.

TI–80

```
PROGRAM:REFLECT
:47Xmin/63→Ymin
:47Xmax/63→Ymax
:Xscl→Yscl
:"X"→Y₂
:DispGraph
:(Xmax−Xmin)/62→I
:Xmin→X
:Lbl A
:Pt-On(Y₁,X)
:X + I→X
:If X > Xmax
:Stop
:Goto A
```

To use this program, enter the function in Y_1 and set a viewing rectangle.

TI–81

```
Prgm3: REFLECT
:2Xmin/3 → Ymin
:2Xmax/3 → Ymax
:Xscl → Yscl
:"X" → Y₂
:DispGraph
:(Xmax-Xmin)/95 → I
:Xmin → X
:Lbl 1
:PT-On(Y₁,X)
:X+I → X
:If X>Xmax
:End
:Goto 1
```

To use this program, enter the function in Y_1 and set a viewing rectangle.

TI–82

```
PROGRAM:REFLECT
:63Xmin/95 → Ymin
:63Xmax/95 → Ymax
:Xscl → Yscl
:"X" → Y₂
:DispGraph
:(Xmax-Xmin)/94 → N
:Xmin → X
:While X≤Xmax
:Pt-On(Y₁,X)
:X+N → X
:End
```

To use this program, enter the function in Y_1 and set a viewing rectangle.

TI–85

```
 PROGRAM:
:63×xMin/127→yMin
:63×xMax/127→yMax
:xScl→yScl
:y2=x
:DispG
:(xMax-xMin)/126→I
:xMin→x
:Lbl A
:PtOn(y1,x)
:x+I→x
:If x>xMax
:Stop
:Goto A
```

Casio fx–6300G

```
REFLECTION:
"-A TO A"◢
"A="?→A:
Range -A,A,1,-2A÷3,2A÷3,1:
-A→B:
Lbl 1:
B→X:
Prog 0:
Ans→Y:
Plot B,Y:
B+A÷24→B:
B≤A⇒Goto 1:
-A→B:
Lbl 2:
B→X:
Prog 0:
Ans→Y:
Plot Y,B:
B+A÷24→B:
B≤A⇒Goto 2:Graph Y=X
```

To use this program, write the function as Prog 0 and set a viewing rectangle.

Sharp EL 9200
Sharp EL 9300

```
Reflection
-----------REAL
Goto top
Label eqtn
Y=X^3+X+1
Return
Label rng
xmin=-10
xmax=10
xstp=(xmax-xmin)/10
ymin=2xmin/3
ymax=2xmax/3
ystp=xstp
Range xmin,xmax,xstp,ymin,ymax,ystp
Return
Label top
Gosub rng
Graph X
step=(xmax-xmin)/(94*2)
X=xmin
Label 1
Gosub eqtn
Plot X,Y
Plot Y,X
X=X+step
If X<=xmax Goto 1
End
```

To use this program, enter a function in X in the third line.

Casio fx–7700G

REFLECTION
"GRAPH -A TO A"
"A="?→A
Range -A,A,1,-2A÷3,2A÷3,1
Graph Y=f₁
-A→B
Lbl 1
B→X
Plot f₁,B
B+A÷32→B
B≤A⇒Goto 1:Graph Y=X

To use this program, enter the function in f_1 and set a viewing rectangle.

Casio fx–9700GE

REFLECTION↵
63Xmin÷127→A↵
63Xmax÷127→B↵
Xscl→C↵
Range ,,,A,B,C↵
(Xmax−Xmin)÷126→I↵
Xmax→M↵
Xmin→D↵
Graph Y=f₁↵
Lbl 1↵
D→X↵
Plot f₁,D↵
D+I→D↵
D≤M⇒Goto 1:Graph Y=X

To use either program, enter a function in f_1 and set a viewing rectangle.

Casio CFX–9800G

REFLECTION↵
63Xmin÷95→A↵
63Xmax÷95→B↵
Xscl→C↵
Range ,,,A,B,C↵
(Xmax−Xmin)÷94→I↵
Xmax→M↵
Xmin→D↵
Graph Y=f₁↵
Lbl 1↵
D→X↵
Plot f₁,D↵
D+I→D↵
D≤M⇒Goto 1:Graph Y=X

PHOTO CREDITS

CHAPTER P

Section P.1 *(page 9)*

7. (a) $1, 4, 6$

(b) $-10, 0, 1, 4, 6$

(c) $-10, -\frac{2}{3}, -\frac{1}{4}, 0, \frac{5}{8}, 1, 4, 6$

(d) $-\sqrt{5}, \sqrt{3}, 2\pi$

9. $-5, -4, -3, -2, -1, 0, 1, 2, 3$

11. $1, 3, 5, 7, 9$

13.

(a)

(b)

15. $2 < 5$

17. $-7 < -2$

19. $-\frac{2}{3} > -\frac{10}{3}$

21. $-1 < 3$ **23.** $-\frac{9}{2} < -2$ **25.** 6

27. 50 **29.** 35 **31.** -3.5 **33.** -25

35. $|-6| > |2|$ **37.** $-|-16.8| = -|16.8|$

39. $-14, 14$ **41.** $\frac{5}{4}, \frac{5}{4}$

43.

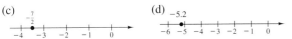

45.

47. $x < 0$ **49.** $p < 225$

51. $-5 < 2$ **53.** $\frac{1}{3} > \frac{1}{4}$

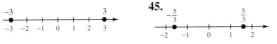

55. $-\frac{5}{8} < \frac{1}{2}$ **57.** $-\frac{5}{3} < -\frac{3}{2}$

59. 19 **61.** 8 **63.** 39 **65.** 18.6

67. -16 **69.** -85 **71.** π **73.** $|-2| = |2|$

75. $-|12.5| < |-25|$ **77.** $-34, 34$

79. $160, 160$ **81.** $\frac{3}{11}, \frac{3}{11}$

83.   **85.**

87. $x \geq 0$ **89.** $2 < z \leq 10$ **91.** True

93. False. $\frac{2}{3}$ is not an integer.

Section P.2 *(page 19)*

7. 45 **9.** 19 **11.** -22 **13.** $\frac{1}{2}$

15. $\frac{1}{24}$ **17.** $\frac{105}{8}$ **19.** -13.2 **21.** $5 + 5 + 5 + 5$

23. $4(-15)$ **25.** -30 **27.** $\frac{1}{2}$ **29.** 6

31. $-\frac{5}{2}$ **33.** $\frac{46}{17}$ **35.** $(-3)(-3)(-3)(-3)$

37. $(-5)^4$ **39.** 16 **41.** -16 **43.** 36 **45.** $-\frac{17}{6}$

47. 4.03 **49.** 145.96 **51.** $\$10,800$ **53.** -19

55. -21 **57.** -20 **59.** $\frac{5}{4}$ **61.** $\frac{1}{10}$ **63.** 60

65. -38.53 **67.** -28 **69.** $(-4) + (-4) + (-4)$

71. $6\left(\frac{1}{4}\right)$ **73.** 48 **75.** 32.13 **77.** $-\frac{6}{13}$

79. -3 **81.** $\frac{2}{5}$ **83.** $\frac{11}{12}$ **85.** $(4)(4)(4)$

87. $\left(-\frac{4}{5}\right)\left(-\frac{4}{5}\right)\left(-\frac{4}{5}\right)\left(-\frac{4}{5}\right)\left(-\frac{4}{5}\right)\left(-\frac{4}{5}\right)$ **89.** $\left(\frac{5}{8}\right)^4$

91. $(-4)^6$ **93.** 256 **95.** -256 **97.** $\frac{64}{125}$

99. 135 **101.** 57 **103.** 72.2 **105.** 161

107. $14,425$ **109.** 171.36 **111.** $\$2533.56$

113. $\frac{17}{180}$

115. (a) $\$5, \$8, -\$5, \$16$

(b) The stock gained $\$24$ in value during the week.

117. 6.125 cubic feet

119. False. The reciprocal of the integer 2 is $\frac{1}{2}$, which is not an integer.

121. True

Section P.3 *(page 28)*

7. Associative Property of Addition

9. Commutative Property of Multiplication

11. Additive Inverse Property

13. Multiplicative Identity Property

15. Distributive Property

17. Associative Property of Multiplication

19. $(3 \cdot 6)y$ **21.** $5 \cdot 6 + 5 \cdot z$ **23.** $x + 8$

25. (a) -10 **27.** (a) $-x - 1$ **29.** $x + (5 - 3)$
(b) $\frac{1}{10}$ (b) $\frac{1}{x + 1}$

31. $(3 \cdot 4)5$ **33.** $(20 \cdot 2) + (20 \cdot 5)$

35. $x \cdot (-2) + 6 \cdot (-2)$ **37.** 28

39. Given equation
Addition Property of Equality
Associative Property of Addition
Additive Inverse Property
Additive Identity Property

41. $a(b + c) = ab + ac$

43. Commutative Property of Addition

45. Associative Property of Multiplication

47. Multiplicative Identity Property

49. Additive Identity Property

51. Multiplicative Inverse Property

53. Additive Inverse Property

55. Commutative Property of Multiplication

57. Associative Property of Addition

59. Associative Property of Addition

61. Associative Property of Addition

63. $(-3)15$ **65.** $-3 \cdot 4 - (-3)x$

67. $-x + 25$ **69.** (a) 16 **71.** (a) $-6z$
(b) $-\frac{1}{16}$ (b) $\frac{1}{6z}$

73. $(3 + 8) - x$ **75.** $(6 \cdot 2)y$

77. $-8 \cdot 3 - (-8)5$ **79.** $5(3x) - 5 \cdot 4$

81. Given equation
Addition Property of Equality
Associative Property of Addition
Additive Inverse Property
Additive Identity Property

83.

$ac = bc, \ c \neq 0$ Given

$\frac{1}{c}(ac) = \frac{1}{c}(bc)$ Multiplication Property of Equality

$\frac{1}{c}(ca) = \frac{1}{c}(cb)$ Commutative Property of Multiplication

$\left(\frac{1}{c} \cdot c\right)a = \left(\frac{1}{c} \cdot c\right)$ Associative Property of Multiplication

$1 \cdot a = 1 \cdot b$ Multiplicative Inverse Property

$a = b$ Multiplicative Identity Property

85. $0.33 **87.** $1.63

89. $4 \odot 7 = 15 \neq 18 = 7 \odot 4$
$3 \odot (4 \odot 7) = 21 \neq 27 = (3 \odot 4) \odot 7$

Mid-Chapter Quiz *(page 31)*

1.

2.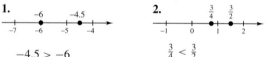

$-4.5 > -6$ $\qquad \frac{3}{4} < \frac{3}{2}$

3. 3.2 **4.** -5.75 **5.** 22 **6.** 3.75

7. 14 **8.** -22 **9.** $\frac{5}{2}$ **10.** $\frac{5}{2}$

11. $\frac{21}{8}$ **12.** $-\frac{3}{8}$ **13.** $\frac{7}{10}$ **14.** $-\frac{27}{8}$

15. 2.89 **16.** 208.51

17. (a) Distributive Property
(b) Additive Inverse Property

18. (a) Associative Property of Addition
(b) Multiplicative Identity Property

19. $3600 **20.** $\frac{7}{24}$

Section P.4 *(page 38)*

7. $10x, 5$　　**9.** $\dfrac{3}{t^2}, \dfrac{-4}{t}, 6$　　**11.** 5

13. Commutative Property of Addition

15. Associative Property of Multiplication

17. Multiplicative Inverse Property

19. Distributive Property

21. Multiplicative Identity Property

23. $(-5x)^4$　　**25.** x^{12}　　**27.** $6x^3y^4$　　**29.** $8x + 18y$

31. $4u^2v^2 + uv$　　**33.** $8x^2 + 4x - 12$　　**35.** $4(x - 2)$

37. $12x - 35$　　**39.** $-2b^2 + 4b - 36$　　**41.** -10

43. (a) 3　　**45.** (a) 0　　**47.** $\frac{1}{2}b(b - 3)$
　　(b) -10　　　　(b) Undefined　　　90

49. $-3y^2, 2y, -8$　　**51.** $x^2, -2.5x, -\dfrac{1}{x}$　　**53.** $-\frac{3}{4}$

55. Associative Property of Multiplication

57. Multiplicative Inverse Property

59. Associative Property of Addition

61. $5x + 5 \cdot 6$　　**63.** $6(x + 1)$　　**65.** $(xy)6$　　**67.** 3

69. 0　　**71.** $3 + (6 - 9)$　　**73.** $(-2x)(-2x)(-2x)$

75. $2x + 4x = 6x$　　**77.** $16x^2$　　**79.** $27y^6$

81. $-54u^5v^3$　　**83.** $7x$　　**85.** $8y$

87. $2\left(\dfrac{1}{x}\right) + 8$　　**89.** $-18y^2 + 3y + 6$　　**91.** $a(b + 2)$

93. $-3y^2 - 7y - 7$　　**95.** $9x^3 - 5$　　**97.** $2y^3 + y^2 + y$

99. (a) 7　　**101.** (a) 0
　　(b) 7　　　　　(b) $\frac{3}{10}$

103. (a) 13　　**105.** (a) 210
　　(b) -36　　　　(b) 140

107. (a)

x	-1	0	1	2	3	4
$2x - 5$	-7	-5	-3	-1	1	3

　　(b) 2
　　(c) $\frac{3}{4}$

109. $1480 million

111. $A = b_1h + \frac{1}{2}(b_2 - b_1)h$
$$= h\left[b_1 + \tfrac{1}{2}(b_2 - b_1)\right]$$
$$= h\left[b_1 + \tfrac{1}{2}b_2 - \tfrac{1}{2}b_1\right]$$
$$= h\left[\tfrac{1}{2}b_1 + \tfrac{1}{2}b_2\right]$$
$$= \frac{1}{2}h(b_1 + b_2) = \frac{h}{2}(b_1 + b_2)$$

113. 1440 square feet

Section P.5 *(page 48)*

7. $8 + n$　　**9.** $15 - 3n$　　**11.** $\dfrac{x}{6}$　　**13.** $\dfrac{3 + 4x}{8}$

15. A number is decreased by 5.

17. The quotient of a number and 2

19. A number is decreased by 2 and the difference is divided by 3.

21. $0.25n$　　**23.** $d = 55t$　　**25.** $0.45y$　　**27.** s^2

29. $n + (n + 1) + (n + 2)$　　**31.** $n + 5$　　**33.** $n - 6$

35. $\frac{1}{3}n$　　**37.** $0.30L$　　**39.** $\dfrac{x + 5}{10}$

41. $3x^2 - 4$　　**43.** $|n - 5|$

45. A number decreased by 2

47. The sum of three times a number x and 2

49. Eight times the difference of a number x and 5

51. The ratio of y to 8

53. The sum of a number and 10, all divided by 3

55. A number x times the sum of x and 7

57. $0.05x$　　**59.** $5m + 10n$　　**61.** $\dfrac{100}{r}$　　**63.** $0.0125I$

65. $0.80L$　　**67.** $8.25 + 0.60q$　　**69.** $69.50 + 32t$

71. $w + 6$　　**73.** Perimeter: $2(2w) + 2w = 6w$
　　　　　　　　　Area: $2w(w) = 2w^2$

75. Perimeter: $6 + 2x + 3 + x + 3 + x = 4x + 12$
　　　Area: $3x + 6x = 9x$

77. $n + 3n = 4n$　　**79.** $(2n + 1) + (2n + 3) = 4n + 4$

81. $l(l - 6) = l^2 - 6l$

83. $-3, 2, 7, 12, 17, 22$　　**85.** a
　　$5, 5, 5, 5, 5$

Review Exercises *(page 53)*

1. $<$ **3.** $<$ **5.** 7.2   **7.** -7.2 **9.** 11

11. 230   **13.** -38 **15.** -4200 **17.** 14

19. 0   **21.** $\frac{11}{21}$ **23.** $\frac{1}{6}$ **25.** $\frac{17}{8}$

27. $-\frac{1}{20}$ **29.** 2   **31.** $\frac{2}{5}$ **33.** 5

35. -216 **37.** $\frac{1}{12}$ **39.** 20

41. Additive Inverse Property

43. Distributive Property

45. Associative Property of Addition

47. Commutative Property of Multiplication

49. Multiplicative Inverse Property

51. $4s - 8t$ **53.** $-3y^2 + 10y$

55. $u - 3v$ **57.** $5x - 10$ **59.** $5x - y$

61. $20u$ **63.** $18b - 15a$ **65.** x^6

67. $-3x^3y^4$ **69.** $-64a^7$ **71.** 6910

73. (a) 0   **75.** (a) 15

    (b) -3 (b) 3

77. $P = 4l - 20$ **79.** $P = 9b + 5$

    $A = l(l - 10)$ $A = 4b^2$

81. 82 billion dollars   **83.** 1 hour   **85.** $644

87. 1   **89.** $200 - 3n$ **91.** $n^2 + 49$ **93.** $\left|\frac{n}{5}\right|$

95. The sum of twice a number y and 7

97. The difference of a number x and 5, all divided by 4

99. $0.18I$ **101.** $l(l - 5)$ **103.** $30p$

Chapter Test *(page 56)*

1. (a) $-\frac{5}{2} > -|-3|$ **2.** 11.9

    (b) $-\frac{2}{3} > -\frac{3}{2}$

3. -150 **4.** $-\frac{1}{2}$ **5.** $\frac{1}{6}$

6. $\frac{4}{27}$ **7.** $-\frac{27}{125}$ **8.** 15

9. (a) Associative Property of Multiplication

    (b) Multiplicative Inverse Property

10. $5(2x) - 5 \cdot 3$ **11.** $3x^4y^3$

12. $-2x^2 + 5x - 1$ **13.** a^2 **14.** $11t + 7$

15. (a) 4   **16.** $5n - 8$

    (b) -12

17. Perimeter: $3.2l$ **18.** $4n + 2$

    Area: $0.6l^2$

19. 16 feet   **20.** 640 cubic feet

CHAPTER 1

Section 1.1 *(page 65)*

7. (a) No   **9.** (a) No

    (b) Yes   (b) Yes

11. No solution   **13.** Identity   **15.** Linear

17. Not linear, because the unknown is in the denominator.

19. 7   **21.** 11

23. Given equation

    Addition Property of Equality

    Additive Inverse Property

    Multiplication Property of Equality

    Multiplicative Inverse Property

25. 4   **27.** $\frac{1}{3}$ **29.** No solution, because $7 \neq 0$.

31. 11   **33.** 1   **35.** $\frac{19}{10}$ **37.** $-\frac{10}{3}$ **39.** 12

41. 3.89   **43.** 125, 126   **45.** -240 **47.** $-\frac{1}{4}$

49. -27 **51.** (a) Yes   **53.** (a) Yes

            (b) No   (b) No

55. Given equation

    Addition Property of Equality

    Additive Inverse Property

    Multiplication Property of Equality

    Multiplicative Inverse Property

57. 7   **59.** $-\frac{2}{3}$ **61.** 2

63. No solution, because $-3 \neq 0$.

65. -2 **67.** 0   **69.** -3 **71.** -2

73. -3 **75.** $\frac{25}{3}$ **77.** 50   **79.** -20

81. 0   **83.** $-\frac{8}{31}$ **85.** 23   **87.** 30.28

89. (a)

t	1	1.5	2	3	4	5
Width	300	240	200	150	120	100
Length	300	360	400	450	480	500
Area	90,000	86,400	80,000	67,500	57,600	50,000

(b) Because the length is t times the width and the perimeter is fixed, as t gets larger the length gets larger and the area gets smaller. The maximum area occurs when the length and width are equal.

91. 2 seconds   **93.**

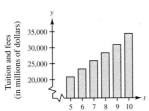

Tuition and fees (in millions of dollars)
Year (5 ↔ 1985)

95. 1976

Section 1.2 *(page 75)*

7. Perimeter $= 3 \times$ length of side
Equation: $129 = 3x$

	Percent	Parts out of 100	Decimal	Fraction
9.	30%	30	0.30	$\frac{3}{10}$
11.	7.5%	7.5	0.075	$\frac{3}{40}$
13.	12.5%	12.5	0.125	$\frac{1}{8}$

15. 87.5   **17.** 12,000   **19.** 33%

21. $420   **23.** 7.53%. Base number is smaller.

25. $\frac{3}{4}$ **27.** $\frac{85}{4}$ **29.** $0.0475   **31.** $0.0695

33. 32-ounce bag   **35.** 12-ounce tube   **37.** 6

39. 4   **41.** $1800   **43.** 3   **45.** $\frac{8}{3}$

47. 14   **49.** -19

51. (A number) $+ 30 = 82$
Equation: $x + 30 = 82$

53. Total bill $=$ charge for parts $+ 35$(number hours worked)
Equation: $380 = 275 + 35x$

55. $41\frac{2}{3}\%$ **57.** 69.36   **59.** 600

61. 350   **63.** 400   **65.** 175%   **67.** $247

69. 15.625%   **71.** 200   **73.** 475

75. 16.5 pounds, 20%   **77.** 1.175×10^{10} pounds

79. Taxes: 7.12%
Wages: 59.35%
Employee Benefits: 10.98%
Misc.: 12.46%
Insurance: 2.37%
Supplies: 2.67%
Utilities: 1.19%
Rent: 3.86%

81. $\frac{1}{8}$ **83.** $\frac{1}{25}$ **85.** $\frac{1}{50}$ **87.** 4   **89.** $\frac{15}{2}$

91. 16   **93.** 48   **95.** 2667   **97.** $\frac{56}{11}$ **99.** $\frac{45}{7}$ feet

Section 1.3 *(page 88)*

7. 10.71 million barrels

	Cost	Selling Price	Markup	Markup Rate
9.	$45.95	$64.33	$18.36	40%
11.	$22,250.00	$26,922.50	$4672.50	21%

	List Price	Sale Price	Discount	Discount Rate
13.	$49.50	$25.74	$23.76	48%
15.	$1145.00	$893.10	$251.90	22%

17. Department store: $223.96   **19.** 9%

21. (a)

Oats x	Corn $500 - x$	Price per Bushel of the Mixture
0	500	$3.00
100	400	$2.74
200	300	$2.48
300	200	$2.22
400	100	$1.96
500	0	$1.70

(b) Decreases
(c) Decreases
(d) Average of the prices of all components

23. Solution 1 = 50 gallons

Solution 2 = 50 gallons

25. Solution 1 = 8 quarts

Solution 2 = 16 quarts

27. 2275 miles   **29.** $\frac{100}{11}$ hours

31. $\frac{2000}{3}$ feet per second   **33.** 17.65 minutes

35. (a) 8 pages per minute

(b) $\frac{15}{4}$ units per hour

37. $R = \dfrac{E}{I}$ **39.** $r = \dfrac{A - P}{Pt}$ **41.** 11.25 inches

43. $2850   **45.** $\frac{11}{4}$ **47.** $\frac{5}{9}$ **49.** $y \le 45$

	Cost	Selling Price	Markup	Markup Rate
51.	$152.00	$250.80	$98.80	65%
53.	$225.00	$416.70	$191.70	85.2%

	List Price	Sale Price	Discount	Discount Rate
55.	$300.00	$111.00	$189.00	63%
57.	$95.00	$33.25	$61.75	65%

59. $54.15   **61.** 18.3%   **63.** 9 minutes, $2.06

65. Tax = $267   **67.** 30 15¢-stamps

Total bill = $4717   40 30¢-stamps

Amount financed = $3717

69. 700   **71.** $\frac{5}{6}$ gallon   **73.** 1440 miles   **75.** $V = 24$

77. 1988, $0.396 per hour per year   **79.** Bus drivers

81. 6.25 meters per second

Mid-Chapter Quiz *(page 93)*

1. 2   **2.** 2   **3.** 2   **4.** Identity

5. $\frac{28}{5}$ **6.** $\frac{12}{7}$ **7.** $\frac{33}{2}$ **8.** 6

9. −0.20   **10.** 1.41   **11.** $\frac{9}{20}$, 45%

12. 200   **13.** $0.1958 per ounce

14. 2000   **15.** Catalog at $1274.95   **16.** 7 hours

17. 25% solution: 40 gallons

50% solution: 10 gallons

18. 1.5 hours, 4.5 hours   **19.** 3.43 hours

20. 169 square inches

Section 1.4 *(page 102)*

7. (a) Solution   **9.** (a) Not a solution

(b) Not a solution   (b) Solution

(c) Solution   (c) Solution

(d) Not a solution   (d) Not a solution

11. d   **13.** a

15. $(-5, 3]$ **17.** $\left(0, \frac{3}{2}\right]$

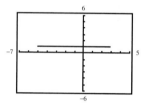   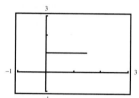

19. $\left(-\frac{15}{4}, -\frac{5}{2}\right)$

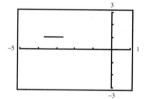

21. $x \le 2$ **23.** $x > 7$

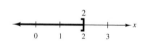   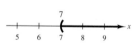

25. $y > 2$ **27.** $-\frac{3}{2} < x < \frac{9}{2}$

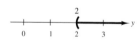   

29. $1 < x < 10$

31. $x \ge 0$ **33.** $n \le -2$ **35.** x is at least $\frac{5}{2}$.

37. $m < 25{,}357$ **39.** $2 \le d \le 8$ **41.** $-\frac{3}{4} > -5$

43. $\pi > -3$ **45.** $\frac{7}{2}$ **47.** −12

49. f   **51.** a   **53.** d

55. $(-\infty, -2]$

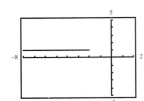

57. $(-2, 4]$

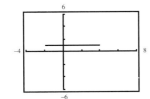

59. $[3, 9]$

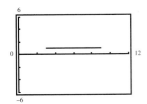

61. $x < \frac{11}{2}$

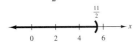

63. $x \le -4$

65. $y > 8$

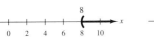

67. $x \ge 7$

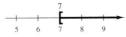

69. $x > -\frac{2}{3}$

71. $x > \frac{9}{2}$

73. $y \le -10$

75. $x \ge -12$

77. $\frac{5}{2} < x < 7$

79. $-\frac{3}{5} < x < -\frac{1}{5}$

81. $z \ge 2$ **83.** y is at least 3.

85. z is more than 0 and no more than π.

87. \$2600   **89.** Miami > New York

91. $12.50 < 8 + 0.75n$ **93.** $x \ge 31$
$n > 6$

95. $0 < t < 6.38$. If a portion of a minute is charged as a full minute, then $0 < t \le 6$.

97. 1983 through 1985   **99.** c

101. $3 \le n \le 9$ **103.** $n > 32$

Section 1.5 *(page 112)*

7. Not a solution   **9.** Solution

11. $x - 10 = 17$ **13.** $4x + 1 = \frac{1}{2}$
$x - 10 = -17$ $4x + 1 = -\frac{1}{2}$

15. $45, -45$ **17.** $11, -14$ **19.** No solution

21. $-\frac{39}{2}, \frac{15}{2}$ **23.** $\frac{3}{2}, -\frac{1}{4}$ **25.** $|2x + 3| = 5$

27. (a) Solution   **29.** (a) Not a solution
(b) Not a solution   (b) Solution
(c) Not a solution   (c) Solution
(d) Solution   (d) Not a solution

31. $-3 < y + 5 < 3$ **33.** $7 - 2h \ge 9$
$7 - 2h \le -9$

35. $-4 < y < 4$ **37.** $-6 < y < 2$

39. $x > 6$ or $x < -6$ **41.** $y \ge 2$ or $y \le -6$

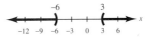

43. $x > 3$ or $x < -6$ **45.** $-5 < x < 35$

47. $z > 110$ or $z < -50$ **49.** $|t - 98.6| < 1$

51. 1.875   **53.** 150%   **55.** 80,000   **57.** 0

59. 21, 11   **61.** $\frac{16}{3}, 16$ **63.** No solution   **65.** $-\frac{11}{5}, \frac{17}{5}$

67. 18.75, -6.25 **69.** $-3, 7$ **71.** $2, -\frac{3}{2}$

73. (a) Not a solution   **75.** (a) Solution
(b) Solution   (b) Not a solution
(c) Solution   (c) Not a solution
(d) Not a solution   (d) Solution

77.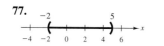

79.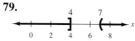

81. $-2 < y < 6$ **83.** $y \geq 4$ or $y \leq -4$

85. $y > 6$ or $y < -2$ **87.** $-7 < x < 7$

89. $-9 \leq y \leq 9$ **91.** $x > \frac{2}{3}$ or $x < -2$

93. $x > 7$ or $x < 3$ **95.** 3   **97.** $-82 \leq x \leq 78$

99. $t \geq \frac{5}{2}$ or $t \leq -\frac{15}{2}$ **101.** $s > 23$ or $s < -17$

103. d   **105.** b   **107.** $|x| \leq 2$ **109.** $|x - 10| < 3$

111. $|x - 19| < 3$ **113.**

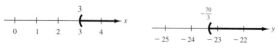

115. $|x| < 3$ **117.** $|x - 5| > 6$

Review Exercises *(page 117)*

1. (a) Not a solution   **3.** (a) Solution   **5.** 2
   (b) Solution   (b) Not a solution

7. 14   **9.** -8.2 **11.** $\frac{16}{3}$ **13.** 2   **15.** $\frac{20}{17}$

17. 3.38   **19.** 33.32   **21.** $87, 0.87, \frac{87}{100}$ **23.** 65

25. 3000   **27.** 125%   **29.** 10.2%, 9.6%, 9.9%, 13.2%

31. $\frac{4}{3}$ **33.** $\frac{7}{2}$ **35.** $1856.25   **37.** 50 feet

39. $27,166.25   **41.** 84.21%   **43.** Department store price

45. 30% solution: $3\frac{1}{3}$ liters   **47.** 6.35 hours
   60% solution: $6\frac{2}{3}$ liters

49. $\frac{18}{7} \approx 2.57$ hours   **51.** $340   **53.** $210,526.32

55. $20,000 at 8.5; $30,000 at 10%

57. 8 inches × 6 inches   **59.** $x = \frac{1}{2}(7y - 4)$

61. $h = \frac{V}{\pi r^2}$ **63.** $-4, 8$ **65.** $3, \frac{1}{2}$

67. $x > 3$ **69.** $y > -\frac{70}{3}$

   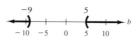

71. $-20 < x \leq 20$ **73.** $-16 < x < -1$

75. $-4 < x < 11$ **77.** $x > 7$ or $x < 1$

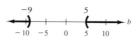

79. $b > 5$ or $b < -9$ **81.** $z \leq 10$

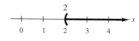

83. $7 \leq y < 14$ **85.** $|x - 3| < 2$

Chapter Test *(page 120)*

1. (a) Not a solution   **2.** 4   **3.** 4   **4.** 24
   (b) Solution

5. $\frac{19}{2}$ **6.** 864   **7.** 150%   **8.** $8000

9. 15-ounce package   **10.** $1466.67

11. $2\frac{1}{2}$ hours   **12.** $33\frac{1}{3}$ liters of 10% solution
                   $67\frac{2}{3}$ liters of 40% solution

13. 40 minutes   **14.** 13, 52   **15.** $2000

16. $x > 2$ **17.** $-7 < x \leq 1$

18. $1 \leq x \leq 5$ **19.** $x > -3$ or $x < -5$

   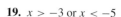

20. $t \geq 8$

CHAPTER 2

Section 2.1 *(page 131)*

7.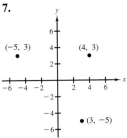

9.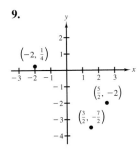

11. A: $(4, -2)$

B: $(-3, -2.5)$

C: $(3, 0.75)$

13.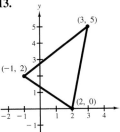

15.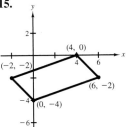

17. Quadrant III   **19.** Quadrants I and II

21. Quadrants II and IV   **23.** $(-5, 2)$ **25.** $(10, 0)$

27.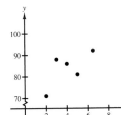

29. (a) Solution

(b) Not a solution

(c) Not a solution

(d) Solution

31.

x	-2	0	2	4	6
$y = 5x - 1$	-11	-1	9	19	29

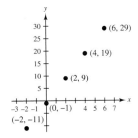

33.

x	-2	0	2	4	6
$y = 4x^2 + x - 2$	12	-2	16	66	148

35.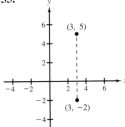

Distance: 7
Vertical line

37.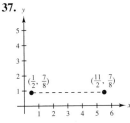

Distance: 5
Horizontal line

39. 5   **41.** 15   **43.** $(x, y) = (10, 2)$

6, 8, 10

45.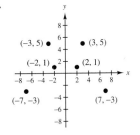

Reflection through the y-axis

47. $-\frac{1}{2} < x < \frac{1}{2}$ **49.** $-6 \leq x \leq 6$ **51.** \$29,018

53.

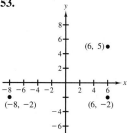

55.

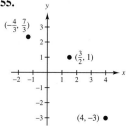

79.

x	-4	$\frac{2}{5}$	4	8	12
$y = -\frac{5}{2}x + 4$	14	3	-6	-16	-26

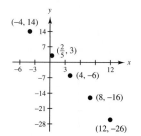

57. A: $(-2, 4)$
 B: $(0, -2)$
 C: $(4, -2)$

59.

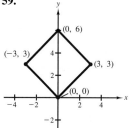

61.

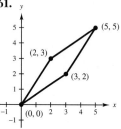

81.

x	100	150	200	250	300
$y = 28x + 3000$	5800	7200	8600	$10,000$	$11,400$

83.

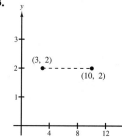

Distance: 7
Horizontal line

85.

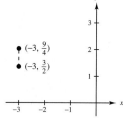

Distance: $\frac{3}{4}$
Vertical line

63. (a) Quadrant IV   **65.** Quadrants II and III
 (b) Quadrant II

67. Quadrants I and III   **69.** $(3, -4)$ **71.** $(-10, -10)$

73.

87. $(x, y) = (4, -4)$ **89.** $\sqrt{61}$ **91.** $\sqrt{29}$
 $8, 7, \sqrt{113}$

93. Not collinear   **95.** Collinear

97. $3 + \sqrt{26} + \sqrt{29} \approx 13.48$

75. (a) Solution   **77.** (a) Not a solution
 (b) Solution   (b) Solution
 (c) Solution   (c) Solution
 (d) Not a solution   (d) Not a solution

99.

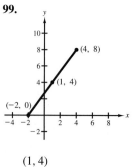

$(1, 4)$

101.

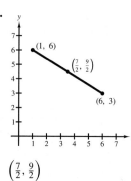

$\left(\frac{7}{2}, \frac{9}{2}\right)$

103. $(-3, -4) \implies (-1, 1)$ **105.** 18.55 feet
$(1, -3) \implies (3, 2)$
$(-2, -1) \implies (0, 4)$

Section 2.2   *(page 142)*

5. e   **7.** f   **9.** d

11.

x	-4	-2	0	2	4
y	11	7	3	-1	-5

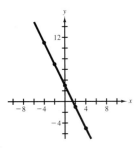

13.

x	± 2	-1	0	2	± 3
y	0	3	4	0	-5

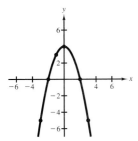

15.

x	-4	-2	0	2	4
y	0	2	4	2	0

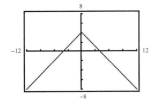

17.

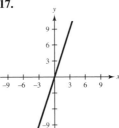

19.

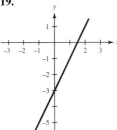

21.

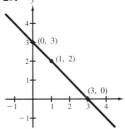

23.

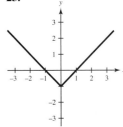

25. $(10, 0), (0, 5)$ **27.** $(\pm 5, 0), (0, -25)$

29.

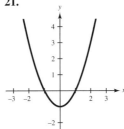

31.

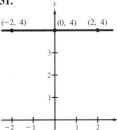

33.

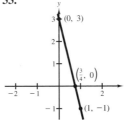

35.

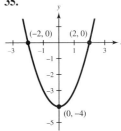

37.

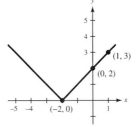

39. (a)

x	0	3	6	9	12
F	0	4	8	12	16

(b)

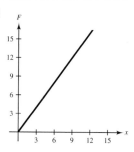

(c) Length doubles

41. a. It is difficult to assess the increase in sales.

43. Multiplicative Inverse Property

45. Distributive Property   **47.** 72 pounds

49.

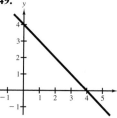

51.

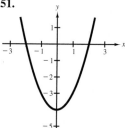

53.

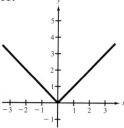

55.

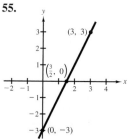

57.

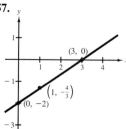

59.

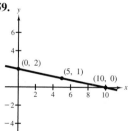

61.

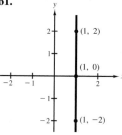

63.

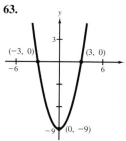

65.

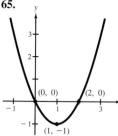

67.

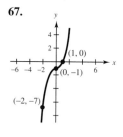

69.

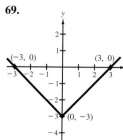

71.

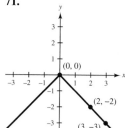

73. $(0, 3)$ **75.** $(2, 0), (0, 2)$

77.

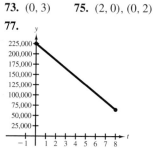

79.

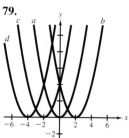

Horizontal translations

Section 2.3 *(page 151)*

5. c   **7.** d   **9.** a

11.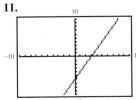

13.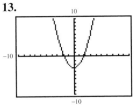

15.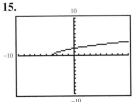

17.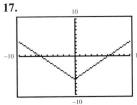

19.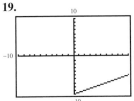

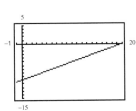

21.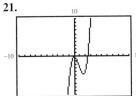

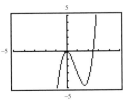

23.

| Xmin = −2 |
| Xmax = 8 |
| Xscl = 1 |
| Ymin = −1 |
| Ymax = 17 |
| Yscl = 1 |

25.

| Xmin = −5 |
| Xmax = 5 |
| Xscl = 1 |
| Ymin = −3 |
| Ymax = 6 |
| Yscl = 1 |

27.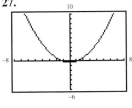

29.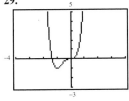

$(0, -0.25)$ $(-1, -1)$

31. Evaluate the expression for values of x near the lowest point on the graph.

33. $(-2.5, 0), (3, 0)$ **35.** $(0.68, 0)$

37. Identical graphs
Distributive Property

39. Identical graphs
Associative Property of Addition

41.

| Xmin = 20 |
| Xmax = 80 |
| Xscl = 5 |
| Ymin = 20 |
| Ymax = 500 |
| Yscl = 50 |

43. $y = 4 - 3x$ **45.** $y = \frac{1}{3}(4 - x^2)$

47.   **49.** $9.35 + 0.75q$

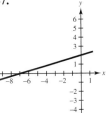

51.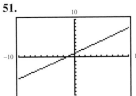

53.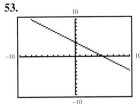

55.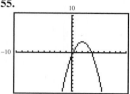

57.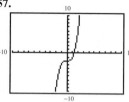

59.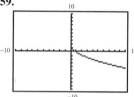

61.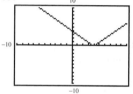

63.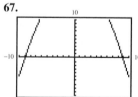

65.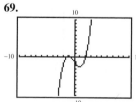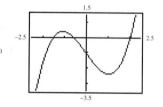

67.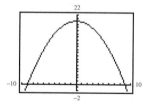

69.

71.

Xmin = −10
Xmax = 40
Xscl = 5
Ymin = −10
Ymax = 100
Yscl = 10

73.

Xmin = −2
Xmax = 10
Xscl = 1
Ymin = −5
Ymax = 5
Yscl = 1

75.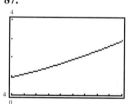

Intersect twice

77.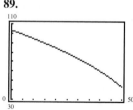

Intersect twice

79. $(-0.65, 0), (4.65, 0)$ **81.** No intercepts

83. $y = 2 - x$

85.

Xmin = 0
Xmax = 50
Xscl = 5
Ymin = −10
Ymax = 100
Yscl = 20

Utilizes the entire calculator display

87.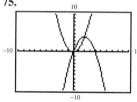

89.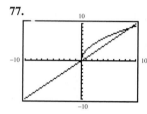

Model: $2.89

Mid-Chapter Quiz *(page 155)*

1.

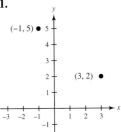

Distance: 5

2.

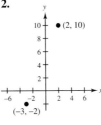

Distance: 13

3. Quadrants I and II   **4.** $(10, -3)$

5. (a) Not a solution   **6.** $(-8, 0), (0, 6)$

    (b) Solution

    (c) Solution

    (d) Solution

7.

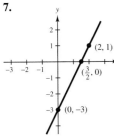

8.

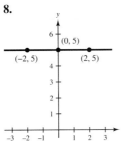

9.

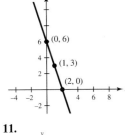

10.

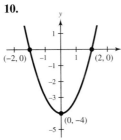

11.

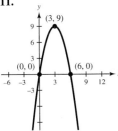

12.

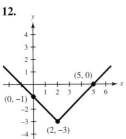

13.

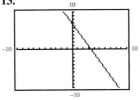

14.

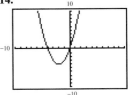

15.

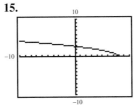

16.

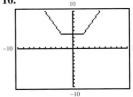

17.

Xmin = −2
Xmax = 8
Xscl = 1
Ymin = −4
Ymax = 24
Yscl = 4

18.

Xmin = −4
Xmax = 4
Xscl = 1
Ymin = −2
Ymax = 12
Yscl = 2

19. $(-1.23, 0), (2.23, 0)$

20. Associative Property of Addition

Section 2.4 *(page 164)*

7. $\frac{2}{3}$ **9.** -2 **11.** Undefined

13.

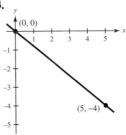

$m = -\frac{4}{5}$; falls

15.

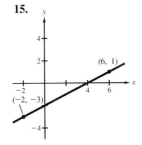

$m = \frac{1}{2}$; rises

17.

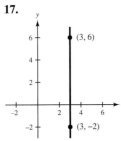

m is undefined; vertical

19.

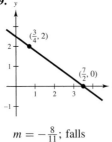

$m = -\frac{8}{11}$; falls

21. $(0, 2), (1, 2)$ **23.** $(4, -1), (5, 2)$

25.

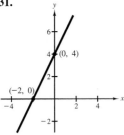

27.

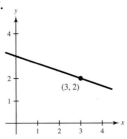

29. Line with slope m_2

31.

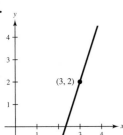

33.

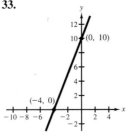

35. $y = 3x - 2$

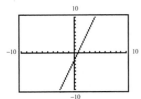

37. $y = -\frac{3}{2}x + 1$

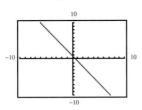

39. Parallel

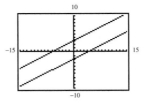

41. Perpendicular

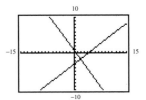

43. (a) $y = 8x + 140$

(b)

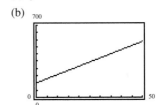

(c) $m = 8$, 8 feet

45. 15   **47.** -6 **49.** 2.4 hours   **51.** (a) L_3

(b) L_2

(c) L_1

53.

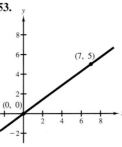

$m = \frac{5}{7}$; rises

55.

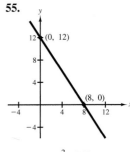

$m = -\frac{3}{2}$; falls

57.

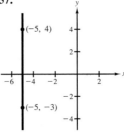

m is undefined; vertical

59.

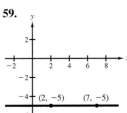

$m = 0$; horizontal

61.
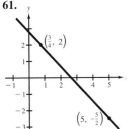
$m = -\frac{18}{17}$; falls

63.
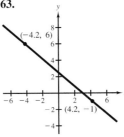
$m = -\frac{5}{6}$; falls

65.
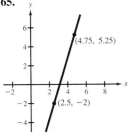
$m = \frac{29}{9}$; rises

67. $x = 1$ **69.** $y = -15$

71. $(1, 2), (2, 1)$ **73.** $(-2, 4), (1, 8)$ **75.** $(4, 0), (4, 1)$

77.
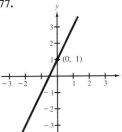

79.
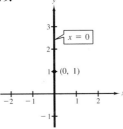

81.
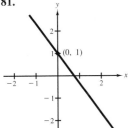

83. $y = -x$
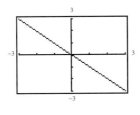

85. $y = \frac{1}{4}x + \frac{1}{2}$
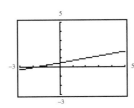

87. $y = -\frac{2}{3}x + 2$
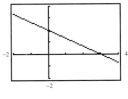

89. $y = 2$
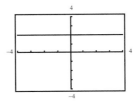

91. Perpendicular

93. Parallel   **95.** y_2 and y_3 are perpendicular.
97. The lines are parallel.   **99.** (a) 1991   **101.** 1989
                                          (b) 1992

Section 2.5 *(page 176)*

7. Domain: $\{-2, 0, 1\}$ **9.** Domain: $\{0, 2, 4, 5, 6\}$
   Range: $\{-1, 0, 1, 4\}$ Range: $\{-3, 0, 5, 8\}$

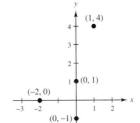
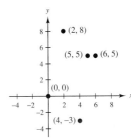

11. $(3, 150), (2, 100), (8, 400), (6, 300), \left(\frac{1}{2}, 25\right)$

13. (1990, Cincinnati), (1991, Minnesota), (1992, Toronto),
   (1993, Toronto)

15. Not a function   **17.** Not a function

19. (a) Function   **21.** Function
   (b) Not a function
   (c) Function
   (d) Function

23. Not a function   **25.** (a) $3(2) + 5 = 11$

(b) $3(-2) + 5 = -1$

(c) $3(k) + 5$

(d) $3(k + 1) + 5 = 3k + 8$

27. (a) 29   **29.** (a) 2   **31.** (a) 2

(b) 11   (b) -2 (b) $\dfrac{2(x - 6)}{x}$

(c) $12a - 2$ (c) 10

(d) $12 + 5$ (d) -8

33. All real values of x such that $x \neq 3$

35. All real values of x such that $x \geq \frac{1}{2}$

37. $P(x) = 4x$ **39.** $V(x) = x(24 - 2x)^2$

$= 4x(12 - x)^2$

41. $1 - 6x$ **43.** $22 - 3x$ **45.** $\frac{1}{4}, \frac{1}{5}, \frac{20}{9} \approx 2.2$ hours

47. Function   **49.** Not a function   **51.** Function

53. Not a function   **55.** Function   **57.** Function

59. Function   **61.** Not a function

63. (a) 8   **65.** (a) 2

(b) $\frac{2}{9}$ (b) 3

(c) $2y^2$ (c) $\sqrt{\dfrac{31}{3}}$

(d) 26   (d) $\sqrt{5(z + 1)}$

67. (a) 0   **69.** (a) -3.84 **71.** (a) 0

(b) $-\frac{3}{2}$ (b) -4.2 (b) $\frac{7}{4}$

(c) $-\frac{5}{2}$ (c) 3

(d) $\dfrac{3(x + 4)}{x - 1}$ (d) 0

73. Domain: $\{0, 2, 4, 6\}$

Range: $\{0, 1, 8, 27\}$

75. Domain: all real numbers r such that $r \geq 0$

Range: all real numbers C such that $C \geq 0$

77. All real values of x

79. All real values of t such that $t \neq 0$ and $t \neq -2$

81. All real values of x such that $x \geq -4$

83. All real values of t **85.** $V(x) = x^3$

87. $C = 1.95x + 8000$

89. (a) 10,680 pounds   (b) 8010 pounds

91. (a) Correct   (b) Not correct

Math Matters   *(page 180)*

At first, it might seem that the answer is for the spider to walk straight down the wall, across the floor, and up the other wall. The total length of this path is 42 feet. There is, however, a shorter path. To see this, imagine that the sides of the room are unfolded as shown in the figure. By imposing a rectangular coordinate system on the unfolded walls, we see that the spider starts at the point $(0, -1)$ and wants to travel to the point $(24, 31)$. The distance between these two points is $\sqrt{24^2 + 32^2} = 40$. Thus, the shortest distance is not the 42-foot path, but the diagonal path, running across five of the six surfaces of the room, as shown in the figure.

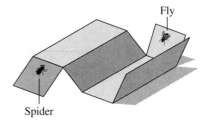

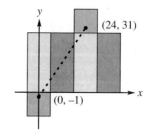

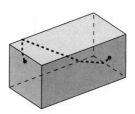

Section 2.6 *(page 176)*

7.

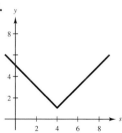

Domain: $-\infty < x < \infty$
Range: $-\infty < y \leq 1$

9.

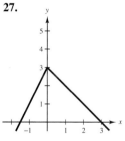

Domain: $2 \leq x < \infty$
Range: $0 \leq y < \infty$

11. Function   **13.** Not a function

15. Function. For each value of y there corresponds one value of x.

17. b   **19.** a

21.

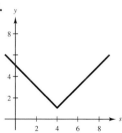

Domain: $-\infty < x < \infty$
Range: $0 \leq y < \infty$

23.

Domain: $-\infty < x < \infty$
Range: $-\infty < y \leq 0$

25.

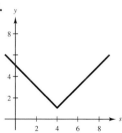

Domain: $-\infty < s < \infty$
Range: $1 \leq y < \infty$

27.

Domain: $-\infty < x < \infty$
Range: $-\infty < y \leq 3$

29. b   **31.** Vertical shift two units upward

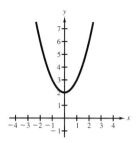

33. Horizontal shift two units to the left

35. Reflection in the x-axis

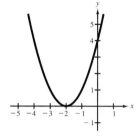 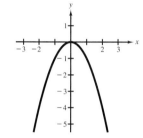

37. $y = -\sqrt{x}$ **39.** $y = \sqrt{x+2}$

41. $y = \sqrt{-x}$ **43.** (a)

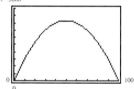

(b) $x = 50$. The figure is a square.

45. -4 **47.** 6   **49.** $0 < t < 23$

51. Function   **53.** Not a function

55.

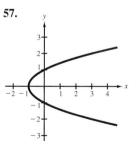

57.

Function

Not a function

59.

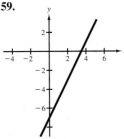

Domain: $-\infty < x < \infty$
Range: $-\infty < y < \infty$

61.

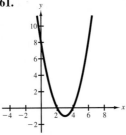

Domain: $-\infty < x < \infty$
Range: $-1 \leq y < \infty$

75.

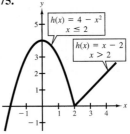

Domain: $-\infty < x < \infty$
Range: $-\infty \leq y < \infty$

77. Vertical shift
three units upward

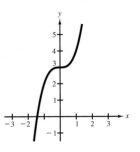

63.

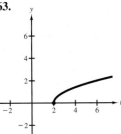

Domain: $2 \leq t < \infty$
Range: $0 \leq y < \infty$

65.

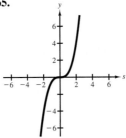

Domain: $-\infty < s < \infty$
Range: $-\infty < y < \infty$

79. Horizontal shift
three units to the right

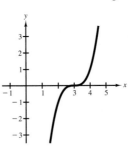

81. Reflection in the x-axis
followed by a horizontal
shift one unit to the right
followed by a vertical
shift two units upward

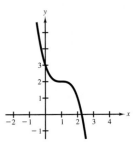

67.

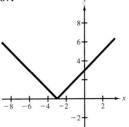

Domain: $-\infty < x < \infty$
Range: $0 \leq y < \infty$

69.

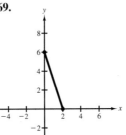

Domain: $0 \leq x \leq 2$
Range: $0 \leq y \leq 6$

83. Horizontal shift
five units to the right

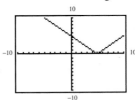

85. Vertical shift
five units downward

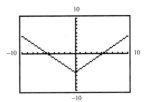

71.

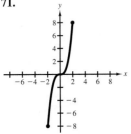

Domain: $-2 \leq x \leq 2$
Range: $-8 \leq y \leq 8$

73.

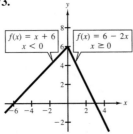

Domain: $-\infty < x < \infty$
Range: $-\infty < y \leq 6$

87. Reflection in
the x-axis

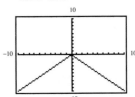

89. $h(x) = (x + 3)^2$

91. $h(x) = -x^2$ **93.** $h(x) = -(x+3)^2$

95. $h(x) = -x^2 + 2$

97. (a)

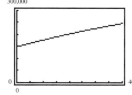

(b) 1970

(c)
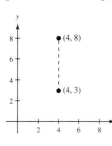

Review Exercises *(page 193)*

1.
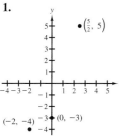

3.
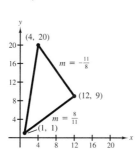

5. Quadrant IV   **7.** Quadrants I and IV

9.
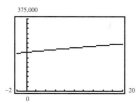

Distance: 5

11.
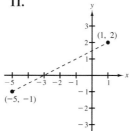

Distance: $3\sqrt{5}$

13. (a) Solution

(b) Not a solution

(c) Not a solution

(d) Solution

15.
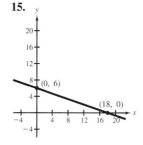

17.
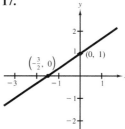

19.
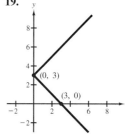

21. $\frac{2}{7}$ **23.** 0   **25.** $-\frac{3}{4}$ **27.** $\frac{3}{2}$ **29.** $(1, -1), (0, 2)$

31. $(7, 6), (11, 11)$ **33.** $(3, 0), (3, 5)$

35. $y = \frac{5}{2}x - 2$ **37.** $y = -\frac{1}{2}x + 1$

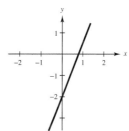

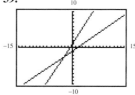

39.
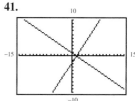

Neither

41.

Perpendicular

43. Not a function   **45.** Function

47. Intercept: $(0, 0)$ **49.** Intercept: $(0, 0), (3, 0)$

Not a function   Function

51. (a) 29
(b) 3
(c) $\dfrac{36 - 5t}{2}$
(d) $4 - \dfrac{5}{2}(x + h)$

53. (a) $2\sqrt{2}$
(b) 0
(c) $\dfrac{2\sqrt{3}}{3}$
(d) $\sqrt{5 - 5z}$

55. (a) -3
(b) 2
(c) 0
(d) -7

57. (a) -2
(b) $\dfrac{2(6 - x)}{x}$

59. All real values of x

61. All real values of x such that $x \le \dfrac{5}{2}$

63. c

65.

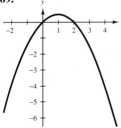

67.

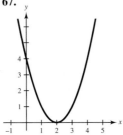

69.

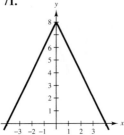

71.

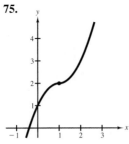

73.

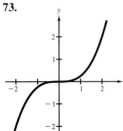

75.

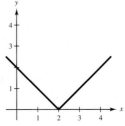

77. Reflection in the x-axis

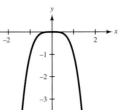

79. Horizontal shift one unit to the right

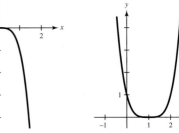

81. (a)

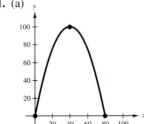

(b) 100 feet
(c) 80 feet

83. (a) $k = \dfrac{1}{8}$
(b) 1953.1 kw

Chapter Test *(page 196)*

1. Quadrant IV

2.

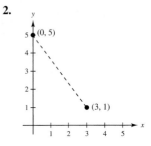

Distance: 5

3. $(-1, 0),\ (0, -3)$

4.

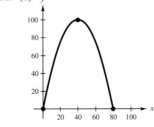

5. $-\frac{2}{3}$ **6.** Undefined

7.

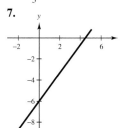

8.

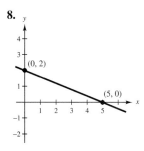

9. $y = -\frac{5}{3}x + 3$ **10.** Not a function

$\frac{3}{5}$

11. Not a function   **12.** Function

13. -2 **14.** 7   **15.** $\dfrac{x+2}{x-1}$

16. All real values of t such that $t \le 9$

17. All real values of x such that $x \ne 4$

18.

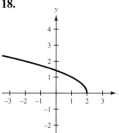

19.

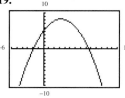

$(0, 5), (-1.94, 0), (8.60, 0)$

20. (a) $y = |x - 2|$

(b) $y = |x| - 2$

(c) $y = 2 - |x|$

CHAPTER 3

Section 3.1   *(page 203)*

7. $3x^2 - x + 2, 2, 3$ **9.** $-3y^4 + 5, 4, -3$

11. Binomial   **13.** Trinomial   **15.** Monomial

17. $7x^2 + 3$ **19.** $-2y^4 - 5y + 4$ **21.** $2x^2 - 3x$

23. 4   **25.** $x^2 - 3x + 2$ **27.** $7x^3 + 2x$

29. $x^2 - 2x + 5$ **31.** $2x^3 - 2x + 3$

33. $3v^2 + 78v + 27$ **35.**

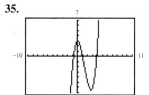

$y_1 = y_2$

37. (a) 16   **39.** (a) 32   **41.** \$15,000   **43.** 3

(b) 0   (b) 28

(c) -16 (c) 16

(d) 16   (d) -4

45. (a)

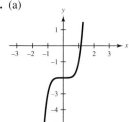

(b)

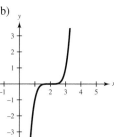

Vertical shift two   Horizontal shift 2
units downward   units to the right

(c)

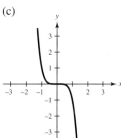

(d)

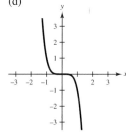

Reflection in   Reflection in
the x-axis   the y-axis

47. $10x - 4, 1, 10$ **49.** $-3x^3 - 2x^2 - 3, 3, -3$

51. $-4, 0, -4$ **53.** $-16t^2 + v_0t, 2, -16$ **55.** $3x^3$

57. $8x^2 - 9$ **59.** 13   **61.** $-3x^2 + x + 3$

63. $6x^2 - 7x + 8$ **65.** $-2x^2 + 15$ **67.** $-2x^4 - x^2 - 9$

69. $2p^2 - 2p - 5$ **71.** $-2y^3$ **73.** $q^2 + 15$

75. $-2x^3 + x^2 + 2x$ **77.** $4x^3 + x + 4$

79. $-4x^3 - 2x + 13$ **81.** $2x^2 + 9x - 11$

83. $7x^3 + 22x^2 + 4$ **85.** $29s + 8$ **87.** $3t^2 + 29$

89. (a) -28 **91.** (a) 50 **93.** Dropped, 100 feet
(b) -16 (b) 146
(c) -16 (c) 114
(d) $-\frac{77}{2}$ (d) 50

95. Thrown downward, 50 feet

97. 224 feet, 216 feet, 176 feet **99.** $36x$

101. (a) 1975: 28.225 gallons; 1985: 25.025 gallons

(b)

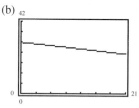

Decreasing

Section 3.2 *(page 213)*

7. $t^{3+4} = t^7$ **9.** $\dfrac{y^4}{5^4} = \dfrac{y^4}{625}$ **11.** $x^{6-4} = x^2$

13. $3^{7-3}x^{5-3} = 81x^2$ **15.** (a) $-3x^8$ (b) $9x^7$

17. (a) $60x^2y^3$ **19.** (a) $\frac{4}{3}m^4n^3$ **21.** (a) $\frac{5}{2}u^4v$
(b) $-120x^2y^4$ (b) $3m^2n^3$ (b) $-\frac{1}{2}u^4v$

23. $16a^3$ **25.** $-10x - 6x^3 + 14x^4$ **27.** $-a^2 + 19a$

29. $2u^3 + 13u^2 + 11u - 20$ **31.** $x^3 - 5x^2 + 10x - 6$

33. $7x^3 + 7x^2 - 33x + 27$ **35.** $u^3 - u^2 + u - 6$

37. $x^2 - 4$ **39.** $x^2 + 10x + 25$

41. $x^2 + 2xy + y^2 + 4x + 4y + 4$ **43.** $8x^2 + 26x$

45. $A = (x + a)(x + b) = x^2 + ax + bx + ab$

47.

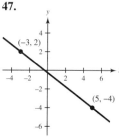

$m = -\frac{3}{4}$

49.

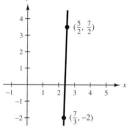

$m = 33$

51. $d = 48t$ **53.** $(-5)^5x^5 = -3125x^5$

55. $3^{4-1}x^{5-2} = 27x^3$ **57.** (a) $-125z^3$ (b) $25z^2$

59. (a) $625z^4$ **61.** (a) $\frac{4}{3}xy$ **63.** (a) $\dfrac{9x^2}{16y^2}$
(b) $-3125z^8$ (b) $\frac{3}{4}xy$
(b) $\dfrac{125u^3}{27v^3}$

65. $20x^3$ **67.** $10y - 2y^2$ **69.** $8x^3 - 12x^2 + 20x$

71. $75x^3 + 30x^2$ **73.** $48y^2 + 32y - 3$

75. $6x^2 + 7xy + 2y^2$ **77.** $2st - 6t^2$

79. $x^4 - 2x^3 - 3x^2 + 8x - 4$ **81.** $t^4 - t^2 + 4t - 4$

83. $-2x^3 + 3x^2 - 1$ **85.** $x^5 - 3x^3 - 4x$ **87.** $4 - 49y^2$

89. $4x^2 - \frac{1}{16}$ **91.** $0.04t^2 - 0.25$ **93.** $x^2 + 20x + 100$

95. $4x^2 - 28xy + 49y^2$ **97.** $u^2 - 2uv + v^2 + 6u - 6v + 9$

99. $12t$ **101.** $x^3 + 9x^2 + 27x + 27$

103. $u^3 + 3u^2v + 3uv^2 + v^3$

105.

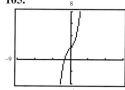

$y_1 = y_2$

107.

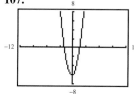

$y_1 = y_2$

109. $3x^2 - 2x + 6$ **111.** (a) $t^2 - 8t + 15$
(b) $h^2 + 2h$

113. (a) $5w$ **115.** $1000 + 2000r + 1000r^2$
(b) $\frac{3}{2}w^2$

117. (a) $x^2 - 1$
(b) $x^3 - 1$
(c) $x^4 - 1, \quad x^5 - 1$

Math Matters *(page 216)*

Section 3.3 *(page 223)*

7. 6   **9.** $14b^2$ **11.** $8(z-1)$ **13.** $7u(3u-2)$

15. $3xy(5y - x + 3)$ **17.** $-(x - 10)$

19. $-(2y^2 + 3y - y)$ **21.** $6x + 5$ **23.** $2x + y$

25. $(y - 3)(2y + 5)$ **27.** $a(a + 6)(1 - a)$

29. $(x + 25)(x + 1)$ **31.** $(a - 4)(a^2 + 2)$

33. $(x + 8)(x - 8)$ **35.** $(2z + y)(2z - y)$

37. $(x + 3)(x - 5)$ **39.** $(x - 2)(x^2 + 2x + 4)$

41. $(x + 4y)(x^2 - 4xy + 16y^2)$ **43.** $3x^2(x + 10)(x - 10)$

45. $2(a + 2)(a - 2)(a^2 + 4)$ **47.** $p = 800 - 0.25x$

49. $\frac{7}{24}x + 8$ **51.** $-30x^2 + 23x + 3$ **53.** $|x| < 5$

55. $3x$ **57.** $6z^2$ **59.** $21(x + 8)^2$ **61.** $2(2u + 5)$

63. $6(4x^2 - 3)$ **65.** $x(2x + 1)$ **67.** Prime

69. $3y(x^2y - 5)$ **71.** $4(7x^2 + 4x - 2)$

73. $x^2(14x^2 + 21x + 9)$ **75.** $-7(2x - 1)$

77. $-(x^2 - 4x - 5)$ **79.** $8t + 9$ **81.** $10y - 3$

83. $(x + 2)(5x - 3)$ **85.** $-(b + 10)(c + 6)(c + 7)$

87. $(y - 6)(y + 2)$ **89.** $(x + 2)(x^2 + 1)$

91. $(z + 3)(z^3 - 2)$ **93.** $(4y + 3)(4y - 3)$

95. $(15 + 3y)(15 - 3y)$ **97.** $\left(u + \frac{1}{4}\right)\left(u - \frac{1}{4}\right)$

99. $[9 + (z + 5)][9 - (z + 5)]$ **101.** $(a + b)(a^2 - ab + b^2)$

103. $(2t - 3)(4t^2 + 6t + 9)$ **105.** $(3u + 1)(9u^2 - 3u + 1)$

107. $2(2 - 5x)(2 + 5x)$ **109.** $(y - 3)(y + 3)(y^2 + 9)$

111. $2(x - 3)(x^2 + 3x + 9)$

113.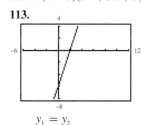

$y_1 = y_2$

115.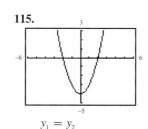

$y_1 = y_2$

117. 6399

119. $x^2(3x + 4) - (3x + 4) = (x - 1)(x + 1)(3x + 4)$

$3x(x^2 - 1) + 4(x^2 - 1) = (x - 1)(x + 1)(3x + 4)$

121. $P(1 + rt)$ **123.** Width $= 45 - l$

125. $\pi(R - r)(R + r)$

Mid-Chapter Quiz *(page 226)*

1. $4, -2$

2. The exponent of $x^{1/2}$ is not an integer.

3. $3t^3 + 3t^2 + 7$ **4.** $7y^2 - 5y$ **5.** $9x^3 - 4x^2 + 1$

6. $2u^2 - u + 1$ **7.** $10n^5$ **8.** $-\frac{3}{4}x$

9. $28y - 21y^2$ **10.** $2z^2 + 3z - 35$ **11.** $36r^2 - 25$

12. $4x^2 - 12x + 9$ **13.** $x^3 + 1$ **14.** $-12v$

15. $7a(4a - 3)$ **16.** $(5 + 2x)(5 - 2x)$

17. $(z + 3)^2(z - 3)$ **18.** $4(y - 2)(y^2 + 2y + 4)$

19. $(5x + 10)(2x + 1)$ $(5x - 10)(2x - 1)$

$(5x + 1)(2x + 10)$ $(5x - 1)(2x - 10)$

$(5x + 2)(2x + 5)$ $(5x - 2)(2x - 5)$

$(5x + 5)(2x + 2)$ $(5x - 5)(2x - 2)$

20. $\frac{1}{2}(x + 2)^2 - \frac{1}{2}x^2 = 2(x + 1)$

Section 3.4 *(page 235)*

7. $(x + 2)^2$ **9.** $(5y - 1)^2$ **11.** ± 18 **13.** ± 12

15. 16   **17.** 9   **19.** $x + 1$ **21.** $y - 5$

23. $(x + 3)(x + 1)$ **25.** $(t - 7)(t + 3)$

27. $(x - 7y)(x + 5y)$ **29.** $\pm 12, \pm 36$ **31.** $8, -16$

33. $5x + 3$ **35.** $(5x + 2)(x + 1)$ **37.** $(2x - 3)(3x - 1)$

39. Prime   **41.** $(2u + v)(3u - 4v)$ **43.** $-(x + 2)(2x - 3)$

45. $(1 + 4x)(1 - 15x)$ **47.** $(x + 2)(3x + 4)$

49. $(3x - 1)(5x - 2)$ **51.** $(15, 13)$ **53.** $(6.84, -1)$

55. $8\frac{3}{4}$ **57.** $(a - 6)^2$ **59.** $(b + 2)^2$

61. $(u + 4v)^2$ **63.** 14   **65.** $x - 6$ **67.** $z - 2$

69. $(x - 3)(x - 2)$ **71.** $(y + 10)(y - 3)$

73. $(x - 8)(x - 12)$ **75.** $(x + 12)(x + 18)$

77. $\pm 9, \pm 11, \pm 19$ **79.** $\pm 4, \pm 20$ **81.** $2, -18$

83. $5a - 3$ **85.** $2y - 9$ **87.** $(3x + 1)(x + 1)$

89. $(2x - 3)(x - 3)$ **91.** $(3b - 1)(2b + 7)$

93. $(4x - 3y)(6x + y)$ **95.** Prime   **97.** $(2x - 1)(3x + 2)$

99. $(3x^3)(x - 4)$ **101.** $2t(5t - 9)(t + 2)$

103. $(3 - z)(9 + z)$ **105.** $2(3x - 1)(9x^2 + 3x + 1)$

107.

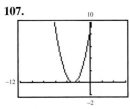

$y_1 = y_2$

109.

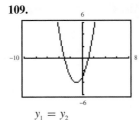

$y_1 = y_2$

111. 2704   **113.** $(x + 3)(x + 1)$

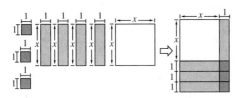

115. c   **117.** b

Section 3.5 *(page 245)*

5. 0, 8   **7.** 2, −4   **9.** $-4, -\frac{1}{2}, 3$ **11.** 0, 5

13. ±5   **15.** −5, 1   **17.** 4   **19.** −2, 5

21. −2, 10   **23.** $-\frac{1}{2}, 7$ **25.** 0, 7, 12   **27.** ±2

29. ±3   **31.** $(-3, 0), (3, 0)$ **33.** $(-1, 0), (3, 0)$

35. 0, 6   **37.** $-4, \frac{3}{2}$

39. (a) length $= 5 - 2x$
width $= 4 - 2x$
height $= x$
volume $=$ (length)(width)(height)
$V = (5 - 2x)(4 - 2x)(x)$

(b) $0, 2, \frac{5}{2}$
$0 < x < 2$

(c)

x	0.25	0.50	0.75	1.00	1.25	1.50	1.75
V	3.94	6	6.56	6	4.69	3	1.31

(d) $\frac{3}{2}$

(e) 0.74

41. 11, 12   **43.** 10   **45.** $-2(4x + 5)$ **47.** $1400

49. 3, −10   **51.** $-\frac{5}{2}, -\frac{1}{3}$ **53.** $0, \frac{3}{2}, -\frac{25}{2}$ **55.** ±4

57. 6, −2   **59.** $0, -\frac{5}{3}$ **61.** 5, −2   **63.** −8

65. $\frac{3}{2}$ **67.** $-\frac{1}{7}, -\frac{1}{2}$ **69.** −12, 6   **71.** $-\frac{7}{2}, 5$

73. −7, 0   **75.** −6, 5   **77.** −10, 2   **79.** $0, -\frac{1}{3}, \frac{1}{2}$

81. ±4, 25   **83.** ±3, −2   **85.** 2, 6   **87.** 0, ±2

89. 15   **91.** 15 feet × 22 feet   **93.** $b = 8, h = 12$

95. 20 inches × 20 inches   **97.** 10 units, 20 units

99. (a) −8, 0, 10, 22, 36, 52

(b) P increases by $2(x + 1)$

(c) 9

101. 0, 1   **103.** (a) $x = -\frac{20}{3}, -2$

(b) $x = -\frac{5}{4}, -\frac{1}{2}$

Review Exercises *(page 251)*

1. A variable cannot be in the denominator of a term.

3. A variable cannot be within absolute value signs.

5. $-4(x - 2) = -4x + 8$ **7.** $(x + 3)^2 = x^2 + 6x + 9$

9. $6x - x^2$ **11.** $-2t^2 + 5t + 10$ **13.** $-9x^3 + 9x - 4$

15. $6y^2 - 2y + 15$ **17.** $-216z^3$ **19.** $4u^7 v^3$

21. $8u^2 v^2$ **23.** $12x^2$ **25.** $-8x^4 - 32x^3$

27. $15x^2 - 11x - 12$ **29.** $4x^3 - 5x + 6$

31. $u^2 - 8u + 7$ **33.** $16x^2 - 56x + 49$

35. $25u^2 - 64$ **37.** $6z^2 - z - 15$

39. $u^2 - v^2 - 6u + 9$ **41.** $3x^2(2 + 5x)$

43. $-14(x + 5)(5x + 23)$ **45.** $(v + 1)(v - 1)(v - 2)$

47. Prime   **49.** $(3a - 10)(3a + 10)$ **51.** $(u - 3)(u + 15)$

53. $(u - 1)(u^2 + u + 1)$ **55.** $(2x + 3)(4x^2 - 6x + 9)$

57. $(x - 20)^2$ **59.** $(2s + 10)^2$ **61.** $(x + 7)(x - 5)$

63. $(3x + 2)(6x + 5)$ **65.** $4a(1 - 4a)(1 + 4a)$

67. $4(2x - 3)(2x - 1)$ **69.** $(2x + 1)(4x^2 - 2x + 1)$

71. $(2u - 7)^2$ **73.** Prime   **75.** Prime

77.   **79.** ±6   **81.** 6, 10

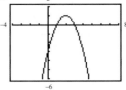

$y_1 = y_2$

83. 0, 3   **85.** 10, −10   **87.** 10, 15   **89.** $-\frac{4}{3}, 2$

91. −4, 9   **93.** 3, 7   **95.** $10p^3 - 20p^4 + 10p^5$

97. (a) $x^2 - y^2$

(b) Resulting rectangle $= (x + y)(x - y)$.

99. 13, 15   **101.** 100 feet × 300 feet

Chapter Test (page 253)

1. Degree: 3; leading coefficient: -5.2

2. The variable appears in the denominator.

3. (a) $6a^2 - 3a$ **4.** (a) $8x^2 - 4x + 10$

  (b) $-2y^2 - 2y$ (b) $11t + 7$

5. (a) $-24u^6v^5$ (b) $60x^3y^2$ **6.** (a) $\frac{1}{8}y^3$ (b) $\frac{27}{2}x^6y^4$

7. (a) $-3x^2 + 12x$ **8.** (a) $3x^2 - 6x + 3$

  (b) $2x^2 + 7xy - 15y^2$ (b) $6s^3 - 17s^2 + 26s - 21$

9. (a) $16x^2 - 24x + 9$ **10.** $6y(3y - 2)$

  (b) $16 - a^2 - 2ab - b^2$

11. $\left(v - \frac{4}{3}\right)\left(v + \frac{4}{3}\right)$ **12.** $(x + 2)(x - 2)(x - 3)$

13. $(3u - 1)^2$ **14.** $2(x - 5)(3x + 2)$

15. $(x + 3)(x^2 - 3x + 9)$ **16.** $1, -5$ **17.** $3, -\frac{4}{3}$

18. $x^2 + 26x$ **19.** 6 centimeters $\times$ 9 centimeters

20. 2 seconds

Cumulative Test: Chapters P–3 (page 254)

1. $-\frac{10}{27}$ **2.** $3n - 8$

3. (a) $8a^8b^7$ **4.** (a) $t^2 - 9t$

  (b) $2x^3 - 11x$ (b) $x^2 - 2xy + y^2 + 4x - 4y + 4$

5. (a) $\frac{3}{2}$ **6.** (a) ± 8 **7.** $1408.75   **8.** $\frac{13}{2}$

  (b) $-\frac{3}{2}$ (b) $-\frac{1}{2}, 3$

9. $x \le -1$ or $x \ge 5$ **10.** $x \ge 103$

11. Function   **12.** $2 \le x < \infty$

13. (a) 4   **14.** $m = \frac{3}{4}$

  (b) $c^2 + 3c$ Distance: 10

15. $(y - 3)^2(y + 3)$ **16.** $(3x + 7)(x - 5)$

17.   **18.**

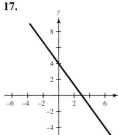   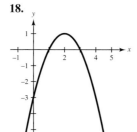

19.   **20.**

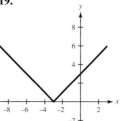   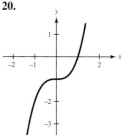

CHAPTER 4

Section 4.1 (page 262)

7. $(-\infty, 8) \cup (8, \infty)$ **9.** $(-\infty, \infty)$

11. $(-\infty, 0) \cup (0, 3) \cup (3, \infty)$

13. (a) 1   **15.** (a) $\frac{25}{22}$

  (b) -8 (b) 0

  (c) Undefined   (c) Undefined

  (d) 0   (d) Undefined

17. $\{1, 2, 3, 4, \ldots\}$ **19.** $x + 3$ **21.** $x + 2$

23. $\frac{x}{5}$ **25.** $\frac{6x}{5y^3}$ **27.** $\frac{3y}{y^2 + 1}$ **29.** $\frac{y - 8}{15}$

31. $\frac{-1}{2x + 3}$ **33.** $\frac{3(m - 2n)}{m + 2n}$

35.

x	-2	-1	0	1	2	3	4
$\dfrac{x^2 - x - 2}{x - 2}$	-1	0	1	2	Undefined	4	5
$x + 1$	-1	0	1	2	3	4	5

37. $\frac{x}{3(x + 3)}$ **39.** $\frac{107.1 + 12.64t + 0.54t^2}{31.6 + 0.51t - 0.14t^2}$

41. $42x^2 - 60x$ **43.** $121 - x^2$ **45.** $\frac{1}{81}y^4$

47. $(-\infty, -4) \cup (-4, \infty)$ **49.** $(-\infty, \infty)$

51. $(-\infty, -4) \cup (-4, 4) \cup (4, \infty)$

53. $(-\infty, -1) \cup (-1, 5) \cup (5, \infty)$ **55.** $(0, \infty)$

57. $(3)(x + 16)^2$ **59.** $(x)(x - 2)$ **61.** $6y$

63. x **65.** $\frac{1}{2}$ **67.** $-\frac{1}{3}$ **69.** $\frac{1}{a + 3}$

71. $\frac{x}{x - 7}$ **73.** $\frac{y(y + 2)}{y + 6}$ **75.** $\frac{5x + 4}{5x + 2}$

77. $\dfrac{3xy + 5}{y^2}$ **79.** $\dfrac{u - 2v}{u - v}$

81. Evaluating both sides when $x = 10$ yields $\frac{3}{2} \neq 9$.

83. $x^n - 2$ **85.** (a) $\dfrac{2500 + 9.25x}{x}$ **87.** π

   (b) $x > 0$

   (c) \$34.25

Section 4.2 *(page 271)*

5. (a) 0   **7.** x^2 **9.** $(-1)(2 + x)$

   (b) Undefined

   (c) $\frac{3}{2}$

   (d) $\frac{1}{24}$

11. $\dfrac{3x}{2}$ **13.** 24   **15.** $-\dfrac{x + 8}{x^2}$ **17.** $\dfrac{(u + v)^2}{(u - v)^2}$

19. $\dfrac{3}{2x}$ **21.** $\dfrac{3}{2(a + b)}$ **23.** $\dfrac{(x + 3)(4x + 1)}{(3x - 1)(x - 1)}$

25. $\dfrac{3x}{10}$ **27.** $\dfrac{x + 4}{3}$ **29.** $\dfrac{1}{4}$

31.   **33.**

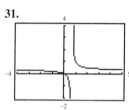   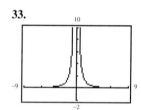

35. $\dfrac{2w^2 + 3w}{6}$ **37.** $3x(x - 7)$ **39.** $(2t + 13)(2t - 13)$

41. \$720   **43.** $(x + 2)^2$ **45.** $u + 1$ **47.** $\dfrac{s^3}{6}$

49. $24u^2$ **51.** $\dfrac{2uv(u + v)}{3(3u + v)}$ **53.** -1

55. $\dfrac{2(r + 2)}{11r}$ **57.** $(u - 2v)(u + 2v)$ **59.** $2t + 5$

61. $\dfrac{(x - 1)(2x + 1)}{(3x - 2)(x + 2)}$ **63.** $\dfrac{3y^2}{2ux^2}$ **65.** $x^4 y(x + 2y)$

67. $-\dfrac{5x}{2}$ **69.** $\dfrac{(x + 2)(x + 3)}{x}$ **71.** $\dfrac{(x + 1)(2x - 5)}{x}$

73. (a) 1/20 minute   **75.** $x/[4(3x - 2)]$

   (b) $x/20$ minutes

   (c) 7/4 minutes

Section 4.3 *(page 280)*

5. $\dfrac{3}{2}$ **7.** $-\dfrac{2}{9}$ **9.** $1, x \neq -4$ **11.** $20x^3$

13. $15x^2(x + 5)$ **15.** $6x(x + 2)(x - 2)$

17. $\dfrac{2n^2(n + 8)}{6n^2(n - 4)}$ **19.** $\dfrac{(x - 8)(x - 5)}{(x + 5)(x - 5)^2}$

   $\dfrac{10(n - 4)}{6n^2(n - 4)}$ $\dfrac{9x(x + 5)}{(x + 5)(x - 5)^2}$

21. $\dfrac{25x - 12}{20}$ **23.** 0   **25.** $\dfrac{5(5x + 22)}{x + 4}$

27. $\dfrac{x^2 + 3x + 9}{x(x^2 - 9)}$ **29.** $\dfrac{5u - 2v}{(u - v)^2}$ **31.** $\dfrac{x}{x - 1}$

33.   **35.** $\dfrac{x}{2(3x + 1)}$

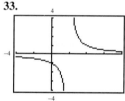

37. $\dfrac{3}{4}$ **39.** $y - x$ **41.** $\dfrac{R_1 R_2}{R_1 + R_2}$

43.

x	-3	-2	-1	0	1	2	3
$\dfrac{\left(1 - \dfrac{1}{x}\right)}{\left(1 - \dfrac{1}{x^2}\right)}$	$\frac{3}{2}$	2	Undef.	Undef.	Undef.	$\frac{2}{3}$	$\frac{3}{4}$
$\dfrac{x}{x + 1}$	$\frac{3}{2}$	2	Undef.	0	$\frac{1}{2}$	$\frac{2}{3}$	$\frac{3}{4}$

45. $54x^{10}$ **47.** 1   **49.** $\dfrac{25}{x^4}$

51. $2500 < 1500 + 0.04x$ **53.** $\dfrac{2 + y}{2}$

   $x > \$25,000$

55. $-\dfrac{3}{a}$ **57.** $\dfrac{x + 6}{3}$ **59.** $-\dfrac{4}{3}$ **61.** $\dfrac{-25}{9}$

63. $36y^3$ **65.** $30x^2(x - 1)$

67. $\dfrac{3v^2}{6v^2(v + 1)}$ **69.** $\dfrac{4x(x - 5)}{(x + 5)^2(x - 5)}$

   $\dfrac{8(v + 1)}{6v^2(v + 1)}$ $\dfrac{(x - 2)(x + 5)}{(x + 5)^2(x - 5)}$

71. $\dfrac{7(a+2)}{a^2}$ **73.** $\dfrac{3(x+2)}{x-8}$ **75.** 1

77. $\dfrac{1}{2x(x-3)}$ **79.** $\dfrac{x^2-7x-15}{(x+3)(x-2)}$ **81.** $\dfrac{x-2}{x(x+1)}$

83. $\dfrac{5(x+1)}{(x+5)(x-5)}$ **85.** $\dfrac{4}{x^2(x^2+1)}$ **87.** $\dfrac{4x}{(x-4)^2}$

89. $\dfrac{y-x}{xy}$ **91.** $\dfrac{2(4x^2+5x-3)}{x^2(x+3)}$ **93.** $-4x-1$

95. $\dfrac{5(x+3)}{2x(5x-2)}$ **97.** $\dfrac{y+1}{y-3}$ **99.** $\dfrac{x(x+6)}{3x^3+10x-30}$

101. $-\dfrac{1}{2(h+2)}$ **103.** $\dfrac{5t}{12}$ **105.** $\dfrac{5x}{24}$

107. $\dfrac{x}{4}, \dfrac{x}{3}, \dfrac{5x}{12}$ **109.** (a) 19.6%

(b) $\dfrac{288(MN-P)}{N(MN+12P)}$

Mid-Chapter Quiz *(page 284)*

1. $(-\infty, 0) \cup (0, 4) \cup (4, \infty)$ **2.** (a) 0

(b) $\dfrac{9}{2}$

(c) Undefined

(d) $\dfrac{8}{9}$

3. $\dfrac{3}{2}y$ **4.** $\dfrac{2u^2}{9v}$ **5.** $-\dfrac{2x+1}{x}$ **6.** $\dfrac{z+3}{2z-1}$

7. $\dfrac{7+3ab}{a}$ **8.** $\dfrac{n^2}{m+n}$ **9.** $\dfrac{t}{2}$ **10.** $\dfrac{5x}{x-2}$

11. $\dfrac{8x}{3(x-1)(x^2+2x-3)}$ **12.** $\dfrac{4(u-v)^2}{5uv}$

13. $-\dfrac{3t}{2}$ **14.** $\dfrac{2(x+1)}{3x}$ **15.** $\dfrac{x(1-3x)}{4(x+5)}$

16. $\dfrac{4x^4+x^3-18x^2+8}{x^2(x^2-4)}$ **17.** $\dfrac{5(2-x)}{4x-15}$

18. $\dfrac{2(x^2+9)}{x+3}$ **19.** (a) $\dfrac{6000+10.50x}{x}$ **20.** $\dfrac{8(x+2)}{(x+4)^2}$

(b) $22.50

Section 4.4 *(page 292)*

7. $-10z^2-6$ **9.** $\dfrac{5}{2}x-4+\dfrac{7}{2}y$ **11.** $x-5$

13. $x+7$ **15.** x^2+4 **17.** $x-4+\dfrac{32}{x+4}$

19. $\dfrac{6}{5}z+\dfrac{41}{25}+\dfrac{41}{25(5z-1)}$ **21.** $x^3-x+\dfrac{x}{x^2+1}$

23. $x+2$ **25.** $x^2-x+4-\dfrac{17}{x+4}$

27. $\dfrac{1}{10}x+\dfrac{41}{50}+\dfrac{291}{250(x-0.2)}$ **29.** $(x-7)(x-8)$

31. $(x^3-3x+1)(x-1)$ **33.**

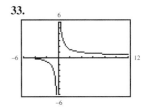

35.

x-Values	Polynomial Values	Divisors	Remainders
-2	-8	$x+2$	-8
-1	-1	$x+1$	-1
0	0	x	0
$\dfrac{1}{2}$	$-\dfrac{9}{8}$	$x-\dfrac{1}{2}$	$-\dfrac{9}{8}$
1	-2	$x-1$	-2
2	0	$x-2$	0

37. $2(x+4)$ **39.** x^2+2x+1 **41.** $16-25z^2$

43. $\dfrac{5}{2}x(2x+9)$ **45.** $3z+5$ **47.** $\dfrac{5}{2}z^2+z-3$

49. $7x^2-2x$ **51.** $4z^2+\dfrac{3}{2}z-1$ **53.** $m^3+2m-\dfrac{7}{m}$

55. $4(x+7)$ **57.** $x+10$ **59.** $y+3$ **61.** $6t-5$

63. $4x-1$ **65.** $x^2-5x+25$ **67.** $2+\dfrac{5}{x+2}$

69. $5x-8+\dfrac{19}{x+2}$ **71.** $4x-\dfrac{25}{3}+\dfrac{35}{3(3x+2)}$

73. $2x^2+x+4+\dfrac{6}{x-3}$

75. $x^5+x^4+x^3+x^2+x+1$ **77.** $x^{2n}+x^n+4$

79. x^3-3x+1 **81.** $5x^2-25x+125-\dfrac{613}{x+5}$

83. $(2a-5)(a+9)$ **85.** $(15x+10)\left(x-\dfrac{4}{5}\right)$

87. $(2t^2+5t-6)(t+5)$ **89.** -8 **91.** x^2-3

Section 4.5 *(page 301)*

5.

x	0	0.5	0.9	0.99	0.999
y	−4	−8	−40	−400	−4000

x	2	1.5	1.1	1.01	1.001
y	4	8	40	400	4000

x	2	5	10	100	1000
y	4	1	0.4444	0.0404	0.0040

7. Domain: $(-\infty, 0) \cup (0, \infty)$
Horizontal asymptote: $y = 0$
Vertical asymptote: $x = 0$

9. Domain: $(-\infty, \infty)$
Horizontal asymptote: $y = 0$
Vertical asymptotes: None

11. Domain: $(-\infty, -1) \cup (-1, 1) \cup (1, \infty)$
Horizontal asymptote: $y = 5$
Vertical asymptotes: $x = -1, x = 1$

13. d   **15.** a

17.

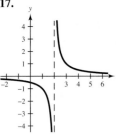

19.

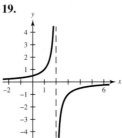

21.

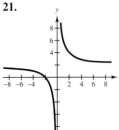

23.

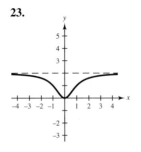

25.

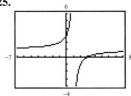

Domain: $(-\infty, 1) \cup (1, \infty)$
Horizontal asymptote: $y = 1$
Vertical asymptote: $x = 1$

27.

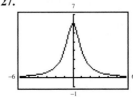

Domain: $(-\infty, \infty)$
Horizontal asymptote: $y = 0$
Vertical asymptotes: None

29.

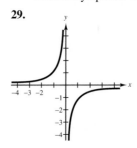

31.

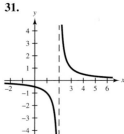

33. (a) $A = \dfrac{2500 + 0.50x}{x}$

(b) \$3, \$0.75

(c)

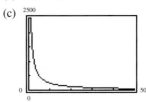

$A = \$0.50$

35. $x \geq 6$ **37.** $1 < x < 5$ **39.** $30 \leq n \leq 150$

41. Domain: $(-\infty, 3) \cup (3, \infty)$
Horizontal asymptote: $y = 2$
Vertical asymptote: $x = 3$

43. Domain: $(-\infty, -3) \cup (-3, 3) \cup (3, \infty)$
Horizontal asymptote: $y = 0$
Vertical asymptotes: $x = -3, x = 3$

45. Domain: $(-\infty, -8) \cup (8, \infty)$
Horizontal asymptote: $y = 1$
Vertical asymptote: $x = -8$

47. Domain: $(-\infty, 0) \cup (0, 1) \cup (1, \infty)$
Horizontal asymptote: $y = 0$
Vertical asymptotes: $t = 0, t = 1$

49. Domain: $(-\infty, \infty)$
Horizontal asymptote: $y = 2$
Vertical asymptotes: None

51.

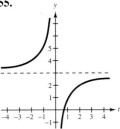

53.

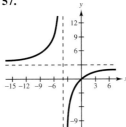

55.

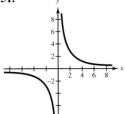

57.

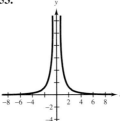

59.

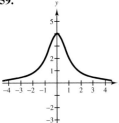

61.

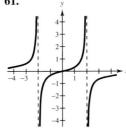

63.

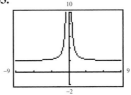

Domain: $(-\infty, 0) \cup (0, \infty)$

65.

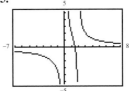

Domain: $(-\infty, 0) \cup (0, 2) \cup (2, \infty)$

67.

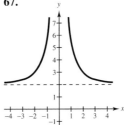

69.

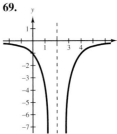

71.

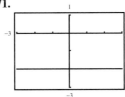

The fraction is not reduced to lowest terms.

73. (a) $C = 0$. The chemical is eliminated from the body.

(b)

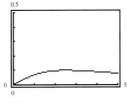

$t \approx 2.5$

75. (a) area = (length)(width)

$$400 = x \cdot y$$

$$y = \frac{400}{x}$$

$$P = 2(x + y)$$

$$P = 2\left(x + \frac{400}{x}\right)$$

(b) $0 < x$

(c)

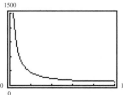

20 units × 20 units

Section 4.6 *(page 311)*

7. (a) Not a solution   **9.** (a) Not a solution

(b) Not a solution   (b) Solution

(c) Not a solution   (c) Solution

(d) Solution   (d) Not a solution

11. $\frac{3}{2}$ **13.** 10   **15.** $\frac{1}{3}$ **17.** 3   **19.** $-\frac{4}{15}$

21. No solution   **23.** ±4   **25.** 0, 2   **27.** 2, 3

29. $(-2, 0)$ **31.** $(-1, 0), (1, 0)$

33.

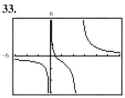

(1, 0)

35.

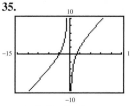

$(-3, 0), (2, 0)$

37. 40 miles per hour   **39.** 12   **41.** $11\frac{1}{4}$ hours

43. (a) 1989

(b) $y = 43.31 - \dfrac{275.25}{9} + \dfrac{654.53}{9^2} \approx 20.8$ billion

(c) $28.66 billion

45. $\frac{5}{2}$ **47.** $-7, 6$ **49.** 24, 26   **51.** $\frac{1}{4}$ **53.** $-\frac{8}{3}$

55. 10   **57.** 61   **59.** $\frac{18}{5}$ **61.** 3   **63.** $-\frac{11}{5}$ **65.** $\frac{4}{3}$

67. No solution   **69.** $\frac{3}{4}$ **71.** 3   **73.** ±6   **75.** $8, -9$

77. 3, 13   **79.** -5 **81.** No solution   **83.** $8, -3$

85.

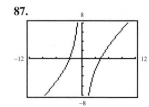

(4, 0)

87.

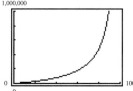

$(-3, 0), (4, 0)$

89. 8   **91.** (a)

(b) 85%

93. 8 miles per hour; 10 miles per hour

95. 4 miles per hour   **97.** 10

99. 3 hours; $\dfrac{15}{8}$ minutes; $\dfrac{5}{3}$ hours

101. 15 hours; $22\frac{1}{2}$ hours

Review Exercises *(page 317)*

1. $(-\infty, 8) \cup (8, \infty)$

3. $(-\infty, 1) \cup (1, 6) \cup (6, \infty)$

5. $\dfrac{2x^3}{5}$ **7.** $\dfrac{b-3}{6(b-4)}$ **9.** -9 **11.** $\dfrac{x}{2(x+5)}$

13. b   **15.** a   **17.** $\dfrac{y}{8x}$ **19.** $12z(z-6)$ **21.** $-\dfrac{1}{4}$

23. $3x^2$ **25.** $\dfrac{125y}{x}$ **27.** $\dfrac{x(x-1)}{x-7}$ **29.** $-\dfrac{7}{9}$

31. $-\dfrac{13}{48}$ **33.** $\dfrac{4x+3}{(x+5)(x-12)}$

35. $\dfrac{5x^3 - 5x^2 - 31x + 13}{(x+2)(x-3)}$ **37.** $\dfrac{x+24}{x(x^2+4)}$

39. $\dfrac{6(x-9)}{(x+3)^2(x-3)}$ **41.** $\dfrac{6(x+5)}{x(x+7)}$ **43.** $\dfrac{3t^2}{5t-2}$

45. $-\dfrac{a^2 - a - 16}{(4a^2 + 16a + 1)(a-4)}$ **47.** $2x^2 - \dfrac{1}{2}$

49. $2x^2 + \dfrac{4}{3}x - \dfrac{8}{9} + \dfrac{10}{9(3x-1)}$ **51.** $x^2 - 2$

53. $x^2 + 5x - 7$ **55.** $x^3 + 3x^2 + 6x + 18 + \dfrac{29}{x-3}$

57.

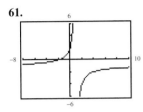

59.

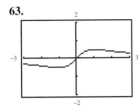

61.

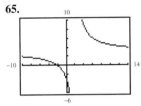

63.

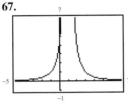

65.

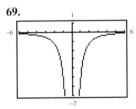

67.

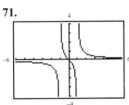

69.

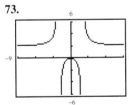

71.

73.

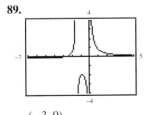

75. -40 **77.** $\frac{3}{2}$

79. 5   **81.** 6, -4,   **83.** 2, -6,   **85.** $-2, 2$
87. $-\frac{9}{5}, 3$ **89.**

$(-3, 0)$

91. 56 miles per hour   **93.** 4

Chapter Test   *(page 320)*

1. $(-\infty, -5) \cup (-5, 5) \cup (5, \infty)$ **2.** $x^3(x + 3)^2(x - 3)$

3. (a) $-\dfrac{1}{3}$ **4.** $\dfrac{5z}{3}$ **5.** $\dfrac{4}{y + 4}$ **6.** $\dfrac{(2x + 3)^2}{x + 1}$

   (b) $\dfrac{2a + 3}{5}$

7. $\dfrac{14y^6}{15}$ **8.** $\dfrac{x^3}{4}$ **9.** $-(3x + 1)$

10. $\dfrac{-2x^2 + 2x + 1}{x + 1}$ **11.** $\dfrac{5x^2 - 15x - 2}{(x - 3)(x + 2)}$

12. $\dfrac{5x^3 + x^2 - 7x - 5}{x^2(x + 1)^2}$ **13.** 4   **14.** $t^2 + 3 - \dfrac{6t - 6}{t^2 - 2}$

15. $2x^3 + 6x^2 + 3x + 9 + \dfrac{20}{x - 3}$

16. (a)   (b)

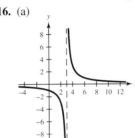 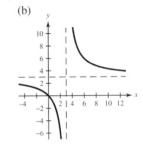

17. 22   **18.** $-1, -\frac{15}{2}$ **19.** No solution
20. $6\frac{2}{3}$ hours, 10 hours

CHAPTER 5

Section 5.1   *(page 328)*

7. $\frac{1}{25}$ **9.** -32 **11.** $\frac{3}{2}$ **13.** 1   **15.** 100,000

17. $\dfrac{1}{16}$ **19.** $\dfrac{16}{15}$ **21.** y^2 **23.** t^2 **25.** $-\dfrac{12}{xy^3}$

27. $\dfrac{y^4}{9x^4}$ **29.** $\dfrac{3x^5}{y^4}$ **31.** $\dfrac{81v^8}{u^6}$ **33.** $\dfrac{ab}{b - a}$

35. 5.75×10^7 **37.** 9.461×10^{15} **39.** 8.99×10^{-5}

41. 28,200,000,000   **43.** 13,000,000

45. 0.00000000048   **47.** 1.3×10^{11} **49.** 6×10^6

51. 3.30×10^8 **53.** 3.33×10^5 **55.** $(x - 1)(x - 2)$

57. $(11x - 5)(x + 1)$ **59.** 7650   **61.** $-\frac{1}{1000}$

63. $-\frac{1}{243}$ **65.** 64   **67.** 1   **69.** 729   **71.** $\frac{1}{64}$

73. $\frac{3}{16}$ **75.** $\frac{64}{121}$ **77.** z^2 **79.** x^6 **81.** a

83. $\frac{1}{4x^4}$ **85.** $\frac{10}{x}$ **87.** $\frac{b^5}{a^5}$ **89.** x^8y^{12}

91. $\frac{v^2}{uv^2+1}$ **93.** $\frac{x^{12}y^8}{16}$ **95.** 3.6×10^6

97. 3.81×10^{-8} **99.** $60,000,000$

101. 0.0000001359 **103.** 6.8×10^5 **105.** 4×10^3

107. 9×10^{15} **109.** 4.70×10^{11} **111.** 3.46×10^{10}

113. 9.3×10^7 **115.** 1.6×10^{-5} years ≈ 8.4 minutes

117. $\$17,159.53$

Section 5.2   *(page 338)*

7. 7 **9.** Cube root   **11.** Square root   **13.** 9

15. $\frac{3}{4}$ **17.** Not a real number   **19.** $-\frac{2}{3}$ **21.** 5

23. 3 **25.** -0.3 **27.** 5 **29.** $\frac{1}{4}$ **31.** $16^{1/2} = 4$

33. $27^{2/3} = 9$ **35.** $\sqrt[4]{256^3} = 64$ **37.** 1.0420

39. 4.3004 **41.** t^2 **43.** 3 **45.** $\frac{1}{2}$ **47.** x^3

49. $x^{1/4}$ **51.** $r = 0.128$ **53.** $[0, \infty)$ **55.** $\frac{5}{56}$

57. 5 **59.** $h + 4$ **61.** 8 **63.** Not a real number

65. 0.4 **67.** 13 **69.** 10 **71.** $-\frac{1}{4}$

73. Not a real number   **75.** Irrational   **77.** Rational

79. 7 **81.** 8 **83.** $\frac{4}{9}$ **85.** $\frac{3}{11}$ **87.** 8.5440

89. -0.1248 **91.** 0.0038 **93.** 66.7213 **95.** $|t|$

97. y^3 **99.** $\frac{4}{9}$ **101.** $\frac{9y^{3/2}}{x^{2/3}}$ **103.** $\frac{3y^2}{4z^{4/3}}$

105. $c^{1/2}$ **107.** $\sqrt[8]{y}$ **109.** $2x^{3/2} - 3x^{1/2}$

111. $1 + 5y$ **113.** $(0, \infty)$

115. 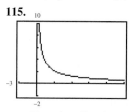   **117.**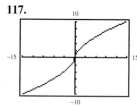

Domain: $(0, \infty)$ Domain: $(-\infty, \infty)$

119. $1, 4, 5, 6, 9, 0$ **121.** 0.026 inch
No

Section 5.3   *(page 346)*

7. $\sqrt[3]{110}$ **9.** $\sqrt{\frac{15}{31}}$ **11.** $3\sqrt{35}$ **13.** $\frac{10}{\sqrt[3]{11}}$

15. $2\sqrt{5}$ **17.** $2\sqrt[3]{3}$ **19.** $\frac{1}{2}\sqrt{15}$ **21.** $\frac{1}{3}\sqrt[5]{15}$

23. $3x^2\sqrt{x}$ **25.** $|x|\sqrt[4]{3y^2}$ **27.** $\frac{2}{y}\sqrt[5]{x^2}$

29. $\frac{4a^2}{|b|}\sqrt{2}$ **31.** $\frac{\sqrt{3}}{3}$ **33.** $\frac{\sqrt[4]{320}}{2}$ **35.** $\frac{\sqrt{y}}{y}$

37. $\frac{\sqrt[3]{18xy^2}}{3y}$ **39.** $2\sqrt{2}$ **41.** $30\sqrt[3]{2}$ **43.** $13\sqrt{y}$

45. 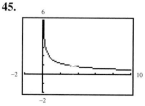   **47.** $3\sqrt{5}$

49. 89.44 cycles per second   **51.** 1 **53.** $4rs^2$

55. Base: 22 inches   **57.** $\sqrt{30}$
Height: 10 inches

59. $\sqrt[5]{\frac{152}{3}}$ **61.** $3\sqrt[4]{11}$ **63.** $\frac{1}{3}\sqrt{35}$ **65.** $3\sqrt{3}$

67. $10\sqrt[4]{3}$ **69.** $\frac{1}{7}\sqrt{15}$ **71.** $\frac{1}{4}\sqrt[3]{35}$ **73.** 0.02

75. $20\sqrt[3]{300}$ **77.** $4y^2\sqrt{3}$ **79.** $xy\sqrt[3]{x}$ **81.** $2xy\sqrt[5]{y}$

83. $\frac{1}{5}\sqrt{13}$ **85.** $\frac{3a\sqrt[3]{2a}}{b^3}$ **87.** $4\sqrt{3}$ **89.** $\frac{3}{2}3\sqrt[3]{2}$

91. $\frac{2\sqrt{x}}{x}$ **93.** $\frac{2\sqrt{3b}}{b^2}$ **95.** $\frac{a^2\sqrt[3]{a^2b}}{b}$

97. $-5\sqrt[4]{7} - 11\sqrt[4]{3}$ **99.** $24\sqrt{2} - 6$ **101.** $12\sqrt{x}$

103. $(10 - z)\sqrt[3]{z}$ **105.** $\frac{2\sqrt{5}}{5}$ **107.** $\frac{9\sqrt{5}}{5}$ **109.** $>$

111. $>$ **113.** $3\sqrt{13}$ **115.** 2.22 seconds   **117.** 1

Math Matters   *(page 349)*

$1,000,000,000$ nanoseconds in a second

1000 millimeters in a meter

1000 watts in a kilowatt

Mid-Chapter Quiz *(page 350)*

1. $-\frac{1}{144}$ **2.** $\frac{64}{27}$ **3.** $\frac{5}{3}$ **4.** -4 **5.** $3t^{3/2}$

6. $\frac{3x^6}{16y^3}$ **7.** $\frac{2}{3u^3}$ **8.** $x^2 + 4$ **9.** (a) 1.34×10^7
                                                (b) 7.5×10^{-4}

10. (a) 8.1×10^{13} **11.** (a) $5\sqrt{6}$ **12.** (a) $3|x|\sqrt{3}$
    (b) 2×10^{-4} (b) $3\sqrt[3]{2}$ (b) $3x\sqrt{x}$

13. (a) $\frac{\sqrt[4]{5}}{2}$ **14.** (a) $\frac{2u\sqrt{10u}}{3}$ **15.** (a) $\frac{\sqrt{6}}{3}$

    (b) $\frac{2\sqrt{6}}{7}$ (b) $\frac{16}{u^2}$ (b) $4\sqrt{3}$

16. (a) $\frac{2\sqrt{5x}}{x}$ **17.** $4\sqrt{2y}$ **18.** $6x\sqrt[3]{5x^2} + 4x\sqrt[3]{5x}$

    (b) $\frac{\sqrt[3]{12a^2}}{2a}$

19. $\sqrt{5^2 + 12^2} = \sqrt{169} = 13$ **20.** $23 + 8\sqrt{2}$

Section 5.4 *(page 355)*

5. $3\sqrt{2}$ **7.** $2\sqrt{5} - \sqrt{15}$ **9.** -1 **11.** $8\sqrt{5} + 24$
13. $y + 4\sqrt{y}$ **15.** $45x - 17\sqrt{x} - 6$ **17.** $2 - 7\sqrt[3]{4}$
19. $\sqrt{3} + 3\sqrt{3}$ **21.** $4 - 3x$ **23.** $2u + \sqrt{2u}$
25. $\sqrt{11} + \sqrt{3}, 8$ **27.** $\sqrt{x} + 3, x - 9$ **29.** $\frac{\sqrt{22} + 2}{3}$

31. $4(3 - \sqrt{7})$ **33.** $\frac{(\sqrt{5} + 1)\, t^{3/2}}{2}$

35. $-\frac{\sqrt{u + v}\,(\sqrt{u - v} + \sqrt{u})}{v}$ **37.** (a) 0 (b) -1

39.   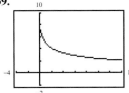   **41.** $192\sqrt{2}$ **43.** 6

45. $-\frac{2}{3}, 5$ **47.** Vertical shift four units upward
49. Reflection in the x-axis   **51.** 4 **53.** $2\sqrt{10} + 8\sqrt{2}$
55. 4 **57.** 4 **59.** $\sqrt{15} + 3\sqrt{3} - 5\sqrt{5} - 15$
61. $2x + 20\sqrt{2x} + 100$ **63.** $9x - 25$ **65.** $x - y$

67. $\sqrt[3]{4x^2} + 10\sqrt[3]{2x} + 25$ **69.** $\frac{1 - 2\sqrt{x}}{3}$

71. $\frac{-1 + \sqrt{3y}}{4}$ **73.** $2 - \sqrt{5}, -1$

75. $\sqrt{2u} + \sqrt{3}, 2u - 3$ **77.** $\frac{6 - \sqrt{2}}{17}$ **79.** $\frac{4\sqrt{7} + 11}{3}$

81. $\frac{(\sqrt{15} + \sqrt{3})\, x}{4}$ **83.** $\frac{(\sqrt{x} - 5)(2\sqrt{x} + 1)}{4x - 1}$

85. (a) $2\sqrt{3} - 4$ **87.**   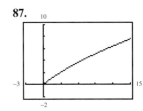   **89.** $\frac{\sqrt{3}}{\sqrt{4}}$
    (b) 0

Section 5.5 *(page 364)*

5. (a) Not a solution   **7.** (a) Not a solution
    (b) Not a solution   (b) Solution
    (c) Not a solution   (c) Not a solution
    (d) Solution   (d) Not a solution

9. 400 **11.** No solution   **13.** $\frac{44}{3}$ **15.** $-\frac{2}{3}$
17. No solution   **19.** 7 **21.** $1, 3$ **23.** 1.407

25. 4.840 **27.** 10 **29.** $\frac{5\sqrt{3}}{2} \approx 4.33$

31.   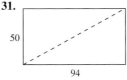   **33.**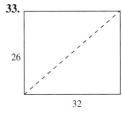

$2\sqrt{2834} \approx 106.47$ feet

$10\sqrt{17} \approx 41.23$ feet

35. 64 feet   **37.** 1.82 feet   **39.** $-\frac{4x}{3}$ **41.** $\frac{2u}{9v^6}$

43. $\frac{4}{9x^4 y^2}$ **45.** The store, at $191.96   **47.** 49

49. 525 **51.** 90 **53.** $\frac{14}{25}$ **55.** No solution
57. 4 **59.** -15 **61.** $4, 12$ **63.** 8 **65.** 1.569
67. 1.978 **69.** $4\sqrt{2} \approx 5.66$ **71.** $5\sqrt{15} \approx 19.36$
73. c   **75.** d   **77.** f   **79.** $8\sqrt{6} \approx 19.596$
81. $40\sqrt{2} \approx 56.57$ feet per second
83. 56.25 feet   **85.** 500 units

87. $r = \sqrt{\dfrac{A}{\pi}}$ **89.** 26,250

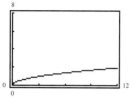

Section 5.6 *(page 374)*

7. $2i$ **9.** $3\sqrt{3}\,i$ **11.** $\sqrt{7}\,i$ **13.** $10i$

15. $3\sqrt{2}\,i$ **17.** -4 **19.** $-\sqrt{15} + \sqrt{10}$

21. $10 + 4i$ **23.** $-14 + 20i$ **25.** $-3 + 49i$

27. -36 **29.** $-40 - 5i$ **31.** $-24 + 2\sqrt{3}\,i$

33. $-2 + 8i, 68$ **35.** $1 - \sqrt{3}\,i, 4$ **37.** $-\frac{24}{53} + \frac{84}{53}i$

39. $-10i$ **41.** $23\sqrt{2}$ **43.** $\frac{4}{5}\sqrt{10}$ **45.** \$1489.65

47. $-12i$ **49.** $\frac{2}{5}i$ **51.** $0.3i$ **53.** $2\sqrt{2}\,i$

55. $-3\sqrt{6}$ **57.** $-2\sqrt{3} - 3$ **59.** $a = 2, b = -3$

61. $a = -4, b = -2\sqrt{2}$ **63.** $-14 - 40i$ **65.** 13

67. $9 - 7i$ **69.** -20 **71.** $-36i$ **73.** $27i$

75. $-65 - 10i$ **77.** $20 - 12i$ **79.** $-14 + 42i$

81. $-7 + 24i$ **83.** 9   **85.** $2 - i, 5$

87. $5 + \sqrt{6}\,i, 31$ **89.** $-10i, 100$ **91.** $2 + 2i$

93. $-\frac{6}{5} + \frac{2}{5}i$ **95.** $\frac{8}{5} - \frac{1}{5}i$ **97.** $1 - \frac{6}{5}i$

99. $-\frac{53}{25} + \frac{29}{25}i$ **101.** (a) Solution

(b) Solution

103. (a) Solution   **105.** $2a$ **107.** $2bi$

(b) Solution

Review Exercises *(page 378)*

1. $\frac{1}{72}$ **3.** $\frac{125}{8}$ **5.** 3.6×10^7 **7.** 500   **9.** $2x$

11. $\dfrac{x^6}{y^8}$ **13.** $\dfrac{1}{t^3}$ **15.** $\dfrac{27}{y^3}$ **17.** 1.2   **19.** $\frac{5}{6}$

21. 12   **23.** 11,414.13   **25.** 10.63   **27.** $49^{1/2} = 7$

29. $\sqrt[3]{216} = 6$ **31.** 81   **33.** 125   **35.** 0.04

37. $x^{7/12}$ **39.** $\dfrac{3}{x^{1/4}y^{2/5}}$ **41.** $6\sqrt{10}$

43. $0.5x^2\sqrt{y}$ **45.** $2ab\sqrt[3]{6b}$ **47.** $\dfrac{\sqrt{30}}{6}$

49. $\dfrac{\sqrt{3x}}{2x}$ **51.** $\dfrac{\sqrt[3]{4x^2}}{x}$ **53.** $\dfrac{7 + \sqrt{7}}{7}$

55. $-24\sqrt{10}$ **57.** $11\sqrt{x}$ **59.** $3 - x$

61. **63.**

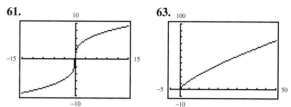

65. 225   **67.** 105   **69.** $-3, -5$ **71.** 5

73. $\frac{3}{32}$ **75.** 1.978   **77.** $4\sqrt{3}\,i$ **79.** $10 - 9\sqrt{3}\,i$

81. $\frac{3}{4} - \sqrt{3}\,i$ **83.** $8 - 3i$ **85.** -90 **87.** 25

89. $59 + 74i$ **91.** $\frac{9}{17} + \frac{2}{17}i$ **93.** $23 + 8\sqrt{2}$

95. 1.37 feet

Chapter Test *(page 380)*

1. (a) $\frac{3}{8}$ **2.** (a) $\frac{1}{9}$ **3.** 3.2×10^{-5}

(b) 3.0×10^{-5} (b) 6

4. 30,400,000   **5.** (a) $\dfrac{3}{5t}$ **6.** (a) $x^{1/3}$

(b) 25

(b) $\dfrac{y^2}{xy^2 + 1}$

7. (a) $\dfrac{4}{3}\sqrt{2}$ **8.** $\dfrac{\sqrt{6}}{2}$ **9.** $-10\sqrt{3x}$

(b) $2\sqrt[3]{3}$

10. $5\sqrt{3x} + 3\sqrt{5}$ **11.** $16 - 8\sqrt{2x} + 2x$

12. $7\sqrt{3}\,(4y + 3)$ **13.** No solution   **14.** 9

15. $2 - 2i$ **16.** $-5 - 12i$ **17.** $-8 + 4i$

18. $13 + 13i$ **19.** $-2 - 5i$ **20.** 100 feet

CHAPTER 6

Section 6.1 *(page 387)*

7. 0, 3   **9.** 6   **11.** 1, 6   **13.** $\frac{5}{3}, 6$ **15.** ± 3

17. ± 8 **19.** $9, -17$ **21.** $\dfrac{-1 \pm 5\sqrt{2}}{2}$ **23.** $\pm 2i$

25. $-6 \pm \frac{11}{3}i$ **27.** $1 \pm 3\sqrt{3}\,i$ **29.** $0, \frac{5}{2}$ **31.** ± 10

33. $5 \pm 10i$ **35.** $-2, 1, 1 \pm \sqrt{3}\,i, -\dfrac{1}{2} \pm \dfrac{\sqrt{3}}{2}i$

37. $4, \frac{25}{4}$ **39.** $\pm 2\sqrt{2}, \pm 2i$ **41.** $\pm\sqrt{15}, \pm\sqrt{5}\,i$

43.

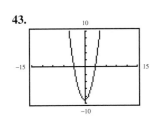

$(-3, 0), (3, 0)$

45.

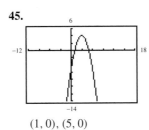

$(1, 0), (5, 0)$

47.

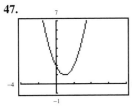

$1 \pm i$

49.

$-3 \pm \sqrt{2}\, i$

51. 1978   **53.** $(4x + 11)(4x - 11)$

55. $(x - 10)(x - 4)$ **57.** 3 miles per hour   **59.** 9, 12

61. $\pm \frac{5}{2}$ **63.** Prime   **65.** $\frac{1}{2}, -\frac{5}{6}$ **67.** ± 8

69. $\pm \frac{4}{5}$ **71.** $\pm \frac{15}{2}$ **73.** 2.5, 3.5   **75.** $2 \pm \sqrt{7}$

77. $\dfrac{3 \pm 7\sqrt{2}}{4}$ **79.** $\pm \sqrt{17}\, i$ **81.** $3 \pm 5i$

83. $\frac{3}{2} \pm \frac{5}{2}i$ **85.** $\frac{2}{3} \pm \frac{1}{3}i$ **87.** $-\dfrac{7}{3} \pm \dfrac{\sqrt{38}}{3}i$

89. ± 30 **91.** $\pm 30i$ **93.** ± 3 **95.** $-5, 15$

97. $5 \pm 10i$ **99.** $\pm 1, \pm 2$ **101.** $\pm \sqrt{2}, \pm \sqrt{3}$

103. $\pm 2, \pm i$ **105.** $\pm 1, \pm \sqrt{5}$ **107.** 4   **109.** $\frac{1}{2}, 1$

111. $-\frac{3}{4}, 2$ **113.** $-8, 27$

115.

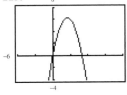

$(0, 0), (5, 0)$

117.

$2 \pm \sqrt{3}\, i$

119. $(x - 5)(x + 2) = x^2 - 3x - 10 = 0$

121. $\left[x - \left(1 + \sqrt{2}\right)\right]\left[x - \left(1 - \sqrt{2}\right)\right] = x^2 - 2x - 1 = 0$
$= x^2 - 2x + 3 = 0$

123. 4 seconds   **125.** 9 seconds   **127.** 1987

129. $0.06 = 6\%$ **131.** $\frac{5}{2}\sqrt{5} \approx 5.590$ seconds

Section 6.2 *(page 395)*

7. 16   **9.** $\frac{25}{4}$ **11.** 0, 6   **13.** $-3, -4$

15. $2 + \sqrt{7} \approx 4.65$ **17.** $-1 + \sqrt{2}\, i \approx -1 + 1.41i$
$\quad\;\; 2 - \sqrt{7} \approx -0.65$ $\quad\;\; -1 - \sqrt{2}\, i \approx -1 - 1.41i$

19. $\dfrac{3 + 2\sqrt{7}}{3} \approx 2.76$ **21.** $\dfrac{-5 + \sqrt{13}}{2} \approx -0.70$

$\quad\;\; \dfrac{3 - 2\sqrt{7}}{3} \approx -0.76$ $\quad\;\; \dfrac{-5 - \sqrt{13}}{2} \approx -4.30$

23. $\dfrac{-4 + \sqrt{10}}{2} \approx -0.42$ **25.** $\dfrac{-5 + \sqrt{17}}{2} \approx -0.44$

$\quad\;\; \dfrac{-4 - \sqrt{10}}{2} \approx -3.58$ $\quad\;\; \dfrac{-5 - \sqrt{17}}{2} \approx -4.56$

27. $1 \pm \sqrt{3}$ **29.** $4 \pm 2\sqrt{2}$

31.

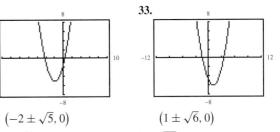

$\left(-2 \pm \sqrt{5}, 0\right)$

33.

$\left(1 \pm \sqrt{6}, 0\right)$

35. $30\sqrt{6} \approx 73.48$ feet   **37.** $2xy\sqrt{2x}$

39. $(|x| - 4)\sqrt{3}$ **41.** 15 minutes   **43.** 100

45. $\frac{9}{25}$ **47.** $\frac{9}{100}$ **49.** $\frac{1}{25}$ **51.** 0, 25

53. 1, 7   **55.** $-6, 4$ **57.** $-3, 6$ **59.** $\frac{3}{2}, 4$

61. $-2 + \sqrt{7} \approx 0.65$ **63.** $2 + \sqrt{3} \approx 3.73$
$\quad\;\; -2 - \sqrt{7} \approx -4.65$ $\quad\;\; 2 - \sqrt{3} \approx 0.27$

65. $5 + 3\sqrt{3} \approx 10.20$ **67.** $-10 + 3\sqrt{10} \approx -0.51$
$\quad\;\; 5 - 3\sqrt{3} \approx -0.20$ $\quad\;\; -10 - 3\sqrt{10} \approx -19.49$

69. $\dfrac{-3 + \sqrt{17}}{2} \approx 0.56$ **71.** $\dfrac{1}{2} + \dfrac{\sqrt{3}}{2}i \approx 0.5 + 0.87i$

$\quad\;\; \dfrac{-3 - \sqrt{17}}{2} \approx -3.56$ $\quad\;\; \dfrac{1}{2} - \dfrac{\sqrt{3}}{2}i \approx 0.5 - 0.87i$

73. $\dfrac{-9 + \sqrt{21}}{6} \approx -0.74$ **75.** $\dfrac{-1 + \sqrt{10}}{2} \approx 1.08$

$\quad\;\; \dfrac{-9 - \sqrt{21}}{6} \approx -2.26$ $\quad\;\; \dfrac{-1 - \sqrt{10}}{2} \approx -2.08$

77. $-1 \pm 2i$ **79.** $\dfrac{7 + \sqrt{57}}{2} \approx 7.27$

$\quad\;\; \dfrac{7 - \sqrt{57}}{2} \approx -0.27$

81. $-8.20, 2.20$ **83.** $-2.30, 1.30$

85. (a) $x^2 + 8x$ **87.** 8 feet × 20 feet

(b) $x^2 + 8x + 16$

(c) $(x + 4)^2$

89. 42 units or 58 units

91. 20 feet × 35 feet or 15 feet × $46\frac{2}{3}$ feet

93. 6.53 feet × 6.53 feet × 3 feet

Section 6.3 *(page 403)*

5. $2x^2 + 2x - 7 = 0$ **7.** $-x^2 + 10x - 5 = 0$

9. $4, 7$ **11.** $-\frac{3}{2}$ **13.** Two distinct irrational solutions

15. Two complex solutions **17.** $1 \pm \sqrt{5}$

19. $-2 \pm \sqrt{3}$ **21.** $-\frac{3}{2} \pm \frac{\sqrt{3}}{2}i$ **23.** $\frac{-1 \pm \sqrt{3}}{3}$

25. $\frac{-1 \pm \sqrt{10}}{5}$ **27.** $-15, 0$ **29.** $-4 \pm 3i$

31.

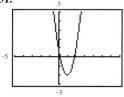

$(0.18, 0), (1.82, 0)$

33.

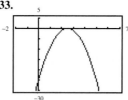

$(2.50, 0)$

35. $\frac{5 \pm \sqrt{185}}{8}$ **37.** $\frac{3 + \sqrt{17}}{2}$ **39.** (a) $c < 9$

(b) $c = 9$

(c) $c > 9$

41. (a) $c < 16$ **43.** 5.1 inches × 11.4 inches

(b) $c = 16$

(c) $c > 16$

45. $x - 9$ **47.** $4t + 12\sqrt{t} + 9$

49. 30% solution: $13\frac{1}{3}$ gallons **51.** $-2, -4$ **53.** $\frac{2}{3}$

60% solution: $6\frac{2}{3}$ gallons

55. $-\frac{1}{2}, \frac{2}{3}$ **57.** Two complex solutions

59. One rational solution **61.** $-3 \pm 2\sqrt{3}$

63. $5 \pm \sqrt{2}$ **65.** $\frac{1 \pm \sqrt{3}}{2}$ **67.** $\frac{-2 \pm \sqrt{10}}{2}$

69. $\frac{1 \pm \sqrt{5}}{5}$ **71.** $\frac{3}{4} \pm \frac{\sqrt{3}}{4}i$ **73.** $\frac{3 \pm \sqrt{13}}{6}$

75. ± 13 **77.** $\frac{9}{5}, \frac{21}{5}$ **79.** $8, 16$ **81.** $0.372, 3.228$

83. $0.200, 99.800$ **85.** (a) 2.5 seconds

(b) $\frac{5 + 5\sqrt{3}}{4} \approx 3.415$ seconds

87.

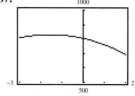

89. 606,870

Mid-Chapter Quiz *(page 406)*

1. ± 6 **2.** $-4, \frac{5}{2}$ **3.** $\pm 2\sqrt{3}$ **4.** $-1, 7$

5. $-5 \pm \sqrt{6}$ **6.** $\frac{-3 \pm \sqrt{19}}{2}$ **7.** $-2 \pm \sqrt{10}$

8. $\frac{3 \pm \sqrt{105}}{12}$ **9.** $-\frac{5}{2} \pm \frac{\sqrt{3}}{2}i$ **10.** $-2, 10$

11. $-3, 10$ **12.** $-2, 5$ **13.** $\frac{3}{2}$

14. $\frac{-5 \pm \sqrt{10}}{3}$ **15.** $-4, 1$ **16.** 2

17.

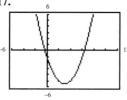

$(-0.32, 0), (6.32, 0)$

18.

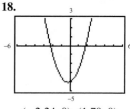

$(-2.24, 0), (1.79, 0)$

19. 50 units **20.** 35 meters × 65 meters

Section 6.4 *(page 413)*

7. 15, 16 **9.** 14, 16 **11.** 108 square inches

13. 440 meters **15.** 8% **17.** 2.59%

19. (a) 89 miles **21.** 100 feet × 125 feet

(b) 178 miles

23. 18 dozen, $1.20 per dozen **25.** 48

27. (a) $d = \sqrt{(3+x)^2 + (4+x)^2}$

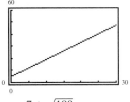

(b) $\dfrac{-7+\sqrt{199}}{2} \approx 3.55$ meters

29. 400 miles per hour   **31.** 9.1 hours, 11.1 hours

33. 3 seconds   **35.** 9.5 seconds   **37.** 30 units

39. $\sqrt{\dfrac{A}{P}} - 1$ **41.** $-10x^2$ **43.** $4x^2 - 60x + 225$

45. \$2300   **47.** 21, 23   **49.** 12, 13

51. 180 square kilometers   **53.** 210 inches

55. 12 inches × 24 inches   **57.** 6%   **59.** 7.5%

61. 5   **63.** 15.86 miles   **65.** 6.8 days, 9.8 days

67. 4.7 seconds   **69.** $\sqrt{\dfrac{S}{4\pi}}$ **71.** $\sqrt{\dfrac{12I - a^2 M}{M}}$

73. 46 miles per hour

75. (a) $b = 20 - a$
  $A = \pi a b$
  $A = \pi a(20 - a)$

(b)

a	4	7	10	13	16
A	201.1	285.9	314.2	285.9	201.1

(c) 7.9, 12.1

(d)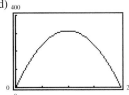

Section 6.5 *(page 425)*

7. $0, \dfrac{5}{2}$ **9.** 1, 3

11. (a) Not a solution   **13.** (a) Not a solution
  (b) Solution   (b) Solution
  (c) Not a solution   (c) Not a solution
  (d) Solution   (d) Not a solution

15. Negative: $(-\infty, 4)$
  Positive: $(4, \infty)$

17. Negative: $(-1, 5)$
  Positive: $(-\infty, -1)$ or $(5, \infty)$

19. $x \geq -3$ **21.** $x < 0$ or $x > 2$

23. $-5 \leq x \leq 2$ **25.** No solution

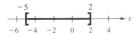

27. $-\infty < x < \infty$ **29.** $-2 < x < 0$ or $x > 2$

   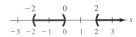

31. 3   **33.** $0, -5$

35. $x > 3$ **37.** $x < 3$ or $x \geq 6$

   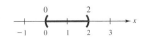

39. $x < -4$ or $x > \dfrac{3}{2}$ **41.** $-5 < x < 2.5$

43. $3 < t < 5$ **45.** $x < \dfrac{3}{2}$ **47.** $1 < x < 5$

49. 10 inches × 15 inches   **51.** $\pm\dfrac{9}{2}$ **53.** $\dfrac{5}{2}$

55. Negative: $(-\infty, 12)$ **57.** Negative: $(0, 4)$
  Positive: $(12, \infty)$ Positive: $(-\infty, 0) \cup (4, \infty)$

59. $x > 8$ **61.** $0 < x < 2$

63. $x \leq -2$ or $x \geq 2$ **65.** $u < -\dfrac{1}{2}$ or $u > 4$

67. $x < 2 - \sqrt{2}$ or $x > 2 + \sqrt{2}$ **69.** No solution

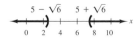

71. $x < 5 - \sqrt{6}$ or $x > 5 + \sqrt{6}$ **73.** All real numbers
except $x = 3$

75. $x < 3$ **77.** $x < 3$

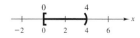   

79. $0 \le x < 4$ **81.** $y \le 3$ or $y > \frac{11}{2}$

83. $4 < x < 7$ **85.** $-3 < x < 7$

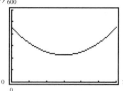

87. $3 < x < 4$ **89.** $0.64 < t < 4.86$ **91.** $12 < l < 16$
93. (a) $A = \pi x^2 + \pi(12 - x)^2, 0 < x < 12$

(b)   (c) $0.74 < x < 2.57$

Review Exercises (page 430)

1. $0, -12$ **3.** $-10, \frac{8}{3}$ **5.** $\pm\frac{1}{2}$ **7.** $-\frac{4}{3}, -\frac{4}{3}$

9. $-9, 10$ **11.** ± 100 **13.** $\pm\frac{3}{2}$ **15.** $-4, 36$

17. $3 \pm 2\sqrt{3}$ **19.** $\frac{3}{2} \pm \frac{\sqrt{3}}{2}i$ **21.** $\frac{-5 \pm \sqrt{19}}{2}$

23. $5, -6$ **25.** $3, -\frac{7}{2}$ **27.** $\frac{10}{3} \pm \frac{5\sqrt{2}}{3}i$

29. $3 \pm 5\sqrt{10}$ **31.** $-7, 12$ **33.** $-2, 20$ **35.** $3 \pm i$

37. $-\frac{1}{2}, -1$ **39.** $\frac{3 \pm \sqrt{17}}{2}$ **41.** $1, 1 \pm \sqrt{6}$

43. $-\frac{2}{3}, 3$ **45.**

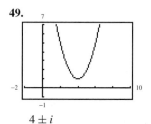

$(0, 0), (7, 0)$

47.   **49.**

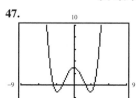

$(\pm 2, 0), (\pm 2\sqrt{3}, 0)$ $4 \pm i$

51. $11, 12$ **53.** (a) 200 feet

(b) Dropped

(c) $\frac{5\sqrt{2}}{2} \approx 3.54$ seconds

55. (a) $P = 2(l + w)$ (b) Substitute $24 - l$ for w.
$48 = 2(l + w)$
$24 = l + w$
$w = 24 - l$

(c)

l	2	4	6	8	10	12	14	16	18
A	44	80	108	128	140	144	140	128	108

57. 15   **59.** 10.7 hours, 13.7 hours   **61.** 48
63. 19 hours, 21 hours   **65.** $(x + 3)(x - 3) = 0$

$x^2 - 9 = 0$

67. $(x + 5)(x + 1) = 0$ **69.** $y = x^2 - 4x + 3$

$x^2 + 6x + 5 = 0$

71. $y = x^2 + 5x + 4$ **73.** $y = \frac{1}{9}x^2 - 1$
75. $x < 3$ **77.** $0 < x < 7$

79. $x \le -2$ or $x \ge 6$

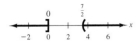

81. $-4 < x < \frac{5}{2}$

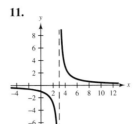

83. $x \le 0$ or $x > \frac{7}{2}$ **85.** 1995

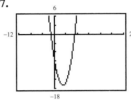

Chapter Test   *(page 433)*

1. $-5, 10$ **2.** $-\frac{3}{8}, 3$ **3.** $1.7, 2.3$

4. $-3 \pm 9i$ **5.** $C = \frac{9}{4}$ **6.** $\frac{3 \pm \sqrt{3}}{2}$

7. -56; two complex solutions   **8.** $\frac{4 \pm \sqrt{7}}{3}$

9. $\frac{2 \pm 3\sqrt{2}}{2}$ **10.** $\frac{7 + \sqrt{61}}{2}$

11. $(x + 4)(x - 5) = x^2 - x - 20 = 0$

12. $[x - (i - 1)][x - (-i - 1)] = x^2 + 2x + 2 = 0$

13. $0 < x < 3$ **14.** $x \le -2$ or $x \ge 6$

15. $2 < x < \frac{11}{4}$ **16.** $-8 \le u < 3$

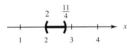

17. $14, 15$ **18.** 12×20 feet   **19.** 50 miles per hour

20. $\frac{\sqrt{10}}{2} \approx 1.58$ seconds

Cumulative Test: Chapters 4–6   *(page 434)*

1. $\frac{x(x + 2)(x + 4)}{9(x - 4)}$ **2.** $\frac{3x + 5}{x(x + 3)}$ **3.** $x + y$

4. $-4 + 3\sqrt{2}\,i$ **5.** $-\frac{2y^6}{3x^4}$ **6.** $t^{1/2}$ **7.** $35\sqrt{5x}$

8. $2x - 6\sqrt{2x} + 9$ **9.** $\sqrt{10} + 2$ **10.** $x^2 - 3x + 9$

11.

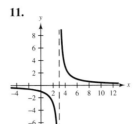

12.

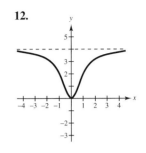

13. 2   **14.** 6   **15.** $5 \pm 5\sqrt{2}\,i$ **16.** $\frac{-3 \pm \sqrt{33}}{3}$

17.

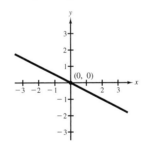

$(-1.12, 0), (7.12, 0)$

18. $y = x^2 - 4x - 12$

19. 1.6×10^7 **20.** $r_2 = \frac{\sqrt{15\,r_1}}{5}$

CHAPTER 7

Section 7.1   *(page 443)*

7. b   **9.** a   **11.** $m = 6, (-4, 3)$ **13.** $m = \frac{3}{2}, (0, 0)$

15. $y = -\frac{1}{2}x$ **17.** $y + 4 = 3x$

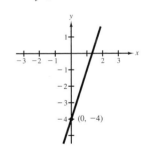

19. $y - \frac{7}{2} = -4(x + 2)$

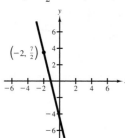

$\left(-2, \frac{7}{2}\right)$

21. $3x - 2y = 0$

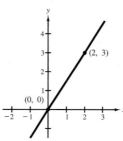

$(2, 3)$

$(0, 0)$

23. $3x + 7y - 15 = 0$

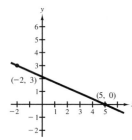

$(-2, 3)$

$(5, 0)$

25. $14x + 6y - 39 = 0$

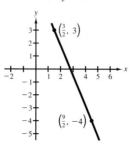

$\left(\frac{3}{2}, 3\right)$

$\left(\frac{9}{2}, -4\right)$

27. $f(x) = \frac{1}{2}x + 3$

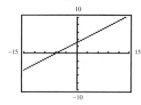

29. $f(x) = 3$

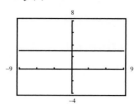

31. (a) $3x - y - 5 = 0$

(b) $x + 3y - 5 = 0$

33. (a) $x + y + 1 = 0$

(b) $x - y + 9 = 0$

35. (a) $y - 2 = 0$

(b) $x + 1 = 0$

37.

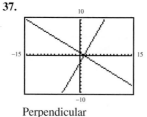

Perpendicular

39.

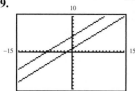

Parallel

41.

x	0	50	100	500	1,000
C	5000	6000	7000	15,000	25,000

43. (a) $P = 1200 - 15x$

(b) 42

(c) 48

45. $5x(1 - 4x)$

47. $(3x - 5)(5x + 3)$

49. (a)

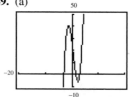

(b) $(2x - 3)(x + 3)(x - 3)$

$= (2x - 3)(x^2 - 9)$

$= 2x^3 - 3x^2 - 18x + 27$

51. $m = \frac{3}{4}, \left(-2, -\frac{5}{8}\right)$ **53.** $m = \frac{2}{3}, (0, -2)$

55. $m = \frac{5}{2}, (0, 12)$ **57.** $y + 6 = \frac{1}{2}(x)$

59. $y - 8 = -2(x + 2)$ **61.** $y = -\frac{2}{3}(x - 5)$

63. $y - 5 = 0(x + 8)$ or $y - 5 = 0$ **65.** $y - \frac{5}{2} = \frac{4}{3}\left(x - \frac{3}{4}\right)$

67. $2x + 5y = 0$ **69.** $5x + 34y - 67 = 0$

71. $2x - 15 = 0$ **73.** $4x + 5y - 11 = 0$

75. (a) $4x - y - 5 = 0$ **77.** (a) $3x + 10y + 22 = 0$

(b) $x + 4y - 31 = 0$ (b) $10x - 3y - 72 = 0$

79. Parallel   **81.** Perpendicular   **83.** Parallel

85. $\frac{x}{3} + \frac{y}{2} = 1$ **87.** $-\frac{6x}{5} - \frac{3y}{7} = 1$

89. $S = 1500 + 0.03M$ **91.** (a) $S = 0.70L$

(b) $94.50

93. (a) $N = 1500 + 60t$ **95.** $C = 0.32t + 2.3$

(b) 2700 $7.1 billion

(c) 1800

97. (a) and (b)   (c) $y = 4x + 19$ (d) 87

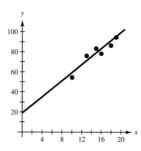

Section 7.2 *(page 452)*

7. b   **9.** d   **11.** f   **13.** (a) Solution

(b) Not a solution

(c) Solution

(d) Solution

15. (a) Solution   **17.**

(b) Not a solution

(c) Not a solution

(d) Solution

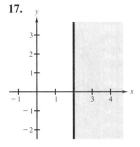

19.

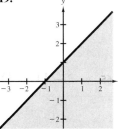

21.

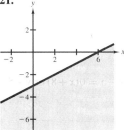

23.

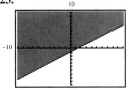

25.

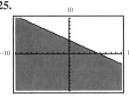

27. $y > x$ **29.** $3x + 4y - 17 > 0$

31. $y < 2$ **33.** $x - 2y < 0$

35.

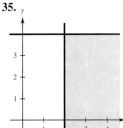

37.

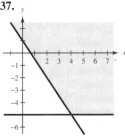

39.

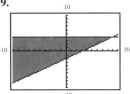

41.

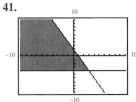

43.

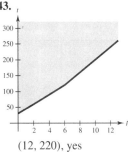

(12, 220), yes

45. $\dfrac{1}{x^3}$ **47.** $\dfrac{9y^2}{4x^2}$

49. $\dfrac{19\sqrt{2}}{2} \approx 13.435$ inches

51. (a) Not a solution   **53.** (a) Not a solution

(b) Not a solution   (b) Solution

(c) Solution   (c) Not a solution

(d) Solution   (d) Solution

55.

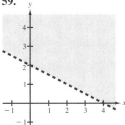

57.

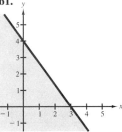

59.

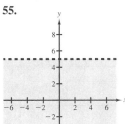

61.

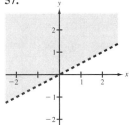

63.

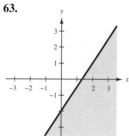

65.

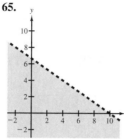

Section 7.3 (*page 460*)

7. e   **9.** b   **11.** d   **13.** Up, $(0, 2)$

15. Up, $(0, -6)$ **17.** $(\pm 5, 0), (0, 25)$ **19.** $(0, 3)$

21. $y = (x - 2)^2 + 3, (2, 3)$

23. $y = -(x - 1)^2 - 6, (1, -6)$

25.

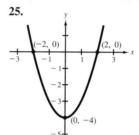

27.

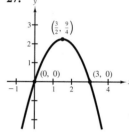

67.

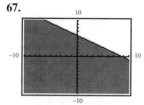

69.

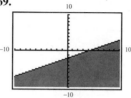

29.

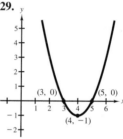

31.

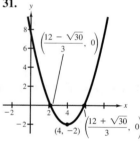

71.

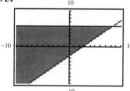

73.

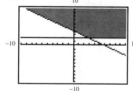

33. $y = (x - 2)^2$ **35.** $y = 4 - (x + 2)^2$

37. $y = -2(x + 3)^2 + 3$

39. (a)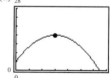

(b) 4 feet

(c) 16 feet

(d) $12 + 8\sqrt{3} \approx 25.9$ feet

75.

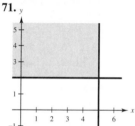

77.

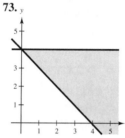

41. (a) Revenue = (number sold) $\cdot$ (price per unit)

$= (100 + x)[90 - x(0.15)]$

$= 9000 + 75x - 0.15x^2$

Cost = (number sold) $\cdot$ (cost per unit)

$= 60(100 + x) = 6000 + 60x$

Profit = Revenue $-$ Cost

$= 3000 + 15x - 0.15x^2$

79. $2x + 2y \le 500$

(Note: x and y cannot be negative.)

81. $30x + 20y \ge 300$

(Note: x and y cannot be negative.)

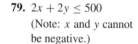

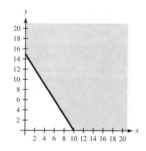

(b)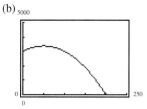

(c) Vertex; (50, 3375); 150 units

(d) Yes for orders up to 150 units

43. $y = \frac{1}{2}x(x - 2)$ **45.** $x < \frac{3}{2}$

47. $1 < x < 5$ **49.** $1, 3$

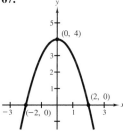

51. $\sqrt{5} \approx 2.236$ seconds   **53.** Down, $(10, 4)$

55. Up, $(3, -9)$ **57.** $(0, 0), (9, 0)$ **59.** $\left(\frac{3}{2}, 0\right), (0, 9)$

61. $y = (x + 3)^2 - 4, (-3, -4)$

63. $y = -(x - 3)^2 - 1, (3, -1)$

65. $y = 2\left(x + \frac{3}{2}\right)^2 - \frac{5}{2}, \left(-\frac{3}{2}, -\frac{5}{2}\right)$

67.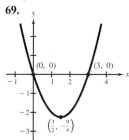

69.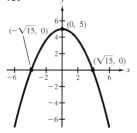

71.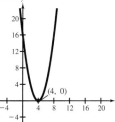

73.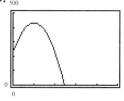

75.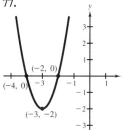

77.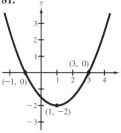

79.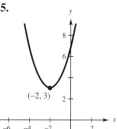

81.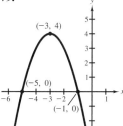

83.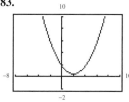

Vertex: $(2, 0.5)$

85.

Vertex: $(-1.9, 4.9)$

87.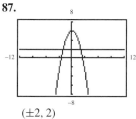

$(\pm 2, 2)$

89.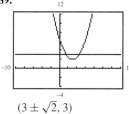

$(3 \pm \sqrt{2}, 3)$

91. $y = x^2 - 4x + 5$ **93.** $y = -x^2 - 6x - 5$

95. $y = x^2 - 4x$ **97.** $y = \frac{1}{2}x^2 - 3x + \frac{13}{2}$

99.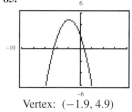

$2000

101. $y = \frac{1}{2500}x^2$

103.

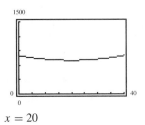

$x = 20$

Math Matters *(page 464)*

(a) $17,500 a year   (b) $12.50 an hour

Mid-Chapter Quiz *(page 465)*

1. $4x - 2y - 3 = 0$ **2.** $3x + 2y - 12 = 0$

3. $6x + 8y - 63 = 0$ **4.** $30x - 10y + 87 = 0$

5. $12x - 11y + 7 = 0$ **6.** $31x + 30y - 24 = 0$

7. $y + 1 = 0$ **8.** $x - 4 = 0$

9. (a) $2x - 3y + 9 = 0$ **10.** (a) Solution
   (b) $3x + 2y - 19 = 0$ (b) Solution
                                       (c) Not a solution
                                       (d) Not a solution

11. $x + 2y \le 11$ **12.** $x - 3y > -5$

13.

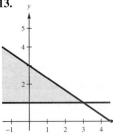

14.

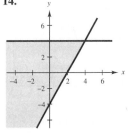

15. $y = (x - 3)^2 - 1$ **16.** $y = -\frac{1}{4}(x - 5)^2 + 4$

17.

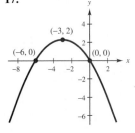

18.

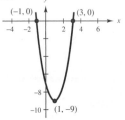

---

19. $V = -1850t + 12{,}400,\ 0 \le t \le 4$ **20.** 55 feet

Section 7.4 *(page 475)*

7. c   **9.** e   **11.** a

13. $x^2 + y^2 = 25$ **15.** $x^2 + y^2 = 29$

17. Center: $(0, 0)$ **19.** Center: $(0, 0)$
   Radius: $r = 4$ Radius: $r = \frac{12}{5}$

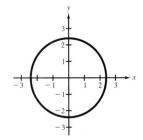

21. $\dfrac{x^2}{16} + \dfrac{y^2}{9} = 1$ **23.** $\dfrac{x^2}{9} + \dfrac{y^2}{16} = 1$ **25.** $\dfrac{x^2}{100} + \dfrac{y^2}{36} = 1$

27.   **29.**

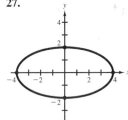

Vertices: $(\pm 4, 0)$ Vertices: $(0, \pm 2)$
Co-vertices: $(0, \pm 2)$ Co-vertices: $(\pm 1, 0)$

31.   **33.**

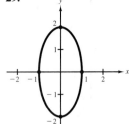

Vertices: $(\pm 3, 0)$ Vertices: $(0, \pm 3)$
Asymptotes: $y = \pm \frac{5}{3}x$ Asymptotes: $y = \pm \frac{3}{5}x$

35. $\dfrac{x^2}{16} - \dfrac{y^2}{64} = 1$ **37.** $\dfrac{y^2}{16} - \dfrac{x^2}{64} = 1$

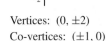
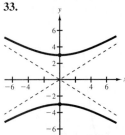
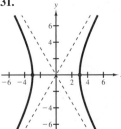
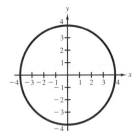

39. Center: (2, 3)
radius: $r = 2$

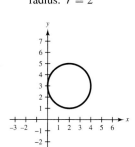

41. $(x - 2)^2 + (y - 1)^2 = 4$

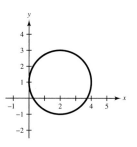

71.

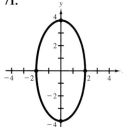

Vertices: $(0, \pm 4)$
Co-vertices: $(\pm 2, 0)$

73.

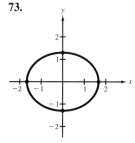

Vertices: $\left(\pm \frac{5}{3}, 0\right)$
Co-vertices: $\left(0, \pm \frac{4}{3}\right)$

43.

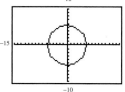

45.

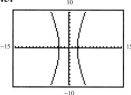

75.

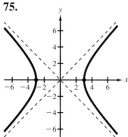

Vertices: $(\pm 3, 0)$
Asymptotes: $y = \pm x$

77.

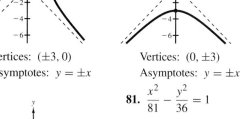

Vertices: $(0, \pm 3)$
Asymptotes: $y = \pm x$

47. $x^2 + y^2 = 4500^2$ **49.** $\sqrt{304} \approx 17.4$ feet

51.

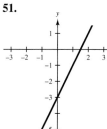

53.

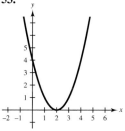

79.

Vertices: $(\pm 4, 0)$
Asymptotes: $y = \pm \frac{1}{2}x$

81. $\dfrac{x^2}{81} - \dfrac{y^2}{36} = 1$

55. $\sqrt{\dfrac{3}{\pi}} \approx 0.9722$ **57.** $x^2 + y^2 = \frac{4}{9}$ **59.** $x^2 + y^2 = 64$

61. Center: (0, 0)
radius: $r = 6$

63. Center: (0, 0)
radius: $r = \frac{1}{2}$

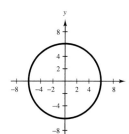

83. $\dfrac{y^2}{1} - \dfrac{x^2}{\frac{1}{4}} = 1$ **85.** Parabola **87.** Ellipse

89. Hyperbola **91.** Circle **93.** Line

95.

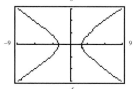

97.

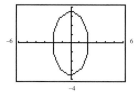

65. $\dfrac{x^2}{4} + \dfrac{y^2}{1} = 1$ **67.** $\dfrac{x^2}{1} + \dfrac{y^2}{4} = 1$ **69.** $\dfrac{x^2}{9} + \dfrac{y^2}{25} = 1$

99. (a) $x^2 + y^2 = 25^2$

$$y^2 = 625 - x^2$$
$$y = \sqrt{625 - x^2}$$
$$\text{width} = 2y = 2\sqrt{625 - x^2}$$
$$\text{Area} = (2x)(2y)$$
$$= 4x\sqrt{625 - x^2}$$

(b)

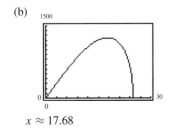

$x \approx 17.68$

Section 7.5 *(page 485)*

5. $I = kV$ **7.** $p = \dfrac{k}{d}$ **9.** $P = \dfrac{k}{V}$

11. The area of a triangle is proportional to the product of the base and height.

13. The volume of a right circular cylinder is proportional to the product of the square of the radius and the height.

15. $s = 5t$ **17.** $F = \dfrac{5}{16}x^2$ **19.** $n = \dfrac{48}{m}$

21. $d = \dfrac{120x^2}{r}$ **23.** 18 pounds   **25.** $208\frac{1}{3}$ feet

27. 3072 watts   **29.** 667 units   **31.** $\frac{1}{4}$

33. 3125 pounds   **35.** $y = \frac{1}{4}(5 - 3x)$

37. $y = \frac{2}{3}(x^2 - 1)$ **39.** 15, 17   **41.** $V = kt$

43. $u = kv^2$ **45.** $P = \dfrac{k}{\sqrt{1+r}}$ **47.** $A = klw$

49. The volume of a sphere varies directly as the cube of the radius.

51. The area of an ellipse varies jointly as the semimajor axis and the semiminor axis.

53. $H = \frac{5}{2}u$ **55.** $g = \dfrac{4}{\sqrt{z}}$ **57.** $F = \dfrac{25}{6}xy$

59. (a) \$4921.25
(b) Price per unit

101. $\dfrac{x^2}{144} + \dfrac{y^2}{64} = 1$

63. 2.44 hours, distance

65. 324 pounds

67. $p = \dfrac{114}{t}$, 17.5%

69.

x	2	4	6	8	10
$y = kx^2$	4	16	36	64	100

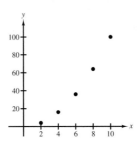

71.

x	2	4	6	8	10
$y = kx^2$	8	32	72	128	200

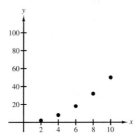

73.

x	2	4	6	8	10
$y = \dfrac{k}{x^2}$	$\frac{1}{2}$	$\frac{1}{8}$	$\frac{1}{18}$	$\frac{1}{32}$	$\frac{1}{50}$

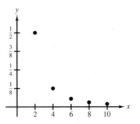

61. 32 feet per second squared

75.

x	2	4	6	8	10
$y = \dfrac{k}{x^2}$	$\dfrac{5}{2}$	$\dfrac{5}{8}$	$\dfrac{5}{18}$	$\dfrac{5}{32}$	$\dfrac{1}{10}$

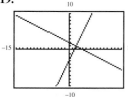

77. $y = \dfrac{4}{x}$ **79.** $y = -\dfrac{3}{10}x$

Review Exercises *(page 491)*

1. $2x - y - 6 = 0$ **3.** $4x + y = 0$

5. $2x + 3y - 17 = 0$ **7.** $x - 7 = 0$

9. $x + 2y + 6 = 0$ **11.** $y - 10 = 0$

13. $9x - 24y - 8 = 0$

15. (a) $3x + y - 1 = 0$ **17.** (a) $x - 12 = 0$
   (b) $x - 3y - 3 = 0$ (b) $y - 1 = 0$

19.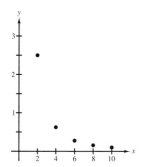

Perpendicular

21.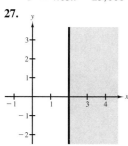

Neither

23. $C = 8.55x + 25,000$ **25.** $y = -x + 3.87$
   $P = 4.05x - 25,000$

27. 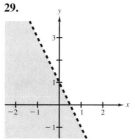   **29.**

31. 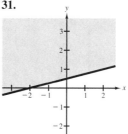   **33.**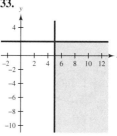

35. 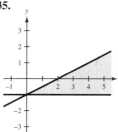   **37.**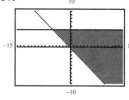

39. x hours at the grocery store   **41.** Parabola
   y hours mowing the lawn
   $8x + 10y \geq 200$

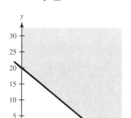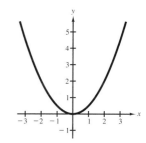

43. Hyperbola   **45.** Parabola

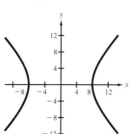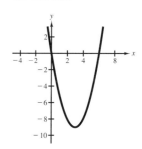

47. Ellipse

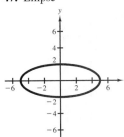

49. Circle

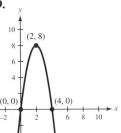

9.

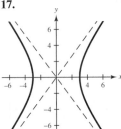

10. $y = \frac{2}{3}(x-3) - 2$

51. $y = \frac{1}{16}x^2 - \frac{5}{8}x + \frac{25}{16}$ **53.** $\frac{x^2}{4} + \frac{y^2}{25} = 1$

55. $x^2 + y^2 = 400$ **57.**

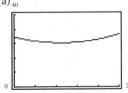

$s = 10$

11. 120 **12.** $x^2 + y^2 = 25$ **13.** $\frac{x^2}{9} + \frac{y^2}{100} = 1$

14. $\frac{x^2}{9} - \frac{y^2}{\frac{9}{4}} = 1$

15.

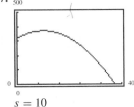

16.

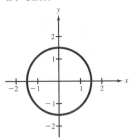

59. (a)

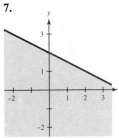

(b) 1981

(c) 32.5 million

17.

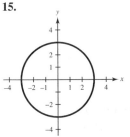

18.

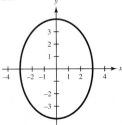

61. c **63.** a **65.** b **67.** $y = 4\sqrt[3]{3}x$

69. $T = \frac{1}{18}rs^2$ **71.** 150 pounds **73.** 945 units

19. $S = \frac{kx^2}{y}$ **20.** $v = \frac{1}{4}\sqrt{u}$

Chapter Test *(page 494)*

1. $2x + y = 0$ **2.** $x - 2y - 55 = 0$ **3.** $y + 1 = 0$

4. $x + 2 = 0$ **5.** $\frac{3}{5}$ **6.** $V = -4000t + 26,000$
$t = 2.5$

7.

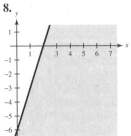

8.

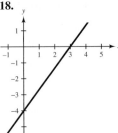

CHAPTER 8

Section 8.1 *(page 503)*

7. (a) Solution **9.** (a) Not a solution
(b) Not a solution (b) Solution

11. (2, 1) **13.** (−2, 4), (1, 1) **15.** (1, 2)

17. $\left(\frac{3}{2}, \frac{3}{2}\right)$ **19.** (0, 5), (−4, −3) **21.** No solution

23. $\left(1, \frac{1}{3}\right)$ **25.** No solution

27.

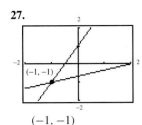

$(-1, -1)$

29.

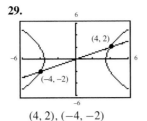

$(4, 2), (-4, -2)$

75.

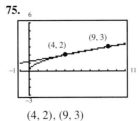

$(4, 2), (9, 3)$

77.

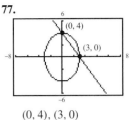

$(0, 4), (3, 0)$

31.

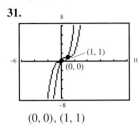

$(0, 0), (1, 1)$

33.

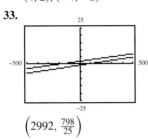

$\left(2992, \frac{798}{25}\right)$

79. $31, 49$ **81.** $20, 32$ **83.** 10 feet $\times$ 15 feet

85. 14 yards $\times$ 20 yards   **87.** \$15,000 at 8%

\$5000 at 9.5%

89. Between the points   **91.** 1987

$\left(-\frac{3}{5}, -\frac{4}{5}\right)$ and $\left(\frac{4}{5}, -\frac{3}{5}\right)$

35. 10,000 units   **37.** $\left(\dfrac{-90 + 96\sqrt{2}}{7}, \dfrac{160 - 96\sqrt{2}}{7}\right)$

39. $2x - y = 0$ **41.** $22x + 16y - 161 = 0$

43. 2.4 hours   **45.** $(4, 3)$ **47.** $\left(4, -\frac{1}{2}\right)$

49. $(-1, 3), (4, 2)$ **51.** $(4, -2)$ **53.** $(7, 2)$

55. $\left(\frac{1}{3}, \frac{17}{3}\right)$ **57.** $(2, 8), (-3, 18)$ **59.** $(2, 5), (-3, 0)$

61. $(10, 30), (20, 15)$ **63.** No solution   **65.** $(0, 2), (3, 1)$

67.

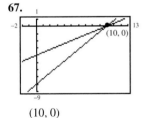

$(10, 0)$

69.

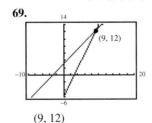

$(9, 12)$

71.

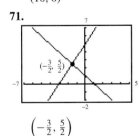

$\left(-\frac{3}{2}, \frac{5}{2}\right)$

73.

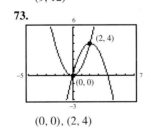

$(0, 0), (2, 4)$

Math Matters   *(page 507)*

1.

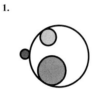

2.

3.

4.

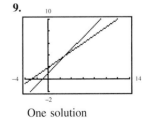

Section 8.2   *(page 514)*

7.

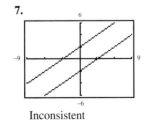

Inconsistent

9.

One solution

11. Inconsistent   **13.** (5, 3)

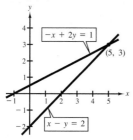

15. (2, −2)   **17.** Inconsistent

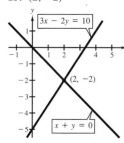   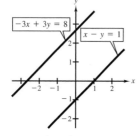

19. Infinitely many solutions   **21.** $\left(\frac{1}{2}, \frac{3}{2}\right)$

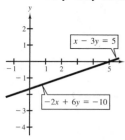   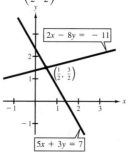

23. (3, 2)   **25.** Inconsistent   **27.** (3000, −2000)

29. (4, 3)   **31.** (2, 7)   **33.** $x + 2y = 0$
$4x + 2y = 9$

35. 121 weeks   **37.** 6.4 hours   **39.** 32 inches, 128 inches

41. Depth: 10 feet; Length of sections: 122 feet, 125 feet

43. $2x + 7y − 45 = 0$ **45.** $y − 3 = 0$

47. 4500 units   **49.** (5, −1)

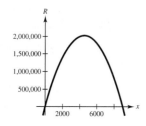

51. (−2, 5)   **53.** (−1, −1)   **55.** (7, −2)   **57.** $\left(\frac{3}{2}, 1\right)$

59. (−2, −1)   **61.** Infinitely many solutions

63. Inconsistent   **65.** $\left(\frac{25}{2}, \frac{1237}{250}\right)$ **67.** (15, 10)

69. (6, 11)   **71.** $k = 4$ **73.** Consistent

75. Consistent   **77.** Inconsistent

79.   **81.**

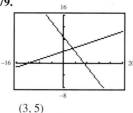   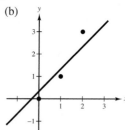

(3, 5)   (1/2, −1)

83. 50 meters × 60 meters   **85.** 5 dimes, 10 quarters

87. 11 nickels, 14 dimes   **89.** Regular unleaded: $1.11
Premium unleaded: $1.22

91. $5.65 variety: 6.1 pounds   **93.** 40% solution: 12 liters
$8.95 variety: 3.9 pounds   65% solution: 8 liters

95. (a) $y = x + \frac{1}{3}$ (b)

Section 8.3   *(page 526)*

 5. (22, −1, −5)   **7.** (14, 3, −1)   **9.** (1, 2, 3)

11. (3, 2, 1)   **13.** Inconsistent   **15.** (1, −1, 2)

17. $\left(\frac{10}{13}, \frac{4}{13}, 0\right)$ **19.** $y = −x^2 + 2x + 5$

21. $x^2 + y^2 − 4x = 0$ **23.** $s = −16t^2 + 144$

25. $s = −16t^2 + 48t$ **27.** Spray X: 20 gallons
Spray Y: 18 gallons
Spray Z: 16 gallons

29. String: 50   **31.** $(−\infty, \infty)$ **33.** $[−4, 4]$
Wind: 20
Percussion: 8

35. 50   **37.** $x − 2y + 3z = 5$ **39.** (a) Not a solution
$− y + 8z = 9$ (b) Solution
$2x − 3z = 0$ (c) Solution
(d) Not a solution

41. $(1, 2, 3)$ **43.** Inconsistent   **45.** $(-4, 8, 5)$

47. $(2, -1, 1)$ **49.** $(0, -4, 5)$ **51.** Inconsistent

53. Inconsistent   **55.** $\left(\frac{1}{4}, \frac{5}{4}, 0\right)$

57. $y = x^2 - 4x + 3$ **59.** $y = 2x - x^2$

61. $x^2 + y^2 - 3x - 2y = 0$

63. $x^2 + y^2 - 2x - 4y - 20 = 0$

65. (a) $y = 0.05t^2 + 0.15t + 6.6$

(b)

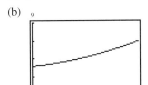

(c) 12 million short tons

67. $x + 2y - z = -4$
$y + 2z = 1$
$3x + y + 3z = 15$

Mid-Chapter Quiz *(page 530)*

1. $(10, 4)$ **2.**

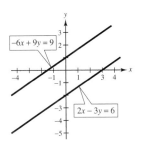

No solution

3.

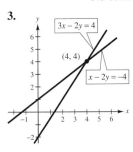

One

4.

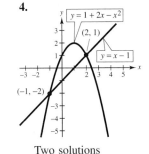

Two solutions

5.

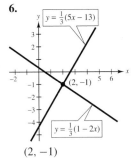

$(4, 2)$

6.

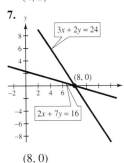

$(2, -1)$

7.

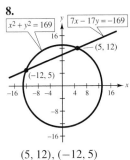

$(8, 0)$

8.

$(5, 12), (-12, 5)$

9. $(5, 2)$ **10.** $(1, 4), (-3, -4)$ **11.** $\left(\frac{90}{13}, \frac{34}{13}\right)$

12. $(5, 10)$ **13.** $(8, 1)$ **14.** $(-2, 4)$

15. $\left(\frac{1}{2}, -\frac{1}{2}, 1\right)$ **16.** $(5, -1, 3)$

17. $x + y = -2$ **18.** $x + y - z = 11$
$2x - y = 32$ $x + 2y - z = 14$
$-2x + y + z = -6$

19. $x + y = 20$ **20.** $y = x^2 + 3x - 2$
$0.2x + 0.5y = 6$
20% solution: $13\frac{1}{3}$ gallons
50% solution: $6\frac{2}{3}$ gallons

Section 8.4 *(page 539)*

7. 3×2 **9.** 2×2 **11.** $\begin{bmatrix} 4 & -5 & -2 \\ -1 & 8 & 10 \end{bmatrix}$

13. $\begin{bmatrix} 1 & 10 & -3 & 2 \\ 5 & -3 & 4 & 0 \\ 2 & 4 & 0 & 6 \end{bmatrix}$ **15.** $4x + 3y = 8$
$x - 2y = 3$

17. $x + 2z = -10$ **19.** $\begin{bmatrix} 1 & 4 & 3 \\ 0 & 2 & -1 \end{bmatrix}$
$3y - z = 5$
$4x + 2y = 3$

21. $\begin{bmatrix} 1 & 2 & 3 \\ 0 & 1 & 2 \end{bmatrix}$ **23.** $\begin{bmatrix} 1 & 1 & -1 & 3 \\ 0 & 1 & -4 & 1 \\ 0 & 0 & 0 & 0 \end{bmatrix}$

25. $x + 5y = 3$
$\quad\quad y = -2$
$\quad (13, -2)$

27. $(1, 1)$ **29.** $(2, -3, 2)$

31. $(2a + 1, 3a + 2, a)$ **33.** Inconsistent

35. 8%: \$800,000 **37.** $y = x^2 + 2x + 4$
9%: \$500,000
12%: \$200,000

39. (a) $y = -\frac{1}{250}x^2 + \frac{3}{5}x + 6$

(b)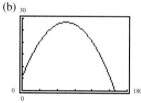

(c) Maximum height: 28.5 feet
Range: 159.4 feet

41. (a) 12 **43.** (a) $\frac{1}{3}$ **45.** $C = 5.75x + 12,000$
(b) $\frac{3}{16}$
(b) $\dfrac{c-6}{c+4}$

47. $\begin{bmatrix} 1 & -2 & \frac{2}{3} \\ 2 & 8 & 15 \end{bmatrix}$ **49.** $\begin{bmatrix} 1 & 1 & 4 & -1 \\ 0 & 5 & -2 & 6 \\ 0 & 3 & 20 & 4 \end{bmatrix}$

$\begin{bmatrix} 1 & 1 & 4 & -1 \\ 0 & 1 & -\frac{2}{5} & \frac{6}{5} \\ 0 & 3 & 20 & 4 \end{bmatrix}$

51. $\begin{bmatrix} 1 & \frac{3}{2} & \frac{1}{4} \\ 0 & 1 & \frac{11}{10} \end{bmatrix}$ **53.** $\begin{bmatrix} 1 & 1 & 0 & 5 \\ 0 & 1 & 2 & 0 \\ 0 & 0 & 1 & -1 \end{bmatrix}$

55. $\begin{bmatrix} 1 & -1 & -1 & 1 \\ 0 & 1 & 6 & 3 \\ 0 & 0 & 1 & \frac{4}{5} \end{bmatrix}$ **57.** $x - 2y = 4$
$\quad\quad\quad y = -3$
$\quad (-2, -3)$

59. $x - y + 2z = 4$ **61.** $\left(\frac{9}{5}, \frac{13}{5}\right)$
$\quad\quad y - z = 2$
$\quad\quad\quad\quad z = -2$
$\quad (8, 0, -2)$

63. Inconsistent **65.** $(1, -1, 2)$ **67.** $(8, 10, 6)$

69. $(-12a - 1, 4a + 1, a)$ **71.** $(-1, -2, 3)$

73. $\left(1, \frac{2}{3}, -1\right)$ **75.** Certificates of deposit: $250,000 - \frac{1}{2}s$
Municipal bonds: $125,000 + \frac{1}{2}s$
Blue-chip stocks: $125,000 - s$
Growth stocks: s

77. $y = \frac{1}{2}x^2 - 3x + \frac{7}{2}$ **79.** 5, 8, 20

81. $\dfrac{2x^2 - 9x}{(x-2)^3} = \dfrac{2}{x-2} - \dfrac{1}{(x-2)^2} - \dfrac{10}{(x-2)^3}$

Section 8.5 *(page 552)*

5. 5 **7.** -24 **9.** -24 **11.** -30 **13.** 0

15. 248 **17.** 5.4 **19.** $(2, -2)$

21. Not possible; $D = 0$ **23.** $(-1, 3, 2)$ **25.** $\left(1, \frac{1}{2}, \frac{3}{2}\right)$

27. $\left(\frac{22}{27}, \frac{22}{9}\right)$ **29.** $\left(\frac{1}{3}, 1, -\frac{2}{3}\right)$ **31.** 16

33. 250 square miles **35.** Collinear

37. $9x + 10y + 3 = 0$ **39.** $y = \frac{1}{2}x^2 - 2x$

41. (a) $y_1 = -0.8t^2 + 28.9t + 393.6$
(b) $y_2 = 26.8t^2 - 35t + 495.3$
(c)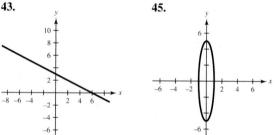

(d) $y = -27.6t^2 + 63.9t - 101.7$

43. 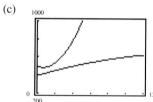 **45.**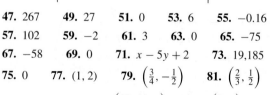

47. 267 **49.** 27 **51.** 0 **53.** 6 **55.** -0.16

57. 102 **59.** -2 **61.** 3 **63.** 0 **65.** -75

67. -58 **69.** 0 **71.** $x - 5y + 2$ **73.** 19,185

75. 0 **77.** $(1, 2)$ **79.** $\left(\frac{3}{4}, -\frac{1}{2}\right)$ **81.** $\left(\frac{2}{3}, \frac{1}{2}\right)$

83. $(1, -2, 1)$ **85.** $\left(\frac{13}{16}, \frac{11}{16}, 0\right)$ **87.** $\left(\frac{1}{2}, 4\right)$

89. $\left(2, \frac{1}{2}, -\frac{1}{2}\right)$ **91.** 7   **93.** $\frac{31}{2}$ **95.** 31 square units

97. Collinear   **99.** Not collinear   **101.** $3x - 5y = 0$

103. $7x - 6y - 28 = 0$ **105.** $y = 2x^2 - 6x + 1$

107. $y = -3x^2 + 2x$ **109.** 1, 6

Review Exercises *(page 558)*

1. $(2, -1)$ **3.** Inconsistent   **5.** $(-10, -5)$

7. $(-1, 5), (-2, 20)$ **9.** $(0, -1), (-1, 0)$

11.
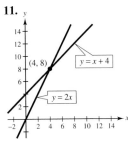
$(4, 8)$

13.
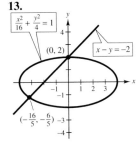
$(0, 2), \left(-\frac{16}{5}, -\frac{6}{5}\right)$

15.
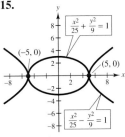
$(\pm 5, 0)$

17. $(3, 4)$

19. $(\pm 3, \pm 4)$ **21.** $(0, 0)$ **23.** $(2, -3, 3)$

25. $(10, -12)$ **27.** $\left(\frac{24}{5}, \frac{22}{5}, -\frac{8}{5}\right)$ **29.** 5

31. -51 **33.** 1   **35.** $(-3, 7)$ **37.** Inconsistent

39. $(2, -3, 3)$ **41.** $x - 2y + 4 = 0$

43. $2x + 6y - 13 = 0$ **45.** 16   **47.** 7

49. $3x + y = -2$ **51.** 16,667 units
$\ 6x + y = 0$

53. 75% solution: 40 gallons
 50% solution: 60 gallons

55. 193.75 miles per hour, 218.75 miles per hour

57. $y = 2x^2 + x - 6$

59. $x^2 + y^2 - 4x + 2y - 4 = 0$

61. (a) $y = -\frac{1}{45}x^2 + \frac{2}{3}x + 11$

(b)
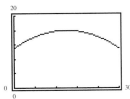

(c) $7\frac{1}{4}$ feet

Chapter Test *(page 433)*

1. $\left(1, \frac{1}{2}\right)$ **2.** $(2, 4)$ **3.** $(2, 6), (6, 2)$ **4.** $(3, 2)$

5. $(-2, 2)$ **6.** $\left(\frac{1}{4}, \frac{1}{3}\right)$ **7.** $(2, -1, 0)$ **8.** $(-1, 3, 3)$

9. $(2, 1, -2)$ **10.** $\left(4, \frac{1}{7}\right)$ **11.** $(5, 4)$

12. $(5, 1, -1)$ **13.** $\left(-\frac{11}{5}, \frac{56}{25}, \frac{32}{25}\right)$

14. Inconsistent   **15.** -62 **16.** $-\frac{24}{5}$
 One solution
 Infinitely many solutions

17. $x + 2y = -1$ **18.** $x + y = 200$
$\ x + y = 2$ $\ 4x - y = 0$
 40 miles, 60 miles

19. $y = 2x^2 - 3x + 4$ **20.** 12

CHAPTER 9

Section 9.1 *(page 503)*

7. 2^{2x-1} **9.** $8e^{3x}$ **11.** (a) $\frac{1}{9}$ **13.** (a) 73.89
 (b) 1 (b) 1.35
 (c) 3 (c) 0.18

15. b   **17.** e   **19.** f

21.
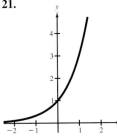

23.
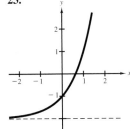

25.

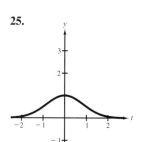

27.

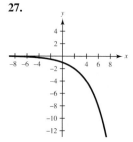

43. $-11x$ **45.** $6x^2 + 9$ **47.** (a) $\dfrac{1}{2\pi}$ **49.** 11.036

(b) $\dfrac{1}{4\pi}$

51. 1.396   **53.** (a) $\frac{1}{5}$ **55.** (a) 500

(b) 5   (b) 250

(c) 125   (c) 56.657

57. (a) 1000   **59.** (a) 0.368   **61.** d   **63.** a   **65.** f

(b) 1628.895   (b) 1

(c) 2653.298   (c) 1.649

29.

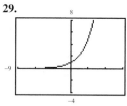

31.

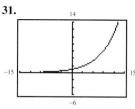

33.

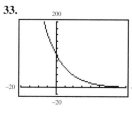

35.

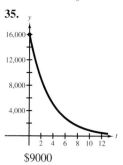

$9000

67.

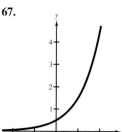

69.

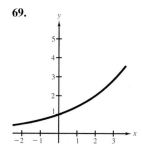

37.

n	1	4	12	365	Continuous
A	$466.10	$487.54	$492.68	$495.23	$495.30

39.

n	1	4	12	365	Continuous
A	$2541.75	$2498.00	$2487.98	$2483.09	$2482.93

71.

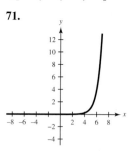

73.

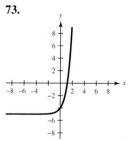

75.

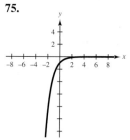

77.

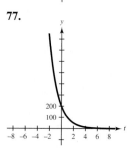

41. (a)

(b)

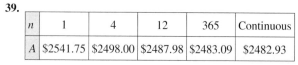

(c) The difference between the functions increases at an increasing rate.

79.

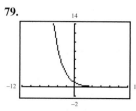

81.

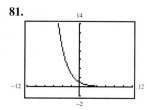

83.

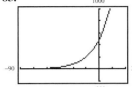

85.

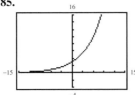

87.

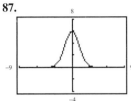

89. (a) 272.184 million

(b) 288.627 million

99.

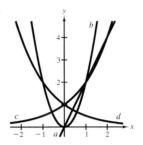

$f(x) = 2^x$ and $f(x) = 2^{-x}$ are exponential.

91. $32.50

93.

n	1	4	12
A	$4734.73	$4870.38	$4902.71

n	365	Continuous
A	$4918.66	$4919.21

95.

n	1	4	12
A	$226,296.28	$259,889.34	$268,503.32

n	365	Continuous
A	$272,841.23	$272,990.75

97.

n	1	4	12
A	$18,429.30	$15,830.43	$15,272.04

n	365	Continuous
A	$15,004.64	$14,995.58

101. (a)

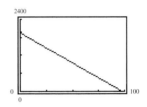

(b)

t	0	25	50	75
h	2000 ft	1450 ft	950 ft	450 ft

Ground level: 97.5 seconds

103. (a)

Year	1987	1988	1989	1990
Price	$99,744	$121,074	$130,035	$131,368

Year	1991	1992
Price	$132,715	$142,537

(b)

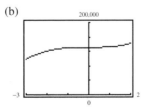

(c) Increasing at a higher rate. Home prices probably will
not increase at a higher rate indefinitely.

Section 9.2   *(page 581)*

7. (a) 13
(b) 16
(c) $x^2 - 3$
(d) $(x - 3)^2$

9. (a) 3
(b) 8
(c) $\sqrt{x + 5}$
(d) $\sqrt{x} + 5$

11. (a) -1
(b) -2
(c) -2

13. (a) -1
(b) 1

15. (a) $[2, \infty)$
(b) $[0, \infty)$

17. $g(f(x)) = 0.02(x - 200{,}000)$
$x > 200{,}000$

19. $f^{-1}(x) = \frac{1}{5}x$ **21.** $f^{-1}(x) = \sqrt[7]{x}$

23. $f(g(x)) = (x - 15) + 15 = x$
$g(f(x)) = (x + 15) - 15 = x$

25. $f(g(x)) = \sqrt[3]{(x^3 - 1) + 1}$
$= \sqrt[3]{x^3} = x$
$g(f(x)) = \left(\sqrt[3]{x + 1}\right)^3 - 1$
$= x + 1 - 1 = x$

27. b   **29.** d   **31.**

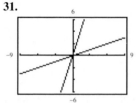

33.

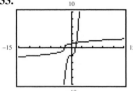

35. $g^{-1}(x) = \frac{1}{4}(3 - x)$

37. $f^{-1}(t) = \sqrt[3]{t + 1}$ **39.** No   **41.** No

43. $x \ge 0$, $f^{-1}(x) = \sqrt[4]{x}$ **45.** $x \ge 2$, $f^{-1}(x) = \sqrt{x} + 2$

47.

x	0	1	3	4
f^{-1}	6	4	2	0

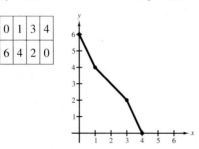

49. True   **51.** $-4, 1$ **53.** $\dfrac{-3 \pm \sqrt{37}}{4}$

55. (a) Down
(b) $(0, 0)$, $(4, 0)$
(c) $(2, 4)$

57. (a) 7
(b) 8
(c) $2x + 1$
(d) $2(x + 1)$

59. (a) 0
(b) 3
(c) $3|x - 1|$
(d) $3|x - 3|$

61. (a) $\frac{1}{4}$
(b) $\frac{1}{3}$
(c) $\dfrac{1}{\sqrt{x} - 3}$
(d) $\dfrac{1}{\sqrt{x - 3}}$

63. (a) 10
(b) 1

65. (a) 0
(b) 10

67. (a) $(-\infty, \infty)$
(b) $(-\infty, \infty)$

69. (a) $(-\infty, \infty)$
(b) $(-\infty, -9) \cup (-9, \infty)$

71. $(C \circ x)(t) = 102t + 300$
Total cost after t hours of production

73. $f^{-1}(x) = \frac{1}{6}x$ **75.** $f^{-1}(x) = x - 10$

77. $f^{-1}(x) = x^3$ **79.** $f^{-1}(x) = \frac{1}{2}(x + 1)$

81. $f(g(x)) = 10\left(\frac{1}{10}x\right) = x$
$g(f(x)) = \frac{1}{10}(10x) = x$

83. $f(g(x)) = 1 - 2\left[\frac{1}{2}(1 - x)\right]$
$= 1 - (1 - x) = x$
$g(f(x)) = \frac{1}{2}[1 - (1 - 2x)]$
$= \frac{1}{2}(2x) = x$

85. $f(g(x)) = 2 - 3\left[\frac{1}{3}(2 - x)\right]$
$= 2 - (2 - x) = x$
$g(f(x)) = \frac{1}{3}[2 - (2 - 3x)]$
$= \frac{1}{3}(3x) = x$

87. $f(g(x)) = \dfrac{1}{\left(\frac{1}{x}\right)} = x$
$g(f(x)) = \dfrac{1}{\left(\frac{1}{x}\right)} = x$

89.
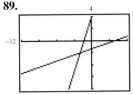

91.
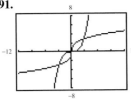

93. $f^{-1}(x) = \frac{1}{8}x$ **95.** $g^{-1}(x) = x - 25$

97. $g^{-1}(t) = -4(t - 2)$ **99.** $h^{-1}(x) = x^2, \ x \geq 0$

101. $g^{-1}(s) = \dfrac{5}{s}$

103. $f^{-1}(x) = \sqrt[3]{x - 1}$ **105.**
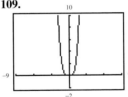
Yes

107.
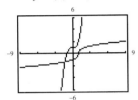
Yes

109.
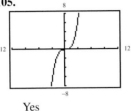
No

111.
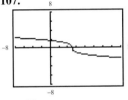
Yes

113.
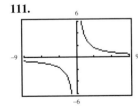
No

115.

x	-4	-2	2	3
f^{-1}	-2	-1	1	3

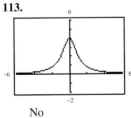

117. (a) $f^{-1}(x) = \frac{1}{2}(3 - x)$

(b) $(f^{-1})^{-1}(x) = 3 - 2x$

119. (a) $y = \frac{20}{13}(x - 9)$ (b) x: hourly wage   (c) 8

y: number of units
produced

121. False. $f(x) = \sqrt{x - 1}$;   domain: $[1, \infty)$

$f^{-1}(x) = x^2 + 1$;   domain: $[0, \infty)$

Section 9.3   *(page 593)*

7. $5^2 = 25$ **9.** $36^{1/2} = 6$ **11.** $\log_3 \frac{1}{9} = -2$

13. $\log_8 4 = \frac{2}{3}$ **15.** 3   **17.** -2

19. There is no power to which 2 can be raised to obtain -3.

21. $\frac{1}{2}$ **23.** 2   **25.** 1   **27.** 1.4914   **29.** -0.2877

31. e   **33.** d   **35.** a   **37.** $f^{-1} = g$

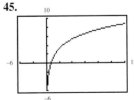

39.
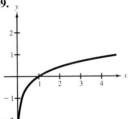

41.
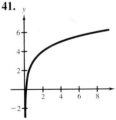

43.
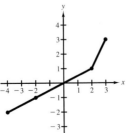

45.
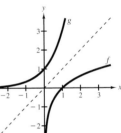

47. 2.3481   **49.** -0.4739

51.

r	0.07	0.08	0.09	0.10	0.11	0.12
t	9.90	8.66	7.70	6.93	6.30	5.78

53.

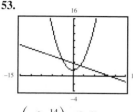

$\left(-3, \frac{14}{3}\right), (2, 3)$

55. $x^2 - 6$

57. Compact discs: 53.3 million   **59.** $6^2 = 36$
Records: 128.2 million

61. $4^{-2} = \frac{1}{16}$ **63.** $8^{2/3} = 4$ **65.** $\log_7 49 = 2$

67. $\log_{25} \frac{1}{5} = -\frac{1}{2}$ **69.** $\log_4 1 = 0$ **71.** 0

73. 1   **75.** 3

77. There is no power to which 4 can be raised to obtain -4.

79. $\frac{3}{4}$ **81.** 4   **83.** 0   **85.** 1   **87.** 0.7335

89. -0.0706 **91.** 3.2189

93. $f^{-1} = g$

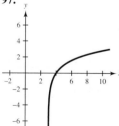

95.

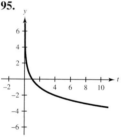

97.

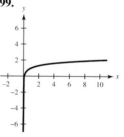

99.

101.

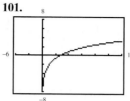

103.

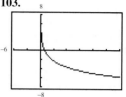

105.

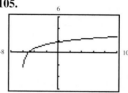

107.

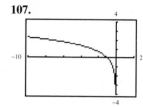

109. 1.7712   **111.** 2.6332   **113.** 1.3481   **115.** -2

117. 53.4 inches   **119.** $0 < x < \infty$

121. $3 \le f(x) \le 4$ **123.** A factor of 10

Math Matters *(page 596)*

79.5%

Mid-Chapter Quiz *(page 597)*

1. (a) $\frac{16}{9}$

(b) 1

(c) $\frac{3}{4}$

(d) $\frac{8\sqrt{3}}{9}$

2. Domain: $(-\infty, \infty)$
Range: $(0, \infty)$

3.

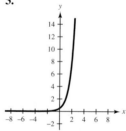

4.

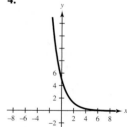

5.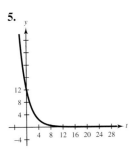

6.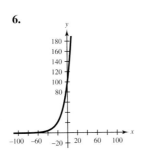

7.

n	1	4	12	365	Continuous
A	\$3185.89	\$3314.90	\$3345.61	\$3360.75	\$3361.27

8. \$2.71   **9.** (a) -19

(b) 125

(c) $2x^3 - 3$

(d) $(2x - 3)^3$

10. $f(g(x)) = 3 - 5\left[\frac{1}{5}(3 - x)\right] = 3 - 3 + x = x$

$g(f(x)) = \frac{1}{5}[3 - (3 - 5x)] = \frac{1}{5}(5x) = x$

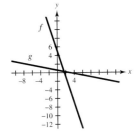

11. $h^{-1}(x) = \frac{1}{10}(x - 3)$ **12.** $g^{-1}(t) = \sqrt[3]{2(t - 2)}$

13. $4^{-2} = \frac{1}{16}$ **14.** $\log_3 81 = 4$ **15.** 3

16. $f^{-1}(x) = g(x)$

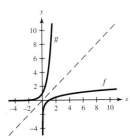

17.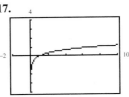

18.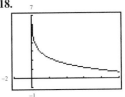

19. $f(x) = \log_5(x - 3) + 1$ **20.** 3.4096

Section 9.4 *(page 603)*

5. 2   **7.** -3 **9.** 0.954   **11.** 1.556

13. 0.778   **15.** $\log_2 3 + \log_2 x$ **17.** $-2\log_5 x$

19. $\ln 5 - \ln(x - 2)$ **21.** $\frac{1}{2}[\ln x + \ln(x + 2)]$

23. $\log_{12} \frac{x}{3}$ **25.** $\log_5(2x)^{-2}$ **27.** $\ln \frac{32y^3}{x}$

29. $\ln 4\sqrt{x}$ **31.** $1 - \log_4 x$ **33.** $1 + \frac{1}{2}\log_5 2$

35. $2 + \ln 3$ **37.**

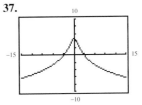

39. Evaluate when $x = 3$ and $y = 2$.

41. $E = 1.4\log_{10} \frac{C_2}{C_1}$ **43.** x^{20} **45.** $\frac{3}{2y^5}$

47. $\frac{1}{6}x^3 y$ **49.** 12   **51.** 2   **53.** 2   **55.** 3

57. 0   **59.** 1   **61.** 1.2925   **63.** 0.2925

65. 0.2500   **67.** 2.7925   **69.** 0

71. $\log_3 11 + \log_3 x$ **73.** $3\ln y$ **75.** $\log_2 z - \log_2 17$

77. $\frac{1}{3}\log_3(x + 1)$ **79.** $\ln 3 + 2\ln x + \ln y$

81. $2\log_2 x - \log_2(x - 3)$ **83.** $\frac{1}{3}[\ln x + \ln(x + 5)]$

85. $\log_2 3x$ **87.** $\log_{10} \frac{4}{x}$ **89.** $\ln b^4$ **91.** $\ln \sqrt[3]{2x + 1}$

93. $\log_3 2\sqrt{y}$ **95.** $\ln \frac{x^2 y^3}{z}$ **97.** $\ln(xy)^4$ **99.** True

101. True   **103.** False   **105.**

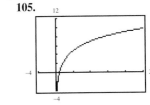

107. $10(\log_{10} I + 16)$, 60 decibels

109. False. 0 is not in the domain of f.

111. False. $f(x - 3) = \ln(x - 3)$

113. False. If $v = u^2$, then $f(v) = \ln u^2 = 2\ln u = 2f(u)$.

Section 9.5 *(page 612)*

5. (a) Not a solution   **7.** (a) Solution

     (b) Solution   (b) Not a solution

9. (a) Not a solution   **11.** 8   **13.** −6   **15.** −7

     (b) Solution

17. 9   **19.** $2x - 1$ **21.** $2x$ **23.** $\frac{3}{2}$ **25.** 1.23

27. 0.59   **29.** 2.48   **31.** 9.73   **33.** 22.63

35. 2187   **37.** 6.52   **39.** 12.18   **41.** 2.46

43.

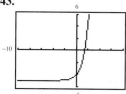

(1.40, 0)

45.

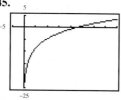

(21.82, 0)

47. 7.70 years   **49.** 262.5°   **51.** $2|x|y\sqrt{6y}$

53. $\dfrac{2\sqrt{3b^3}}{a^2}$ **55.** $5395.40   **57.** 5   **59.** 8

61. 3   **63.** −3   **65.** 18   **67.** $\frac{22}{5}$ **69.** 81

71. 500,000   **73.** 5.49   **75.** 0.86   **77.** 3.28

79. 2.49   **81.** 0.39   **83.** 6.80   **85.** 0.80

87. 7.39   **89.** 21.82   **91.** 8.99   **93.** 1

95. 1000   **97.** 7.91   **99.** 8.17   **101.** −0.78

103. 31,622,771.60   **105.** ±20.09   **107.** 5

109.

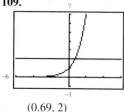

(0.69, 2)

111.

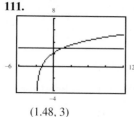

(1.48, 3)

113. 3.16×10^{-9} **115.** 205

Section 9.6 *(page 621)*

7. 9%   **9.** 8%   **11.** 8.75 years   **13.** 6.60 years

15. Continuous   **17.** Quarterly   **19.** 8.33%

21. 7.23%   **23.** $1652.99   **25.** $626.46

27. $k = \frac{1}{2}\ln\frac{8}{3} \approx 0.4904$ **29.** $k = -\dfrac{\ln 2}{3} \approx -0.2310$

31. $y = 10.1e^{0.0072t}$ **33.** $y = 4.6e^{0.0432t}$

     11.1 million   8.1 million

35. (a) k is larger in Exercise 33, because the population of Dhaka is increasing faster than the population of Los Angeles.

     (b) k corresponds to r; k gives the annual percentage rate of growth.

37. 3.2595 grams   **39.** 63 times as great   **41.** 7.04

43. $\dfrac{9}{2(x + 3)}$ **45.** $\dfrac{x^2 + 2x - 13}{x(x - 2)}$ **47.** $\dfrac{7x}{9}, \dfrac{19x}{18}$

49. 7%   **51.** 6%   **53.** 7%   **55.** 14.21 years

57. 9.99 years   **59.** 9.24 years   **61.** 6.136%

63. 8.300%   **65.** 7.788%   **67.** No   **69.** $3080.15

71. $951.23   **73.** $67,667.64   **75.** $5,496.57

77. $320,250.81   **79.** Total deposits: $7200.00

                       Total interest: $10,529.42

81. 0.8517 grams   **83.** $y = 6.5e^{0.0019t}$

                       6.7 million

85. $y = 21.6e^{0.0320t}$

     32.7 million

87. (a)

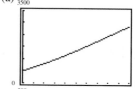

     (b) 1000

     (c) 2642

     (d) 5.88 years

89. $9281.25

91. (a)

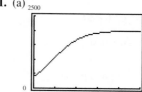

     (b) 1300 units

     (c) 3 years

     (d) 2000 units

93. The one in Chile is 63 times as great.   **95.** 10^7 times

97. (a)

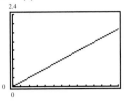

$E = 0.146t + 0.007$

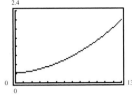

$E = 0.009t^2 + 0.018t + 0.325$

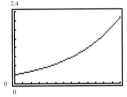

$E = 0.301(1.165)^t$

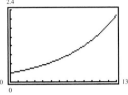

$E = 0.301e^{0.153t}$

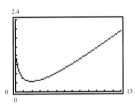

$E = 0.201 + 0.228t - 0.444 \ln t$

(b) $E = 0.146t + 0.007$.
   Sum of Squares: 0.1725

$E = 0.009t^2 + 0.018t + 0.325$.
   Sum of Squares: 0.0076

$E = 0.301(1.165)^t$.
   Sum of Squares: 0.0315

$E = 0.301e^{0.153t}$.
   Sum of Squares: 0.0313

$E = 0.201 + 0.228t - 0.444 \ln t$.
   Sum of Squares: 0.0209

Quadratic

Review Exercises *(page 628)*

1. (a) $\frac{1}{8}$ **3.** (a) 2.718   **5.** (a) 0
   (b) 2   (b) 0.351   (b) 3
   (c) 4   (c) 0.135   (c) -0.631

7. (a) 1   **9.** (a) -6 **11.** d   **13.** a   **15.** c
   (b) -1.099 (b) 0
   (c) 2.303   (c) 22.5

17.

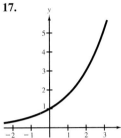

19.

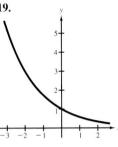

21.

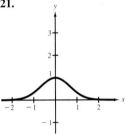

23.

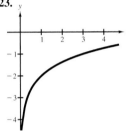

25.

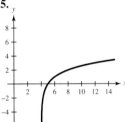

27.

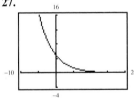

29.

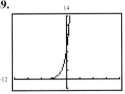

31.

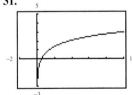

33.

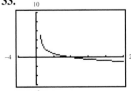

35. $(f \circ g)(x) = x^2 + 2$
        $(g \circ f)(x) = (x + 2)^2$

37. $(f \circ g)(x) = x$ **39.** (a) $[2, \infty)$
$(g \circ f)(x) = x$ (b) $[4, \infty)$

41.

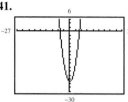

Not one-to-one

43.

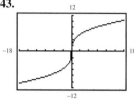

One-to-one

45. $f^{-1}(x) = 4x$ **47.** $h^{-1}(x) = x^2, x \geq 0$

49. Not one-to-one   **51.** $\log_4 64 = 3$ **53.** $e^1 = e$

55. 3   **57.** -2 **59.** 7   **61.** 0

63. $\log_4 6 + 4 \log_4 x$ **65.** $\frac{1}{5} \log_5 (x + 2)$

67. $\ln(x + 2) - \ln(x - 2)$ **69.** $\log_4 \dfrac{x}{10}$

71. $4 + \ln x^8$ **73.** $\ln \left(\dfrac{3}{2x} \right)^2$

75.

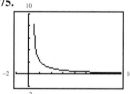

77. False.
$\log_2 4x = 2 + \log_2 x$

79. True   **81.** True   **83.** 1.79588   **85.** -0.43068

87. 1.02931   **89.** 1.585   **91.** 2.132   **93.** 6

95. 1   **97.** 125   **99.** 5.66   **101.** 6.23

103. 32.99   **105.** 15.81   **107.** 1408.10   **109.** 0.32

111.

n	1	4	12
A	\$3806.13	\$4009.59	\$4058.25

n	365	Continuous
A	\$4082.26	\$4083.08

113.

n	1	4	12
A	\$67,275.00	\$72,095.68	\$73,280.74

n	365	Continuous
A	\$73,870.32	\$73,890.56

115.

n	1	4	12
P	\$2301.55	\$2103.50	\$2059.87

n	365	Continuous
P	\$2038.82	\$2038.11

117. 4.6 years   **119.** 150 units

121.

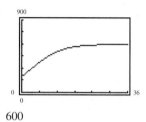

Chapter Test   *(page 631)*

1. $f(-1) = 81$
$f(0) = 54$
$f\left(\frac{1}{2}\right) = 18\sqrt{6} \approx 44.09$
$f(2) = 24$

2.

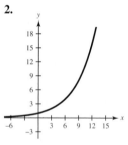

3. $5^3 = 125$ **4.** $\log_4 \frac{1}{16} = -2$ **5.** $\frac{1}{3}$

6. $g = f^{-1}$ **7.** $\log_4 5 + 2 \log_4 x - \frac{1}{2} \log_4 y$

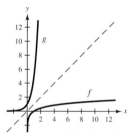

8. $\ln \dfrac{x}{y^4}$ **9.** $3 + \log_5 6$ **10.** 3   **11.** 0.973

12. 13.733   **13.** 15.516   **14.** (a) \$8012.78
                                          (b) \$8110.40

15. $10,806.08   **16.** 7%   **17.** $8469.14

18. 600   **19.** 1141   **20.** 4.4 years

Cumulative Test: Chapters 7–9 *(page 632)*

1. $3x - 4y + 12 = 0$ **2.**

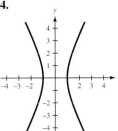

3.

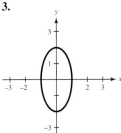

4.

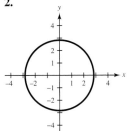

5.

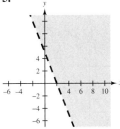

6. $y = \frac{2}{3}(x - 3)^2 - 2$

7. 11 feet   **8.** 128 feet   **9.** $(2, 1)$ **10.** $(3, -2)$

11. $(5, 4)$ **12.** $\left(-\frac{1}{5}, -\frac{22}{5}\right)$ **13.** 8, 12, 24

14.

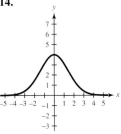

15. -2

16. Reflections in the line $y = x$ because
$f^{-1}(x) = g(x)$.

17. $\log_2\left(\dfrac{x^3 y^3}{z}\right)$ **18.** (a) 3   **19.** 8.329%

(b) 12.18

(c) 18.01

(d) 0.87

20. 15.40 years

CHAPTER 10

Section 10.1 *(page 640)*

7. 2, 4, 6, 8, 10   **9.** $-\frac{1}{2}, \frac{1}{4}, -\frac{1}{8}, \frac{1}{16}, -\frac{1}{32}$

11. $\frac{2}{5}, \frac{4}{8}, \frac{6}{11}, \frac{8}{14}, \frac{10}{17}$ **13.** 2, 2, $\frac{4}{3}, \frac{2}{3}, \frac{4}{15}$

15. c   **17.** b   **19.**

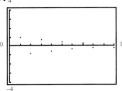

21. -72 **23.** $\frac{1}{702}$ **25.** $n(n + 1)$

27. 63   **29.** 77   **31.** 7.283   **33.** 54

35. $\displaystyle\sum_{k=1}^{5} 2k$ **37.** $\displaystyle\sum_{k=1}^{20} \frac{4}{k+3}$ **39.** $\displaystyle\sum_{k=1}^{20} \frac{2k}{k+3}$

41. (a) $502.92, $505.85, $508.80, $511.77, $514.75,
$517.76, $520.78, $523.82

(b) $2019.37   (c)

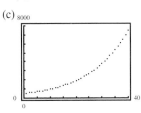

(d) Yes. Investment earning compound interest increases
at an increasing rate.

43. $x^2 + 20x + 100$ **45.** $\dfrac{1}{a^8}$ **47.** $x - y + 1 = 0$

49. $-2, 4, -6, 8, -10$ **51.** $\frac{1}{2}, \frac{1}{4}, \frac{1}{8}, \frac{1}{16}, \frac{1}{32}$

53. $\frac{1}{2}, \frac{1}{3}, \frac{1}{4}, \frac{1}{5}, \frac{1}{6}$ **55.** $-1, \frac{1}{4}, -\frac{1}{9}, \frac{1}{16}, -\frac{1}{25}$

57. $\frac{9}{2}, \frac{19}{4}, \frac{39}{8}, \frac{79}{16}, \frac{159}{32}$ **59.** 0, 4, 0, 4, 0

61. 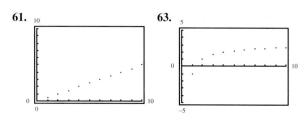   **63.**

79. 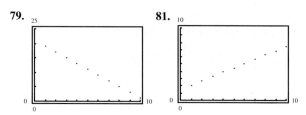   **81.**

65. 5   **67.** $\frac{1}{132}$ **69.** $\frac{1}{n+1}$ **71.** $2n$ **73.** 35

75. $\frac{47}{60}$ **77.** -48 **79.** $\frac{8}{9}$ **81.** $\frac{182}{243}$ **83.** 112

85. 852   **87.** 6.579   **89.** $\sum_{k=1}^{5} k$ **91.** $\sum_{k=1}^{10} \frac{1}{2k}$

93. $\sum_{k=1}^{20} \frac{1}{k^2}$ **95.** $\sum_{k=0}^{9} \frac{(-1)^k}{3^k}$ **97.** $\sum_{k=1}^{11} \frac{k}{k+1}$ **99.** $\sum_{k=0}^{6} k!$

101. 3   **103.** 12.5   **105.** $a_5 = 108°, a_6 = 120°$
$a_5 + 2a_6 = 348° < 360°$

107. $25.7°, 45°, 60°, 72°, 81.8°$

83. $3 + \frac{1}{2}(n-1)$ **85.** $64 - 8(n-1)$

87. $8 + 4(n-1)$ **89.** $50 - 10(n-1)$

91. $9, 6, 3, 0, -3$ **93.** $-10, -4, 2, 8, 14$

95. a   **97.** e   **99.** c   **101.** 210   **103.** 2700

105. 62,625   **107.** 19,500   **109.** 522   **111.** 900

113. 12,200   **115.** 243   **117.** 2850   **119.** 2550

121. \$246,000   **123.** \$25.42

125. (a) 4, 9, 16, 25, 36   (b) 49

(c) $\sum_{k=1}^{n} [1 + 2(k-1)] = n^2$

Section 10.2 *(page 648)*

7. 12   **9.** $-\frac{5}{4}$ **11.** Arithmetic, $d = -2$

13. Arithmetic, $d = -\frac{1}{2}$ **15.** Arithmetic, $d = 4$

17. Not arithmetic   **19.** 40, 35, 30, 25, 20

21. $\frac{8}{5}, \frac{11}{5}, \frac{14}{5}, \frac{17}{5}, 4$ **23.** 25, 28, 31, 34, 37

25. d   **27.** c   **29.** $3 + \frac{3}{2}(n-1)$

31. $28 - 4(n-1)$ **33.** $5 + 2(n-1)$ **35.** 275

37. 9000   **39.** 1850   **41.** 23   **43.** 4914

45. 1024 feet   **47.** $\left(-\frac{1}{2}, \frac{1}{4}\right), (2, 4)$ **49.** (1, 2, 1)

51. Adults: 816   **53.** 3   **55.** -6 **57.** $\frac{2}{3}$
Children: 384

59. Arithmetic, $d = 2$ **61.** Arithmetic, $d = \frac{3}{2}$

63. Not arithmetic   **65.** Arithmetic, $d = \frac{1}{6}$

67. Arithmetic, $d = 0.8$ **69.** Not arithmetic

71. 7, 10, 13, 16, 19   **73.** 6, 4, 2, 0, -2

75. $\frac{3}{2}, 4, \frac{13}{2}, 9, \frac{23}{2}$ **77.** 4, $\frac{15}{4}, \frac{7}{2}, \frac{13}{4}, 3$

Section 10.3 *(page 657)*

7. -3 **9.** π **11.** Geometric, $r = 2$

13. Not geometric   **15.** $4, -2, 1, -\frac{1}{2}, \frac{1}{4}$

17. 20, 21.40, 22.90, 24.50, 26.22   **19.** -0.0061

21. 1486.02   **23.** $25(4)^{n-1}$ **25.** $\left(\frac{3}{2}\right)^{n-1}$ **27.** b

29. a   **31.**   **33.** -5460

35. 6.67   **37.** 399.93   **39.** 13,120   **41.** 48.00

43. (a) $250{,}000(0.75)^n$ **45.** \$75,715.32

(b) \$59,326.17

(c) The first year

47. 70.875 square inches   **49.** -200 **51.** -60

53. 58   **55.** 3   **57.** $\frac{1}{3}$ **59.** $-\frac{3}{2}$ **61.** e

63. Not geometric   **65.** Geometric, $r = \frac{1}{2}$

67. Geometric, $r = -\frac{2}{3}$ **69.** Geometric, $r = 1.02$

71. 4, 8, 16, 32, 64   **73.** 6, 2, $\frac{2}{3}, \frac{2}{9}, \frac{2}{27}$

75. $1, -\frac{1}{2}, \frac{1}{4}, -\frac{1}{8}, \frac{1}{16}$

77. 1000, 1010, 1020.10, 1030.30, 1040.60

79. $\frac{3}{256}$ **81.** $48\sqrt{2}$ **83.** $\frac{81}{256}$ **85.** $\frac{243}{32}$

87. $2(3)^{n-1}$ **89.** 2^{n-1} **91.** $4\left(-\frac{1}{2}\right)^{n-1}$

93. $8\left(\frac{1}{4}\right)^{n-1}$ **95.** 1023   **97.** 772.478

99. 2.250   **101.** 6300.250   **103.** $-14{,}762$

105. 16.000   **107.** 152.095   **109.** 110.357

111. (a) $P(0.999)^n$ (b) 69.4%

(c)
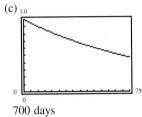
700 days

113. \$3,623,993.23   **115.** \$19,496.56   **117.** \$105,428.44

119. (a) \$53,687.09   (b) \$107,374.18

121. 72.969 square inches

Mid-Chapter Quiz *(page 662)*

1. $32, 8, 2, \frac{1}{2}, \frac{1}{8}$ **2.** $-\frac{3}{5}, 3, -\frac{81}{7}, \frac{81}{2}, -135$

3. 100   **4.** 40   **5.** 87   **6.** -32

7. $\sum\limits_{k=1}^{20} \frac{2}{3k}$ **8.** $\sum\limits_{k=1}^{25} \frac{(-1)^{k-1}}{k^3}$ **9.** $\frac{1}{2}$ **10.** -6

11. $20 - 3(n-1)$ **12.** $32\left(-\frac{1}{4}\right)^{n-1}$ **13.** 4075

14. 9030   **15.** 25.947   **16.** 18,392.796   **17.** -0.026

18. a_n: upper graph   **19.** $5.5°$ **20.** Arithmetic
    b_n: lower graph

Section 10.4 *(page 668)*

7. 15   **9.** 1   **11.** 593,775

13. 2,598,960   **15.** 455   **17.** 53,130

19. 1, 9, 36, 84, 126, 126, 84, 36, 9, 1

21. $x^6 + 18x^5 + 135x^4 + 540x^3 + 1215x^2 + 1458x + 729$

23. $u^3 - 6u^2v + 12uv^2 - 8v^3$

25. $x^8 + 8x^7y + 28x^6y^2 + 56x^5y^3 + 70x^4y^4 + 56x^3y^5 + 28x^2y^6 + 8xy^7 + y^8$

27. $x^6 - 12x^5 + 60x^4 - 160x^3 + 240x^2 - 192x + 64$

29. 120   **31.** -1365

33. $\frac{1}{32} + \frac{5}{32} + \frac{10}{32} + \frac{10}{32} + \frac{5}{32} + \frac{1}{32}$

35. $\frac{1}{256} + \frac{12}{256} + \frac{54}{256} + \frac{108}{256} + \frac{81}{256}$

37. 1.17   **39.** 510,568.79   **41.** $4^3 = 64$ **43.** $e^0 = 1$

45.   **47.** 252   **49.** 1

$t = 1.72$

51. 1225   **53.** 12,650   **55.** 792

57. 22,451,004,309,013,280   **59.** 5,200,300

61. 10   **63.** 792   **65.** 1   **67.** 35   **69.** 70

71. $x^5 + 5x^4 + 10x^3 + 10x^2 + 5x + 1$

73. $x^6 - 24x^5 + 240x^4 - 1280x^3 + 3840x^2 - 6144x + 4096$

75. $x^4 + 4x^3y + 6x^2y^2 + 4xy^3 + y^4$

77. $x^5 - 5x^4y + 10x^3y^2 - 10x^2y^3 + 5xy^4 - y^5$

79. $a^3 + 6a^2 + 12a + 8$

81. $16x^4 - 32x^3 + 24x^2 - 8x + 1$

83. $64y^6 + 192y^5z + 240y^4z^2 + 160y^3z^3 + 60y^2z^4 + 12yz^5 + z^6$

85. -120 **87.** 1760   **89.** 54

91. 1, 3, 6, 10, 15
The difference between consecutive determinants increases by 1.

Section 10.5 *(page 677)*

5. 5   **7.** 9   **9.** 8   **11.** 6   **13.** 72

15. $xyz, xzy, yxz, yzx, zxy, zyx$ **17.** 120   **19.** 40,320

21. $\{A, B\}, \{A, C\}, \{A, D\}, \{A, E\}, \{A, F\}, \{B, C\}, \{B, D\}, \{B, E\}, \{B, F\}, \{C, D\}, \{C, E\}, \{C, F\}, \{D, E\}, \{D, F\}, \{E, F\}$

23. (a) 3   (b) 6   **25.** 56   **27.** (a) 56
    (c) 15   (d) 28   (b) 56
    (e) 45   (f) 66   (c) 8

29. 41   **31.** 69   **33.** $2x - 3y + 9 = 0$

35. 10   **37.** 8   **39.** 6   **41.** 7   **43.** 6

45. 6,760,000   **47.** AB   BA   **49.** 24   **51.** 5040

AC   CA

AD   DA

BC   CB

BD   DB

CD   DC

53. $\{A, B, C\}, \{A, B, D\}, \{A, B, E\}, \{A, C, D\}, \{A, C, E\},$
$\{A, D, E\}, \{B, C, D\}, \{B, C, E\}, \{B, D, E\}, \{C, D, E\}$

55. 260   **57.** 142,506   **59.** 3003

61. (a) 70   (b) 16   **63.** (a) 5   (b) 9   (c) 20

Section 10.6 *(page 685)*

5. $\{A, B, C, D, E, \ldots, X, Y, Z\}$

7. $\{AB, AC, AD, AE, BC, BD, BE, CD, CE, DE\}$

9. 0.65   **11.** $\frac{3}{8}$ **13.** $\frac{7}{8}$ **15.** $\frac{9}{10}$

17. (a) $\frac{1}{20}$ (b) $\frac{2}{5}$ (c) $\frac{1}{2}$ (d) $\frac{1}{4}$

19. $\frac{1917}{6565}, \frac{4648}{6565}$ **21.** $\frac{1}{24}$ **23.** $\frac{1}{45}$ **25.** $\frac{1}{54,145}$

27. g is a vertical shift of f four units downward.

29. g is a reflection of f in the x-axis.

31.

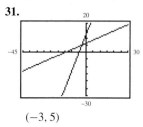

$(-3, 5)$

33. $\{ABC, ACB, BAC, BCA, CAB, CBA\}$

35. $\{WWW, WWL, WLW, WLL, LWW, LWL, LLW, LLL\}$

37. 0.18   **39.** $\frac{1}{2}$ **41.** $\frac{3}{13}$ **43.** $\frac{1}{6}$ **45.** $\frac{5}{6}$

47. 0.158   **49.** 0.733   **51.** 0.022   **53.** 0.455

55. (a) $\frac{1}{5}$ (b) $\frac{1}{3}$ (c) 1

57.

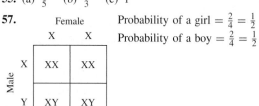

Probability of a girl $= \frac{2}{4} = \frac{1}{2}$

Probability of a boy $= \frac{2}{4} = \frac{1}{2}$

59. $\frac{14}{65}$ **61.** $\frac{1}{100,000}$ **63.** $\frac{1}{45}$ **65.** (a) $\frac{6}{11}$ (b) $\frac{5}{11}$

Review Exercises *(page 691)*

1. $\sum_{k=1}^{4}(5k - 3)$ **3.** $\sum_{n=1}^{6}\frac{1}{3n}$ **5.** 380

7. $n(n - 1)(n - 2)$ **9.** 127, 122, 117, 112, 107

11. $\frac{5}{4}, 2, \frac{11}{4}, \frac{7}{2}, \frac{17}{4}$ **13.** -2.5 **15.** 10, 30, 90, 270, 810

17. $100, -50, 25, -\frac{25}{2}, \frac{25}{4}$ **19.** $\frac{3}{2}$ **21.** $10 + 4(n - 1)$

23. $1000 - 50(n - 1)$ **25.** $\left(-\frac{2}{3}\right)^{n-1}$ **27.** $24(2)^{n-1}$

29. $12\left(-\frac{1}{2}\right)^{n-1}$ **31.** a   **33.** b   **35.** d

37.   **39.**

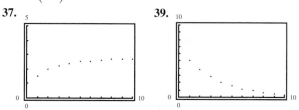

41. 28   **43.** $\frac{4}{5}$ **45.** 486   **47.** 1262.5   **49.** 8190

51. -1.928 **53.** 19.842   **55.** 116,169.538

57. 5100   **59.** 462   **61.** (a) $85,000(1.012)^t$

(b) 154,328

63. 56   **65.** 1   **67.** 91,390

69. $x^{10} + 10x^9 + 45x^8 + 120x^7 + 210x^6 + 252x^5 + 210x^4 +$
$120x^3 + 45x^2 + 10x + 1$

71. $y^6 - 12y^5 + 60y^4 - 160y^3 + 240y^2 - 192y + 64$

73. $x^8 - 4x^7 + 7x^6 - 7x^5 + \frac{35}{8}x^4 - \frac{7}{4}x^3 + \frac{7}{16}x^2 - \frac{1}{16}x + \frac{1}{256}$

75. $-61,236$ **77.** 8   **79.** 3003   **81.** $\frac{1}{3}$

83. $\frac{1}{24}$ **85.** 0.346

Chapter Test *(page 631)*

1. $1, -\frac{2}{3}, \frac{4}{9}, -\frac{8}{27}, \frac{16}{81}$ **2.** 35   **3.** -45 **4.** $\sum_{k=1}^{12}\frac{2}{3k + 1}$

5. 12, 16, 20, 24, 28   **6.** $5000 - 100(n - 1)$

7. 3825   **8.** $-\frac{3}{2}$ **9.** $4\left(\frac{1}{2}\right)^{n-1}$ **10.** 1020

11. $\frac{3069}{1024}$ **12.** \$47,868.33   **13.** 1140

14. $x^5 - 10x^4 + 40x^3 - 80x^2 + 80x - 32$ **15.** 56

16. 26,000   **17.** 12,650   **18.** 0.25   **19.** $\frac{3}{26}$ **20.** $\frac{1}{6}$

APPENDIX A

Section A.1 *(page A5)*

1. Statement **3.** Open statement **5.** Open statement

7. Open statement **9.** Nonstatement

11. Open statement **13.** (a) True **15.** (a) True
(b) False (b) True

17. (a) False **19.** (a) True
(b) False (b) False

21. (a) The sun is not shining.
(b) It is not hot.
(c) The sun is shining and it is hot.
(d) The sun is shining or it is hot.

23. (a) Lions are not mammals.
(b) Lions are not carnivorous.
(c) Lions are mammals and lions are carnivorous.
(d) Lions are mammals or lions are carnivorous.

25. (a) The sun is not shining and it is hot.
(b) The sun is not shining or it is hot.
(c) The sun is shining and it is not hot.
(d) The sun is shining or it is not hot.

27. (a) Lions are not mammals and lions are carnivorous.
(b) Lions are not mammals or lions are carnivorous.
(c) Lions are mammals and lions are not carnivorous.
(d) Lions are mammals or lions are not carnivorous.

29. $p \wedge \sim q$ **31.** $\sim p \vee q$ **33.** $\sim p \vee \sim q$

35. $\sim p \wedge q$ **37.** The bus is blue.

39. x is not equal to 4. **41.** The earth is flat.

43.

p	q	$\sim p$	$\sim p \wedge q$
T	T	F	F
T	F	F	F
F	T	T	T
F	F	T	F

45.

p	q	$\sim p$	$\sim q$	$\sim p \vee \sim q$
T	T	F	F	F
T	F	F	T	T
F	T	T	F	T
F	F	T	T	T

47.

p	q	$\sim q$	$p \vee \sim q$
T	T	F	T
T	F	T	T
F	T	F	F
F	F	T	T

49. Not logically equivalent **51.** Logically equivalent

53. Logically equivalent **55.** Not logically equivalent

57. Logically equivalent **59.** Not a tautology

61. A tautology

63.

p	q	$\sim p$	$\sim q$	$p \wedge q$	$\sim(p \wedge q)$	$\sim p \vee \sim q$
T	T	F	F	T	F	F
T	F	F	T	F	T	T
F	T	T	F	F	T	T
F	F	T	T	F	T	T

Identical columns: $\sim(p \wedge q)$ and $\sim p \vee \sim q$

Section A.2 *(page A13)*

1. (a) If the engine is running, then the engine is wasting gasoline.
(b) If the engine is wasting gasoline, then the engine is running.
(c) If the engine is not wasting gasoline, then the engine is not running.
(d) If the engine is running, then the engine is not wasting gasoline.

3. (a) If the integer is even, then it is divisible by 2.
(b) If it is divisible by 2, then the integer is even.
(c) If it is not divisible by 2, then the integer is not even.
(d) If the integer is even, then it is not divisible by 2.

5. $q \rightarrow p$ **7.** $p \rightarrow q$ **9.** $p \rightarrow q$ **11.** True

13. True   **15.** False   **17.** True   **19.** True

21. Converse:
If you can see the eclipse, then the sky is clear.

Inverse:
If the sky is not clear, then you cannot see the eclipse.

Contrapositive:
If you cannot see the eclipse, then the sky is not clear.

23. Converse:
If the deficit increases, then taxes were raised.

Inverse:
If taxes are not raised, then the deficit will not increase.

Contrapositive:
If the deficit does not increase, then taxes were not raised.

25. Converse:
It is necessary to apply for the visa to have a birth certificate.

Inverse:
It is not necessary to have a birth certificate to not apply for the visa.

Contrapositive:
It is not necessary to apply for the visa to not have a birth certificate.

27. Paul is not a junior and not a senior.

29. The temperature will increase and the metal rod will not expand.

31. We will go to the ocean and the weather forecast is not good.

33. No students are in extracurricular activities.

35. Some contact sports are not dangerous.

37. Some children are allowed at the concert.

39. None of the $20 bills is counterfeit.

41.

p	q	$\sim q$	$p \rightarrow \sim q$	$\sim(p \rightarrow \sim q)$
T	T	F	F	T
T	F	T	T	F
F	T	F	T	F
F	F	T	T	T

43.

p	q	$q \rightarrow p$	$\sim(q \rightarrow p)$	$\sim(q \rightarrow p) \wedge q$
T	T	T	F	F
T	F	T	F	F
F	T	F	T	T
F	F	T	F	F

45.

p	q	$\sim p$	$p \vee q$	$(p \vee q) \wedge (\sim p)$
T	T	F	T	F
T	F	F	T	F
F	T	T	T	T
F	F	T	F	F

$[(p \vee q) \wedge (\sim p)] \rightarrow q$
T
T
T
T

47.

p	q	$\sim p$	$\sim q$	$p \leftrightarrow (\sim q)$	$(p \leftrightarrow \sim q) \rightarrow \sim p$
T	T	F	F	F	T
T	F	F	T	T	F
F	T	T	F	T	T
F	F	T	T	F	T

49.

p	q	$\sim p$	$\sim q$	$q \rightarrow p$	$\sim p \rightarrow \sim q$
T	T	F	F	T	T
T	F	F	T	T	T
F	T	T	F	F	F
F	F	T	T	T	T

└─ Identical ─┘

51.

p	q	$\sim q$	$p \rightarrow q$	$\sim(p \rightarrow q)$	$p \bigwedge \sim q$
T	T	F	T	F	F
T	F	T	F	T	T
F	T	F	T	F	F
F	F	T	T	F	F

$\longleftarrow$ Identical $\longrightarrow$

53.

p	q	$\sim p$	$\sim q$	$p \rightarrow q$	$(p \rightarrow q) \bigvee \sim q$
T	T	F	F	T	T
T	F	F	T	F	T
F	T	T	F	T	T
F	F	T	T	T	T

$p \bigvee \sim p$
T
T
T
T

$\longleftarrow$ Identical $\longrightarrow$

55.

p	q	$\sim p$	$\sim p \bigwedge q$	$p \rightarrow (\sim p \bigwedge q)$
T	T	F	F	F
T	F	F	F	F
F	T	T	T	T
F	F	T	F	T

$\longleftarrow$ Identical $\longrightarrow$

57. (c)   **59.** (a)   **61.**

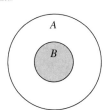

63.

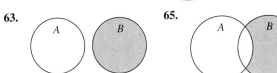

65.

67.

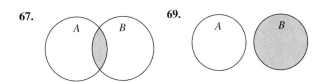

69.

71. (a) Statement does not follow
    (b) Statement follows

73. (a) Statement does not follow
    (b) Statement does not follow

Section A.3 *(page A22)*

1.

p	q	$\sim p$	$\sim q$	$p \rightarrow \sim q$	$(p \rightarrow \sim q) \bigwedge q$
T	T	F	F	F	F
T	F	F	T	T	F
F	T	T	F	T	T
F	F	T	T	T	F

$[(p \rightarrow \sim q) \bigwedge q] \rightarrow \sim p$
T
T
T
T

3.

p	q	$\sim p$	$p \bigvee q$	$(p \bigvee q) \bigwedge \sim p$
T	T	F	T	F
T	F	F	T	F
F	T	T	T	T
F	F	T	F	F

$[(p \bigvee q) \bigwedge \sim p] \rightarrow q$
T
T
T
T

5.

p	q	$\sim p$	$\sim q$	$\sim p \to q$	$(\sim p \to q) \bigwedge p$
T	T	F	F	T	T
T	F	F	T	T	T
F	T	T	F	T	F
F	F	T	T	F	F

$[(\sim p \to q) \bigwedge p] \to \sim q$
F
T
T
T

7.

p	q	$p \bigvee q$	$(p \bigvee q) \bigwedge q$	$[(p \bigvee q) \bigwedge q] \to p$
T	T	T	T	T
T	F	T	F	T
F	T	T	T	F
F	F	F	F	T

9. Valid   **11.** Invalid   **13.** Valid   **15.** Valid

17. Invalid   **19.** Valid   **21.** Invalid

23. (b)   **25.** (c)   **27.** (b)   **29.** (c)

31. Valid

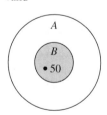

A: All numbers divisible by five
B: All numbers divisible by ten

33. Invalid

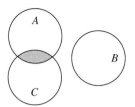

A: People eligible to vote
B: People under the age of 18
C: College students

35. Let p represent the statement "Sue drives to work," let q represent "Sue will stop at the grocery store," and let r represent "She'll buy milk."

First write:

> Premise #1: $p \to q$
> Premise #2: $q \to r$
> Premise #3: p

Reorder the premises:

> Premise #3: p
> Premise #1: $p \to q$
> Premise #2: $q \to r$
> Conclusion: r

Then we can conclude r. That is, "Sue will get milk."

37. Let p represent "This is a good product," let q represent "We will buy it," and let r represent "The product was made by XYZ Corporation."

First write:

> Premise #1: $p \to q$
> Premise #2: $r \bigvee \sim q$
> Premise #3: $\sim r$

Note that $p \to q \equiv \sim q \to \sim p$, and reorder the premises:

> Premise #2: $r \bigvee \sim q$
> Premise #3: $\sim r$
> (Conclusion from Premise #2, Premise #3 : $\sim q$)
> Premise #1: $\sim q \to \sim p$
> Conclusion: $\sim p$

Then we can conclude $\sim p$. That is, "It is not a good product."

Appendix B *(page A35)*

1.

Stems	Leaves
7	0 5 5 5 7 7 8 8 8
8	1 1 1 1 2 3 4 5 5 5 5 7 8 9 9 9
9	0 2 8
10	0 0

3.

Stems	Leaves
5	2 5 9
6	2 3 6 6 7
7	0 1 2 3 4 7 8 8 9
8	0 1 3 4 5 7 9
9	0 0 2 3 3 3 5 6 8 9
10	0 0

5. Frequency Distribution

Interval	Tally
[15, 22)	⦀⦀ |||
[22, 29)	⦀⦀	
[29, 36)	⦀⦀
[36, 43)	||||
[43, 50)	⦀⦀ ||

Histogram

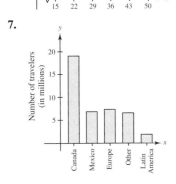

7.

9. 150 million tons, 195 million tons

11. Total waste, recycled waste

13. Total waste equals the sum of the other three quantities.

15.

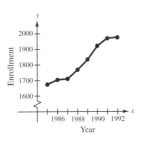

17.

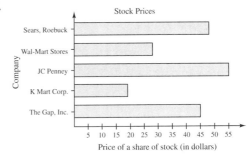

19.

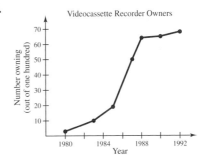

21. Positive correlation   **23.** Yes

25. Negative correlation   **27.** Positive correlation

29.

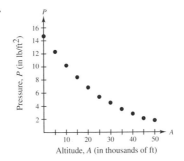

31. 2.45 pounds per square inch

33.

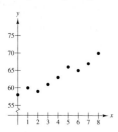

35. $y = 57.49 + 1.43x$; 71.8

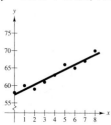

37.

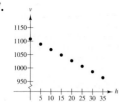

39. $v = 1117.3 - 4.1h$; 1006.6

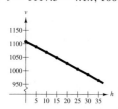

41. $y = -2.179x + 22.964$

43. $y = 2.378x + 23.546$

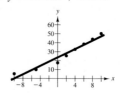

45. $S = 384.1 + 21.2x$; \$479,500

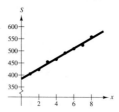

INDEX OF APPLICATIONS

INDEX

A

Absolute value, 7
  equations, 106
  inequality, 109
  of real number, 7
Addition
  of complex numbers, 371
  of fractions, 12
  of rational expressions with like
    denominators, 274
  of real numbers, 11
Addition Property of Equality, 25
Addition and Subtraction Properties, 96
Additive Identity Property, 23
Additive Inverse Property, 23
Algebraic
  expression, 32
  inequalities, 94
Approximately equal to, 3
Area formula, 85
Area of a triangle, 549
Argument, A16
Arithmetic sequence, 634, 644
  nth partial sum of, 646
  nth term of, 645
Associative Property of Addition, 23
Associative Property of Multiplication, 23
Asymptotes, 296, 470
Augmented matrix, 531

B

Back-substitute, 497
Bar graph, A29
Base, 15
Base number, 70
Biconditional, A11
Binomial, 198, 663
  coefficients, 663
  square of, 211
Binomial Theorem, 663
Bounded intervals on the real number line, 94
Branches, 295, 470
Break-even point, 501

C

Cancellation law for fractions, 258
Cancellation Property of Addition, 25
Cancellation Property of Multiplication, 25
Cancellation rule for fractions, 258
Cartesian plane, 122
Center, 466, 468
Central rectangle, 470
Change of base formula, 591
Characteristics of a function, 170
Circle, 466
Coefficient, 32, 198
Coefficient matrix, 531
Collinear, 130
Combination of n elements taken m at a time, 675
Common difference, 644
Common formulas, 85
Common logarithmic function, 588
Common ratio, 652
Commutative Property of Addition, 23
Commutative Property of Multiplication, 23
Complete the square, 391, 392
Completely factored, 222
Complex conjugate, 372
Complex fractions, 270, 278
Complex number, 370
  imaginary part, 371
  real part, 371
Composition, 574
  of two functions, 574
Compound interest, 567
  formulas for, 568
Compound statements, A2
Conclusion, A8, A16
Condense a logarithmic expression, 600
Conditional equation, 58
Conic sections, 466
Conjugate, 352
Conjunction, A3
Consistent, 510
Constant of proportionality, 479
Constant term, 32, 198

D

Definition
  of absolute value of a real number, 7
  of an arithmetic sequence, 644
  of a complex number, 370
  of composition of two functions, 574
  of the determinant of a 2×2 matrix, 544
  of exponential function, 562
  of factorial, 636
  of a function, 169
  of geometric sequence, 652
  of intercepts, 139
  of infinite sequences, 634
  of the inverse of a function, 577
  of linear equation, 60
  of the logarithmic function, 586
  of nth root of a number, 331
  of permutation, 673
  of a polynomial in x, 198
  of a quadratic equation in x, 239
  of ratio, 72
  of rational exponents, 334
  of rational expression, 256
  of a relation, 168
  of row-echelon form of a matrix, 534
  of sigma notation, 638
  of the slope of a line, 156
  of zero exponents and negative exponents, 322
Degree of a polynomial, 198
DeMorgan's Laws, A5

Continuous compounding, 567
Contrapositive, A10
Converse, A10
Coordinate of point, 122
Coordinate system, 122
Correlation
  negative, A32
  positive, A32
Co-vertices, 468
Cramer's Rule, 547
Critical number, 419, 423
Cubing, 15

Graphing Linear Equations

A **linear equation in two variables** is an equation of first degree in both variables.

The **graph of an equation** is the set of all points in the rectangular coordinate system whose coordinates are solutions of the equation.

The **graph of a linear equation in two variables** is a straight line.

To find the **x-intercepts** (where the graph intersects the x-axis), let y be zero and solve the equation for x.

To find the **y-intercepts** (where the graph intersects the y-axis), let x be zero and solve the equation for y.

Slope, $m = \dfrac{y_2 - y_1}{x_2 - x_1} = \dfrac{\text{Change in } y}{\text{Change in } x}$.

Slope-intercept form of the equation of a line: $y = mx + b$, m is the slope, and $(0, b)$ is the y-intercept.

Point-slope form of the equation of a line: $y - y_1 = m(x - x_1)$, m is the slope, and (x_1, y_1) is a point on the line.

Graphs of Systems of Linear Equations

The **solution** of a system of linear equations is an ordered pair (a, b) that satisfies each of the equations. A system of linear equations may have no solution, exactly one solution, or infinitely many solutions.

Graphs of Quadratic Equations

Standard form of a quadratic equation:
$ax^2 + bx + c = 0$, a, b, and c are real numbers with $a \neq 0$. A quadratic equation can be solved by factoring, extracting square roots, completing the square, or using the Quadratic Formula.

Extracting square roots: If $u^2 = d$, where $d > 0$, then $u = \pm\sqrt{d}$.

Quadratic Formula: $x = \dfrac{-b \pm \sqrt{b^2 - 4ac}}{2a}$

Discriminant: $b^2 - 4ac$
If $b^2 - 4ac > 0$, then the equation has two real number solutions.
If $b^2 - 4ac = 0$, then the equation has one (repeated) real number solution.
If $b^2 - 4ac < 0$, then the equation has no real number solutions.

Vertical line
Undefined slope

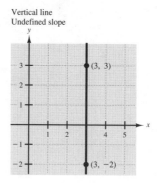

Line falls
Negative slope

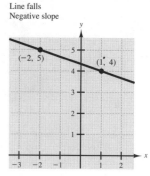

Horizontal line
Zero slope

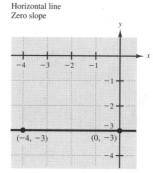

Line rises
Positive slope

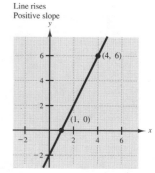

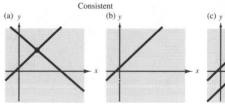

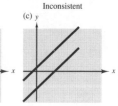

(a) Two lines that intersect at a single point
(b) Two lines that coincide with infinitely many points of intersection
(c) Two parallel lines with no point of intersection

The graph of $y = ax^2 + bx + c$, $a \neq 0$, is called a **parabola** which opens up if $a > 0$ and opens down if $a < 0$.

Parabola Opens Up   Parabola Opens Down

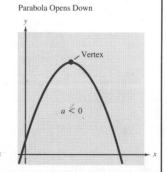